Electromagnetic Concepts and Applications

Electromagnetic Concepts and Applications

Third Edition

S. V. MARSHALL

Professor of Electrical Engineering
University of Missouri-Rolla

G. G. SKITEK

Professor Emeritus of Electrical Engineering
University of Missouri-Rolla

Prentice Hall
Englewood Cliffs, New Jersey 07632

Library of Congress Cataloging-in-Publication Data

MARSHALL, S. V. (STANLEY V.)

Electromagnetic concepts and applications/S. V. Marshall, G. G. Skitek.

p. cm.
Includes index.
ISBN 0-13-250960-1
1. Electromagnetic theory. I. Skitek, G. G. (Gabriel G.)
II. Title.
QC670.M344 1990 89-4020
530.1'41—dc20 CIP

Editorial/production supervision and
interior design: Jennifer Wenzel
Cover design: Wanda Lubelska Design
Manufacturing buyer: Bob Anderson

Printed in the United States of America

10 9 8 7 6 5 4 3 2 1

ISBN 0-13-250960-1

Prentice-Hall International (UK) Limited, *London*
Prentice-Hall of Australia Pty. Limited, *Sydney*
Prentice-Hall Canada Inc., *Toronto*
Prentice-Hall Hispanoamericana, S.A., *Mexico*
Prentice-Hall of India Private Limited, *New Delhi*
Prentice-Hall of Japan, Inc., *Tokyo*
Simon & Schuster Asia Pte. Ltd., *Singapore*
Editora Prentice-Hall do Brasil, Ltda., *Rio de Janeiro*

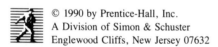

To our students,
who tolerated the use of our manuscript in its many stages of iteration and assisted in finding errors and making constructive comments.

Contents

Preface

This text evolved from notes used in teaching a two-semester junior-level electromagnetics course at the University of Missouri–Rolla. This third edition reflects feedback from our UMR students as well as the many users of the first and second editions. The problem set has been renewed and expanded, but the most significant changes are the addition of a section on the solution of Laplace's equation in cylindrical coordinates in Chapter 6, and the rearrangement of material in Chapter 12. A section on lossy line analysis has also been added to Chapter 12.

Our goals remain essentially the same as for the first edition, and they are being met by the following techniques:

1. The material presented starts with electrostatics and magnetostatics, then progresses to time-varying electric and magnetic fields.
2. There are a large number of example problems, which relate to the material in the section where the examples appear.
3. At the end of most sections, there are related drill problems (relatively simple ones), with answers shown at the back of the book.
4. More challenging problems are found at the end of each chapter, sequenced with the material covered in that chapter.
5. Review questions, and citations as to where to find the answers to those questions, are found at the end of each chapter.
6. An introduction is found at the start of each chapter for the purpose of tying the previously presented material to that found in the upcoming chapter. Highlights of the material found in the chapter are discussed in the order of their appearance.

7. As a pedagogical aid, several related terms (concepts) of an equation have been set equal to each other and displayed on one line with arrows identifying the concepts they express. A good example of this is found in Chapter 3, equations (3.5-11) and (3.7-1).
8. Boldface letters are not used to represent vectors in the text, since they cannot be easily reproduced in the classroom and by the student in the solution of problems, etc. A bar over a letter is used to represent a vector quantity.

The study of electrostatics begins in Chapter 2 with the Coulomb force law between two point charges. It is an experimental fact and may be viewed as the field of charge Q_1 acting to produce a force on charge Q_2, or the field of Q_2 acting to produce a force on Q_1. This charge-field-charge viewpoint versus the action-at-a distance viewpoint is further discussed at the beginning of Chapter 9. The electric field from a line of charge or from a sheet of charge is derived by using the point charge field concept as the basic building block.

The first Maxwell equation, i.e., $\nabla \cdot \overline{D} = \rho_v$ is introduced in Chapter 3, where it logically evolves from the defining equation for divergence. The remaining Maxwell equations for static fields do not formally appear until they are summarized in Chapter 8, after the concept of curl has been introduced and the divergence of the magnetic field is shown to be zero because of the non-isolatable nature of magnetic poles. Thus, rather than starting with Maxwell's equations as though they were inspired, we introduce them as they naturally occur in the development of the material.

The concepts of electric flux, electric flux density, Gauss's law, and the divergence theorem are covered by the end of Chapter 3. The concepts developed in Chapter 4 are potential difference, potential gradient (and its relation to the electric field), and energy density at a point. Material properties, such as conductivity and dielectric constant, are discussed in Chapter 5, followed by the concept of capacitance.

Chapter 6 is devoted to the uniqueness theorem and to the solution of Laplace's and Poisson's equations for simple boundary conditions. Graphical, numerical, and image solution techniques are introduced in Chapter 7, which completes the study of electrostatics.

Magnetostatics is covered in Chapters 8 and 9. Chapter 8 is concerned mainly with magnetic fields resulting from current and with the concept of vector magnetic potential. The Biot-Savart law, Ampere's law, toroidal and solenoidal fields, and a summary of Maxwell's equations for static fields are found in Chapter 8. Chapter 9 deals with magnetic materials, magnetic circuits, and inductance.

The study of time-varying fields and their applications begins in Chapter 10 with Faraday's induction law, another experimental fact. The first application is the transformer. Following a plausibility argument for the existence of displacement current, Maxwell's equations for time-varying fields are then completed by the incorporation of Faraday's law and displacement current. From this point, Maxwell's equations provide a convenient introduction for the derivation of wave equations, waveguide equations, and radiation fields, for applications in transmission lines, waveguides, and antennas. Plane wave propagation, Poynting's vector, reflections, standing waves, and skin effect are treated in Chapter 11.

Chapter 12, on transmission lines, is fairly lengthy compared to most texts, and it is somewhat of a departure since voltage and current are used instead of the $\overline{E}$ and $\overline{H}$

fields. We feel that transmission lines and transmission line analogies are fundamental to all electrical engineers, whether they are working at dc, 60 Hz, or at optical frequencies, and whether they are involved with analog or with digital circuits.

For sinusoidal analysis, we do not feel as some do that the Smith chart has been made obsolete by the hand-held calculator and the personal computer. While these may make possible greater accuracy, there is no substitute for the display of information afforded by the Smith chart. Transients on transmission lines has been treated by using step function changes in sources and impedances for pure resistance circuits.

To keep the book at an introductory level, circular waveguides have been omitted from Chapter 13, but there is a descriptive section on optical fibers. The detailed analysis of rectangular waveguides includes calculation of wall losses.

Chapter 14, on antennas, begins with the concept of gain and beamwidth, then proceeds to the classical treatment of the elemental dipole and half-wave dipole. The aperture antenna is shown to be a limiting case of an array antenna, and the effect of tapering an antenna is discussed in qualitative terms. A brief qualitative coverage of some aperture antennas includes the metal lens, the Luneberg lens, the cassegrain antenna, and the four-horn feed used for a monopulse radar antenna. This chapter closes with a section on the radar equation.

This book is intended primarily for juniors in electrical engineering, but it is hoped that engineers who have been out of school for a few years will find this a useful volume for review, especially preparatory to taking more advanced courses in electromagnetics. Normally, the student taking this course has had two years of mathematics through differential equations, with possibly some vector analysis, although the latter is not essential since the necessary skill in vectors may be developed as the course progresses. In this era of heavy emphasis on research, the instructor available to teach this course may not have sufficient time to prepare lectures or may not have a good background in electromagnetics. With these thoughts in mind, this book has been written to be as helpful to the instructor as possible, as well as being a self-study book for the student. The average or above average student can learn from this book with a minimum of instruction.

SI units and the names of those units are used throughout.

Section numbering is sequential within each chapter and the drill problems are identified with the section that precedes them; e.g., Problem 1.8-6 is the sixth problem found at the end of Section 1.8. Section 1.7-1 is the first subsection of Section 1.7, etc. Equations are sequential in each section, beginning with (1). When an equation is referred to in the reading material, a (3) means equation 3 in that section, a (1.6-3) means equation 3 in Section 1.6, a notation used if the reader is reading within a section other than Section 1.6. Figure numbers are also sequential within each chapter; e.g., Fig. 4–1, Fig. 4–2 are the first and second figures in Chapter 4.

For the instructor, a solutions manual is available that has more detailed solutions to the section problems and also includes solutions to the end-of-chapter problems.

For students with a good math background, the instructor may wish to skip Chapter 1, but it has been our experience that most students will benefit from two or three class periods studying this chapter. All the material in this book can be covered in two semesters, but if the instructor prefers to cover some sections more slowly, or if it is necessary to spend more than a week on Chapter 1, then Sections 5.8-2, 5.8-4, 9.7-1, 9.7-2, 9.7-4, 10.9, and 14.7 may be omitted. After teaching this material for several semesters, we recommend the following schedule for a two-semester course, based on three class meetings per week for 15 weeks for each semester:

FIRST SEMESTER

Chapter	Number of Class Meetings
1	3
2	4
3	3
4	3
5	6
6	5
7	5
8	6
9	6
Exams	4
	45

SECOND SEMESTER

Chapter	Number of Class Meetings
10	5
11	9
12	12
13	9
14	6
Exams	4
	45

For a one-semester course that includes material on time-varying fields, we suggest the omission of Sections 5.8-4, 6.5, 6.6, 7.4, 7.5, 7.6, 7.7, 8.10, 9.7-1, and the inclusion of Chapter 10 (except possibly Section 10.9), and Sections 11.1, 11.2, and 11.3. The following schedule might then be appropriate:

ONE-SEMESTER COURSE

Chapter	Number of Class Meetings
1	3
2	4
3	3
4	3
5	5
6	4
7	3
8	5
9	4
10	4
11	3
Exams	4
	45

S. V. Marshall
G. G. Skitek

Electromagnetic Concepts and Applications

1

Vector Analysis

1.1 INTRODUCTION

The use of vectors and vector analysis can greatly simplify the mathematics used in expressing and manipulating the laws and theorems of electric and magnetic fields. This simplification is possible because vector analysis is a mathematical shorthand that greatly reduces the number of equations needed by using special vector operators and operations on scalar and vector quantities.

To many students, vector analysis may not be totally new. The amount of time spent on this chapter should depend on the amount and quality of the student's previous exposure to vector analysis.

In this text, as in others, it is very important that the student become familiar with the symbols used. The authors are well aware that change of symbols from text to text and course to course contributes considerably to learning inefficiency and frustration. This being the case, great care has been taken in the selection of symbols used in this text. The SI system of units is used throughout. See Tables A-1 and A-2 of Appendix A for a complete list of symbols and their SI units. SI prefixes are given in Table A-3.

1.2 SCALARS

> A *scalar* is a quantity, such as temperature or energy, having only magnitude.

Symbolically, a scalar is represented by either lower- or uppercase letters. The student should be familiar with scalars since they are the real numbers learned in elementary education. All the laws of algebra of real numbers apply to scalars.

Other examples of scalars are time, spatial volume, potential difference, temperature, distance, and coordinate variables. These scalars are symbolically represented by t, υ, V, T, ℓ, and (x, y, z), respectively.

1.3 VECTORS

A *vector* is a quantity, such as velocity or force, having magnitude and direction.

It is important that we learn how a vector is represented *graphically* and *symbolically*. Graphically, a vector is represented by an arrow whose length represents its magnitude, and the arrowhead indicates its direction, as shown in Fig. 1-1(a).

Symbolically, a vector is represented by placing a bar over the letter symbol used for a given quantity. This symbolic representation is placed at the arrowhead end of the

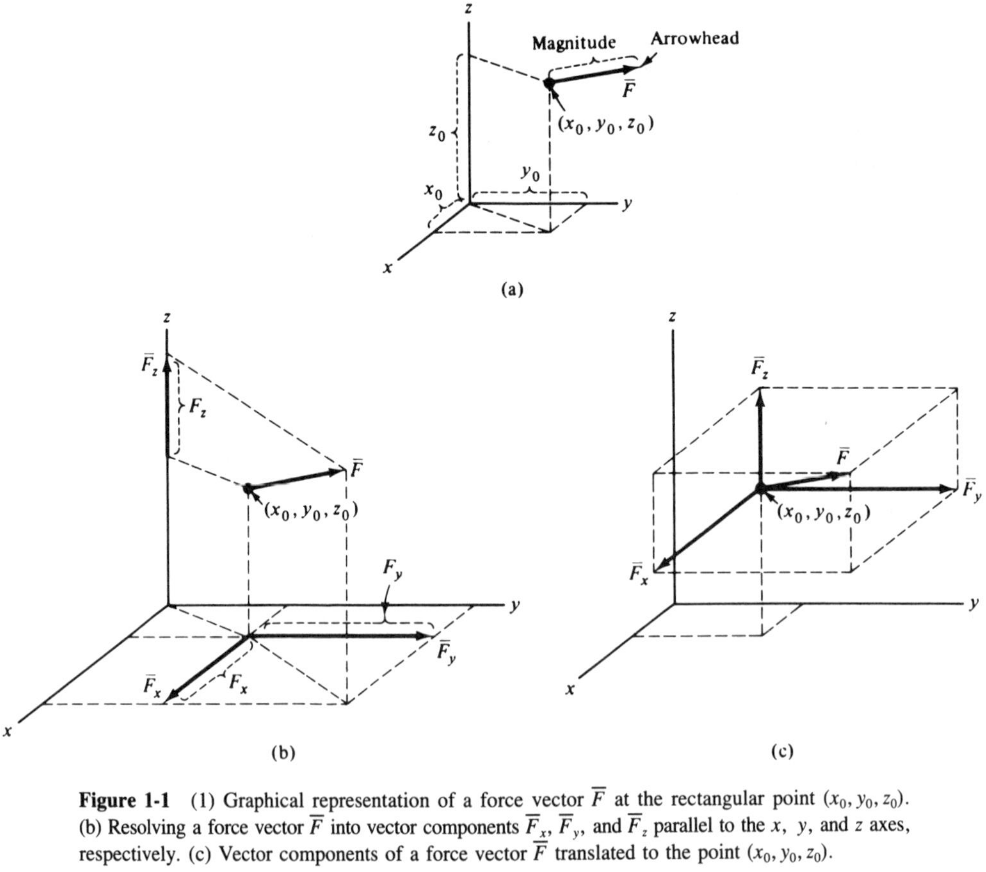

Figure 1-1 (1) Graphical representation of a force vector $\overline{F}$ at the rectangular point (x_0, y_0, z_0). (b) Resolving a force vector $\overline{F}$ into vector components $\overline{F}_x$, $\overline{F}_y$, and $\overline{F}_z$ parallel to the x, y, and z axes, respectively. (c) Vector components of a force vector $\overline{F}$ translated to the point (x_0, y_0, z_0).

graphical representation of the vector. The vector $\overline{F}$ could logically be used to represent a force vector. Thus, Fig. 1-1(a) could represent a force vector $\overline{F}$, graphically and symbolically at the point $P(x_0, y_0, z_0)$ in rectangular space.

The vector $\overline{F}$ can be resolved into vector components parallel to the x, y, and z directions of a right-hand rectangular coordinate system, as shown in Fig. 1-1(b). In this coordinate system, the thumb points in the $+z$ direction when the fingers of the right hand rotate from $+x$ to $+y$ axes. The student should be able to graphically construct the vector components $\overline{F}_x$, $\overline{F}_y$, and $\overline{F}_z$ of the vector $\overline{F}$. Figure 1-1(c) shows the vector compo-ponents $\overline{F}_x$, $\overline{F}_y$, and $\overline{F}_z$ translated to the point $P(x_0, y_0, z_0)$. Thus, we have resolved vector $\overline{F}$ into *three mutually perpendicular vectors*. From Fig. 1-1(c), we can write

$$\boxed{\overline{F} = \overline{F}_x + \overline{F}_y + \overline{F}_z} \tag{1}$$

This equation will be graphically proven when we discuss vector addition. Other examples of a vector are velocity, space-directed distance, electric field intensity, and magnetic field intensity. These quantities are symbolically represented by $\overline{v}$, $\overline{\ell}$, $\overline{E}$, and $\overline{H}$, respectively.

1.4 UNIT VECTORS

The vector projection $\overline{F}$ parallel to the x axis has been designated by $\overline{F}_x$. This vector can be expressed as

$$\boxed{\overline{F}_x = \hat{x}F_x} \tag{1}$$

where $\hat{x}$ is a *unit vector,* magnitude one, parallel to and in the direction of the $+x$ axis, and F_x is the scalar projection of $\overline{F}$ parallel to the x axis. It should be noted that F_x can be a positive or negative scalar quantity, depending on the direction of $\overline{F}$. Thus, (1.3-1) can be expressed as

$$\boxed{\overline{F} = \hat{x}F_x + \hat{y}F_y + \hat{z}F_z} \tag{2}$$

where $\hat{x}$, $\hat{y}$, and $\hat{z}$ are unit vectors in the $+x$, $+y$, and $+z$ axis directions, respectively. Unit vectors will be designated by lowercase letters topped by a *caret* (ˆ), also referred to as a *hat*. The scalars F_x, F_y, and F_z are scalar projections of $\overline{F}$ parallel to the x, y, and z axes, respectively, as shown in Fig. 1-1(b).

The left-hand side of (2) can also be expressed in unit vector form

$$\boxed{\overline{F} = \hat{a}_{\overline{F}}F} \tag{3}$$

where $\hat{a}_{\overline{F}}$ is a unit vector in the direction of $\overline{F}$, and F is a positive scalar equal to the absolute value of $\overline{F}$; thus, we can write $F = |\overline{F}|$. It should be noted that $F_x \neq |\overline{F}_x|$ in

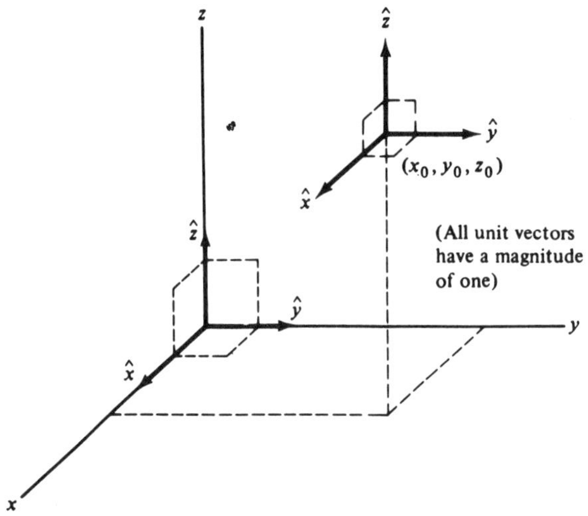

Figure 1-2 Graphical display of unit vectors $\hat{x}$, $\hat{y}$, and $\hat{z}$ at the origin and point (x_0, y_0, z_0) of a rectangular coordinate system.

general. This is due to the fact that $\overline{F}_x$ may not be in the direction of $\hat{x}$. If $\overline{F}_x$ is in the $\hat{x}$ direction, F_x will be positive; otherwise, F_x will be negative.

Unit vectors $\hat{x}$, $\hat{y}$, and $\hat{z}$ are shown in Fig. 1-2, at the origin and at the point $P(x_0, y_0, z_0)$. It should be noted that the unit vectors $\hat{x}$, $\hat{y}$, and $\hat{z}$, in the rectangular coordinate system, are not functions of x, y, or z, and thus they are called *constant vectors*. Not all unit vectors are constant vectors. We shall find in Secs. 1.9 and 1.10, that unit vectors in both cylindrical and spherical coordinate systems are functions of coordinate variables; this must be taken into account when they are integrated and differentiated with respect to coordinate variables. From (2) and (3), the magnitude of $\overline{F}$ becomes $|\overline{F}| = F = (F_x^2 + F_y^2 + F_z^2)^{1/2}$, since F_x, F_y, and F_z are along mutually perpendicular axes.

Example 1

Express the vector distance $\overline{R}_P$ from the origin to the rectangular point $P(2, 4, 3)$ in component form, using unit vectors. This vector is commonly called the *position vector* of the point $P(2, 4, 3)$.

Solution. The vector distance $\overline{R}_P$, from the origin to the point $P(2, 4, 3)$, is shown in Fig. 1-3. The projections R_x, R_y, and R_z are 2, 4, and 3, respectively. Thus, the vector $\overline{R}_P$ can be expressed as

$$\overline{R}_P = \hat{x}R_x + \hat{y}R_y + \hat{z}R_z = (\hat{x}2 + \hat{y}4 + \hat{z}3) \qquad \text{(m)}$$

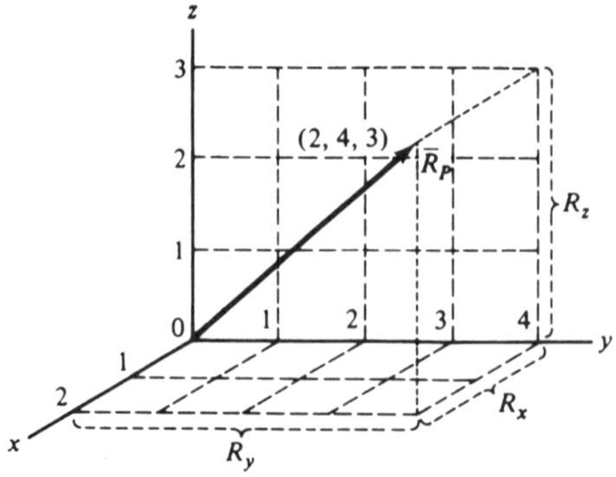

Figure 1-3 Position vector $\overline{R}_P$ from the origin to the point (2, 4, 3) (see Example 1).

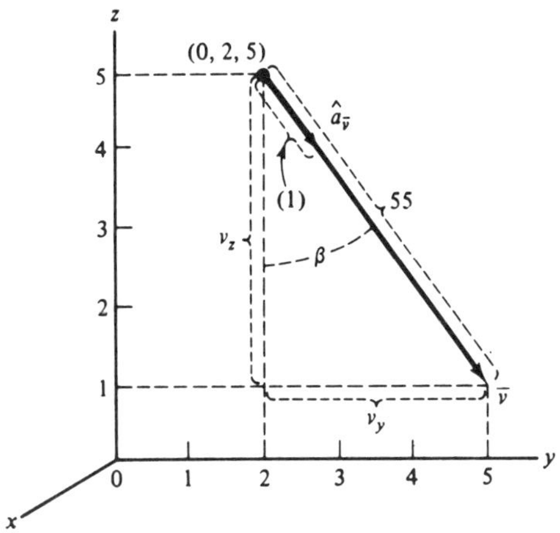

Figure 1-4 The velocity vector $\bar{v} = \hat{a}_{\bar{v}}$ 55(m s^{-1}) at the point (0, 2, 5) (see Example 2). Note that $\hat{a}_{\bar{v}}$ is not drawn to the scale of $|\bar{v}|$.

Example 2

Express the vector velocity $\bar{v} = \hat{a}_{\bar{v}}$ 55 (m s^{-1}) at $P(0, 2, 5)$, shown in Fig. 1-4 to be in the y-z plane, in component form, using unit vectors.

Solution. The vector velocity $\bar{v}$ can, in general, be expressed as $\bar{v} = \hat{x}v_x + \hat{y}v_y + \hat{z}v_z$. From Fig. 1-4, $v_x = 0$, $v_y = 55 \sin\beta$, $v_z = -55\cos\beta$, $\cos\beta = \frac{4}{5}$, and $\sin\beta = \frac{3}{5}$. Thus,

$$\bar{v} = \hat{x}(0) + \hat{y}(55)\left(\tfrac{3}{5}\right) + \hat{z}(-55)\left(\tfrac{4}{5}\right) = [\hat{y}(33) - \hat{z}(44)] \qquad (\text{m } s^{-1})$$

Problem 1.4-1 Find the distance and vector distance in component form from the origin to the point $(4, -3, 5)$ in rectangular coordinates, using the unit vectors $\hat{x}$, $\hat{y}$, and $\hat{z}$.

Problem 1.4-2 Find the unit vector $\hat{a}_{\bar{R}_p}$ for the vector in Prob. 1.4-1

Problem 1.4-3 Find the unit vector in the direction of the vector $\bar{F} = -\hat{x}4 + \hat{y}2 + \hat{z}2$.

1.5 LAWS OF ALGEBRA

The algebra of scalar numbers is based on a group of laws commonly called the *laws of algebra*. In our study of the interaction of vectors with scalars and with other vectors, only three laws will be found useful from the entire group of laws of scalar algebra. These three laws of scalar algebra, as they apply to scalar addition and multiplication, are summarized below.

LAW	ADDITION	MULTIPLICATION
Commutative	$a + b = b + a$	$ab = ba$
Associative	$(a + b) + c = a + (b + c)$	$(ab)c = a(bc)$
Distributive		$a(b + c) = ab + ac$

We will find that not all of the laws of scalar algebra above apply to all mathematical operations involving vectors.

1.6 VECTOR ADDITION AND SUBTRACTION

Graphically, the sum of two vectors $\overline{A}$ and $\overline{B}$ is a vector $\overline{C}$ that begins at the start of vector $\overline{A}$ and ends at the arrowhead of vector $\overline{B}$, when the vector $\overline{B}$ is positioned so that its start is joined to the arrowhead of vector $\overline{A}$, as shown in Fig. 1-5. Symbolically, the sum is written as $\overline{A} + \overline{B} = \overline{C}$.

If we express $\overline{A}$ and $\overline{B}$ in component forms, we have

$$\boxed{\overline{A} + \overline{B} = (\hat{x}A_x + \hat{y}A_y + \hat{z}A_z) + (\hat{x}B_x + \hat{y}B_y + \hat{z}B_z) = \overline{C}} \tag{1}$$

Adding vector components parallel to the x, y, and z directions, we obtain

$$\boxed{\overline{A} + \overline{B} = \hat{x}(A_x + B_x) + \hat{y}(A_y + B_y) + \hat{z}(A_z + B_z) = \overline{C}} \tag{2}$$

From the expansion of $\overline{C}$, in component form in (2), we obtain

$$C_x = (A_x + B_x), \qquad C_y = (A_y + B_y), \qquad C_z = (A_z + B_z)\,. \tag{3}$$

Equations (2) and (3) can be proven graphically. This proof is left as an exercise for the student.

Subtraction of two vectors, $\overline{A} - \overline{B}$, can be looked upon as addition of $\overline{A} + (-\overline{B})$. The negative of vector $\overline{B}$ can be graphically expressed by drawing the arrowhead on the opposite end of the arrow that represents the vector $+\overline{B}$. With the aid of (2), we have

$$\boxed{\overline{A} - \overline{B} = \overline{A} + (-\overline{B}) = \hat{x}(A_x - B_x) + \hat{y}(A_y - B_y) + \hat{z}(A_z - B_z)} \tag{4}$$

If the vectors $\overline{A}$ and $\overline{B}$, shown in Fig. 1-5, are drawn in the plane of the page, the sum and difference of these two vectors can be shown as the diagonals of a parallelogram, as shown in Fig. 1-6. The principles of addition and subtraction of two vectors can be easily extended to more than two vectors. It should be noted that the commutative and associative laws of scalar algebra also apply to vector addition and subtraction. This can easily be proven graphically (see Probs. 1.6-2 and 1.6-3).

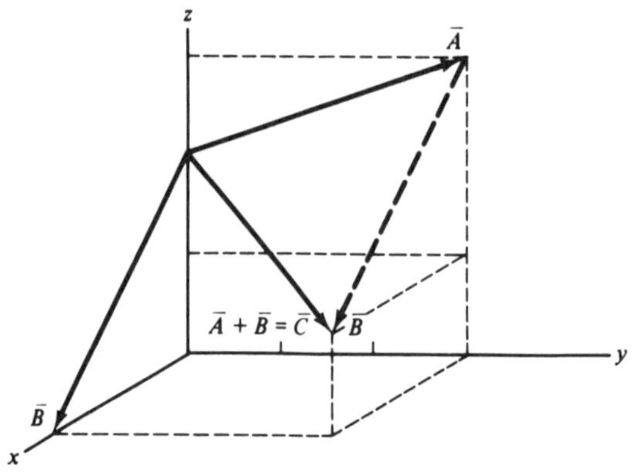

Figure 1-5 Graphical addition of two vectors $\overline{A}$ and $\overline{B}$ at the point $(0, 0, 2)$ to give vector $\overline{A} + \overline{B} = \overline{C}$.

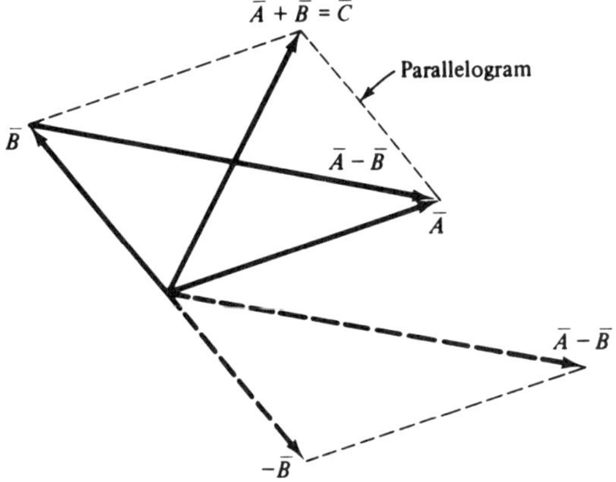

Figure 1-6 The vector sum $\overline{A} + \overline{B}$ and the vector difference $\overline{A} - \overline{B}$ are diagonals of a parallelogram.

Example 3

Two force vectors, $\overline{F}_1$ and $\overline{F}_2$, act at $P(1, 2, 0)$. If $\overline{F}_1 = (\hat{x}10 + \hat{y}5 + \hat{z}3)$ (N) and $\overline{F}_2 = (-\hat{x}5 - \hat{y}8)$ (N), find the resultant force $\overline{F}_r$.

Solution

$$\overline{F}_r = \overline{F}_1 + \overline{F}_2 = \hat{x}(F_{1x} + F_{2x}) + \hat{y}(F_{1y} + F_{2y}) + \hat{z}(F_{1z} + F_{2z})$$

$$= \hat{x}[10 + (-5)] + \hat{y}[5 + (-8)] + \hat{z}(3 + 0)$$

$$= [\hat{x}(5) + \hat{y}(-3) + \hat{z}(3)] \quad \text{(N)}$$

Example 4

Find the distance vector from $P_1(1, 3, 2)$ to the point $P_2(2, 6, -2)$ through the use of position vectors as defined in Example 1.

Solution. From Fig. 1-7, it can be seen that the distance vector $\overline{R}_{12}$, from a general point $P_1(x_1, y_1, z_1)$ to a general point $P_2(x_2, y_2, z_2)$ is $\overline{R}_{12} = \overline{R}_{P_2} - \overline{R}_{P_1}$, where $\overline{R}_{P_2}$ and $\overline{R}_{P_1}$ are posi-

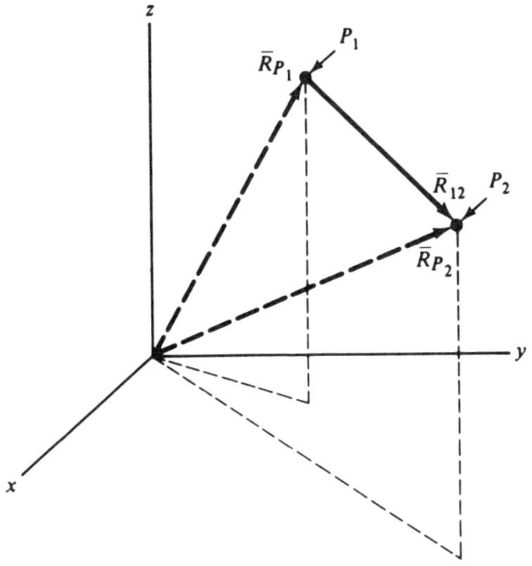

Figure 1-7 The distance vector from point P_1 to point P_2 is found to be $\overline{R}_{12} = \overline{R}_{P_2} - \overline{R}_{P_1}$ (see Example 4).

tion vectors to the points P_2 and P_1, respectively. If we expand the position vectors in general component form, we have

$$\begin{aligned}\overline{R}_{12} &= (\hat{x}x_2 + \hat{y}y_2 + \hat{z}z_2) - (\hat{x}x_1 + \hat{y}y_1 + \hat{z}z_1)\\ &= \hat{x}(x_2 - x_1) + \hat{y}(y_2 - y_1) + \hat{z}(z_2 - z_1)\end{aligned} \tag{5}$$

Substituting the coordinate point values, we have

$$\begin{aligned}\overline{R}_{12} &= \hat{x}(2 - 1) + \hat{y}(6 - 3) + \hat{z}(-2 - (2))\\ &= [\hat{x}(1) + \hat{y}(3) + \hat{z}(-4)] \quad \text{(m)}\end{aligned}$$

Problem 1.6-1 Find the resultant of two forces, $\overline{F}_1 = -\hat{x}6 + \hat{y}4 + \hat{z}4$ (N) and $\overline{F}_2 = -\hat{x}7 + \hat{y}2 + \hat{z}2$ (N).

Problem 1.6-2 Show graphically that the commutative law of algebra also applies to addition of two vectors $\overline{A}$ and $\overline{B}$.

Problem 1.6-3 Show graphically that the associative law of algebra applies to addition of three vectors.

1.7 POSITION VECTORS, DISTANCE VECTORS, FIELDS, AND FIELD VECTORS

Now that some basic concepts of vectors have been mastered, we shall introduce three types of vectors that are frequently found in the study of electromagnetics. These three vectors are (1) *position vectors,* (2) *distance vectors,* and (3) *field vectors*.

1.7-1 Position Vectors

> A *position vector* is defined as the directed distance from the origin to a coordinate point in space.

Thus, the position vector can be used to describe the spatial position of a particle or other physical quantity. In Example 1, the position vector $\overline{R}_P$ of the rectangular point $P(2, 4, 3)$ was found to be $\overline{R}_P = \hat{x}2 + \hat{y}4 + \hat{z}3$. Later, we will find that the position vector is most easily expressed in spherical coordinates since the direction of $\overline{R}_P$ is always in the radial unit vector direction $\hat{r}_s$ of the spherical coordinate system.

1.7-2 Distance Vectors

> A *distance vector* is defined as the directed distance between two coordinate points in space.

In Example 4, the distance vector $\overline{R}_{12}$ from some point P_1 to another point P_2, was found to be $\overline{R}_{12} = \hat{x}(x_2 - x_1) + \hat{y}(y_2 - y_1) + \hat{z}(z_2 - z_1)$, where x_2, y_2, z_2 are the coordinates of the point P_2 and x_1, y_1, z_1 are the coordinates of the point P_1. The distance vector will also be used to express directed differential distances such as $\overline{d\ell}$, directed in a given direction along a specific path. Note that the position vector is a special case of a distance vector.

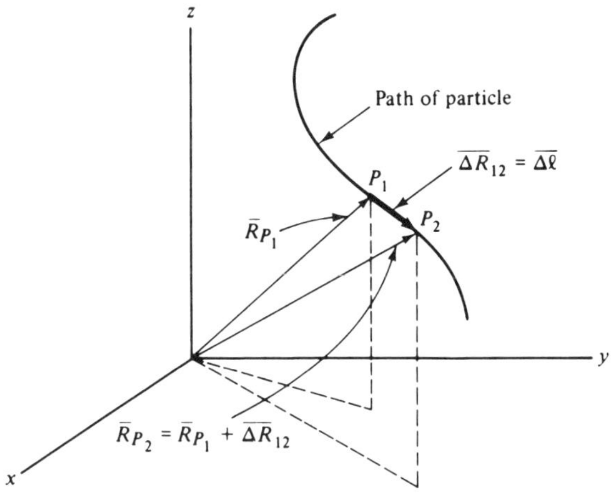

Figure 1-8 Vector change in position vector of a particle in space from P_1 to P_2.

If the position vector $\overline{R}_P$ is used to denote the position of a particle in space as it moves from point $P_1(x_1, y_1, z_1)$ to point $P_2(x_2, y_2, z_2)$, in a time Δt seconds, we can obtain the vector distance $\overline{\Delta R}_{12}$, as shown in Fig. 1-8, that the particle moved by forming

$$\boxed{\overline{\Delta R}_{12} = \overline{R}_{P_2}(t + \Delta t) - \overline{R}_{P_1}(t)} \tag{1}$$

where the position vector of the particle $\overline{R}_P$ is a function of the variables x, y, z, and t. Equation (1) is a distance vector, the distance that the particle traveled in Δt seconds. This distance vector is also a function of x, y, z, and t. If we divide $\overline{\Delta R}_{12}$ by Δt, we obtain the average velocity of the particle as it moves from $P_1(x_1, y_1, z_1)$ to $P_2(x_2, y_2, z_2)$.

$$\boxed{\overline{v}_{\text{ave}} = \frac{\overline{\Delta R}_{12}}{\Delta t} = \frac{\overline{R}_{P_2}(t + \Delta t) - \overline{R}_{P_1}(t)}{\Delta t}} \tag{2}$$

Now, if we form the limit of (2) as $\Delta t \to 0$, we obtain the velocity of the particle at point P_1,

$$\boxed{\overline{v} = \lim_{\Delta t \to 0} \left[\frac{\overline{R}_{P_2}(t + \Delta t) - \overline{R}_{P_1}(t)}{\Delta t}\right] = \frac{\overline{dR_P}}{dt}} \tag{3}$$

where $\overline{dR_P}$ is the differential vector change in the position vector $\overline{R}_P$, and $\overline{dR_P}/\overline{dt}$ is the time rate of change of the position vector $\overline{R}_P$. Note that, through the limit process, the distance vector $\overline{\Delta R}_{12}$ between two points becomes $\overline{dR}_{12}$ defined at a point, as is $\overline{v}$.

The foregoing velocity vector of a particle will, in general, change in magnitude and direction from point to point over a given region of space and thus leads us to the concept of a *field* and *field vectors*.

1.7-3 Fields and Field Vectors

> A *field* is defined as the mathematical specification, in terms of position variables and time, of a physical quantity, such as temperature, in a given region.

Fields can be *scalar fields* or *vector fields,* depending on the physical quantity involved. A good example of a scalar field would be the temperature distribution in a block of unevenly heated steel as it is allowed to cool. On cooling, the temperature in the block will not only vary spatially, i.e., from point to point, but also with time. For this scalar-field example, the mathematical form for the scalar field could well be

$$T(x, y, z, t) = 10e^{-0.2t}y^2z \cos 4x \qquad (°\text{K}) \tag{4}$$

Now, if the physical quantity is a vector, we have a vector field that differs from a scalar field in that the physical quantity has direction as well as magnitude that may vary from point to point. Examples of a vector field could be the velocity field (distribution) of the particle in space previously discussed, a gravitational force field, an electric field intensity, etc. The mathematical expression for the magnetic field intensity $\overline{H}(x, y, z, t)$ in a waveguide is found to be

$$\overline{H}(x, y, z, t) = \left[\hat{x}10^{-4} \sin\left(\frac{\pi x}{a}\right) \cos(\omega t - 0.02z) + \hat{z}\frac{10^{-4}}{2} \cos\left(\frac{\pi x}{a}\right) \sin(\omega t - 0.02z)\right] \qquad (\text{A m}^{-1}) \tag{5}$$

It can be seen that the magnitude and direction of this vector field will vary with x, z, and t.

It is possible for a field, scalar or vector, not to vary with any of the coordinate variables or with time. If the field does not vary with any of the coordinate variables, in a given region, the field is said to be *uniform* in that region and is thus called a *uniform field*. A field that does not vary with time is called a *constant field*. The earth's gravitational field, over a small region of space, is uniform and constant.

The physical quantity that is specified in our definition of a field is called a *field quantity*. Thus, T in (4) and $\overline{H}$ in (5) are examples of scalar and vector field quantities, respectively. A non-field type of vector is called a *discrete vector* and is defined only at one point in space. The force vectors $\overline{F}_1$ and $\overline{F}_2$ in Example 3 are good examples of discrete vectors.

The mathematical specification of a vector field quantity at a general point $P(x, y, z)$, whose magnitude is equal to the distance between the origin and $P(x, y, z)$ and whose direction is radial, becomes $\hat{a}_{\text{rad}}(x^2 + y^2 + z^2)^{1/2}$. A close examination of this specification of a vector field quantity will show that it is the position vector $\overline{R}_{P(x,y,z)}$ at the point $P(x, y, z)$ and that it is truly a vector field. This field concept of a position vector will be used in Chapter 2 to specify other field vectors such as the vector force field about a point charge.

In the study of electromagnetics, vector field quantities may be added or subtracted only if they are at the same point. Discrete vectors at different locations can be added or subtracted if the physical configuration of the problem allows the operation to lend a useful physical result.

Example 5

A force vector field is found to have the mathematical expression $\overline{F} = (\hat{x}5x^2 + \hat{y}2y^2 - \hat{z}4z^2)e^{-t}$ (N) in a region about the origin. (a) Evaluate the vector field $\overline{F}$ at the rectangular points $P_1(1, 1, 0)$, $P_2(1, 2, 0)$ and $P_3(2, 1, 0)$, when $t = 0$. (b) Plot the vector field $\overline{F}$ in the $z = 0$ plane at the three points, using a scale of $\frac{1}{2}$ cm (actual distance measured on page) to represent 1 N.

Solution. (a) $\overline{F}_1(x, y, z, t)\,|_{P_1} = (\hat{x}5x^2 + \hat{y}2y^2 - \hat{z}4z^2)e^{-t}\Big|_{\substack{x=1\\y=1\\z=0\\t=0}} = (\hat{x}5 + \hat{y}2) \quad \text{(N)}$

$$\overline{F}_2(x, y, z, t)\,|_{P_2} = (\hat{x}5 + \hat{y}8) \qquad \text{(N)}$$
$$\overline{F}_3(x, y, z, t)\,|_{P_3} = (\hat{x}20 + \hat{y}2) \qquad \text{(N)}$$

(b) See Fig. 1-9. Note that the scalar projections of $\overline{F}$ are measured in newtons by the newton scale provided and not the scale of the coordinate axes marked in meters.

Problem 1.7-1 The position of an object is given by $\overline{R} = \hat{x}t + \hat{y}t^2$ (m) where t is the time measured in seconds. Characterize the type of force field that gives rise to this vector.

Problem 1.7-2 Give the position vector that describes the location of point $P_2(0, 2 + 3t, 3)$ from point $P_1(1, 5, 6)$.

Problem 1.7-3 The electric field in a rectangular waveguide that accompanies the magnetic field given by (5) is given by

$$\overline{E}(x, y, z) = -\hat{y}\left[0.05 \sin\left(\frac{\pi x}{a}\right) \cos(\omega t - 0.02z)\right] \qquad (\text{V m}^{-1})$$

(a) Evaluate and plot to scale $\overline{E}$ at the following rectangular points: $P_1(0, 0, 0)$, $P_2(a/4, 0, 0)$, $P_3(a/2, 0, 0)$, $P_4(3a/4, 0, 0)$, and $P_5(a, 0, 0)$, at $t = 0$. (b) Evaluate at the point $(a/2, 0, \pi/0.04)$, at $t = 0, \pi/4\omega, \pi/2\omega, \pi/\omega$.

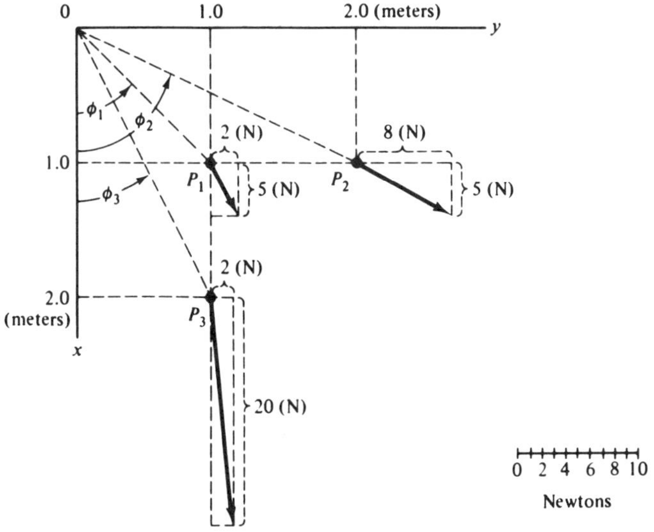

Figure 1-9 Plot of $\overline{F}$ at three points in the $z = 0$ plane (see Example 5). *Note:* The meter and newton scales are different.

1.8 VECTOR MULTIPLICATION

In the development of electromagnetic theory and laws, scientists have devised mathematical operations involving vectors to greatly reduce work and to formulate shorthand notation for use in mathematically expressing theory and laws.

1.8-1 Scalar Times a Vector

If we multiply a vector by a scalar m,

$$m\overline{A} = m\hat{a}_{\overline{A}}|\overline{A}| = m(\hat{x}A_x + \hat{y}A_y + \hat{z}A_z) \tag{1}$$

we can see by (1) that the magnitude of the vector is increased by a factor of m, but its direction is unchanged. Each of the scalar components of $\overline{A}$ are also increased by a factor of m, to yield

$$m\overline{A} = \hat{x}mA_x + \hat{y}mA_y + \hat{z}mA_z \tag{2}$$

Note that the distributive law applies in (2).

1.8-2 Dot Product (Scalar Product)

The dot product is a special multiplication, between two vectors, defined as

$$\boxed{\overline{A} \bullet \overline{B} \triangleq |\overline{A}||\overline{B}| \cos \beta_{\overline{A}}^{\overline{B}} \quad \text{(smallest)}} \tag{3}$$

where β is the smallest angle between vectors $\overline{A}$ and $\overline{B}$, as shown in Fig. 1-10. The symbol $\triangleq$, found in (3), is used in this text to mean "equal to by definition," and thus the equation involving this symbol will be referred to as the *defining equation*. This multiplication between two vectors takes its name from the *dot* between $\overline{A}$ and $\overline{B}$ used in the mathematical symbolism of this operation. Some texts call this multiplication *scalar product* since the result yields a *scalar* [see (3)]. One must admit that the left side of (3) is a shorthand form of the right side.

One of many uses of the dot product is found in expressing the work (energy) expended when a constant force $\overline{F}$ (N) acts over a distance $\overline{\ell}$ (m):

$$\text{work} = |\overline{F}||\overline{\ell}| \cos \beta = \overline{F} \bullet \overline{\ell} \qquad \text{(J)} \tag{4}$$

where β is the smallest angle between the force vector $\overline{F}$ and the distance vector $\overline{\ell}$.

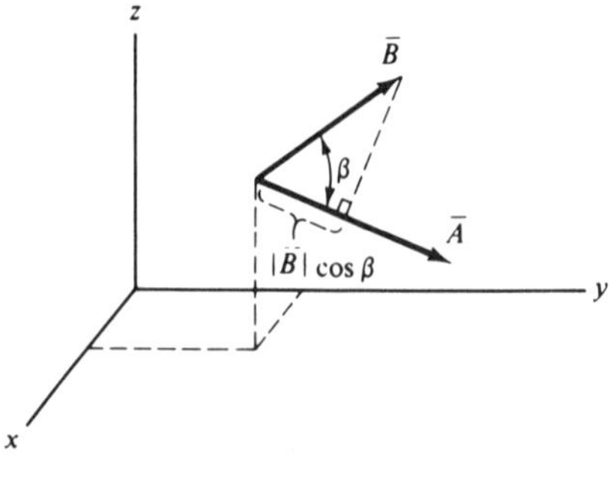

Figure 1-10 The angle β is the smallest angle between $\overline{A}$ and $\overline{B}$ in the dot product, and $|\overline{B}| \cos \beta$ equals the scalar projection of $\overline{B}$ onto $\overline{A}$.

From the defining equation (3), $\overline{B} \bullet \overline{A} = |\overline{B}||\overline{A}| \cos \beta$. This indicates that the commutative law applies to the dot product:

$$\boxed{\overline{A} \bullet \overline{B} = \overline{B} \bullet \overline{A}} \tag{5}$$

Not readily obvious, but readily proven graphically, is the distributive law:

$$\boxed{\overline{A} \bullet (\overline{B} + \overline{C}) = \overline{A} \bullet \overline{B} + \overline{A} \bullet \overline{C}} \tag{6}$$

Equation (6) can be extended to

$$\begin{aligned}(\overline{A}_1 + \overline{A}_2 + \ldots + \overline{A}_n) \bullet (\overline{B}_1 + \overline{B}_2 + \ldots + \overline{B}_m) \\ = \overline{A}_1 \bullet \overline{B}_1 + \overline{A}_1 \bullet \overline{B}_2 + \ldots + \overline{A}_1 \bullet \overline{B}_m + \ldots \\ + \overline{A}_n \bullet \overline{B}_1 + \overline{A}_n \bullet \overline{B}_2 + \ldots + \overline{A}_n \bullet \overline{B}_m\end{aligned} \tag{7}$$

If we use (3) to evaluate dot products between the unit vectors $\hat{x}$, $\hat{y}$, and $\hat{z}$, we obtain

$$\boxed{\begin{aligned}\hat{x} \bullet \hat{x} = \hat{y} \bullet \hat{y} = \hat{z} \bullet \hat{z} = 1 \\ \hat{x} \bullet \hat{y} = \hat{x} \bullet \hat{z} = \hat{y} \bullet \hat{z} = 0\end{aligned}} \tag{8}$$

Dot products between unit vectors of rectangular, cylindrical, and spherical coordinate systems are found in Table B-1 of Appendix B.

In (3), let us expand $\overline{A}$ and $\overline{B}$ in component form and apply the distributive law (7) to obtain

$$\begin{aligned}\overline{A} \bullet \overline{B} &= (\hat{x}A_x + \hat{y}A_y + \hat{z}A_z) \bullet (\hat{x}B_x + \hat{y}B_y + \hat{z}B_z) \\ &= (\hat{x} \bullet \hat{x})A_xB_x + (\hat{x} \bullet \hat{y})A_xB_y + (\hat{x} \bullet \hat{z})A_xB_z + (\hat{y} \bullet \hat{x})A_yB_x \\ &\quad + (\hat{y} \bullet \hat{y})A_yB_y + (\hat{y} \bullet \hat{z})A_yB_z + (\hat{z} \bullet \hat{x})A_zB_x + (\hat{z} \bullet \hat{y})A_zB_y \\ &\quad + (\hat{z} \bullet \hat{z})A_zB_z\end{aligned} \tag{9}$$

Through the use of (8), eq. (9) reduces to

$$\boxed{\overline{A} \bullet \overline{B} = A_xB_x + A_yB_y + A_zB_z} \tag{10}$$

From Fig. 1-10, we can see that $|\overline{B}| \cos \beta$ is equal to the scalar projection of $\overline{B}$ onto the direction of $\overline{A}$. Using (3), we obtain this projection to be

$$\boxed{|\overline{B}| \cos \beta = \frac{\overline{A} \bullet \overline{B}}{|\overline{A}|}} \tag{11}$$

The scalar projection of $\overline{A}$ onto the direction of $\overline{B}$ can also be found from (3):

$$|\overline{A}| \cos \beta = \frac{\overline{A} \bullet \overline{B}}{|\overline{B}|} \tag{12}$$

If in (11) we let $\overline{A} = \hat{x}$, we obtain the scalar projection of $\overline{B}$ onto the direction of $\hat{x}$:

$$|\overline{B}| \cos \beta = \frac{\hat{x} \bullet \overline{B}}{(1)} = \overline{B} \bullet \hat{x} \tag{13}$$

Equation (13) can be extended to the general form

$$\overline{B} \bullet \hat{a} = |\overline{B}| \cos \beta \tag{14}$$

where $|\overline{B}| \cos \beta$ is the scalar projection of $\overline{B}$ onto the direction of $\hat{a}$. The dot product between any two unit vectors $\hat{a}$ and $\hat{b}$ equals, by (3),

$$\hat{a} \bullet \hat{b} = (1)(1) \cos \beta = \cos \beta \tag{15}$$

In terms of scalar projections (15) can be interpreted as the projection of $\hat{a}$ onto the direction of $\hat{b}$ or the projection of $\hat{b}$ onto the direction of $\hat{a}$.

The angle β between $\overline{A}$ and $\overline{B}$ can be found from (3) to equal

$$\beta = \cos^{-1}\left[\frac{\overline{A} \bullet \overline{B}}{|\overline{A}||\overline{B}|}\right] \tag{16}$$

Example 6

A uniform vector force field $\overline{F} = (\hat{x}5 + \hat{y}2 - \hat{z}2)$ (N) acts on a particle that experiences a vector distance $\overline{\ell} = (\hat{x}3 + \hat{y}3 - \hat{z}4)$ (m). Find the work done by the force field.

Solution. From (4), we have

$$W = \overline{F} \bullet \overline{\ell} = (\hat{x}5 + \hat{y}2 - \hat{z}2) \bullet (\hat{x}3 + \hat{y}3 - \hat{z}4) = (15 + 6 + 8) = 29 \quad \text{(J)}$$

Example 7

Find the scalar projection of $\overline{A} = \hat{x}10 + \hat{y}2 + \hat{z}2$ onto the direction of $\overline{B} = \hat{x}4 - \hat{z}3$.

Solution. From (12), we have

$$\text{projection} = |\overline{A}| \cos \beta = \frac{\overline{A} \bullet \overline{B}}{|\overline{B}|} = \frac{A_x B_x + A_y B_y + A_z B_z}{(B_x^2 + B_y^2 + B_z^2)^{1/2}} = \frac{34}{5} = 6.8$$

Example 8

Find the unit vector $\hat{a}_{\overline{A}}$ in the direction of the vector $\overline{A} = \hat{x}5 + \hat{y}3 - \hat{z}6$.

Solution. From (1), we have

$$\boxed{\frac{1}{|\overline{A}|}\overline{A} = \frac{1}{|\overline{A}|}\hat{a}_{\overline{A}}|\overline{A}| = \hat{a}_{\overline{A}}} \tag{17}$$

$$\hat{a}_{\overline{A}} = \frac{\overline{A}}{|\overline{A}|} = \frac{\hat{x}5 + \hat{y}3 - \hat{z}6}{[5^2 + 3^2 + (-6)^2]^{1/2}} = \hat{x}\frac{5}{\sqrt{70}} + \hat{y}\frac{3}{\sqrt{70}} - \hat{z}\frac{6}{\sqrt{70}}$$

Problem 1.8-1 Use of the dot product to show that the two vectors $\overline{A} = \hat{x}2 - \hat{y}4 + \hat{z}6$ and $\overline{B} = \hat{x}2 + \hat{y}4 + \hat{z}2$ are perpendicular.

Problem 1.8-2 Show that $\overline{C} = \hat{x}4 - \hat{y} - \hat{z}2$ is perpendicular to both $\overline{A}$ and $\overline{B}$ of Prob. 1.8-1. Are there other vectors that are mutually perpendicular to both $\overline{A}$ and $\overline{B}$? How many?

Problem 1.8-3 Show graphically that the distributive law $\overline{A} \cdot (\overline{B} + \overline{C}) = (\overline{A} \cdot \overline{B}) + (\overline{A} \cdot \overline{C})$ applies to the dot product.

Problem 1.8-4 Find the dot product of $\overline{A} = \hat{x}6 \sin 10t + \hat{y}6 \cos 10t$ and $\overline{B} = \hat{y}5 \cos \cdot (10t + \pi/4)$ at (a) $t = 0$; (b) $t = 0.1$ s.

1.8-3 Cross Product (Vector Product)

The cross product is another multiplication between two vectors, defined as

$$\boxed{\overline{A} \times \overline{B} \triangleq \hat{a}_n|\overline{A}|\,|\overline{B}| \sin \beta\angle_{\overline{A}}^{\overline{B}} \quad \text{(smallest)}} \tag{18}$$

where β is the smallest angle between $\overline{A}$ and $\overline{B}$, and $\hat{a}_n$ is a unit vector normal to the plane containing both $\overline{A}$ and $\overline{B}$ and having the direction of the right thumb when the fingers of the right hand rotate from $\overline{A}$ (first vector) to $\overline{B}$ (second vector). The cross product is illustrated in Fig. 1-11. This multiplication between two vectors takes its

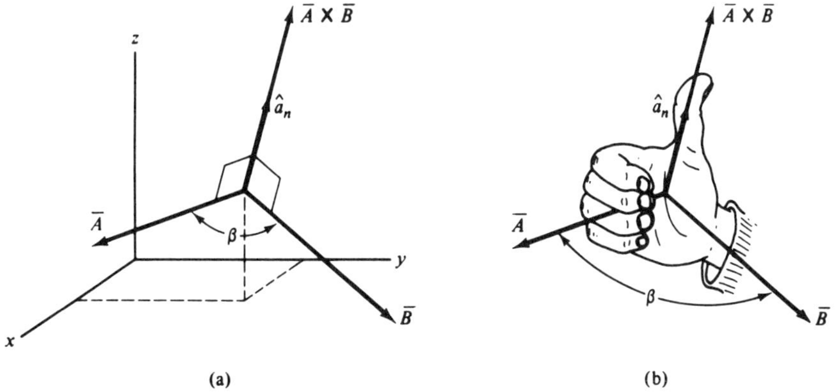

Figure 1-11 (a) Directions of $\overline{A} \times \overline{B}$ and $\hat{a}_n$. (b) Right-hand concept to find $\hat{a}_n$.

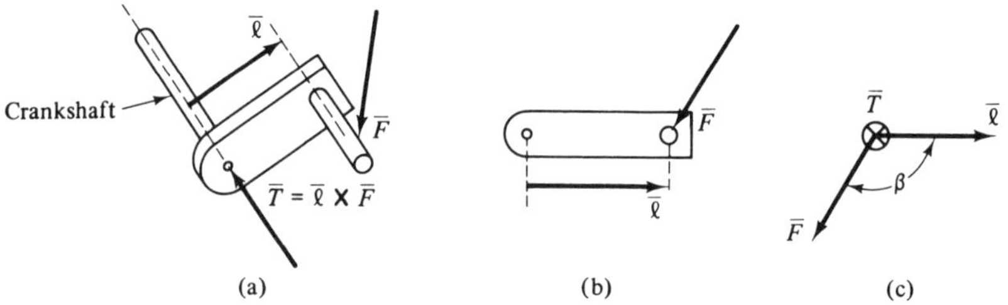

Figure 1-12 Torque produced on a crank by a force $\overline{F}$ acting a distance ℓ from the crankshaft: (a) a three-dimensional view; (b) the vectors $\overline{F}$ and ℓ drawn in the plane of the page; (c) the direction of $\overline{T}$ is into the page along the crankshaft. Direction indicated by $\oplus$.

name from the *cross* between $\overline{A}$ and $\overline{B}$ in the mathematical symbolism of this operation. Some texts call this multiplication a *vector product* since the result yields a *vector* [see (18)].

An example of this multiplication can be found in the evaluation of the vector torque produced by a vector force acting to produce rotation, as shown in Fig. 1-12. The torque $\overline{T}$ produced by the vector force $\overline{F}$, acting on the crank handle at a perpendicular distance $\overline{\ell}$ from the crankshaft axis is

$$\overline{T} = \overline{\ell} \times \overline{F} = \hat{a}_n|\overline{\ell}|\,|\overline{F}| \sin \beta \qquad (\text{N} \cdot \text{m}) \tag{19}$$

Note that $\overline{T}$ is along the crankshaft in Fig. 1-12.

From the definition of $\hat{a}_n$ in (18), it can be seen that

$$\boxed{\overline{A} \times \overline{B} = -\overline{B} \times \overline{A}} \tag{20}$$

and thus the commutative law does not apply to the cross product. It can be shown that the distributive law holds:

$$\boxed{\overline{A} \times (\overline{B} + \overline{C}) = \overline{A} \times \overline{B} + \overline{A} \times \overline{C}} \tag{21}$$

Equation (21) can be extended to

$$\begin{aligned}(\overline{A}_1 + \overline{A}_2 + \ldots + \overline{A}_n) \times (\overline{B}_1 + \overline{B}_2 + \ldots + \overline{B}_m) \\ = \overline{A}_1 \times \overline{B}_1 + \overline{A}_1 \times \overline{B}_2 + \ldots \overline{A}_1 \times \overline{B}_m \\ + \ldots \overline{A}_n \times \overline{B}_1 + \overline{A}_n \times \overline{B}_2 + \ldots + \overline{A}_n \times \overline{B}_m\end{aligned} \tag{22}$$

From (18), the cross products between the unit vectors $\hat{x}$, $\hat{y}$, and $\hat{z}$ are

$$\boxed{\begin{array}{lll} \hat{x} \times \hat{x} = \hat{y} \times \hat{y} = \hat{z} \times \hat{z} = 0 & & \\ \hat{x} \times \hat{y} = \hat{z}, & \hat{z} \times \hat{x} = \hat{y}, & \hat{y} \times \hat{z} = \hat{x} \\ \hat{y} \times \hat{x} = -\hat{z}, & \hat{x} \times \hat{z} = -\hat{y}, & \hat{z} \times \hat{y} = -\hat{x} \end{array}} \tag{23}$$

In (18), let us expand $\overline{A}$ and $\overline{B}$ in component form and apply (22) to obtain

$$\begin{aligned}\overline{A} \times \overline{B} &= (\hat{x}A_x + \hat{y}A_y + \hat{z}A_z) \times (\hat{x}B_x + \hat{y}B_y + \hat{z}B_z) \\ &= \hat{x} \times \hat{x}A_xB_x + \hat{x} \times \hat{y}A_xB_y + \hat{x} \times \hat{z}A_xB_z \\ &\quad + \hat{y} \times \hat{x}A_yB_x + \hat{y} \times \hat{y}A_yB_y + \hat{y} \times \hat{z}A_yB_z \\ &\quad + \hat{z} \times \hat{x}A_zB_x + \hat{z} \times \hat{y}A_zB_y + \hat{z} \times \hat{z}A_zB_z \end{aligned} \tag{24}$$

Through the use of (23), eq. (24) reduces to

$$\boxed{\overline{A} \times \overline{B} = \hat{x}(A_yB_z - A_zB_y) + \hat{y}(A_zB_x - A_xB_z) + \hat{z}(A_xB_y - A_yB_x)} \tag{25}$$

Equation (25) can also be expressed in a determinant form:

$$\boxed{\overline{A} \times \overline{B} = \begin{vmatrix} \hat{x} & \hat{y} & \hat{z} \\ A_x & A_y & A_z \\ B_x & B_y & B_z \end{vmatrix}} \tag{26}$$

Example 9

The vector force $\overline{F}$ on a conductor of length $\overline{\ell}$ carrying a current I in a uniform vector magnetic field $\overline{B}$ is equal to $\overline{F} = I\overline{\ell} \times \overline{B}$ (N). If $I = 10$ A, $\overline{\ell} = (\hat{x}2 + \hat{y}3)$ (m), and $\overline{B} = (\hat{y}10^{-2} + \hat{z}10^{-3})$ tesla (T), find $\overline{F}$ in newtons. The direction $\overline{\ell}$ is that of I.

Solution

$$\begin{aligned}\overline{F} = I\overline{\ell} \times \overline{B} &= 10[(\hat{x}2 + \hat{y}3) \times (\hat{y}10^{-2} + \hat{z}10^{-3})] \\ &= 10[\hat{x} \times \hat{y}2(10^{-2}) + \hat{x} \times \hat{z}2(10^{-3}) + \hat{y} \times \hat{y}3(10^{-2}) + \hat{y} \times \hat{z}3(10^{-3})] \\ &= 10[\hat{z}2(10^{-2}) - \hat{y}2(10^{-3}) + 0 + \hat{x}3(10^{-3})] \\ &= [\hat{x}3(10^{-2}) - \hat{y}2(10^{-2}) + \hat{z}2(10^{-1})] \quad \text{(N)}\end{aligned}$$

Example 10

Find the angle β between $\overline{A} = \hat{x}4 - \hat{y}2 - \hat{z}3$ and $\overline{B} = \hat{x}3 - \hat{y}4$, using the cross product.

Solution. From (18),

$$|\overline{A} \times \overline{B}| = |\overline{A}||\overline{B}| \sin \beta$$

Solving for β, we have

$$\boxed{\beta = \sin^{-1}\left[\frac{|\overline{A} \times \overline{B}|}{|\overline{A}||\overline{B}|}\right]}$$

$$\beta = \sin^{-1}\left[\frac{|\hat{x}(A_yB_z - A_zB_y) + \hat{y}(A_zB_x - A_xB_z) + \hat{z}(A_xB_y - A_yB_x)|}{(A_x^2 + A_y^2 + A_z^2)^{1/2}(B_x^2 + B_y^2 + B_z^2)^{1/2}}\right]$$

$$= \sin^{-1}\left[\frac{[(0 - 12)^2 + (-9 - 0)^2 + (-16 - (-6))^2]^{1/2}}{(16 + 4 + 9)^{1/2}(9 + 16)^{1/2}}\right]$$

$$= \sin^{-1}\left[\frac{(144 + 81 + 100)^{1/2}}{(29)^{1/2}(25)^{1/2}}\right] = \sin^{-1}\left[\frac{18.03}{26.9}\right] = 42°$$

Problem 1.8-5 Is

$$\begin{vmatrix} \hat{x} & \hat{y} & \hat{z} \\ A_x & A_y & A_z \\ B_x & B_y & B_z \end{vmatrix} = \begin{vmatrix} \hat{x} & \hat{y} & \hat{z} \\ B_x & B_y & B_z \\ A_x & A_y & A_z \end{vmatrix} = \begin{vmatrix} \hat{x} & \hat{z} & \hat{y} \\ B_x & B_z & B_y \\ A_x & A_z & A_y \end{vmatrix} ?$$

Explain.

Problem 1.8-6 Find the unit vector $\hat{a}_n$ in the direction of $\overline{B} \times \overline{A}$ when $\overline{A} = \hat{x} + \hat{y}2 + \hat{z}3$, and $\overline{B} = \hat{x}3 + \hat{y}2 + \hat{z}$.

Problem 1.8-7 Find the vector torque $\overline{T}$ when a vector force $\overline{F} = \hat{x}3 + \hat{y}4$ (N) is applied to a crank handle located a vector distance $\overline{\ell} = +\hat{x}5$ (m) from the crankshaft, as shown in Fig. 1-12.

1.8-4 Triple Scalar Product

When three vectors are multiplied as

$$\boxed{\overline{A} \times \overline{B} \bullet \overline{C}} \tag{27}$$

it can be seen that the only defined sequence of multiplication is

$$\boxed{\overline{A} \times \overline{B} \bullet \overline{C} = (\overline{A} \times \overline{B}) \bullet C} \tag{28}$$

since $\overline{A} \times (\overline{B} \bullet \overline{C})$ is not defined. Thus, the associative law does not hold for the triple scalar product. The result of (28) is a scalar quantity that can be displayed in a determinant form:

$$\boxed{(\overline{A} \times \overline{B}) \bullet \overline{C} = \begin{vmatrix} A_x & A_y & A_z \\ B_x & B_y & B_z \\ C_x & C_y & C_z \end{vmatrix}} \tag{29}$$

The triple scalar product seems to be quite strange since equality exists if the cyclic order (*ABCABC* . . .) is preserved and no matter where the cross and dot are placed; thus,

$$\begin{aligned}(\overline{A} \times \overline{B}) \bullet \overline{C} &= \overline{A} \bullet (\overline{B} \times \overline{C}) = \overline{B} \bullet (\overline{C} \times \overline{A}) \\ &= (\overline{B} \times \overline{C}) \bullet \overline{A} = \overline{C} \bullet (\overline{A} \times \overline{B}) = (\overline{C} \times \overline{A}) \bullet \overline{B}\end{aligned} \tag{30}$$

Equation (30) can be readily proven through the use of the determinant form, eq. (29).

1.8-5 Triple Vector Product

When three vectors are multiplied as

$$\boxed{(\overline{A} \times \overline{B}) \times \overline{C}} \tag{31}$$

it can be seen that the result is a vector. If we multiply $(\overline{B} \times \overline{C})$ first and then cross $\overline{A}$ with this, we can show that, in general,

$$\boxed{(\overline{A} \times \overline{B}) \times \overline{C} \neq \overline{A} \times (\overline{B} \times \overline{C})} \tag{32}$$

Inequality in (32) indicates that the associative law does not hold for the triple vector product. This product (31) can also be written as

$$\boxed{(\overline{A} \times \overline{B}) \times \overline{C} = \overline{B}(\overline{C} \bullet \overline{A}) - \overline{A}(\overline{C} \bullet \overline{B})} \tag{33}$$

Due to the inequality in (32), it can be seen that it is important to know which multiplication is to be carried out first. In the development of theory and proofs in electromagnetics, two of the vectors will be crossed first and then this result crossed with another vector. Thus, the parentheses must be placed about the first multiplication to be performed.

Problem 1.8-8 By expanding $\overline{A}$, $\overline{B}$, and $\overline{C}$ in component form, prove that

$$(\overline{A} \times \overline{B}) \times \overline{C} = \overline{B}(\overline{C} \bullet \overline{A}) - \overline{A}(\overline{C} \bullet \overline{B}) \tag{33}$$

Problem 1.8-9 Through the use of the determinant for the triple scalar product (29) show that

$$(\overline{A} \times \overline{B}) \bullet \overline{C} = \overline{A} \bullet (\overline{B} \times \overline{C}) = (\overline{C} \times \overline{A}) \bullet \overline{B}$$

1.9 CYLINDRICAL COORDINATE SYSTEM

In a cylindrical coordinate system, a point in space is uniquely defined by three variables, r_c, ϕ, and z, as seen in Fig. 1-13(a). The ranges on the variables are $r_c \geqq 0$, $0 \leqq \phi < 2\pi$, and $-\infty < z < +\infty$. The unit vectors are $\hat{r}_c$, $\hat{\phi}$, and $\hat{z}$, in mutually perpendicular directions, as shown in Fig. 1-13(b). A right-hand cylindrical coordinate system exists if the thumb points in the direction of $\hat{z}$ when the fingers of the right hand rotate from $\hat{r}_c$ to $\hat{\phi}$ directions. The ordered sequence of variables is r_c, ϕ, and z.

A vector force $\overline{F}$, discrete or field, will have projections onto the $\hat{r}_c$, $\hat{\phi}$, and $\hat{z}$ directions as suggested in Fig. 1-13(c). This vector, at $P(r_c, \phi, z)$, is thus expressed in cylindrical component form as

$$\boxed{\overline{F} = \hat{r}_c F_{r_c} + \hat{\phi} F_\phi + \hat{z} F_z} \tag{1}$$

where F_{r_c}, F_ϕ, and F_z are scalar projections of $\overline{F}$ onto $\hat{r}_c$, $\hat{\phi}$, and $\hat{z}$ directions, respectively. The magnitude of $\overline{F}$ is equal to $|\overline{F}| = F = (F_{r_c}^2 + F_\phi^2 + F_z^2)^{1/2}$.

If we keep r_c constant and allow ϕ and z to vary, a cylindrical surface will be generated by the point as shown in Fig. 1-13(d). In the same manner, surfaces will be generated when ϕ = constant and when z = constant. The intersection of these three surfaces thus locates the point $P(r_c, \phi, z)$.

A line, surface, and volume will be generated when a single variable, two variables, and three variables, respectively, are varied. When these changes are differential,

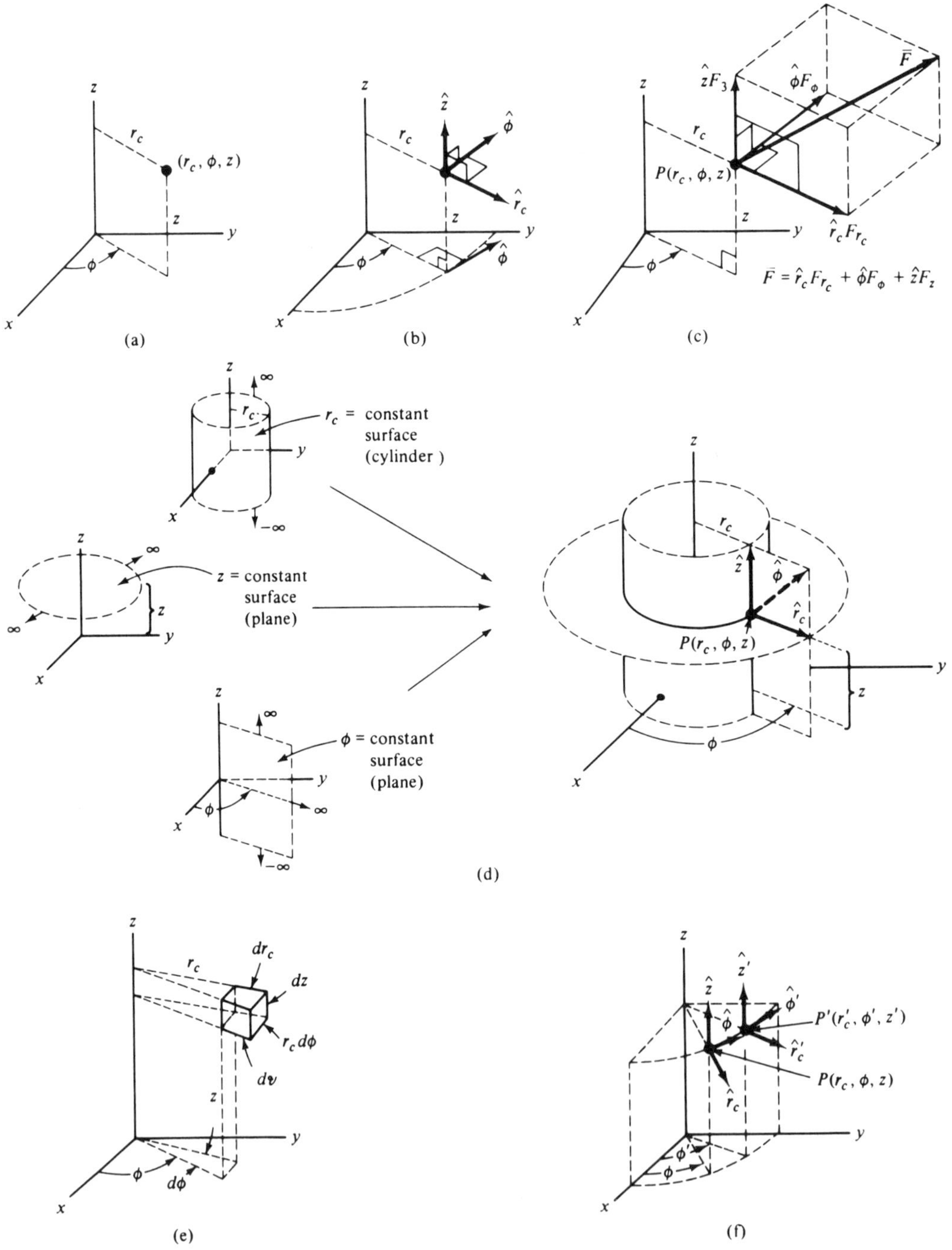

Figure 1-13 Cylindrical coordinate system: (a) location of a point; (b) unit vectors; (c) vector $\overline{F}$ in space; (d) constant r_c, z, ϕ surfaces; (e) differential volume; (f) unit vectors at two points.

as shown in Fig. 1-13(e), we generate the following differential lines, differential surfaces, and differential volume:

LINE	SURFACE ds	VOLUME dv
dr_c	$(dr_c)(r_c\,d\phi)$	
$(r_c\,d\phi)$	$(dr_c)(dz)$	$(dr_c)(r_c\,d\phi)(dz)$
dz	$(dz)(r_c\,d\phi)$	

(2)

The differential vector path length $\overline{d\ell}$ over a general distance equals

$$\overline{d\ell} = \hat{r}_c\,dr_c + \hat{\phi}r_c\,d\phi + \hat{z}\,dz \tag{3}$$

A close examination of the unit vectors at two different points, as shown in Fig. 1-13(f), shows that the unit vectors $\hat{r}_c$ and $\hat{\phi}$ are functions of ϕ, while $\hat{z}$ is still a constant vector. The unit vectors $\hat{r}_c$ and $\hat{\phi}$ can be related to rectangular unit vectors as

$$\hat{r}_c = \hat{r}_c(\phi) = \hat{x}\cos\phi + \hat{y}\sin\phi \tag{4}$$

$$\hat{\phi} = \hat{\phi}(\phi) = \hat{x}(-\sin\phi) + \hat{y}\cos\phi \tag{5}$$

Thus, the angle ϕ must be known if we expect to specify $\hat{r}_c$ and $\hat{\phi}$ since they are functions of ϕ. It should be noted again that $\hat{x}$, $\hat{y}$, and $\hat{z}$ are constant unit vectors.

A position vector can be expressed by only two components in the cylindrical coordinate system. From Fig. 1-14, it can be seen that the projection of a position vector in the $\hat{\phi}$ direction is zero; thus,

$$\overline{R}_{p(\text{cyl})} = \hat{r}_c r_c + \hat{z}z \tag{6}$$

Example 11

Show graphically that $\hat{r}_c = \hat{x}\cos\phi + \hat{y}\sin\phi$ at some general point $P(r_c, \phi, z)$.

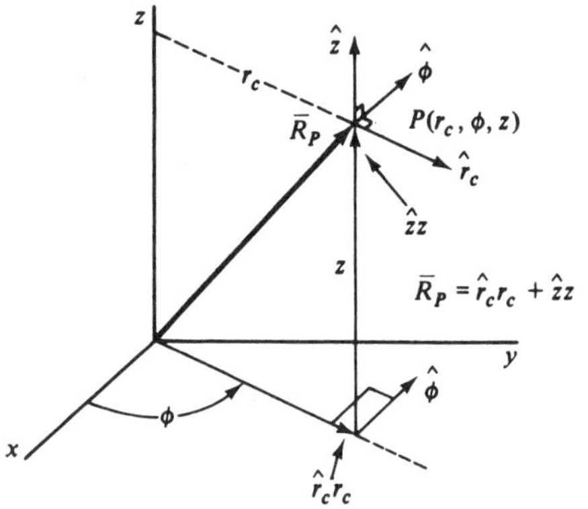

Figure 1-14 Position vector $\overline{R}_P$ in the cylindrical coordinate system. *Note:* The position vector has no projection onto the $\hat{\phi}$ direction in the cylindrical coordinate system.

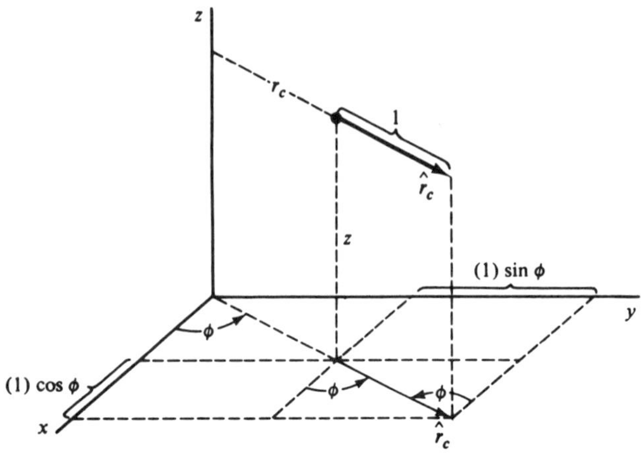

Figure 1-15 Projection of $\hat{r}_c$ onto the x and y axes (see Example 11).

Solution. From Fig. 1-15, the projections of $\hat{r}_c$ onto the x and y axes are $\cos\phi$ and $\sin\phi$, respectively; thus, $\hat{r}_c = \hat{x}\cos\phi + \hat{y}\sin\phi$.

Example 12

A velocity vector field $\overline{v} = (\hat{x}5x^2 + \hat{y}2y^2 - \hat{z}4z^2)$ (m s^{-1}) is found at $P_{\text{rec}}(1, 1, 0)$. Find the the projection of $\overline{v}$ onto the $\hat{r}_c$ direction.

Solution. Let us solve this problem graphically and then through the use of the dot product. Let us first evaluate $\overline{v}\,|_{(1,1,0)}$ and then sketch the vector, as shown in Fig. 1-16(a). Thus, Thus,

$$\overline{v}\,|_{(1,1,0)} = (\hat{x}5 + \hat{y}2) \qquad (\text{m } s^{-1})$$

We can more readily see the projection of $\overline{v}$ onto the $\hat{r}_c$ direction from Fig. 1-16(b), where the x-y plane is in the plane of the page. The projection of $\overline{v}$ onto $\hat{r}_c$ is equal to the sum of the projections of the components of $\overline{v}$ onto $\hat{r}_c$. From Fig. 1-16(b), we have

$$\text{projection} = v_{r_c} = (5\cos\phi + 2\sin\phi)\,|_{\phi=\pi/4} = 4.95 \text{ m } s^{-1}$$

Now, if we form

$$\overline{v} \bullet \hat{r}_c\,|_{\phi=\pi/4} = v_{r_c} = (\hat{x}5 + \hat{y}2) \bullet (\hat{x}\cos\phi + \hat{y}\sin\phi)\,|_{\phi=\pi/4}$$
$$\text{projection} = (5\cos\phi + 2\sin\phi)\,|_{\phi=\pi/4} = 4.95 \ (\text{m } s^{-1})$$

Note that ϕ is the angle location of the point where $\overline{v}$ is evaluated and not the angle that $\overline{v}$ makes with the x axis.

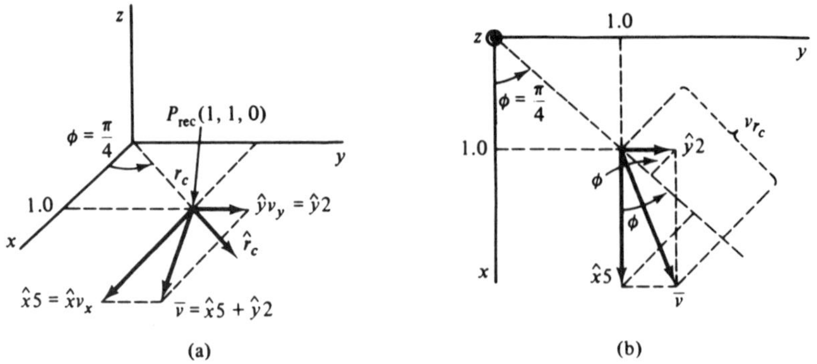

Figure 1-16 (a) Display of $\overline{v}$. (b) View of $\overline{v}$ in the x-y plane drawn in the plane of the page. The z axis is out of the page and is noted by the circled dot ⊙. See Example 12.

Problem 1.9-1 Show graphically that $\hat{\phi} = \hat{x}(-\sin\phi) + \hat{y}\cos\phi$ at some general point $P(r_c, \phi, z)$.

Problem 1.9-2 Evaluate at the point $P_{\text{cyl}}(2, \pi/6, 5)$ the following: (a) $\overline{A} \times \overline{B}$; (b) $|\overline{B}|$; (c) the smaller angle between $\overline{A}$ and $\overline{B}$, where $\overline{A} = \hat{r}_c 3r_c^2 \sin\phi + \hat{\phi}4r_c\cos\phi + \hat{z}r_c$ and $\overline{B} = \hat{r}_c 4r_c \sin\phi + \hat{z}z^3\cos\phi$.

1.10 SPHERICAL COORDINATE SYSTEM

In a spherical coordinate system, a point in space is uniquely defined by three variables, r_s, θ, and ϕ, as seen in Fig. 1-17(a). The ranges on the variables are $r_s \geqq 0$, $0 \leqq \theta \leqq \pi$, and $0 \leqq \phi \leqq 2\pi$. The unit vectors are $\hat{r}_s$, $\hat{\theta}$, and $\hat{\phi}$ in mutually perpendicular directions, as shown in Fig. 1-17(b). A right-hand spherical coordinate system exists if the thumb points in the direction of $\hat{\phi}$ when the fingers of the right hand rotate from $\hat{r}_s$ to $\hat{\theta}$ directions. The ordered sequence of variables is r_s, θ, ϕ.

A vector force $\overline{F}$, discrete or field, will have projections onto the $\hat{r}_s$, $\hat{\theta}$, and $\hat{\phi}$ directions, as shown in Fig. 1-17(c). In the spherical coordinate system, $\overline{F}$ is expressed in spherical component form as

$$\boxed{\overline{F} = \hat{r}_s F_{r_s} + \hat{\theta}F_\theta + \hat{\phi}F_\phi} \tag{1}$$

where F_{r_s}, F_θ, and F_ϕ are scalar projections of $\overline{F}$ onto $\hat{r}_s$, $\hat{\theta}$, and $\hat{\phi}$ directions, respectively. It should be noted that ϕ and $\hat{\phi}$ are defined exactly as in the cylindrical coordinate system. The unit vector $\hat{r}_s$ and r_s are quite different from $\hat{r}_c$ and r_c in cylindrical coordinates. In this part of the text, we shall be careful to indicate the difference through the use of the s subscript in spherical and c in the cylindrical coordinate systems.

If we keep r_s constant and allow θ and ϕ to vary, a spherical surface will be generated by the point, as shown in Fig. 1-17(d). In the same manner, conical and planar surfaces will be generated when θ = constant and ϕ = constant, respectively. The intersection of these three surfaces thus locates the point $P(r_s, \theta, \phi)$.

The differential line lengths, differential surfaces, and differential volume are found to be [see Fig. 1-17(e)]

LINE	SURFACE ds	VOLUME dv
dr_s	$(dr_s)(r_s \sin\theta\, d\phi)$	$(dr_s)(r_s d\theta)(r_s \sin\theta\, d\phi)$
$r_s \sin\theta\, d\phi$	$(dr_s)(r_s d\theta)$	
$r_s d\theta$	$(r_s d\theta)(r_s \sin\theta\, d\phi)$	

(2)

The differential vector length $\overline{d\ell}$ over a general distance equals

$$\boxed{\overline{d\ell} = \hat{r}_s dr_s + \hat{\theta}r_s d\theta + \hat{\phi}r_s \sin\theta(d\phi)} \tag{3}$$

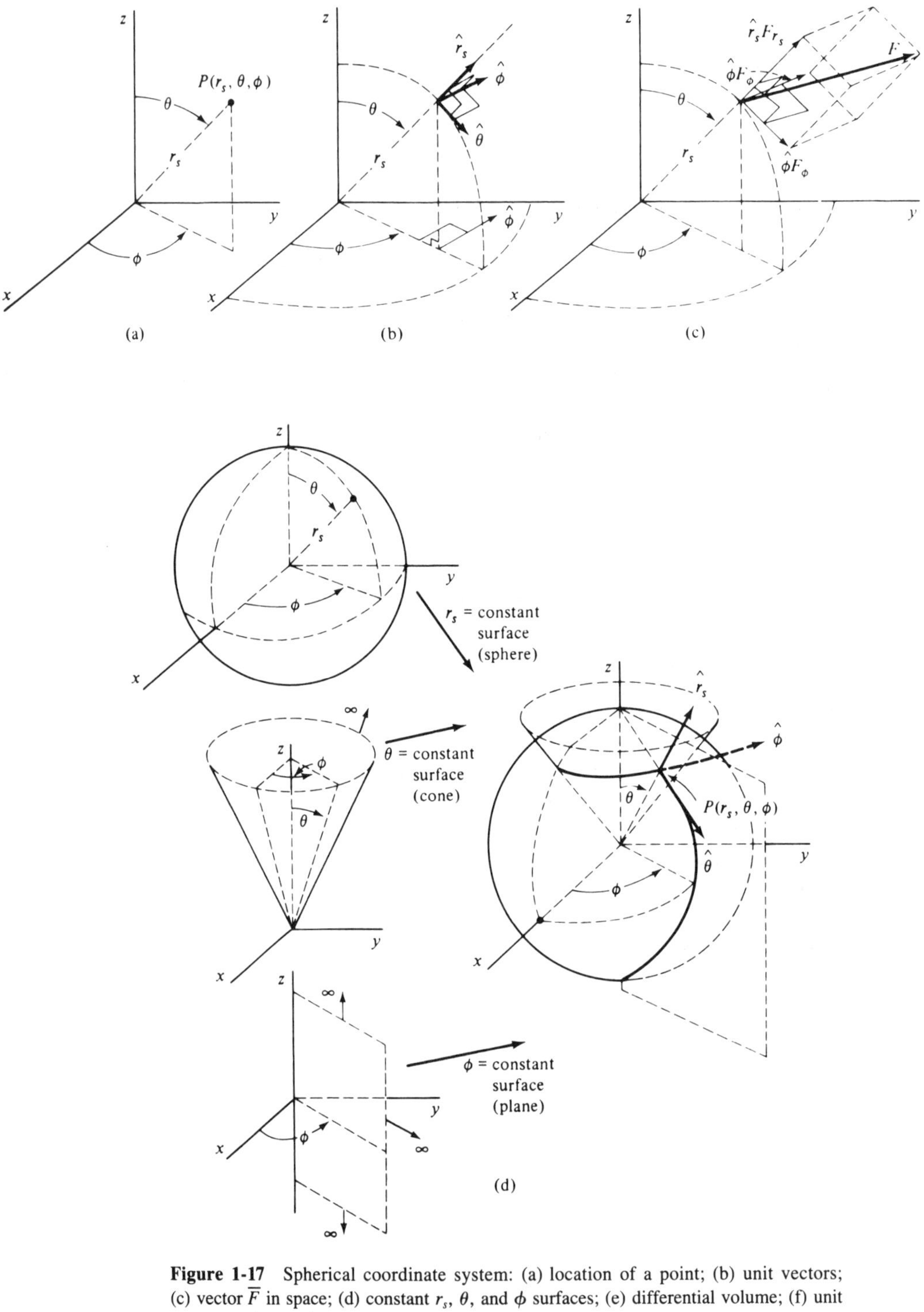

Figure 1-17 Spherical coordinate system: (a) location of a point; (b) unit vectors; (c) vector $\overline{F}$ in space; (d) constant r_s, θ, and ϕ surfaces; (e) differential volume; (f) unit vectors at three points.

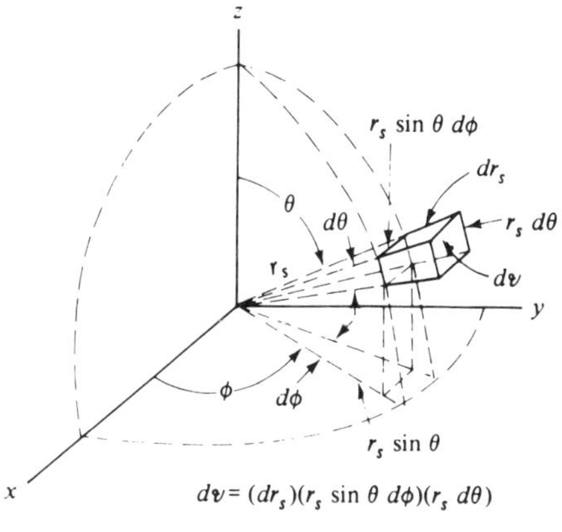

(e)

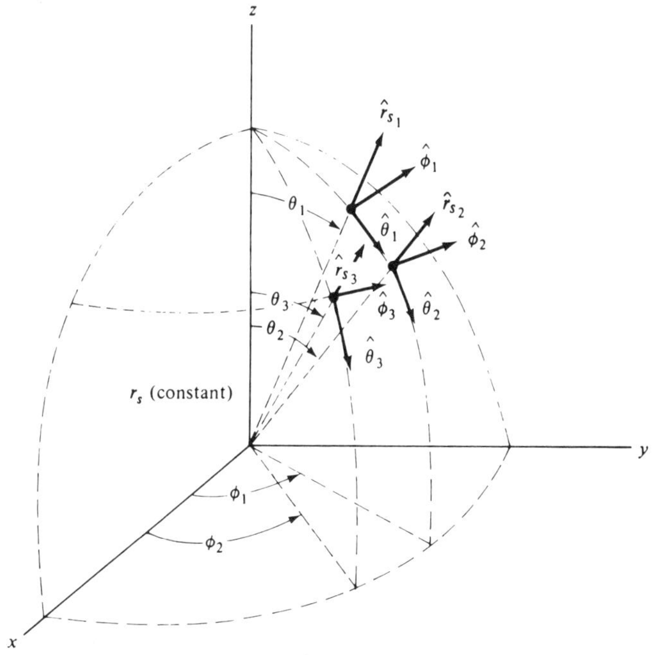

(f)

Figure 1-17 *(cont.)*

From Fig. 1-17(f), it can be seen that $\hat{r}_s$ and $\hat{\theta}$ are functions of θ and ϕ, while $\hat{\phi}$ is a function of ϕ only. The unit vectors $\hat{r}_s$, $\hat{\theta}$, and $\hat{\phi}$ can be expressed as

$$\hat{r}_s = \hat{r}_s(\theta, \phi) = \hat{x} \sin\theta \cos\phi + \hat{y} \sin\theta \sin\phi + \hat{z} \cos\theta \quad (4)$$
$$\hat{\theta} = \hat{\theta}(\theta, \phi) = \hat{x} \cos\theta \cos\phi + \hat{y} \cos\theta \sin\phi + \hat{z}(-\sin\theta) \quad (5)$$
$$\hat{\phi} = \hat{\phi}(\phi) = \hat{x}(-\sin\phi) + \hat{y} \cos\phi \quad (1.9\text{-}5)$$

Thus, the angles θ and ϕ must be known if we expect to specify $\hat{r}_s$ and $\hat{\theta}$, while the ϕ must be known to specify $\hat{\phi}$.

A position vector can be expressed by only one component in the spherical coordinate system. From Fig. 1-17(b), it can be seen that

$$\boxed{\overline{R}_{P(\text{sph})} = \hat{r}_s r_s} \tag{6}$$

Cylindrical and spherical coordinate systems greatly simplify the mathematics when the boundary or boundaries of a problem coincide with the constant surface or surfaces found in the coordinate systems; for example, cones, spheres, and planes.

In our study of the cylindrical and spherical coordinate systems, the angles ϕ and θ were measured from the x and z axes, respectively. It is for this reason that one should always locate the right-hand rectangular coordinate axes before attempting to locate a point in the cylindrical or spherical coordinate systems. By now it should be apparent that the right-hand coordinate systems are defined through an ordered sequence of variables, called *cyclic order*. This cyclic order of variables is $xyzxyzx\ldots$ in the rectangular system, $r_c\phi z r_c \phi z r_c\ldots$ in the cylindrical system, and $r_s\theta\phi r_s\theta\phi r_s\ldots$ in the spherical system. The right-hand system designation comes from the axes directions being dictated by the right hand as the fingers turn, in cyclic order, from the $+x$ axis to the $+y$ axis, while the thumb will point in the $+z$ axis direction in the rectangular coordinate system. This also applies to the cylindrical and spherical systems. Note that the axes, or unit vector directions, are mutually perpendicular in each of the three systems discussed.

The right-hand systems can also be defined by the cross product of the coordinate system unit vectors taken in cyclic order; thus,

$$\boxed{\begin{array}{llll} \hat{x} \times \hat{y} = \hat{z} & \text{or} & \hat{y} \times \hat{z} = \hat{x} \text{ etc.} & \text{(rectangular)} \\ \hat{r}_c \times \hat{\phi} = \hat{z} & \text{or} & \hat{\phi} \times \hat{z} = \hat{r}_c \text{ etc.} & \text{(cylindrical)} \\ \hat{r}_s \times \hat{\theta} = \hat{\phi} & \text{or} & \hat{\theta} \times \hat{\phi} = \hat{r}_s \text{ etc.} & \text{(spherical)} \end{array}} \tag{7}$$

Point designations also involve cyclic order as $P_{\text{rec}}(x, y, z)$, $P_{\text{cyl}}(r_c, \phi, z)$, and $P_{\text{sph}}(r_s, \theta, \phi)$ in the rectangular, cylindrical, and spherical systems, respectively.

In the transformation of vectors from one coordinate system to another, and in many other applications, we will have need for dot products between unit vectors and variable relationships between coordinate systems. Table B-1 of Appendix B contains dot products between unit vectors, while Table B-2 gives the variable relationships between variables in the rectangular, cylindrical, and spherical coordinate systems.

Example 13

Evaluate $\hat{r}_s \cdot \hat{x}$: (a) graphically; (b) through the use of eq. (4).

Solution. (a) from Fig. 1-18, it can be seen that the projection of $\hat{r}_s$ onto the x-y plane is $\sin\theta$, which projects $\sin\theta\cos\phi$ onto the $\hat{x}$ direction. Since $\hat{r}_s \cdot \hat{x}$ equals the projection of $\hat{r}_s$ onto the $\hat{x}$ direction, $\hat{r}_s \cdot \hat{x} = \sin\theta\cos\phi$.

(b) From eq. (4), we have

$$\hat{r}_s \cdot \hat{x} = (\hat{x}\sin\theta\cos\phi + \hat{y}\sin\theta\sin\phi + \hat{z}\cos\theta) \cdot \hat{x} = \sin\theta\cos\phi$$

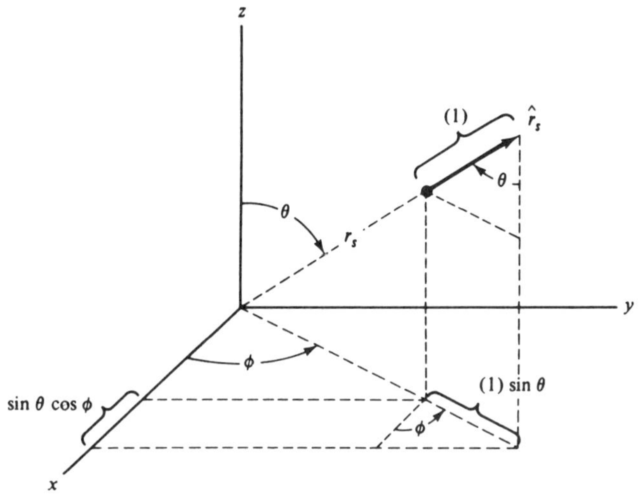

Figure 1-18 Projection of the unit vector $\hat{r}_s$ onto the $\hat{x}$ direction (see Example 13).

Example 14

Find the position vector $\overline{R}_p$ in spherical coordinates of the cylindrical point $P_{\text{cyl}}(3, \pi/4, 3)$.

Solution. From Fig. 1-19, it can be seen that $r_s = \sqrt{(3)^2 + (3)^2} = \sqrt{18}$. Thus, from eq. (6) we have

$$\overline{R}_P = \hat{r}_s r_s = \hat{r}_s \sqrt{18}\,\Big|_{\substack{\theta=\pi/4 \\ \phi=\pi/4}}$$

where $\theta = \tan^{-1}(r_c/z) = \pi/4$.

Problem 1.10-1 Evaluate $\hat{\theta} \cdot \hat{y}$: (a) graphically; (b) through the use of eq. (5).

Problem 1.10-2 In a given region, the vector field is found to be $\overline{E} = (10/r_s^3)(\hat{r}_s 2 \cos\theta + \hat{\theta} \sin\theta)$. (a) Evaluate the field at the point $P_{\text{sph}}(3, \pi/2, 0)$, and at $(3, 0, 0)$. (b) Find the unit vector in the direction of $\overline{E}$ in part (a).

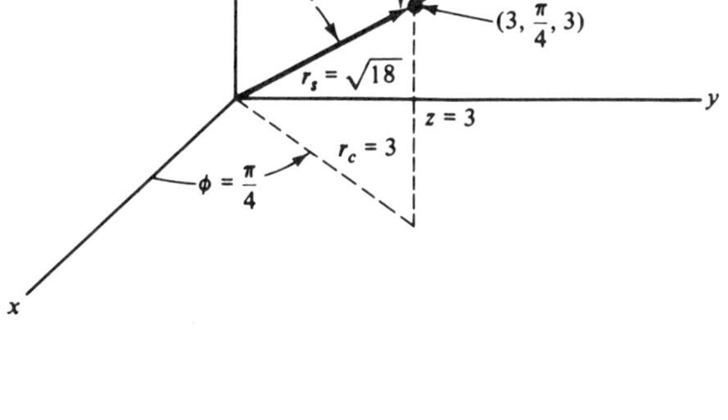

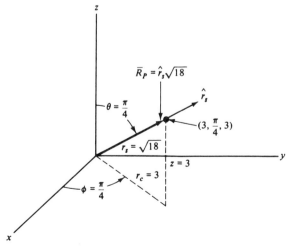

Figure 1-19 Position vector of the cylindrical point $P_{\text{cyl}}(3, \pi/4, 3)$ (see Example 14).

1.11 SURFACE VECTORS

In evaluating scalar quantities such as flux and current, we will find it convenient to express surface as a vector.

> A *surface vector* has the magnitude of the physical surface area it represents and the direction of a unit vector normal to the surface.

An example of a differential surface vector is shown in Fig. 1-20, where r_s is constant. The unit normal vector $\hat{a}_n$ equals $\hat{r}_s$, and the physical differential surface ds equals $r_s^2 \sin\theta\, d\theta\, d\phi$. Here $\overline{ds}$ would represent an outward surface area on the sphere.

Example 15

Evaluate $\oint_s \overline{D} \bullet \overline{ds}$ on the surface of a sphere, when $r_s = 0.2$ and $\overline{D} = \hat{r}_s/r_s^2$. The notation $\oint_s$ indicates a *closed integral*. The sphere is centered at the origin.

Solution

$$\oint_s \overline{D} \bullet \overline{ds} = \int_0^{\phi=2\pi} \int_0^{\theta=\pi} \left(\frac{\hat{r}_s}{r_s^2}\right) \bullet (\hat{r}_s r_s^2 \sin\theta\, d\theta\, d\phi)$$

$$= \int_0^{\phi=2\pi} \int_0^{\theta=\pi} \sin\theta\, d\theta\, d\phi = 4\pi$$

Problem 1.11-1 Evaluate $\int_s \overline{B} \bullet \overline{ds}$ on the surface of a hollow (no area at ends) cylinder, when $\overline{B} = (\hat{r}_c/r_c^2)10^{-2} + \hat{z}r_c^2 z$, radius of cylinder is 2 m, and the cylinder is 2 m in length.

Problem 1.11-2 Evaluate the integral $\oint_s \overline{J} \bullet \overline{ds}$ over the surface of a sphere, centered at the origin, where $r_s = 1$ and $\overline{J} = \hat{z}5|z|r_s^2$.

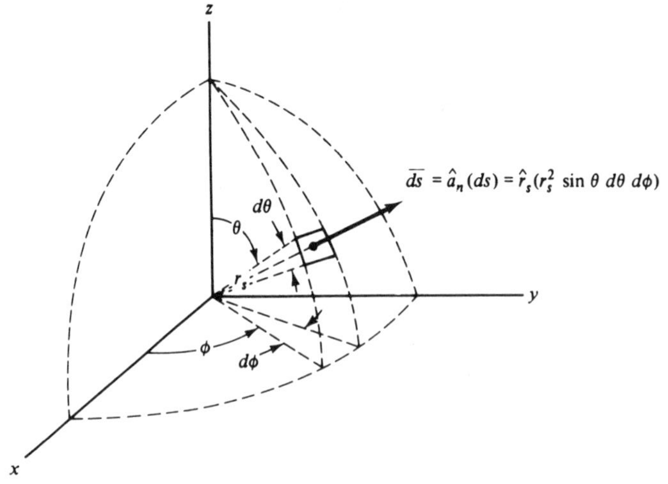

Figure 1-20 Differential surface vector on the surface of a sphere.

1.12 ∇ (del) OPERATOR

To reduce the length of certain equations found in electromagnetics we shall define, in rectangular coordinates, an operator ∇ called *del* or *nabla*.

$$\nabla \triangleq \hat{x}\frac{\partial}{\partial x} + \hat{y}\frac{\partial}{\partial y} + \hat{z}\frac{\partial}{\partial z} \tag{1}$$

The ∇ operator is a vector operator that has no physical meaning or vector direction by itself. We will later find the following operations involving the ∇ operator with scalars and vectors:

$$\text{(Gradient of } f) \triangleq \nabla f = \hat{x}\frac{\partial f}{\partial x} + \hat{y}\frac{\partial f}{\partial y} + \hat{z}\frac{\partial f}{\partial z} \tag{2}$$

$$\text{(Divergence of } \overline{A}) \triangleq \nabla \bullet \overline{A} = \frac{\partial A_x}{\partial x} + \frac{\partial A_y}{\partial y} + \frac{\partial A_z}{\partial z} \tag{3}$$

$$\text{(Curl of } \overline{A}) \triangleq \nabla \times \overline{A} = \hat{x}\left(\frac{\partial A_z}{\partial y} - \frac{\partial A_y}{\partial z}\right) + \hat{y}\left(\frac{\partial A_x}{\partial z} - \frac{\partial A_z}{\partial x}\right) + \hat{z}\left(\frac{\partial A_y}{\partial x} - \frac{\partial A_x}{\partial y}\right) \tag{4}$$

$$\text{(Laplacian of } f) \triangleq \nabla \bullet \nabla f \triangleq \nabla^2 f = \frac{\partial^2 f}{\partial x^2} + \frac{\partial^2 f}{\partial y^2} + \frac{\partial^2 f}{\partial z^2} \tag{5}$$

It should be noted that some of the ∇ operations yield scalars while others yield vectors. In (5), $\nabla \bullet \nabla = \nabla^2$ is called the *Laplacian operator* and is also used to operate on a vector.

$$\text{(Laplacian of } \overline{A}) = \nabla^2 \overline{A} = \hat{x}\nabla^2 A_x + \hat{y}\nabla^2 A_y + \hat{z}\nabla^2 A_z \tag{6}$$

It should be noted that the expanded form of (6) applies only to the rectangular coordinate system.

The gradient, divergence, curl, and Laplacian in the rectangular, cylindrical, and spherical coordinate systems are shown on the inside of the back cover of this book. Algebraic vector identities and those involving the ∇ operator can be found in Table B-4.

The student will, in due time, become acquainted with the meaning of gradient, divergence, curl, and Laplacian.

Problem 1.12-1 If $\overline{E} = -\nabla V$ (V m^{-1}), find $\overline{E}$ when $V = k \cos \theta / r_s^2$ (V). ($\overline{E}$ is the electric field in volts per meter, and V is the electric potential in volts.)

Problem 1.12-2 Find the divergence of $\overline{D}$ at $P_{\text{rec}}(2, 2, 3)$ when $\overline{D} = \hat{x}(xyz) + \hat{y}(x^2yz) + \hat{z}(xy^2z)$.

Problem 1.12-3 Find the divergence of $\overline{E}$ at (1, 2, 3) when $\overline{E} = \hat{x}x^3 + \hat{y}y^2 + \hat{z}z$.

1.13 TRANSFORMATION OF VECTORS

There will be many occasions when it becomes desirable to transform a vector from one coordinate system to another. For a vector to be expressed in a single coordinate system, it must satisfy the following three requirements:

1. Contain unit vectors only of a single coordinate system.
2. Contain scalar projections only onto the unit vector directions of a single coordinate system.
3. Contain variables only of a single coordinate system.

Examples of vectors in the rectangular, cylindrical, and spherical coordinate systems, expressed in general functional form, are:

$$\overline{A}_{\text{rec}} = \hat{x}A_x(x, y, z) + \hat{y}A_y(x, y, z) + \hat{z}A_z(x, y, z) \tag{1}$$

$$\overline{A}_{\text{cyl}} = \hat{r}_c A_{r_c}(r_c, \phi, z) + \hat{\phi}A_\phi(r_c, \phi, z) + \hat{z}A_z(r_c, \phi, z) \tag{2}$$

$$\overline{A}_{\text{sph}} = \hat{r}_s A_{r_s}(r_s, \theta, \phi) + \hat{\theta}A_\theta(r_s, \theta, \phi) + \hat{\phi}A_\phi(r_s, \theta, \phi) \tag{3}$$

In some of the previous work in this chapter, we have written vectors in mixed coordinate form, e.g., $\hat{r}_c = \hat{x}\cos\phi + \hat{y}\sin\phi$. We have seen that this mixed form of expressing a vector has definite advantages for some problems.

From the requirement placed on a vector expressed in a single coordinate system, it is clear that there are three distinct steps to be taken in transforming a vector; they are as follows:

Step 1. Write the general vector expression, using vectors of the new coordinate system [see (1), (2), or (3)].

Step 2. Evaluate the scalar projections onto unit vector directions of the new coordinate system.

Step 3. Change the variables from old to the new coordinate system.

When the vector to be transformed has only one or two components and is simple in form, we can "smell out" through graphical means the new vector without going through the highly organized procedure specified above.

Relationships of scalar projections of a vector between rectangular, cylindrical, and spherical coordinate systems are tabulated in Table B-3. Through the use of eqs. (1), (2), and (3), and Tables B-1, B-2, and B-3, a vector can be transformed between any of the three coordinate systems.

Example 16

Transform the rectangular vector $\overline{A}_{\text{rec}} = \hat{x}x^2y + \hat{y}y^2z + \hat{z}x^2z$ to a cylindrical vector $\overline{A}_{\text{cyl}}$. Note that $\overline{A}_{\text{rec}} = \overline{A}_{\text{cyl}}$.

Solution. Let us first express $\overline{A}_{\text{rec}}$ in a general symbolic form $\overline{A}_{\text{rec}} = \hat{x}A_x + \hat{y}A_y + \hat{z}A_z$ and apply the three steps outlined above.

Step 1

$$\overline{A}_{\text{cyl}} = \hat{r}_c A_{r_c} + \hat{\phi}A_\phi + \hat{z}A_z \tag{4}$$

Step 2

$$\begin{aligned} A_{r_c} &= \overline{A}_{\text{rec}} \bullet \hat{r}_c = (\hat{x}A_x + \hat{y}A_y + \hat{z}A_z) \bullet (\hat{x}\cos\phi + \hat{y}\sin\phi) \\ &= A_x \cos\phi + A_y \sin\phi \end{aligned} \tag{5}$$

$$\begin{aligned} A_\phi &= \overline{A}_{\text{rec}} \bullet \hat{\phi} = (\hat{x}A_x + \hat{y}A_y + \hat{z}A_z) \bullet [\hat{x}(-\sin\phi) + \hat{y}\cos\phi] \\ &= -A_x \sin\phi + A_y \cos\phi \end{aligned} \tag{6}$$

$$A_z = \overline{A}_{\text{rec}} \bullet \hat{z} = (\hat{x}A_x + \hat{y}A_y + \hat{z}A_z) \bullet (\hat{z}) = A_z \tag{7}$$

Now, substitute $A_x = x^2y$, $A_y = y^2z$, and $A_z = x^2z$ into the foregoing expressions for A_{r_c}, A_ϕ, and A_z; thus,

$$A_{r_c} = x^2y\cos\phi + y^2z\sin\phi \tag{8}$$

$$A_\phi = -x^2y\sin\phi + y^2z\cos\phi \tag{9}$$

$$A_z = x^2z \tag{10}$$

Step 3 Change variables in (8), (9), and (10) from x, y, and z to r_c, ϕ, z. Thus, from Table B-2, we have

$$A_{r_c} = (r_c^2\cos^2\phi)(r_c\sin\phi)(\cos\phi) + (r_c^2\sin^2\phi)(z)\sin\phi \tag{11}$$

$$A_\phi = -(r_c^2\cos^2\phi)(r_c\sin\phi)(\sin\phi) + (r_c^2\sin^2\phi)(z)\cos\phi \tag{12}$$

$$A_z = (r_c^2\cos^2\phi)z \tag{13}$$

Reducing (11), (12), and (13) and substituting into (4), we have

$$\begin{aligned} \overline{A}_{\text{cyl}} &= \hat{r}_c(r_c^3\cos^3\phi\sin\phi + r_c^2z\sin^3\phi) \\ &\quad - \hat{\phi}(r_c^3\cos^2\phi\sin^2\phi - r_c^2z\sin^2\phi\cos\phi) + \hat{z}(r_c^2\cos^2\phi)z \end{aligned} \tag{14}$$

Problem 1.13.1 Transform $\overline{A}_{\text{rec}} = \hat{x}x + \hat{y}y + \hat{z}z$ to: (a) $\overline{A}_{\text{cyl}}$; (b) $\overline{A}_{\text{sph}}$ by any means.

Problem 1.13.2 Transform $\overline{A}_{\text{cyl}} = \hat{r}_c r_c^2 + \hat{\phi}\cos\phi$ to: (a) $\overline{A}_{\text{rec}}$; (b) $\overline{A}_{\text{sph}}$.

REVIEW QUESTIONS

1. What is a scalar quantity? *Sec. 1.2*
2. What is a vector quantity? *Sec. 1.3*
3. Are unit vectors mutually perpendicular in the rectangular coordinate system? *Sec. 1.4*
4. Are unit vectors $\hat{x}$, $\hat{y}$, and $\hat{z}$ functions of x, y, z? *Sec. 1.4*
5. What laws of algebra relate to vectors? *Sec. 1.5*
6. A position vector is most easily expressed in what coordinate system? *Eq. (1.10-6), Sec. 1.7-1*
7. What is a vector field? *Sec. 1.7-3*
8. What is the defining equation of the dot product? *Eq. (1.8-3)*
9. What is the defining equation of the cross product? *Eq. (1.8-18)*
10. What is the direction of $\overline{A} \times \overline{B}$? *Sec. 1.8-3*
11. Is the dot product commutative? *Eq. (1.8-5)*
12. Is the cross product commutative? *Eq. (1.8-20)*
13. The unit vectors $\hat{r}_c$ and $\hat{\phi}$ in the cylindrical coordinates are functions of what cylindrical variables? *Eqs. (1.9-4) and (1.9-5)*
14. A position vector in cylindrical coordinates can be expressed by how many components? *Eq. (1.9-6)*
15. The unit vectors $\hat{r}_s$ and $\hat{\theta}$ are functions of what variable? *Eqs. (1.10-4) and (1.9-5)*

16. What is a surface vector? *Sec. 1-11*
17. What is the defining equation for the $\overline{V}$ operator? *Eq. (1.12-1)*
18. In operator form, what is the gradient of f? *Eq. (1.12-2)*
19. How would you express $\hat{r}_c$ in mixed coordinate form? *Sec. 1.13*
20. What three steps are required in vector transformation? *Sec. 1.13*

PROBLEMS

1-1. By expansion in rectangular component form, prove that

$$\overline{A} \times (\overline{B} \times \overline{C}) = (\overline{A} \cdot \overline{C})\overline{B} - (\overline{A} \cdot \overline{B})\overline{C}$$

1-2. Given two vectors, $\overline{A} = \hat{x} - \hat{y}2 + \hat{z}$ and $\overline{B} = \hat{x} - \hat{y}2 + \hat{z}$, find: (a) $\overline{A} \cdot \overline{B}$; (b) $\overline{A} \times \overline{B}$; (c) $\overline{B} \times \overline{A}$.

1-3. At the point $P_{cyl}(2, \pi/3, 4)$, evaluate: (a) $\overline{A} \times \overline{B}$; (b) $|\overline{A}|$; (c) the angle β between $\overline{A}$ and $\overline{B}$, where $\overline{A} = \hat{r}_2 2r_c \cos^2\phi + \phi 3z \sin\phi$ and $\overline{B} = \hat{r}_c 3r_c^2 \sin\phi + \hat{z}z^2 \cos\phi$.

1-4. Given two vectors, $\overline{A} = \hat{x}2 + \hat{y} + \hat{z}5$ and $\overline{B} = \hat{r}_s 4$, with $\phi = 90°$, $\theta = 45°$, find (a) $\overline{A} + \overline{B}$; (b) $\overline{A} \cdot \overline{B}$; (c) $\overline{A} \times \overline{B}$. Express your results in rectangular coordinates.

1-5. Find $\overline{A} \times \overline{B}$ and the unit vector in the direction of $\overline{A} \times \overline{B}$ when $\overline{A} = \hat{r}_c 4 + \hat{\phi}2$ and $\overline{B} = \hat{r}_c 4 - \hat{\phi}2$ in cylindrical coordinates.

1-6. In a given region, the vector field is found to be $\overline{E} = (10/r_s^3)(\hat{r}_s 2 \cos\theta \sin\phi + \hat{\theta} \sin\theta \cos\phi + \hat{\phi}\, r_s)$. (a) Evaluate the field at the point $P_{sph}(3, \pi/3, \pi/4)$. (b) Find the unit vector in the direction of **E** in part (a).

1-7. In cylindrical coordinates, $P = P_{cyl}(1, 90°, 1)$ and $Q = Q_{cyl}(0, 0°, 1)$. (a) Determine the length of line connecting P and Q. (b) Find the unit vector from P toward Q.

1-8. A certain field is described by the vector $\overline{E} = \hat{r}_c 5 + \hat{z}2$ in cylindrical coordinates. Find: (a) $|\overline{E}|$; (b) the unit vector in the direction of $\overline{E}$; (c) $\overline{E}$ in rectangular coordinates at $\phi = 135°$.

1-9. Given the vectors $\overline{A}$, $\overline{B}$, and $\overline{C}$, where $\overline{A} = \hat{x} + \hat{y} + \hat{z}$, $\overline{B} = \hat{x} + \hat{y}3 + \hat{z}2$, and $\overline{C} = \hat{x}3 + \hat{y}2 + \hat{z}$, find $\overline{A} \times \overline{B} \cdot \overline{C}$, $\overline{A} \cdot \overline{B} \times \overline{C}$, $\overline{C} \cdot \overline{A} \times \overline{B}$, and $\overline{A} \cdot \overline{C} \times \overline{B}$. Are they the same?

1-10. Show through component expansion and differentiation, in the rectangular coordinate system, that the divergence of the curl of any vector A equals zero; i.e., $\nabla \cdot (\nabla \times \overline{A}) = 0$.

1-11. Show that $\nabla \times \nabla f = 0$ for any scalar $f = f(x, y, z)$ by component expansion and differentiation required.

1-12. Express the position vector $\overline{R}_p$ for the point $P_{rec}(2, 2, 2\sqrt{2})$ in: (a) rectangular form; (b) cylindrical form; (c) spherical form.

1-13. Evaluate $\oint_s \overline{A} \cdot \overline{ds}$ over a sphere of constant r_s when

$$\overline{A} = \hat{r}_s \frac{10}{r_s^2} + \hat{\theta}\frac{10}{r_s^2}$$

1-14. Find the dot product of two vector fields, $\overline{A} = (\hat{x}10xy^2 + \hat{y}20x^2y)e^{-t}$ and $\overline{B} = (\hat{x}5zx + \hat{y}2zx^2)e^{-t}$, at the point $(3, 1, 2, t = 1)$.

1-15. Use integral tables to evaluate the following integral for: (a) $m = n$; (b) $m \neq n$. m and n are integers.

$$\int_0^{2\pi} \sin(mx)\sin(nx)\,dx$$

What can you say about the "dot product" of $f = \sin(mx)$ and $g = \sin(nx)$?

2

Charge, Charge Densities, Coulomb's Law, and Electric Field Intensity

2.1 INTRODUCTION

The entire universe is composed of matter that has electric charge, positive and negative, at the root of its structure. The atom contains the smallest electric charges known to date, the negatively charged electron and the positively charged proton.

Stationary and moving electric charges produce forces on other stationary and moving electric charges. To aid in the study of these forces, a concept called *force field* has been devised. This force field is called an *electric field, magnetic field,* or *electromagnetic field.* When charges are stationary, the resulting force field is an electric field or an *electrostatic field,* whereas charges moving with constant velocity give rise to a magnetic field that is called a *magnetostatic field.* Accelerated charges produce an *electromagnetic field* that consists of related time-varying electric and magnetic fields. An ever-present example of an electromagnetic field is the radio signal radiated by a radio station antenna due to the presence of accelerated charges in it. The room that you are now in contains millions of electromagnetic fields due to TV stations, radio stations, electromagnetic noises, CB signals, radiation from radio stars in outer space, radar communication, and many, many more.

Electric and magnetic fields are also found in and about all electric circuits, transistors, capacitors, inductors, electric motors, generators, relays, TV picture tubes, solar cells, etc. As a matter of fact, science had devised the concept of fields well in advance of the concepts of circuit theory, which itself is based on field theory. With this brief introduction, it is obvious that electric and magnetic fields are fundamental to the study of electrical and electronics engineering.

Our study in this chapter will begin with stationary electric charges and the electrostatic force field that they produce in a vacuum or free space (absence of material). Coulomb's force law will be used to develop the concept of an electric field. Vector concepts introduced in Chapter 1 will be used extensively throughout this text.

2.2 ELECTRIC CHARGES AND ELECTRIC CHARGE DENSITIES

As stated in Sec. 2.1, the smallest unit of electric charge is found on the negatively charged electron and the positively charged proton. Thus, electric charge must exist in multiples of the magnitude of the charge found on the electron,

$$\boxed{q_e = -1.602 \times 10^{-19}}$$

(C; coulombs). A negative charge of 1 C would represent about 6×10^{18} electrons. The coulomb is an extremely large unit of charge, as we shall discover later. In the real world, electric charges are found: (1) at a point, (2) on a line, (3) on a surface, and (4) within a volume, or as any combination of these distributions.

The concept of a point charge is used when the dimensions of an electric charge distribution are very small compared to the distance to neighboring electric charges. Thus, an aggregate of electric charges on a pinhead can be considered a point charge in a problem where the nearest electric charge is a meter away. In general, we think of a point charge as occupying a very small physical space. Shortly, we shall introduce an experiment of Charles Augustin de Coulomb, who used charged pith balls which he assumed to be point charges, to establish Coulomb's force law.

An electric charge aggregate along a thin line is referred to as a *line charge*. A small length $\Delta\ell$ on the line would thus contain a charge ΔQ. A line charge density ρ_ℓ along a line of charge is defined as

$$\boxed{\rho_\ell \triangleq \lim_{\Delta\ell \to 0} \left[\frac{\Delta Q}{\Delta \ell}\right] \qquad (\text{C m}^{-1}) \quad (\text{aap})} \tag{1}$$

Figure 2-1(a) illustrates this concept. Equation (1), due to the limit process, is true "at a point" (aap). This notation will be used occasionally to remind the student that equations such as (1) are defined *at a point*. The charge ΔQ in the limit becomes dQ, a point charge, and equals $\rho_\ell \, d\ell$ (C).

An electric charge aggregate on a surface is called a *surface charge*. The charge on a differential surface ds is dQ and can be viewed as a point charge. A surface charge density is defined as

$$\boxed{\rho_s \triangleq \lim_{\Delta s \to 0} \left[\frac{\Delta Q}{\Delta s}\right] \qquad (\text{C m}^{-2}) \quad (\text{aap})} \tag{2}$$

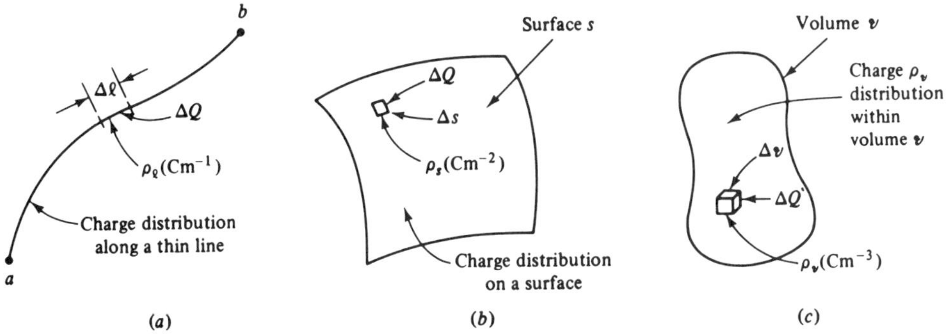

Figure 2-1 Geometry used in defining charge densities: (a) line charge density ρ_ℓ; (b) surface charge density ρ_s; (c) volume charge density ρ_v.

where ρ_s is the surface charge density in (C m^{-2}) on the surface. Figure 2-1(b) illustrates this concept.

An electric charge aggregate within a volume is called a *volume charge*. This volume charge can be viewed as a cloud of charged particles. A volume charge density ρ_v is defined as

$$\boxed{\rho_v \triangleq \lim_{\Delta v \to 0} \left[\frac{\Delta Q}{\Delta v}\right] \qquad (\text{C m}^{-3}) \quad (\text{aap})} \tag{3}$$

Figure 2-1(c) illustrates this concept.

The total charge on a given length of line of charge, surface of charge, and volume of charge can be found by integration to become

$$\boxed{Q_\ell = \int_\ell \rho_\ell \, d\ell, \qquad Q_s = \int_s \rho_s \, ds, \qquad Q_v = \int_v \rho_v \, dv} \tag{4}$$

Note that, in general, ρ_ℓ, ρ_s, ρ_v are functions of position and thus will vary with the variables of the coordinate system used. In (4), $\int_s$ indicates a double integral over a surface s and $\int_v$ indicates a triple integral over a volume v.

Example 1

A uniform, spherical volume charge distribution contains a total charge of 10^{-8} C. If the radius of this spherical volume is 2×10^{-2} m, find ρ_v.

Solution. From (4), $Q_v = \int_v \rho_v \, dv$. Since the charge distribution is uniform,

$$Q_v = \rho_v \int_v dv = \rho_v \left(\frac{4}{3}\pi r_s^3\right)\bigg|_{r_s = 2 \times 10^{-2}}$$

Therefore,

$$\rho_v = \frac{Q_v}{\frac{4}{3}\pi(2 \times 10^{-2})^3} = \frac{10^{-8}}{\frac{4}{3}\pi(8 \times 10^{-6})} = 2.98 \times 10^{-4} \ (\text{C m}^{-3})$$

Problem 2.2-1 Find the total charge enclosed by a sphere of radius a, centered at the origin, when: (a) $\rho_v = k_1/r_s^2$; (b) $\rho_v = k_2/r_s$; (c) $\rho_v = k_3$; (d) $\rho_v = k_4 r_s$ (C m^{-3})

Problem 2.2-2 Find the total charge per unit length enclosed by a cylinder of radius a whose axis is the z axis, and where $\rho_v = kr_c^2|\sin\phi|$ (C m^{-3}).

2.3 FORCE BETWEEN POINT CHARGES

It is highly appropriate that our study of electrostatics, static electricity, begin with the first reported experiment of Coulomb in 1785.[1] These experiments are considered the organized starting point of electrostatics. The results of these experiments are expressed in Coulomb's law, which states that the force F between two stationary point charges Q_1 and Q_2 is proportional to the product of the charges and inversely proportional to the square of the distance R between them; i.e.,

$$\boxed{F = k\frac{Q_1 Q_2}{R^2} \qquad \text{(N)}} \tag{1}$$

In Coulomb's experiments charges were placed on small pith balls, considered point charges, and the force was measured by a delicate torsion balance that he invented. The force F is along the line connecting the two charges. It is a force of repulsion if the charges are of like sign and a force of attraction if the charges are of unlike sign. The k in (1) is a constant of proportionality whose value depends on the system of units. In vector notation form, and using the International System of units, abbreviated SI, (1) becomes

$$\boxed{\overline{F}_2 = \left(\frac{1}{4\pi\epsilon_0}\right)\frac{Q_1 Q_2}{R_{12}^2}\hat{a}_{R12} \qquad \text{(N)}} \tag{2}$$

where $\overline{F}_2$ = force on point charge Q_2 due to point charge Q_1 (N)
Q_1 = charge at location (1) (C)
Q_2 = charge at location (2) (C)
R_{12} = distance between the location of Q_1 and Q_2 (m)
$\hat{a}_{R_{12}}$ = unit vector along a straight line from Q_1 to Q_2 locations (from Q_1, the source of the force, to the location of the force $\overline{F}_2$, at Q_2)
ϵ_0 = permittivity of free space (F m^{-1}; farads per meter)

Figure 2-2 illustrates the Coulomb force $\overline{F}_2$ on charge Q_2 due to charge Q_1. One can readily show that (2), the vector form of Coulomb's law (1), gives the correct magnitude and direction for like or unlike point charges. It should be noted that Q_1 and Q_2 take on the sign of the charge. In SI units, the constant of proportionality in (1) becomes

$$\boxed{k = \frac{1}{4\pi\epsilon_0}}$$

[1] Charles A. de Coulomb, *History Royal Academie Sciences* (France), 1785, pp. 569 and 579.

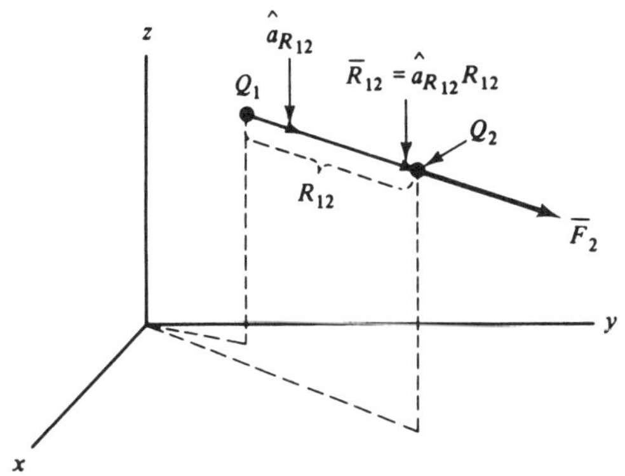

Figure 2-2 Coulomb's force $\overline{F}_2$ on Q_2 is in the direction of $\hat{a}_{R_{12}}$ when the charges have like signs.

where ϵ_0, the permittivity of free space, is

$$\boxed{\epsilon_0 = 8.854 \times 10^{-12} \cong \frac{1}{36\pi} 10^{-9} \qquad (\text{F m}^{-1})}$$

Equation (2) can be also written in the form

$$\boxed{\overline{F}_2 = \left(\frac{1}{4\pi\epsilon_0}\right) \frac{Q_1 Q_2 \overline{R}_{12}}{R_{12}^3} \qquad (\text{N})} \tag{3}$$

since $\hat{a}_{R_{12}} = \overline{R}_{12}/R_{12}$, where $\overline{R}_{12}$ is the distance vector from Q_1 to Q_2 locations, as shown in Fig. 2-2.

The force $\overline{F}_1$ on Q_1 due to Q_2, through the use of (2), becomes

$$\boxed{\overline{F}_1 = \frac{Q_2 Q_1 \hat{a}_{R_{21}}}{4\pi\epsilon_0 R_{21}^2} = -\overline{F}_2 \qquad (\text{N})} \tag{4}$$

since $\hat{a}_{R_{21}} = -\hat{a}_{R_{12}}$.

Coulomb's law is based on point charges and linearity exists; thus, when charge Q_1 is increased by a factor m, the force $\overline{F}_2$ is also increased by the same factor m. If more than two point charges are present, the force on one of these charges is equal to the vector sum of the forces on that charge due to each of the other charges acting alone. Thus, we can say that superposition can be extended to the summation of vector forces due to point charges. This concept will become quite important when we calculate the electric field intensity due to a system of charges in Sec. 2.4. Through the use of (2) we will find the force between two charges of 1 coulomb, separated 1 meter, to equal approximately 9×10^9 N or approximately 1 million tons. This shows that the coulomb is a large unit of charge.

Example 2

Find the force $\overline{F}_2$, in vacuum, on a point charge $Q_2 = 10^{-6}$ C due to a point charge $Q_1 = 2 \times 10^{-5}$ C when Q_2 is at the rectangular point $P_2(2, 4, 5)$ and Q_1 is at the rectangular point $P_1(0, 1, 2)$.

Solution. From (2),

$$\overline{F}_2 = \frac{Q_1 Q_2}{4\pi\epsilon_0 R_{12}^2}\hat{a}_{R12}$$

From Sec. 1.7-2,

$$\begin{aligned}\overline{R}_{12} &= \hat{x}(x_2 - x_1) + \hat{y}(y_2 - y_1) + \hat{z}(z_2 - z_1)\\ &= \hat{x}(2 - 0) + \hat{y}(4 - 1) + \hat{z}(5 - 2)\\ &= \hat{x}2 + \hat{y}3 + \hat{z}3\end{aligned}$$

Now

$$R_{12} = (2^2 + 3^2 + 3^2)^{1/2} = 4.69$$

From (1.8-17),

$$\hat{a}_{R_{12}} = \frac{\overline{R}_{12}}{R_{12}} = \frac{\hat{x}2 + \hat{y}3 + \hat{z}3}{4.69}$$

Substitute into the $\overline{F}_2$ equation:

$$\begin{aligned}\overline{F}_2 &= \frac{(2 \times 10^{-5})(10^{-6})}{4\pi\left(\dfrac{1}{36\pi} \times 10^{-9}\right)(4.69)^2}\left(\frac{\hat{x}2 + \hat{y}3 + \hat{z}3}{4.69}\right)\\ &= [\hat{x}(3.49) + \hat{y}(5.23) + \hat{z}(5.23)] \qquad \text{(m N)}\end{aligned}$$

Problem 2.3-1 A positive charge Q_1 of 10^{-9} C is located on the y axis at $y = 3$, and a charge Q_2 of -10^{-9} C is located on the y axis at $y = -3$. Find the total force on a small positive test charge Q_t located at: (a) (0, 10, 0); (b) (10, 0, 0).

Problem 2.3-2 Repeat Prob. 2.3-1(b) with $Q_2 = 10^{-9}$ C instead.

Problem 2.3-3 Find the force on a charge Q_3 of 10^{-7} C located at the origin due to charge Q_1 of 10^{-9} C at $(-1, -1, 1)$ and charge Q_2 of 10^{-9} C at $(-1, -1, -1)$.

Problem 2.3-4 Find the force on Q_2 in Prob. 2.3-3 due to Q_1 and Q_3.

2.4 ELECTRIC FIELD INTENSITY OF POINT CHARGES

If we fix in space a point electric charge Q_1 and find the force $\overline{F}_2$ on another point electric charge Q_2 at some variable point $P(x, y, z)$, the force $\overline{F}_2$ will be a function of the variables x, y, z and thus can be called a *vector force field*. Through this force field concept we can explain the "force action at a distance" between two point electric charges. Note that our only detector of this phenomenon is the electric point charge.

> The vector force field $\overline{F}_t$, acting on a positive electric point test charge Q_t, divided by Q_t, is defined as the vector electric field intensity $\overline{E}_t$ (or just $\overline{E}$) at the location of Q_t.

From (2.3-2) it can be seen that the vector electric field intensity $\overline{E}_t$ (or just $\overline{E}$) becomes

$$\boxed{\overline{E}_t \triangleq \frac{\overline{F}_t}{Q_t} = \frac{Q_1}{4\pi\epsilon_0 R_{1t}^2}\hat{a}_{R_{1t}} \qquad (\text{N C}^{-1}) \quad (\text{aap})} \tag{1}$$

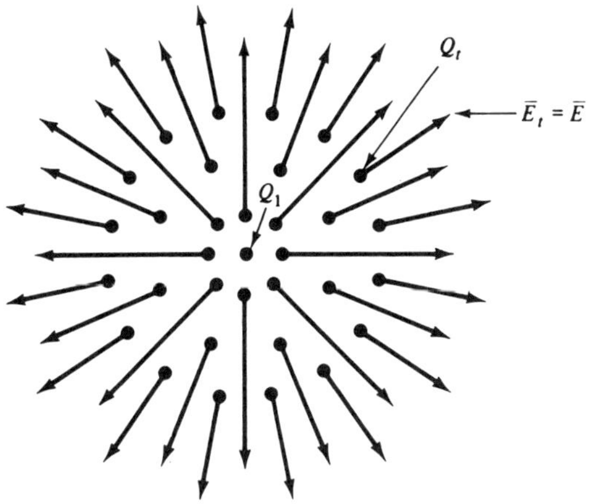

Figure 2-3 Graphical representation of the electric field intensity vector by a system of $\overline{E}_t$ vectors about a fixed point charge Q_1, as Q_t (test charge) is positioned at various points about Q_1. Charges Q_1 and Q_t here are of like signs.

where the quantities $\overline{F}_t$, R_{1t}, and $\hat{a}_{R_{1t}}$ are as shown in Fig. 2-2, with the 2 subscript replaced by t. The direction of $\overline{E}_t$ is the same as that of $\overline{F}_t$. Later, we show that an equivalent unit for the electric field intensity is the volt per meter (V m^{-1}). Note that $\overline{E}_t$ is defined at a point, the location of Q_t. Figure 2-3 illustrates the vector electric field intensity $\overline{E}_t$ by a system of vectors about a point charge Q_1.

From the word definition of the electric field intensity and (1), it can be seen that the magnitude of $\overline{E}_t$ is independent of the magnitude of the test charge Q_t. This is due to the linearity of Coulomb's law and that the electric field intensity is defined as a ratio of force and charge. If we allow the test charge Q_t to be equal to $+1$ C, the magnitude and direction of $\overline{E}_t$ will be that of $\overline{F}_t$ force on $+1$ C. This can be proven by (1). From the definition of $\overline{E}_t$, the force on a charge Q_t located in an $\overline{E}_t$ field is $\overline{F}_t = \overline{E}_t Q_t$ (N). In cases where Q_1 is not a point charge but a charge distribution on a conducting surface, it is possible that a point test charge Q_t, of large charge magnitude, will redistribute Q_1 and thus change the physical charge system. This dilemma can be circumvented by reducing Q_t to ΔQ_t and defining $\overline{E}_t$ as

$$\boxed{\overline{E}_t \triangleq \lim_{\Delta Q_t \to 0} \left[\frac{\Delta \overline{F}_t}{\Delta Q_t}\right] = \frac{Q_1}{4\pi\epsilon_0 R_{1t}^2}\hat{a}_{R_{1t}} \qquad (\text{V m}^{-1}) \quad (\text{aap})} \tag{2}$$

where ΔQ_t is a small positive test charge. Equations (1) and (2) give the same values for $\overline{E}_t$ (or just $\overline{E}$), the vector electric field intensity.

From Fig. 2-3 it can be seen that the plot of a system of $\overline{E}_t$ vectors gives us a visual picture of this mysterious force between electric charges. Later, we will improve our visual picture of the force field about a charge when we introduce electric flux lines and equipotential surfaces. With these two new concepts, we shall complete our visual picture of a force field about a charge Q_1. By this visual picture, called a *field map*, we shall be able to visually ascertain the direction and magnitude of the force on a point charge Q_t when placed in the vicinity of Q_1.

The electric field intensity $\overline{E}$ (the subscript t has been dropped) of n point charges at a point in space is equal to the vector sum of the electric field intensities due to each

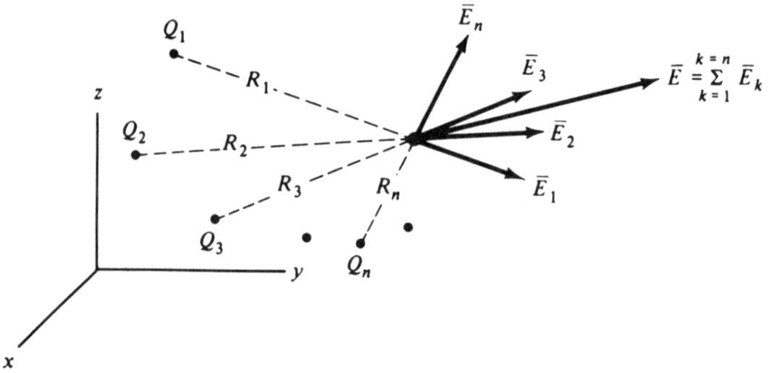

Figure 2-4 The resultant field intensity $\overline{E}$ of n point charges is the vector sum of the electric field intensities of each charge acting alone.

charge acting alone, as shown in Fig. 2-4. Thus,

$$\boxed{\overline{E} = \overline{E}_1 + \overline{E}_2 + \ldots + \overline{E}_k + \ldots + \overline{E}_n = \sum_{k=1}^{k=n} \overline{E}_k \qquad (\text{N C}^{-1})} \tag{3}$$

The superposition in (3) can be justified by the validity of superposition in Coulomb's force law (Sec. 2-3). Through the use of (1), eq. (3) becomes

$$\begin{aligned} \overline{E} &= \frac{Q_1}{4\pi\epsilon_0 R_1^2}\hat{a}_{R_1} + \frac{Q_2}{4\pi\epsilon_0 R_2^2}\hat{a}_{R_2} + \ldots + \frac{Q_k}{4\pi\epsilon_0 R_k^2}\hat{a}_{R_k} + \ldots + \frac{Q_n}{4\pi\epsilon_0 R_n^2}\hat{a}_{R_n} \\ &= \sum_{k=1}^{k=n} \frac{Q_k}{4\pi\epsilon_0 R_k^2}\hat{a}_{R_k} \qquad (\text{V m}^{-1}) \end{aligned} \tag{4}$$

Note that the basic contribution from a point charge in (4) is of the form

$$\boxed{\overline{E}_k = \frac{Q_k}{4\pi\epsilon_0 R_k^2}\hat{a}_{R_k} \qquad (\text{V m}^{-1})} \qquad \text{(building block)} \tag{5}$$

Equation (5) will be referred to as the *building block* for the $\overline{E}$ of a point charge.

Example 3

Find the electric field intensity $\overline{E}$ at $P_{\text{rec}}(2, 4, 5)$ due to a point charge $Q = 2 \times 10^{-5}$ C, located at $P_{\text{rec}}(0, 1, 2)$, in vacuum.

Solution. From (1) or (5),

$$\overline{E} = \frac{Q\hat{a}_R}{4\pi\epsilon_0 R^2}$$

where R is the distance from Q, the source of the field, to the location of the field $\overline{E}$, and $\hat{a}_R$ is the unit vector directed from Q to the field point $\overline{E}$. From Sec. 1.7-2, the vector distance $\overline{R}$ from Q to the field point becomes

$$\overline{R} = \hat{x}(2 - 0) + \hat{y}(4 - 1) + \hat{z}(5 - 2) = \hat{x}2 + \hat{y}3 + \hat{z}3$$

Now,

$$R = (2^2 + 3^2 + 3^2)^{1/2} = 4.69$$

From (1.8-17),

$$\hat{a}_R = \frac{\overline{R}}{R} = \frac{\hat{x}2 + \hat{y}3 + \hat{z}3}{4.69}$$

Substitute into the $\overline{E}$ equation above:

$$\overline{E} = \frac{2 \times 10^{-5}}{4\pi(1/36\pi \times 10^{-9})(4.69)^2}\left(\frac{\hat{x}2 + \hat{y}3 + \hat{z}3}{4.69}\right)$$

$$= [\hat{x}(3.49) + \hat{y}(5.23) + \hat{z}(5.23)] \qquad (\text{kV m}^{-1})$$

Problem 2.4-1 Find the total electric field at the origin from a cluster of charges, each of charge Q, located at $(1, 0, 1)$, $(-1, 0, 1)$, $(0, 1, 1)$, and $(0, -1, 1)$.

Problem 2.4-2 Point charges $Q_1 = 10^{-4}$ C and $Q_2 = -3 \times 10^{-5}$ C are located at $P_{\text{rec}}(0, 0, 5)$ and $P_{\text{rec}}(0, 0, -8)$, respectively. Find the location on the z axis, excluding infinity, where $\overline{E} = 0$.

Problem 2.4-3 Find the total electric field at the origin due to a -10^{-8} C charge located at $P_{\text{rec}}(4, 0, 2)$ and a 0.5×10^{-8} C charge at $P_{\text{rec}}(0, 4, 4)$.

2.5 ELECTRIC FIELD INTENSITY OF A LINE OF CHARGE

To evaluate the electric field intensity from systems of charges such as line charge, surface charge, and volume charge we shall apply the building block eq. (2.4-5) to a point charge on the charge system and sum the vector contributions over the charge distribution of the system.

Figure 2-5(a) illustrates this technique for evaluating the $\overline{E}$ of a general line of charge. The point charge on the line is established at some general location by selecting a $d\ell$ length of line that contains a charge $dQ = \rho_\ell \, d\ell$ (C). Through the use of the building block (2.4-5), the differential electric field intensity $\overline{dE}$ becomes

$$\overline{dE} = \frac{dQ\,\hat{a}_R}{4\pi\epsilon_0 R^2} = \frac{\rho_\ell \, d\ell \, \hat{a}_R}{4\pi\epsilon_0 R^2} \qquad (\text{V m}^{-1}) \tag{1}$$

Summing all contributions from a to b, we form a vector integral to obtain $\overline{E}$; thus,

$$\overline{E} = \int_a^b \overline{dE} = \int_a^b \frac{\rho_\ell \, d\ell \, \hat{a}_R}{4\pi\epsilon_0 R^2} \qquad (\text{V m}^{-1}) \tag{2}$$

where the subscripts in $\hat{a}_R$, $\overline{E}$, and R have been dropped, and the unit vector $\hat{a}_R$ is directed from the source point (dQ) to the field point ($\overline{E}$). If we assign primed variables to locate points on the line of charge and unprimed variables to locate the field point $\overline{E}$, (2) will take on the following functional form in the rectangular coordinate system:

$$\overline{E}(x, y, z) = \int_{a'}^{b'} \frac{\rho_\ell(x', y', z')\, d\ell'}{4\pi\epsilon_0 R^2(x', y', z', x, y, z)} \hat{a}_R(x', y', z', x, y, z) \qquad (\text{V m}^{-1}) \tag{3}$$

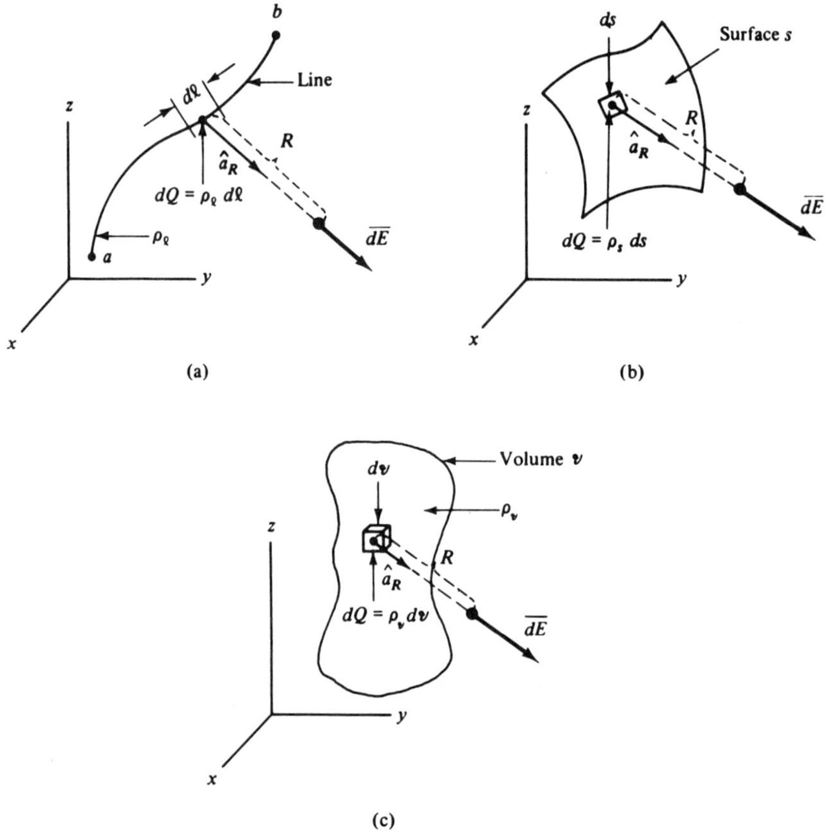

Figure 2-5 Graphical formulation for evaluating the differential electric field intensity $\overline{dE}$: (a) line of charge; (b) surface of charge; (c) volume of charge.

Note that $\overline{E}$ is a function of the unprimed variables since through the definite integral, with respect to the primed variables, the primed variables are eliminated.

Example 4

Find the electric field intensity about the finite line charge of uniform ρ_ℓ distribution along the z axis, as shown in Fig. 2-6.

Solution. The differential electric field intensity $\overline{dE}$ can be found through (1); thus,

$$\overline{dE} = \frac{\rho_\ell \, d\ell}{4\pi\epsilon_0 R^2}\hat{a}_R$$

Let us use the primed variables to locate points on the line of charge and the unprimed variables to locate the electric field point. Thus, point charge dQ is located at $P_{\text{cyl}}(r_c', \phi', z')$. With the aid of the graphical construction in Fig. 2-6, we have

$$R = (r_c^2 + (z' - z)^2)^{1/2}$$

$$\overline{R} = \hat{r}_c r_c - \hat{z}(z' - z) = \hat{a}_R R$$

$$\hat{a}_R = \frac{\overline{R}}{R} = \frac{\hat{r}_c r_c - \hat{z}(z' - z)}{(r_c^2 + (z' - z)^2)^{1/2}}$$

$$d\ell = dz'$$

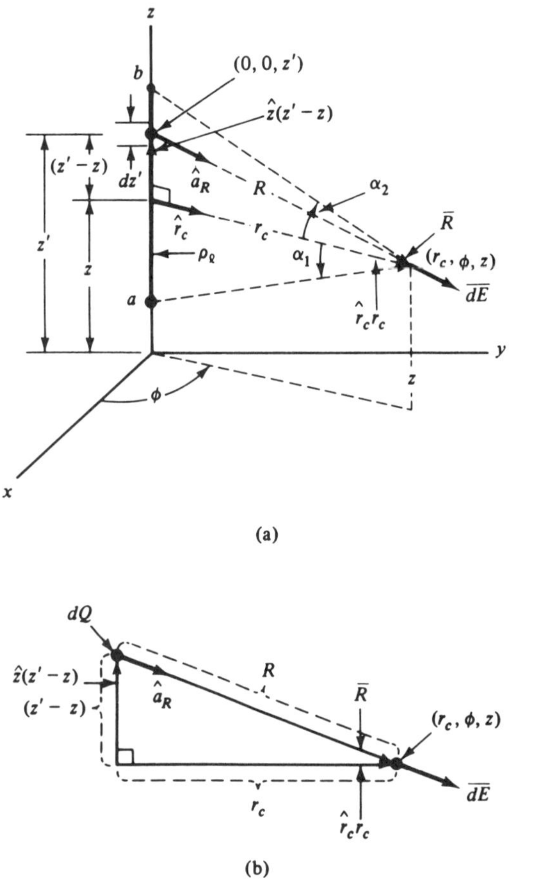

Figure 2-6 (a) Graphical construction to evaluate $\overline{dE}$ and thus $\overline{E}$ about a finite length of line charge of uniform ρ_ℓ (see Example 4). (b) Isolated view for evaluating $\hat{a}_R$.

Substituting into the $\overline{dE}$ expression yields

$$\overline{dE} = \frac{\rho_\ell \, dz'}{4\pi\epsilon_0(r_c^2 + (z' - z)^2)} \left[\frac{\hat{r}_c r_c - \hat{z}(z' - z)}{(r_c^2 + (z' - z)^2)^{1/2}} \right]$$

Now

$$\overline{E} = \int_a^b \overline{dE} = \frac{\rho_\ell}{4\pi\epsilon_0} \int_a^b \frac{\hat{r}_c r_c - \hat{z}(z' - z)}{(r_c^2 + (z' - z)^2)^{3/2}} \, dz'$$

Over the range of the integral above, the only variable is z', while $\hat{r}_c$ is a constant, since the point of $\overline{E}$ is fixed for the integration over the line from a to b. The $\overline{E}$ integral can be rewritten as

$$\overline{E} = \frac{\rho_\ell}{4\pi\epsilon_0} \left\{ \hat{r}_c r_c \int_a^b \frac{dz'}{(r_c^2 + (z' - z)^2)^{3/2}} - \hat{z} \int_a^b \frac{(z' - z)\, dz'}{(r_c^2 + (z' - z)^2)^{3/2}} \right\}$$

The integrals found in the expression above are of the forms

$$\boxed{\int \frac{dx}{(c^2 + x^2)^{3/2}} = \frac{x}{c^2(c^2 + x^2)^{1/2}}} \tag{4}$$

$$\boxed{\int \frac{x\,dx}{(c^2 + x^2)^{3/2}} = \frac{-1}{(c^2 + x^2)^{1/2}}} \tag{5}$$

Through the use of (4) and (5), $\overline{E}$ becomes

$$\boxed{\begin{aligned}\overline{E} &= \frac{\rho_\ell}{4\pi\epsilon_0}\left\{\hat{r}_c r_c \frac{z' - z}{r_c^2(r_c^2 + (z' - z)^2)^{1/2}}\Bigg|_a^b + \hat{z}\frac{1}{(r_c^2 + (z' - z)^2)^{1/2}}\Bigg|_a^b\right\}\\ &= \frac{\rho_\ell}{4\pi\epsilon_0}\left\{\frac{\hat{r}_c}{r_c}\left[\frac{b - z}{(r_c^2 + (b - z)^2)^{1/2}} - \frac{a - z}{(r_c^2 + (a - z)^2)^{1/2}}\right]\right.\\ &\quad \left. + \hat{z}\left[\frac{1}{(r_c^2 + (b - z)^2)^{1/2}} - \frac{1}{(r_c^2 + (a - z)^2)^{1/2}}\right]\right\} \quad (\text{V m}^{-1})\end{aligned}} \tag{6}$$

In terms of α_1 and α_2 [see Fig. 2-6(a)], eq. (6) becomes

$$\boxed{\overline{E} = \frac{\rho_\ell}{4\pi\epsilon_0}\left\{\frac{\hat{r}_c}{r_c}(\sin\alpha_2 + \sin\alpha_1) + \frac{\hat{z}}{r_c}(\cos\alpha_2 - \cos\alpha_1)\right\} \quad (\text{V m}^{-1})} \tag{7}$$

Example 5

Find the electric field intensity $\overline{E}$ about an infinite line of charge of uniform ρ_ℓ distributed along the z axis.

Solution. In (6), let $b = +\infty$, $a = -\infty$ to obtain

$$\overline{E} = \frac{\rho_\ell}{2\pi\epsilon_0 r_c}\hat{r}_c \quad (\text{V m}^{-1})$$

Also, let $\alpha_1 = \pi/2$ and $\alpha_2 = \pi/2$ in (7) to obtain

$$\boxed{\overline{E} = \frac{\rho_\ell}{2\pi\epsilon_0 r_c}\hat{r}_c \quad (\text{V m}^{-1})} \tag{8}$$

It should be noted that only the radial component exists for the infinite line of charge of uniform ρ_ℓ.

Problem 2.5-1 Find the electric field intensity $\overline{E}$ on the axis of a circular ring of uniform charge ρ_ℓ and radius a. Let the axis of the ring be along the z axis.

Problem 2.5-2 Find the electric field at $P_{\text{rec}}(0, 0, 1.5)$ due to two infinite and parallel line charges of uniform ρ_ℓ. Let one $\rho_\ell = 10^{-6}$ C m^{-1} line be located at $y = 2$ m and another $\rho_\ell = -10^{-6}$ C m^{-1} line be located at $y = -2$ m, with both lines parallel to the x axis and in the $z = 0$ plane.

Problem 2.5-3 (a) At what distance along the z axis is the electric field from the charged ring of Prob. 2.5-1 a maximum? (b) What is the magnitude of this field?

2.6 ELECTRIC FIELD INTENSITY OF A SURFACE OF CHARGE

For a surface charge distribution, as shown in Fig. 2-5(b), the point charge is formulated by the charge on a differential surface ds. The differential point charge on ds is $dQ = \rho_s\, ds$ (C). Through the use of the building block (2.4-5), the differential electric field intensity $\overline{dE}$ becomes

$$\overline{dE} = \frac{dQ\,\hat{a}_R}{4\pi\epsilon_0 R^2} = \frac{\rho_s\, ds}{4\pi\epsilon_0 R^2}\hat{a}_R \qquad (\text{V m}^{-1}) \quad (\text{aap}) \tag{1}$$

The total electric field intensity $\overline{E}$ of the surface charge distribution is obtained by integrating (1) over the surface s; thus,

$$\overline{E} = \int_s \overline{dE} = \int_s \frac{\rho_s\, ds}{4\pi\epsilon_0 R^2}\hat{a}_R \qquad (\text{V m}^{-1}) \quad (\text{aap}) \tag{2}$$

where $\int_s$ denotes a surface integral and thus a double integral. Equation (2) will take on a functional dependence similar to (2.5-3) in the rectangular coordinate system.

Example 6

Through the use of cylindrical coordinates, find the electric field intensity about an infinite sheet of uniform charge distribution ρ_s, as shown in Fig. 2-7.

Solution. Locate the infinite sheet in the $z = 0$ plane and the electric field point on the $+z$ axis at the point $P_{\text{cyl}}(0, 0, z)$. The differential electric field intensity $\overline{dE}$ can be found through the use of (1); thus,

$$\overline{dE} = \frac{\rho_s\, ds}{4\pi\epsilon_0 R^2}\hat{a}_R$$

With the aid of the graphical construction in Fig. 2-7 and the use of primed variables to locate source points on the surface, we have

$$ds = ds' = (dr_c')(r_c'\, d\phi')$$
$$R = (r_c'^2 + z^2)^{1/2}$$
$$\overline{R} = \hat{z}z - \hat{r}_c' r_c' = \hat{a}_R R$$
$$\hat{a}_R = \frac{\hat{z}z - \hat{r}_c' r_c'}{(r_c'^2 + z^2)^{1/2}}$$

Note, that $\hat{r}_c'$ is a primed unit vector at the source point. Substituting into the $\overline{dE}$ expression, we have

$$\overline{dE} = \frac{\rho_s(r_c'\, dr_c'\, d\phi')}{4\pi\epsilon_0(r_c'^2 + z^2)}\left[\frac{\hat{z}z - \hat{r}_c' r_c'}{(r_c'^2 + z^2)^{1/2}}\right]$$

Now,

$$\overline{E} = \int_s \overline{dE} = \int_0^{r_c'=\infty} \int_0^{\phi'=2\pi} \frac{\rho_s(r_c'\, d\phi'\, dr_c')}{4\pi\epsilon_0(r_c'^2 + z^2)^{3/2}}(\hat{z}z - \hat{r}_c' r_c') \qquad (\text{V m}^{-1}) \tag{3}$$

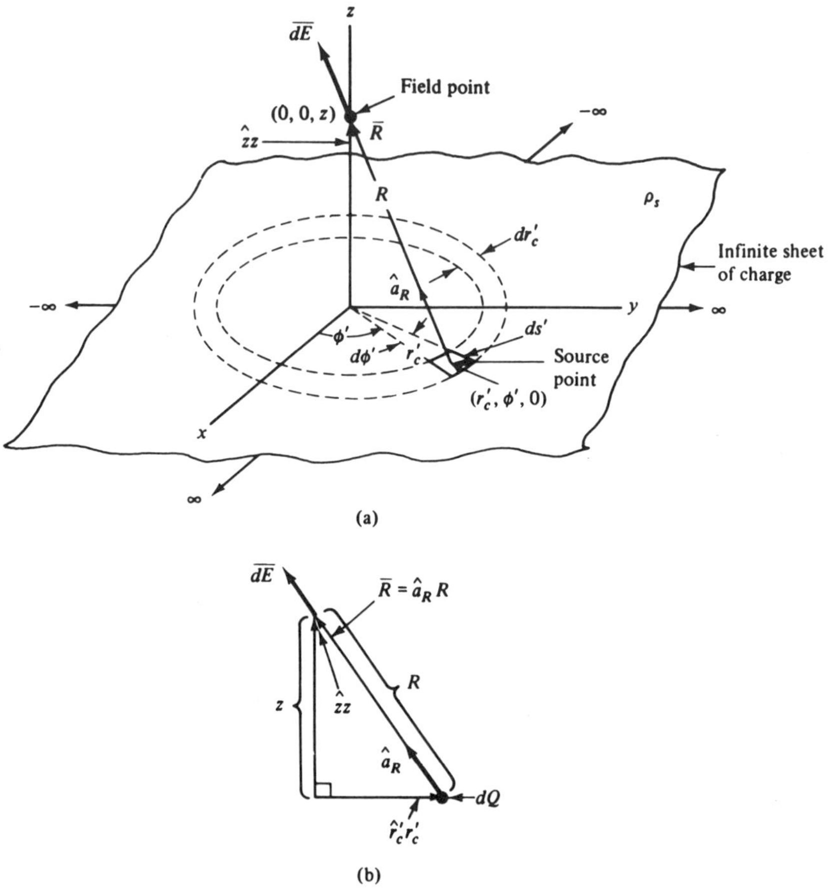

Figure 2-7 (a) Graphical construction to evaluate the $\overline{dE}$ and thus $\overline{E}$ about an infinite sheet of uniform charge distribution (see Example 6). (b) Isolated view for evaluating $\hat{a}_R$.

Over the range of integration $0 \leqq \phi' < 2\pi$ and $0 \leqq r_c' \leqq \infty$, the only variables in (3) are r_c', ϕ', and $\hat{r}_c'$. From (1.9-4), we can express

$$\hat{r}_c' = \hat{x}\cos\phi' + \hat{y}\sin\phi' \tag{1.9-4}$$

Thus, (3) can be rewritten as

$$\overline{E} = \frac{\rho_s}{4\pi\epsilon_0}\left[\hat{z}z\int_0^{\infty}\int_0^{2\pi}\frac{r_c'\,d\phi'\,dr_c'}{(r_c'^2+z^2)^{3/2}}\right.$$
$$\left. - \left(\hat{x}\int_0^{\infty}\int_0^{2\pi}\frac{r_c'^2\cos\phi'\,d\phi'\,dr_c'}{(r_c'^2+z^2)^{3/2}} + \hat{y}\int_0^{\infty}\int_0^{2\pi}\frac{r_c'^2\sin\phi'\,d\phi'\,dr_c'}{(r_c'^2+z^2)^{3/2}}\right)\right] \tag{4}$$

In (4) the last two integrals will become zero due to the integration with respect to ϕ', from 0 to 2π; thus, we have

$$\overline{E} = \frac{\rho_s}{4\pi\epsilon_0}\hat{z}z\int_0^{\infty}\int_0^{2\pi}\frac{r_c'\,d\phi'\,dr_c'}{(r_c'^2+z^2)^{3/2}} \tag{5}$$

Integrating with respect to ϕ' and using (2.5-5) we obtain

$$\overline{E} = \frac{\rho_s}{4\pi\epsilon_0}\hat{z}z(2\pi)\left[\frac{-1}{(r_c'^2 + z^2)^{1/2}}\right]\Bigg|_0^\infty \tag{6}$$

$$\boxed{\overline{E} = \frac{\rho_s}{2\epsilon_0}\hat{z} \qquad (\text{V m}^{-1})} \tag{7}$$

Equation (7) gives the electric field intensity above an infinite sheet of uniform ρ_s distribution. In this development the $+z$ axis, on which $\overline{E}$ was found, was taken at the center of the infinite sheet. Since the center of an infinite sheet is at any finite point, we can say that (7) is true for any point in the upper half-space, $z > 0$. For the lower half space, $z < 0$, we will find that $\overline{E}$ is in the $-\hat{z}$ direction. Note that $\overline{E}$ above and below the infinite surface is independent of z.

Example 7

Rewrite (4) to apply to: (a) a disc of charge; (b) an annular ring of charge; (c) a thin ring of charge.

Solution. All of these charge configurations are shown in Fig. 2-8, in terms of bound values on ϕ' and r_c'. Equation (4) can be used to find the $\overline{E}$ on the $+z$ axis in all these cases if the limits on the integrals are changed to read

$$\overline{E} = \frac{\rho_s}{4\pi\epsilon_0}\left[\hat{z}z\int_{r_1'}^{r_2'}\int_{\phi_1'}^{\phi_2'}\frac{r_c\,d\phi'\,dr_c'}{(r_c'^2 + z^2)^{3/2}} - \left(\hat{x}\int_{r_1'}^{r_2'}\int_{\phi_1'}^{\phi_2'}\frac{r_c'^2\cos\phi'\,d\phi'\,dr_c'}{(r_c'^2 + z^2)^{3/2}} + \hat{y}\int_{r_1'}^{r_2'}\int_{\phi_1'}^{\phi_2'}\frac{r_c'^2\sin\phi'\,d\phi'\,dr_c'}{(r_c'^2 + z^2)^{3/2}}\right)\right] \tag{8}$$

Equation (8) will reduce to the following:

(a) For a disc of charge [Fig. 2-8(a)], $\phi_1' = 0$, $\phi_2' = 2\pi$, $r_1' = 0$, $r_2' = b$,

$$\overline{E} = \frac{\rho_s}{4\pi\epsilon_0}\hat{z}z\int_0^b\int_0^{2\pi}\frac{r_c'\,d\phi'\,dr_c'}{(r_c'^2 + z^2)^{3/2}}$$

$$= \frac{\rho_s}{4\pi\epsilon_0}\hat{z}z(2\pi)\left[\frac{-1}{(r_c'^2 + z^2)^{1/2}}\right]\Bigg|_0^b$$

$$\boxed{\overline{E} = \frac{\rho_s\hat{z}z}{2\epsilon_0}\left[\frac{1}{z} - \frac{1}{(b^2 + z^2)^{1/2}}\right] \qquad (\text{V m}^{-1})} \tag{9}$$

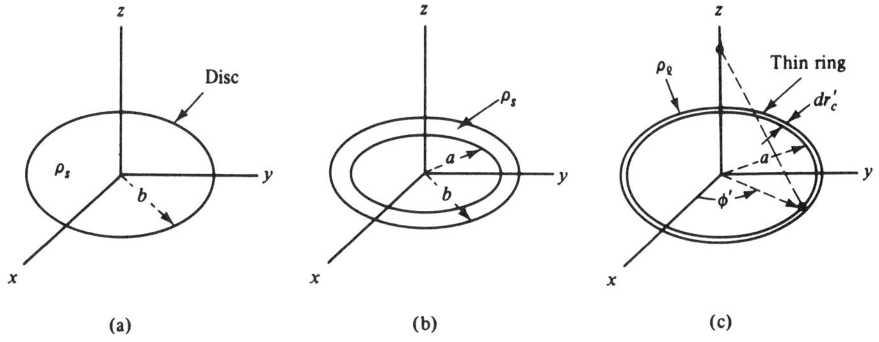

Figure 2-8 Charge configurations for use in Example 7: (a) disc of uniform ρ_s distribution; (b) annular ring of uniform ρ_s distribution; (c) thin ring of uniform ρ_ℓ distribution.

(b) For an annular ring of charge [Fig. 2-8(b)], $\phi_1' = 0$, $\phi_2' = 2\pi$, $r_1' = a$, $r_2' = b$,

$$\overline{E} = \frac{\rho_s}{4\pi\epsilon_0}\hat{z}z\int_a^b\int_0^{2\pi}\frac{r_c'\,d\phi'\,dr_c'}{(r_c'^2 + z^2)^{3/2}}$$

$$\overline{E} = \frac{\rho_s\hat{z}z}{2\epsilon_0}\left[\frac{1}{(a^2 + z^2)^{1/2}} - \frac{1}{(b^2 + z^2)^{1/2}}\right] \quad (\text{V m}^{-1}) \tag{10}$$

(c) For a thin ring of charge [Fig. 2-8(c)], $\phi_1' = 0$, $\phi_2' = 2\pi$, $r_1' = a$, $r_2' = a + dr_c'$,

$$\overline{E} = \frac{\rho_s}{4\pi\epsilon_0}\hat{z}z\int_a^{a+dr_c'}\int_0^{2\pi}\frac{r_c'\,d\phi'\,dr_c'}{(r_c'^2 + z^2)^{3/2}}$$

The $\int_a^{a+dr_c'}$ can be dropped since we are summing over a differential distance dr_c'. Note that $r_c' = a$ and we have

$$\overline{E} = \frac{\rho_s}{2\epsilon_0}\hat{z}z\left[\frac{a\,dr_c'}{(a^2 + z^2)^{3/2}}\right] \quad (\text{V m}^{-1}) \tag{11}$$

The charge dQ on ds' and on the arc length $a\,d\phi'$ should be the same; thus,

$$dQ = \rho_s\,ds' = \rho_\ell\,d\ell' = \rho_s(a\,d\phi')(dr_c') = \rho_\ell(a\,d\phi')$$

Solving for ρ_ℓ, we have

$$\rho_\ell = \rho_s\,dr_c' \tag{12}$$

Thus, (11) becomes

$$\overline{E} = \frac{\hat{z}z\rho_\ell}{2\epsilon_0}\left[\frac{a}{(a^2 + z^2)^{3/2}}\right] \quad (\text{V m}^{-1}) \tag{13}$$

It should be noted that for Examples 6 and 7 the work could have been reduced considerably by taking advantage of symmetry to eliminate all but the $\hat{z}$ component of $\overline{E}$ along the z axis.

Problem 2.6-1 A semi-infinite sheet of charge density ρ_s is described by $-\infty < x < 0$, $-\infty < y < \infty$ in the $z = 0$ plane. Calculate the component of electric field normal to the sheet at a distance a directly above the edge at $x = 0$. Compare your results with (7).

Problem 2.6-2 Find the expression for the electric field intensity between two parallel sheets of charge, separated d meters and of infinite extent. Assume uniform $+\rho_s$ on one of the sheets, uniform $-\rho_s$ on the other, and use the results found in (7).

2.7 ELECTRIC FIELD INTENSITY OF A VOLUME OF CHARGE

For a volume charge distribution, as shown in Fig. 2-5(c), the point charge is formulated by the differential volume dv. The differential point charge within dv is $dQ = \rho_v\,dv$ (C). Through the use of the building block (2.4-5), the differential electric field intensity $d\overline{E}$ becomes

$$\boxed{d\overline{E} = \frac{dQ\,\hat{a}_R}{4\pi\epsilon_0 R^2} = \frac{\rho_v\,dv}{4\pi\epsilon_0 R_2}\hat{a}_R \qquad (\text{V m}^{-1}) \quad (\text{aap})} \tag{1}$$

The $\overline{E}$ due to the volume charge distribution is obtained by integrating (1) over the volume v; thus,

$$\boxed{\overline{E} = \int_v d\overline{E} = \int_v \frac{\rho_v\,dv}{4\pi\epsilon_0 R^2}\hat{a}_R \qquad (\text{V m}^{-1}) \quad (\text{aap})} \tag{2}$$

where $\int_v$ denotes a volume integral and is thus a triple integral. The functional dependence of (2) is similar to (2.5-3).

Example 8

Find the electric field intensity about a sphere of uniform ρ_v distribution and $r_s = r_0$ (m).

Solution. Figure 2-9 shows the sphere of uniform ρ_v distribution along the essential graphical construction required to find $\overline{dE}$ from a point charge dQ within the differential

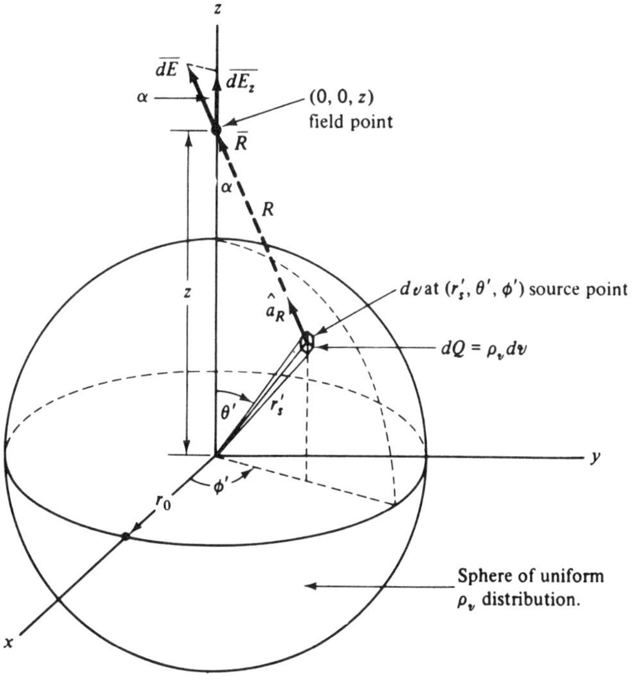

Figure 2-9 Graphical formulation for finding $\overline{dE}$ and thus $\overline{E}$ about a sphere of uniform ρ_v distribution (see Example 8).

volume dv. Spherical coordinate variables will be used since the problem has spherical symmetry. From (1.10-2) and (1) we have

$$\overline{dE} = \frac{\rho_v\, dv}{4\pi\epsilon_0 R^2}\hat{a}_R = \frac{\rho_v(r_s'^2 \sin\theta'\, d\theta'\, d\phi'\, dr_s')}{4\pi\epsilon_0 R^2}\hat{a}_R \tag{3}$$

If we vary ϕ' from 0 to 2π and hold all other quantities fixed in (3), we will generate a ring of charge that will produce only a $\hat{z}$-directed (radial) component of electric field intensity due to symmetry; thus,

$$\overline{dE}_z = \overline{dE}\cos\alpha = \hat{z}\frac{\rho_v(r_s'^2 \sin\theta'\, d\theta'\, d\phi'\, dr_s')}{4\pi\epsilon_0 R^2}\cos\alpha \tag{4}$$

To obtain $\overline{E}_z$, we must perform three integrations. Integration of (4) with respect to ϕ' from 0 to 2π will yield a factor of 2π. To perform the remaining two integrations, it is convenient to use r_s' and R as independent variables.[2] To express (4) in terms of these two variables, we use the cosine law to obtain

$$r_s'^2 = z^2 + R^2 - 2zR\cos\alpha \tag{5}$$

and

$$R^2 = z^2 + r_s'^2 - 2zr_s'\cos\theta' \tag{6}$$

Solving (5) and (6) for $\cos\alpha$ and $\cos\theta'$, respectively, we have

$$\cos\alpha = \frac{z^2 + R^2 - r_s'^2}{2zR} \tag{7}$$

$$\cos\theta' = \frac{z^2 + r_s'^2 - R^2}{2zr_s'} \tag{8}$$

A close examination of (4) and Fig. 2-9 will show that z and r_s' will be constant when integration with respect to θ' is performed. The terms $\sin\theta'\, d\theta'$ in (4) can be eliminated by differentiating (8), with z and r_s' held constant, to yield

$$\sin\theta'\, d\theta' = \frac{R\, dR}{zr_s'} \tag{9}$$

where the range on θ' is $0 \leqq \theta' \leqq \pi$ and on R is $(z - r_s') \leqq R \leqq (z + r_s')$.

Substituting (7) and (9) into (4) and integrating to obtain

$$\overline{E}_z = \frac{\rho_v \hat{z}}{8\pi\epsilon_0 z^2}\int_0^{2\pi}\int_{r_s'=0}^{r_s'=r_0}\int_{R=z-r_s'}^{R=z+r_s'} r_s'\left(1 + \frac{z^2 - r_s'^2}{R^2}\right) dR\, dr_s'\, d\phi' \tag{10}$$

Integrating yields,

$$\overline{E}_z = \frac{\hat{z}}{4\pi\epsilon_0}\left(\frac{\frac{4}{3}\pi r_0^3 \rho_v}{z^2}\right) \tag{11}$$

where $\frac{4}{3}\pi r_0^3\rho_v$ is the charge Q within the sphere of radius r_0. Let us rewrite (11) in the form

$$\boxed{\overline{E} = \frac{Q}{4\pi\epsilon_0 r_s^2}\hat{r}_s \qquad (\text{V m}^{-1})} \tag{12}$$

where $\hat{z}$ was changed to $\hat{r}_s$ and z to r_s due to the symmetry of the charge distribution. Equation (12) indicates that the electric field intensity of a volume charge of uniform ρ_v distribution for $r_s > r_0$ is identical to that produced by a point charge Q, equal to that found within the sphere, located at the origin.

[2] See Dale R. Corson and Paul Lorrain, *Introduction to Electromagnetic Fields and Waves*, W. H. Freeman and Company, Publishers, San Francisco, 1962, pp. 38–40.

Example 8 should clearly indicate the mathematical complexity encountered in the solution for $\overline{E}$ in volume charge distribution problems. Later, we shall develop Gauss's law and show that Example 8 will be greatly simplified through its use.

Problem 2.7-1 A spherical charge distribution has a uniform volume charge density described by $\rho_v = 10^{-12}$ C m^{-3} for $r_s < 10^{-3}$ m. Find the electric field at a distance of 3 cm from the center of the sphere.

REVIEW QUESTIONS

1. What is at the root of all matter? *Sec. 2.1*
2. What is a force field? *Sec. 2.1*
3. In addition to point charges, what three other charge distributions exist? *Sec. 2.2*
4. Define charge densities ρ_ℓ, ρ_s, and ρ_v. *Eqs. (2.2-1), (2.2-2), (2.2-3)*
5. Were point charges used by Coulomb in his experiment? *Sec. 2.3*
6. Does the vector form of Coulomb's force law give the correct magnitude and direction of force between like and unlike point charges? *Sec. 2.3, Eq. (2.3-2)*
7. Does superposition apply to Coulomb's force law when more than two point charges are present? *Sec. 2.3*
8. In your own words, what is the definition of the electric field intensity? *Sec. 2.4*
9. What is the defining equation for the vector electric field intensity due to a point charge? *Eq. (2.4-1)*
10. In the defining equation for the vector electric field intensity, is the magnitude of the electric field intensity independent of the magnitude of the positive test charge Q_t? *Sec. 2.4*
11. If a +1 C test charge is used to find the electric field intensity, will the magnitude and direction of the electric field intensity be the same as the vector force on the +1 C? *Sec. 2.4*
12. If the magnitude of the test charge affects the charge distribution of a system of charge, how can the electric field intensity be defined? *Eq. (2.4-2)*
13. Can superposition be applied in finding the resultant $\overline{E}$ from more than two point charges? *Sec. 2.4, Eq. (2.4-3)*
14. What is the building block equation? *Eq. (2.4-5)*
15. The unit vector $\hat{a}_R$, in general, is a function of how many variables? *Eq. (2.5-3)*
16. Are different sets of variables required to locate charge points and field points? *Eq. (2.5-3)*
17. What is the expression for $\overline{E}$ above and below an infinite sheet of uniform ρ_s distribution? *Eq. (2.6-7)*
18. What is the direction of $\overline{E}$ along the axis of a ring of uniform ρ_ℓ distribution? *Eq. (2.6-11)*
19. Can an infinite sheet of uniform ρ_s be built up through the use of thin rings of uniform ρ_ℓ distribution? *Example 7, Sec. 2.6*
20. What is the direction of $\overline{E}$ from a sphere of uniform ρ_v? *Example 8, Sec. 2.7*

PROBLEMS

2-1. (a) Find the charge within a sphere of radius 0.03 m when the charge density is given by

$$\rho_v = \frac{2 \times 10^{-3} r_s^2 \cos^2 \phi}{\sin \theta} \qquad (\text{C m}^{-3})$$

(b) What is the average charge density in the sphere?

2-2. Find the electric field along the axis of a square loop of line charge ρ_ℓ if each side of the loop is a (m) in length. Compare this result with that of Prob. 2.5-1.

2-3. Find the force on a point charge $Q = 10^{-5}$ C on the axis of a square loop of uniform line charge $\rho_\ell = 3 \times 10^{-3}$ C m^{-1} in the $z = 0$ plane. Assume that each leg of the loop is 2.5 m long.

2-4. Prove that superposition can be used in finding the electric field intensity $\overline{E}$ due to n point charges by first finding the Coulomb force on a test charge and then finding the force/Q_t that leads to $\overline{E}$.

2-5. Find the repulsive force between two parallel rings of charge, ρ_ℓ (C m^{-1}) each, if the radius of each ring is a (m) and the separation between rings is d (m), where $a \gg d$.

2-6. Find the force per unit length between two parallel infinite line charges of ρ_ℓ (C m^{-1}) each, separated by a distance d (m).

2-7. Find the pressure, i.e., force per unit area, tending to separate two discs of surface charge density ρ_s (C m^{-2}) each, of radius a and separation d, where $a \gg d$.

2-8. (a) Rework Prob. 2.5-2 with $\rho_\ell = 2$ nC m^{-1} for both lines. (b) Plot $|\overline{E}|$ on the z axis as a function of z for $0 < z < 10$ m.

2-9. Two parallel charged plates each have an area of 1 m^2. The charge density on one plate is $700\epsilon_0$ and on the other, $-700\epsilon_0$. (a) If the separation between plates is 1 mm, find the acceleration on an electron that is injected between the plates. Assume that $Q_e = -1.6 \times 10^{-19}$ C and $m_e = 9.1 \times 10^{-31}$ kg. (b) If the electron started from rest at the negative plate, how long would it take to reach the positive plate?

2-10. (a) Find the electric field at a distance a (m) on the positive x axis for a semi-infinite line charge extending from $-\infty$ to 0 on the x axis. (b) Find both E_x and E_z for the same line charge but from a point a (m) on a positive z axis.

2-11. Find the $\overline{E}$ inside a thin spherical shell of uniform charge distribution ρ_s, as shown in Fig. 2-10. [*Hint:* Construct a ring of charge at a general location, establish that $\overline{dE}$ due to the ring is along the axis of the ring due to symmetry, change variable θ' to variable R, and integrate with respect to ϕ' and R. See Example 8].

2-12. Derive the expression for $\overline{E}$ along the z axis of an infinite sheet of uniform $\rho_s = 10^{-6}$ C m^{-2}, as shown in Fig. 2-11. Note that the infinite sheet contains a hole of a (m) radius. [*Hint:* Use the results of an infinite sheet of uniform charge distribution and a disc of radius a (m) and uniform charge distribution $-\rho_s$.]

2-13. Repeat Prob. 2-11 for a point outside the shell charge.

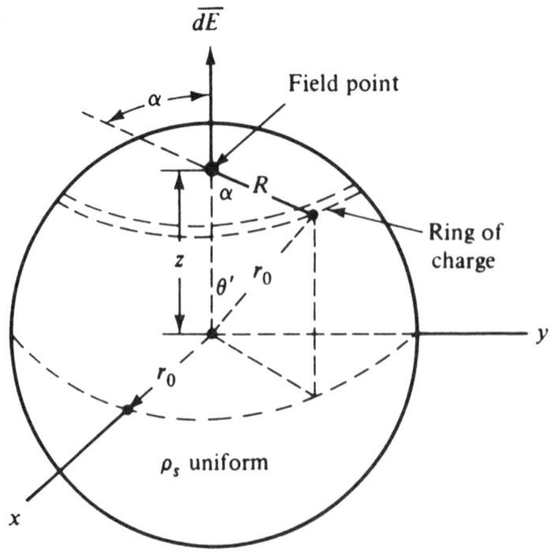

Figure 2-10 Graphical formulation for finding $\overline{E}$ inside a thin spherical shell of uniform distribution found in Prob. 2-11.

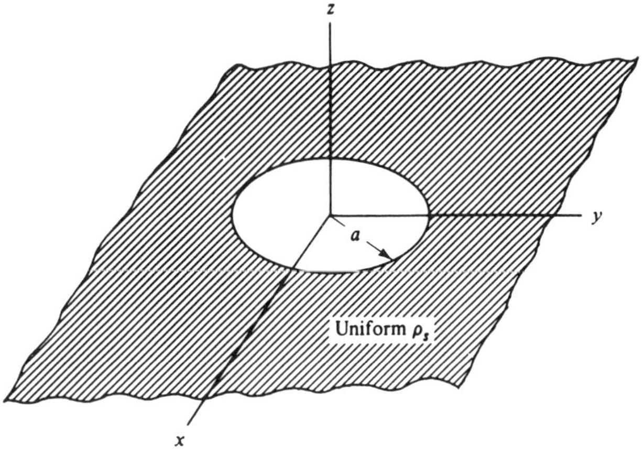

Figure 2-11 Graphical formulation for finding $\overline{E}$ above an infinite sheet of charge containing a hole of radius a (see Prob. 2-12).

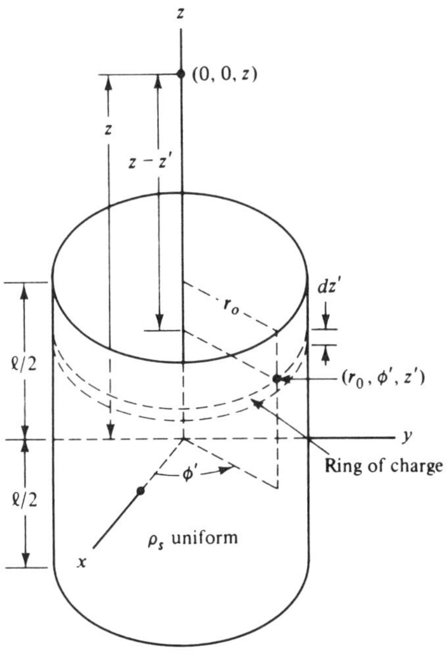

Figure 2-12 Graphical formulation for finding $\overline{E}$ along the axis of a uniform cylindrical surface charge distribution (see Prob. 2-17).

2-14. (a) From the results of Prob. 2-13, show that the field outside a uniformly charged spherical shell, like that outside a solid sphere of charge as found in (2.7-12), is

$$E_r = \frac{Q}{4\pi\epsilon_0 r_s^2}, \qquad r_s > r_0$$

the sphere radius, and thus is the same as if all the charge were concentrated at the origin. (b) Further show that the electric field normal to the surface on the outside of the sphere is ρ_s/ϵ_0 rather than $\rho_s/2\epsilon_0$, as it is for a single infinite flat sheet of charge density ρ_s. Comment on this difference.

2-15. A non-uniformly charged spherical shell is located with the equator in the $z = 0$ plane. Find the charge distribution $\rho_s\ (a, \theta, \phi)$ such that the electric field everywhere inside the charged shell is $\hat{z}(Q/3\pi a^2 \epsilon_0)$, where a is the radius of the sphere, Q is the total charge on the hemisphere below the $z = 0$ plane, and $-Q$ is the total charge above the $z = 0$ plane.

2-16. Two concentric shells of uniform ρ_s (C m^{-2}) distribution are centered at the origin. The radius of the inner shell is a (m) and has a $Q_{\text{total}} = 10^{-5}$ C; the outer shell has a radius of b (m) and a $Q_{\text{total}} = -10^{-5}$ C. Find the $\overline{E}$ field for: (a) $a > r_s > 0$; (b) $b > r_s > a$; (c) $r_s > b$.

2-17. Find the $\overline{E}$ along the $+z$ axis due to a uniform cylindrical surface charge distribution ρ_s when the axis of the cylinder is along the $+z$ axis. Let the cylinder be ℓ (m) long, radius r_0 (m), and centered at the origin. [*Hint:* Form a ring charge of thickness dz at a general location on the cylinder, convert ρ_ℓ to ρ_s, and integrate with respect to z, as shown in Fig. 2-12.]

2-18. A solid conducting sphere is placed in what had previously been a uniform field E_0 V m^{-1}. Assuming that the "induced" charge distribution on the surface of the sphere is $\rho_s = \rho_{s\,\text{max}} \cos\theta$, where the $\theta = 0$ direction is parallel to the field E_0, and assuming that the field inside the conducting sphere is zero, find the maximum field just outside the sphere and normal to the spherical surface.

3

Electric Flux, Electric Flux Density, Gauss's Law, Divergence, and Divergence Theorem

3.1 INTRODUCTION

In Chapter 2, the concept of a vector force field acting on a point charge was used to define the electric field intensity $\overline{E}$. The plot of this vector force field (Fig. 2-3) did to a certain extent give us a visual picture of the mysterious force between electric charges. In this chapter, we shall improve on this visual picture through the introduction of the electric flux lines and their density.

The concept of electric flux line density will lead us to Gauss's law, divergence, and the divergence theorem. Through the use of Gauss's law, we will be able to readily solve many problems possessing charge symmetry. In this chapter, as in Chapter 2, we shall assume free space.

3.2 ELECTRIC FLUX

About 1837, Michael Faraday performed several basic experiments in electrostatics that have had a direct bearing on our present concept of electric flux or electric flux lines. We shall use the electric flux concept to improve our visual picture of the vector force field about charges. This concept is illustrated by two-dimensional plots in Fig. 3-1. Figure 3-1(a) shows the plot of the vector force field $\overline{E}$ (electric field intensity) at discrete points about a $+Q$ (C) point charge, whereas Fig. 3-1(b) shows a system of lines about the same point charge. We shall call these lines *electric flux lines* or, simply, *electric flux*. Figure 3-1(c) extends the flux concept to two point charges: one $+Q$ (C) and the other $-Q$ (C). The electric flux concept is based on the following rules:

1. Electric flux begins on positive charges and terminates on negative charges.
2. Flux is in the same direction as the electric field $\overline{E}$.
3. Flux density is proportional to the magnitude of $\overline{E}$.
4. In the SI system of units, the total flux emanating from a charge of Q (C) is Q (C). A single line will emanate from 1 C of charge.

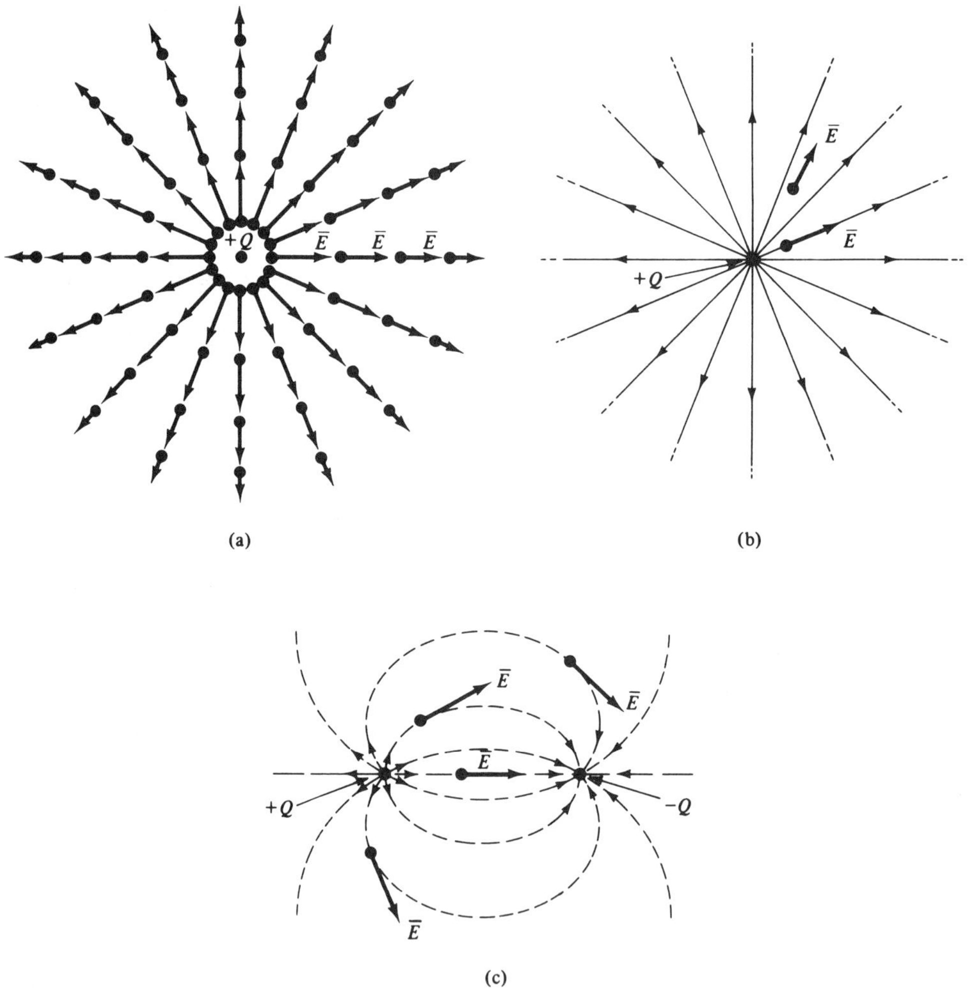

Figure 3-1 Development of the flux line concept: (a) plot of the vector force field $\overline{E}$ at discrete points in two-dimensional space about a $+Q$ (C) point charge; (b) plot of electric flux lines about a $+Q$ point charge; (c) plot of electric flux lines about two point charges. The actual plots must be drawn in three-dimensional space.

In Fig. 3-1(b) and (c), $\overline{E}$ has been drawn to show the relationship to the electric flux concept. Thus, it can be seen that we have tremendously improved, over Fig. 3-1(a), the visual picture of the mysterious forces between charges. From Fig. 3-1(b), we can obtain at a glance the following information:[1]

1. The $\overline{E}$ is radial and outward (direction of electric flux).
2. The magnitude of $\overline{E}$ is the same at a fixed radius (electric flux density is the same).
3. The magnitude of $\overline{E}$ decreases with distance from the charge (electric flux density decreases with distance from the charge).
4. The $\overline{E}$ is symmetrical about the $+Q$ (C) charge.

In the list of rules for the construction of electric flux lines, rule 4 states that the number of electric flux lines emanating from a $+Q$ (C) charge is equal to Q in the SI units. If we let Ψ_E symbolize the number of electric lines or coulombs of flux, we may write

$$\boxed{\Psi_E = Q \qquad \text{(lines or C)}} \tag{1}$$

where Ψ_E is in coulombs of electric flux, emanating from charge Q (C), and Q is the charge in coulombs.

Let us apply (1) or rule 4 when $Q = 1$ C and $Q = 1000$ C to obtain the flux plots of Fig. 3-2. Figure 3-2(a) illustrates the case when $Q = 1$ C. It can be seen that symmetry is lost, and the plot does not convey the correct information. Figure 3-2(b) illustrates the case when $Q = 1000$ C. It can be seen that too much work is involved in drawing a thousand flux lines, and that the flux plot tends to become a solid black blob. This dilemma of too few lines of flux, when $Q = 1$ C, and too many lines of flux, when $Q = 1000$ C, can be averted by negating rule 4 in making a flux plot, but allowing (1), that stems from rule 4, to hold in quantitative formulations. Thus, the number of flux lines used in a flux plot should be sufficient to give a clear visual picture of the force field about the charge or system of charges. If symmetry exists in the electric field intensity, the electric flux plot should clearly indicate this property. All the foregoing concepts of electric flux lines and flux plots can be extended to any vector field.

Example 1

Using the first three rules of the electric flux concept and the expression for $\overline{E}$, construct the electric flux plot for an infinite line of uniform charge $+\rho_\ell$ (C m^{-1}).

Solution. From (2.5-13),

$$\overline{E} = \hat{r}_c \frac{\rho_\ell}{2\pi\epsilon_0 r_c}$$

we note that $\overline{E}$ is radial and that $|\overline{E}|$ varies as $1/r_c$. Thus, the electric flux lines are radial and their density will decrease with r_c. The electric flux plot is shown in Fig. 3-3(a) in three-quarter space only. Figure 3-3(b) shows the plot in the z = constant plane, and the plot of $|\overline{E}|$ versus r_c is shown in Fig. 3-3(c).

[1] Here we assume that a charge of $-Q$ (C) is distributed uniformly at infinity.

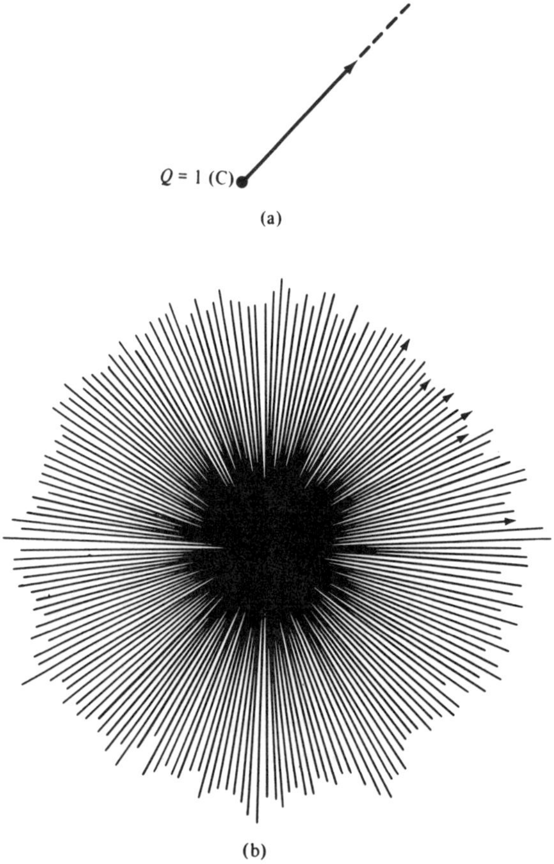

Figure 3-2 (a) Plot of one electric flux line about a +1 C point charge. Symmetry is lost. (b) Plot of 1000 flux lines about a +1000 C point charge. Too much work is involved and a poor flux plot results.

Example 2

Repeat Example 1 for an infinite sheet of uniform $+\rho_s$ (C m^{-2}) distribution.

Solution. From (2.6-7),

$$\overline{E} = \hat{z}\frac{\rho_s}{2\epsilon_0}$$

we note that $\overline{E}$ is in the $\hat{z}$ direction, perpendicular to the sheet of charge, and $|\overline{E}|$ is constant above and below the sheet. The electric flux plot is shown in Fig. 3-4.

Problem 3.2-1 Starting with the results of Example 2, construct the flux plot between two infinite sheets of uniform but opposite charge densities.

Problem 3.2-2 Construct the electric flux plot for the field between two oppositely charged concentric cylinders of radii r_a and r_b with length L, where $r_a < r_b$. Let Q be the total charge on the outer cylinder and $-Q$ be the charge on the inner cylinder. Find the charge densities on each.

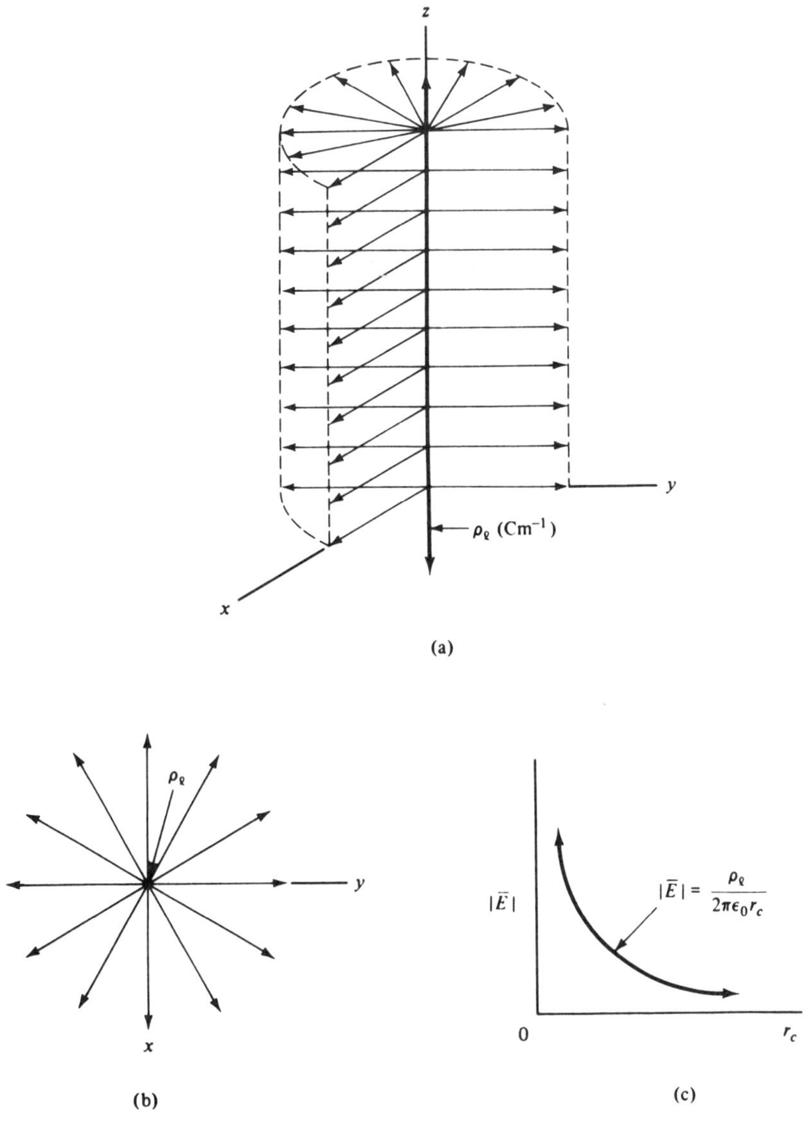

Figure 3-3 (a) Electric flux plot about an infinite line charge of uniform ρ_ℓ (C m^{-1}) in three-quarter space only. (b) Electric flux plot in a z = constant plane. (c) Plot of $|\bar{E}|$ versus r_c.

Problem 3.2-3 Construct the flux plot for the field between two concentric and oppositely charged spheres of radii r_a and r_b, where $r_a < r_b$. Find the surface charge densities on each if there are Q (C) on the inner sphere and $-Q$ (C) on the outer sphere.

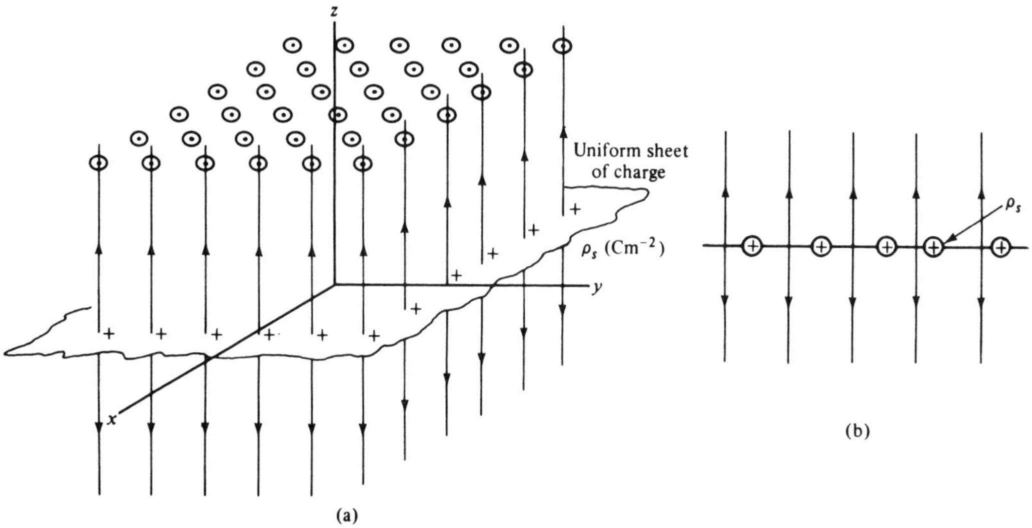

Figure 3-4 (a) Electric flux plot about an infinite sheet of uniform $+\rho_s$ (C m^{-2}). (b) Electric flux plot in an x = constant or y = constant planes.

3.3 ELECTRIC FLUX DENSITY VECTOR

In free space, the electric flux density vector $\overline{D}$ is defined to be in the same direction as the electric flux lines (same as $\overline{E}$) and to have magnitude

$$|\overline{D}| \triangleq \lim_{\Delta s \to 0} \left[\frac{\Delta \Psi_E}{\Delta s} \right] \quad \text{(lines m}^{-2}\text{)} \quad \text{(aap)} \tag{1}$$

where $\Delta \Psi_E$ equals the number of electric lines of flux that are perpendicular to the surface Δs, as shown in Fig. 3-5(a).

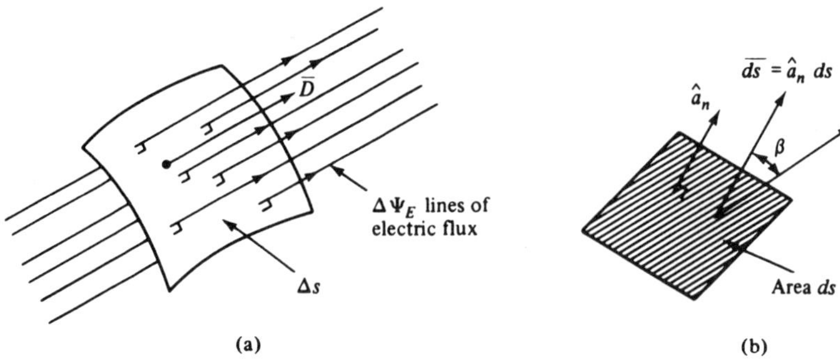

Figure 3-5 (a) Electric lines of flux through a surface Δs. (b) The flux, $d\Psi_E$, through the surface ds is equal to $d\Psi_E = \overline{D} \cdot \overline{ds}$. Note that $d\Psi_E = 0$ when $\beta = \pi/2$.

In Fig. 3-5(b), the $d\Psi_E$ through a surface ds can be expressed as

$$d\Psi_E = \overline{D} \bullet \overline{ds} \qquad \text{(lines or C)} \tag{2}$$

where $\overline{ds}$ is a differential surface vector. The vector $\overline{ds} = \hat{a}_n\, ds$, where $\hat{a}_n$ is a unit vector perpendicular to the physical surface and ds is the surface area in (m^2). The surface vector concept will be used extensively in this text. Note that the flux $d\Psi_E$ will be maximum when $\overline{ds}$ and $\overline{D}$ are in the same direction.

If we locate a point charge $+Q$ (C) at the origin, as shown in Fig. 3-6, and construct a concentric sphere of radius r_s, we can evaluate $\overline{D}$ by dividing (3.2-1) by the surface area of the sphere; thus,

$$\overline{D} = \hat{r}_s \frac{\Psi_E}{4\pi r_s^2} = \hat{r}_s \frac{Q}{4\pi r_s^2} \qquad (\text{lines m}^{-2} \text{ or C m}^{-2}) \tag{3}$$

The expression for $\overline{E}$ on the surface at r_s, from (2.4-5), is

$$\overline{E} = \hat{r}_s \frac{Q}{4\pi \epsilon_0 r_s^2} \qquad (\text{V m}^{-1}) \quad \text{(free space)} \tag{2.4-5}$$

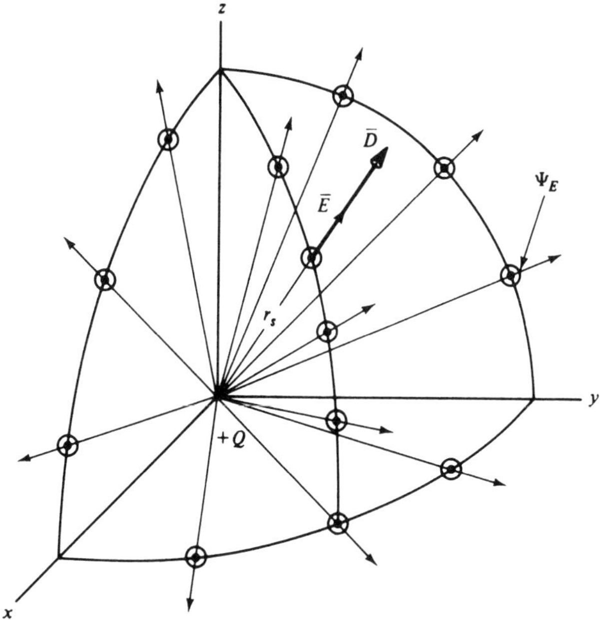

Figure 3-6 The flux density vector due to a point charge Q, in free space, at a radius of r_s, is equal to $Q/4\pi r_s^2$ (lines m^{-2}). Note that only the $+x$, $+y$, $+z$ space is shown.

From (3) and (2.4-5) we see that

$$\overline{D} = \epsilon_0 \overline{E} \quad \text{(free space)} \quad \text{(aap)} \tag{4}$$

Equation (4) was derived using a point charge Q, but its validity can be extended to a general charge distribution. For a general charge distribution,

$$\overline{E} = \int_v \frac{\rho_v \, dv}{4\pi\epsilon_0 R^2} \hat{a}_R \quad \text{(free space)} \quad \text{(aap)} \tag{5}$$

at a given point in space. Now, through the use of (3), good for a point charge, the $\overline{D}$ at the same given point becomes

$$\overline{D} = \int_v \frac{\rho_v \, dv}{4\pi R^2} \hat{a}_R \qquad \text{(lines m}^{-2}\text{)} \quad \text{(aap)} \tag{6}$$

for the same general charge distribution. From (5) and (6), it can be seen that (4) applies to any general charge distribution in free space.

From the famous ice pail (electric induction experiments of Michael Faraday, Ψ_E and thus $\overline{D}$ are independent of the dielectric media in which Q is embedded. This is not true for $\overline{E}$ and thus (2.4-5), (4), and (5) are only true for free space. The relationship (4) between $\overline{D}$ and $\overline{E}$ will become slightly more complicated when we consider media other than free space. From (4) we deduce that the field plot of $\overline{D}$ is identical to that of $\overline{E}$.

Example 3

Find the number of lines of electric flux emanating from a point charge of Q (C) at the origin by finding the electric lines of flux through an imaginary concentric sphere of radius r_s (m).

Solution. From Fig. 3-7 and (2), the differential flux through ds is equal to

$$d\Psi_E = \overline{D} \cdot \overline{ds} \qquad \text{(lines)}$$

Thus, $\Psi_E = \oint_s \overline{D}_s \cdot \overline{ds}$, where $\overline{D}_s$ is evaluated on the closed surface of the imaginary sphere of radius r_s. From $\overline{D}_s = \epsilon_0 \overline{E}_s$, $\overline{E}_s = \hat{r}_s(Q/4\pi\epsilon_0 r_s^2)$, and $\overline{ds} = \hat{r}_s r_s^2 \sin\theta \, d\theta \, d\phi$, we have

$$\Psi_E = \oint_s \left(\epsilon_0 \hat{r}_s \frac{Q}{4\pi\epsilon_0 r_s^2} \right) \cdot (\hat{r}_s r_s^2 \sin\theta \, d\theta \, d\phi)$$

$$= \int_0^{2\pi} \int_0^{\pi} \frac{Q}{4\pi} \sin\theta \, d\theta \, d\phi = Q \qquad \text{(lines or C)}$$

Note that $\Psi_E = Q$ (lines), where Q is the charge enclosed by the closed surface s. The solution is obvious from (3.2-1) and is independent of the closed surface selected.

Example 4

Find the electric flux Ψ_E that passes through the surface shown in Fig. 3-8, when $\overline{D} = (\hat{x}y + \hat{y}x)10^{-2}$ C m^{-2}.

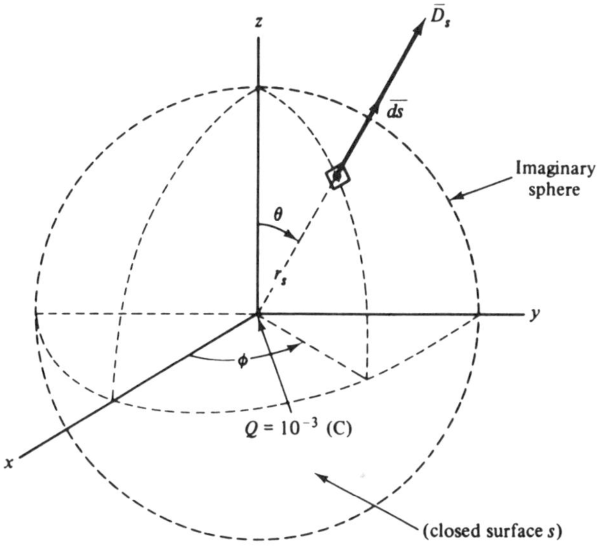

Figure 3-7 Graphical display for finding the flux density vector $\bar{D}$ on an imaginary, closed, and concentric surface of radius r_s (m).

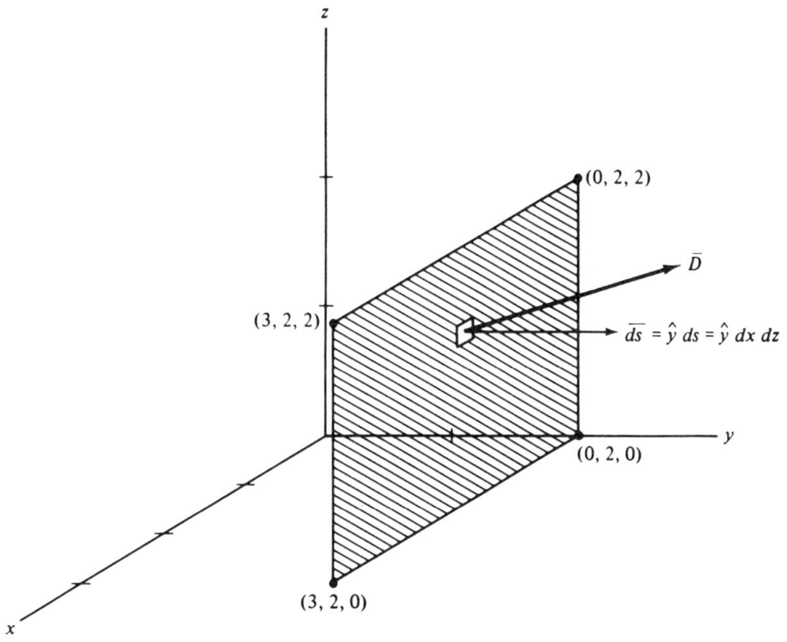

Figure 3-8 Graphical display for finding flux through a surface in Example 4.

Solution. From Fig. 3-8 and (3.3-2),

$$d\Psi_E = \overline{D} \bullet \overline{ds} = (\hat{x}y + \hat{y}x)10^{-2} \bullet (\hat{y}\,dx\,dz) = x10^{-2}\,dx\,dz$$

$$\Psi_E = \int_0^2 \int_0^3 x10^{-2}\,dx\,dz = 9 \times 10^{-2} \text{ lines}$$

Problem 3.3-1 Find the total electric flux Ψ_E between two parallel uniformly but oppositely charged plates if the area of each plate is 1 m^2, ρ_s on the one plate is 2×10^{-8} C m^{-2}, and ρ_s on the other plate is -2×10^{-8} C m^{-2}.

Problem 3.3-2 Find the total electric flux between cylinders of Prob. 3.2-2 if $L = 1$ m, $r_a = 1$ mm, and $r_b = 1$ cm, if the charge density on the outer cylinder is -5×10^{-8} C m^{-2}.

Problem 3.3-3 Find the total electric flux between the concentric spheres of Prob. 3.2-3 if ρ_s on the inner sphere is 6×10^{-7} C m^{-2}, $r_a = 0.7$ cm, and $r_b = 3$ cm.

Problem 3.3-4 Find the total electric flux that passes outwardly through the surface of a cube, 0.2 m on an edge, centered at the origin, when the flux density is $\overline{D} = \hat{y}|y| + \hat{x}x^2 + \hat{z}z$ (μC m^{-2}). Let the sides of the cube be parallel to the rectangular axes.

3.4 GAUSS'S LAW

From rule 4 of the electric flux concept we were able to write $\Psi_E = Q$ [eq. (3.2-1)]. A statement of this equation is that the total electric flux emanating from a charge $+Q$ (C) is equal to Q (C) in the SI units. We can restate this by saying that the total electric flux passing through any closed imaginary surface, enclosing the charge Q (C), is equal to Q (C) in the SI units. Note that Q is enclosed by the closed surface. For this reason we will refer to Q as Q enclosed, or just Q_{en}. A graphical representation of this last statement, $\Psi_E = Q_{\text{en}}$, is shown in Fig. 3-9. The total flux Ψ_E is thus equal to

$$\Psi_E = \oint_s d\Psi_E = \oint_s \overline{D}_s \bullet \overline{ds} = Q_{\text{en}} \qquad \text{(C)} \tag{1}$$

where $\oint_s$ indicates a double integral over the closed surface s. The mathematical form, from (1),

$$\boxed{\oint_s \overline{D}_s \bullet \overline{ds} = Q_{\text{en}} \qquad \text{(C)}} \tag{2}$$

is named *Gauss's law* after Karl Friedrich Gauss, a great mathematician. The Q_{en} enclosed by surface s, due to a ρ_v distribution, becomes

$$\boxed{Q_{\text{en}} = \int_v \rho_v\,dv \qquad \text{(C)}} \tag{3}$$

where the v is the volume enclosed by the closed surface s. In (2), the $\overline{D}_s$ is evaluated on the closed surface s. The charge enclosed Q_{en} can be a net charge due to any combination of positive and negative charge distributions. Any charge not enclosed will pass the same number of Ψ_E lines into and out of the closed surface s.

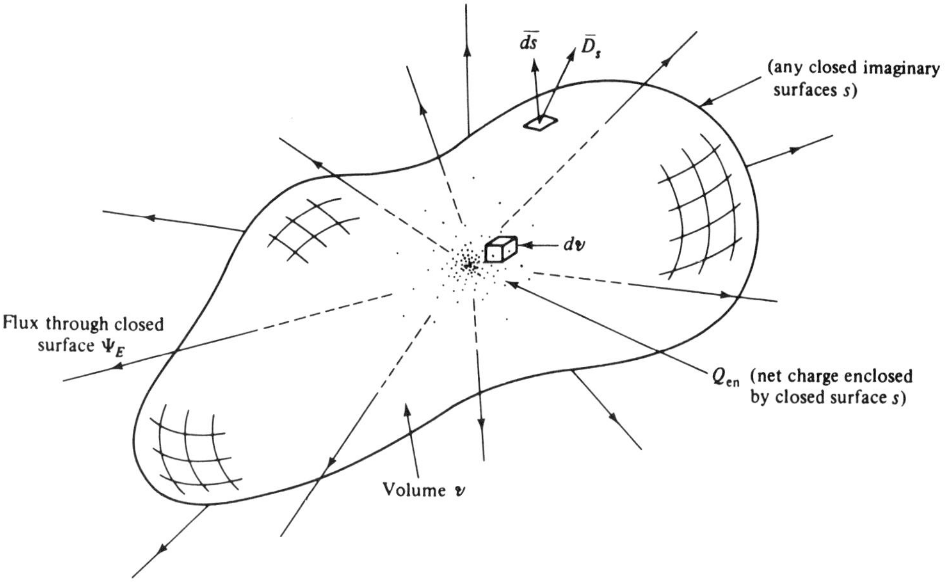

Figure 3-9 The total electric lines of flux Ψ_E passing through any closed imaginary surface enclosing the charge Q_{en} (C) is equal to Q_{en} (lines).

Gauss's law was used in Prob. 3.3-2 to find the charge enclosed by a closed surface. Another application of Gauss's law is in finding $\overline{D}$ or $\overline{E}$ of symmetrical charge distributions as a point charge, spherical shell of uniform charge, sphere of uniform charge, infinite line of uniform charge, infinite cylindrical surface of uniform charge, infinite sheet of uniform charge, etc.

Example 5

Through the use of Gauss's law (2), find the $\overline{D}$ and $\overline{E}$ about a point charge Q.

Solution. From (2),

$$\oint_s \overline{D}_s \cdot \overline{ds} = Q_{en}$$

Construct a spherical closed surface concentric about the point charge Q, as shown in Fig. 3-10. This surface we will call a *Gaussian surface*. From symmetry, $\overline{D}_s = \hat{r}_s D_{r_s}$ on the surface at some radius r_s; also, $\overline{ds} = \hat{r}_s\, ds$. Substituting into the above, we have

$$\oint_s \overline{D}_s \cdot \overline{ds} = \oint_s \hat{r}_s D_{r_s} \cdot \hat{r}_s\, ds = \oint_s D_{r_s}\, ds = Q_{en}$$

Now, at a fixed radius, D_{r_s} is constant over the range of $\oint_s$ and thus can be removed from under the integral sign to obtain

$$D_{r_s} \oint_s ds = Q_{en} = Q$$

The integral $\oint_s ds$ equals $4\pi r_s^2$, the Gaussian surface area. Thus,

$$D_{r_s}(4\pi r_s^2) = Q$$

and

$$D_{r_s} = \frac{Q}{4\pi r_s^2}$$

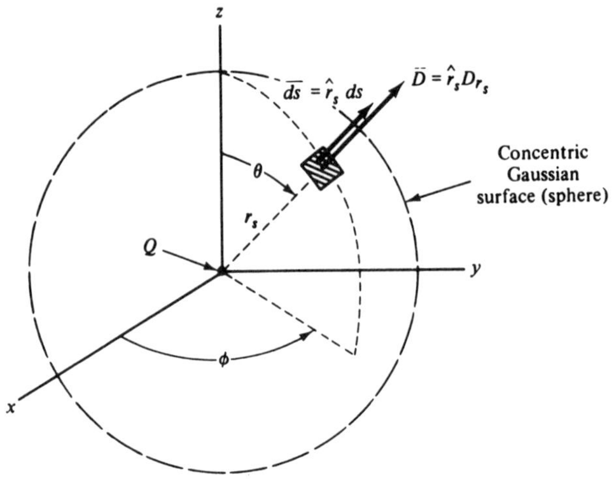

Figure 3-10 Graphical display for finding $\overline{D}$ and $\overline{E}$ about a point charge Q in Example 5 through the use of Gauss's law.

Therefore,

$$\overline{D} = \hat{r}_s \frac{Q}{4\pi r_s^2} \qquad (\text{C m}^{-2} \text{ or lines m}^{-2}) \tag{4}$$

and

$$\overline{E} = \frac{\overline{D}}{\epsilon_0} = \hat{r}_s \frac{Q}{4\pi\epsilon_0 r_s^2} \qquad (\text{V m}^{-1}) \tag{5}$$

Close examination will show that the key to finding $\overline{D}$ through the use of Gauss's law is the ability to remove D_{r_s} from under the integral sign. This can be accomplished in problems of symmetrical charge distributions.

Table 3-1 contains the expressions for $\overline{D}$ about several spherically symmetrical charge distributions. These expressions were obtained with ease, through application of Gauss's law. If the total charge, enclosed with $r_s = b$ in each case, is made equal to Q, the expressions for $\overline{D}$ will have identical form for $r_s \geqq b$. Table 3-2 contains the expressions for $\overline{D}$ about several cylindrically symmetrical charge distributions.

These expressions were also obtained through the application of Gauss's law. If the charge per unit length, enclosed within $r_c = b$ in each case, is made equal to ρ_ℓ, the expressions for $\overline{D}$ will have identical form for $r_c \geqq b$. In fact, we can say that an infinite number of different spherically symmetrical charge distributions will produce the same $\overline{D}$ for $r_s \geqq b$ when they enclose the same Q within $r_s = b$. This also applies to cylindrically symmetrical charge distributions.

Example 6

Through the use of Gauss's law, find $\overline{D}$ and $\overline{E}$ about an infinite length line charge of uniform ρ_ℓ.

Solution. From (2),

$$\oint_s \overline{D}_s \bullet \overline{ds} = Q_{\text{en}}$$

TABLE 3-1 EXPRESSIONS AND PLOTS FOR $\overline{D}$ ABOUT SEVERAL SPHERICALLY SYMMETRICAL CHARGE DISTRIBUTIONS; IN EACH CASE, THE TOTAL CHARGE IS Q (C)

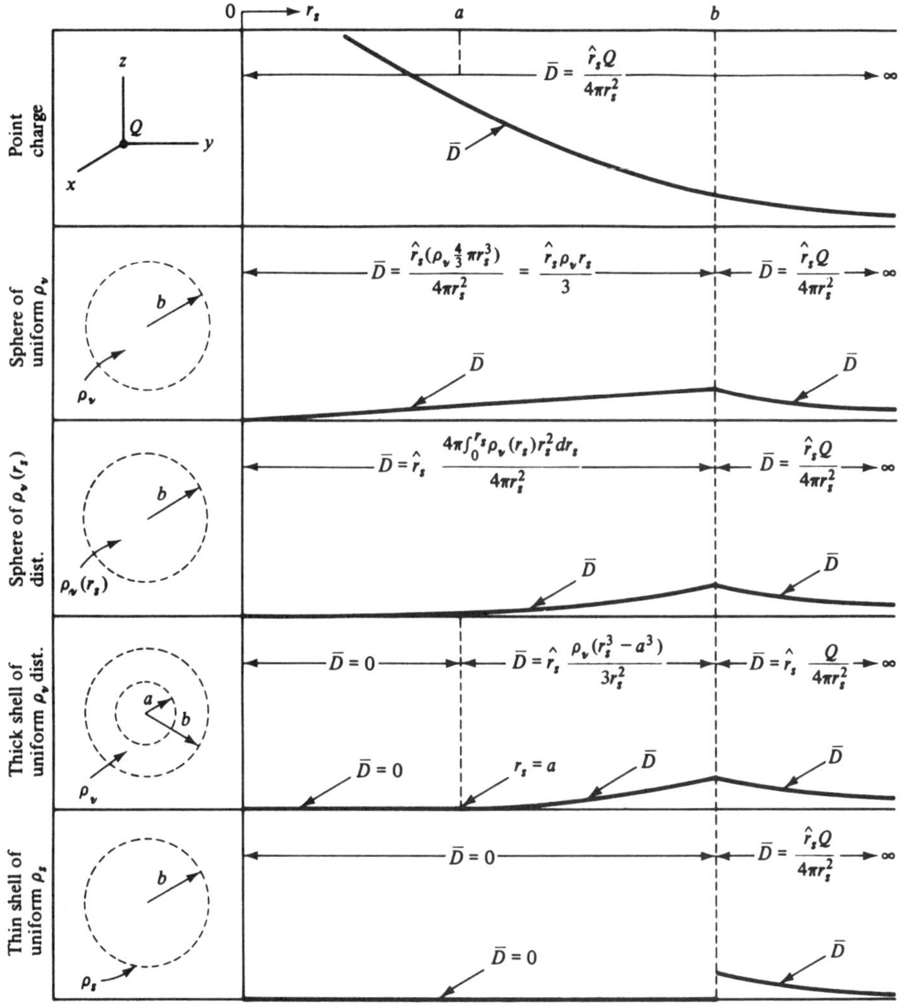

Construct a Gaussian closed surface concentric about the uniform line charge as shown in Fig. 3-11. Note that the closed surface resembles a can of radius r_c (m). The closed surface integral must be expressed as

$$\oint_s \overline{D}_s \cdot \overline{ds} = \int_{s(\text{top})} \overline{D}_s \cdot \overline{ds} + \int_{s(\text{bottom})} \overline{D}_s \cdot \overline{ds} + \int_{s(\text{side})} \overline{D}_s \cdot \overline{ds} \tag{6}$$

where $\overline{D}_s$ and $\overline{ds}$ are evaluated on the respective surfaces. Through the use of Coulomb's force law, it can be shown that $\overline{E}$, and thus $\overline{D}$, are in the radial direction and do not vary

TABLE 3-2 EXPRESSIONS AND PLOTS FOR $\overline{D}$ ABOUT SEVERAL CYLINDRICALLY SYMMETRICAL CHARGE DISTRIBUTIONS; IN EACH CASE, THE TOTAL CHARGE PER UNIT LENGTH EQUALS ρ_ℓ (C m^{-1})

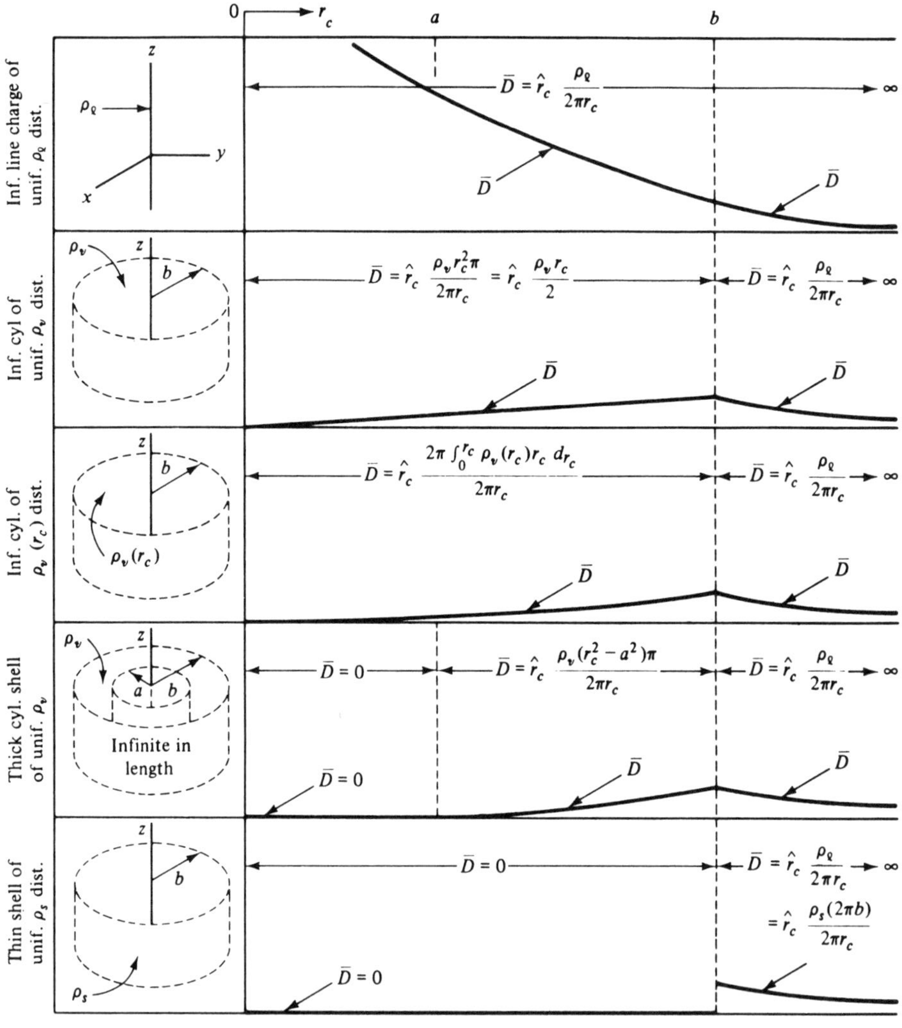

with z. Thus, (6) becomes

$$\oint_s \overline{D}_s \bullet \overline{ds} = \int_{s(\text{top})} (\hat{r}_c D_{r_c}) \bullet (\hat{z}\, ds) + \int_{s(\text{bottom})} (\hat{r}_c D_{r_c}) \bullet (-\hat{z}\, ds)$$

$$+ \int_{s(\text{side})} (\hat{r}_c D_{r_c}) \bullet (\hat{r}_c\, ds) \tag{7}$$

In (7), the integrals over the top and bottom surfaces are zero since $\overline{D}$ and $\overline{ds}$ are perpendicular. In the integral over the side surface, D_{r_c} is constant and can be taken out from under the integral sign to give

$$\oint_s \overline{D}_s \bullet \overline{ds} = D_{r_c} \int_{s(\text{side})} ds = D_{r_c}(2\pi r_c\, \Delta\ell) = Q_{\text{en}} \tag{8}$$

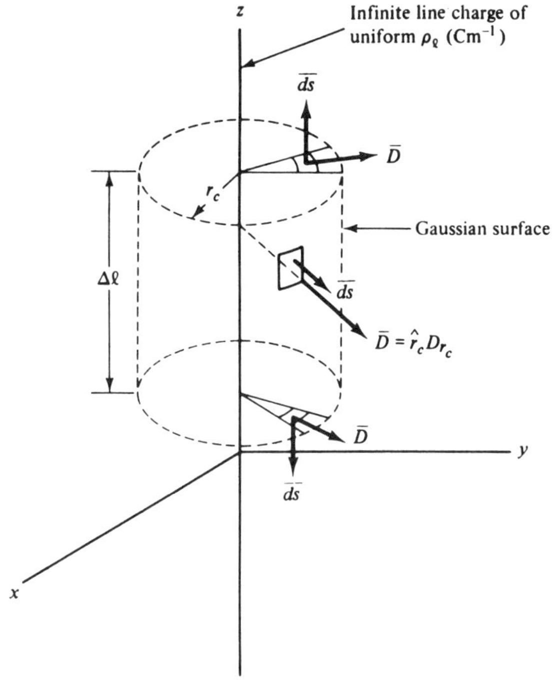

Figure 3-11 Graphical display for finding $\overline{D}$ and $\overline{E}$ about an infinite and uniform line charge through the use of Gauss's law in Example 6.

In (8), $Q_{en} = \rho_\ell \Delta\ell$ and upon substitution we obtain

$$D_{r_c}(2\pi r_c \Delta\ell) = \rho_\ell \Delta\ell$$

Solving for D_{r_c} we obtain

$$D_{r_c} = \frac{\rho_\ell}{2\pi r_c}$$

$$\overline{D} = \hat{r}_c \frac{\rho_\ell}{2\pi r_c} \qquad (\text{C m}^{-2})$$

and

$$\overline{E} = \frac{\overline{D}}{\epsilon_0} = \frac{\hat{r}_c \rho_\ell}{2\pi\epsilon_0 r_c} \qquad (\text{V m}^{-1})$$

This agrees with (2.5-8).

Example 7

Through the use of Gauss's law, find $\overline{D}$ and $\overline{E}$ inside and outside a thin spherical shell of uniform ρ_s and of radius r_0 (m), as shown in Fig. 3-12.

Solution. From (2),

$$\oint_s \overline{D}_s \bullet \overline{ds} = Q_{en}$$

For $\overline{D}$ and $\overline{E}$ inside the shell, $r_s < r_0$, we construct a concentric Gaussian spherical surface of $r_s < r_0$, as shown in Fig. 3-12. Let us generate the thin shell of charge by means of an infinite number of thin rings of charge and then generate each ring by means of an infinite number of point charges, two of which are shown in Fig. 3-12. It can be seen that two radially opposite point charges, on the ring, will only contribute to a radial $\overline{E}$. Extension of this concept to all radially opposite point charges on a given ring and to all the infinite rings of charge will show that $\overline{E}$ and thus $\overline{D}$ are radial inside and outside of the shell of charge. Due

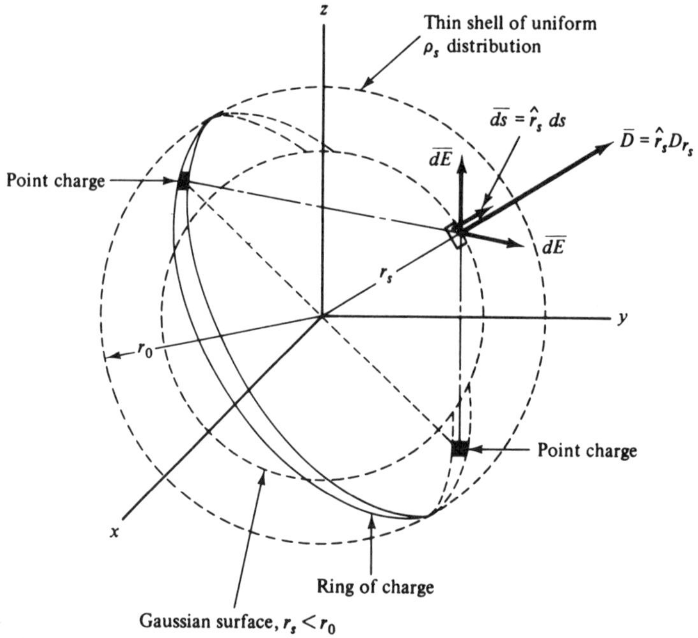

Figure 3-12 Graphical construction for finding $\overline{D}$ and $\overline{E}$ inside a thin shell of uniform charge distribution through the use of Gauss's law in Example 7.

to symmetry D_s is constant on a constant r_s surface. Substituting these facts into Gauss's law expression, we have

$$\oint_s \overline{D}_s \cdot \overline{ds} = \oint_s (\hat{r}_s D_{r_s}) \cdot (\hat{r}_s\, ds) = D_{r_s} \oint_s ds = D_{r_s}(4\pi r_s^2) = Q_{\text{en}} \tag{9}$$

The Q_{en} by the Gaussian surface is zero; thus, from (9) we have

$$D_{r_s}(4\pi r_s^2) = 0 \tag{10}$$

This equation leads to $D_{r_s} = 0$. Thus, $\overline{D}$ and $\overline{E}$ are zero for $r_s < r_0$.

To find $\overline{D}$ and $\overline{E}$ outside the shell, $r_s > r_0$, we construct a concentric spherical Gaussian surface of $r_s > r_0$. Substituting into Gauss's law expression, we obtain (9) except now Q_{en} becomes the total charge on the shell of radius r_0 (m). Thus, from (9), we obtain

$$D_{r_s}(4\pi r_s^2) = Q_{\text{en}} = \rho_s(4\pi r_0^2) \tag{11}$$

Solving for D_{r_s}, we have

$$D_{r_s} = \frac{\rho_s r_0^2}{r_s^2}$$

Thus,

$$\overline{D} = \hat{r}_s \frac{\rho_s r_0^2}{r_s^2} \qquad (\text{C m}^{-2}) \tag{12}$$

and

$$\overline{E} = \hat{r}_s \frac{\rho_s r_0^2}{\epsilon_0 r_s^2} \qquad (\text{V m}^{-1}) \tag{13}$$

If in (11) we let $\rho_s(4\pi r_0^2) = Q$, the expression for $\overline{D}$ becomes

$$\overline{D} = \hat{r}_s \frac{Q}{4\pi r_s^2} \qquad (\text{C m}^{-2}) \tag{14}$$

Equation (14) has the same form as (4) and thus we find, for $r_s > r_0$, that the $\overline{D}$ field due to a thin shell of uniform charge is the same as if the total charge were located at a point at the center of the sphere.

Problem 3.4-1 Through the use of Gauss's law, find the expression for $\overline{D}$ and $\overline{E}$ inside and outside a sphere of r_0 radius and of uniform ρ_v (C m^{-3}) distribution. Compare the results for $\overline{E}$ outside the sphere with (2.7-11) and (2.7-12).

Problem 3.4-2 Through the use of Gauss's law, find the expression for $\overline{D}$ and $\overline{E}$ inside and outside an infinite cylinder of r_0 radius and of uniform ρ_v (C m^{-3}) distribution.

Problem 3.4-3 Through the use of Gauss's law, find the expression for $\overline{D}$ and $\overline{E}$ above and below an infinite sheet of uniform ρ_s (C m^{-2}) distribution. Locate the infinite sheet in the $z = 0$ plane, and compare the results with (2.6-7).

3.5 DIVERGENCE

Let us apply Gauss's law to a point $P(x_0, y_0, z_0)$ in a $\overline{D}$ field that has been produced by some system of charge distribution. A portion of this $\overline{D}$ field is shown in Fig. 3-13(a) with a small cubical Gaussian surface enclosing at its center the point $P(x_0, y_0, z_0)$, where $\overline{D} = \overline{D}_0 = \hat{x}D_{x0} + \hat{y}D_{y0} + \hat{z}D_{z0}$. This cubical surface encloses the volume $\Delta v = \Delta x\, \Delta y\, \Delta z$, which later will be allowed to shrink to zero and thus to the point $P(x_0, y_0, z_0)$. Through the application of Gauss's law to a point we will not evaluate $\overline{D}_0$, as in Sec. 3.4, but we will find a partial differential equation that will relate components of the vector $\overline{D}$ to another physical quantity at the point $P(x_0, y_0, z_0)$.

An expanded sketch of the small cube is shown in Fig. 3-13(b). To apply Gauss's law on each of the six surfaces, we must express the closed surface integral as a sum of six integrals as

$$\oint_s \overline{D} \cdot \overline{ds} = \int_{s(\text{front})} \overline{D}_s \cdot \overline{ds} + \int_{s(\text{back})} \overline{D}_s \cdot \overline{ds} + \int_{s(\text{right})} \overline{D}_s \cdot \overline{ds}$$
$$+ \int_{s(\text{left})} \overline{D}_s \cdot \overline{ds} + \int_{s(\text{top})} \overline{D}_s \cdot \overline{ds} + \int_{s(\text{bottom})} \overline{D}_s \cdot \overline{ds} = Q_{\text{en}} \tag{1}$$

From (1), it can be seen that only the component of $\overline{D}_s$ that is normal to the surface, on each of the surfaces, contributes to the surface integral. Since each of the surfaces will be later shrunk to zero, we will assume that the normal components of $\overline{D}_s$ are uniform over their respective surfaces. Incorporating these ideas into the first two surface integrals found in (1), we have

$$\int_{s(\text{front})} \overline{D}_s \cdot \overline{ds} + \int_{s(\text{back})} \overline{D}_s \cdot \overline{ds} \simeq \overline{D}_{\text{front}} \cdot \overline{\Delta s}_{\text{front}} + \overline{D}_{\text{back}} \cdot \overline{\Delta s}_{\text{back}}$$
$$\simeq \hat{x}D_{x(\text{front})} \cdot \hat{x}\,\Delta y\, \Delta z + \hat{x}D_{x(\text{back})} \cdot (-\hat{x}\,\Delta y\, \Delta z)$$
$$\simeq D_{x(\text{front})}\,\Delta y\, \Delta z - D_{x(\text{back})}\,\Delta y\, \Delta z$$
$$\simeq (D_{x(\text{front})} - D_{x(\text{back})})\,\Delta z\, \Delta y \tag{2}$$

The normal components $D_{x(\text{front})}$ and $D_{x(\text{back})}$ can be found through the use of the first two terms of the Taylor expansion for D_x about the point $P(x_0, y_0, z_0)$.
Thus, we have

$$D_{x(\text{front})} \simeq D_{x0} + \frac{\partial D_x}{\partial x}\left(\frac{\Delta x}{2}\right) \tag{3}$$

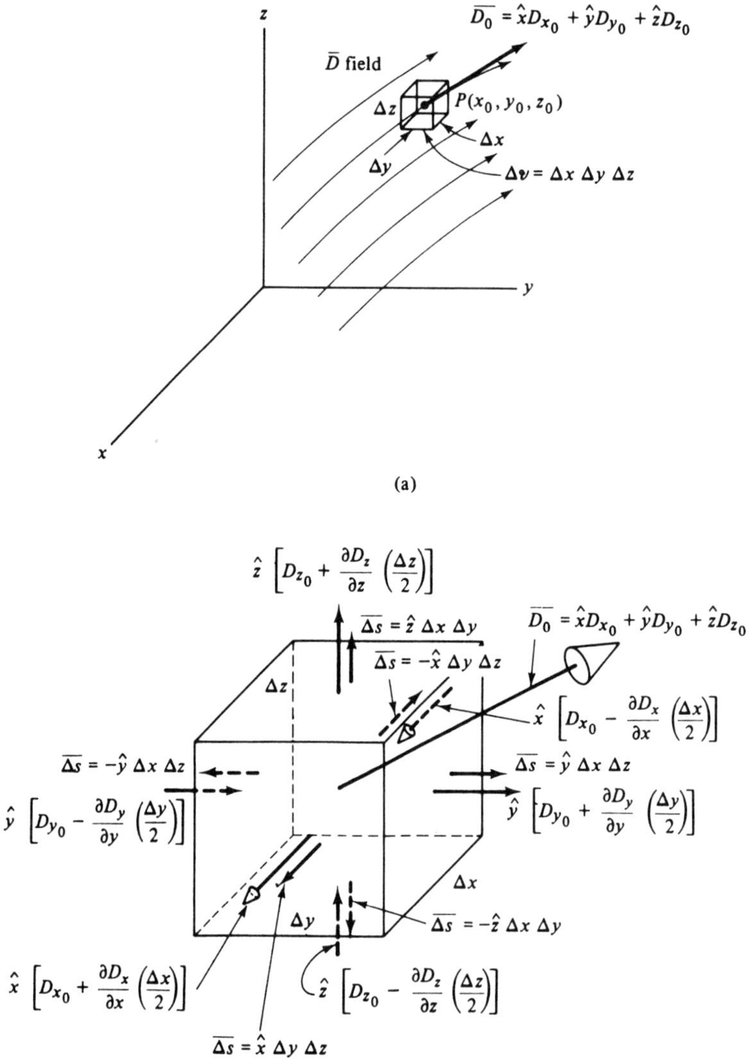

Figure 3-13 Graphical display for finding divergence of $\overline{D}$ at $P(x_0, y_0, z_0)$.

$$D_{x(\text{back})} \simeq D_{x_0} + \frac{\partial D_x}{\partial x}\left(-\frac{\Delta x}{2}\right) \tag{4}$$

Substituting (3) and (4) into (2), we have

$$\int_{s(\text{front})} \overline{D_s} \bullet \overline{ds} + \int_{s(\text{back})} \overline{D_s} \bullet \overline{ds} \simeq \frac{\partial D_x}{\partial x}\,\Delta x\,\Delta y\,\Delta z \tag{5}$$

Following the same procedure, we find that

$$\int_{s(\text{right})} \overline{D_s} \bullet \overline{ds} + \int_{s(\text{left})} \overline{D_s} \bullet \overline{ds} \simeq \frac{\partial D_y}{\partial y}\,\Delta x\,\Delta y\,\Delta z \tag{6}$$

and

$$\int_{s(\text{top})} \overline{D}_s \cdot \overline{ds} + \int_{s(\text{bottom})} \overline{D}_s \cdot \overline{ds} \simeq \frac{\partial D_z}{\partial z} \Delta x \, \Delta y \, \Delta z \tag{7}$$

Substituting (5), (6), and (7) into (1) yields

$$\oint_s \overline{D} \cdot \overline{ds} \simeq \left(\frac{\partial D_x}{\partial x} + \frac{\partial D_y}{\partial y} + \frac{\partial D_z}{\partial z} \right) \Delta x \, \Delta y \, \Delta z \simeq Q_{\text{en}} \tag{8}$$

Let us divide (8) by $\Delta v = \Delta x \, \Delta y \, \Delta z$ and then take the limit as $\Delta v \to 0$; thus,

$$\lim_{\Delta v \to 0} \left[\frac{\oint_s \overline{D} \cdot \overline{ds}}{\Delta v} \right] = \left(\frac{\partial D_x}{\partial x} + \frac{\partial D_y}{\partial y} + \frac{\partial D_z}{\partial z} \right) = \lim_{\Delta v \to 0} \left[\frac{Q_{\text{en}}}{\Delta v} \right] \tag{9}$$

Note that taking the limit as $\Delta v \to 0$ justifies the use of only the first two terms of the Taylor expansion and the changing of the approximation signs to equality signs. The left term of (9) is defined as the divergence of $\overline{D}$; thus,

$$\text{divergence of } \overline{D} \triangleq \lim_{\Delta v \to 0} \left[\frac{\oint_s \overline{D}_s \cdot \overline{ds}}{\Delta v} \right] \quad (\text{lines m}^{-3} \text{ or C m}^{-3}) \quad (\text{aap}) \tag{10}$$

A physical significance of divergence of $\overline{D}$ can be obtained from (10) by stating what is implied

> The divergence of $\overline{D}$ equals the net flux of the vector $\overline{D}$ that flows outwardly through a closed surface s per unit volume (enclosed by $\oint_s$) as the volume goes to zero.

The last term in (9) equals ρ_v (C m^{-3}) at the point $P(x_0, y_0, z_0)$. From (1.12-3) and the center term of (9), we can symbolize divergence of $\overline{D}$ as $\nabla \cdot \overline{D}$. Incorporating these concepts with (9), we have

$$\underbrace{\text{Div } \overline{D}}_{\text{Concept}} = \underbrace{\nabla \cdot \overline{D}}_{\text{Vector analysis compact symbolization}} \triangleq \underbrace{\lim_{\Delta v \to 0} \left[\frac{\oint_s \overline{D}_s \cdot \overline{ds}}{\Delta v} \right]}_{\text{Defining equation}} = \underbrace{\left(\frac{\partial D_x}{\partial x} + \frac{\partial D_y}{\partial y} + \frac{\partial D_z}{\partial z} \right)}_{\text{Mathematical relationship resulting from application of defining equation in the rectangular coordinate system}} = \rho_v \quad \underbrace{(\text{C m}^{-3}) \quad (\text{aap})}_{\text{Physical quantity at the point } P(x_0, y_0, z_0)} \tag{11}$$

Thus, in rectangular coordinates

$$\text{Div } \overline{D} = \nabla \cdot \overline{D} = \frac{\partial D_x}{\partial x} + \frac{\partial D_y}{\partial y} + \frac{\partial D_z}{\partial z} \quad (\text{C m}^{-3}) \quad (\text{aap}) \tag{12}$$

From the last two terms of (11), the divergence concept gives us a partial differential equation that relates rates of change of the $\overline{D}$ component to the charge density ρ_v at a point. This equation is expressed in compact form as

$$\boxed{\nabla \bullet \overline{D} = \rho_v \qquad (\text{C m}^{-3}) \quad (\text{aap})} \tag{13}$$

Equation (13) is frequently referred to as the first of *Maxwell's equations* for static electric and steady magnetic fields. It is very important to note that the $\overline{D}$ in (13) is not solely due to the ρ_v at the point of $\overline{D}$. As stated at the start of this section, the $\overline{D}$ field at point $P(x_0, y_0, z_0)$ has been produced by some system of charge distribution which could be external to the point $P(x_0, y_0, z_0)$.

From (3.2-1) and Gauss's law, it can be readily seen that if a positive ρ_v exists at a point we must have flux lines leaving a closed surface, about ρ_v, that gives rise to a positive divergence of $\overline{D}$. This would indicate a source of flux lines. In the same manner, a negative ρ_v would indicate a sink of flux lines since flux lines would flow into this point.

The concept of divergence can be applied to any vector $\overline{A}$, and the divergence of $\overline{A}$ would be defined by (10) with $\overline{D}$ replaced by $\overline{A}$. Thus, $\oint_s \overline{A} \bullet \overline{ds}$ would equal the flux of $\overline{A}$ through the closed surface s. From $\nabla \bullet \overline{D} = \rho_v$ and $\overline{D} = \epsilon_0 \overline{E}$ we have

$$\boxed{\nabla \bullet \overline{E} = \frac{\rho_v}{\epsilon_0} \qquad (\text{V m}^{-2}) \quad (\text{aap})} \tag{14}$$

If we use cylindrical or spherical coordinates in (10), the divergence of $\overline{D}$ will then be expressed in cylindrical and spherical coordinates as

$$\text{Div } \overline{D} = \nabla \bullet \overline{D} = \frac{1}{r_c}\left[\frac{\partial}{\partial r_c}(r_c D_{r_c})\right] + \frac{1}{r_c}\frac{\partial D_\phi}{\partial \phi} + \frac{\partial D_z}{\partial z} \quad (\text{cylindrical}) \tag{15}$$

$$\text{Div } \overline{D} = \nabla \bullet \overline{D} = \frac{1}{r_s^2}\left[\frac{\partial}{\partial r_s}(r_s^2 D_{r_s})\right] + \frac{1}{r_s \sin\theta}\left[\frac{\partial}{\partial \theta}(D_\theta \sin\theta)\right] + \frac{1}{r_s \sin\theta}\frac{\partial D_\phi}{\partial \phi} \quad (\text{spherical}) \tag{16}$$

where $\nabla \bullet \overline{D}$ here is used to indicate divergence of $\overline{D}$ only in symbolic form. Equations (12), (15), and (16) are also given, for convenient reference, on the inside of the back cover.

Example 8

In rectangular coordinates, obtain the expanded form for $\nabla \bullet \overline{A}$.

Solution. Using (1.12-1) for ∇ and the expanded form for $\overline{A}$, we have

$$\nabla \bullet \overline{A} = \left(\hat{x}\frac{\partial}{\partial x} + \hat{y}\frac{\partial}{\partial y} + \hat{z}\frac{\partial}{\partial z}\right) \bullet (\hat{x}A_x + \hat{y}A_y + \hat{z}A_z) = \frac{\partial A_x}{\partial x} + \frac{\partial A_y}{\partial y} + \frac{\partial A_z}{\partial z}$$

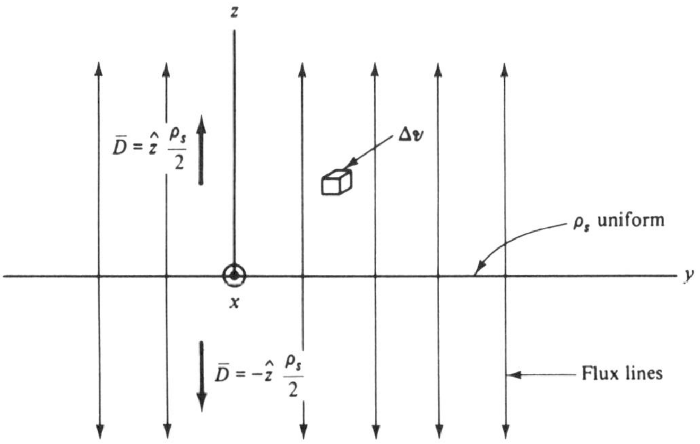

Figure 3-14 Two-dimensional flux plot of $\overline{D}$ about an infinite sheet of uniform ρ_s distribution for Example 9.

Example 9

Above an infinite sheet of uniform ρ_s distribution in the $z = 0$ plane, $\overline{D} = \hat{z}(\rho_s/2)$ (C m^{-2}). Evaluate $\nabla \cdot \overline{D}$ above the sheet, and from the flux plot discern the correctness of the answer.

Solution. From (12),

$$\nabla \cdot \overline{D} = \frac{\partial D_x}{\partial x} + \frac{\partial D_y}{\partial y} + \frac{\partial D_z}{\partial z} = \frac{\partial}{\partial z}\left(\frac{\rho_s}{2}\right) = 0$$

Thus, from (13),

$$\nabla \cdot \overline{D} = \rho_v = 0$$

A flux plot of $\overline{D}$ is shown in Fig. 3-14. If we locate a small cube within the $\overline{D}$ field, as we have in Fig. 3-13, we will note that the net flux emanating through the surface of the cube is zero. Thus, no charge is present within the cube which leads to $\rho_v = 0$ as the volume shrinks to zero.

Example 10

For a thick spherical shell of uniform ρ_v distribution, as shown in Fig. 3-15(a), the $\overline{D}$ expressions are

$$\overline{D} = \hat{r}_s \frac{\rho_v(r_s^3 - a^3)}{3r_s^2} \qquad \text{for } a \leqq r_s \leqq b$$

and

$$\overline{D} = \hat{r}_s \frac{\rho_v(b^3 - a^3)}{3r_s^2} \qquad \text{for } r_s \geqq b$$

Find the $\nabla \cdot \overline{D}$ in the two regions, and from the flux plot discern the correctness of the answers.

Solution. From (16), for $a \leqq r_s \leqq b$,

$$\nabla \cdot \overline{D} = \frac{1}{r_s^2}\frac{\partial}{\partial r_s}\left[r_s^2 \frac{\rho_v(r_s^3 - a^3)}{3r_s^2}\right]$$

$$= \frac{1}{r_s^2}\left[\rho_v \frac{3r_s^2}{3}\right] = \rho_v \qquad (\text{C m}^{-3})$$

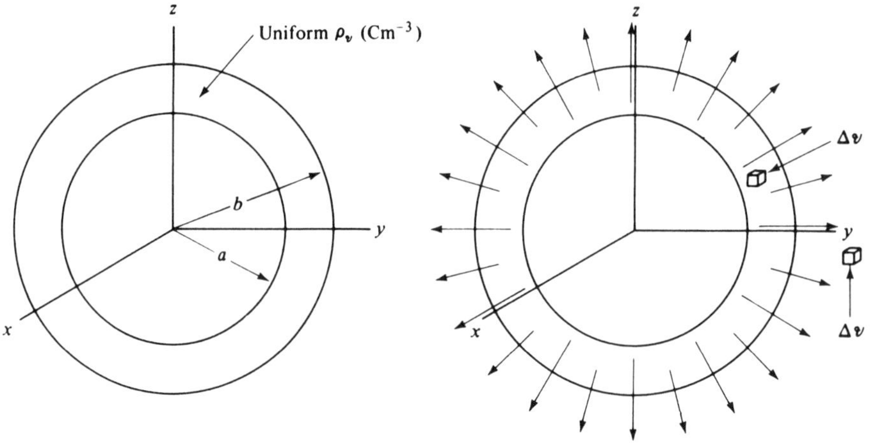

Figure 3-15 Thick spherical shell of uniform ρ_v (C m^{-3}) distribution for Example 10: (a) graphical display of ρ_v only in the region $a \leqq r_s \leqq b$; (b) flux plot and locations of sampling Δv's. Note the flux lines start within ρ_v distribution.

From (16), for $r_s \geqq b$,

$$\nabla \cdot \overline{D} = \frac{1}{r_s^2}\frac{\partial}{\partial r_s}\left[r_s^2 \frac{\rho_v(b^3 - a^3)}{3r_s^2}\right] = 0$$

Thus, from (13), we note that our solutions in the two regions are correct. A flux plot of $\overline{D}$ is shown in Fig. 3-15(b). If we locate a small cube in the range $a < r_s < b$, we will note that some net flux will emanate from the cube since some flux lines start within the cube due to ρ_v distribution within this range. A cube placed in the range $r_s > b$ will have zero net flux emanating through its surface; thus $\rho_v = 0$ in this range.

Problem 3.5-1 Evaluate $\nabla \cdot \overline{D}$ for the following $\overline{D}$ fields: (a) $\hat{x}Kx^2$; (b) $\hat{x}K(x + y^2)$; (c) $\hat{x}Kxz + \hat{y}Kxz$; (d) $\hat{r}_s Kr_s^2$; (e) $\hat{r}_s Kr_s^{-3}$; (f) $\hat{r}_s K \sin\theta$; (g) $\hat{r}_s K\phi$; (h) $\hat{r}_c Kr_c^2$; (i) $\hat{\phi}(\sin\phi/r_s)/\cos\theta$.

Problem 3.5-2 Find the ρ_v at the origin for parts (a), (c), (e), and (f) of Prob. 3.5-1.

Problem 3.5-3 Find the ρ_v on the positive y axis for a charge density described by $\overline{D} = \hat{r}_c K_1 \sin\phi + \hat{z}K_2 z$.

3.6 DIVERGENCE THEOREM

Let us apply Gauss's law to a region of a $\overline{D}$ field, as suggested in Fig. 3-16, to obtain

$$\oint_s \overline{D}_s \cdot \overline{ds} = Q_{\text{en}}$$

Now, let

$$Q_{\text{en}} = \int_v \rho_v \, dv$$

and through the use of (3.5-13),

$$\rho_v = \nabla \cdot \overline{D}$$

we obtain

$$\oint_s \overline{D}_s \bullet \overline{ds} = Q_{en} = \int_v \rho_v \, dv = \int_v \nabla \bullet \overline{D} \, dv \tag{1}$$

Equating the first and last terms of (1), we have

$$\boxed{\oint_s \overline{D}_s \bullet \overline{ds} = \int_v \nabla \bullet \overline{D} \, dv} \tag{2}$$

Equation (2) is a mathematical statement of the *divergence theorem*. The divergence theorem relates a closed surface integral to a volume integral involving the same vector. It should be noted that the closed surface s encloses the volume v, as shown in Fig. 3-16. The divergence theorem will be used quite frequently in the study of electromagnetics.

In our development of the divergence theorem, we selected $\overline{D}$ as the vector field since we are more familiar with it and its flux. The divergence theorem can be extended to any vector $\overline{A}$ field by replacing $\overline{D}$ by $\overline{A}$ in (2).

The divergence theorem is an obvious extension of the basic concept $\Psi_E = Q$. In (2) the closed surface integral yields Ψ_E while the volume integral yields Q_{en} within volume v, enclosed by the surface s. It should be pointed out that, in (2), the mathematical function $\overline{D}$ must be "well-behaved" within v and on the surface s. By "well-behaved" we mean that $\overline{D}$ and $\nabla \bullet \overline{D}$ are continuous and defined (not infinite). For a good discussion of the divergence theorem when $\overline{D}$ is not "well-behaved," see the text by Johnk.[2]

Example 11

Evaluate both sides of the divergence theorem equation when $\overline{D} = \hat{x}10^{-2}x^2y$ (C m^{-2}) and the Gaussian surface is a cube measuring 1 m on each side, as shown in Fig. 3-17.

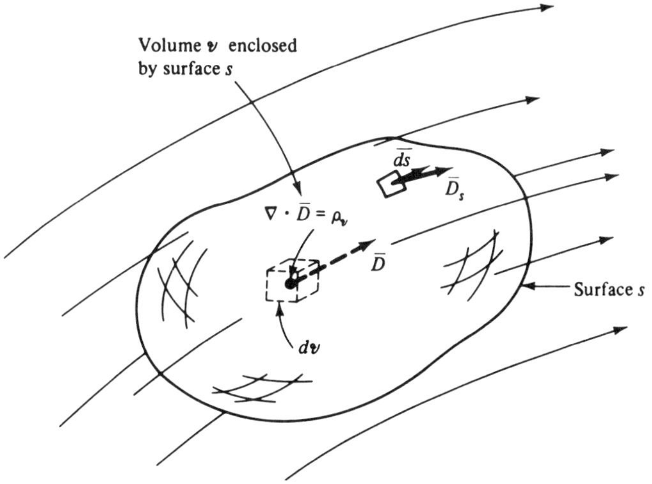

Figure 3-16 Graphical display of quantities involved in the development of the divergence theorem.

[2] C. T. A. Johnk, *Engineering Electromagnetic Fields and Waves*, John Wiley & Sons, Inc., New York, 1975, pp. 73–76.

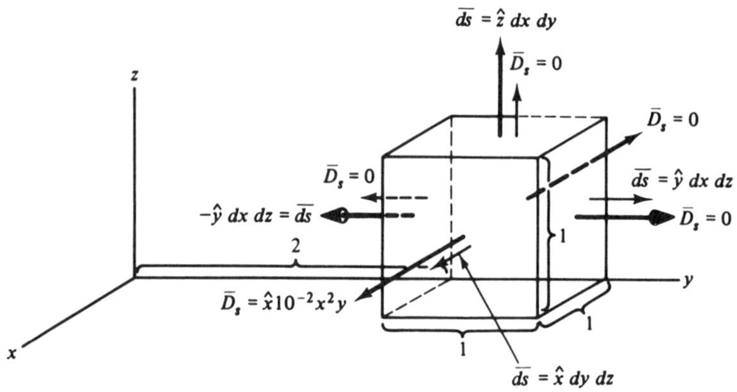

Figure 3-17 Graphical display for evaluating both sides of the divergence theorem in Example 11.

Solution. Divergence theorem,

$$\oint_s \overline{D}_s \bullet \overline{ds} = \int_v \nabla \bullet \overline{D}\, dv$$

Let us evaluate the left side:

$$\oint_s \overline{D}_s \bullet \overline{ds} = \int_{s(\text{top})} + \int_{s(\text{bottom})} + \int_{s(\text{right})} + \int_{s(\text{left})} + \int_{s(\text{front})} + \int_{s(\text{back})}$$

From the values of $\overline{D}_s$ and $\overline{ds}$ shown in Fig. 3-17, all but one of the surface integrals equal zero; thus,

$$\oint_s \overline{D}_s \bullet \overline{ds} = \int_{s(\text{front})} (\hat{x}10^{-2}y) \bullet (\hat{x}\, dy\, dz) = \int_0^1 \int_2^3 10^{-2} y\, dy\, dz = 0.025 \text{ C} \tag{3}$$

where $x = 1$ on this surface.

Let us now evaluate the right side of divergence theorem:

$$\int_v \nabla \bullet \overline{D}\, dv$$

where

$$\nabla \bullet \overline{D} = \frac{\partial D_x}{\partial x} + \frac{\partial D_y}{\partial y} + \frac{\partial D_z}{\partial z} = \frac{\partial}{\partial x}(10^{-2}x^2y) = 0.02xy$$

Substituting into the above, we have

$$\int_v \nabla \bullet \overline{D}\, dv = \int_0^1 \int_2^3 \int_0^1 0.02xy\, dx\, dy\, dz = 0.025 \text{ C} \tag{4}$$

Thus, from (3) and (4), we see that both sides of the divergence theorem give the same results. From Gauss's law it can be shown that the charge enclosed by the cube is 0.025 C.

Problem 3.6-1 Evaluate both the surface integral and the volume integral of (3.6-2) for the following functions and boundaries: (a) $\overline{D} = \hat{r}_s K r_s^2$ for a sphere of radius b (m) centered at the origin; (b) $\overline{D} = \hat{r}_s K r_s^{-2}$ for a sphere of radius b (m) centered at the origin; (c) $\overline{D} = \hat{x}Kx^2 + \hat{z}Kxz$, for a cube, 2 m on each edge, centered at the origin; (d) $\overline{D} = \hat{r}_c K r_c$ for a cylinder of radius a (m) and length L (m), centered at the origin.

Problem 3.6-2 Evaluate Prob. 3.6-1(c) with the cube centered at (0, 1, 1).

3.7 SUMMARY

This chapter was started with the concept $\Psi_E = Q$. Based on this concept, Gauss's law, divergence of $\overline{D}$, and the divergence theorem were developed. This sequence of development can be expressed in equation form as

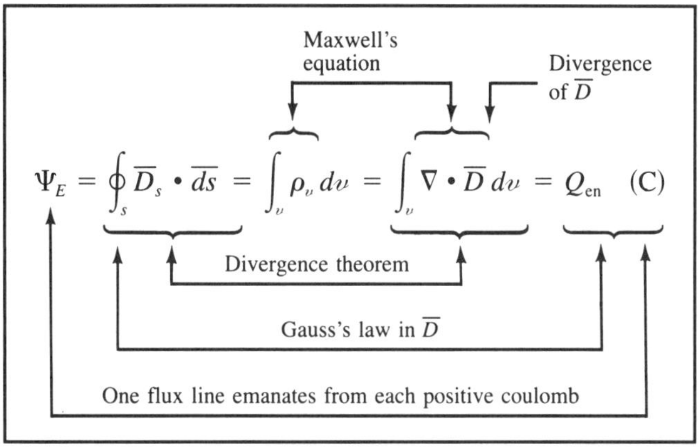

$$\Psi_E = \oint_s \overline{D}_s \bullet \overline{ds} = \int_v \rho_v \, dv = \int_v \nabla \bullet \overline{D} \, dv = Q_{en} \quad \text{(C)} \qquad (1)$$

Another sequence of equality stems from the divergence of $\overline{D}$ concept (3.5-11).

$$\text{Div } \overline{D} = \nabla \bullet \overline{D} \triangleq \lim_{\Delta v \to 0} \left[\frac{\oint_s \overline{D}_s \bullet \overline{ds}}{\Delta v} \right] = \frac{\partial D_x}{\partial x} + \frac{\partial D_y}{\partial y} + \frac{\partial D_z}{\partial z} = \rho_v \quad \text{(aap)} \qquad (3.5\text{-}11)$$

Equations (1) and (3.5-11) summarize quite well the mathematical relationships between concepts developed in this chapter and thus should be an aid to the student.

REVIEW QUESTIONS

1. Do electric lines of flux actually exist? *Sec. 3.2*
2. Are the electric lines of flux in the same direction as $\overline{E}$? *Sec. 3.2*
3. In the SI units, how many electric lines of flux emanate from a charge of +1 C? *Sec. 3.2, Eq. (3.2-1)*
4. In free space, is $\overline{D}$ in the same direction as $\overline{E}$? *Sec. 3.3, Eq. (3.3-1)*
5. Is the differential surface vector $\overline{ds}$ perpendicular to the physical differential surface ds? *Sec. 3.3*
6. What is the mathematical formulation $\oint_s \overline{D}_s \bullet \overline{ds} = Q_{en}$ called? *Eq. (3.4-2)*
7. Is Gauss's law based on $\Psi_E = Q_{en}$? *Eq. (3.4-1)*
8. In symmetrical charge distributions can we use Gauss's law to find $\overline{D}$ and $\overline{E}$? *Sec. 3.4, Example 5*
9. What is the key in finding $\overline{D}$ through the use of Gauss's law? *Sec. 3.4, Example 5*
10. What are $\overline{D}$ and $\overline{E}$ equal to inside a thin spherical shell of uniform ρ_s distribution? *Example 7*
11. What is the defining equation for divergence of $\overline{D}$? *Eq. (3.5-10)*
12. What is the physical significance of divergence of $\overline{D}$? *Sec. 3.5*
13. What is the compact vector analysis symbolization for divergence of $\overline{D}$? *Eq. (3.5-11)*
14. What is the mathematical relationship in rectan-

gular form, resulting from the application of the defining equation for divergence? *Eqs. (3.5-9), (3.5-11), (3.5-12)*

15. The divergence of $\overline{D}$, at a point, is equal to what physical quantity located at the same point? *Eqs. (3.5-11), (3.5-13)*
16. At a certain point $\nabla \cdot \overline{D} = -10^{-3}$ C m^{-3}. What is ρ_v equal to at the same point? *Eq. (3.5-13)*
17. Can the concept of divergence of a vector be applied to any vector field $\overline{A}$? *Sec. 3.5*
18. Does the $\nabla \cdot \overline{D}$ in cylindrical and spherical coordinate systems have the same mathematical form as in the rectangular coordinate system? *Eqs. (3.5-12), (3.5-15), (3.5-16)*
19. Is there a difference in mathematical form between Gauss's law and divergence theorem? *Eqs. (3.4-7), (3.6-2)*
20. Is there a difference between the divergence of $\overline{D}$ and divergence theorem? *Eqs. (3.5-10), (3.6-2)*

PROBLEMS

3-1. Using Gauss's law, find the net outward electric flux from a spherical surface of radius r_s enclosing a flux density given by $\overline{D} = (K/r_s^3)(\hat{r}_s 2 \cos\theta + \hat{\theta}\sin\theta)$ C m^{-2}. (b) Find the total flux from the upper hemisphere only, $0 < \theta < 90°$.

3-2. Suppose that a charged sphere of charge Q and of radius r_a is placed concentric with and insulated from a larger conducting sphere of radius r_b, which is uncharged. (The outer sphere could consist of two hemispheres which can be placed around the inner sphere.) Find the $\overline{D}$ field for: (a) $r_s < r_a$; (b) $r_a < r_s < r_b$; (c) $r_s > r_b$.

3-3. The electric field from a particular spherical surface is found to be $\overline{E} = \hat{r}_s(K \sin\theta/\epsilon_0 r_s^2)$ V m^{-1} outside the sphere. (a) Does the sphere contain a net charge, or is it neutral? (b) Assuming that all the charge is on the surface of the sphere, find the surface charge density ρ_s.

3-4. For a spherical surface having a surface charge density $\rho_s = 10^{-12}$ C m^{-2}, and for the Q contained on that surface to be equal to 10^{-13} C, find the angular boundaries (limits) of ϕ if $15° < \theta < 20°$.

3-5. Find the solid angle, i.e., $\int_{\phi_1}^{\phi_2}\int_{\theta_1}^{\theta_2} \sin\theta\, d\theta\, d\phi$ for (a) $0 < \phi < 2\pi$, $0 < \theta < \pi$; (b) $0 < \phi < 2\pi$, $0 < \theta < \pi/2$; (c) $0 < \phi < \pi$, $0 < \theta < \pi/2$; (d) $0 < \phi < \pi/2$, $\pi/8 < \theta < \pi/4$.

3-6. Suppose that we have two concentric spheres of radii r_a and r_b, where $r_b > r_a$, and that the inner sphere has a positive charge of Q (C). (a) Sketch the flux pattern and calculate $\overline{D}$ for $r_s < r_a$, $r_a < r_s < r_b$, $r_s > r_b$ for a charge of $-Q$ (C) on the outer sphere. (b) Repeat for a charge of $+Q$ on the outer sphere.

3-7. Located midway between two oppositely charged parallel plates is a third "uncharged" plate parallel to the other two. Using the "rules for electric flux" discussed in Sec. 3.2, sketch the flux lines between the two oppositely charged plates and discuss what happens on the uncharged plate; i.e., does it become charged? Assume perfect insulation between all plates and between all plates and the rest of the world.

3-8. Let us modify the situation described in Prob. 3-7. Let the negative plate be plate A, the positive plate, plate B, and the middle plate, plate C. Initially, we have the conditions as stated in Prob. 3-7. Next, we momentarily make an electrical connection between A and C. Sketch the new flux pattern that now exists between the plates.

3-9. Find the axial $\overline{E}$ and $\overline{D}$ fields from a uniformly charged disc whose radius is 2 cm and whose $\rho_s = 10^{-7}$ C m^{-2}. Plot the field for distances of 1 cm to 1 m from the disc. Compare the fields at 1 m to that from a point charge of 1.257×10^{-12} C at a distance of 1 m.

3-10. The electric field from a charged sphere of 10 cm radius is 20 kV m^{-1} at a distance of 20 cm from the center of the sphere. Assuming uniform charge distribution on the surface of the sphere, find: (a) the $\overline{D}$ field; (b) the total charge on the sphere.

3-11. Three concentric spheres, $r_s = a$, $r_s = b$, $r_s = c$, carry uniform charges of $Q_a = 3$ nC, $Q_b = -1$ nC, and $Q_c = -2$ nC. Through use of Gauss's law, find $\overline{D}$ in all regions and make a graph of $|\overline{D}|$ versus r_s. Assume that $a = 2$ mm, $b = 3$ mm, and $c = 4$ mm.

3-12. A beam of focused electrons traveling at constant velocity has a charge density that is uniform along its axis, but varies as $e^{-10^4 r_c}$ from the center of the beam. Evaluate $\overline{D}$ and $\overline{E}$ at a radius of 2 mm.

3-13. Two coaxial cylinders of radii r_a and r_b, were $r_a < r_b$, have charges of $-Q$ and Q, respectively, and the cylinders are of unit length. For $r_c > r_b$, find the following: (a) E_{r_c}; (b) E_{r_c} with the inner cylinder removed; (c) E_{r_c} when the two cylinders have been momentarily shorted together and then the inner cylinder removed.

3-14. Using the same two cylinders as in Prob. 3-13, begin with the inner cylinder charged with Q (C) but with the outer cylinder uncharged. Now find the E_{r_c} field repeating the steps as in Prob. 3-13.

3-15. Indicate from which of the following one can find the $\overline{D}$ or $\overline{E}$ fields by simple application of Gauss's law: (a) $\rho_v = r_s^{-2} \cos\theta \cos\phi$ C m^{-3}; (b) $\rho_s = a^{-2} \cos\theta \cos\phi$; (c) $\rho_s = a^{-2}$; (d) $\rho_v = K/(r_s^2 \sin\theta)$. Discuss.

3-16. Find the electric field for a charge density distribution that is constant in x and y but varies as $z^{-1/2}$.

3-17. Between two parallel sheets of opposite and uniform ρ_s distribution, the $\overline{D}$ field is found to be $\hat{z} \times 10^{-3}$ C m^{-2}. Deduce the following: (a) general orientation of the surface; (b) magnitude of ρ_s on the surfaces; (c) relationship of the normal scalar component of $\overline{D}$ at the surface to the surface charge density at the same point.

3.18. Find the ρ_v distribution within the following $\overline{E}$ fields: (a) $\hat{r}_s r_s$ V m^{-1}; (b) $\hat{r}_s r_s^{-2}$ V m^{-1}; (c) $\hat{r}_s r_s^2$ V m^{-1}; (d) $\hat{r}_c r_c^{-1}$ V m^{-1}; (e) $\hat{r}_c r_c$ V m^{-1}.

4

Energy Exchanges, Potential Difference, Gradient and Energy in a System of Charges

4.1 INTRODUCTION

A system of parallel surfaces of charge or any other system of charge distribution can be built up by expending energy to separate and position charges. The energy may come from the sun, heat, chemicals, and food by means of the following converters; solar cell, thermopile, battery, and human muscles, respectively. In the case of surfaces of charge on two parallel conducting plates, the energy in the charge distribution manifests itself through a force of attraction that exists between the plates and a force field between the plates that led us to the electric field intensity concept $\overline{E}$ in Chapter 2. Within this force field, energy must be expended to move a point charge from one location to another. This energy requirement will lead us to the potential difference and potential field concepts that we shall introduce in this chapter. The energy required to build up a system of charges and the concept of energy density at a point will also be developed.

We shall also discover that the electric field intensity $\overline{E}$ can be found through simple differentiation of the scalar potential field function. In most cases, we will find that it is much easier to evaluate the scalar potential field function from which we can evaluate $\overline{E}$, than to find $\overline{E}$ directly. In cases of symmetrical charge distributions, using Gauss's law to find $\overline{E}$ is still the simplest method, if we can find a Gaussian surface that will allow us to remove the field function from under the integral sign. Thus, by the end of this chapter, we will have three methods for finding the $\overline{E}$ or $\overline{D}$ fields due to a system of known charge distribution.

4.2 ENERGY EXCHANGES IN THE ELECTRIC FIELD OF A SYSTEM OF CHARGES

In Chapter 2, the electric field concept was developed from the force field about a system of charge distribution. The electric field intensity $\overline{E}$ was defined as the force on a unit positive point charge. Thus, the force on a point charge Q, located in an electric field $\overline{E}$, is equal to $Q\overline{E}$ (N). This force is due to the system of charge distribution that in turn gives rise to $\overline{E}$. Thus, we denote this force as $\overline{F}_{\overline{E}}$, where the subscript $\overline{E}$ indicates that the force is due to $\overline{E}$. To reduce "sign" confusion, we will assume that Q is a positive point charge; thus, the directions of $\overline{F}_{\overline{E}}$ and $\overline{E}$ will be the same, as shown in Fig. 4-1(a). If we release Q and allow it to travel under the influence of $\overline{E}$, its initial movement $\overline{d\ell}$ will be in the $\overline{E}$ direction. From particle mechanics we know that the charge Q will acquire kinetic energy from $\overline{E}$. This energy comes from the electric field $\overline{E}$, and the energy of the system of charges will be reduced by this amount. The incremental energy released by the electric field $\overline{E}$ during the initial movement $\overline{d\ell}$ of a point charge Q is

$$\boxed{dW_{\overline{E}} = \overline{F}_{\overline{E}} \cdot \overline{d\ell} = Q\overline{E} \cdot \overline{d\ell} \qquad \text{(J)}} \tag{1}$$

Equation (1) can be viewed as the amount of work done by the electric field $\overline{E}$. Examples of this type of energy exchange can be found in a cathode ray tube and in high frequency oscillators such as the magnetron and klystron. In the magnetron and klystron, this type of energy exchange occurs when electrons are injected into the cavity regions when the field is of such a polarity as to increase the kinetic energy of the electrons; thus, work is done by the electric field.

Let us now move, by some externally applied force $\overline{F}_{\text{app}}$, the charge Q against the direction of $\overline{E}$ as suggested in Fig. 4-1(b). In this case, $\overline{F}_{\text{app}} = -\overline{E}Q$ and the differential work performed by the force $\overline{F}_{\text{app}}$ becomes

$$\boxed{dW = \overline{F}_{\text{app}} \cdot \overline{d\ell} = -Q\overline{E} \cdot \overline{d\ell} \qquad \text{(J)}} \tag{2}$$

We see that (1) and (2) are opposite in algebraic sign, meaning that positive work done by the field [Eq. (1) positive] corresponds to negative work by an applied external force. Positive work is done by the field when (1) is positive, and positive work is done by a force acting in opposition to the electric field when (2) is positive. When the kinetic en-

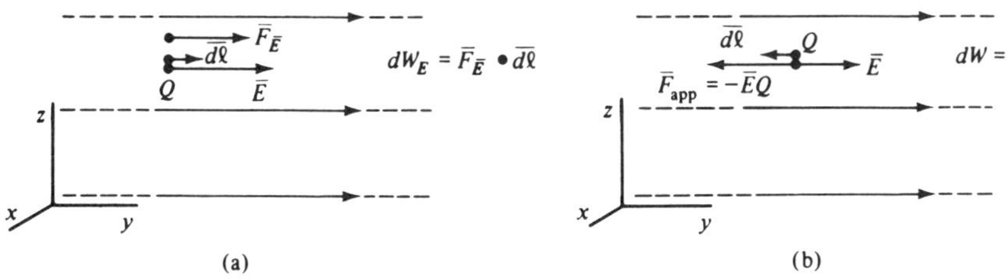

Figure 4-1 Graphical display of forces on a charge Q: (a) when the charge is released in an $\overline{E}$ field; (b) when the charge is forced by an externally applied force $\overline{F}_{\text{app}}$.

ergy of either a negatively or positively charged particle is increased by the field, that is referred to as positive work by the field. A very common non-electrical example of energy exchange is found in moving a mass in a gravitational field.

Equations (1) and (2) can both be generalized to $\overline{d\ell}$ being in any direction. When $\overline{d\ell}$ is perpendicular to $\overline{E}$, we will note that (1) gives zero, and thus the work done by the field will be zero, and from (2), the work done by an outside source is also equal to zero.

The work done by an external source in moving a point charge Q from some point b to a point a, through the use of (2), becomes

$$W = \int_b^a (-Q\overline{E} \bullet \overline{d\ell}) \qquad \text{(J)} \tag{3}$$

where the integral from b to a is a line integral over a specified path from point b to point a. The line integral will be fully discussed in Sec. 4.3, where we define potential difference.

Example 1

A point charge $Q = 5\ \mu\text{C}$ is released with zero initial velocity in an electric field $\overline{E} = \hat{x}2x + \hat{y}y^3 - \hat{z}z$. Find the approximate energy exchange from the field to the kinetic energy of the charge when the charge travels a distance of 10 μm at the point $(2, 1, 2)$.

Solution. The initial movement of the charge will be in the $\overline{E}$ direction. From (1),

$$\Delta W_{\overline{E}} \cong Q\overline{E} \bullet \overline{\Delta\ell} = Q|\overline{E}||\overline{\Delta\ell}|$$
$$= 5(10^{-6})[(2x)^2 + (y^3)^2 + z^2)]^{1/2}|_{(2,1,2)}(10^{-5}) = 229 \text{ pJ}$$

Problem 4.2-1 Find the approximate energy required to move a charge $Q = 3\ \mu\text{C}$ a distance $\overline{\Delta\ell} = (\hat{x}2 + \hat{y}3 - \hat{z}4)$ (μm) in a field $\overline{E} = (-\hat{x}x^2 + \hat{y}y^2 - \hat{z}2z)$ (V m^{-1}) at: (a) the point $(2, 2, 1)$; (b) the point $(2, 3, 0)$.

Problem 4.2-2 Find the energy required to move an electron from a positively charged plate to a negatively charged plate if the separation between plates is 1 cm and the voltage between plates is 1000 V. Express the answer in joules and in electron volts.

4.3 POTENTIAL DIFFERENCE (POTENTIAL) ABOUT A POINT CHARGE

> The potential difference between points a and b, V_{ab}, is defined as the work in joules that we must expend in moving a unit positive point charge from point b to a in an electric field $\overline{E}$.

The above word definition of potential difference can also be expressed in a defining equation form as

$$V_{ab} \triangleq \int_b^a (-\overline{E} \bullet \overline{d\ell}) \qquad (\text{N C}^{-1}\text{ m, J C}^{-1}\text{, V}) \tag{1}$$

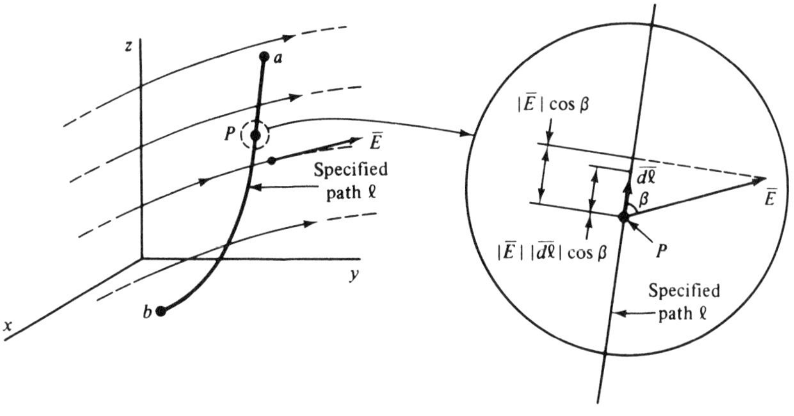

Figure 4-2 Graphical display of the line integral of $\int_b^a \overline{E} \cdot \overline{d\ell}$: (a) a specified path from point b to point a; (b) magnified view of the point P and formulation of $\overline{E} \cdot \overline{d\ell}$.

The integrand of (1) is (4.2-2), work done in moving a +1-C point charge a distance $\overline{d\ell}$, by some externally applied force. In this case, we are the applicators of the external force. Equation (1) is a line integral over a specified path from point b to a. Figure 4-2(a) shows a specified path ℓ, from point b to a, over which we shall evaluate $\int_b^a \overline{E} \cdot \overline{d\ell}$. A general point P on this path has been selected, and a magnified view of this point is shown in Fig. 4-2(b). The integrand $\overline{E} \cdot \overline{d\ell}$ is the dot product between the field vector $\overline{E}$ and the path length $\overline{d\ell}$ at point P. The integrand at point P thus becomes $|\overline{E}|\,|\overline{d\ell}| \cos\beta$ as shown. Thus, the integral $\int_b^a \overline{E} \cdot \overline{d\ell}$ is the summation of scalar quantities such as $|\overline{E}|\,|\overline{d\ell}| \cos\beta$ from all points along the specified path ℓ, from point b to point a. The dependency of the value of the integral on the path taken from b to a is determined by the vector field function. We will show later that, if the vector field function is an electric field $\overline{E}$ from a static charge distribution, the line integrals (4.2-3) and (1) will be independent of the path taken from b to a. In evaluating line integrals, the differential distance $\overline{d\ell}$ is always positive, even though the path is directed in a decreasing coordinate value. The expressions for $\overline{d\ell}$ will take on one of the following forms, depending on the coordinate system selected:

$$\begin{aligned} \overline{d\ell} &= \hat{x}\,dx + \hat{y}\,dy + \hat{z}\,dz \quad \text{(rectangular)} \\ \overline{d\ell} &= \hat{r}_c\,dr_c + \hat{\phi} r_c\,d\phi + \hat{z}\,dz \quad \text{(cylindrical)} \\ \overline{d\ell} &= \hat{r}_s\,dr_s + \hat{\theta} r_s\,d\theta + \hat{\phi} r_s \sin\theta\,d\phi \quad \text{(spherical)} \end{aligned} \tag{2}$$

Also, the lower limit b, in (1), is the initial point while the upper limit a is the final point.

From the definition of potential difference, it should be noted that we are dealing with the potential energy level of point a with respect to that of point b. If V_{ab} is positive, point a is at a higher potential energy level than point b. Commonly, the potential difference is referred to as the *potential of point a*. This usage must be backed up with a statement of the reference point b since the potential of point a depends on the reference selected. In practice, when potential difference or potential measurements are made, the reference or "ground" is usually the chassis upon which the electric circuits are constructed or some convenient common tie point. When the reference is taken to be at infinity, the potential is referred to as the *absolute potential*. It should be noted that the absolute potential is a potential difference with reference point b at infinity.

To obtain a better understanding of potential difference, references of potential, and the use of the line integral (1), we will find the potential differences about a positive point charge at the origin, as shown in Fig. 4-3. Let us first find V_{ab} by integrating (1) over a general path ℓ from point $b(r_b, \theta_b, \phi_b)$ to point $a(r_a, \theta_a, \phi_a)$ to obtain

$$V_{ab} = \int_b^a (-\overline{E} \cdot \overline{d\ell}) = \int_b^a \left(-\frac{Q\hat{r}_s}{4\pi\epsilon_0 r_s^2}\right) \cdot (\hat{r}_s\, dr_s + \hat{\theta} r_s\, d\theta + \hat{\phi} r_s \sin\theta\, d\hat{\phi})$$

$$= \frac{Q}{4\pi\epsilon_0}\int_b^a \left(-\frac{dr_s}{r_s^2}\right) = \frac{Q}{4\pi\epsilon_0}\left(\frac{1}{r_s}\right)\Bigg|_{r_b}^{r_a}$$

$$\boxed{V_{ab} = \frac{Q}{4\pi\epsilon_0}\left(\frac{1}{r_a} - \frac{1}{r_b}\right) \quad \text{(V)} \quad \text{(over path } \ell)} \tag{3}$$

From (3), it can be seen that V_{ab} from a static point charge depends on the end point coordinates r_a and r_b and is thus independent of the path taken from point b to point a. This is due to the fact that $\overline{E}$ has a radial component only, thus yielding zero contribution to the amount of work we do in carrying a +1 C charge in the $\hat{\theta}$ and $\hat{\phi}$ directions.

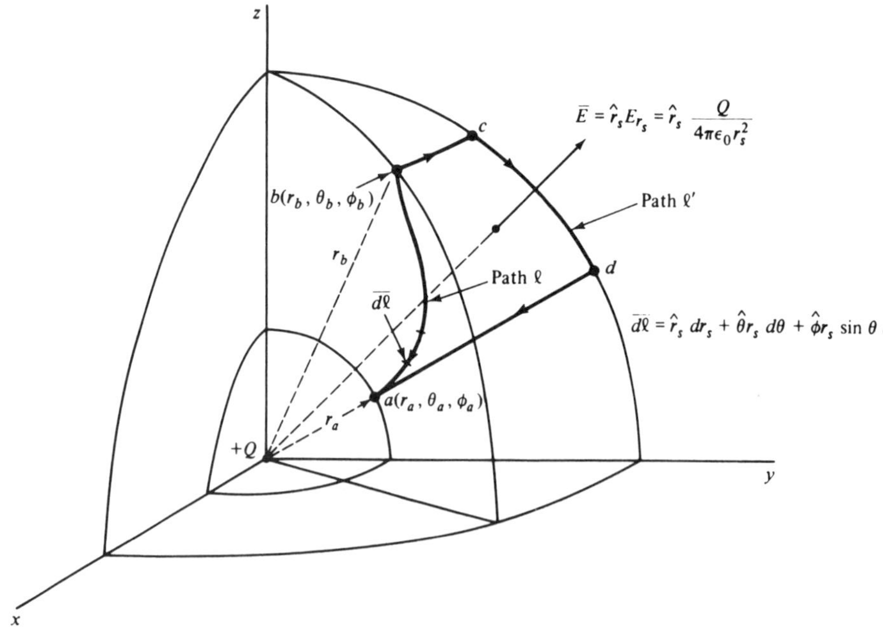

Figure 4-3 Graphical display of the paths taken between points b and a in evaluating the potential V_{ab} about a point charge at the origin.

This path independence is also illustrated in Fig. 4-3 when we integrate (1) over the path ℓ'. Over this path we have

$$V_{ab} = \int_b^a (-\overline{E} \bullet \overline{d\ell'}) = \int_b^c + \int_c^d + \int_d^a = 0 + 0 + \int_d^a = \int_b^a \tag{4}$$

where the integrals from b to c and from c to d are zero since $\overline{E}$ and $\overline{d\ell'}$ along those paths are perpendicular. It is clear from (3) and (4) that

$$\boxed{\int_{b(\text{path } \ell)}^{a} (-\overline{E} \bullet \overline{d\ell}) = \int_{b(\text{path } \ell')}^{a} (-\overline{E} \bullet \overline{d\ell'})} \tag{5}$$

where path ℓ and ℓ' are any paths between point b and point a. An $\overline{E}$ field whose integral $\int_b^a(-\overline{E} \bullet \overline{d\ell})$ is independent of path will also yield, from (5),

$$\int_b^a (-\overline{E} \bullet \overline{d\ell}) - \int_b^a (-\overline{E} \bullet \overline{d\ell'}) = 0 = \int_b^a (-\overline{E} \bullet \overline{d\ell}) + \int_a^b (-\overline{E} \bullet \overline{d\ell'})$$

$$= \oint_\ell (-\overline{E} \bullet \overline{d\ell}) \tag{6}$$

where $\oint_\ell (-\overline{E} \bullet \overline{d\ell})$ is a closed loop integral from point b about any path and then back to b. Equation (6) indicates that zero amount of work is done by us if we carry a +1 C about any closed path. This conservation of energy property of the static $\overline{E}$ field prompts us to call this field a *conservative field*. The conservative field property of a static point charge can be extended to the field from a general static charge distribution since any static charge distribution can be built up from point charges and the resultant field is the superposition of fields from these point charges. When $\overline{E}$ is a time-varying field, nonstatic, we will find $\overline{E}$ to be nonconservative.

Let us now return to (3) where, from our definition of potential difference, r_b is the reference point for the potential difference V_{ab}. Let us rewrite (3) in the form

$$\boxed{V_{a(\text{ref})} = \frac{Q}{4\pi\epsilon_0}\left[\frac{1}{r_a} - \frac{1}{r_{\text{ref}}}\right] \quad (\text{V})} \tag{7}$$

If we allow r_{ref} to become infinite, (7) becomes

$$\boxed{V_{a(\infty)} = \frac{Q}{4\pi\epsilon_0}\left(\frac{1}{r_a}\right) \quad (\text{V})} \tag{8}$$

where $V_{a(\infty)}$ is the potential difference between point $r_{\text{ref}} = \infty$ and point a. The potential difference $V_{a(\infty)}$ is also called the *absolute potential* or just the *potential* of point a, referenced to infinity. Equation (8) indicates that any point on a sphere of constant radius, $r_s = r_a$, will have the same potential. Surfaces of the same potential will be called *equipotential surfaces*. Through the use of (8), we can write

$$\boxed{V_{b(\infty)} = \frac{Q}{4\pi\epsilon_0}\left(\frac{1}{r_b}\right) \quad (\text{V})} \tag{9}$$

where $V_{b(\infty)}$ is the potential of point $b = r_b$ with reference at infinity. From (3), (8), and (9) we have

$$\boxed{V_{ab} = \frac{Q}{4\pi\epsilon_0}\left[\frac{1}{r_a} - \frac{1}{r_b}\right] = \frac{Q}{4\pi\epsilon_0}\left(\frac{1}{r_a}\right) - \frac{Q}{4\pi\epsilon_0}\left(\frac{1}{r_b}\right) = V_{a(\infty)} - V_{b(\infty)}} \tag{10}$$

Thus, from (10), it can be seen that we can evaluate the potential difference V_{ab} by taking the difference of potentials found at points a and b, if they are referenced to the same point: infinity in this case.

If we let $r_a = r_{\text{ref}}$ in (7), we find that the potential at the reference point is equal to zero. This is consistent with our definition of potential difference. Equation (7) could have been expressed in the more general form

$$\boxed{V_{a(\text{ref})} = \frac{Q}{4\pi\epsilon_0}\left(\frac{1}{r_a}\right) + C \qquad (\text{V})} \tag{11}$$

where C is a constant that determines the references for our potential. From (7), $C = 0$ when $r_{\text{ref}} = \infty$, to yield (8) while from (3), $C = -(Q/4\pi\epsilon_0)\,(1/r_b)$ when $r_{\text{ref}} = r_b$. Thus, through the use of (11), a constant C can be evaluated from the knowledge of the potential at one point. This point can be the reference point; thus, from (11),

$$C = -\frac{Q}{4\pi\epsilon_0 r_{\text{ref}}} \tag{12}$$

The reference radius r_{ref}, where the potential equals zero, can be found from (12) when C is known.

Figure 4-4 shows a plot of equipotential surfaces and electric flux lines about a positive point charge Q at the origin. Two potential scales, one based on $r_{\text{ref}} = r_0$ and the other based on $r_{\text{ref}} = \infty$, are shown. The potential differences V_{ab}, between two equipotential surfaces, is found to be 15 V in both cases. From Fig. 4-4, it should be noted that equipotential surfaces are perpendicular to electric flux lines.

Example 2

Find V_{ab} by integrating over path ℓ' of Fig. 4-5 in an electric field $\overline{E} = -\hat{x}y - \hat{y}x$ when the points a and b are at $(4, 6, 0)$ and $(2, 2, 0)$, respectively.

Solution. Equation (1) must be expressed by two integrals over the path ℓ' as

$$V_{ab} = \int_b^a (-\overline{E} \cdot \overline{d\ell'}) = \int_b^c (-\overline{E} \cdot \overline{d\ell'}) + \int_c^a (-\overline{E} \cdot \overline{d\ell'})$$

In rectangular coordinates, $\overline{d\ell'} = \hat{x}\,dx + \hat{y}\,dy + \hat{z}\,dz$; thus,

$$V_{ab} = \int_b^c (-1)(-\hat{x}y - \hat{y}x) \cdot (\hat{x}\,dx + \hat{y}\,dy + \hat{z}\,dz) + \int_c^a (-1)(-\hat{x}y - \hat{y}x) \cdot (\hat{x}\,dx + \hat{y}\,dy + \hat{z}\,dz)$$

Over the path from b to c, we find that $dx = dz = 0$ and $x = 2$. Over the path from c to a, we find that $dy = dz = 0$ and $y = 6$. Thus, the equation above reduces to

$$V_{ab} = \int_b^c 2\,dy + \int_c^a 6\,dx = \int_2^6 2\,dy + \int_2^4 6\,dx = 20 \text{ V}$$

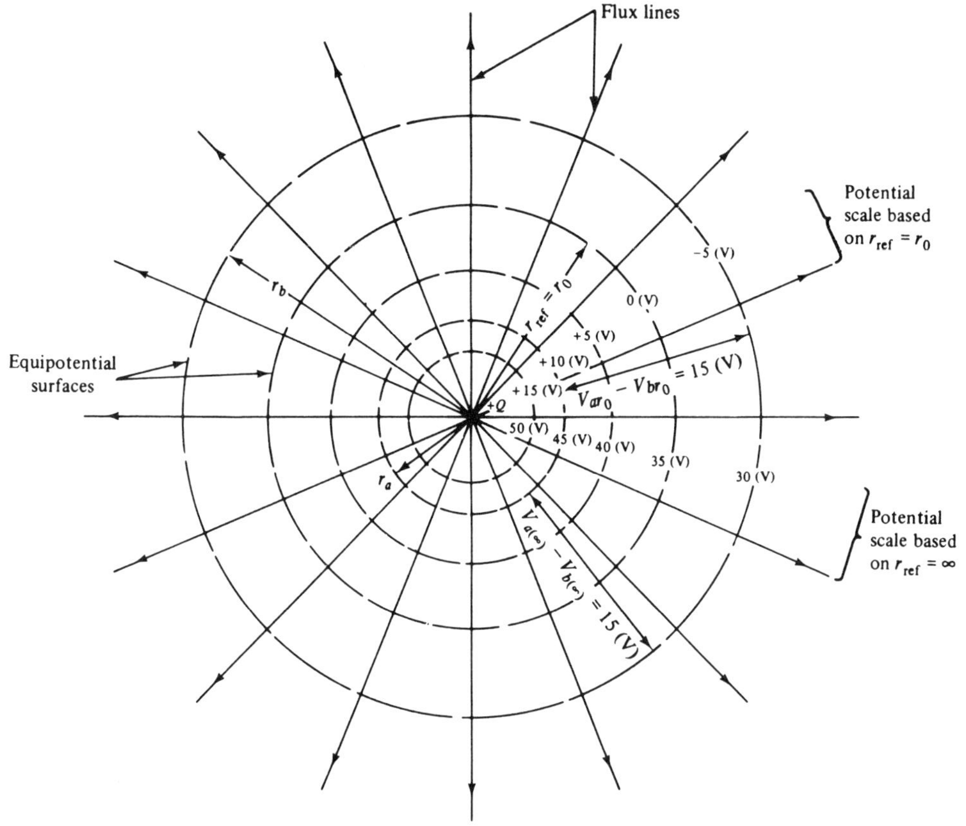

Figure 4-4 Plot of electric flux lines and equipotential surfaces about a positive point charge Q. Two potential scales, one based on $r_{ref} = r_0$ and the other based on $r_{ref} = \infty$ are shown. The potential difference V_{ab} between two equipotential surfaces is found to equal 15 V in both cases.

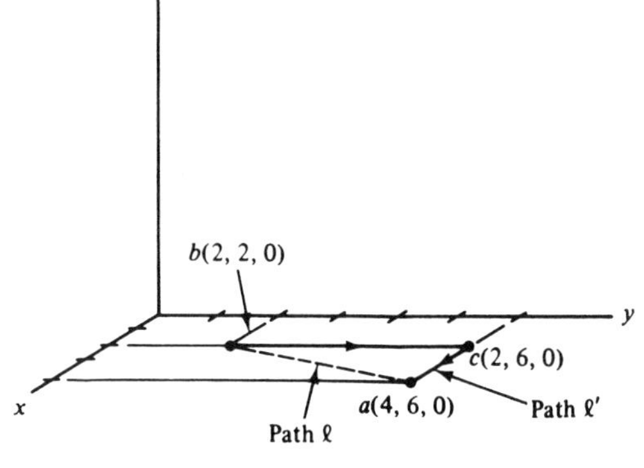

Figure 4-5 Graphical display of paths ℓ' and ℓ for Examples 2 and 3.

Example 3

Repeat Example 2 over the path ℓ, a straight-line path from b to a.

Solution. From (1), we have

$$V_{ab} = \int_b^a (-\overline{E} \bullet \overline{d\ell}) = \int_b^a (-1)(-\hat{x}y - \hat{y}x) \bullet (\hat{x}\,dx + \hat{y}\,dy + \hat{z}\,dz)$$

Over path ℓ, we find that $dz = 0$, $y = 2x - 2$, and $dy = 2\,dx$; thus,

$$V_{ab} = \int_b^a (y\,dx + x\,dy) = \int_2^6 y\left(\frac{dy}{2}\right) + \int_2^4 x(2\,dx) = 20 \text{ V}$$

The above equation could have been expressed in terms of y only as

$$V_{ab} = \int_2^6 \left[y\left(\frac{dy}{2}\right) + \left(\frac{y}{2} + 1\right)dy\right] = 20 \text{ V}$$

From Examples 2 and 3 it can be seen that the field $\overline{E}$ is a conservative field and that it requires 20 J of energy to carry a +1 C of charge from point b to point a.

Example 4

Through the use of (11), find the expression for V_{ab}, about a point charge $Q = 1$ C, when the reference point is $r_b = r_{\text{ref}} = 5$ m.

Solution. Solve for C in (11) when we set $V_{a(\text{ref})}\big|_{r\,\text{ref}=r_b=5} = 0$ thus,

$$V_{a(\text{ref})}\Bigg|_{r\,\text{ref}=r_b=5} = 0 = \left[\frac{Q}{4\pi\epsilon_0}\left(\frac{1}{r_a}\right) + C\right]\Bigg|_{r\,\text{ref}=r_b=5}$$

$$\therefore\quad C = -\frac{Q}{4\pi\epsilon_0 5}$$

Substituting back into (11), we have

$$V_{ab} = \frac{1}{4\pi\epsilon_0}\left(\frac{1}{r_a} - \frac{1}{5}\right) \qquad \text{(V)}$$

The constant C could also have been found from (12) as could the expression for V_{ab} by simply letting $r_{\text{ref}} = 5$ in (7).

Problem 4.3-1 Evaluate the potential V_{ab} in an electrostatic field $\overline{E} = \hat{x}2x^2 + \hat{y}y + \hat{z}z^3$ when the path of integration is: (a) a series of straight lines from $b(0, 3, -3)$ to $(1, 3, -3)$ to $(1, 3, 0)$ to $(0, 3, 0)$ to $a(0, 0, 0)$; (b) a straight path from b to a.

Problem 4.3-2 Repeat Prob. 4.3-1 for $\overline{E} = \hat{x}2x^2z + \hat{y}y + \hat{z}z^3$.

Problem 4.3-3 A two-dimensional E field is described by $\overline{E} = \hat{y}100 \sin(\pi x/b)$ (V m^{-1}) for $0 \le x \le b$. Evaluate the line integral $\oint \overline{E} \bullet \overline{d\ell}$ over: (a) the path from $(0, 0)$ to $(b/2, 0)$ to $(b/2, 1)$ to $(0, 1)$ to $(0, 0)$; (b) the path from $(b/4, 0)$ to $(3b/4, 0)$ to $(3b/4, 1)$ to $(b/4, 1)$ to $(b/4, 0)$.

4.4 POTENTIAL DIFFERENCE (POTENTIAL) ABOUT A CHARGE SYSTEM

In this section, we extend the potential field concept about a point charge, discussed in Sec. 4.3, to a system of point charges. For the system of finite point charges shown in Fig. 4-6, the absolute potential at point a, from (4.3-1), becomes

$$V_{a(\infty)} = \int_\infty^a (-\overline{E} \bullet \overline{d\ell}) = \int_\infty^a (-\overline{E}_1 \bullet \overline{d\ell}) + \int_\infty^a (-\overline{E}_2 \bullet \overline{d\ell}) + \int_\infty^a (-\overline{E}_3 \bullet \overline{d\ell}) \qquad (1)$$

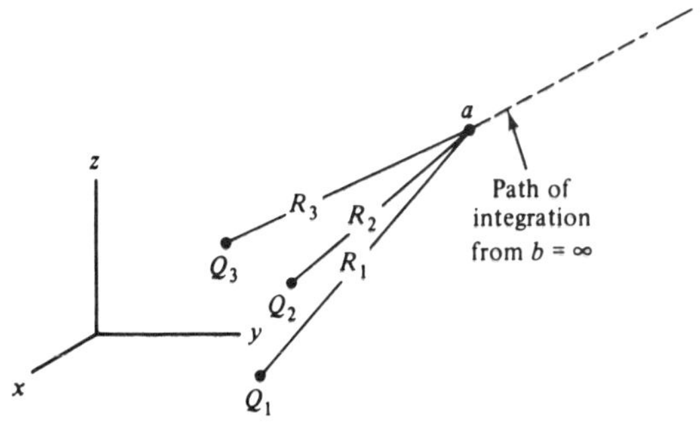

Figure 4-6 Graphical display for finding the absolute potential at point a due to a system of three point charges.

where $\overline{E}_1$, $\overline{E}_2$, and $\overline{E}_3$ are the electric fields due to the point charges Q_1, Q_2, and Q_3 acting alone. Through the use of (4.3-8), the last three integrals in (1) become

$$V_{a(\infty)} = \frac{Q_1}{4\pi\epsilon_0 R_1} + \frac{Q_2}{4\pi\epsilon_0 R_2} + \frac{Q_3}{4\pi\epsilon_0 R_3} = \sum_{i=1}^{i=3} \frac{Q_i}{4\pi\epsilon_0 R_i} \tag{2}$$

where R_1, R_2, and R_3 are the distances from Q_1, Q_2, and Q_3 to the point a, respectively. Equation (2) can be rewritten as

$$\boxed{V_{a(\infty)} = (V_1 + V_2 + V_3) \qquad \text{(V)}} \tag{3}$$

where V_1, V_2, and V_3 are the absolute potentials at point a due to Q_1, Q_2, and Q_3 acting alone.

In Sec. 4.3 we have shown that the $\overline{E}$ field from a point charge is conservative; thus, from (1) we can extend the conservative concept to the resultant field $\overline{E}$ from a system of charges.

If in (2), we let $Q_i \to dQ$ and increase the number of point charges to infinity we have

$$V_{a(\infty)} = \int_{\text{(charge)}} \frac{dQ}{4\pi\epsilon_0 R} = \int_{\text{(charge)}} dV \tag{4}$$

where R is the distance from dQ to the point a, and the integral is carried out over the charge in the system. The integrand of (4) is thus the contribution from a differential point charge dQ to the absolute potential $V_{a(\infty)}$ at point a. A building block for potential difference stems from (4) in the form

$$\boxed{dV = \frac{dQ}{4\pi\epsilon_0 R} \quad \text{(building block)} \quad \text{(V)}} \tag{5}$$

Equation (4) takes on the following forms for line, surface, and volume charge distributions

$$V_{a(\infty)} = \int_\ell \frac{\rho_\ell \, d\ell}{4\pi\epsilon_0 R} \tag{6}$$

$$V_{a(\infty)} = \int_s \frac{\rho_s \, ds}{4\pi\epsilon_0 R} \tag{7}$$

$$V_{a(\infty)} = \int_v \frac{\rho_v \, dv}{4\pi\epsilon_0 R} \tag{8}$$

Let us rewrite (1) and (8) in functional form, using primed variables for charge (source) locations and unprimed variables for V field locations; thus,

$$V_{a(\infty)} = \int_{\infty}^{a} (-\overline{E} \cdot \overline{d\ell}) = V_{a(\infty)}(x, y, z)$$

$$= \int_{\infty}^{a} [-\overline{E}(x, y, z) \cdot \overline{d\ell}(x, y, z)] \tag{9}$$

and

$$V_{a(\infty)} = \int_{v'} \frac{\rho_v' \, dv'}{4\pi\epsilon_0 R} = V_{a(\infty)}(x, y, z)$$

$$= \int_{v'} \frac{\rho_v(x', y', z') \, dv'(x', y', z')}{4\pi\epsilon_0 R(x', y', z', x, y, z)} \tag{10}$$

where the integral in (9) is evaluated with respect to x, y, and z while the integral in (10) is evaluated with respect to x', y', and z'. From (9) and (10) it can be seen that the absolute potential field is a function of x, y, and z and is single-valued due to the conservative nature of the static $\overline{E}$ field. It should be noted that $\overline{d\ell}$ in (9) is a differential vector distance along a path from ∞ to a in space while the $d\ell$ in (6) is the differential length along a line of charge.

We now have two distinct methods, (9) and (10), to find the absolute potential. One method, (9), requires knowledge of the $\overline{E}$ while the other, (10), requires the knowledge of the charge distribution.

Example 5

Through the use of (9), obtain the expression for the absolute potential along the z axis of a ring of uniform ρ_ℓ (C m^{-1}) and whose radius is r_0, as shown in Fig. 4-7.

Solution. From (2.6-13), the $\overline{E}$ along the z axis,

$$\overline{E} = \hat{z}\frac{zr_0\rho_\ell}{2\epsilon_0(r_0^2 + z^2)^{3/2}} \tag{2.6-13}$$

where r_0 replaced the radius a. From (4.3-1),

$$V_{a(\infty)} = V_{z(\infty)} = \int_{\infty}^{z} (-\overline{E} \cdot \overline{d\ell}) \tag{11}$$

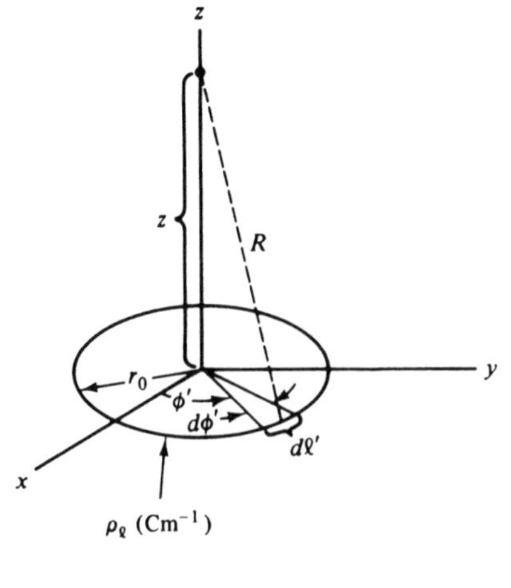

Figure 4-7 Graphical display for finding the potential along the $+z$ axis of a ring of charge in Examples 5 and 6.

Since (11) will be integrated along the z axis, $dx = dy = 0$ in the expression for $\overline{d\ell}$ and thus $\overline{d\ell} = \hat{z}\,dz$. Equation (11) becomes

$$V_{z(\infty)} = \int_{\infty}^{z} -\frac{\hat{z}zr_0\rho_\ell}{2\epsilon_0(r_0^2 + z^2)^{3/2}} \bullet (\hat{z}\,dz)$$

$$= -\frac{r_0\rho_\ell}{2\epsilon_0}\int_{\infty}^{z} \frac{z\,dz}{(r_0^2 + z^2)^{3/2}}$$

$$= \left.\frac{r_0\rho_\ell}{2\epsilon_0(r_0^2 + z^2)^{1/2}}\right|_{\infty}^{z} = \frac{r_0\rho_\ell}{2\epsilon_0(r_0^2 + z^2)^{1/2}} \quad \text{(V)} \qquad (12)$$

Example 6

Repeat Example 5 through the use of (4.4-6).

Solution. From (4.4-6),

$$V_{a(\infty)} = V_{z(\infty)} = \int_{\text{ring}} \frac{\rho_\ell\,d\ell'}{4\pi\epsilon_0 R} \qquad (13)$$

where $d\ell'$ is along the line of charge, and R is the distance from the location of the point charge $\rho_\ell\,d\ell'$ to the point $a = z$ on the z axis where the absolute potential is to be found. Thus, $R = (r_0^2 + z^2)^{1/2}$ and $d\ell' = r_0\,d\phi'$. Equation (13) becomes

$$V_{z(\infty)} = \int_0^{2\pi} \frac{\rho_\ell r_0\,d\phi'}{4\pi\epsilon_0(r_0^2 + z^2)^{1/2}} = \frac{r_0\rho_\ell}{2\epsilon_0(r_0^2 + z^2)^{1/2}} \quad \text{(V)} \qquad (14)$$

Note that eqs. (12) and (14) agree. In the next section, we will show that a simple differentiation of (12) or (14) will yield the $\overline{E}$ field along the z axis [eqs. (2.6-13)].

Problem 4.4-1 Obtain the absolute potential field $V_a(\infty)$ inside and outside a thin shell, centered at the origin, of uniform ρ_s C m^{-2} and of radius r_0. Assume Q to be the total charge on the thin shell and use previously derived expressions for $\overline{E}$. Obtain the potential at $a = r_0/2$, r_0, $2r_0$, $4r_0$, $8r_0$.

Problem 4.4-2 In Example 5, find the potential difference along the $+z$ axis when the reference point is specified at the point $z_{\text{ref}} = 100$ m.

4.5 GRADIENT OF POTENTIAL DIFFERENCE

From the defining equation for potential difference V_{ab},

$$\boxed{V_{ab} \triangleq \int_b^a (-\overline{E} \bullet \overline{d\ell}) \quad \text{(V)}} \qquad (4.3\text{-}1)$$

where $(-\overline{E} \bullet \overline{d\ell})$ is the differential potential difference over $\overline{d\ell}$ distance along the path from point b to point a. Thus,

$$\boxed{dV = -\overline{E} \bullet \overline{d\ell} = -|\overline{E}|\,|\overline{d\ell}| \cos\beta = -E\,d\ell \cos\beta} \qquad (1)$$

From (1), form

$$\frac{dV}{d\ell} = -E\cos\beta \qquad (2)$$

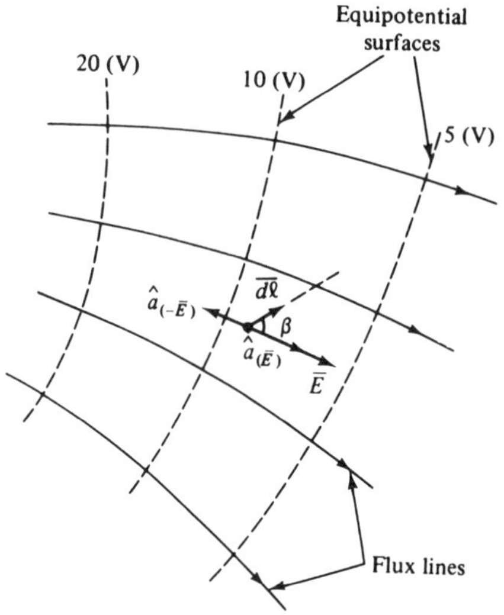

Figure 4-8 Graphical construction for determining the gradient of the potential difference.

where $dV/d\ell$ is the derivative of potential difference with respect to a direction $\overline{d\ell}$. This derivative is a directional derivative whose value depends on $\overline{d\ell}$ and thus on β, as shown in Fig. 4-8. If $\overline{d\ell} = d\ell\hat{a}_{(\overline{E})}$, in the same direction as $\overline{E}$, $\beta = 0$ and $dV/d\ell = -E$, as expected, since the potential decreases in the direction of $\overline{E}$ as indicated along the equipotential surfaces. Now, if $\overline{d\ell} = \hat{a}_{(-\overline{E})}\,d\ell$, opposite to the direction of $\overline{E}$, $\beta = \pi$ and

$$\boxed{\left.\frac{dV}{d\ell}\right|_{\overline{d\ell}=\hat{a}_{(-\overline{E})}d\ell} = \left.\frac{dV}{d\ell}\right|_{\max} = -E\cos\beta\Big|_{\beta=\pi} = E} \tag{3}$$

Thus, we see that the maximum derivative is in a direction opposite to $\overline{E}$ and is equal to the magnitude of $\overline{E}$. Let us write (3) in a vector form by multiplying both sides by $\hat{a}_{(-\overline{E})}$, we have

$$\boxed{\hat{a}_{(-\overline{E})}\left.\frac{dV}{d\ell}\right|_{\max} = \hat{a}_{(-\overline{E})}E = -\hat{a}_{(-\overline{E})}E = -\overline{E}} \tag{4}$$

where the first term is a vector quantity defined as the *gradient* of V; thus,

$$\boxed{\hat{a}_{(-\overline{E})}\left.\frac{dV}{d\ell}\right|_{\max} \triangleq \text{gradient of } V = -\overline{E} \quad \text{(aap)}} \tag{5}$$

The defining expression for the gradient of V is quite cumbersome to use in the form found in (5).

In (1), let us expand $\overline{d\ell}$ and $\overline{E}$ in component form to obtain

$$dV = -(\hat{x}E_x + \hat{y}E_y + \hat{z}E_z) \bullet (\hat{x}\,dx + \hat{y}\,dy + \hat{z}\,dz) \tag{6}$$

Now let $\overline{d\ell} = \hat{x}\,dx$ and $dy = dz = 0$; thus, (6) becomes

$$dV = -(\overline{E}) \bullet (\hat{x}\,dx) = -E_x\,dx \tag{7}$$

From (7),

$$\boxed{\left.\frac{dV}{dx}\right|_{\substack{y=\text{constant}\\ z=\text{constant}}} \triangleq \frac{\partial V}{\partial x} = -E_x} \tag{8}$$

Now, if we follow the same procedure for $\overline{d\ell} = \hat{y}\,dy$ and $\overline{d\ell} = z\,dz$, we obtain

$$\boxed{\left.\frac{dV}{dy}\right|_{\substack{x=\text{constant}\\ z=\text{constant}}} \triangleq \frac{\partial V}{\partial y} = -E_y} \tag{9}$$

and

$$\boxed{\left.\frac{dV}{dz}\right|_{\substack{x=\text{constant}\\ y=\text{constant}}} \triangleq \frac{\partial V}{\partial z} = -E_z} \tag{10}$$

Expanding $\overline{E}$ in component form and through the use of (8), (9), and (10), we have

$$\boxed{\overline{E} = (\hat{x}E_x + \hat{y}E_y + \hat{z}E_z) = -\left(\hat{x}\frac{\partial V}{\partial x} + \hat{y}\frac{\partial V}{\partial y} + \hat{z}\frac{\partial V}{\partial z}\right) \qquad (\text{V m}^{-1})} \tag{11}$$

Comparing (5) and (11) leads to[1]

$$\boxed{\text{gradient of } V = \hat{x}\frac{\partial V}{\partial x} + \hat{y}\frac{\partial V}{\partial y} + \hat{z}\frac{\partial V}{\partial z}} \tag{12}$$

Equation (12) is a more useful expression for the gradient. It should be noted that we have dropped subscripts on the potential function V. The potential function V is still the potential difference function V_{ab}.

Through the use of the ∇ operator (1.12-1), (1.12-2), and (12) we can express (5) as

$$\boxed{\overline{E} = -\nabla V \quad (\text{aap})} \tag{13}$$

where ∇V is the gradient of V and is the shorthand form for (12). Equation (13) is thus a partial differential equation that allows us to readily solve for $\overline{E}$ through simple differentiation on the scalar potential function V.

[1] Grad V or ∇V may be concisely defined by $dV = \nabla V \bullet \overline{d\ell}$, where in cartesian coordinates,

$$dV = \frac{\partial V}{\partial x}dx + \frac{\partial V}{\partial y}dy + \frac{\partial V}{\partial z}dz \qquad \text{and} \qquad \overline{d\ell} = \hat{x}\,dx + \hat{y}\,dy + \hat{z}\,dz$$

The direction of the greatest rate of change of V is in the direction of ∇V, as $\nabla V \bullet \overline{d\ell}$ is then a maximum.

The expanded expressions for ∇V in rectangular, cylindrical, and spherical coordinate systems are given below and also inside the back cover.

$$\nabla V = \hat{x}\frac{\partial V}{\partial x} + \hat{y}\frac{\partial V}{\partial y} + \hat{z}\frac{\partial V}{\partial z} \quad \text{(rectangular)} \tag{14}$$

$$\nabla V = \hat{r}_c\frac{\partial V}{\partial r_c} + \hat{\phi}\frac{1}{r_c}\frac{\partial V}{\partial \phi} + \hat{z}\frac{\partial V}{\partial z} \quad \text{(cylindrical)} \tag{15}$$

$$\nabla V = \hat{r}_s\frac{\partial V}{\partial r_s} + \hat{\theta}\frac{1}{r_s}\frac{\partial V}{\partial \theta} + \hat{\phi}\frac{1}{r_s \sin\theta}\frac{\partial V}{\partial \phi} \quad \text{(spherical)} \tag{16}$$

Example 7

Through the use of $\overline{E} = -\nabla V$, find the $\overline{E}$ field about a point charge at the origin, using the following expressions for potential difference:

(a) $V_{ab} = \dfrac{Q}{4\pi\epsilon_0}\left(\dfrac{1}{r_a} - \dfrac{1}{r_b}\right)$ [eq. (4.3-3)]

(b) $V_{a(\infty)} = \dfrac{Q}{4\pi\epsilon_0}\left(\dfrac{1}{r_a}\right)$ [eq. (4.3-8)]

(c) $V_{a(\text{ref})} = \dfrac{Q}{4\pi\epsilon_0}\left(\dfrac{1}{r_a}\right) + C$ [eq. (4.3-11)]

Solution. Since the potentials about a point charge have been expressed in spherical r_s, we must use the spherical expression for ∇V; thus, $-\nabla V$ reduces to

$$\overline{E} = -\nabla V = -\hat{r}_s\frac{\partial V}{\partial r_s}$$

since V is not a function of θ and ϕ. In the expressions for potential difference, r_b is viewed as a constant point while r_a is the variable point. Thus, we will change from r_a to r_s to more clearly show the spherical variable dependence.

(a)
$$\overline{E} = -\hat{r}_s\frac{\partial}{\partial r_s}\left[\frac{Q}{4\pi\epsilon_0}\left(\frac{1}{r_s} - \frac{1}{r_b}\right)\right] = \hat{r}_s\frac{Q}{4\pi\epsilon_0 r_s^2} \tag{17}$$

(b)
$$\overline{E} = -\hat{r}_s\frac{\partial}{\partial r_s}\left[\frac{Q}{4\pi\epsilon_0}\left(\frac{1}{r_s}\right)\right] = \hat{r}_s\frac{Q}{4\pi\epsilon_0 r_s^2} \tag{18}$$

(c)
$$\overline{E} = -\hat{r}_s\frac{\partial}{\partial r_s}\left[\frac{Q}{4\pi\epsilon_0}\left(\frac{1}{r_s}\right) + C\right] = \hat{r}_s\frac{Q}{4\pi\epsilon_0 r_s^2} \tag{19}$$

From (17), (18), and (19) it can be seen that the $\overline{E}$ is the same no matter where the reference of potential difference is taken.

Example 8

Through the use of $\overline{E} = -\nabla V$, show that $\int_b^a (-\overline{E} \cdot \overline{d\ell})$ is independent of path taken from point b to point a, and thus prove that $\overline{E}$ is a conservative field.

Solution. Substitute $\overline{E} = -\nabla V$ and expand in component form to obtain

$$\begin{aligned}\int_b^a (-\overline{E} \cdot \overline{d\ell}) &= \int_b^a (\nabla V) \cdot \overline{d\ell} \\ &= \int_b^a \left(\hat{x}\frac{\partial V}{\partial x} + \hat{y}\frac{\partial V}{\partial y} + \hat{z}\frac{\partial V}{\partial z}\right) \cdot (\hat{x}\,dx + \hat{y}\,dy + \hat{z}\,dz) \\ &= \int_b^a \left(\frac{\partial V}{\partial x}dx + \frac{\partial V}{\partial y}dy + \frac{\partial V}{\partial z}dz\right) = \int_b^a dV = V_a - V_b\end{aligned} \tag{20}$$

Equation (20) indicates $\int_b^a (-\overline{E} \cdot \overline{d\ell})$ equals the difference between potentials at the end points of the path and thus is independent of the path from b to a. The path independency is based on the fact that $\overline{E} = -\nabla V$.

Example 9

The expression for the potential difference between r_b (reference) and a variable radius r_c, about an infinite line of uniform ρ_ℓ along the z axis, in cylindrical coordinates is

$$V_{cb} = \frac{-\rho_\ell}{2\pi\epsilon_0} \ln\left(\frac{r_c}{r_b}\right)$$

Find the expression for $\overline{E}$ through the use of the gradient concept.

Solution. From (13) and (15),

$$\overline{E} = -\nabla V = -\left(\hat{r}_c \frac{\partial V}{\partial r_c}\right)$$

since V_{cb} is not a function of ϕ or z.

$$\therefore \quad \overline{E} = -\hat{r}_c \frac{\partial}{\partial r_c}\left(-\frac{\rho_\ell}{2\pi\epsilon_0} \ln \frac{r_c}{r_b}\right) = \hat{r}_c \frac{\rho_\ell}{2\pi\epsilon_0}\left(\frac{1}{r_c}\right) \tag{21}$$

This checks with (2.5-13).

Example 10

Show that electric flux lines are perpendicular to equipotential surfaces.

Solution. Let $d\ell$ in (1) be along an equipotential surface; thus, $dV = 0$ and (1) becomes

$$0 = -E\,d\ell \cos\beta$$

From the equation above, for non-zero $d\ell$ and E, it can be seen that the angle β between $d\ell$ and E must equal $\pi/2$. Since $d\ell$ was chosen along an equipotential surface and flux lines are in the same direction as $\overline{E}$, the electric lines of flux are perpendicular to equipotential surfaces.

Example 11

For the electric dipole shown in Fig. 4-9(a), find for $r_s \gg d$: (a) the absolute potential (potential) at point $P(r_s, \theta, \phi)$; (b) the $\overline{E}$ field through the use of the gradient concept.

Solution. (a) The sum of the absolute potentials from $+Q$ and $-Q$ becomes

$$V = V_{+Q} + V_{-Q} = \frac{Q}{4\pi\epsilon_0 R_1} + \frac{-Q}{4\pi\epsilon_0 R_2} = \frac{Q}{4\pi\epsilon_0}\left(\frac{R_2 - R_1}{R_1 R_2}\right) \tag{22}$$

For $r_s \gg d$, $R_2 - R_1 \cong d \cos\theta$, and $R_1 R_2$ in the denominator of (22) can be set equal to r_s^2, thus; we have

$$V \cong \frac{Qd \cos\theta}{4\pi\epsilon_0 r_s^2} \tag{23}$$

If we define the electric dipole moment as

$$\boxed{\overline{p} \triangleq Q\overline{d} \qquad (\text{C} \cdot \text{m})} \tag{24}$$

where $\overline{d}$ is from $-Q$ to $+Q$, we can rewrite (23) as

$$V \cong \frac{\hat{r}_s \cdot \overline{p}}{4\pi\epsilon_0 r_s^2} \tag{25}$$

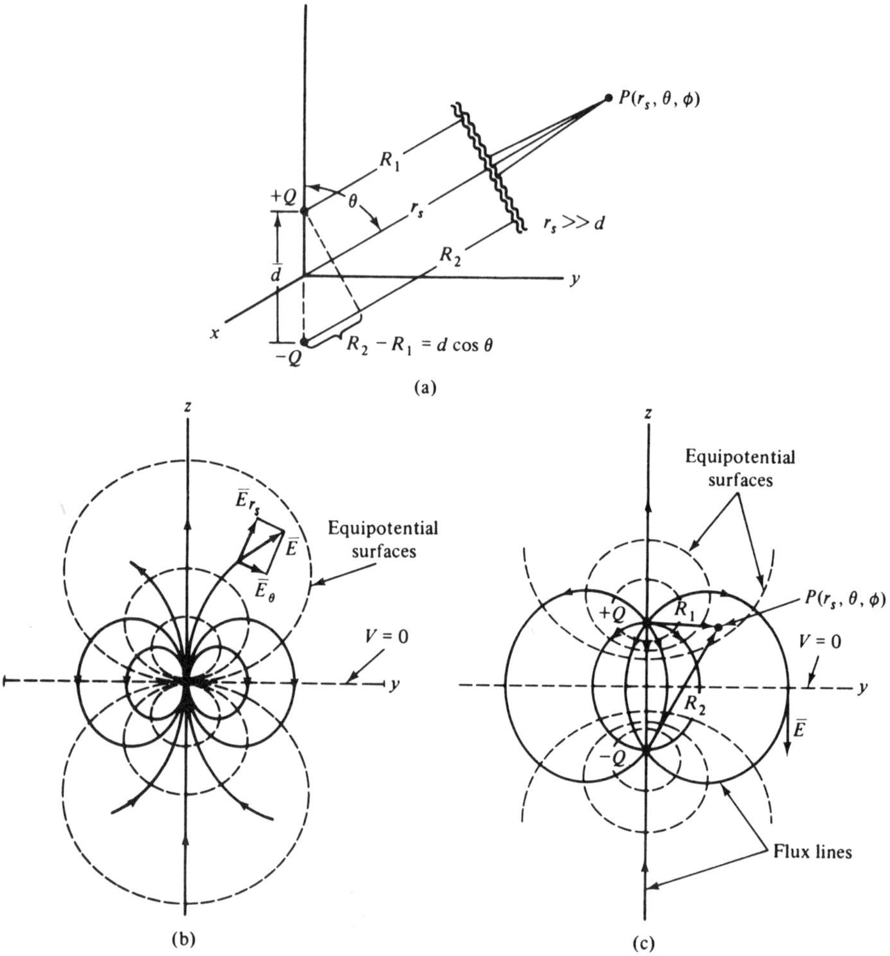

Figure 4-9 (a) Graphical construction for finding V and $\overline{E}$ about an electric dipole located on the z axis. (b) Two-dimensional plot of equipotential surfaces and flux lines about an electric dipole for $r_s \gg d$. (c) Two-dimensional plot of equipotential surfaces and flux lines near the electric dipole.

(b) The $\overline{E}$ field can be found from (23) through the gradient concept in spherical coordinates,

$$\overline{E} = -\nabla V = -\left(\hat{r}_s \frac{\partial V}{\partial r_s} + \frac{\hat{\theta}}{r_s}\frac{\partial V}{\partial \theta} + \frac{\hat{\phi}}{r_s \sin\theta}\frac{\partial V}{\partial \phi}\right) \tag{16}$$

From (23), we find that V is a function of r_s and θ only. Thus, (16) becomes

$$\boxed{\overline{E} \simeq -\left(-\hat{r}_s \frac{Qd\cos\theta}{2\pi\epsilon_0 r_s^3} - \hat{\theta}\frac{Qd\sin\theta}{4\pi\epsilon_0 r_s^3}\right) = \frac{Qd}{4\pi\epsilon_0 r_s^3}(\hat{r}_s 2\cos\theta + \hat{\theta}\sin\theta)} \tag{26}$$

The electric dipole has been selected as an example problem for two reasons: first, it illustrates the use of the gradient concept to find $\overline{E}$ from V; second, we will find the electric dipole a very important concept when we replace free space by material that consists of atoms and molecules.

A two-dimensional plot of equipotential surfaces and flux lines about an electric dipole, for $r_s \gg d$, is shown in Fig. 4-9(b). This plot checks with the results found in (23) and (26). Figure 4-9(c) shows a two-dimensional plot of equipotential surfaces and flux lines very near the electric dipole. Equations (23) and (26) do not hold for this case since the inequality $r_s \gg d$ does not hold. Note that, as proven in Example 10, the equipotential surfaces are perpendicular to the electric lines of flux in Fig. 4-9(b) and (c). A three-dimensional plot of the equipotential surfaces and flux lines can be obtained by revolving Fig. 4-9(b) and (c) about the z axis.

Problem 4.5-1 Find the potential along the axis of a ring of charge density ρ_ℓ and radius b (see Example 6), and from this potential function and (13), find the $\overline{E}$ field. Compare this result with that of (2.6-13).

Problem 4.5-2 Sketch the two-dimensional field $\overline{F} = \hat{x}10y$ defined for $0 \leq y \leq 2$. Is $\overline{F}$ a conservative field? Demonstrate by integrating $\overline{F} \cdot \overline{d\ell}$ over the paths from (0, 0) to (0, 2) to (2, 2) to (2, 0) to (0, 0).

Problem 4.5-3 The potential field between the two conductors of a coaxial cable is $V(r_c) = (\rho_\ell/2\pi\epsilon_0) \ln (r_a/r_c)$ (V), where $r_a \leq r_c \leq r_b$. (a) Where is the reference potential for $V(r_c)$ in this case? (b) If the total potential difference between conductors is 10 V, find r_c where $V(r_c) = -5$ V. Express r_c in terms of r_a and r_b. (c) Find r_c when $V(r_c) = -2V$.

4.6 ENERGY EXPENDED TO BUILD UP A SYSTEM OF CHARGES

In Sec. 4.2 we discussed the energy exchanges between the field of a system of charges and a point charge Q. In this section we shall start with empty space and calculate the energy required to build up a system of $n = 4$ point charges, as shown in Fig. 4-10. We

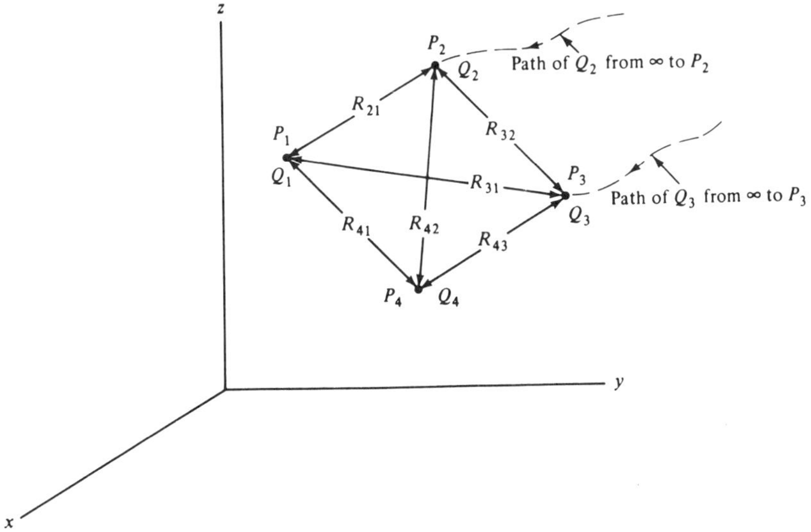

Figure 4-10 Graphical construction for finding the energy expended in building up a system of four point charges.

will assume that all charges are initially located at infinity and that we build up the system by bringing from infinity one charge at a time.

No work is done in bringing point charge Q_1 from infinity to P_1 since the $\overline{E}$ field that Q_1 encounters is zero. The story is quite different when we position Q_2 at P_2. Now, with Q_1 at P_1 an electric field exists in space and the energy expended to position Q_2 is

$$\boxed{W_2 = V_{21} Q_2 = \frac{Q_1}{4\pi\epsilon_0 R_{21}} Q_2 \qquad \text{(J)}} \tag{1}$$

where V_{21} is the absolute potential at P_2 due to Q_1 alone, and R_{21} is the distance between P_1 and P_2. The validity of (1) stems from the fact that V_{21}, from the definition of potential difference, equals the energy expended in bringing a $+1$ C charge from infinity to P_2; thus, $V_{21} Q_2$ equals the energy expended in bringing Q_2 from infinity to P_2.

The energy expended to position Q_3 at P_3 equals

$$W_3 = V_{31} Q_3 + V_{32} Q_3 = \frac{Q_1}{4\pi\epsilon_0 R_{31}} Q_3 + \frac{Q_2}{4\pi\epsilon_0 R_{32}} Q_3 \tag{2}$$

where V_{31} and V_{32} are the absolute potentials at P_3 due to Q_1 and Q_2, acting alone, respectively. Now, to position Q_4 we have

$$W_4 = V_{41} Q_4 + V_{42} Q_4 + V_{43} Q_4 \tag{3}$$

The total energy to position the four charges is the sum of (1), (2), and (3); thus,

$$\boxed{W = V_{21} Q_2 + V_{31} Q_3 + V_{32} Q_3 + V_{41} Q_4 + V_{42} Q_4 + V_{43} Q_4} \tag{4}$$

The subscripts in (1) and (2) can be interchanged and the equality will still exist; e.g., (2) becomes

$$W_3 = V_{13} Q_1 + V_{23} Q_2 = \frac{Q_3}{4\pi\epsilon_0 R_{13}} Q_1 + \frac{Q_3}{4\pi\epsilon_0 R_{23}} Q_2 \tag{5}$$

Let us now interchange the subscripts in (4) to obtain

$$\boxed{W = V_{12} Q_1 + V_{13} Q_1 + V_{23} Q_2 + V_{14} Q_1 + V_{24} Q_2 + V_{34} Q_3} \tag{6}$$

Now, add (4) and (6) and divide by two to obtain

$$\begin{aligned} W = \tfrac{1}{2}[&(V_{12} + V_{13} + V_{14})Q_1 + (V_{21} + V_{23} + V_{24})Q_2 \\ &+ (V_{31} + V_{32} + V_{34})Q_3 + (V_{41} + V_{42} + V_{43})Q_4] \end{aligned} \tag{7}$$

Note that (7) does not contain such terms as $V_{11} Q_1$, $V_{22} Q_2$, $V_{33} Q_3$, and $V_{44} Q_4$. The term $V_{11} Q_1$ implies the product of the absolute potential at P_1 due to Q_1. For a point charge, V_{11} is infinite and the product $V_{11} Q_1$ is equal to infinity. The term $V_{11} Q_1$ is called the *self-energy of the point charge,* and is the energy required to build up the point charge from smaller charges. The same notions apply to the other self-energy terms, $V_{22} Q_2$, $V_{33} Q_3$, and $V_{44} Q_4$. Thus, (7) does not include the energy required to build up point charges Q_1, Q_2, Q_3, and Q_4.

The term $(V_{12} + V_{13} + V_{14})$ is the sum of absolute potentials at P_1 due to Q_2, Q_3, and Q_4 acting alone. We shall denote this potential as V_1. Equation (7) can be rewritten as

$$\boxed{W = \tfrac{1}{2}(V_1 Q_1 + V_2 Q_2 + V_3 Q_3 + V_4 Q_4) \qquad \text{(J)}} \tag{8}$$

Extending (8) to n point charges, we have

$$\boxed{W = \sum_{k=1}^{n} \tfrac{1}{2} V_k Q_k \qquad \text{(J)}} \tag{9}$$

where Q_k is the charge brought from infinity to P_k, and V_k is the absolute potential at P_k due to all charges except Q_k.

For a ρ_v distribution of charge within a volume v, Q_k becomes $\rho_v\,dv$ and (9) becomes

$$\boxed{W = \int_v \tfrac{1}{2} V \rho_v\,dv \qquad \text{(J)}} \tag{10}$$

where V is the absolute potential, at the position of ρ_v, due to all charges of the system including $\rho_v\,dv$. Note that the definition of V in (10) differs from that in (8) or (9).

If we use (10) in a region of a continuous charge distribution, the absolute potential will be a continuous function of position and the contribution due to $\rho_v\,dv$, at the point where V is found, will approach zero in the limit. Now, if we apply (10) to a system of point charges, we will find that $V = V_{11} = \infty$ and $\rho_v = \infty$ at the location of the point charge. Thus, due to the changed definition of V in (10) from that in (9), we find that (10) will involve the self-energies of the point charges.

For line and surface charge distributions, (9) can be expressed as

$$\boxed{W = \int_\ell \tfrac{1}{2} V \rho_\ell\,d\ell \qquad \text{(J)}} \tag{11}$$

and

$$\boxed{W = \int_s \tfrac{1}{2} V \rho_s\,d_s \qquad \text{(J)}} \tag{12}$$

where V is defined as in (10).

Let us write $\rho_v = \nabla \cdot \overline{D}$ in (10) to obtain

$$W = \int_v \tfrac{1}{2} V(\nabla \cdot \overline{D})\,dv \tag{13}$$

Now, through the use of the identity $\nabla \bullet (f\overline{A}) = f(\nabla \bullet \overline{A}) + \overline{A} \bullet (\nabla f)$, where f is any scalar and $\overline{A}$ is any vector, (13) becomes

$$W = \tfrac{1}{2}\int_v \nabla \bullet (V\overline{D})\, dv - \tfrac{1}{2}\int_v \overline{D} \bullet (\nabla V)\, dv \tag{14}$$

In (14) we have set $f = V$ and $\overline{A} = \overline{D}$. Applying the divergence theorem to the first integral of (14), we have

$$W = \tfrac{1}{2}\oint_s (V\overline{D}) \bullet \overline{ds} - \tfrac{1}{2}\int_v \overline{D} \bullet (\nabla V)\, dv \tag{15}$$

Let us now, in (15) allow v to include all space; thus, s will be the closed surface at infinity. For a system of charge distribution of finite size, we will find that the closed surface integral becomes zero. This is due to the fact that as r_s becomes infinite V decreases at least as rapidly as $1/r_s$, $\overline{D}$ decreases at least as rapidly as $1/r_s^2$, and $\overline{ds}$ increases as r_s^2. Thus, the integrand decreases at least as rapidly as $1/r_s$ and becomes zero when $r_s = \infty$. Let us now replace ∇V by $-\overline{E}$ in the remaining integral of (15) to obtain

$$\boxed{W = \int_v \tfrac{1}{2}\overline{D} \bullet \overline{E}\, dv = \int_v \tfrac{1}{2}\epsilon_0 E^2\, dv \qquad \text{(J)}} \tag{16}$$

where the integral is taken over all space.

We now have two integrals, (10) and (16), for finding the energy expended in building up a system of charges. Equation (10) is in terms of absolute potential and charge while (16) is in terms of the field produced by the system of charges. From (10), we can easily infer that the energy expended to build up a system of charges is stored in the charges while, from (16), we can just as readily say that the energy is stored in the $\overline{E}$ field. The question of where the energy is stored has not been resolved to date. In many applications of electromagnetic radiation problems, we will find it convincing that the energy is stored in the field. This leads to the concept of energy density in a field.

From the integrand of (16),

$$\boxed{dW_E = \tfrac{1}{2}\overline{D} \bullet \overline{E}\, dv \qquad \text{(J)}} \tag{17}$$

where the subscript E has been added to remind us that we view the energy to be stored in the electric field. Dividing (17) by dv, we obtain

$$\boxed{\frac{dW_E}{dv} = \tfrac{1}{2}\overline{D} \bullet \overline{E} = \tfrac{1}{2}\epsilon_0 E^2 \qquad (\text{J m}^{-3}) \quad \text{(aap)}} \tag{18}$$

Equation (18) is useful to obtain the total energy expended to build up a system of charges by integrating over all space, where $\overline{E}$, due to a system of charges, is found.

It can easily be shown that (18) represents the energy density between two parallel oppositely charged plates, each of area s, that have been separated a distance ℓ, where the initial separation had been zero (see Prob. 4.6-3). From a circuit's point of view, assuming that the volume $s\ell$ between plates is the variable, we have

$$\frac{dW}{dv} = \frac{d}{dv}\left(\frac{CV^2}{2}\right)$$
$$= \frac{d}{d(s\ell)}\left(\frac{\epsilon_0 s(E\ell)^2}{2\ell}\right)$$
$$= \frac{d}{d(s\ell)}\left(\frac{\epsilon_0 s\ell E^2}{2}\right)$$
$$= \frac{\epsilon_0 E^2}{2} \qquad (\text{J m}^{-3})$$

where $C = \epsilon_0 s/\ell$ for the capacitance of a parallel-plate capacitor.

The force between two parallel oppositely charged plates may be obtained by assuming that s and E remain constant, and that the separation is a variable. Then the energy increase caused by increasing the separation by a distance $d\ell$ is

$$dW = \tfrac{1}{2}\epsilon_0 E^2 s\, d\ell \tag{19}$$

and since $dW = \text{force} \cdot d\ell$,

$$\boxed{F = \frac{dW}{d\ell} = \tfrac{1}{2}\epsilon_0 E^2 s \qquad (\text{N})} \tag{20}$$

for the force between plates. Equation (20) may also be written as

$$F = \frac{1}{2}\epsilon_0\left(\frac{D}{\epsilon_0}\right)^2 s = \frac{\tfrac{1}{2}D^2 s}{\epsilon_0} \qquad (\text{N}) \tag{21}$$

It will be seen in Chapter 9 [eq. (9.11-8)] that the magnetic force between two magnetic pole faces is analogous to the electric force between two charged plates.

Example 12

A spherical charge distribution of radius r_a and uniform ρ_v is centered at the origin. If the total charge is Q: (a) find the energy expended to build up the charge system through the use of (10); (b) repeat part (a) through the use of (16); (c) let $r_a = 0$ and thus find the energy to assemble a point charge Q; (d) find the energy expended to bring from infinity the first point charge Q in building up a system of point charges. Assume that the point charge Q was assembled at infinity.

Solution. (a) Equation (10) must be integrated only over the volume of radius r_a since ρ_v is zero for $r_s > r_a$. Thus, the absolute potential $V = V_{rs(\infty)}$ must also be evaluated over the same range from the knowledge of $\overline{E}_{r_s>r_a}$ and $\overline{E}_{r_s<r_a}$. From (4.3-1), we have

$$V_{rs(\infty)} = \int_{\infty}^{r_a}(-\overline{E}_{r_s>r_a} \bullet \overline{d\ell}) + \int_{r_a}^{r_s}(-\overline{E}_{(r_s<r_a)} \bullet \overline{d\ell})$$

Substituting for $\overline{E}_{r_s>r_a}$ and $\overline{E}_{(r_s<r_a)}$, we have

$$V_{rs(\infty)} = \int_{\infty}^{r_a}\left(-\frac{Q\hat{r}_s}{4\pi\epsilon_0 r_s^2} \bullet \hat{r}_s\, dr_s\right) + \int_{r_a}^{r_s}\left(-\frac{Qr_s\hat{r}_s}{4\pi\epsilon_0 r_a^3} \bullet \hat{r}_s\, dr_s\right)$$
$$= \frac{Q}{4\pi\epsilon_0 r_a} - \frac{Q}{8\pi\epsilon_0 r_a^3}(r_s^2 - r_a^2) \quad (\text{for } r_s \leqq r_a) \tag{22}$$

Substituting (22) into (10), we have

$$W = \int_v \rho_v \frac{V\,dv}{2} = \int_v \frac{1}{2}\left(\frac{Q}{\frac{4}{3}\pi r_a^3}\right)\left[\frac{Q}{4\pi\epsilon_0 r_a} - \frac{Q}{8\pi\epsilon_0 r_a^3}(r_s^2 - r_a^2)\right]r_s^2 \sin\theta\,dr_s\,d\theta\,d\phi$$

$$= \frac{3Q^2}{20\pi\epsilon_0 r_a} \quad (\text{J}) \tag{23}$$

(b) From (16),

$$W = \int_v \tfrac{1}{2}\epsilon_0 E^2\,dv = \int_{v(r_s > r_a)} \tfrac{1}{2}\epsilon_0 E_{(r_s > r_a)}^2\,dv + \int_{v(r_s < r_a)} \tfrac{1}{2}\epsilon_0 E_{(r_s < r_a)}^2\,dv$$

$$= \int_{v(r_s > r_a)} \tfrac{1}{2}\epsilon_0\left(\frac{Q}{4\pi\epsilon_0 r_s^2}\right)^2 (r_s^2 \sin\theta\,dr_s\,d\theta\,d\phi)$$

$$+ \int_{v(r_s < r_a)} \tfrac{1}{2}\epsilon_0\left(\frac{Qr_s}{4\pi\epsilon_0 r_a^3}\right)^2 (r_s^2 \sin\theta\,dr_s\,d\theta\,d\phi)$$

$$= \frac{Q^2}{8\pi\epsilon_0 r_a} + \frac{Q^2}{40\pi\epsilon_0 r_a} = \frac{3Q^2}{20\pi\epsilon_0 r_a} \quad (\text{J}) \tag{24}$$

Note that (23) and (24) are the same.

(c) Now, let $r_a = 0$ in (23) or (24) to obtain $W = \infty$ (J). This is the energy required to build the sphere of charge as $r_a \to 0$. Thus, we can say it is the energy to assemble a point charge of Q (C). Note that $\rho_v = \infty$ for a point charge.

(d) The energy to bring the first point charge of Q (C) from infinity to the origin is equal to zero. Here we assume that the point charge Q was assembled at infinity with infinite amount of energy as found in part (c).

Problem 4.6-1   Through the use of (9), find the energy required to bring two point charges $Q_1 = 15\ \mu\text{C}$ and $Q_2 = -10\ \mu\text{C}$ from infinity. Place Q_1 at the origin and Q_2 at $P_{\text{rec}}(0, 5, 0)$.

Problem 4.6-2   A thin spherical shell of radius r_a and uniform ρ_s (C m^{-2}) distribution is centered at the origin. If the total charge is Q: (a) find the energy expended to build up the charge system through the use of (12); (b) repeat part (a) through the use of (16); (c) let $r_a = 0$ and find the energy required to assemble a point charge of Q (C); (d) plot the absolute potential versus r_s for parts (a) and (c).

Problem 4.6-3   Find the energy density between two large parallel sheets of uniform ρ_s (C m^{-2}) distribution when the sheets are separated d (m) with $+\rho_s$ on one sheet and $-\rho_s$ on the other. What is the relation between this energy density and the work required to separate the plates if the spacing were initially zero?

REVIEW QUESTIONS

1. Can energy be exchanged from an electric field to a charge? *Sec. 4.2*

2. Can kinetic energy of a charge be transferred to an electric field? *Sec. 4.2*

3. What is the physical significance of $dW = -Q\overline{E}\cdot\overline{d\ell}$? *Eq. (4.2-2)*

4. In $W = \int_b^a (-Q\overline{E}\cdot\overline{d\ell})$, what is the integral called? *Eq. (4.2-3)*

5. The $\int_b^a (-\overline{E}\cdot\overline{d\ell})$ is the defining equation for what physical quantity? *Eq. (4.3-1)*

6. In words, what is the definition for the potential difference V_{ab}? *Sec. 4.3*

7. In a line integral what is the general expression for $\overline{d\ell}$ in rectangular coordinates? *Eq. (4.3-2)*

8. Must the potential at a point always have a reference? *Sec. 4.3*

9. Is $\overline{E}$ a conservative field when $\int_b^a (-\overline{E}\cdot\overline{d\ell})$ is independent of the path taken from b to a? *Sec. 4.3*

10. What name is given to the potential difference whose reference is at infinity? *Eq. (4.3-8)*
11. Is $\overline{E}$ a conservative field when $\oint_\ell \overline{E} \cdot \overline{d\ell} = 0$? *Eq. (4.3-6)*
12. When subtracting potentials at two different points must the potentials at each of the points have the same reference? *Sec. 4.3, Eq. (4.3-10)*
13. What is the physical significance of $dV = dQ/4\pi\epsilon_0 R$? *Eq. (4.4-5)*
14. What is the defining expression for the gradient of V? *Eq. (4.5-5)*
15. What is the expression for the ∇ operation? *Eq. (1.12.1)*
16. What expression connects $\overline{E}$ to V? *Eq. (4.5-13)*
17. How can it be shown that electric flux lines and equipotential surfaces are perpendicular? *Example 10*
18. How is the electric dipole moment defined? *Eq. (4.5-24)*
19. In building up a system of point charges, why is no work done in bringing in the first point charge? *Sec. 4.6*
20. How much energy is required to bring two point charges from infinity? *Eq. (4.6-1)*
21. What is the expression for the energy required to assemble a ρ_v charge distribution within a volume v? *Eq. (4.6-10)*
22. The energy to build up a charge distribution manifests itself in the electric field about the charges. What is the expression for the energy in terms of the $\overline{E}$ field? *Eq. (4.6-16)*
23. What is the expression for the energy density in a field $\overline{E}$? *Eq. (4.6-18)*
24. How much energy is required to assemble a point charge? *Example 12*

PROBLEMS

4-1. An electron is released at the negative plate of a charged pair of parallel plates with zero initial velocity. The field between the plates is 1200 V m^{-1}, and the spacing between plates is 1 cm. Find the acceleration on the electron, its final velocity, and the kinetic energy in joules and in electron volts, upon reaching the positive plate. How much is the total stored energy between plates reduced by the transfer of the electron?

4-2. A charge $Q = 8\ \mu$C is moved by an external force in a field $\overline{E} = \hat{x}x + \hat{y}2y$. If the charge moves a distance $\overline{d\ell} = -\hat{x} + \hat{z}\ \mu$m, find the approximate energy exchanged and indicate if the energy is given to or taken from the $\overline{E}$ field.

4-3. A charge is moved in the presence of an electric field. If $\overline{E} \cdot \overline{d\ell}$ is negative, is energy to move the charge supplied by the field $\overline{E}$, or does it come from an outside source?

4-4. For the configuration of Prob. 4-1, what is the ratio of final kinetic energy compared to the gain in potential energy due to the gravitational field if the path of the electron is vertical?

4-5. Consider the charged parallel plates of Prob. 4-1. Since opposite charges attract, the two oppositely charged plates are attracted to each other. If the spacing between plates were allowed to decrease to 0.5 cm with no change in charge density, find the new values of: (a) potential between plates; (b) $\overline{E}$ field between plates; (c) the proportionate change in stored energy, i.e., the percentage increase or decrease.

4-6. Find the potential difference between two parallel line charges of ρ_ℓ (C m^{-1}) and $-\rho_\ell$ (C m^{-1}) separated a distance d (m). [*Hint:* Assume a radius a (m) for each line, and integrate only from a to $d - a$.]

4-7. The voltage between two parallel wires is 1000 V. The radius of each wire is 1 mm, and the separation is 1 m. (a) Find the ρ_ℓ (C m^{-1}) on each wire. (b) Find the $\overline{E}$ field between wires along a straight line joining the two. Use results from Prob. 4-6 if available, and observe the hint given in Prob. 4-6.

4-8. Repeat Prob. 4.4-1 for a cylinder instead of a sphere, and let the reference potential be 0 V at $r_c = 10r_0$ instead of at infinity.

4-9. The difference in potential at 20 m from a point charge and the potential at 30 m from the same charge is 50 V. Find the value of the point charge, and at what r_s is the potential equal to 25 V?

4-10. The potential between two concentric cylinders is 65 V. The radius of the small cylinder is 1 cm and of the larger cylinder, 3 cm. Find the line charge density in (C m^{-1}) on each if the inner cylinder is positive.

4-11. Find V_{ab} by integrating over the path ℓ' of Fig. 4-5 in an electric field $\overline{E} = -\hat{x}2yx - \hat{y}x^2$ when the points a and b are at $(4, 6, 0)$ and $(2, 2, 0)$, respectively.

4-12. Repeat Prob. 4-10 with the radius of the larger cylinder 6 cm instead of 3 cm.

4-13. For the electric dipole shown in Fig. 4-9(a), find the exact absolute potential along the $+z$ axis and compare with the approximate values obtained, using the expression assuming $r_s \gg d$ at the following rectangular points: (a) $P_1(0, 0, 0)$; (b) $P_2(0, 0, d/4)$; (c) $P_3(0, 0, 2d)$; (d) $P_4(0, 0, 10d)$. Assume that $Q = 1\ \mu$C and $d = 0.25$ m.

4-14. An $\overline{E}$ field, $\overline{E} = \hat{x}y^2$ (V m^{-1}) exists between a plane at $x = 1$ m and another plane at $x = 2$ m. Are the planes each at a constant potential everywhere in the plane? Explain, using line integrals.

4-15. For the electric dipole shown in Fig. 4-9(a), d is 1 mm, $|\overline{E}| = 1$ V m^{-1} at $r_s = 10$ cm, and $\theta = 90°$. Find $\overline{E}$ at $r_s = 20$ cm and $\theta = 0°$.

4-16. There is a potential difference of 12 V between two concentric spheres of radii 1 cm and 10 cm. Find the $\overline{E}$ field between spheres. Assuming that the inner sphere is positive, find $V(r_s)$ for $0.01\ \text{m} \le r_s \le 0.1$ m, using infinity as a reference.

4-17. Repeat Prob. 4-16 with concentric cylinders instead of spheres, and use as a zero reference the potential at $r_c = r_a$.

4-18. Find the absolute potential along the axis of a disc in the $z = 0$ plane, and centered at the origin, when $\rho_s = r_c'$ (C m^{-2}), $0 \leqq r_c' \leqq b$. [*Hint:* Use the building block concept.]

4-19. Find $\overline{E}$, $\overline{D}$, and ρ_v for the following potential difference fields: (a) $V = 10x^2$; (b) $V = 2r_c \sin\phi$; (c) $V = (5/r_s)\cos\theta$.

4-20. Find the potential along the axis of a disc of charge density ρ_s and radius b, and from this potential function and (4.5-13), find the $\overline{E}$ field. Compare this result with that of (2.6-9).

4-21. The ratio r_b/r_a for a particular coaxial cable is 3. Find the potential difference between conductors if the radial electric field is 3000 V m^{-1} at the surface of the inner conductor of 1 cm radius.

4-22. The electric field between two coaxial cylinders is 500 V m^{-1} at the inside surface of the outer conductor. Find the potential difference between conductors if the radii are 2 cm and 5 cm.

4-23. The expression for the potential difference above an infinite sheet of uniform ρ_s, in the $z = 0$ plane and with reference at $z = 0$, is found to be $V_{z0} = -(3z/\epsilon_0)$ (V). Find: (a) $\overline{E}$ through the use of the gradient concept for $+z$; (b) ρ_s on the sheet.

4-24. For the electric dipole of Example 11, find the exact expression for E at $r_s = 0$ and compare with results obtained by using (4.5-26).

4-25. An air dielectric parallel plate capacitor is charged to 200 V. If the area of each plate is 5 cm^2 and the spacing between plates is 0.1 mm, find: (a) the charge density ρ_s (C m^{-2}) on each plate; (b) the force pulling the plates together; (c) the total energy stored in the volume between plates.

4-26. A thin spherical shell of radius r_a and uniform $\rho_s = 10^{-7}$ C m^{-2} is centered at the origin. If the total charge is $Q = 1\ \mu$C, find: (a) r_a; (b) the absolute potential for $r_s \leqq r_a$; (c) the absolute potential for $r_s \geqq r_a$. Plot V for $0 \leqq r \leqq 4r_a$.

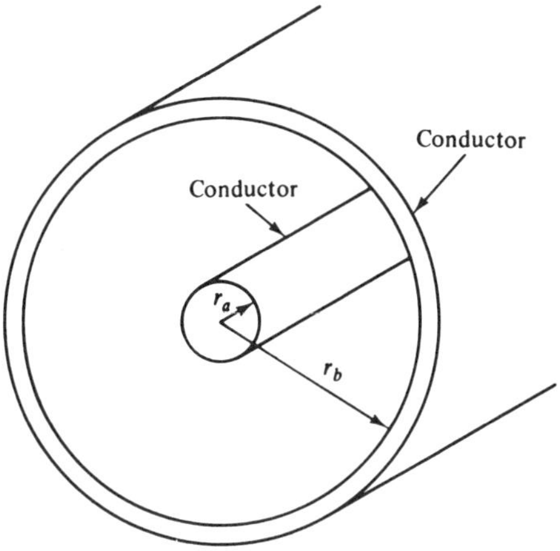

Figure 4-11 Graphical display of a coaxial cable for Prob. 4-27.

4-27. A coaxial cable, shown in Fig. 4-11, has $-\rho_\ell$ on the surface at r_a and $+\rho_\ell$ at the surface at r_b. (a) Find the energy stored per meter length through the use of (4.6-16). (b) Show that the result of part (a) will reduce to $W_E = \frac{1}{2}\rho_\ell V_{ab}$. (c) Find the ρ_s at r_a. (d) Find the ρ_s at r_b.

4-28. The capacitance per meter length of the coaxial cable of Prob. 4-27 is defined as $C \triangleq \rho_\ell / V_{ab}$ (F m^{-1}). Show that

(a) $$C = \frac{\rho_\ell^2}{2W_E} \quad (\text{F m}^{-1})$$

(b) $$C = \frac{2\pi\epsilon_0}{\ln(r_b/r_a)} \quad (\text{F m}^{-1})$$

(c) $$W_E = \tfrac{1}{2}CV_{ab}^2 \quad (\text{J m}^{-1})$$

5

Moving Charges (Current), Conductors, Ohm's Law, Resistance, Semiconductors, Dielectrics, and Capacitance

5.1 INTRODUCTION

In Chapters 2, 3, and 4 our discussion was restricted to static (stationary) charges in free space (vacuum). Under these conditions we found electrostatic $\overline{E}$, $\overline{D}$, and V fields. In this chapter we will allow the charges to move with a constant velocity and thus introduce the concept of *current*. The concept of current will lead us to conducting media whose prominent characteristic is that of conducting electric charge. It should be noted that a constant (steady) current will give rise to a constant magnetic field that will be covered in Chapter 8.

With the introduction of the conducting media, our next logical discussion will be that of dielectric media whose prominent characteristic is that of electric polarization—formation of electric dipoles—within the media. The effect of this electric polarization of the media on the $\overline{E}$ and $\overline{D}$ fields will be studied with the aid of the electric dipole introduced in Chapter 4.

The introduction of conducting media (conductors) will lead us to the conductivity and resistance concepts while the dielectric media (insulators) will lead us to the dielectric constant and capacitance concepts. The concepts of resistance and capacitance are called *circuit concepts* since they are very useful in electronic circuit analysis.

5.2 MOVING CHARGES (CURRENT)

In this section we study moving charges or current in free space (vacuum). Later, the current concept will be extended to metal conductors and semiconductors.

Let us assume that we have, in vacuum, a long cylinder of ρ_v distribution, as shown in Fig. 5-1(a). The element of charge ΔQ, found in the volume $\Delta v = \Delta s\, \Delta x$, is

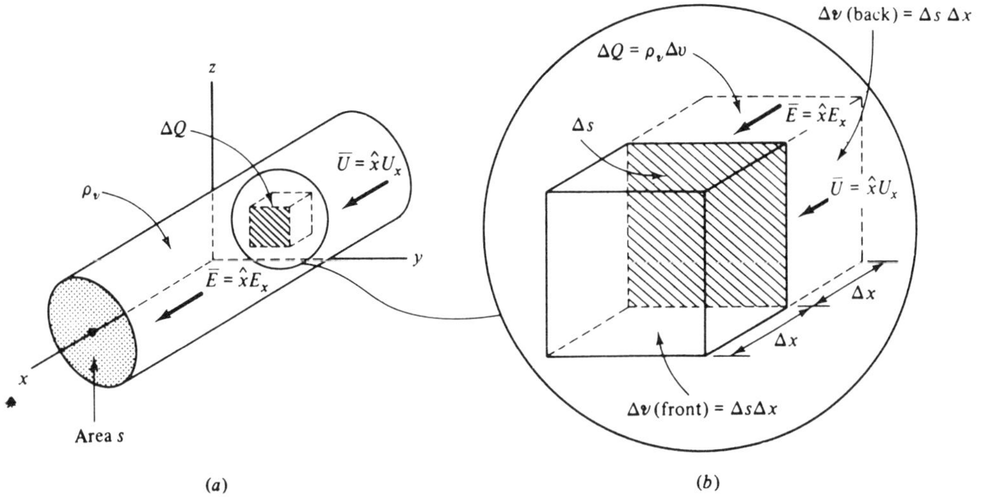

Figure 5-1 Graphical display for the current concept: (a) a long cylinder of ρ_v subjected to an $\overline{E} = \hat{x}E_x$ field; (b) an exploded view of the ΔQ that travels in the $\overline{U} = \hat{x}U_x$ direction when subjected to the $\overline{E}$ field.

assumed to move with a velocity $\overline{U} = \hat{x}U_x$ when subjected to the applied field $\overline{E} = \hat{x}E_x$. This is shown in an exploded view in Fig. 5-1(b). At some time t_0 let ΔQ be located in the back volume Δv (back) and at $t_0 + \Delta t$ let it be found in the front volume Δv (front). Thus, within the time Δt the charge $\Delta Q = \rho_v \Delta v$ has passed through the cross-hatched surface Δs. Let us form

$$\frac{\Delta Q}{\Delta t} = \frac{\rho_v \Delta v}{\Delta t} = \rho_v \frac{\Delta x \, \Delta s}{\Delta t} = \rho_v \frac{\Delta x}{\Delta t} \Delta s \tag{1}$$

The limit of (1) as $\Delta t \to 0$ leads us to the definition of the current ΔI that flows through the surface Δs; thus,

$$\boxed{\Delta I \triangleq \lim_{\Delta t \to 0} \left[\frac{\Delta Q}{\Delta t}\right] = \lim_{\Delta t \to 0} \left[\rho_v \frac{\Delta x}{\Delta t} \Delta s\right] = \rho_v U_x \Delta s \qquad \text{(A)}} \tag{2}$$

where the unit of ΔI is the ampere (A).

> Current is defined as the movement of charge through a given surface and is equal to the coulombs per second through that surface.

Let us now divide (2) by Δs and take the limit as $\Delta s \to 0$ to obtain the defining equation for the x-directed current density J_x (A m^{-2}); thus,

$$\boxed{J_x \triangleq \lim_{\Delta s \to 0} \left[\frac{\Delta I}{\Delta s}\right] = \rho_v U_x \qquad (\text{A m}^{-2})} \tag{3}$$

Equation (3) can be written in general vector form as

$$\boxed{\overline{J} = \rho_v \overline{U} \qquad (\text{A m}^{-2})} \tag{4}$$

It should be noted that the current density is a vector quantity. If ρ_v is positive and $\overline{E} = \hat{x}E_x$, the current density is $\overline{J} = \hat{x}J_x$. Now, if ρ_v is negative and $\overline{E} = \hat{x}E_x$, we have $\overline{J} = \hat{x}J_x$. This comes about from (4) since the sign on the ρ_v and $\overline{U}$ terms are both negative. From this simple exercise we note that $\overline{J}$ will have the same direction for $+\rho_v$ moving in the $+x$ direction and $-\rho_v$ moving in the $-x$ direction. The current ΔI and the current density $\overline{J} = \rho_v \overline{U}$ are called the *convection current* and *convection current density*, respectively. Let us emphasize again that $\overline{U}$ is the velocity of the charges whose density is ρ_v (C m^{-3}). If both positive and negative charges are present, (4) can be rewritten as follows:

$$\boxed{\overline{J} = \rho_{v+}\overline{U}_+ + \rho_{v-}\overline{U}_-} \tag{5}$$

The total current I flowing through the large perpendicular cross section of area s of Fig. 5-1 can be found through

$$\boxed{I = \int_s \overline{J} \cdot \overline{ds} \qquad (\text{A})} \tag{6}$$

Equation (6) can be extended to a general case to give the current I that flows through any surface s.

At this point of development it is logical to consider the *principle of conservation of charge*. This principle states that charge can neither be created nor destroyed. Let us locate, within a ρ_v distribution, a closed surface s that encloses a volume $\mathcal{v}$. The charge enclosed by the surface s will be denoted as Q_{en} and in terms of ρ_v, found within the volume, becomes

$$Q_{\text{en}} = \int_{\mathcal{v}} \rho_v \, d\mathcal{v} \tag{7}$$

The charge Q_{en} can be increased or decreased by charge flowing through s, inwardly or outwardly. This movement of charge through the surface s gives rise to a current flow through the surface s. Thus, the time rate of decrease of Q_{en} must be substantiated by a current I_s that flows outwardly through s. For a fixed surface s, $Q_{\text{en}} = Q_{\text{en}}(t)$ and the current I_s becomes

$$\boxed{I_s = -\frac{\partial Q_{\text{en}}}{\partial t} = \oint_s \overline{J} \cdot \overline{ds}} \tag{8}$$

where the last term was obtained through the use of (6) with $\overline{ds}$ directed outwardly.

Combining (7) and (8), we obtain

$$\oint_s \overline{J} \cdot \overline{ds} = -\frac{\partial}{\partial t}\left(\int_v \rho_v \, dv\right) = -\int_v \frac{\partial \rho_v}{\partial t} \, dv \tag{9}$$

Equations (8) and (9) state that the time rate of decrease of charge within volume v must equal the net current that flows outwardly through the closed surface s. The accounting of charge in (8) and (9) thus expresses in integral form the conservation of electric charge.

Through the use of the divergence theorem, (9) becomes

$$\int_v (\nabla \cdot \overline{J}) \, dv = -\int_v \frac{\partial \rho_v}{\partial t} \, dv \tag{10}$$

Now, let $v \to 0$ and obtain

$$\nabla \cdot \overline{J} = -\frac{\partial \rho_v}{\partial t} \quad \text{(A m}^{-3}) \tag{11}$$

Equation (11) is the point form of the conservation of charge. Many authors call (11) the *point form of the continuity of current equation*. From the divergence concept, (11) states that current (flux of $\overline{J}$) per unit volume emanating from a point equals the time rate decrease of volume charge density at the same point.

By the introduction of the principle of conservation of charge, we have strayed from static field ideas since (8), (9), and (10) are functions of time. Equation (11) will be used when we discuss the concept of relaxation time in our study of conductors.

Example 1

For a ρ_v distribution, in a long cylinder, moving with a velocity $\overline{U} = \hat{z}5z$ (m s^{-1}), as shown in Fig. 5-2, find: (a) $\rho_v|_{z_1=2}$ when $I|_{z_1=2} = \pi$ (μA); (b) $\rho_v|_{z_2=4}$ when $I|_{z_2=4} = \pi$ (μA); (c) $\rho_v(z)$ when $I(z) = \pi$ (μA); (d) $\overline{J}$ (A m^{-2}).

Solution.   (a) Assume that I and $\overline{J}$ are uniform over the cross section of the cylinder. From (6) and (4), in functional form, we have

$$I(z) = \int_s \overline{J}(z) \cdot \overline{ds} = \int_s \rho_v(z)\overline{U}(z) \cdot \overline{ds} \tag{12}$$

Since $\rho_v(z)$ and $\overline{U}(z)$ do not vary with x and y, and $\overline{U}(z)$ is parallel to $\overline{ds}$, (12) reduces to

$$I(z) = \rho_v(z)U_z(z)\int_s ds = \rho_v(z)U_z(z)\pi r_c^2 \tag{13}$$

Solving for $\rho_v(z)$ and substituting values, we have

$$\rho_v(z)\Big|_{z_1=2} = \frac{I(z)}{U_z(z)\pi r_c^2}\Big|_{z_1=2} = \frac{\pi 10^{-6}}{10\pi(0.2)^2} = 2.5 \ \mu\text{C m}^{-3} \tag{14}$$

(b) Evaluate (14) at $z_2 = 4$ to obtain

$$\rho_v(z)\Big|_{z_2=4} = \frac{\pi 10^{-6}}{20\pi(0.2)^2} = 1.25 \ \mu\text{C m}^{-3} \tag{15}$$

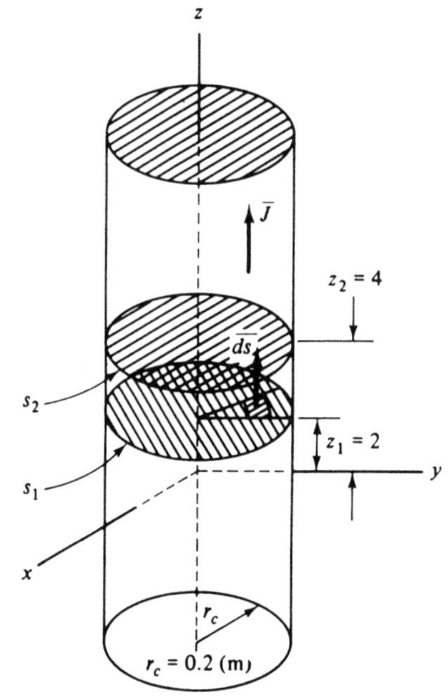

Figure 5-2 Graphical display for finding ρ_v in a long cylinder of charge distribution for Example 1.

(c) From (14),

$$\rho_v(z) = \frac{\pi 10^{-6}}{5z\pi(0.2)^2} = 5z^{-1} \qquad (\mu\text{C m}^{-3})$$

(d) From (4),

$$\overline{J} = \rho_v \overline{U} = \frac{5}{z}\hat{z}5z = \hat{z}25 \qquad (\mu\text{A m}^{-2}) \tag{16}$$

Example 2

Evaluate $\oint_s \overline{J} \cdot \overline{ds}$ over the portion of the cylinder of Example 1 bounded by $r_c = 0.2$ and the surfaces at z_1 and z_2, as shown in Fig. 5-2.

Solution

$$\oint_s \overline{J} \cdot \overline{ds} = \int_{s(\text{at } r_c=0.2)} \overline{J} \cdot \overline{ds} + \int_{s1} \overline{J} \cdot \overline{ds} + \int_{s2} \overline{J} \cdot \overline{ds}$$

$$= \int_{s(\text{at } r_c=0.2)} (\hat{z}25 \times 10^{-6}) \cdot (\hat{r}_c\, ds) + \int_{s1} (\hat{z}25 \times 10^{-6}) \cdot (-\hat{z}\, ds)$$

$$+ \int_{s2} (\hat{z}25 \times 10^{-6}) \cdot (\hat{z}\, ds)$$

$$= 0 - 25\pi \times 10^{-6}(0.2)^2 + 25\pi \times 10^{-6}(0.2)^2 = 0 \text{ A}$$

Thus, the total current flowing outward through the closed surface s is zero.

Problem 5.2-1 From the divergence formulas (3.5-9), (3.5-15), (3.5-16), or those given on the inside back cover of this book, show that $\nabla \cdot \overline{J} = 0$, and thus $\partial\rho_v/\partial t = 0$, for the following current densities: (a) $\overline{J} = \hat{x}K$; (b) $\overline{J} = \hat{r}_c(K/r_c)$; (c) $\overline{J} = \hat{r}_s(K/r_s^2)$ (A m^{-2}); (d) $\overline{J} = \hat{\phi}K$.

Problem 5.2-2 In a solid cylindrical conductor with uniform ρ_v over its cross section, the current density is described by $\overline{J} = \hat{z}K \cos(\omega t - \beta z)$ (A m^{-2}). Is $\nabla \cdot \overline{J} = 0$? If not, find $\nabla \cdot \overline{J}$.

5.3 CONDUCTORS AND CONDUCTIVITY

A conducting material (medium) has electric charge conduction as its prominent characteristic. A metal conductor has enormous electric charge conduction through the presence of a large number of outer-orbit electrons that are free to move about in the lattice structures of the material. These free electrons, with zero external $\overline{E}$ applied within the conductor, move in random directions and with varying velocities to produce zero net current through any surface in the conductor. The direction of the electron motion is changed in the process of collisions with the thermally excited lattices. Now, if we apply an electric field the electrons will still have motion in random directions but will drift slowly in the $-\overline{E}$ direction with an average velocity called *drift velocity* $\overline{U}_d$. The motion of electrons in the $-\overline{E}$ direction with a velocity $\overline{U}_d$ gives rise to a conduction current in the conductor. The drift velocity $\overline{U}_d$ of the electrons is related to the externally applied $\overline{E}$ field through

$$\boxed{\overline{U}_d = -\mu_e \overline{E}} \tag{1}$$

where μ_e is a positive quantity called the *electron mobility* with units of ($m^2\ V^{-1}\ s^{-1}$).

Through the use of (5.2-4), we obtain

$$\boxed{\overline{J} = \rho_{ve}(-\mu_e \overline{E}) = -\rho_{ve}\mu_e \overline{E} \qquad (\text{A m}^{-2})} \tag{2}$$

where ρ_{ve} is the free-electron charge density and thus is a negative quantity. The positive ρ_v stemming from protons in the lattice structures contribute zero to $\overline{J}$ since, though they vibrate about a fixed point, they do not have a net $\overline{U}$, influenced by $\overline{E}$, as do the more agile free electrons. It should be noted that, although the electrons move about, the material remains neutral (net $\rho_v = 0$) on the macroscopic scale (volume containing thousands of atoms). The term $-\rho_{ve}\mu_e$ is called *conductivity* σ and has the units of siemens per meter ($S\ m^{-1}$). Equation (2) becomes

$$\boxed{\overline{J} = \sigma \overline{E} \qquad (\text{A m}^{-2})} \tag{3}$$

Equation (3) can be called the *point form* of Ohm's law that we will find useful in electromagnetic field theory. The more common electric circuit theory form of Ohm's law will be derived in Sec. 5-4 from the point form.

The conductivity σ is thus a measure of the free-electron conductive properties of a conductor. In good conductors its value decreases with temperature since increased thermal energy of the lattice structure increases lattice vibration, thus increasing the collision possibility with the moving free electrons. Material possessing low values of conductivity are called *insulators*. Table 5-1 lists several materials and their conductivity.

The reciprocal of σ is called *resistivity* and is generally symbolized by ρ with no subscript and has the units of ($\Omega \cdot m$). Care should be taken not to confuse this with the symbols used for charge density ρ_e, ρ_s, and ρ_v.

Example 3

Given for copper $\sigma = 5.8 \times 10^7$, $\mu_e = 0.0032$, and $E = 1\ V\ m^{-1}$, find for a circular copper conductor of $s = 1\ cm^2$ cross-sectional area: (a) J; (b) I; (c) ρ_{ve}; (d) U_d.

Solution. (a) From (3),

$$J = \sigma E = (5.8 \times 10^7)(1) = 58 \text{ MA m}^{-2}$$

(b) $I = \int_s \overline{J} \cdot \overline{ds} = Js = (5.8 \times 10^7)(10^{-4}) = 5.8 \text{ kA}$

(c) From $\sigma = -\rho_{ve}\mu_e$, we have

$$\rho_{ve} = -\frac{\sigma}{\mu_e} = -\frac{5.8 \times 10^7}{0.0032} = -18 \text{ GC m}^{-3}$$

(d) From (1),

$$U_d = -\mu_e E = -(0.0032)(1) = -0.32 \text{ cm s}^{-1}$$

where the minus sign indicates that U_d is in the opposite direction to $\overline{E}$. From these calculations it can be seen that the drift velocity of free electrons is very small, about one-third centimeter per second, even though a large current of 5.8 kA flows through a square centimeter cross section.

Example 4

From the ρ_{ve} found in Example 3 and the charge on an electron $q_e = -1.602 \times 10^{-19}$ C, find the free electron density n_e in a copper conductor.

Solution. From $\rho_{ve} = q_e n_e$, we have

$$n_e = \frac{\rho_{ve}}{q_e} = \frac{-18 \times 10^9}{-1.602 \times 10^{-19}} = 1.124 \times 10^{29} \text{ m}^{-3}$$

With 10^{29} m^{-3} free electrons, it can readily be seen that copper is a good conductor of charge.

Problem 5.3-1 Find the current that would flow in the z direction of a cylindrical copper conductor, of configuration shown in Fig. 5-2, when $\overline{E} = \hat{z}5$ (V m^{-1}).

TABLE 5-1 REPRESENTATIVE VALUES FOR CONDUCTIVITY OF SEVERAL MATERIALS AT ROOM TEMPERATURE, RANGING FROM GOOD CONDUCTORS TO GOOD INSULATORS

Material	σ (S m^{-1})	Classification
Silver	6.17×10^7	Conductors
Copper	5.8×10^7	
Aluminum	3.82×10^7	
Brass	2.56×10^7	
Tungsten	1.83×10^7	
Nickel	1.45×10^7	
Iron	1.03×10^7	
Nichrome	0.1×10^7	
Mercury	1.0×10^6	
Graphite	$\sim 3.0 \times 10^4$	
Sea water	~ 4.0	
Intrinsic germanium	~ 2.2	Intrinsic semiconductors
Ferrite	$\sim 1.0 \times 10^{-2}$	
Intrinsic silicon	$\sim 0.44 \times 10^{-3}$	
Distilled water	$\sim 1.0 \times 10^{-4}$	Insulators
Bakelite	$\sim 1.0 \times 10^{-9}$	
Glass	$\sim 1.0 \times 10^{-12}$	
Mica	$\sim 1.0 \times 10^{-15}$	
Fused quartz	$\sim 1.0 \times 10^{-17}$	

Problem 5-3.2 In a certain conductor of 10^{-4} m^2 cross section, it is known that $n_e = 10^{26}$ m^{-3}, $E = 0.1$ V m^{-1}, and $I = 0.5$ kA. Find: (a) ρ_{ve}; (b) U_d; (c) μ_e; (d) J; (e) σ.

5.4 OHM'S LAW AND RESISTANCE FROM FIELD THEORY

In Sec. 5.2 we found charge motion in free space under the influence of a static $\overline{E}$ field to give rise to the convection current concept. A close examination of the charge motion will indicate that the velocity $\overline{U}$ of the charge ΔQ will increase indefinitely with time since in free space there exists nothing to impede the charge. Now, in Sec. 5.3 we found that, in a good conductor, the free-electron drift velocity $\overline{U}_d$ is very small and does not increase indefinitely as in the free space condition. The small drift velocity in a good conductor is due to the damping forces which result from collision with the vibrating lattice structure. Such collisions result in energy exchange from the kinetic energy of the electrons to thermal energy of the lattice structure. It should be noted that the exchanged kinetic energy of the electron has the applied $\overline{E}$ field as its source.[1] The increased thermal energy of the lattice structure will manifest itself in the temperature rise of the material.

The damping forces or resistance to free electron motion in a conductor give rise to a concept that is logically called *resistance*. The resistance of a conducting material will now be derived, starting with the point form of Ohm's law (5.3-3). Let us consider a finite length, ℓ, portion of a conductor of uniform cross section as shown in Fig. 5-3(a), with an $\overline{E} = \hat{x}E_x$. By applying (5.2-6) and (5.3-3) the current I, through the cross sections, becomes

$$I = \int_s \overline{J} \bullet \overline{ds} = \int_s \sigma\overline{E} \bullet \overline{ds} = \sigma E_x s \qquad \text{(A)} \qquad (1)$$

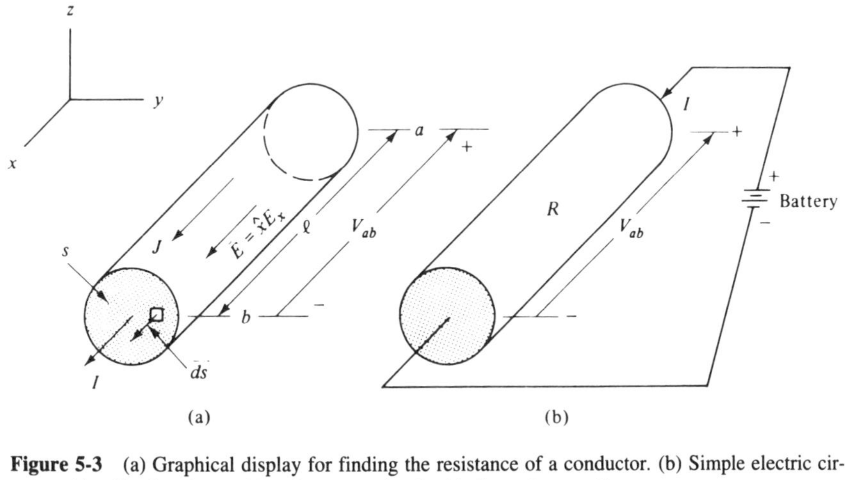

Figure 5-3 (a) Graphical display for finding the resistance of a conductor. (b) Simple electric circuit to identify directions of I and V_{ab} associated with the resistance R.

[1] M. A. Plonus, *Applied Electromagnetics*, McGraw-Hill Book Company, New York, 1978. See pp. 54–69 for a good discussion on conduction in metallic conductors.

where $\overline{ds}$ is taken in the $\overline{J}$ direction, and $\overline{E}$ is assumed uniform over s and ℓ. Now, from (4.3-1), V_{ab} becomes

$$V_{ab} = \int_b^a (-\overline{E} \bullet \overline{d\ell}) = E_x \ell \qquad \text{(V)} \tag{2}$$

Solving for E_x in (1) and substituting into (2), we have

$$V_{ab} = \left(\frac{I}{\sigma s}\right)\ell = I\left(\frac{\ell}{\sigma s}\right) \qquad \text{(V)} \tag{3}$$

From (3), let us define resistance R as

$$\boxed{R \triangleq \frac{V_{ab}}{I} \qquad (\text{V A}^{-1} \text{ or } \Omega)} \tag{4}$$

thus

$$\boxed{R = \frac{\ell}{\sigma s} \qquad (\Omega)} \tag{5}$$

the resistance in ohms to free-electron flow or to the current I of a conductor of uniform cross section s, length ℓ, and uniform $\overline{E}$ (thus uniform $\overline{J}$).

From (4), the formulation

$$\boxed{V_{ab} = IR \qquad \text{(V)}} \tag{6}$$

is called *Ohm's law* and is termed as a circuit concept since it is defined in terms of V_{ab}, and I, which are circuit concepts used extensively in electric circuit analysis. Let us place this length of conductor in a closed electric circuit, as shown in Fig. 5-3(b). In a circuit it will be called a *circuit element* R or a *resistor* of resistance R (Ω). The $\overline{E}$ field in the conductor is now due to the battery, and any thermal energy loss due to free-electron collisions must be supplied by this source of energy. In Fig. 5-3, the positive sign has been used with V_{ab} to denote that the potential is higher at that point than at the negative sign. The arrow on the line connecting points b and a also indicates the same notion, thus they are redundant ideas and only one notation need be used in circuit analysis.

From the defining equation (4), for the resistance R, and the directions of the quantities in Fig. 5-3, it can readily be seen that R is the ratio of the V_{ab}, between two equipotential surfaces, to I that flows into the positive surface or terminal. The concept of resistance can be extended to conducting material of non-uniform cross section and non-uniform $\overline{E}$ or $\overline{J}$ through the use of integral forms of (1) and (2).

$$\boxed{R = \frac{V_{ab}}{I} = \frac{-\int_b^a (\overline{E} \bullet \overline{d\ell})}{\int_s \sigma\overline{E} \bullet \overline{ds}} = \frac{\int_b^a \left(-\frac{\overline{J}}{\sigma} \bullet \overline{d\ell}\right)}{\int_s \overline{J} \bullet \overline{ds}}} \tag{7}$$

where $\overline{E}$ or $\overline{J}$ in the denominator are integrated over any cross section s while $\overline{E}$ or $\overline{J}$ in the numerator are integrated over the length ℓ between two equipotential surfaces. To reduce confusion in signs in (7), it might be wise to place absolute magnitude signs about the integrals.

It may be more convenient to solve directly for the resistance R or the conductance G between two surfaces from

$$R = \int_{\substack{\text{path}\\\text{between}\\\text{surfaces}}} dR = \int \frac{d\ell}{\sigma s} = \frac{1}{G} \tag{8}$$

or

$$G = \int_{\substack{\text{surface}\\\text{area}}} dG = \int \frac{\sigma\, ds}{\ell} = \frac{1}{R} \tag{9}$$

Example 5

Find the resistance between the $\phi = 0$ and $\phi = \pi/2$ surfaces of the truncated wedge section shown in Fig. 5-4 when $\sigma = 4 \times 10^7$ S m^{-1} and $\overline{E} = -\hat{\phi}/r_c$ (V m^{-1}).

Solution. First, let us find V_{ab} between the faces at $\phi = 0$ and $\phi = \pi/2$; thus,

$$V_{ab} = \int_b^a (-\overline{E} \bullet \overline{d\ell}) = \int_b^a -\left(-\frac{\hat{\phi}}{r_c}\right) \bullet (\hat{\phi} r_c\, d\phi) = \int_0^{\pi/2} d\phi = \frac{\pi}{2} \quad \text{(V)} \tag{10}$$

It can be seen that V_{ab} is a constant between the two faces. Now, the current I through face a or any cross section can be found from (1):

$$I = \int_s \sigma\overline{E} \bullet \overline{ds} = \int_{0.1}^{1}\int_0^1 \left(-\sigma\frac{\hat{\phi}}{r_c}\right) \bullet (-\hat{\phi}\, dz\, dr_c) = \sigma \ln 10 \quad \text{(A)} \tag{11}$$

Dividing (10) by (11), we have

$$R = \frac{V_{ab}}{I} = \frac{\pi/2}{\sigma \ln 10} = 0.01705\ \mu\Omega$$

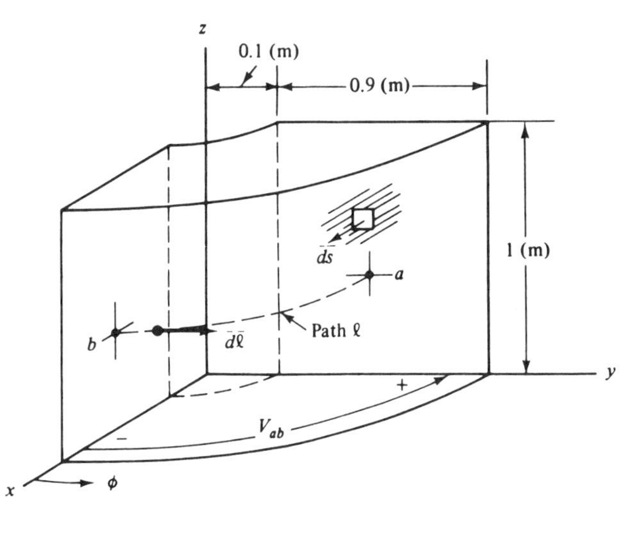

Figure 5-4 Graphical construction for Example 5 to find the resistance of a truncated wedge of metal between faces at $\phi = 0$ and $\phi = \pi/2$.

We could also calculate G directly from (9),

$$G = \int_{z=0}^{1}\int_{r=0.1}^{1} \frac{\sigma\, dr\, dz}{r(\pi/2)} = \frac{2\sigma \ln r}{\pi}\bigg|_{0.1}^{1}$$

$$= \frac{2\sigma \ln 10}{\pi} \quad \text{(S)}$$

and

$$R = \frac{1}{G} = \frac{\pi}{2\sigma \ln 10} \quad (\Omega)$$

Problem 5.4-1 Find the resistance of 150 ft of No. 33 gauge copper wire. The diameter of No. 33 gauge copper wire is 0.0071 in.

Problem 5.4-2 For the truncated conducting wedge of Example 5, find the resistance between faces at $z = 0$ and $z = c$ (m), when the inner radius is r_a and the outer radius is r_b. Calculate the resistance first by using (1) and (2), and then by using (9).

Problem 5.4-3 For a truncated wedge like that of Fig. 5-4, find the resistance between the curved face at radius of r_a and that at a radius of r_b, where $r_b > r_a$, and the length in the z direction is c (m). Calculate the resistance first by using (1) and (2), and then by using (8).

5.5 CONDUCTOR PROPERTIES UNDER STATIC CONDITIONS

Static condition exists in a conductor when all current due to moving charges is zero. This does not mean that all thermal vibrations of the free electrons and those of the lattice structure have been reduced to zero.

To aid us in understanding what happens to excess free charge in a conductor, we will introduce charge at some interior point and study the decay of ρ_v at this point. Let us start with the conservation of charge expression (5.2-11):

$$\boxed{\nabla \cdot \bar{J} = -\frac{\partial \rho_v}{\partial t}}$$

Through the use of $\sigma\bar{E} = \bar{J}$ (5.3-3), the equation above will become

$$\nabla \cdot \bar{E} = -\frac{1}{\sigma}\left(\frac{\partial \rho_v}{\partial t}\right) \tag{1}$$

Now, from $\nabla \cdot \bar{E} = \rho_v/\epsilon_0$, (1) becomes a simple partial differential equation

$$\boxed{\frac{\partial \rho_v}{\partial t} + \frac{\sigma}{\epsilon_0}\rho_v = 0} \tag{2}$$

Equation (2) can be written in ordinary differential form and solved for ρ_v to obtain

$$\boxed{\rho_v = \rho_v(t) = \rho_{v0}e^{-(\sigma/\epsilon_0)t} = \rho_{v0}e^{-t/t_r}} \tag{3}$$

where ρ_{v0} is the charge density at $t = 0$, and t_r is called the *relaxation time constant* that is equal to the time it takes for the ρ_v to decay to $(1/e)$th of its initial value. From (3), we have

$$t_r = \frac{\epsilon_0}{\sigma} \quad (s) \tag{4}$$

For a good conductor such as copper, $\epsilon_0 = 8.85 \times 10^{-12}$, $\sigma = 5.8 \times 10^7$, we find $t_r \cong 1.5 \times 10^{-19}$ (s). For good insulators, t_r will be in the range of hours and days due to their low σ. Thus, a charge placed in a good insulator will not disperse as rapidly as in a good conductor. The decay of ρ_v at a point is associated with charge movement toward the conductor surface until all the free charge that was introduced internally to the conductor appears on the surface. In a good conductor, this process takes place in a time interval equal to a few time constants. During this process current will flow and a non-static condition exists within the conductor.

From the above discussion, it can be seen that under static conditions the following conductor properties exist within the conductor:

1. The net $\rho_v = 0$ inside the conductor. This takes into account all protons and electrons found in the atoms in addition to any other charge.
2. The net $\overline{E} = 0$ inside the conductor; otherwise current would flow. This is the macroscopic $\overline{E}$ in the metal and not the microscopic $\overline{E}$ found near the nucleus of the atom.
3. The conductor is an equipotential body since $-\nabla V = \overline{E} = 0$ inside the conductor.
4. Charge density ρ_s, if present, is found only on the surface.
5. The tangent component of $\overline{E}$, on the surface of the conductor, equals zero since $\sigma E_{\tan} = J_{\tan}$ must equal zero.
6. The normal component of $\overline{E}$, on the surface of the conductor, is non-zero if ρ_s exists on the surface.

If a slab of neutral conductor is immersed in an applied field $\overline{E}_a$, we will find that within an extremely short period of time, several relaxation time constants, free charges will be forced to the surface and the net $\overline{E}$ inside the slab will become zero. The charge on the surface will be surface charge ρ_s and will distribute itself so as to produce a net $\overline{E} = 0$ inside the slab. Thus, the surface charge ρ_s must induce an $\overline{E}_i$ inside the conductor so that $\overline{E} = \overline{E}_a + \overline{E}_i = 0$.

Now, if a charge Q is placed in a neutral slab of conductor, we will find that charge will very quickly find itself on the surface in the form of a ρ_s, so distributed as to produce a net $\overline{E} = 0$ within the conductor.

Example 6

Show graphically the charge distribution and flux plot inside and outside a parallel-faced slab conductor when immersed in a uniform field $\overline{E}_a$ under static conditions.

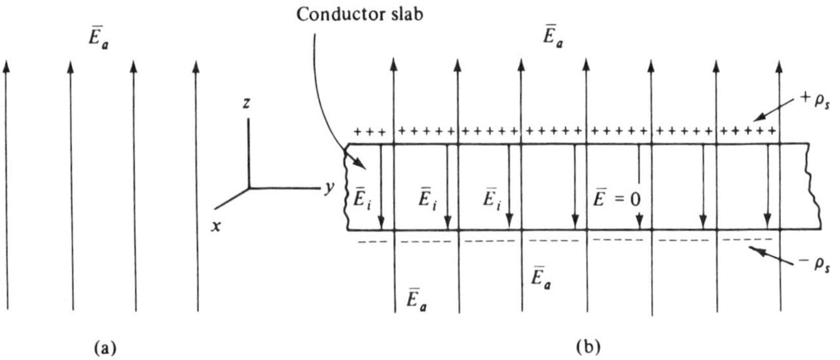

Figure 5-5 Graphical constuction for Example 6: (a) the uniform $\overline{E}_a$ field only; (b) the resultant ρ_s and $\overline{E}$ inside a conductor slab. Inside the slab $\overline{E} = \overline{E}_a + \overline{E}_i = 0$.

Solution. Figure 5-5(b) illustrates surface charges on the top and bottom surfaces of the slab that was caused by $\overline{E}_a$. These surface charges give rise to an induced $\overline{E}_i$, and this $\overline{E}_i$ becomes equal to $-\overline{E}_a$ in the conductor to make the net $\overline{E} = 0$ as required in static condition. Note that here $\overline{E}_a$ outside the slab remains unchanged.

If an elongated conducting object has its axis parallel to the electric field, charges will accumulate at the ends as shown for the prolate spheroid in Fig. 5-6. The internal field due to these charges will cancel the $\overline{E}_a$ field such that the net field inside the

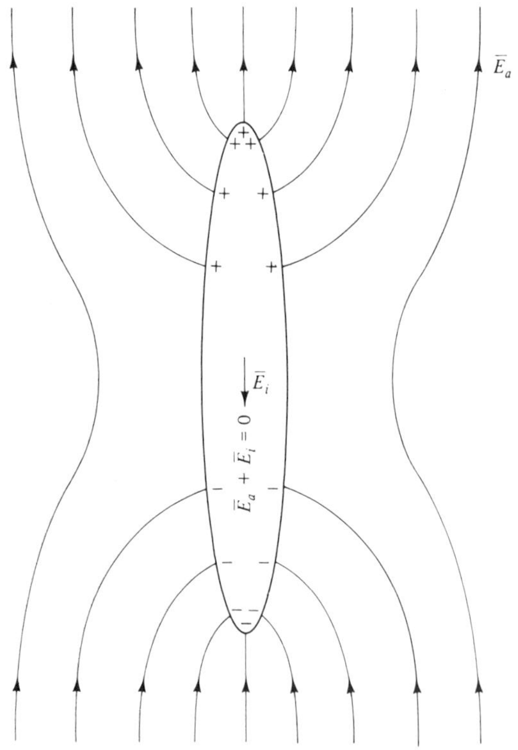

Figure 5-6 Conducting prolate spheroid in an electric field.

spheroid is zero. The external field is enhanced at the poles of the conducting surface. For a sphere, the enhancement factor is exactly 3. For a more elongated shape, the enhancement is greater. If the field is alternating, the charge distribution also alternates, and current flows between the poles of the object. This is the mechanism by which current is induced in objects (or people) near high-voltage-power-line fields or near radio antennas.

Problem 5.5-1 For a flat conducting slab oriented perpendicular to the $\overline{E}$ field as shown in Fig. 5-5(b), the $\overline{E}_a$ field is $\hat{z}7$ (V m^{-1}). Find: (a) $\overline{D}_a$; (b) ρ_s on each plate.

Problem 5.5-2 A conducting sphere is placed in an $\overline{E}_a$ field. Assuming that the enhancement factor at the poles of the sphere is 3, show that for $|\rho_s| = 3\epsilon_0 E_a \cos\theta$ (C m^{-2}) on either hemisphere, the net $\overline{E}$ field inside the sphere is zero. Assume that θ is measured from a pole.

Problem 5.5-3 A total free charge of -10^{-2} C is placed on a neutral, thin, and parallel-faced conductor slab. If each face of the slab has an area of 100 m^2 and the charge distributes itself uniformly on the two faces only, obtain under static conditions: (a) $\overline{D}$ and $\overline{E}$ inside and outside the slab; (b) a flux plot of $\overline{E}$ inside and outside the slab.

5.6 CONDUCTOR-FREE SPACE BOUNDARY CONDITIONS

Let us now relate in mathematical form some of the conductor properties found in Sec. 5.5, under static condition, to the fields in the free space just across the conductor boundary, as shown in Fig. 5-7. The boundary condition on the normal field components can be readily found by applying Gauss's law to a small cylinder centered at the boundary as $\Delta h \to 0$. From Gauss's law,

$$\oint_s \overline{D} \cdot \overline{ds} = Q_{\text{en}} = \int_{\text{top}} \overline{D}_1 \cdot \overline{ds} + \int_{\text{bottom}} \overline{D}_2 \cdot \overline{ds} + \int_{\text{side}} \overline{D} \cdot \overline{ds}$$

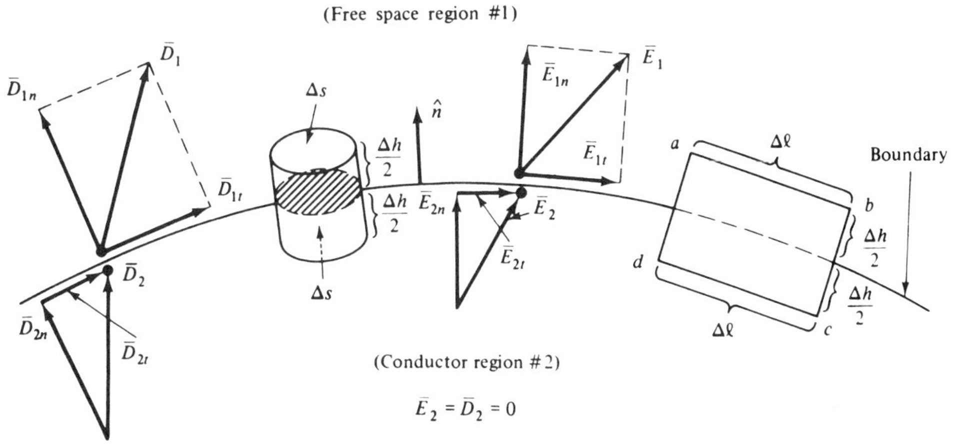

Figure 5-7 Graphical display of $\overline{D}$'s and $\overline{E}$'s, closed path and closed surface, about a conductor-free space boundary for determining static boundary conditions as $\Delta h \to 0$.

The integral over the bottom surface of the cylinder is zero since $\overline{D}_2 = \overline{E}_2\epsilon_0 = 0$ in the conductor while the integral over the side is zero since $\Delta h \rightarrow 0$; thus,

$$\int_{\text{top}} \overline{D}_1 \bullet \overline{ds} = Q_{\text{en}} = D_{1n}\,\Delta s \tag{1}$$

where $\overline{ds}$ is taken outwardly and $\Delta h \rightarrow 0$. The Q_{en} by the cylinder is the free charge on the conductor boundary and equals

$$Q_{\text{en}} = \rho_s\,\Delta s \tag{2}$$

Combining (1) and (2), we have

$$\boxed{D_{1n} = \rho_s} \tag{3}$$

The boundary condition relating the tangent field components can be readily obtained through the use of the conservative property of the electrostatic fields, $\oint_\ell \overline{E} \bullet \overline{d\ell} = 0$, about a small closed loop as $\Delta h \rightarrow 0$. The closed loop must be broken up into four parts as

$$\oint_\ell \overline{E} \bullet \overline{d\ell} = \int_a^b \overline{E}_1 \bullet \overline{d\ell} + \int_b^c \overline{E} \bullet \overline{d\ell} + \int_c^d \overline{E}_2 \bullet \overline{d\ell} + \int_d^a \overline{E} \bullet \overline{d\ell} = 0$$

The $\int_b^c$ and $\int_d^a$ are zero since $\Delta h \rightarrow 0$ while $\int_c^d$ is zero since $\overline{E}_2 = 0$; thus, we have remaining

$$0 = \int_a^b \overline{E}_1 \bullet \overline{d\ell} = E_{1t}\,\Delta\ell \tag{4}$$

From (4),

$$\boxed{E_{1t} = 0} \tag{5}$$

From (3), (5), and the relationship $\overline{D} = \epsilon_0 \overline{E}$, the boundary conditions for the conductor-free space boundary, under static conditions, can be summarized as

$$\boxed{D_{1n} = E_{1n}\epsilon_0 = \rho_s} \tag{6}$$

and

$$\boxed{E_{1t} = \frac{D_{1t}}{\epsilon_0} = 0} \tag{7}$$

with $D_{2n} = E_{2n} = D_{2t} = E_{2t} = 0$ in the conductor.

Equation (6) can be expressed also in vector form as

$$\boxed{\overline{D}_1 = \overline{E}_1\epsilon_0 = \hat{n}\rho_s} \tag{8}$$

where $\hat{n}$ is a unit vector directed normally outward from the conducting media, as shown in Fig. 5-7.

Example 7

A parallel-faced, infinite width, conducting slab is found to have $\hat{n} = \hat{z}$, uniform $\rho_s = 10\ \mu\text{C m}^{-2}$ on one face and $\hat{n} = -\hat{z}$, uniform $\rho_s = -10\ \mu\text{C m}^{-2}$ on the other face. Find: (a) $\overline{D}$ in free space at both faces; (b) $\overline{D}_i$ inside the slab due to the two surface charges only; (c) $\overline{D}_a$ in which the slab is immersed.

Solution. Figure 5-5 will be used in the solution.

(a) From (8), $\overline{D}\,|_{\text{top}} = \hat{n}\rho_s = \hat{z}10\ (\mu\text{C m}^{-2})$ and $\overline{D}\,|_{\text{bottom}} = (-\hat{z})\,(-10) = \hat{z}10\ (\mu\text{C m}^{-2})$ in free space just off the conductor surfaces.

(b) From two parallel sheets of uniform and opposite ρ_s we have for the $\overline{D}_i$ in the conductor,

$$\overline{D}_i = \overline{D}_{2(\rho_s\text{'s only})} = -\hat{z}\rho_s \qquad (\mu\text{C m}^{-2})$$

Note that the $\overline{D}_i$ outside the conductor equals zero.

(c) Now, for the net $\overline{D}_2$ (in the conductor) to be zero,

$$\overline{D}_2 = \overline{D}_i + \overline{D}_a$$

and

$$\overline{D}_a = -\overline{D}_i = \hat{z}\rho_s \qquad (\mu\text{C m}^{-2})$$

also

$$\overline{E}_a = \frac{\overline{D}_a}{\epsilon_0} = \hat{z}\frac{\rho_s}{\epsilon_0} \qquad (\text{V m}^{-1})$$

in the conductor. From parts (a) and (b) we note that $\overline{D}$ just off the conductor must be $\overline{D}_a$ since $\overline{D}_i$, due to the surface charges, equals zero outside the conductor. Since $\overline{D}$ just off the conductor is uniform due to uniform ρ_s, we can deduce that $\overline{D}_a = \hat{z}\rho_s$ and is uniform in the entire free space region. We have also found that $\overline{D}_a = \hat{z}\rho_s$ inside the conductor. Thus, $\overline{D}_a = \hat{z}\rho_s$ and is uniform in all space (free space and conductor).

Problem 5.6-1 At a conductor-free space boundary, $\overline{E} = \hat{r}_s 5\ (\text{V m}^{-2})$. Find: (a) the orientation of the boundary; (b) ρ_s at the boundary; (c) $\overline{D}$ at the boundary.

Problem 5.6-2 At a point on a conductor-free space boundary, $\overline{E} = \hat{x}2 + \hat{y}3\ \text{V m}^{-1}$. Find: (a) ρ_s at the point on the boundary; (b) $\overline{D}$ at the boundary.

5.7 SEMICONDUCTORS

All materials are classified on the basis of their value of conductivity σ as conductors, semiconductors, or insulators. From Table 5-1 it can be seen that σ varies over a range of about 25 orders of magnitude from that of silver, a good conductor, to fused quartz, a good insulator. A display of this overall range and defined subranges of σ for conductors, semiconductors, and insulators can be found in Fig. 5-8.

A middle range of conductivity, from about 10^{-3} to about 1 S m^{-1}, exists where the materials are called *intrinsic* (pure) *semiconductors,* for they are neither conductors nor insulators. Two intrinsic semiconductors of importance in electrical engineering are intrinsic (pure) germanium and intrinsic (pure) silicon. Their importance is due to their property of increase in conductivity with temperature and through addition of impurities in a process called *doping*. The increase in conductivity with temperature property is found in a device called a *thermistor* while the controlled increase in conductivity through doping property is found in semiconductor diodes and transistors.

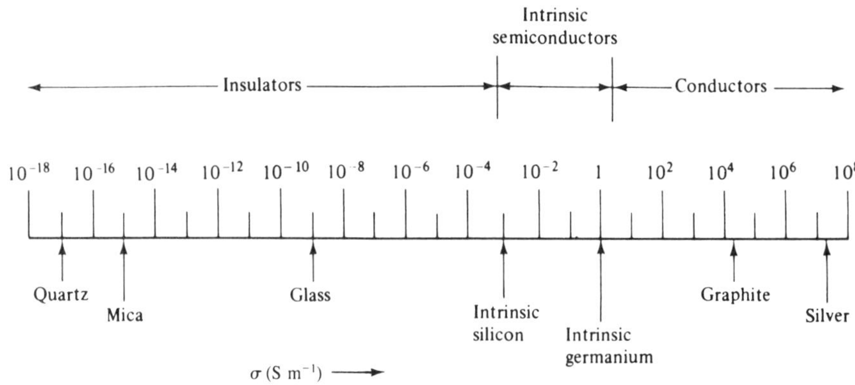

Figure 5-8 Graphical display of range of conductivity from good conductors to good insulators and defining sub-ranges for conductors, semiconductors, and insulators.

In a good conductor, the free electrons are the current carriers while in an intrinsic (pure) semiconductor two current carriers are present—free electrons and bound electrons (bound positive holes). The lattice structure of intrinsic germanium or silicon has four covalent bonds between each of its atoms. At temperatures near absolute zero, the outer shell electrons are held tightly within the covalent bonds and thermal vibration of the lattice structure is zero. Under this condition, there are no free carriers and the material is an insulator. As the temperature is increased, thermal vibration of the lattice structure is increased with the result that some electrons gain sufficient thermal energy to break away from the covalent bond and become free electrons in random motion similar to free electrons in a conductor. Each electron that leaves a covalent bond creates a bond vacancy called a *hole*. These holes also move randomly throughout the lattice since electrons from other covalent bonds can make a transition into a hole position, and thus the hole replaces the initial position of the covalent bond electron. It should be noted that the covalent bond electron is not a free electron but a bound electron that has insufficient thermal energy to become a free electron. With zero applied $\overline{E}$ within the semiconductor, the net current due to free electrons and holes (bound electrons) is zero since the drift velocity of each is zero. Now, with an applied $\overline{E}$ the free electrons will drift with a $\overline{U}_e$ in the $-\overline{E}$ direction while the holes drift with $\overline{U}_h$ in the $\overline{E}$ direction. Thus, the total conduction current is due to two carriers: free electrons and covalent-bound electrons that can be visualized as covalent-bound holes. Since each free electron creates a covalent-bound hole, the free-electron density is equal to the covalent-bound hole density. Many semiconductor properties can be explained by treating the hole as if it had positive charge equal to $|q_e|$ and a mobility of μ_h.

With two current carriers, the expression for the conduction current density becomes

$$\boxed{\overline{J} = (-\rho_{ve}\mu_e + \rho_{vh}\mu_h)\overline{E} = \sigma\overline{E} \qquad (\text{A m}^{-2})} \tag{1}$$

From (1), σ becomes

$$\boxed{\sigma = (-\rho_{ve}\mu_e + \rho_{vh}\mu_h) \qquad (\text{S m}^{-1})} \tag{2}$$

TABLE 5-2 SOME IMPORTANT PARAMETERS OF GERMANIUM AND SILICON AT 300° K

Parameter	Symbol (units)	Intrinsic germanium	Intrinsic silicon
Free-electron mobility	μ_e (m^2 V^{-1}s^{-1})	0.39	0.135
Hole mobility	μ_h (m^2 V^{-1}s^{-1})	0.19	0.048
Free-electron density	η_e (m^{-3})	2.4×10^{19}	1.5×10^{16}
Hole density	η_h (m^{-3})	2.4×10^{19}	1.5×10^{16}
Conductivity	σ (S m^{-1})	2.2	0.44×10^{-3}

Table 5-2 lists some of the important parameters of germanium and silicon.[2]

With increasing temperature, the mobilities μ_e and μ_h decrease, but the free-electron density increases more rapidly to increase the conductivity.

For a more complete coverage of semiconductors, doping, and semiconductor devices, the student is referred to texts on solid-state electronics.[3]

Example 8

Calculate the σ of germanium, using (2) and the parameters found in Table 5-2.

Solution. From (2),

$$\sigma = (-\rho_{ve}\mu_e + \rho_{vh}\mu_h) = (-n_e q_e \mu_e + n_h|q_e|\mu_h)$$

For $n_e = n_h$,

$$\sigma = (\mu_e + \mu_h)n_e|q_e|$$
$$= (0.39 + 0.19)(2.4 \times 10^{19})(1.6 \times 10^{-19}) = 2.2 \text{ S m}^{-1}$$

Note that the intrinsic germanium is a poor conductor as compared to copper.

Problem 5.7-1 Calculate the drift velocity U_d in a semiconductor when $E = 1$ V m^{-1} for: (a) a free electron in germanium; (b) a hole in germanium; (c) a free electron in silicon; (d) a hole in silicon.

5.8 DIELECTRICS (INSULATORS)

In Sec. 5.3 we found that electrical charge conduction in the form of free electrons was the prominent characteristic of materials called *conductors*. Good conductors are listed at the top of Table 5-1 while poor conductors or insulators are shown at the bottom of the table. Insulators are commonly referred to as *dielectrics;* thus, this usage will be used in this text. The prominent characteristic of dielectric materials is *polarization;* i.e., the formation of electric dipoles.

5.8-1 Polarization in Dielectrics

When a dielectric material is placed in an electric field, we find that a microscopic displacement takes place between the average equilibrium positions of the positive and negative bound charges of atoms and molecules. This displacement gives rise to an electric dipole moment $\overline{p}$ discussed in Example 4-11. These electric dipoles have an effect on the electric field in which the dielectric material is placed.

[2] E. M. Conwell, "Properties of Silicon and Germanium II," *Proc. IRE,* June 1958, p. 1281.

[3] M. F. Uman, *Introduction to the Physics of Electronics,* Prentice-Hall, Inc., Englewood Cliffs, N.J., 1974.

There are three basic polarization mechanisms that produce polarization; they are:

1. Electronic polarization exists in an atom when, upon the application of an electric field, the center of the cloud of electrons is displaced relative to the center of the nucleus.
2. Ionic polarization exists in a molecule having ionic bonds that can be viewed as a build up of positive and negative ions. An applied electric field will displace the positive ions relative to the negative ions. Sodium chloride (NaCl) is a good example of this type of polarization.
3. Orientational polarization exists in a material whose molecules, because of structure, possess permanent dipole moments that are randomly oriented in the absence of an applied electric field but tend to orient themselves in the direction of an applied electric field. This type of material is also called *polar material,* water (H_2O) being a good example.

Dielectric material polarized through orientational polarization is called *polar material* while material polarized through electronic or ionic polarization is called *non-polar material*. It should be noted that in the process of polarization the charges forming the electric dipoles are bound charges. Graphical representation of a polarization mechanism on a microscopic base is shown in Fig. 5-9. Figure 5-9(a) illustrates electronic polarization while Fig. 5-9(b) illustrates orientational polarization for zero and

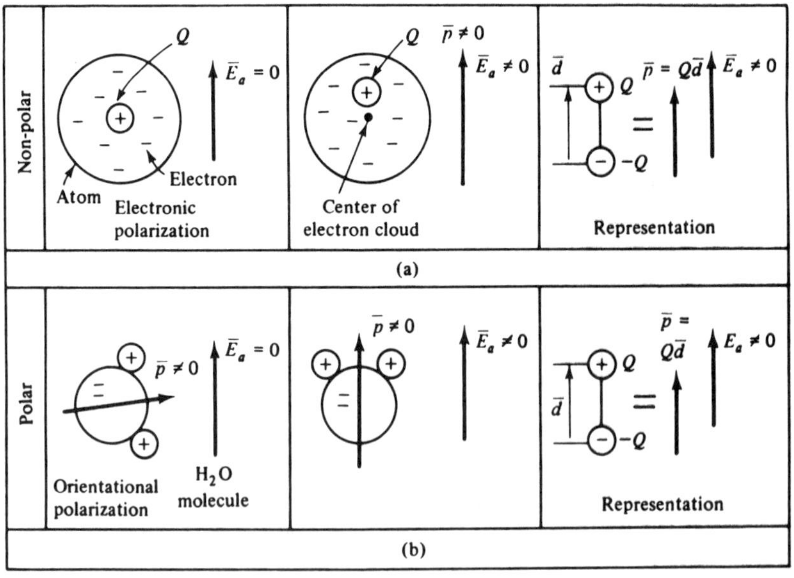

Figure 5-9 Graphical representation of a polarization mechanism on a microscopic base: (a) electronic (non-polar) polarization; (b) orientational (polar) polarization.

non-zero applied electric field $\overline{E}_a$. The resulting microscopic dipole moments are defined by (4.5-24)

$$\boxed{\overline{p} = Q\overline{d} \qquad \text{(C m)}} \tag{1}$$

where $\overline{d}$ is the vector displacement from the center of the negative bound charge accumulation to the center of the positive bound charge accumulation.

The microscopic dipole moment defined in (1) cannot be mathematically applied to a slab of dielectric material since we must know the spatial location of each atom or molecule comprising the material. To overcome this problem, we will define an electric polarization vector $\overline{P}$ on a macroscopic base (involving thousands of atoms or molecules) as the net dipole moment per unit volume given by

$$\boxed{\overline{P} \triangleq \lim_{\Delta\upsilon \to 0}\left[\frac{1}{\Delta\upsilon}\sum_{i=1}^{n\Delta\upsilon}\overline{p}_i\right] \qquad \text{(C m}^{-2}\text{)}} \tag{2}$$

where lim $\Delta\upsilon \to 0$ is a macroscopic limit, and n is the dipole density (dipoles per cubic meter). Within the dielectric material, the function $\overline{P}$ is a continuous function while $\overline{p}$ (if we knew the exact location of each atom or molecule) would be a discrete function having values only at discrete points. Figure 5-10 illustrates the polarization mechanism on a macroscopic base that leads to the polarization $\overline{P}$ function. It should be noted that all $\overline{p}$'s were assumed to be equal and aligned with $\overline{E}_a$ for non-zero $\overline{E}_a$ conditions. This assumption leads to

$$\boxed{\overline{P} = n\overline{p} = nQ\overline{d} \qquad \text{(C m}^{-2}\text{)}} \tag{3}$$

In the case of the orientational (polar) polarization, it should be noted that $\overline{p} \neq 0$ but $\overline{P} = 0$ due to random orientation of $\overline{p}$ when $\overline{E}_a = 0$.

The torque on the electric dipole moment $\overline{p}$ that causes the alignment of electric dipole with $\overline{E}_a$ is

$$\boxed{\overline{T} = \overline{p} \times \overline{E}_a \qquad \text{(N m)}} \tag{4}$$

In a dielectric material, the dipoles may not completely align with $\overline{E}_a$ due to Coulomb restraining forces.

With the use of microscopic and macroscopic concepts, we will, in a qualitative way, show what happens inside and on the surface of a dielectric slab when placed in an applied $\overline{E}_a$ field. Figure 5-11 shows, in a very simplified manner, how the dipoles align themselves in a slab of dielectric material when placed in a uniform $\overline{E}_a$. From this illustration, two very important conclusions can be drawn: first, a bound positive surface charge density $\rho_{sb(+)}$ is found on the top surface while a bound negative surface charge density $\rho_{sb(-)}$ is found on the lower surface; second, a bound volume charge density $\rho_{\upsilon b}$ is equal to zero inside the slab. The $\rho_{\upsilon b}$ is zero since all the $\overline{p}$'s are equal, and thus

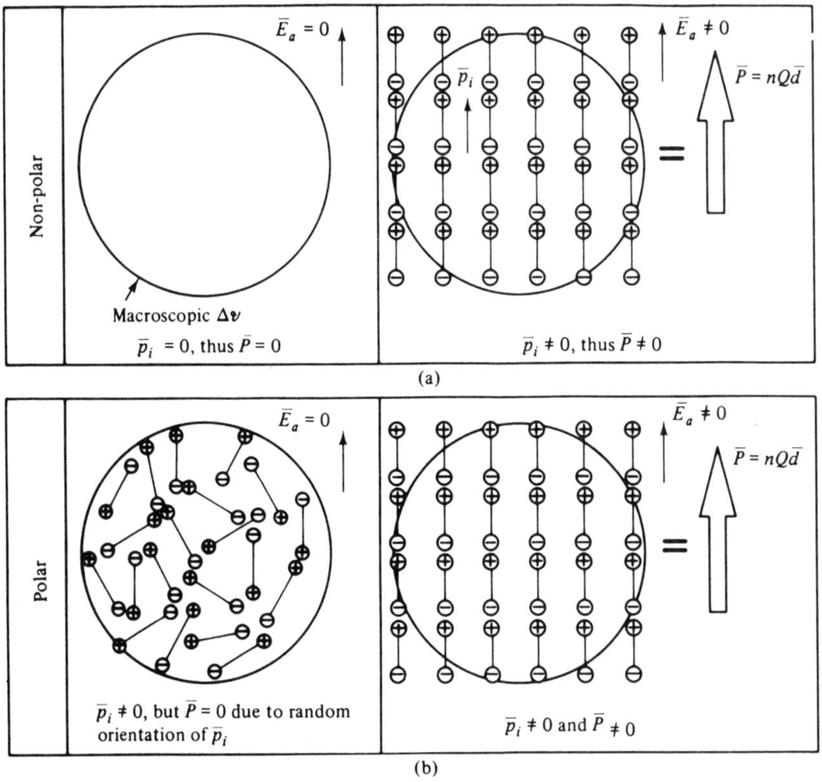

Figure 5-10 Graphical representation of polarization mechanism on a macroscopic base: (a) electronic (non-polar) polarization; (b) orientational (polar) polarization.

within a macroscopic volume Δv the negative and positive charges cancel. It should be noted that ρ_{sb} and ρ_{vb} are macroscopic quantities.

Figure 5-12 illustrates a polarization mechanism that gives rise to a non-zero ρ_{vb} within a dielectric slab subjected to a non-uniform $\bar{E}_a$ field. The $\bar{E}_a$ shown is assumed to increase with z. This increase in $\bar{P}$ is accounted for by the double and triple charge dipoles shown. Now, within a macroscopic Δv it can be seen that a $\rho_{vb(-)}$ exists when $\bar{P}$ increases with z. For a decreasing $\bar{P}$ with z, it can readily be shown that a $\rho_{vb(+)}$ exists within a dielectric slab. If we start with a neutral slab of dielectric, we will find that after polarization the slab will still be neutral since polarization separates bound charges to produce bound charge on the surfaces and within the dielectric. Thus, the integral of ρ_{sb} over the surfaces plus the integral of ρ_{vb} over the inside volume should give zero net charge. Quantitative expressions for ρ_{sb} and ρ_{vb} will be developed in the following section.

5.8-2 Bound Charge Densities ρ_{sb} and ρ_{vb}

The bound charge densities ρ_{sb} and ρ_{vb} can be found through the analysis of the amount of bound charge that is found within a volume v inside a dielectric, due to polarization, as shown in Fig. 5-13(a). Let us consider a volume dv on the surface s that is in the

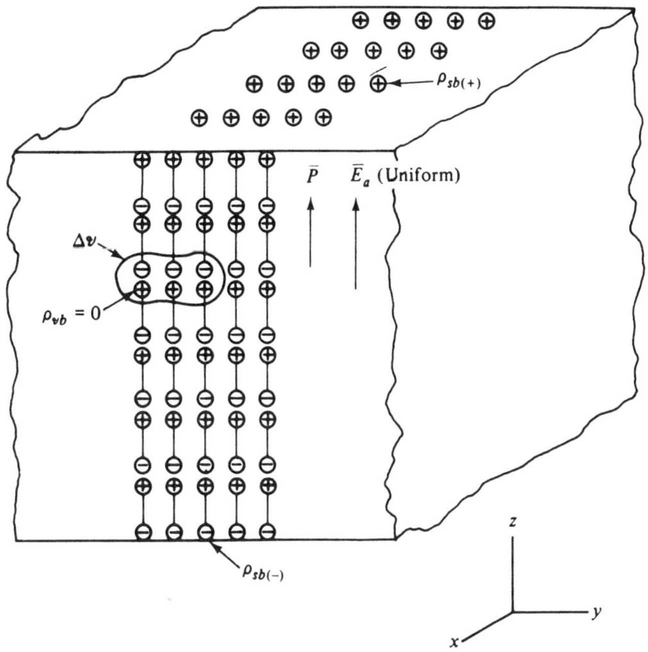

Figure 5-11 Graphical illustration of the polarization mechanism that leads to bound surface charge density ρ_{sb} on the surfaces and $\rho_{vb} = 0$ within a dielectric slab that is subjected to a uniform $\overline{E}_a$ field.

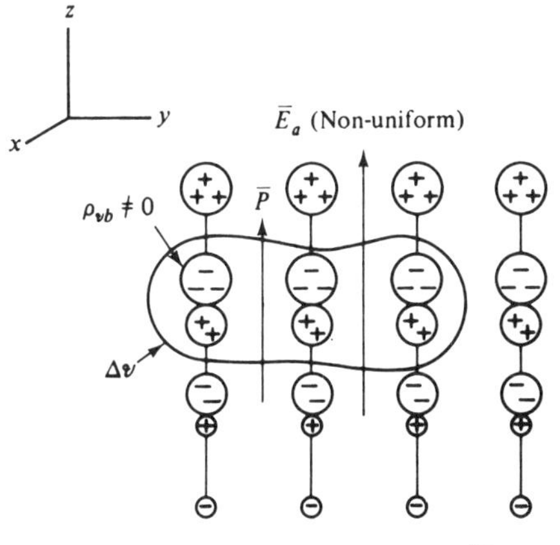

Figure 5-12 Graphical illustration of the polarization mechanism that gives rise to a non-zero ρ_{vb} within a dielectric slab subjected to a non-uniform $\overline{E}_a$ field (assume $\overline{E}_a$ increasing with z only).

form of a parallelepiped formed by length $\overline{d}$ and surface ds, as shown in an expanded view in Fig. 5-13(b). Half of the dv is above the surface ds while the other half is just below. In our analysis, we shall assume a non-polar type polarization and that all microscopic $\overline{p}$'s are equal and aligned in the direction of $\overline{E}_a$ upon polarization. Thus, when $\overline{E}_a = 0$, $\overline{p} = 0$ within dv. When $\overline{E}_a \neq 0$, molecules in the two half volumes of dv will be polarized and dipoles of strength $\overline{p} = Q\overline{d}$ will be formed. Each molecule in the half volume v (below ds) will cause a bound $+Q$ to flow through ds upward when $\overline{E}_a$ is

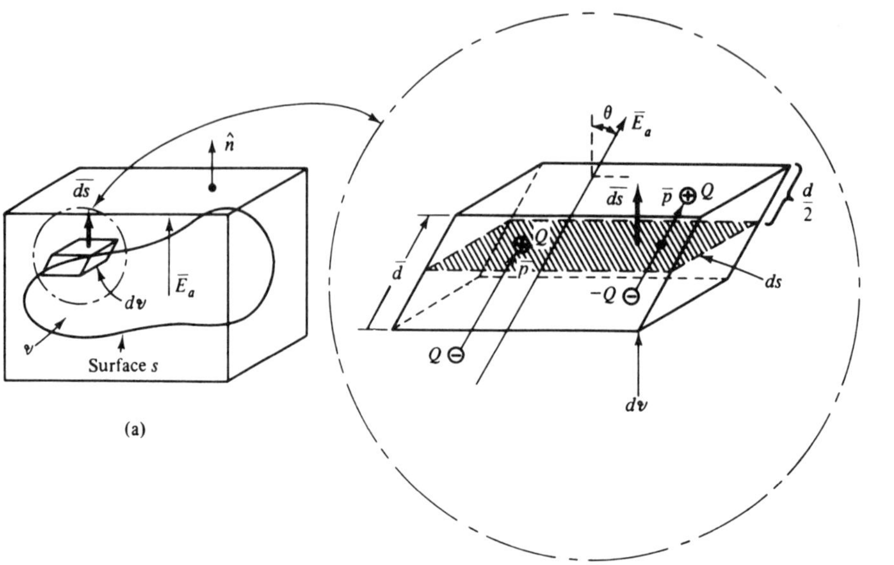

Figure 5-13 Graphical display of bound charge flowing through a surface s within a polarized dielectric: (a) volume within the slab of dielectric; (b) enlarged view of differential volume on the surface s.

applied and a dipole is formed. Now each molecule in the half volume outside v (above ds) will cause a bound $-Q$ to flow through ds downward. If we assume a molecular density of n (molecules/m^3), the bound charge that flows through ds upward becomes

$$dQ_{b(\text{up})} = Qn\frac{dv}{2} = Qn\frac{\overline{d} \bullet \overline{ds}}{2} \tag{5}$$

while the bound charge that flows through ds downward becomes

$$dQ_{b(\text{down})} = -Qn\frac{dv}{2} = -Qn\frac{\overline{d} \bullet \overline{ds}}{2} \tag{6}$$

Since $dQ_{b(\text{up})}$ and $dQ_{b(\text{down})}$ produce equal effects, the net bound positive charge flowing upward through ds, from (5), (6), and (3), becomes

$$\boxed{dQ_b = Qn\overline{d} \bullet \overline{ds} = \overline{P} \bullet \overline{ds}} \tag{7}$$

If we now expand v to be the volume of the entire dielectric slab, the surface s will be on the surface of the dielectric, and thus dQ_{sb} will be a bound charge on the surface.

Rewriting (7), we have

$$dQ_{sb} = \overline{P} \bullet \overline{ds} = \overline{P} \bullet \hat{n}\, ds \tag{8}$$

where $\hat{n}$ is the outward unit vector on the surface.

Now form, from (8),

$$\boxed{\frac{dQ_{sb}}{ds} = \overline{P} \bullet \hat{n} = \rho_{sb} \qquad (\text{C m}^{-2})} \tag{9}$$

where ρ_{sb} is our bound surface charge density on the surface of a dielectric, as found in Sec. 5.8-1.

Let us now return to our original volume υ within the dielectric. From (7), the total bound charge that flows outward through s from volume υ becomes

$$\boxed{Q_{sb} = \oint_s \overline{P} \bullet \overline{ds} \qquad (\text{C})} \tag{10}$$

From (10), the bound charge $Q_{\upsilon b}$ that remains within υ becomes

$$\boxed{Q_{\upsilon b} = -Q_{sb} = -\oint_s \overline{P} \bullet \overline{ds} \qquad (\text{C})} \tag{11}$$

Let us apply the divergence theorem to (11) and obtain

$$Q_{\upsilon b} = -\int_\upsilon \nabla \bullet \overline{P}\, d\upsilon = \int_\upsilon (-\nabla \bullet \overline{P})\, d\upsilon \tag{12}$$

From (12), we note that $-\nabla \bullet \overline{P}$ must have the units of (C m^{-3}), and thus we can write

$$Q_{\upsilon b} = \int_\upsilon \rho_{\upsilon b}\, d\upsilon \tag{13}$$

where $\rho_{\upsilon b}$ is the bound volume charge density discussed in Sec. 5.8-1. From (12) and (13),

$$\boxed{\rho_{\upsilon b} = -\nabla \bullet \overline{P} \qquad (\text{C m}^{-3})} \tag{14}$$

It should be noted that $Q_{sb} + Q_{\upsilon b} = 0$; thus, from (10) and (11),

$$\boxed{\oint_s \overline{P} \bullet \overline{ds} + \left[-\int_\upsilon \nabla \bullet \overline{P}\, d\upsilon\right] = 0} \tag{15}$$

Equation (15) can be proven through the use of the divergence theorem and can be extended to include the entire dielectric slab.

Example 9

Find the ρ_{sb} and $\rho_{\upsilon b}$ for the dielectric cube shown in Fig. 5-14 when $\overline{P} = \hat{z}10^{-2}$ (C m^{-2}) inside the cube.

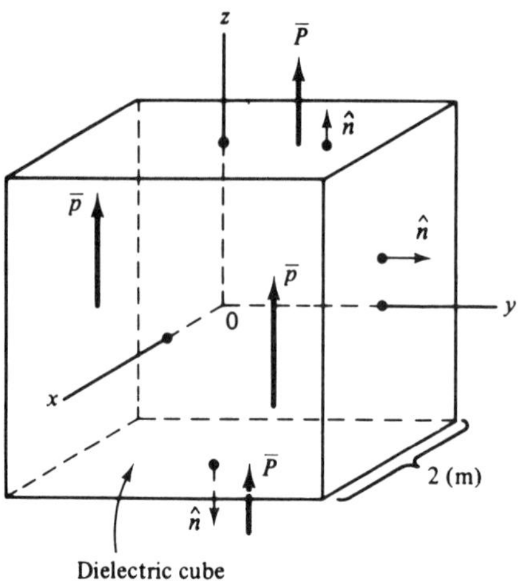

Figure 5-14 Graphical display for finding ρ_{sb} and ρ_{vb} for a dielectric cube for Example 9.

Solution. From (9), we find that $\rho_{sb(\text{top})}$ on the top surface is

$$\rho_{sb(\text{top})} = \overline{P} \bullet \hat{n} = (\hat{z}10^{-2}) \bullet (\hat{z}) = 10^{-2} \text{ C m}^{-2}$$

Also,

$$\rho_{sb(\text{bot})} = \overline{P} \bullet \hat{n} = (\hat{z}10^{-2}) \bullet (-\hat{z}) = -10^{-2} \text{ C m}^{-2}$$

On all other surfaces, $\overline{P} \bullet \hat{n} = 0$ and thus $\rho_{sb} = 0$. From (14), we find that

$$\rho_{vb} = -\nabla \bullet \overline{P} = -\left(\frac{\partial P_x}{\partial x} + \frac{\partial P_y}{\partial y} + \frac{\partial P_z}{\partial z}\right) = 0$$

Example 10

Show that the total bound charge associated with the cube of Example 9 is zero.

Solution. From (8),

$$Q_{sb} = \oint_s \overline{P} \bullet \overline{ds} = \int_{s(\text{top})} (\overline{P} \bullet \hat{n})\, ds + \int_{s(\text{bot})} (\overline{P} \bullet \hat{n})\, ds$$

$$= \int_{s(\text{top})} 10^{-2}\, dx\, dy + \int_{s(\text{bot})} -10^{-2}\, dx\, dy$$

$$= 4 \times 10^{-2} + (-4 \times 10^{-2}) = 0$$

Since $-\nabla \bullet \overline{P} = \rho_{vb} = 0$, we find that $Q_{vb} = 0$. Thus, $Q_{sb} + Q_{vb} = 0$.

Problem 5.8-1 For the cube of Example 9, find ρ_{sb} on all faces and ρ_{vb} inside the cube when: (a) $\overline{P} = \hat{x}x \text{ C m}^{-2}$; (b) $\overline{P} = \hat{x}(x^2 + 2) \text{ C m}^{-2}$ inside the cube.

Problem 5.8-2 Find the total bound surface charge Q_{sb} and the total bound volume charge Q_{vb} for Prob. 5.8-1 (a) and (b).

5.8-3 Effect of Polarization on Fields

Due to polarization in a dielectric we have seen the formation of a bound surface charge density ρ_{sb} and a bound volume charge density ρ_{vb}. These bound charges interact with free charges as though they themselves were free charges. Thus, to take into account these bound charges on the $\overline{E}$ field, we must add ρ_{vb} to ρ_v (the free charge density) in (3.5-14) to obtain

$$\boxed{\epsilon_0 \nabla \bullet \overline{E} = (\rho_v + \rho_{vb})} \qquad (16)$$

when $\overline{E}$ is the resultant field.

Through the use of (14), we have

$$\epsilon_0 \nabla \cdot \overline{E} = \rho_v - \nabla \cdot \overline{P} \tag{17}$$

Rearranging (17), we have

$$\epsilon_0 \nabla \cdot \overline{E} + \nabla \cdot \overline{P} = \nabla \cdot (\epsilon_0 \overline{E} + \overline{P}) = \rho_v \tag{18}$$

Let us now, in (18), set

$$\boxed{\overline{D} = (\epsilon_0 \overline{E} + \overline{P}) \qquad (\mathrm{C\ m^{-2}})} \tag{19}$$

to form

$$\boxed{\nabla \cdot \overline{D} = \rho_v} \tag{20}$$

It should be noted that (20) is independent of bound charge densities and thus independent of material. The expression for $\overline{D}$, found in (19), reduces to (3.3-4) for free space since $\overline{P} = 0$.

The polarization vector $\overline{P}$ can be eliminated from the equations above through the relationship between $\overline{P}$ and $\overline{E}$. In an isotropic material, $\overline{E}$ and $\overline{P}$ are in the same direction and we shall basically limit our discussion to this type of material. In non-isotropic (anisotropic) materials such as crystals, $\overline{E}$ and $\overline{P}$ are not in the same direction since the periodic nature of crystalline materials allows polarization most easily along crystal axes.

Let us express $\overline{P}$ in terms of $\overline{E}$ as

$$\boxed{\overline{P} = \chi_e \epsilon_0 \overline{E}} \tag{21}$$

where χ_e (chi) is the electric susceptibility of the material, and $\overline{E}$ is the resultant $\overline{E}$ at the location of $\overline{P}$. Substituting (21) into (19), we have

$$\overline{D} = (\epsilon_0 \overline{E} + \epsilon_0 \chi_e \overline{E}) = \epsilon_0 (1 + \chi_e) \overline{E} \tag{22}$$

Now, define

$$\boxed{\epsilon_r = (1 + \chi_e)} \tag{23}$$

to obtain

$$\boxed{\overline{D} = \epsilon_0 \epsilon_r \overline{E} = \epsilon \overline{E}} \tag{24}$$

where

$$\boxed{\epsilon = \epsilon_0 \epsilon_r} \tag{25}$$

TABLE 5-3 REPRESENTATIVE VALUES FOR THE RELATIVE PERMITTIVITY ϵ_r OF SEVERAL DIELECTRIC MATERIALS AT LOW FREQUENCY

Material	ϵ_r	Material	ϵ_r
Air	1.0006	Polyethylene	2.26
Bakelite	4.8	Polystyrene	2.5
Glass	6.0	Quartz	3.8
Lucite	3.2	Soil (dry)	3.0
Nylon	3.6	Teflon	2.1
Plexiglas	3.45	Water	80

In (25), ϵ is the permittivity and ϵ_r is the relative permittivity, or dielectric constant, of the material. Thus, we have eliminated $\overline{P}$, which is now reflected through a change of permittivity from ϵ_0 (free space) to $\epsilon = \epsilon_0\epsilon_r$. Table 5-3 contains representative values of relative permittivity of several dielectric materials.

Polarization can also be created by embedding small metal spheres, rods, or slivers in a dielectric. An applied field polarizes the metal objects as shown in Fig. 5-6, and the effect is similar to ionic or molecular polarization. Such materials are called *artificial dielectrics* and find application in radio-frequency lenses for antennas. See Sec. 14.6.

In (19), we have an expression that relates the resultant fields $\overline{D}$, $\overline{E}$, and the polarization vector $\overline{P}$ at a point. This expression was developed through the addition of ρ_{vb} in (16). Another method for finding the fields is to start with the potential produced by dipoles in a polarized block of material, as shown in Fig. 5-15.[4] Through a slight modi-

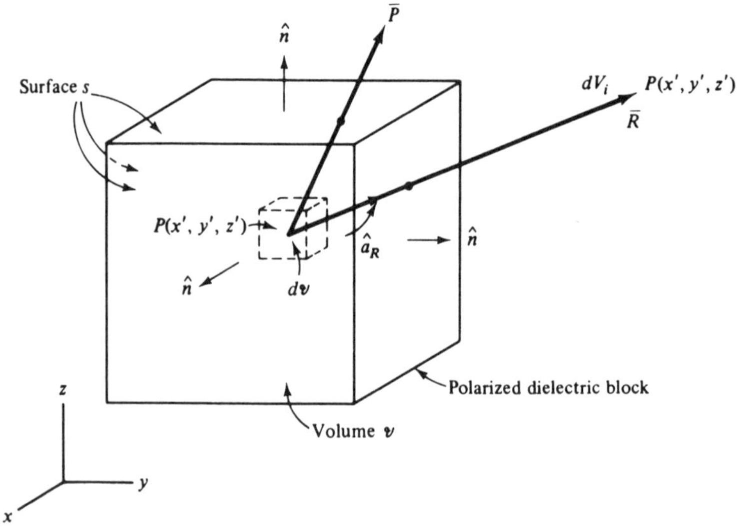

Figure 5-15 Block of polarized dielectric material with $\overline{P}$ dipole moment per unit volume. An induced potential dV_i is produced at point $P(x, y, z)$ by the dipoles at point $P(x', y', z')$.

[4] See Dale R. Corson and Paul Lorrain, *Introduction to Electromagnetic Fields and Waves*, W. H. Freeman and Company, Publishers, San Francisco, 1962, pp. 84–100.

fication of (4.5-25), we find the expression for the induced potential dV_i, due only to the dipoles in the polarized block, to be

$$dV_i = \frac{\overline{P} \bullet \hat{a}_R \, dv}{4\pi\epsilon_0 R^2} \tag{26}$$

where $\overline{P}\, dv$ is the macroscopic dipole moment. Through mathematical manipulation, the solution for V_i in (26) becomes

$$V_i = \oint_s \frac{\overline{P} \bullet \overline{ds}}{4\pi\epsilon_0 R} + \int_v \frac{(-\nabla \bullet \overline{P})\, dv}{4\pi\epsilon_0 R} \tag{27}$$

where the first integral is taken over the closed surface s of the block while the second integral is over the volume enclosed by s. It should be noted that, in (27), $\overline{P} \bullet \overline{ds} = \overline{P} \bullet \hat{n}\, ds = \rho_{sb}\, ds$ and $(-\nabla \bullet \overline{P})\, dv = \rho_{vb}\, dv$. From $-\nabla V_i = \overline{E}_i$, we have

$$\overline{E}_i = \oint_s \frac{\rho_{sb}\, ds\, \hat{a}_R}{4\pi\epsilon_0 R^2} + \int_v \frac{\rho_{vb}\, dv\, \hat{a}_R}{4\pi\epsilon_0 R^2} \tag{28}$$

where $\overline{E}_i$ is the induced electric field due only to polarization. Thus, the induced field $\overline{E}_i$, due to polarization only, can be obtained by treating ρ_{sb} and ρ_{vb} as if they were free charge densities. The resultant $\overline{E}$ is

$$\overline{E} = \overline{E}_a + \overline{E}_i \tag{29}$$

where $\overline{E}_a$ is the applied field with no dielectric present.

Let us solve for resultant $\overline{E}$ in (19) to obtain

$$\overline{E} = \frac{\overline{D}}{\epsilon_0} - \frac{\overline{P}}{\epsilon_0} \tag{30}$$

From (29) and (30), it follows that

$$\frac{\overline{D}}{\epsilon_0} = (\overline{E}_a + \overline{E}_i) \quad \text{(aap)} \tag{31}$$

at a point outside of the dielectric, where $\overline{P} = 0$. In symmetrical $\overline{P}$ distributions, $\overline{E}_i$ outside the dielectric may also be equal to zero and thus

$$\frac{\overline{D}}{\epsilon_0} = \overline{E}_a \quad \text{(aap)} \tag{32}$$

From (29) and (30), in general, we find inside the dielectric

$$\boxed{\frac{\overline{D}}{\epsilon_0} - \frac{\overline{P}}{\epsilon_0} = \overline{E}_a + \overline{E}_i} \tag{33}$$

If (32) is true outside and inside the dielectric, from (33) we see that (see Example 11, where $\overline{E}_i$ exists only in region 2)

$$\boxed{-\frac{\overline{P}}{\epsilon_0} = \overline{E}_i} \tag{34}$$

Example 11

An ungrounded spherical configuration of concentric spherical dielectric shells, enclosed by a conductor shell, is shown in Fig. 5-16. Regions 1 and 3 are free space, region 2 is a dielectric whose relative permittivity equals ϵ_r, and region 4 is a conductor. If a charge Q is placed at the center, find: (a) $\overline{D}$ in all regions through the use of Gauss's law; (b) $\overline{E}$ in all regions through the use of $\overline{E} = \overline{D}/\epsilon$; (c) $\overline{P}$ in all regions through the use of $\overline{P} = \overline{D} - \epsilon_0 \overline{E}$; (d) ρ_s on the conductor surfaces; (e) ρ_{sb} on the dielectric surfaces; (f) ρ_{vb} within the dielectric.

Solution. (a) From Gauss's law $\oint_s \overline{D} \bullet \overline{ds} = Q_{en}$, where Q_{en} is the free charge enclosed by the Gaussian surface. It should be noted that any bound charges, due to polarization, that are enclosed by the Gaussian surface are not included in Q_{en}. Applying Gauss's law to region 1, we have

$$\oint_s \overline{D}_1 \bullet \overline{ds} = D_{1r} \oint_s ds = D_{1r}(4\pi r_s^2) = Q$$

Thus,

$$\overline{D}_1 = \frac{Q\hat{r}_s}{4\pi r_s^2} \qquad (0 < r_s < a) \tag{35}$$

Repeating the same procedure in regions 2 and 3 yields

$$\overline{D}_2 = \frac{Q\hat{r}_s}{4\pi r_s^2} \qquad (a \leqq r_s \leqq b) \tag{36}$$

$$\overline{D}_3 = \frac{Q\hat{r}_s}{4\pi r_s^2} \qquad (b \leqq r_s \leqq c) \tag{37}$$

In region 4, $\overline{D}_4 = 0$ since we are in a conductor under static conditions. Let us apply Gauss's law in region 4 to obtain

$$\oint_s \overline{D}_4 \bullet \overline{ds} = D_{4r} \oint_s ds = Q + Q' = Q_{en}$$

where Q' is the free charge induced on the conductor surface. Since $\overline{D}_4 = 0$, $Q' = -Q$. Now, if $-Q$ is induced at $r_s = c$, then Q must be induced at $r_s = d$. If we apply Gauss's law to region 5, we find $Q_{en} = Q$ and

$$\overline{D}_5 = \frac{Q\hat{r}_s}{4\pi r_s^2} \qquad (d \leqq r_s) \tag{38}$$

From (35), (36), (37), and (38) it should be noted that $\overline{D}$ has the same functional form and is independent of any bound charges. The values of $\overline{D}$ are tabulated in Fig. 5-16.

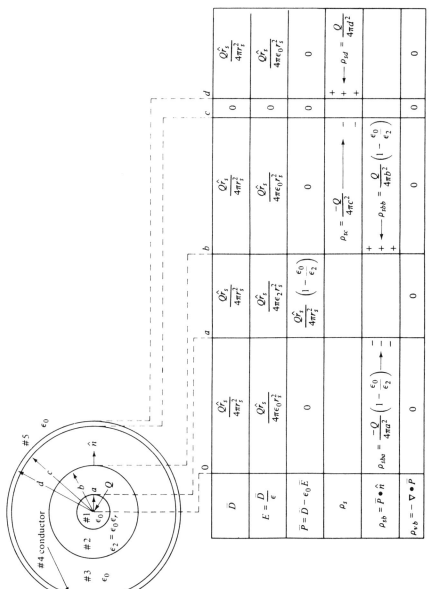

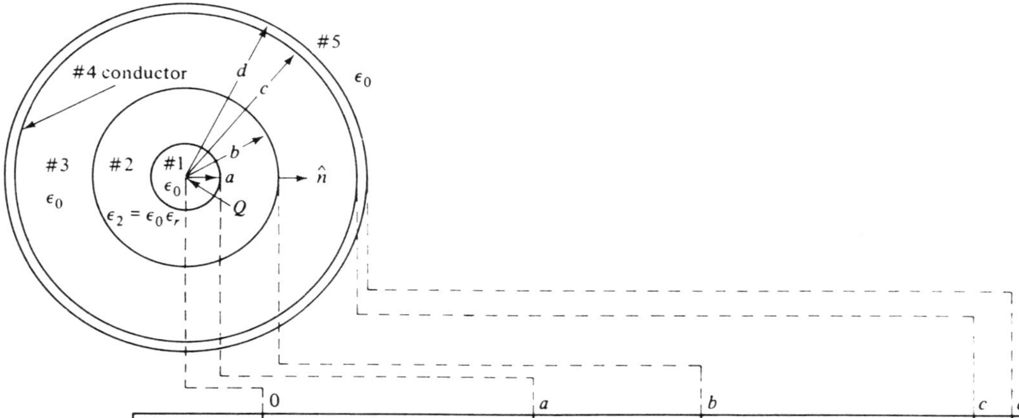

	0 to a	a to b	b to c	c to d	beyond d
$\bar{D}$	$\frac{Q\hat{r}_s}{4\pi r_s^2}$	$\frac{Q\hat{r}_s}{4\pi r_s^2}$	$\frac{Q\hat{r}_s}{4\pi r_s^2}$	0	$\frac{Q\hat{r}_s}{4\pi r_s^2}$
$E = \frac{\bar{D}}{\epsilon}$	$\frac{Q\hat{r}_s}{4\pi\epsilon_0 r_s^2}$	$\frac{Q\hat{r}_s}{4\pi\epsilon_2 r_s^2}$	$\frac{Q\hat{r}_s}{4\pi\epsilon_0 r_s^2}$	0	$\frac{Q\hat{r}_s}{4\pi\epsilon_0 r_s^2}$
$\bar{P} = \bar{D} - \epsilon_0 \bar{E}$	0	$\frac{Q\hat{r}_s}{4\pi r_s^2}\left(1 - \frac{\epsilon_0}{\epsilon_2}\right)$	0	0	0
ρ_s			$\rho_{sc} = \frac{-Q}{4\pi c^2}$ → $-$ $-$		$+$ $+$ $+$ ← $\rho_{sd} = \frac{Q}{4\pi d^2}$
$\rho_{sb} = \bar{P} \bullet \hat{n}$	$\rho_{sba} = \frac{-Q}{4\pi a^2}\left(1 - \frac{\epsilon_0}{\epsilon_2}\right)$ → $-$ $-$ $-$		$+$ $+$ $+$ ← $\rho_{sbb} = \frac{Q}{4\pi b^2}\left(1 - \frac{\epsilon_0}{\epsilon_2}\right)$		
$\rho_{vb} = -\nabla \bullet \bar{P}$	0	0	0	0	0

Figure 5-16 Graphical display of an ungrounded spherical configuration containing a free charge Q at its center and concentric spherical shells of dielectrics all enclosed by a conductor shell, for Example 11.

(b) The expressions for $\overline{E}$ are found through $\overline{E} = \overline{D}/\epsilon$. The values of $\overline{E}$ in all regions are found tabulated in Fig. 5-16.

(c) The expressions for $\overline{P}$ are found through $\overline{P} = \overline{D} - \epsilon_0 \overline{E}$. From the values for $\overline{D}$ and $\overline{E}$ found previously, we obtain the values for $\overline{P}$ tabulated in Fig. 5-16.

(d) The free surface charge density ρ_s is found only on the conductor surface. Since the charge at $r_s = c$ is $-Q$, we find that

$$\rho_{sc} = \frac{-Q}{4\pi c^2} \qquad (\text{C m}^{-2}) \tag{39}$$

At $r_s = d$ we find that

$$\rho_{sd} = \frac{Q}{4\pi d^2} \qquad (\text{C m}^{-2}) \tag{40}$$

(e) The bound surface charge density is found only on the surfaces of region 2. From $\overline{P} \cdot \hat{n} = \rho_{sb}$ we have at $r_s = a$,

$$\rho_{sba} = \overline{P} \cdot \hat{n} = \frac{\hat{r}_s Q}{4\pi a^2}\left(1 - \frac{\epsilon_0}{\epsilon_2}\right) \cdot (-\hat{r}_s) = \frac{-Q}{4\pi a^2}\left(1 - \frac{\epsilon_0}{\epsilon_2}\right) \qquad (\text{C m}^{-2}) \tag{41}$$

and at $r_s = b$ we have

$$\rho_{sbb} = \frac{Q}{4\pi b^2}\left(1 - \frac{\epsilon_0}{\epsilon_2}\right) \qquad (\text{C m}^{-2}) \tag{42}$$

(f) The bound volume charge density $\rho_{vb} = -\nabla \cdot \overline{P}$. The $\nabla \cdot \overline{P}$ in region 2 equals zero; thus, $\rho_{vb} = 0$. Flux plots of $\overline{D}$, $\epsilon_0 \overline{E}$, and $\overline{P}$ are shown in Fig. 5-17. Note from (34) that $\overline{E}_i$ is found only in region 2 due to the symmetry of the problem.

Problem 5.8-3 Using the results found in Example 11 for ρ_{sb} and ρ_{vb}, find in regions 1, 2, 3, and 5 of Fig. 5-16: (a) $\overline{E}_i$ (due to bound charges) through the use of Gauss's law in $\overline{E}_i$, $\oint_s \overline{E}_i \cdot \overline{ds} = Q_{\text{ben}}/\epsilon_0$; (b) $\overline{E}_a$ through the use of Gauss's law in $\overline{E}_a$, $\oint_s \overline{E}_a \cdot \overline{ds} = Q_{\text{en}}/\epsilon_0 (Q_{\text{en}}$ is free charge); (c) $\overline{E}$ through the use of (29) and results found in parts (a) and (b); (d) $\overline{D}$ through the use of (24) and compare with the results found in Example 11.

Problem 5.8-4 If, in Example 11, $\chi_e = 4$ in region 2, by what factor is the field reduced in the dielectric, from the case when $\chi_e = 0$?

5.8-4 Linearity, Homogeneity, and Isotropy

A dielectric material is considered linear when, at a given point, ϵ is a constant. Thus, a plot of $|\overline{D}|$ versus $|\overline{E}|$, from (24), will be a straight line. If the permittivity, at a given point, is a function of $|\overline{E}|$, $\epsilon = \epsilon(|\overline{E}|)$, the material is non-linear and a plot of $|\overline{D}|$ versus $|\overline{E}|$ will not be a straight line.

A dielectric material is homogeneous when ϵ does not vary from point to point; thus, ϵ is not a function of position. If the permittivity varies with position, $\epsilon = \epsilon(x, y, z)$, then the material is non-homogeneous.

As stated in Sec. 5.8-3, $\overline{P}$ and $\overline{E}$ are in the same direction in an isotropic material. From (21), $P_x = \epsilon_0 \chi_e E_x$, $P_y = \epsilon_0 \chi_e E_y$, and $P_z = \epsilon_0 \chi_e E_z$, where the χ_e's are the same. Thus, from (23), (24), and (25) we have $D_x = \epsilon E_x$, $D_y = \epsilon E_y$, and $D_z = \epsilon E_z$. In a non-isotropic (anisotropic) medium, we find that

$$\begin{aligned} P_x &= \chi_{e11}\epsilon_0 E_x + \chi_{e12}\epsilon_0 E_y + \chi_{e13}\epsilon_0 E_z \\ P_y &= \chi_{e21}\epsilon_0 E_x + \chi_{e22}\epsilon_0 E_y + \chi_{e23}\epsilon_0 E_z \\ P_z &= \chi_{e31}\epsilon_0 E_x + \chi_{e32}\epsilon_0 E_y + \chi_{e33}\epsilon_0 E_z \end{aligned} \tag{43}$$

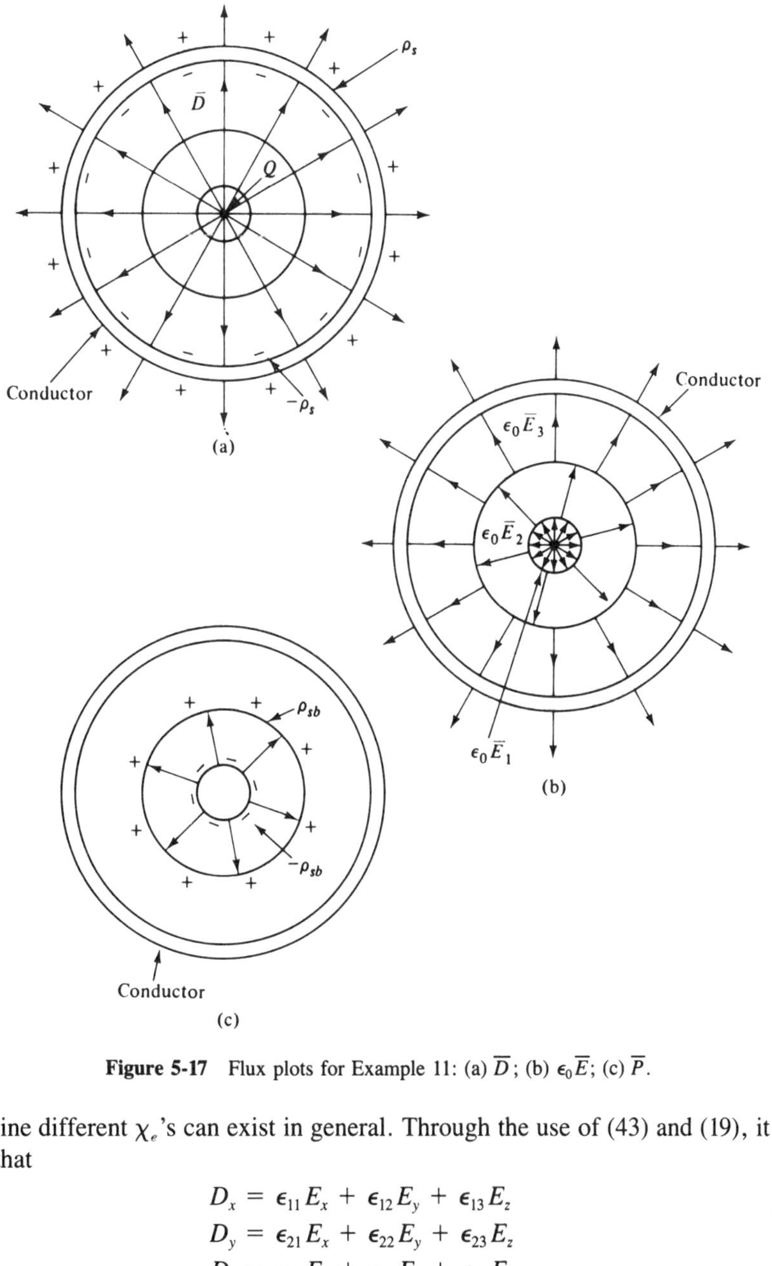

Figure 5-17 Flux plots for Example 11: (a) $\overline{D}$; (b) $\epsilon_0\overline{E}$; (c) $\overline{P}$.

where nine different χ_e's can exist in general. Through the use of (43) and (19), it can be shown that

$$\begin{aligned} D_x &= \epsilon_{11} E_x + \epsilon_{12} E_y + \epsilon_{13} E_z \\ D_y &= \epsilon_{21} E_x + \epsilon_{22} E_y + \epsilon_{23} E_z \\ D_z &= \epsilon_{31} E_x + \epsilon_{32} E_y + \epsilon_{33} E_z \end{aligned} \tag{44}$$

where nine different ϵ's may exist. In matrix form, (44) can be written as

$$\begin{bmatrix} D_x \\ D_y \\ D_z \end{bmatrix} = \begin{bmatrix} \epsilon_{11} & \epsilon_{12} & \epsilon_{13} \\ \epsilon_{21} & \epsilon_{22} & \epsilon_{23} \\ \epsilon_{31} & \epsilon_{32} & \epsilon_{33} \end{bmatrix} \begin{bmatrix} E_x \\ E_y \\ E_z \end{bmatrix} \tag{45}$$

where ϵ is now a matrix of nine terms in general. Equation (45) can be written in vector-matrix form.

$$\boxed{[\overline{D}] = [\epsilon][\overline{E}]} \tag{46}$$

If $\epsilon_{11} = \epsilon_{22} = \epsilon_{33} = \epsilon$ and all other terms in $[\epsilon]$ are zero, we have an isotropic material and (46) can be written as

$$[\overline{D}] = \epsilon[\overline{E}] \tag{47}$$

The permittivity of a dielectric can also vary with frequency and temperature. The dependency on frequency becomes an important factor at high frequencies. In this text, we will basically consider linear, homogeneous, and isotropic dielectric media.

Problem 5.8-5 Through the use of (19) and (43), derive (44).

5.9 DIELECTRIC-DIELECTRIC BOUNDARY CONDITIONS

Let us now find the relationship between $\overline{E}$'s and $\overline{D}$'s at a boundary between two dielectrics, as shown in Fig. 5-18. The procedure that we shall use is identical to that used in Sec. 5.6 to find the boundary conditions at a conductor-free space boundary. The boundary conditions on the normal field components can be found by applying Gauss's law to a small cylinder centered at the boundary as $\Delta h \to 0$. From Gauss's law,

$$\oint_s \overline{D} \bullet \overline{ds} = Q_{\text{en}} \cong D_{1n}\,\Delta s - D_{2n}\,\Delta s = \rho_s\,\Delta s \tag{1}$$

where ρ_s is the free surface charge density placed at the boundary. From (1), we have

$$\boxed{D_{1n} - D_{2n} = \rho_s \qquad (\text{C m}^{-2})} \tag{2}$$

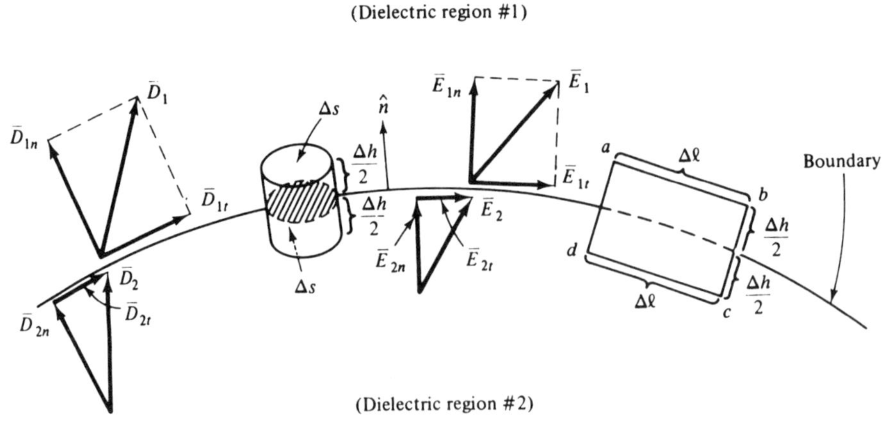

Figure 5-18 Graphical display of $\overline{D}$'s, and $\overline{E}$'s, closed path and closed surface, about a dielectric boundary for determining static boundary conditions as $\Delta h \to 0$.

The boundary condition relating the tangent field components can be obtained through $\oint_\ell \overline{E} \bullet \overline{d\ell} = 0$ about the closed loop shown as $\Delta h \to 0$. Thus,

$$\oint_\ell \overline{E} \bullet \overline{d\ell} = \int_a^b \overline{E}_1 \bullet \overline{d\ell} + \int_b^c \overline{E} \bullet \overline{d\ell} + \int_c^d \overline{E}_2 \bullet \overline{d\ell} + \int_d^a \overline{E} \bullet \overline{d\ell} = 0 \tag{3}$$

The $\int_b^c$ and $\int_d^a$ are zero since $\Delta h \to 0$. Therefore, (3) becomes

$$E_{1t}\Delta\ell - E_{2t}\Delta\ell = 0$$

Thus,

$$\boxed{E_{1t} = E_{2t} \qquad (\text{V m}^{-1})} \tag{4}$$

From (2), (4) and the relationship $\overline{D} = \epsilon\overline{E}$, we obtain

$$\boxed{\epsilon_1 E_{1n} - \epsilon_2 E_{2n} = \rho_s} \tag{5}$$

and

$$\boxed{\frac{D_{1t}}{\epsilon_1} = \frac{D_{2t}}{\epsilon_2}} \tag{6}$$

Equations (2), (4), (5), and (6) make up the set of boundary conditions at a dielectric-dielectric boundary.

From $\nabla \bullet \overline{P} = -\rho_{vb}$ and Gauss's law, it can be shown that

$$P_{1n} - P_{2n} = -\rho_{sb} \qquad (\text{C m}^{-2}) \tag{7}$$

Example 12

A thin disc-shaped dielectric, with $\epsilon_r = 10$, is placed perpendicular to a uniform field $\overline{D} = \hat{z}3$ (nC m^{-2}). Find at the center of the disc: (a) $\overline{D}$; (b) $\overline{E}$; (c) $\overline{P}$; (d) ρ_{sb}. Assume that $\rho_s = 0$ on the disc surface.

Solution. Let region 1 be the region containing a uniform $\overline{D}$ field of ϵ_0 permittivity and region 2 be the disc of $\epsilon_2 = \epsilon_0\epsilon_r = 10\epsilon_0$. From the statement of the problem the faces of the disc are perpendicular to the $\hat{z}$ direction; thus, $D_{1n} = 3$ nC m^{-2}.

(a) From (2), at the top face we have

$$D_{1n} = D_{2n} = 3 \text{ nC m}^{-2}$$

or

$$\overline{D}_2 = \hat{z}3 \qquad (\text{nC m}^{-2}) \tag{8}$$

Since the disc is thin and has parallel faces, we can assume that $\overline{D}_2$ found above is the same from top to bottom of the disc.

(b) From (5), at the top face we have

$$E_{2n} = E_{1n}\frac{\epsilon_1}{\epsilon_2} = \left(\frac{D_{1n}}{\epsilon_1}\right)\frac{\epsilon_1}{\epsilon_2}$$

$$= \left(\frac{D_{1n}}{\epsilon_0}\right)\frac{\epsilon_0}{10\epsilon_0} = \frac{0.3}{\epsilon_0} \qquad (\text{nV m}^{-1})$$

or

$$\overline{E}_2 = \hat{z}\frac{0.3}{\epsilon_0} \qquad (\text{nV m}^{-1}) \tag{9}$$

(c) From (5.8-19), inside the disc we have

$$\overline{P}_2 = \overline{D}_2 - \epsilon_0 \overline{E}_2 = \hat{z}3 - \hat{z}\epsilon_0\left(\frac{0.3}{\epsilon_0}\right) = \hat{z}2.7 \qquad (\text{nC m}^{-2}) \tag{10}$$

(d) From (5.8-9), at the top face we have

$$\rho_{sb} = \overline{P}_2 \bullet \hat{n} = (\hat{z}2.7) \bullet \hat{z} = 2.7 \text{ nC m}^{-2} \tag{11}$$

At the lower face $\rho_{sb} = -2.7$ nC m^{-2}.

Problem 5.9-1 In the region $y > 0$, we find $\epsilon_{r1} = 4$, and in the region $y < 0$, we find $\epsilon_{r2} = 3$. If $\overline{E}_1 = \hat{z}10$ V m^{-1} at the boundary, find: (a) $\overline{D}_1$; (b) $\overline{E}_2$; (c) $\overline{D}_2$; (d) $\overline{P}_1$; (e) $\overline{P}_2$; (f) ρ_{sb}. Assume that $\rho_s = 0$ at the boundary.

Problem 5.9-2 A boundary between two dielectrics is found at the $x = 0$ plane. If material 1 exists for $x > 0$ with $\epsilon_{r1} = 4$ and material 2 exists for $x < 0$ with $\epsilon_{r2} = 7$, and when $\overline{E}_1 = (\hat{x}2 + \hat{y}3 - \hat{z}6)$ (at boundary), find: (a) $\overline{D}_1$; (b) $\overline{P}_1$; (c) $\overline{E}_2$; (d) $\overline{D}_2$; (e) $\overline{P}_2$; (f) ρ_{sb}. Assume that $\rho_s = 0$ at the boundary.

Problem 5.9-3 Repeat Prob. 5.9-1 for $\overline{E}_1 = \hat{y}5$ V m^{-1}.

5.10 CAPACITANCE

In electric circuit analysis, the capacitor is found to be an important circuit element. A generalized two-conductor capacitor is formed by two conductors, as shown in Fig. 5-19. Charge of equal and opposite polarity will accumulate on the conductors when connected to a battery.

> The capacitance of a two-conductor capacitor is defined as the ratio of the magnitude of charge on one of the conductors to the magnitude of the potential difference between the two conductors.

Thus, the mathematical definition of capacitance becomes

$$C \triangleq \frac{|Q|}{|V_{ab}|} \qquad (\text{C V}^{-1} \text{ or F}) \tag{1}$$

where V_{ab} is the potential difference between any point on conductor b and any point on conductor a. It should be noted that under static conditions each conductor is at a constant potential. The unit of capacitance is the farad (F) and is found from (1) to be 1 coulomb per volt (C V^{-1}).

Equation (1) can be written in terms of the surface charge density ρ_s, and $\overline{E}$ between the conductors as

$$C = \frac{\left|\int_{S_a} \rho_s \, ds\right|}{\left|-\int_b^a \overline{E} \bullet \overline{d\ell}\right|} \qquad (\text{F}) \tag{2}$$

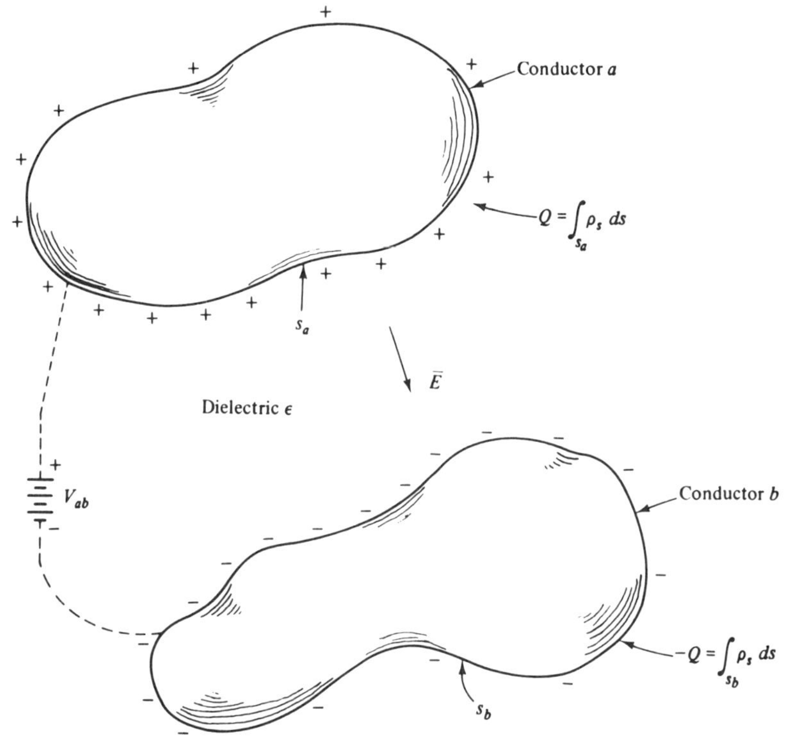

Figure 5-19 Two-conductor generalized capacitor with equal but opposite charge on the conductors.

From boundary conditions at the conductor surface, we find that $\rho_s = \epsilon E_n = \epsilon \overline{E} \cdot \hat{n}$. Now $\rho_s \, ds = \epsilon \overline{E} \cdot \hat{n} \, ds = \epsilon \overline{E} \cdot \overline{ds}$, and (2) becomes

$$\boxed{C = \frac{\left| \int_{s_a} \epsilon \overline{E} \cdot \overline{ds} \right|}{\left| -\int_b^a \overline{E} \cdot \overline{d\ell} \right|} \qquad \text{(F)}} \tag{3}$$

where the $\overline{E}$ in the numerator is that found just off the conductor surface, and the $\overline{E}$ in the denominator is that found between the conductors.

Through the use of (4.6-12), the energy W required to build up the two-conductor capacitor charge system becomes

$$W = \int_s \tfrac{1}{2} V \rho_s \, ds = \int_{s_a} \tfrac{1}{2} V_a \rho_{s_a} \, ds_a + \int_{s_b} \tfrac{1}{2} V_b \rho_{s_b} \, ds_b$$

$$= \frac{V_a}{2} \int_{s_a} \rho_{s_a} \, ds_a + \frac{V_b}{2} \int_{s_b} \rho_{s_b} \, ds_b = \frac{V_a Q}{2} - \frac{V_b Q}{2} = \frac{V_{ab} Q}{2} \qquad \text{(J)} \tag{4}$$

where the integrals are performed only over the surfaces of the two conductors. Through the use of (1) and (4), we can express the energy W as

$$\boxed{W = \tfrac{1}{2}CV_{ab}^2 \qquad \text{(J)}} \tag{5}$$

The capacitance can also be defined in terms of the charge Q and W by solving for V_{ab} in (4) and substituting into (1) to obtain

$$\boxed{C \triangleq \frac{Q^2}{2W} \qquad \text{(F)}} \tag{6}$$

It should be noted that W can also be viewed as the energy stored in the $\overline{E}$ field between the two conductors.

From (3) it can be seen that the capacitance will be a function only of the geometry of the capacitor when the dielectric is linear (ϵ is not a function of $\overline{E}$ or V_{ab}). Voltage-controllable capacitors possess dielectrics whose ϵ is a function of V_{ab}, and thus the capacitance is a function of the applied potential.

Example 13

Find the capacitance of the parallel-plate capacitor shown in Fig. 5-20, when uniform ρ_s but of opposite polarity is found on the conductor plates. Assume zero $\overline{E}$ field fringing at the edges.

Solution. From (3) and $\overline{E}|_{\text{cond.}} = \overline{E}$ (between conductors) $= -\hat{z}\rho_s/\epsilon$, we have

$$C = \frac{\left|\int_{s_a} \epsilon\overline{E} \bullet \overline{ds}\right|}{\left|-\int_b^a \overline{E} \bullet \overline{d\ell}\right|} = \frac{\left|\int_{\text{top}} \epsilon\left(\frac{-\hat{z}\rho_s}{\epsilon}\right) \bullet (-\hat{z}\,ds)\right|}{\left|-\int_0^\ell \left(-\hat{z}\frac{\rho_s}{\epsilon}\right) \bullet (\hat{z}\,dz)\right|}$$

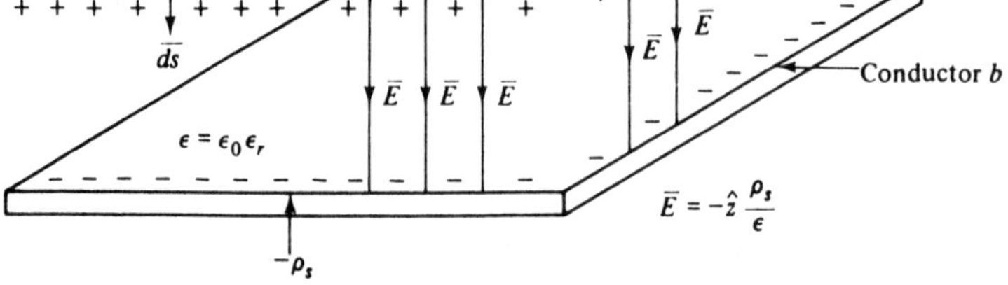

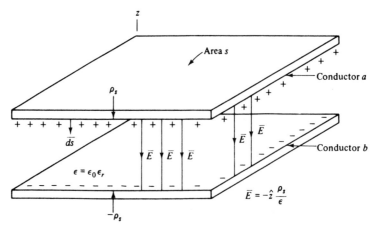

Figure 5-20 Graphical display of two-conductor parallel-plate capacitor of uniform ρ_s for Example 13.

Thus,

$$\boxed{C = \frac{\rho_s s}{\rho_s \ell/\epsilon} = \frac{\epsilon s}{\ell} \qquad \text{(F)}} \tag{7}$$

Comparing (7) and (1), we note that $\rho_s s = Q$ and $\rho_s \ell/\epsilon = V_{ab}$.

Noting the analogy with G from Sec. 5.4[5], we may also use

$$C = \iint_{\substack{\text{surface}\\ \text{area}}} \epsilon \frac{ds}{\ell} \tag{8}$$

or

$$\frac{1}{C} = \int_{\text{path}} \frac{d\ell}{\epsilon s} \tag{9}$$

depending on the particular geometry involved.

Example 14

Find the capacitance of an L length coaxial capacitor shown in Fig. 5-21. Assume zero $\overline{E}$ field fringing at the ends and uniform ρ_ℓ along the conductors.

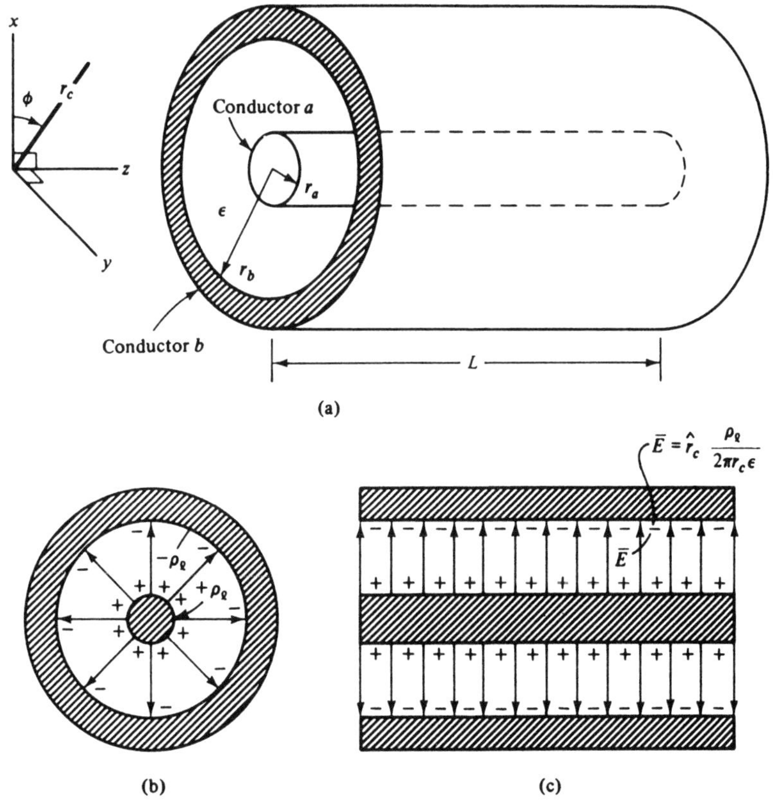

Figure 5-21 (a) Graphical display of a two-conductor cylindrical capacitor. (b) End view of $\overline{E}$ flux lines. (c) $\overline{E}$ flux lines along the length of the capacitor.

[5] This analogy will be discussed further in Secs. 7.5 and 7.6.

Solution. Through the application of Gauss's law for $r_a \leqq r_c \leqq r_b$, we obtain for the $\overline{E}$ field,

$$\boxed{\overline{E} = \hat{r}_c \frac{\rho_\ell}{2\pi r_c \epsilon}} \tag{10}$$

where the form is found to be the same as about an infinite line charge in (4.6-24) or in Table 3-2. The potential difference V_{ab} becomes

$$V_{ab} = -\int_b^a \overline{E} \bullet \overline{d\ell} = -\int_b^a \left(\hat{r}_c \frac{\rho_\ell}{2\pi r_c \epsilon}\right) \bullet (\hat{r}_c\, dr_c) = \frac{\rho_\ell}{2\pi\epsilon} \ln\left(\frac{r_b}{r_a}\right) \tag{11}$$

Through the use of (1) and (11), we obtain for the capacitance of length L,

$$\boxed{C = \frac{|Q|}{|V_{ab}|} = \frac{|\rho_\ell \mathrm{L}|}{|(\rho_\ell/2\pi\epsilon)\ln(r_b/r_a)|} = \frac{2\pi\epsilon \mathrm{L}}{\ln(r_b/r_a)} \quad \text{(F)}} \tag{12}$$

The same results could have been obtained through the use of (2) and (3). The capacitance is found to be independent of ρ_ℓ or V_{ab} since ϵ was assumed to be constant (independent of $\overline{E}$ as required in a linear dielectric).

Solving by use of (9), we have

$$\frac{1}{C} = \int_{r_a}^{r_b} \frac{dr}{\epsilon \mathrm{L} 2\pi r} = \frac{1}{2\pi\epsilon \mathrm{L}} \ln\left(\frac{r_b}{r_a}\right)$$

or

$$C = \frac{2\pi\epsilon \mathrm{L}}{\ln(r_b/r_a)}$$

which agrees with (12).

Example 15

Find the capacitance of the parallel-plate capacitor shown in Fig. 5-22, where two different parallel slabs of dielectric are used. Assume uniform ρ_s distribution and neglect $\overline{E}$ field fringing at the edges.

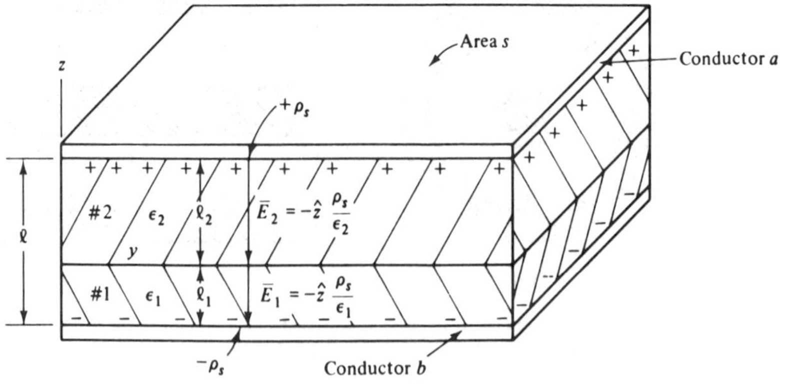

Figure 5-22 Graphical display of a parallel-plate capacitor, using two different dielectric slabs in series for Example 15.

Solution. From the boundary conditions at the upper or lower conductor, we find that $\overline{D} = -\hat{z}\rho_s$ at the two conductors. The $\overline{D}$ in both dielectrics is found to be uniform and equal to $\overline{D} = -\hat{z}\rho_s$. The potential difference V_{ab} is

$$V_{ab} = -\int_b^a \overline{E} \bullet \overline{d\ell} = -\left[\int_0^{\ell_1} \overline{E}_1 \bullet \overline{d\ell} + \int_{\ell_1}^{\ell_1+\ell_2} \overline{E}_2 \bullet \overline{d\ell}\right]$$

$$= -\left[\int_0^{\ell_1} \left(\frac{-\hat{z}\rho_s}{\epsilon_1}\right) \bullet (\hat{z}\, dz) + \int_{\ell_1}^{\ell_1+\ell_2} \left(\frac{-\hat{z}\rho_s}{\epsilon_2}\right) \bullet (\hat{z}\, dz)\right]$$

$$= \left(\frac{\rho_s \ell_1}{\epsilon_1} + \frac{\rho_s \ell_2}{\epsilon_2}\right) \quad \text{(V)} \qquad (13)$$

From (1) and (13), we have

$$C = \frac{|Q|}{|V_{ab}|} = \frac{|\rho_s s|}{\left|\dfrac{\rho_s \ell_1}{\epsilon_1} + \dfrac{\rho_s \ell_2}{\epsilon_2}\right|} = \frac{s}{\dfrac{\ell_1}{\epsilon_1} + \dfrac{\ell_2}{\epsilon_2}} \quad \text{(F)} \qquad (14)$$

Let us arrange (14) in the form

$$\boxed{C = \frac{1}{\dfrac{\ell_1}{\epsilon_1 s} + \dfrac{\ell_2}{\epsilon_2 s}} = \frac{1}{\dfrac{1}{\epsilon_1 s/\ell_1} + \dfrac{1}{\epsilon_2 s/\ell_2}} \quad \text{(F)}} \qquad (15)$$

The expressions $\epsilon_1 s/\ell_1$, and $\epsilon_2 s/\ell_2$ can be recognized as the capacitance of the lower and upper dielectric regions, respectively. Thus, we can write

$$\boxed{C = \frac{1}{1/C_1 + 1/C_2} = \frac{C_1 C_2}{C_1 + C_2} \quad \text{(F)}} \qquad (16)$$

where $C_1 = \epsilon_1 s/\ell_1$ and $C_2 = \epsilon_2 s/\ell_2$. The same results would have been obtained through the use of (2) or (3).

Problem 5.10-1 Use (9) to calculate the capacitance of two concentric metallic spheres separated by a dielectric. Let r_b be the inner radius of the outer sphere and r_a be the outer radius of the inner sphere.

Problem 5.10-2 Find the expression for the capacitance per meter length of a cylindrical capacitor whose cross section is the same as that of the spherical configuration found in Fig. 5-16, with a conducting cylinder placed at the $r_c = a$ position, $a = 0.01$ m, $b = 0.03$ m, $c = 0.04$ m, and $\epsilon_2 = 6\epsilon_0$.

Problem 5.10-3 Find the capacitance of a parallel-plate capacitor whose plates are separated 1.5 cm and whose surface area equals 2 cm^2. Assume air dielectric and neglect $\overline{E}$ field fringing.

Problem 5.10-4 Solve Example 13 through the use of (6).

REVIEW QUESTIONS

1. What is electric current? *Sec. 5.2*
2. Is the current density defined at a point (a point function)? *Eq. (5.2-3)*
3. State the point form expression for the conservation of charge. *Eq. (5.2-11)*
4. Is the drift velocity equal to $-\mu_e \overline{E}$? *Eq. (5.3-1)*
5. What is the point form of Ohm's law? *Eq. (5.3-3)*
6. What is relaxation time? *Sec. 5.5, Eq. (5.5-4)*
7. Under static condition, what is $\overline{E}$ inside a conductor? *Sec. 5.5*
8. Under static condition, what is the normal component of $\overline{D}$ equal to just above the conductor surface? *Eq. (5.6-3)*
9. What is a bond vacancy created by an electron that leaves a covalent bond called? *Sec. 5.7*
10. What are the two current carriers found in a semiconductor? *Sec. 5.7*
11. What are the three basic polarization mechanisms that produce polarization in a dielectric? *Sec. 5.8-1*
12. What is the definition of the polarization vector $\overline{P}$? *Eq. (5.8-2)*
13. What physical quantity is $\overline{P} \bullet \hat{n}$ equal to? *Eq. (5.8-9)*
14. In a dielectric $-\nabla \bullet \overline{P}$ is equal to what physical quantity? *Eq. (5.8-14)*
15. What is $\epsilon_0 \overline{E} + \overline{P}$ equal to? *Eq. (5.8-19)*
16. What is $\nabla \bullet \overline{D}$ equal to in general? *Eq. (5.8-20)*
17. In $\oint_s \overline{D} \bullet \overline{ds} = Q_{en}$, what charges are included in Q_{en}? *Example 11*
18. What is a linear, homogeneous, and isotropic dielectric? *Sec. 5.8-4*
19. At a dielectric-dielectric boundary, what is the relationship between the normal components of $\overline{D}$ and free surface charge density ρ_s? *Eq. (5.9-2)*
20. What is the definition of capacitance of a two-conductor capacitor? *Eq. (5.10-1)*
21. What is the definition of capacitance in terms of energy W expended in building up the charges on the two-conductor capacitor? *Eq. (5.10-6)*
22. What is the expression for the capacitance of a parallel-plate capacitor? *Eq. (5.10-7)*
23. In a capacitor containing a linear dielectric, is the capacitance independent of charge and potential difference between conductors? *Example 14*

PROBLEMS

5-1. In a cylindrical distribution of ρ_v along the z axis, it is found that $\overline{J} = \hat{z}10^{-3}r_c^2$ A m^{-2}. Find: (a) the current through a perpendicular cross section of $r_c = 0.3$ m; (b) $\nabla \cdot \overline{J}$ within the range of $\overline{J}$; (c) $\partial\rho_v/\partial t$; (d) ρ_v if $\overline{U} = \hat{z}10^{-5}$ m s^{-1}.

5-2. The charge within a spherical volume of 2 m radius changes linearly from 10^{-9} C to 10^{-10} C in 10^{-6} s. What is the total convection current that flows outwardly?

5-3. In an aluminum conductor, $\mu_e = 1.4 \times 10^{-4}$, we find that $\overline{J} = \hat{x}(y^2 + z^2)$ (A m^{-2}) over a given range. Find: (a) $\overline{E}$; (b) $\overline{U}_d$; (c) ρ_{ve}.

5-4. A sample of conducting material having a uniform cross section of 0.008 (m^2) and a length of 2 (m) is found to conduct a uniform current of 9 (mA) when a potential difference of 0.05 (μV) is applied to the two ends. Find: (a) E; (b) R; (c) σ; (d) U_d. [Assume that μ_e = 0.003 (m^2 V$^{-1}s^{-1}$).]

5-5. A truncated cone of slightly conductive material, $\sigma = 10^3$ S m^{-1}, has a radius of 1 cm at its base, and a radius of 4 cm at its top. The height is 1 m. Assuming that highly conductive end plates are used for electrodes, find: (a) the resistance between the base and the top of the cone; (b) the radius of a 1-m-length uniform-cross-sectional cylinder that has the same resistance between ends.

5-6. Show that the resistance between the base and the top of a truncated cone of conductivity σ (S m^{-1}) is given by $R = h/\sigma\pi r_1 r_2$ (Ω), where h is the height of the truncated cone, r_1 is the radius of the base, and r_2 is the radius of the top of the cone.

5-7. Repeat Prob. 5-5 for a flared shape whose radius in centimeters is given by $r_c = (y + 1)^2$, where y is the distance in meters measured from the end having 1-cm radius.

5-8. Find the resistance between cylindrical electrodes of a coaxial resistor 1 m in length, whose radii are 1 cm and 5 cm. The conductivity of the material between electrodes is 10^7 S m^{-1}.

5-9. The terminals of a 1-m-long coaxial resistor consists of two concentric conducting cylinders with radii of 1 cm and 5 cm. In the range 1 cm $\leq r_c \leq$ 3 cm, $\sigma_1 = 10^7$ S m^{-1}, and in the range 3 cm $\leqq r_c \leqq$ 5 cm, $\sigma_2 = 0.5 \times 10^7$ S m^{-1}. Find: (a) the total resistance between terminals; (b) the ratio $E(r_c = 1\text{ cm})/E(r_c = 5\text{ cm})$.

5-10. A cylindrical test sample of germanium has a uniform radius $r_c = 2$ mm and a length of 3 cm. If the ends are coated with a thin layer of high-conductivity material and a voltage of 10 V is applied between the ends, find: (a) $\overline{E}$; (b) σ; (c) $\overline{J}$; (d) R, end to end; (e) I; (f) U_e; (g) U_h.

5-11. Calculate the σ of silicon, using (5.7-2) and parameters given in Table 5-2.

5-12. In a sample of dielectric, $\overline{P} = \hat{z}(5 \times 10^{-14})$ (C m^{-2}) at a point where $\overline{E} = \hat{z}3$ (V m^{-1}). Find: (a) χ_e; (b) ϵ_r; (c) ϵ; (d) $\overline{D}$.

5-13. Starting with Maxwell's first equation $\nabla \bullet \overline{D} = \rho_v$ (free charge density), show that Gauss's law in $\overline{D}$ becomes

$$\oint_s \overline{D} \bullet \overline{ds} = Q_{\text{en(free charge)}}$$

5-14. Starting with $\epsilon_0 \nabla \bullet \overline{E} = \rho_v + \rho_{vb}$, eq. (5.8-16), show that Gauss's law in $\overline{E}$ becomes

$$\oint_s \overline{E} \bullet \overline{ds} = \frac{Q_{\text{en}} + Q_{\text{ben}}}{\epsilon_0}$$

where Q_{en} is the free charge enclosed and Q_{ben} is the total bound charge enclosed.

5-15. The plates of a parallel-plate capacitor are 10 mm by 10 mm and are separated by a 0.1-mm thickness of a dielectric, $\epsilon_r = 2.4$. For an applied voltage of 100 V, find: (a) the charge on the plates; (b) the percent of the charge that is due to the polarization of the dielectric.

5-16. Show that at a dielectric-dielectric boundary

$$\hat{n}_{12} \bullet (\overline{P}_1 - \overline{P}_2) = -\rho_{sb(\text{total})} = P_{1n} - P_{2n}$$

5-17. An infinite slab of polystyrene of 0.1 m thickness is placed in an applied field $\overline{E}_a = \hat{z}2$ (μV m^{-1}) so that the faces are perpendicular to $\overline{E}_a$ with one boundary at $z = 0$ and the other at $z = 0.1$ m. If $\epsilon = 4\epsilon_0$, find: (a) $\overline{E}$, $\overline{D}$, and $\overline{P}$ for $z < 0$; (b) $\overline{E}$, $\overline{D}$, and $\overline{P}$ for $0 < z < 0.1$; (c) $\overline{E}$, $\overline{D}$, and $\overline{P}$ for $z > 0.1$; (d) $\overline{E}_i$ for $0 < z < 0.1$

5-18. A capacitor, whose dielectric has an ϵ_r of 4, has a capacitance of 300 pF. If the dielectric is removed and replaced by a material having a conductivity of 10^{-3} S m^{-1}, what is the resistance between terminals?

5-19. If 100 V is applied to the parallel-plate capacitor of Fig. 5-23, find: (a) E_1; (b) E_2; (c) C_{total}; (d) D_1; (e) D_2; (f) P_1; (g) P_2.

5-20. Capacitor $C_1 = 0.1\ \mu$F is charged to 600 V, and capacitor $C_2 = 0.2\ \mu$F is charged to 400 V. The two capacitors are then connected in parallel with the +600-V terminal of C_1 connected to the +400-V terminal of C_2. Find: (a) V across the two paralleled capacitors; (b) the total charge still stored on the plates; (c) the initial stored energy and the final stored energy.

5-21. Repeat Example 11 for a cylindrical configuration of the same dimensions, and display results in a manner similar to those found in Fig. 5-16.

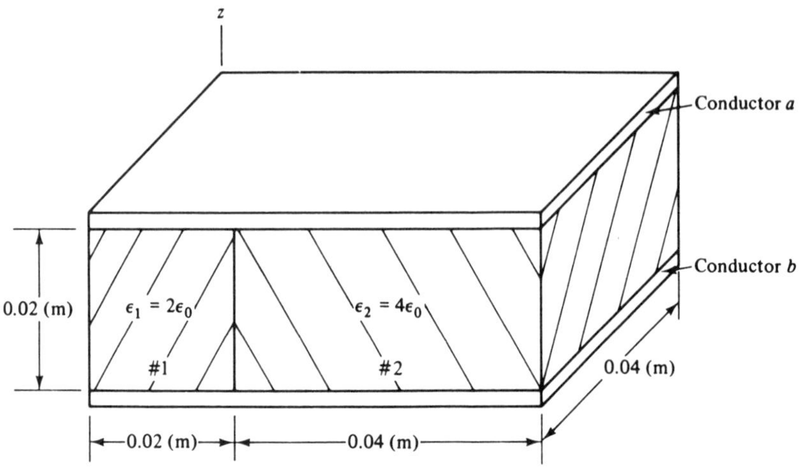

Figure 5-23 Graphical display of a parallel-plate capacitor with two different dielectric slabs in parallel for Prob. 5-19.

5-22. A coaxial capacitor, 1 m in length, has a capacitance of 100 pF. Another capacitor using the same dielectric material is 2 m long, and the radii are each twice as large as for the first capacitor. What is the capacitance of the second capacitor?

5-23. Find Q and V after each of the following steps:

1. A parallel-plate capacitor of capacitance C_0 (F), with a dielectric between the plates, has been charged to V_0 (V). The capacitor is then disconnected from the charging source.
2. The dielectric is now removed from between the plates.
3. The capacitor is momentarily reconnected to the charging source of voltage V_0, then disconnected.
4. The dielectric is now replaced between the plates.

5-24. A parallel-plate capacitor, $C = 1\ \mu$F, has been charged to 3000 V, then disconnected from the charging source. If the dielectric constant ϵ_r is 5, how much stored energy would be removed by removing the dielectric?

5-25. For a coaxial capacitor like that shown in Fig. 5-21(a), having a uniform dielectric between the conducting cylinders, the electric field is greatest just outside and normal to the surface of the innermost conductor. However, it is theoretically possible to have a constant E field from r_a to r_b if the dielectric constant ϵ_r varies in a certain manner with respect to r_c. If $\epsilon_r = 2$ at $r_c = r_a$, with $r_a = 1$ cm and $r_b = 4$ cm, find: (a) E_r for 100 V applied if $E_r =$ constant between r_a and r_b; (b) ϵ_r at $r_c = r_b$.

5-26. A coaxial cable, $r_a = 0.8$ cm, $r_b = 3$ cm, has two concentric layers of dielectric between conductors. The innermost layer has an ϵ_r of 6.0 for 0.8 cm $< r_c <$ 1.0 cm and an ϵ_r of 3.0 for 1.0 cm $< r_c <$ 3.0 cm. Find the maximum E field in each dielectric for an applied voltage of 12.5 kV.

6

Uniqueness Theorem, Solution of Laplace's Equation, and Solution of Poisson's Equation

6.1 INTRODUCTION

In the previous chapters, we have obtained electrostatic field solutions for $\overline{E}$ and V basically by three methods:

1. Through the integration of $\rho_v \, dv \hat{a}_R / 4\pi\epsilon_0 R^2$ [eq. (2.7-2)] over the known charge distribution, we were able to obtain $\overline{E}$ in free space. Then, through integration of $-\overline{E} \cdot \overline{d\ell}$ from some reference point b to point a, we were able to obtain the potential difference V_{ab} [eq. (4.3-1)].
2. Through the use of Gauss's law [eq. (3.4-7)] we were able to obtain $\overline{D}$, and thus $\overline{E}$, since $\overline{D} = \epsilon_0 \overline{E}$ in free space. Since $\nabla \cdot \overline{D} = \rho_v$ [eq. (5.8-20)] holds for free space as well as dielectrics, Gauss's law also applies to dielectric regions and $\overline{E}$ can be found through the use of $\overline{D} = \epsilon \overline{E}$. It should be noted that the charge distribution had to be known and possess symmetry. Then, through integration of $-\overline{E} \cdot \overline{d\ell}$ from some reference point b to point a, we were able to obtain the potential difference V_{ab} [eq. (4.3-1)].
3. Through the integration of $\rho_v \, dv / 4\pi\epsilon_0 R$ [eq. (4.4-8)] over the known charge distribution, we were able to obtain the absolute potential difference in free space. Then, through the use of $\overline{E} = -\nabla V$ [eq. (4.5-13)] we were able to find $\overline{E}$ in free space. In problems containing bound charges due to dielectrics [eq. (5.8-27)] indicates that bound as well as free charges must be considered. Then, through the use of $\overline{E} = -\nabla V$, $\overline{E}$ can be found. In this case, the distribution of all bound and free charges must be known.

Upon close examination of the three above solution methods, we will note that the charge distribution must be known or assumed. In realistic electrostatic problems, one seldom knows the charge distribution; thus, all the solution methods introduced up to this point have a limited use. The knowledge of the charge distribution can be circumvented through analytical solution of Laplace's and Poisson's equations that will be presented in this chapter. The exact solution will be obtained when we solve these equations, subject to the boundary conditions of the problem.

Other solution methods, classified as graphical, experimental, analog, and numerical, will be presented in Chapter 7. We shall find that some of these solution methods are based on Laplace's equation and will not require the knowledge of the distribution of charge. Some of these solution methods will yield approximate solutions.

6.2 LAPLACE'S AND POISSON'S EQUATIONS

A partial differential equation in V can be obtained through the use of (5.8-20), (5.8-24), and (4.5-13). It should be noted that all three equations apply to dielectric media as well as free space. Let us substitute $\overline{D} = \epsilon\overline{E}$ into (5.8-20) to obtain

$$\nabla \cdot \overline{D} = \nabla \cdot (\epsilon\overline{E}) = \rho_v \tag{1}$$

From $\overline{E} = -\nabla V$ [eq. (4.5-13)], eq. (1) becomes

$$\nabla \cdot [\epsilon(-\nabla V)] = \rho_v \tag{2}$$

If ϵ is uniform, as found in a homogeneous medium, (2) can be written as

$$\nabla \cdot \nabla V = -\frac{\rho_v}{\epsilon} \tag{3}$$

From (1.12-5), $\nabla \cdot \nabla = \nabla^2$ and (3) can be written as

$$\boxed{\nabla^2 V = -\frac{\rho_v}{\epsilon} \qquad (\mathrm{V\ m^{-2}})} \tag{4}$$

Equation (4) is called *Poisson's equation* and applies to a homogeneous media. When the free charge density $\rho_v = 0$, (4) will reduce to

$$\boxed{\nabla^2 V = 0 \qquad (\mathrm{V\ m^{-2}})} \tag{5}$$

and is called *Laplace's equation for homogeneous media*. As noted in Sec. 1.12, ∇^2 is called the *Laplacian operator* and $\nabla^2 V$ is the *Laplacian of V*.

The expanded expressions for Laplace's equation in rectangular, cylindrical, and spherical coordinate systems become

Rectangular:

$$\nabla^2 V = \frac{\partial^2 V}{\partial x^2} + \frac{\partial^2 V}{\partial y^2} + \frac{\partial^2 V}{\partial z^2} = 0 \tag{6}$$

Cylindrical:

$$\nabla^2 V = \frac{1}{r_c}\frac{\partial}{\partial r_c}\left(r_c \frac{\partial V}{\partial r_c}\right) + \frac{1}{r_c^2}\frac{\partial^2 V}{\partial \phi^2} + \frac{\partial^2 V}{\partial z^2} = 0 \tag{7}$$

Spherical:

$$\nabla^2 V = \frac{1}{r_s^2}\frac{\partial}{\partial r_s}\left(r_s^2 \frac{\partial V}{\partial r_s}\right) + \frac{1}{r_s^2 \sin\theta}\frac{\partial}{\partial \theta}\left(\sin\theta \frac{\partial V}{\partial \theta}\right) + \frac{1}{r_s^2 \sin^2\theta}\frac{\partial^2 V}{\partial \phi^2} = 0 \tag{8}$$

Equations (7) and (8) are obtained through the use of the divergence and gradient expressions in the cylindrical and spherical coordinate systems, respectively. The Laplacian of f (a scalar field function) in the three coordinate systems is given on the inside of the back cover.

Example 1

Equation (2) is Poisson's equation for a non-homogeneous media. If we allow $\rho_v = 0$ in (2), we obtain

$$\nabla \bullet [\epsilon(\nabla V)] = 0 \tag{9}$$

which is Laplace's equation for a non-homogeneous media. For (9) to reduce to (5), what must $\nabla\epsilon$ equal?

Solution. From Table B-4, we find that $\nabla \bullet (f\overline{A}) = f(\nabla \bullet \overline{A}) + \overline{A} \bullet \nabla f$. Now, let $\epsilon = f$ and $\nabla V = \overline{A}$ to obtain

$$\nabla \bullet [\epsilon(\nabla V)] = \epsilon \nabla \bullet (\nabla V) + \nabla V \bullet \nabla \epsilon = \epsilon \nabla^2 V + \nabla V \bullet \nabla \epsilon = 0$$

To reduce to $\nabla^2 V = 0$, we note that $\nabla \epsilon = 0$ and thus in rectangular coordinates

$$\frac{\partial \epsilon}{\partial x} = \frac{\partial \epsilon}{\partial y} = \frac{\partial \epsilon}{\partial z} = 0$$

This forces ϵ to be uniform as required in a homogeneous medium.

Example 2

The potential field, referenced at r_b, about an infinite length cylinder found along the z axis and of radius r_0 is

$$V = \frac{10^{-2}}{2\pi\epsilon} \ln\left(\frac{r_b}{r_c}\right) \quad (r_0 \leqq r_c \leqq r_b) \qquad \text{(V)}$$

Show that $\rho_v = 0$ in the specified region and that Laplace's equation is satisfied.

Solution. From (4) and (7), we have

$$\begin{aligned}
\nabla^2 V = \frac{-\rho_v}{\epsilon} &= \frac{1}{r_c}\frac{\partial}{\partial r_c}\left(r_c \frac{\partial V}{\partial r_c}\right) + \frac{1}{r_c^2}\frac{\partial^2 V}{\partial \phi^2} + \frac{\partial^2 V}{\partial z^2} \\
&= \frac{1}{r_c}\frac{\partial}{\partial r_c}\left\{r_c \frac{\partial}{\partial r_c}\left[\frac{10^{-2}}{2\pi\epsilon}\ln\left(\frac{r_b}{r_c}\right)\right]\right\} + 0 + 0 \\
&= \frac{1}{r_c}\frac{\partial}{\partial r_c}\left[r_c \frac{10^{-2}}{2\pi\epsilon}\left(-\frac{1}{r_c}\right)\right] = 0
\end{aligned}$$

Thus, $\rho_v = 0$ and Laplace's equation $\nabla^2 V = 0$ is satisfied.

Problem 6.2-1 Between two conducting and parallel planes, the potential function is found to be $V = 20x$ (V) for $0 \leqq x \leqq 3$. In the specified region show that: (a) $\rho_v = 0$; (b) Laplace's equation is satisfied.

Problem 6.2-2 For the following potential fields determine ρ_v, and if Laplace's equation is satisfied: (a) $V = 20x^2$; (b) $V = K_1/r_s$; (c) $V = K_2 r_s$; (d) $V = (K_3 \cos\theta)/r_s^2$; (e) $V = K_4 r_c$.

Problem 6.2-3 Derive (7) and (8) through the use of the divergence and gradient expressions in the cylindrical and spherical coordinate systems, respectively.

6.3 UNIQUENESS THEOREM

In Sec. 6.2, we found several V fields that satisfy Laplace's equation. In fact, we can find an infinite number of V fields that will satisfy Laplace's equation. This situation might indicate to us that we may be able to obtain several different solutions to a particular electrostatic problem, each of which satisfies Laplace's equation. This dilemma does not exist since the V solution of a particular electrostatic problem must not only satisfy Laplace's equation, but, in addition, it must satisfy the values of V at the boundaries of the problem. Each electrostatic problem will have its own set of physical boundaries and known boundary potentials. In a problem containing two infinite and parallel conductors, one conductor in the $z = 0$ plane at $V = 0$ (V) and the other in the $z = d$ plane at $V = V_0$ (V), we will later find that the V field solution between the conductors is $V = V_0 z/d$ (V). This solution will satisfy Laplace's equation and the known boundary potentials at $z = 0$ and $z = d$. Now, the V field solution $V = V_0\,(z + 1)/d$ will satisfy Laplace's equation but will not give the known boundary potentials and thus is not a solution of our particular electrostatic problem. Thus, $V_0 z/d$ is the only solution (unique solution) of our particular problem.

We will now prove that for a V solution of a particular electrostatic problem to be unique, it must satisfy Laplace's equation and the potentials on the boundaries. The requirements of a unique solution are specified by the uniqueness theorem. To prove the uniqueness theorem, we will assume, for a general electrostatic problem as shown in Fig. 6-1, two solutions of Laplace's equation, V_1 and V_2, that are general functions of some coordinate system within the volume v. Thus, $\nabla^2 V_1 = 0$ and $\nabla^2 V_2 = 0$ from which $\nabla^2(V_1 - V_2) = 0$. Let each solution satisfy the given potentials at the boundary, i.e., boundary conditions. Therefore, $V_{1b} = V_{2b}$ at each point on the boundary s_1 and s_2 that encloses the volume v, or $V_{1b} - V_{2b} = 0$. From Table B-4 we obtain the identity

$$\nabla \cdot (f\overline{A}) = f(\nabla \cdot \overline{A}) + \overline{A} \cdot (\nabla f) \tag{1}$$

where f is any scalar function, and $\overline{A}$ is any vector function. In (1), let $f = (V_1 - V_2)$ and $\overline{A} = \nabla(V_1 - V_2)$ to obtain

$$\nabla \cdot [(V_1 - V_2)\nabla(V_1 - V_2)] = (V_1 - V_2)\nabla \cdot \nabla(V_1 - V_2) + \nabla(V_1 - V_2) \cdot \nabla(V_1 - V_2) \tag{2}$$

Let us now integrate (2) over the volume v and simplify to obtain

$$\int_v \nabla \cdot [(V_1 - V_2)\nabla(V_1 - V_2)]\,dv = \int_v (V_1 - V_2)\nabla^2(V_1 - V_2)\,dv + \int_v |\nabla(V_1 - V_2)|^2\,dv \tag{3}$$

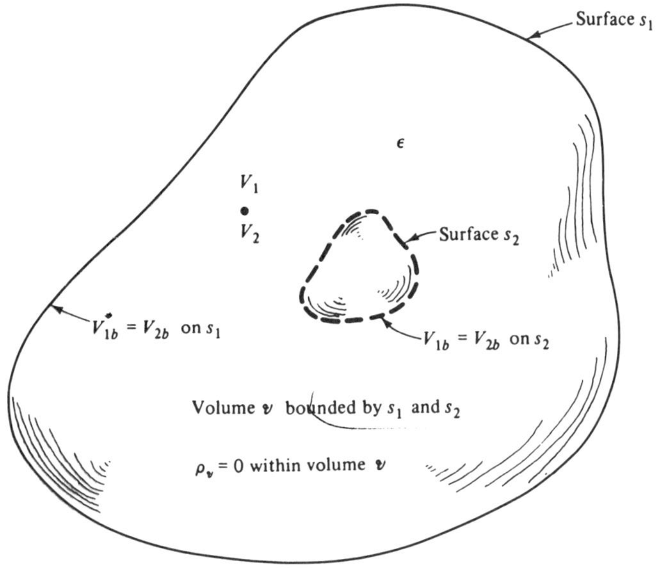

Figure 6-1 Graphical display of a general electrostatic problem for the proof of the uniqueness theorem.

If the divergence theorem is applied to the first integral of (3), we have

$$\oint_s (V_1 - V_2)|_b \nabla(V_1 - V_2)|_b \bullet \overline{ds}$$

where V_1, V_2, and $\nabla(V_1 - V_2)$ must be evaluated at the boundary surface s that consists of s_1 and s_2. This integral can be written as

$$\oint_s (V_{1b} - V_{2b})\left(\frac{\partial V_1}{\partial n} - \frac{\partial V_2}{\partial n}\right)\bigg|_b ds$$

where $\partial V_1/\partial n$ and $\partial V_2/\partial n$ are partial derivatives normal to the boundary. Thus, (3) can be written as

$$\oint_s (V_{1b} - V_{2b})\left(\frac{\partial V_1}{\partial n} - \frac{\partial V_2}{\partial n}\right)\bigg|_b ds = \int_v (V_1 - V_2)\nabla^2(V_1 - V_2)\,dv + \int_v \left|\nabla(V_1 - V_2)\right|^2 dv \tag{4}$$

The first and second integrals are zero since we have hypothesized that $(V_{1b} - V_{2b}) = 0$ and $\nabla^2(V_1 - V_2) = 0$. Equation (4) thus reduces to

$$\int_v \left|\nabla(V_1 - V_2)\right|^2 dv = 0 \tag{5}$$

Since the quantity $|\nabla(V_1 - V_2)|^2$ is always positive and the integral equals zero, we deduce that $|\nabla(V_1 - V_2)|^2 = 0$ and thus $\nabla(V_1 - V_2) = 0$. For $\nabla(V_1 - V_2)$ to equal zero everywhere in volume v, then $(V_1 - V_2)$ cannot vary with coordinate variables and thus $V_1 - V_2 =$ constant. The constant can be evaluated by finding $V_1 - V_2$ at the boundary point, to obtain

$$(V_1 - V_2)|_b = V_{1b} - V_{2b} = 0 = \text{constant} \tag{6}$$

From (6) we see that the constant equals zero, and thus $(V_1 - V_2) = 0$ everywhere within the volume v and on the surface $s = s_1 + s_2$. We have thus proven that V_1 and V_2 are identical solutions if both satisfy Laplace's equation and they have the same potential values at the boundary.

The first integral in (4) will be zero also when $(\partial V_1/\partial n - \partial V_2/\partial n)|_b = 0$. Through the use of this condition, we will find that V_1 and V_2 will be unique within a constant value. Now, if the first integral of (4) is formulated in two parts, one over the surface where $V_{1b} - V_{2b} = 0$ and the other over the remaining surface where $(\partial V_1/\partial n - \partial V_2/\partial n)|_b = 0$, we also will find that $V_1 = V_2$.

The above proof of the uniqueness theorem also applies to Poisson's equation, for if $\nabla^2 V_1 = -\rho_v/\epsilon$ and $\nabla^2 V_2 = -\rho_v/\epsilon$, then $\nabla^2(V_1 - V_2) = 0$ as in the case where $\rho_v = 0$, and the proof is identical from this point.

From the proof of the uniqueness theorem, we can state three problem formulations:

> *Dirichlet Problem.* The potential solution V is unique if it satisfies Laplace's equation and the known V_b on the closed surface s.
>
> *Neumann Problem.* The potential solution V is unique, within a constant, if it satisfies Laplace's equation and the known $\partial V/\partial n|_b$ on the closed surface s.
>
> *Mixed Boundary-Value Problem.* The potential V is unique if it satisfies Laplace's equation, the known V_b on part of surface s, and the known $\partial V/\partial n|_b$ on the remaining closed surface s.

In each of the problem formulations above, the $\overline{E}$ and $\overline{D}$ will also be unique within the volume v and on the closed surface s. In our solutions of Laplace's and Poisson's equations, we will involve ourselves only with the Dirichlet and Neumann problem formulations.

Example 3

Two infinite length concentric and conducting cylinders of radii r_a and r_b are located on the z axis, as in Fig. 5-21. The potential $V = 30$ V at r_a and $V = 0$ V at r_b when $\rho_v = 0$ for $r_a < r_c < r_b$. Show that $V = 30 \ln(r_b/r_c)/\ln(r_b/r_a)$, for $r_a \leqq r_c \leqq r_b$ is a unique solution for the given particular electrostatic problem. Note that r_c is a variable.

Solution. Since $\rho_v = 0$ in the range $r_a < r_c < r_b$, Laplace's equation must be satisfied within this range as well as the given boundary potentials at r_b and r_a. In cylindrical coordinates,

$$\nabla^2 V = \frac{1}{r_c}\frac{\partial}{\partial r_c}\left(r_c\frac{\partial V}{\partial r_c}\right) + \frac{1}{r_c^2}\frac{\partial^2 V}{\partial \phi^2} + \frac{\partial^2 V}{\partial z^2}$$

$$\therefore \quad \nabla^2 V = \frac{1}{r_c}\frac{\partial}{\partial r_c}\left\{r_c\frac{\partial}{\partial r_c}\left[\frac{30\ln(r_b/r_c)}{\ln(r_b/r_a)}\right]\right\}$$

$$= \frac{1}{r_c}\frac{\partial}{\partial r_c}\left\{r_c\frac{-30}{r_c\ln(r_b/r_a)}\right\} = 0 \qquad (7)$$

Now

$$V\Big|_{r_b} = \frac{30 \ln (r_b/r_c)}{\ln (r_b/r_a)}\Bigg|_{r_c=r_b} = 0 \tag{8}$$

Also,

$$V\Big|_{r_a} = \frac{30 \ln (r_b/r_c)}{\ln (r_b/r_a)}\Bigg|_{r_c=r_a} = 30 \tag{9}$$

Thus, from (7), (8), and (9), we can see that Laplace's equation and potentials at the boundaries are satisfied. The uniqueness theorem thus tells us that the given solution is unique.

Problem 6.3-1 Is

$$V = \frac{30 \ln (r_b/r_c)}{\ln (r_b/r_a)} + 20 \ln\left(\frac{r_b}{r_c}\right)$$

a solution to the electrostatic problem in Example 3? Explain.

Problem 6.3-2 Can $V = K/r_c$, $r_a \le r_c \le r_b$, be a potential function for the potential between two concentric cylinders having a dielectric between the surfaces at $r_c = r_a$ and at $r_c = r_b$?

6.4 SOLUTION OF LAPLACE'S EQUATION IN ONE VARIABLE

Now that we are familiar with uniqueness theorem, we shall embark on the solution of Laplace's equation in one variable. The solution of Laplace's equation in one variable can be readily obtained by direct integration. In the next section we will try our hand at a two-variable solution only in the rectangular coordinate system.

Example 4

Two infinite and parallel conducting planes are separated d (m), with one of the conductors in the $z = 0$ plane at $V = 0$ (V) and the other in the $z = d$ plane at $V = V_0$ (V). Assume $\rho_v = 0$ and $\epsilon = 2\epsilon_0$ between the conductors. Find: (a) V in the range $0 < z < d$; (b) $\overline{E}$ between the conductors; (c) $\overline{D}$ between the conductors; (d) D_n on the conductors; (e) ρ_s on the conductors; (f) capacitance per square meter.

Solution. (a) Since $\rho_v = 0$, Laplace's equation applies in the region between the conductors. Now, due to rectangular symmetry in the stated problem, we will solve Laplace's equation in rectangular form; thus,

$$\nabla^2 V = \frac{\partial^2 V}{\partial x^2} + \frac{\partial^2 V}{\partial y^2} + \frac{\partial^2 V}{\partial z^2} = 0 \tag{1}$$

From the problem specification, we note that V will be a function of z only, $V = V(z)$; thus, Laplace's equation will reduce to

$$\nabla^2 V = \frac{\partial^2 V}{\partial z^2} = 0 \tag{2}$$

The partial derivative can be replaced by an ordinary derivative since V is a function of z only; thus,

$$\frac{d^2 V}{dz^2} = \frac{d}{dz}\left(\frac{dV}{dz}\right) = 0 \tag{3}$$

Integrating twice, we obtain

$$\frac{dV}{dz} = A \tag{4}$$

and

$$V = Az + B \tag{5}$$

where A and B are constants of integration that must be evaluated using the given potential values at the boundaries, $V = 0$ at $z = 0$ and $V = V_0$ at $z = d$. Applying these potential boundary values (boundary conditions), we have

$$V\big|_{z=0} = 0 = B \tag{6}$$

and

$$V\big|_{z=d} = V_0 = Ad \tag{7}$$

Thus, from (6) and (7), $B = 0$ and $A = V_0/d$. Equation (5) thus becomes

$$V = \frac{V_0 z}{d} \quad (0 \leqq z \leqq d) \qquad \text{(V)} \tag{8}$$

It should be noted that (5) is a general solution for V for problems involving conductors in $z =$ constant planes. Equation (8) is a unique solution for the stated problem and its boundary conditions.

(b)
$$E = -\nabla V = -\left(\hat{x}\frac{\partial V}{\partial x} + \hat{y}\frac{\partial V}{\partial y} + \hat{z}\frac{\partial V}{\partial z}\right)$$
$$= -\hat{z}\frac{\partial}{\partial z}\left(\frac{V_0 z}{d}\right) = -\hat{z}\frac{V_0}{d} \qquad (\text{V m}^{-1}) \tag{9}$$

(c)
$$\overline{D} = \epsilon\overline{E} = 2\epsilon_0\left(-\hat{z}\frac{V_0}{d}\right) = -\hat{z}\left(\frac{2\epsilon_0 V_0}{d}\right) \qquad (\text{C m}^{-2}) \tag{10}$$

(d)
$$D_n\big|_{z=0} = (\overline{D} \bullet \hat{z})\big|_{z=0} = -\frac{2\epsilon_0 V_0}{d} \qquad (\text{C m}^{-2}) \tag{11}$$

$$D_n\big|_{z=d} = [\overline{D} \bullet (-\hat{z})]\big|_{z=d} = \frac{2\epsilon_0 V_0}{d} \qquad (\text{C m}^{-2}) \tag{12}$$

(e)
$$\rho_s\big|_{z=0} = D_n\big|_{z=0} = -\frac{2\epsilon_0 V_0}{d} \qquad (\text{C m}^{-2}) \tag{13}$$

$$\rho_s\big|_{z=d} = D_n\big|_{z=d} = \frac{2\epsilon_0 V_0}{d} \qquad (\text{C m}^{-2}) \tag{14}$$

(f)
$$\text{C/m}^2 = \frac{|Q/m^2|}{|V_{ab}|} = \frac{|\rho_s|}{V_0} = \frac{2\epsilon_0(V_0/d)}{V_0} = \frac{\epsilon}{d} \qquad (\text{F m}^{-2}) \tag{15}$$

where C m^{-2} is the capacitance per square meter.

If the conductors were located in the $x =$ constant planes or in the $y =$ constant planes, the solutions would involve $\partial^2 V/\partial x^2 = 0$ and $\partial^2 V/\partial y^2 = 0$, respectively, and would parallel those of Example 4.

Example 5

Two infinite length, concentric, and conducting cylinders of radii r_a and r_b are located on the z axis, as in Fig. 5-21. If $\epsilon = 3\epsilon_0$, $\rho_v = 0$ between the cylinders, $V = V_0(V)$ at r_a, $V = 0$ at r_b, and $r_b > r_a$ find: (a) V in the range $r_a < r_c < r_b$; (b) $\overline{E}$ between the conductors; (c) $\overline{D}$ between the conductors; (d) D_n on the conductors; (e) ρ_s on both conductors; (f) ρ_ℓ on both conductors; (g) capacitance per meter length.

Solution. (a) In the range $r_a < r_c < r_b$ we find $\rho_v = 0$; thus, Laplace's equation applies. The cylindrical coordinate system will be used since the problem has cylindrical symmetry; thus,

$$\nabla^2 V = \frac{1}{r_c}\frac{\partial}{\partial r_c}\left(r_c\frac{\partial V}{\partial r_c}\right) + \frac{1}{r_c^2}\frac{\partial^2 V}{\partial \phi^2} + \frac{\partial^2 V}{\partial z^2} = 0 \tag{16}$$

From the problem we note that V will be a function of r_c only, $V = V(r_c)$; thus, Laplace's equation will reduce to

$$\nabla^2 V = \frac{1}{r_c}\frac{\partial}{\partial r_c}\left(r_c\frac{\partial V}{\partial r_c}\right) = 0 \tag{17}$$

Excluding $r_c = 0$, multiplying by r_c, and writing in ordinary derivative form, (17) becomes

$$\frac{d}{dr_c}\left(r_c\frac{dV}{dr_c}\right) = 0 \tag{18}$$

Integrating (18), we have

$$r_c\frac{dV}{dr_c} = A \tag{19}$$

Dividing by r_c and integrating again, we get

$$V = A\,\ln(r_c) + B \tag{20}$$

Applying the given boundary conditions, we have

$$V\big|_{ra} = V_0 = A\,\ln(r_a) + B \tag{21}$$

and

$$V\big|_{rb} = 0 = A\,\ln(r_b) + B \tag{22}$$

Solving for A and B in (21) and (22), we have

$$A = \frac{V_0}{\ln(r_a/r_b)} \tag{23}$$

and

$$B = \frac{-V_0\,\ln(r_b)}{\ln(r_a/r_b)} \tag{24}$$

Equation (20) thus becomes

$$V = V_0\frac{\ln(r_b/r_c)}{\ln(r_b/r_a)} \quad (r_a \leqq r_c \leqq r_b) \quad \text{(V)} \tag{25}$$

(b) $$\overline{E} = -\nabla V = -\hat{r}_c\frac{\partial V}{\partial r_c} = \frac{V_0\hat{r}_c}{r_c\,\ln(r_b/r_a)} \quad (\text{V m}^{-1}) \tag{26}$$

(c) $$\overline{D} = \epsilon\overline{E} = \frac{\epsilon V_0\hat{r}_c}{r_c\,\ln(r_b/r_a)} \quad (\text{C m}^{-2}) \tag{27}$$

(d) $$D_n\big|_{ra} = (\overline{D}\bullet\hat{r}_c)\big|_{ra} = \frac{\epsilon V_0}{r_c\,\ln(r_b/r_a)} \quad (\text{C m}^{-2}) \tag{28}$$

$$D_n\big|_{rb} = [\overline{D}\bullet(-\hat{r}_c)]\big|_{rb} = \frac{-\epsilon V_0}{r_b\,\ln(r_b/r_a)} \quad (\text{C m}^{-2}) \tag{29}$$

(e) $$\rho_s\big|_{ra} = D_n\big|_{ra} = \frac{\epsilon V_0}{r_a\,\ln(r_b/r_a)} \quad (\text{C m}^{-2}) \tag{30}$$

$$\rho_s\big|_{rb} = D_n\big|_{rb} = -\frac{\epsilon V_0}{r_b\,\ln(r_b/r_a)} \quad (\text{C m}^{-2}) \tag{31}$$

(f) $$\rho_\ell\big|_{ra} = \rho_s\big|_{ra}(2\pi r_a) = \frac{2\pi\epsilon V_0}{\ln(r_b/r_a)} \quad (\text{C m}^{-1}) \tag{32}$$

$$\rho_\ell\big|_{rb} = \rho_s\big|_{rb}(2\pi r_b) = -\frac{2\pi\epsilon V_0}{\ln(r_b/r_a)} \quad (\text{C m}^{-1}) \tag{33}$$

(g) $$\text{C/m} = \frac{|\rho_\ell|}{|V_{ab}|} = \frac{|\rho_\ell|}{V_0} = \frac{2\pi\epsilon}{\ln(r_b/r_a)} \quad (\text{F m}^{-1}) \tag{34}$$

which agrees with (5.10-10) for L = 1.

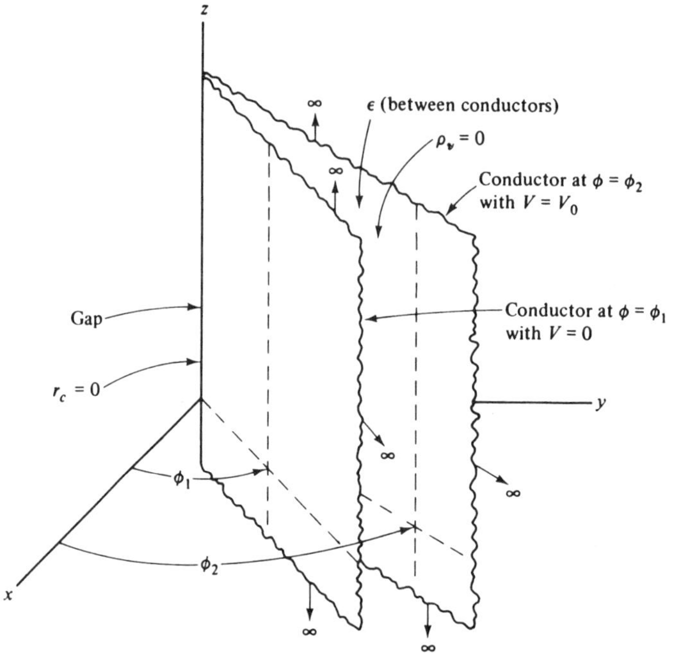

Figure 6-2 Graphical display of two infinite conductor planes, one in the $\phi = \phi_1$ (constant) plane and the other in the $\phi = \phi_2$ (constant) plane, for finding the potential $V = V(\phi)$ between two conductors in Example 6. A very small gap exists between the two conductors at $r_c = 0$, along the z axis.

An example problem where V is a function of ϕ only, $V = V(\phi)$, is shown in Fig. 6-2.

Example 6

Two infinite conductors form a wedge region, $\phi_1 \leqq \phi \leqq \phi_2$, as shown in Fig. 6-2. If this region is characterized by $\rho_v = 0$ and $\epsilon = \epsilon_0$ while $V = 0$ at $\phi = \phi_1$ and $V = V_0$ at $\phi = \phi_2$, find: (a) V in the range $\phi_1 < \phi < \phi_2$; (b) $\overline{E}$ between the conductors; (c) $\overline{D}$ between conductors; (d) D_n on both conductors; (e) ρ_s on both conductors.

Solution. (a) Since the problem has cylindrical symmetry and V is a function of ϕ only, $V = V(\phi)$, Laplace's equation reduces to

$$\nabla^2 V = \frac{1}{r_c^2}\frac{\partial^2 V}{\partial \phi^2} = 0 \tag{35}$$

Excluding $r_c = 0$, multiplying by r_c^2, and writing in ordinary derivative form, we have

$$\frac{d^2V}{d\phi^2} = 0 \tag{36}$$

Integrating twice yields

$$V = A\phi + B \tag{37}$$

Solving for A and B through the application of boundary conditions yields

$$V = \frac{V_0(\phi - \phi_1)}{\phi_2 - \phi_1} \quad (\phi_1 \leqq \phi \leqq \phi_2) \qquad \text{(V)} \tag{38}$$

(b) $$\overline{E} = -\nabla V = -\frac{\hat{\phi}}{r_c}\frac{\partial V}{\partial \phi} = -\hat{\phi}\frac{V_0}{r_c(\phi_2 - \phi_1)} \quad (\text{V m}^{-1}) \tag{39}$$

(c) $$\overline{D} = \epsilon\overline{E} = -\hat{\phi}\frac{\epsilon V_0}{r_c(\phi_2 - \phi_1)} \quad (\text{C m}^{-2}) \tag{40}$$

(d) $$D_n|_{\phi=\phi_1} = (\overline{D} \cdot \hat{\phi})|_{\phi=\phi_1} = -\frac{\epsilon V_0}{r_c(\phi_2 - \phi_1)} \quad (\text{C m}^{-2}) \tag{41}$$

$$D_n|_{\phi=\phi_2} = [\overline{D} \cdot (-\hat{\phi})]|_{\phi=\phi_2} = \frac{\epsilon V_0}{r_c(\phi_2 - \phi_1)} \quad (\text{C m}^{-2}) \tag{42}$$

(e) $$\rho_s|_{\phi=\phi_1} = D_n|_{\phi=\phi_1} = -\frac{\epsilon V_0}{r_c(\phi_2 - \phi_1)} \quad (\text{C m}^{-2}) \tag{43}$$

$$\rho_s|_{\phi=\phi_2} = D_n|_{\phi=\phi_2} = \frac{\epsilon V_0}{r_c(\phi_2 - \phi_1)} \quad (\text{C m}^{-2}) \tag{44}$$

From $\rho_\ell = \int_0^1 \int_{r_c=0}^{r_c=\infty} \rho_s|_{\phi=\phi_2} dr_c\, dz$, we find that $\rho_\ell = \infty$ (C m^{-1}), and thus the capacitance per unit length along the z direction will be infinite since r_c ranges from 0 to ∞. It should be noted that ρ_s becomes infinite as $r_c \to 0$. The only other remaining cylindrical variable is z.

Problems involving V as a function of z only have already been discussed in Example 4.

In the spherical coordinate system, we can generate problems where $V = V(r_s)$, $V = V(\theta)$, and $V = V(\phi)$. In Example 6, we have covered the case when $V = V(\phi)$; thus, we have only two remaining solutions of one variable to investigate.

Example 7

Two concentric conducting shells are found centered at the origin. The outer radius of the inner shell is r_a, and the inner radius of the outer shell is r_b. If the region between the two shells is characterized by $\rho_v = 0$, ϵ, $V = 0$ at r_a, and $V = V_0$ at r_b, find V between the two shells, $r_a < r_s < r_b$.

Solution. Since the problem has spherical symmetry and V is a function of r_s only, $V = V(r_s)$, Laplace's equation reduces to

$$\nabla^2 V = \frac{1}{r_s^2}\frac{\partial}{\partial r_s}\left(r_s^2\frac{\partial V}{\partial r_s}\right) = 0 \tag{45}$$

Excluding $r_s = 0$, multiplying by r_s^2, and writing (45) in ordinary derivative form, we have

$$\frac{d}{dr_s}\left(r_s^2\frac{dV}{dr_s}\right) = 0 \tag{46}$$

Integrating twice yields

$$V = -\frac{A}{r_s} + B \tag{47}$$

Solving for A and B through the application of boundary conditions yields

$$V = \frac{V_0 r_b(r_s - r_a)}{r_s(r_b - r_a)} \quad (r_a \leqq r_s \leqq r_b) \quad (\text{V}) \tag{48}$$

Our last single-variable solution of Laplace's equation, in spherical coordinates, is graphically displayed in Fig. 6-3, where V is found to be a function of θ only.

Example 8

Consider two infinite and conducting cones, both on the z axis, one in the $\theta = \theta_1$ (constant) cone and the other in the $\theta = \theta_2$ (constant) cone. If the region between is characterized by ϵ, $\rho_v = 0$, $V = 0$ at θ_1, and $V = V_0$ at θ_2, find V between the two cones, $\theta_1 < \theta < \theta_2$.

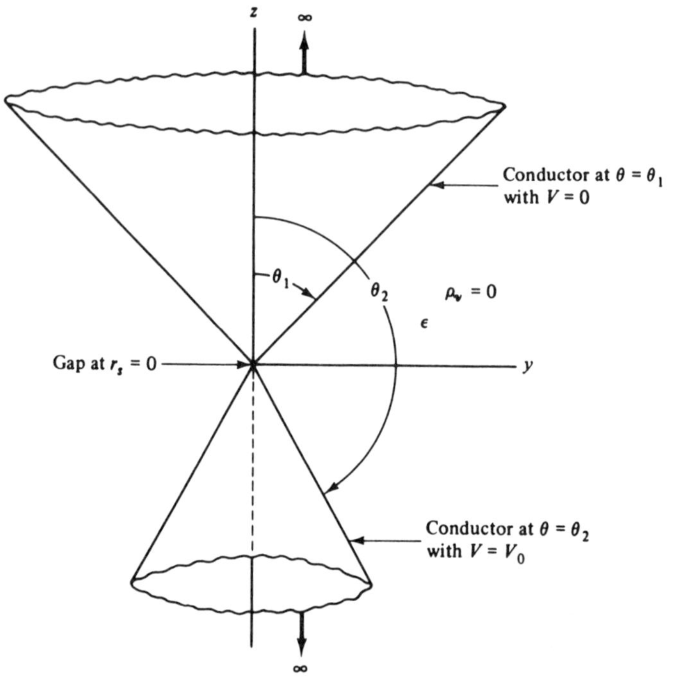

Figure 6-3 Graphical display of two infinite and conducting cones, both on the z axis, one in the $\theta = \theta_1$ (constant) cone and the other in the $\theta = \theta_2$ (constant) cone, for finding the potential $V = V(\theta)$ between the two cones in Example 8. A very small gap exists at the cone vertex at $r_s = 0$.

Solution. Since the problem has spherical symmetry and V is a function of θ only, $V = V(\theta)$, Laplace's equation reduces to

$$\nabla^2 V = \frac{1}{r_s^2 \sin\theta}\frac{\partial}{\partial\theta}\left(\sin\theta\frac{\partial V}{\partial\theta}\right) = 0 \tag{49}$$

Excluding $r_s = 0$ and $\theta = 0$ or π, and multiplying by $r_s^2 \sin\theta$, we obtain

$$\frac{d}{d\theta}\left(\sin\theta\frac{dV}{d\theta}\right) = 0 \tag{50}$$

Integrating once, we obtain

$$\sin\theta\frac{dV}{d\theta} = A$$

and

$$\frac{dV}{d\theta} = \frac{A}{\sin\theta} \tag{51}$$

Integrating again, we obtain

$$V = \int \frac{A}{\sin\theta}\,d\theta + B \tag{52}$$

From a table of integrals,

$$\int \frac{d\theta}{\sin\theta} = \ln\left(\tan\frac{\theta}{2}\right) \tag{53}$$

Thus, (52) becomes

$$V = A \ln\left(\tan\frac{\theta}{2}\right) + B \tag{54}$$

Applying boundary conditions to solve for A and B, we have

$$V = V_0 \frac{\ln\left[\dfrac{\tan(\theta/2)}{\tan(\theta_1/2)}\right]}{\ln\left[\dfrac{\tan(\theta_2/2)}{\tan(\theta_1/2)}\right]} \quad (\theta_1 \leqq \theta \leqq \theta_2) \qquad \text{(V)} \tag{55}$$

The $\overline{E}$ between the cones can be found from $\overline{E} = -\nabla V$; thus,

$$\overline{E} = -\frac{\hat{\theta}}{r_s}\left(\frac{\partial V}{\partial \theta}\right)$$

From (51), $\partial V/\partial\theta = A/\sin\theta$; thus,

$$\overline{E} = -\hat{\theta}\frac{A}{r_s \sin\theta} = -\hat{\theta}\frac{V_0}{r_s \sin\theta \ln\left[\dfrac{\tan(\theta_2/2)}{\tan(\theta_1/2)}\right]} \tag{56}$$

The details of this development are left as a problem at the end of this section for the student (Prob. 6.4-4).

We will conclude single-variable solutions of Laplace's equation by a solution of an electrostatic problem that contains two different dielectrics, as found in the capacitor of Fig. 5-22. In such a problem a potential is found in each dielectric region, and the two sets of constants in the general solutions are evaluated by applying boundary conditions on V at all boundaries and on normal components of $\overline{D}$ at the dielectric-dielectric boundary. This method can be extended to any number of dielectrics, with the amount of work growing extremely rapidly when more than two dielectrics are involved.

Example 9

Through the use of Laplace's equation, find the expressions for V_1 and V_2, in regions 1 and 2, respectively, as found in the two-dielectric capacitor of Fig. 5-22. Let $V = V_0$ at $z = \ell$ and $V = 0$ at $z = 0$. Assume that ℓ is very small compared to the width and depth so that we can assume infinite plates and infinite dielectric slabs.

Solution. In region 1, we have

$$\nabla^2 V_1 = \frac{\partial^2 V_1}{\partial z^2} = 0 \tag{57}$$

Thus,

$$V_1 = A_1 z + A_2 \tag{58}$$

In region 2, we have

$$\nabla^2 V_2 = \frac{\partial^2 V_2}{\partial z^2} = 0 \tag{59}$$

Thus,

$$V_2 = A_3 z + A_4 \tag{60}$$

Now solve for A_1, A_2, A_3, and A_4 through the application of the following boundary conditions:

$$V_1\big|_{z=0} = 0 \tag{61}$$

$$V_2\big|_{z=\ell} = V_0 \tag{62}$$

$$V_1\big|_{z=\ell_1} = V_2\big|_{z=\ell_1} \tag{63}$$

$$-\epsilon_1\frac{\partial V_1}{\partial z}\bigg|_{z=\ell_1} = -\epsilon_2\frac{\partial V_2}{\partial z}\bigg|_{z=\ell_1} \tag{64}$$

Equation (64) states the boundary on the normal components of $\overline{D}$ at the dielectric-dielectric boundary, $D_{1n}\big|_{z=\ell_1} = D_{2n}\big|_{z=\ell_1}$. From (61), (62), (63), and (64), we have

$$0 = A_1(0) + A_2 \tag{65}$$

$$V_0 = A_3\ell + A_4 \tag{66}$$

$$A_1\ell_1 + A_2 = A_3\ell_1 + A_4 \tag{67}$$

$$-\epsilon_1 A_1 = -\epsilon_2 A_3 \tag{68}$$

Solving for A_1, A_2, A_3, and A_4, we find that $A_2 = 0$ and

$$A_1 = \frac{V_0\epsilon_2}{\epsilon_2\ell_1 + \epsilon_1\ell_2}, \qquad A_3 = \frac{V_0\epsilon_1}{\epsilon_2\ell_1 + \epsilon_1\ell_2}, \qquad A_4 = V_0 - \frac{V_0\epsilon_1\ell}{\epsilon_2\ell_1 + \epsilon_1\ell_2}$$

Thus, V_1 and V_2 become

$$V_1 = \left(\frac{V_0\epsilon_2}{\epsilon_2\ell_1 + \epsilon_1\ell_2}\right)z \quad (0 \leqq z \leqq \ell_1) \tag{69}$$

$$V_2 = \left(\frac{V_0\epsilon_1}{\epsilon_2\ell_1 + \epsilon_1\ell_2}\right)z + V_0 - \left(\frac{V_0\epsilon_1\ell}{\epsilon_2\ell_1 + \epsilon_1\ell_2}\right) \quad (\ell_1 \leqq z \leqq \ell) \tag{70}$$

The solution set, V_1 and V_2, satisfies Laplace's equation and the boundary conditions; thus, it represents a unique solution set for the given problem.

Problem 6.4-1 A parallel-plate capacitor has a separation of 0.5 mm between plates and a potential difference between plates of 50 V. If $\rho_v = 0$ and $\epsilon = 4\epsilon_0$ between plates, find: (a) $V(y)$ $(0 \leqq y \leqq 0.5\text{ mm})$ if the plate at $y = 0$ is 25 V and at $y = 0.5$ mm is 75 V; (b) $\overline{E}$ between plates; (c) capacitance per square meter.

Problem 6.4-2 Two infinite length, concentric, and conducting cylinders of radii $r_a = 0.02$ m and $r_b = 0.05$ m are located with axes on the z axis. If $\epsilon = 4\epsilon_0$, $\rho_v = 0$ between the cylinders, $V = 50$ V at r_a, $V = 100$ V at r_b, find: (a) V in the range $0.02 \leq r_c \leq 0.05$; (b) $\overline{E}$; (c) $\overline{D}$; (d) ρ_s at r_b; (e) capacitance per meter length.

Problem 6.4-3 Two infinite and radial planes are separated by a small gap along the z axis. One of the planes is in the $\phi = 0$ plane at $V = 100$ V while the other is in the $\phi = \pi/2$ plane at $V = 200$ V. If $\epsilon = 2\epsilon_0$ and $\rho_v = 0$ between the planes, find: (a) V in the range $0 \leqq \phi \leqq \pi/2$; (b) $\overline{E}$; (c) $\overline{D}$; (d) ρ_s at $r_c = 2$ m on the plane at $\phi = \pi/2$.

Problem 6.4-4 Solve for the constants A and B in (54) of Example 8 and obtain: (a) eq. (55); (b) eq. (56); (c) expression for ρ_s on $\theta = \theta_1$, (constant) cone.

Problem 6.4-5 Using eqs. (4.3-1) and (4.5-13), and the boundary conditions, derive eqs. (69) and (70) of Example 9 through use of definite integrals.

Problem 6.4-6 For the problem of Example 9, using (58) and (60) show that: (a) $\overline{E}_1 = -\hat{z}A_1$; (b) $\overline{E}_2 = -\hat{z}A_3$; (c) $\rho_s\big|_{z=\ell} = \epsilon_2 A_3$; (d) $\rho_s\big|_{z=0} = -\epsilon_1 A_1$; (e) $C = \epsilon_2 A_3 s/V_0 = \epsilon_2\epsilon_1 s/(\epsilon_2\ell_1 + \epsilon_1\ell_2)$, which can be placed in the form

$$C = \frac{1}{1/C_1 + 1/C_2}$$

[eq. (5.10-14)].

6.5 SOLUTION OF LAPLACE'S EQUATION IN TWO VARIABLES—RECTANGULAR HARMONICS

We will now extend our solution of Laplace's equation in two variables. To keep the mathematics within our reach, we shall work only in the rectangular coordinate system in this section. We will look briefly at cylindrical coordinates in the next section, but in general, for solutions of Laplace's equation in three variables and those of two and three variables in the cylindrical and spherical coordinate systems, the student is referred to more advanced texts in electromagnetics.[1,2]

The method of solution that we shall use to solve Laplace's equation in two variables is called *separation of variables* or *product solution*.

Example 10

A trough of infinite length is constructed of thin conducting sheets whose cross section is shown in Fig. 6-4. A conducting lid is placed at $x = a$, with small gaps between the lid and the trough. If $\rho_v = 0$ within the trough, $V = V_0$ on the lid, and $V = 0$ on the three trough walls, find the expression for V within the trough.

Solution. From Fig. 6-4 it can be seen that V is a function of x and y only since the cross section is uniform and the trough is infinite in the z direction; thus, $V = V(x, y)$. Laplace's equation, in rectangular coordinates, will thus reduce to

$$\nabla^2 V = \frac{\partial^2 V}{\partial x^2} + \frac{\partial^2 V}{\partial y^2} = 0 \tag{1}$$

We now assume that the potential V can be expressed as a product of two functions

$$\boxed{V = XY} \tag{2}$$

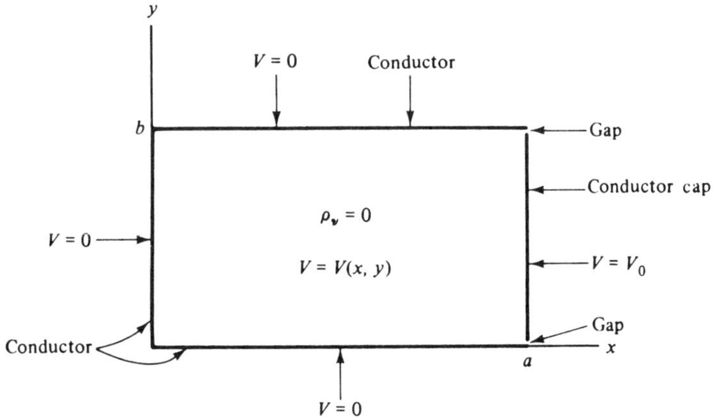

Figure 6-4 Infinite length conducting trough of uniform cross section, with an applied conducting cap at $x = a$ for Example 10.

[1] E. Weber, *Electromagnetic Fields,* Vol. 1, John Wiley & Sons, Inc., New York, 1950.

[2] W. R. Smythe, *Static and Dynamic Electricity,* 3rd ed., McGraw-Hill Book Company, New York, 1968.

where X is a function of x only, and Y is a function of y only. Let us substitute this assumed form of solution into (1) and obtain

$$Y\frac{\partial^2 X}{\partial x^2} + X\frac{\partial^2 Y}{\partial y^2} = 0 \tag{3}$$

Dividing (3) by XY and changing to ordinary derivatives, we have

$$\frac{1}{X}\frac{d^2X}{dx^2} + \frac{1}{Y}\frac{d^2Y}{dy^2} = 0 \tag{4}$$

where $(1/X)\ (d^2X/dx^2)$ seems to be a function of x or equal to a constant, and $(1/Y) \cdot (d^2Y/dy^2)$ seems to be a function of y or equal to a constant. Now, since (4) at this point of development must be true for any x and y values, we deduce that

$$\boxed{\frac{1}{X}\frac{d^2X}{dx^2} = A_1^2 \quad \text{(constant)}} \tag{5}$$

$$\boxed{\frac{1}{Y}\frac{d^2Y}{dy^2} = A_2^2 \quad \text{(constant)}} \tag{6}$$

Thus, (4) can be written as

$$\boxed{A_1^2 + A_2^2 = 0} \tag{7}$$

Equation (7) is commonly referred to as the *separation equation*. We now see that the solution of (1) can be obtained by solving two ordinary differential equations, (5) and (6), and the separation equation (7). Once we have X and Y, $V = XY$ from (2). From (7), $-A_1^2 = A_2^2$; thus, (5) and (6) can be written as[3]

$$\frac{d^2X}{dx^2} - A_1^2X = 0 \tag{8}$$

$$\frac{d^2Y}{dy^2} + A_1^2Y = 0 \tag{9}$$

The solutions of (8) and (9) are

$$X = B_1 \cosh A_1x + B_2 \sinh A_1x \tag{10}$$

or

$$X = B_1'e^{-A_1x} + B_2'e^{A_1x} \tag{11}$$

and

$$Y = B_3 \cos A_1y + B_4 \sin A_1y \tag{12}$$

or

$$Y = B_3'e^{-jA_1y} + B_4'e^{+jA_1y} \tag{13}$$

[3]We could also write $d^2X/dx^2 + A_1x = 0$ and $d^2Y/dy^2 - A_1y = 0$, but for the boundary conditions of Example 10, (8) and (9) turn out to be far more convenient.

Any combination of solutions from (10) or (11) and (12) or (13) can be used as a solution $V = XY$. Let us select the solutions (10) and (12) since they are the most convenient form to satisfy the boundary conditions of the problem; thus,

$$\begin{aligned} V = XY &= V(x, y) \\ &= (B_1 \cosh A_1 x + B_2 \sinh A_1 x)(B_3 \cos A_1 y + B_4 \sin A_1 y) \end{aligned} \tag{14}$$

Equation (14) is the general solution of our partial differential equation of second order and two variables, (1).

The constants B_1, B_2, B_3, and B_4 will now be evaluated from the boundary conditions on our particular problem. They are

$$\begin{aligned} V &= 0, & x &= 0, & (0 &\leqq y \leqq b) & \text{(I)} \\ V &= V_0, & x &= a, & (0 &< y < b) & \text{(II)} \\ V &= 0, & y &= 0, & (0 &\leqq x \leqq a) & \text{(III)} \\ V &= 0, & y &= b, & (0 &\leqq x \leqq a) & \text{(IV)} \end{aligned}$$

From boundary condition (III) and eq. (14), we note that B_3 must be zero. Thus,

$$V = V(x, y) = (B_1 \cosh A_1 x + B_2 \sinh A_1 x)(B_4 \sin A_1 y) \tag{15}$$

From boundary condition (IV) and Eq. (15), we find that

$$\sin A_1 y \,|_{y=b} = 0 \tag{16}$$

Thus,

$$A_1 y \,|_{y=b} = m\pi \tag{17}$$

where $m = 0, 1, 2, \ldots,$ and

$$A_1 = \frac{m\pi}{b} \tag{18}$$

Now (15) can be expressed as

$$V = V(x, y) = \left[B_1 \cosh\left(\frac{m\pi}{b}x\right) + B_2 \sinh\left(\frac{m\pi}{b}x\right)\right]\left[B_4 \sin\left(\frac{m\pi}{b}y\right)\right] \tag{19}$$

Boundary condition (I) and Eq. (19) require that $B_1 = 0$; thus,

$$V = V(x, y) = B \sinh\left(\frac{m\pi}{b}x\right) \sin\left(\frac{m\pi}{b}y\right) \tag{20}$$

where $B = B_2 B_4$.

Our last boundary condition (II) at $x = a$ is displayed in Fig. 6-5(a) as it varies with y. For mathematical reasons that shall unfold shortly, we have extended the boundary condition beyond our problem range, $0 \leqq y \leqq b$, to form a periodic function $f(y)$. From (20), we find that $V(a, y)$ becomes

$$V(a, y) = B \sinh\left(\frac{m\pi}{b}a\right) \sin\left(\frac{m\pi}{b}y\right) = C_m \sin\left(\frac{m\pi}{b}y\right) \tag{21}$$

where $C_m = B \sinh(m\pi a/b)$. Now $V(a, y)$ must equal V_0 in the range $0 < y < b$ to satisfy the boundary condition at $x = a$. The solution (21) for values of $m = 1$, 2, and 3 is plotted in Fig. 6-5(b) between the walls of the trough and beyond. It can be seen that a single solution when $m = 1$, 2, 3, etc., cannot satisfy the boundary condition at $x = a$. To construct the rectangular form of $V = V_0$ or $f(y)$, we need an infinite number of solutions of the type found in (21). Due to symmetry about $y = b/2$, it can be seen that m must be odd, $m = 1$, 3, 5, etc. The solution for $V(x, y)$ must be an infinite series and (20) becomes

$$V(x, y) = \sum_{m=1}^{m=\infty} B \sinh\left(\frac{m\pi}{b}x\right) \sin\left(\frac{m\pi}{b}y\right) \tag{22}$$

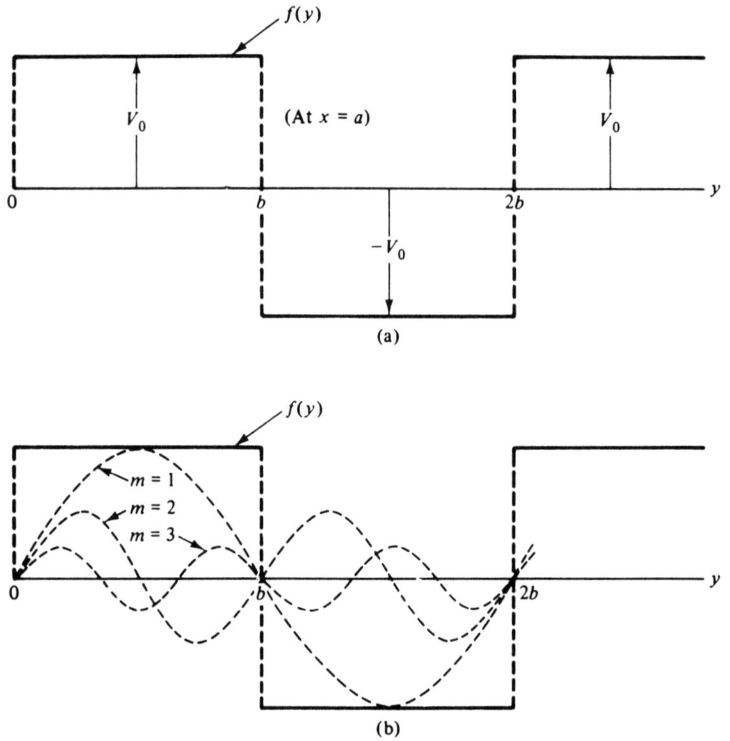

Figure 6-5 (a) Graph of the $V(a, y)$ boundary condition at $x = a$ versus y for Example 10. (b) Display of solutions between the walls of the trough when $m = 1, 2,$ and 3.

From this, $V(a, y)$ becomes

$$V(a,y) = \sum_{m=1}^{m=\infty} B \sinh\left(\frac{m\pi}{b}a\right) \sin\left(\frac{m\pi}{b}y\right) = \sum_{m=1}^{m=\infty} C_m \sin\left(\frac{m\pi}{b}y\right) \tag{23}$$

Equation (23) is an infinite Fourier sine series, and the constant C_m must be evaluated by the Fourier series method that requires $V(a, y)$ to be a periodic function of y. If we extend our boundary condition beyond $y = b$ and form a periodic function $f(y)$, a rectangular waveform, we can evaluate C_m. Thus,

$$f(y) = \sum_{m=1}^{m=\infty} C_m \sin\left(\frac{m\pi}{b}y\right) \tag{24}$$

From Fourier series methods, we have

$$C_m = \frac{2}{(\text{period})} \int_0^{\text{period}} f(y) \sin\left(\frac{m\pi}{b}y\right) dy \tag{25}$$

where the period is $2b$, as seen in Fig. 6-5. Thus,

$$C_m = \frac{2}{2b}\left[\int_0^b V_0 \sin\left(\frac{m\pi}{b}y\right) dy + \int_b^{2b} (-V_0) \sin\left(\frac{m\pi}{b}y\right) dy\right] \tag{26}$$

Integrating (26), we have

$$C_m = \frac{4V_0}{m\pi} \text{ for } m \text{ (odd)}, \qquad C_m = 0 \text{ for } m \text{ (even)}$$

From $C_m = B \sinh\left(\frac{m\pi}{b}a\right)$,

$$B = \frac{C_m}{\sinh\left(\frac{m\pi}{b}a\right)} = \frac{4V_0}{m\pi \sinh\left(\frac{m\pi}{b}a\right)} \tag{27}$$

Substituting (27) into (22), we have our solution

$$V(x, y) = \sum_{m=1}^{m=\infty} \frac{4V_0}{m\pi \sinh\left(\frac{m\pi}{b}a\right)} \sinh\left(\frac{m\pi}{b}x\right) \sin\left(\frac{m\pi}{b}y\right) \tag{28}$$

where m is odd. Since (28) satisfies Laplace's equation within the trough and all the potential boundary condition, it is a unique solution of our trough problem. Equation (28) is an infinite series whose amplitudes converge quite rapidly, and only a few terms need to be summed to obtain a high degree of accuracy.

Problem 6.5-1 For the trough of Example 10, when $a = b$ and $V_0 = 100$ V, calculate V through the use of (28) at: (a) $x = a/2$ and $y = b/2$; (b) $x = a/4$ and $y = b/4$; (c) $x = 3a/4$ and $y = b/4$.

Problem 6.5-2 For the trough of Fig. 6-4, the lid potential of V_0 is replaced by a potential distribution that resembles a triangular form. The potential is found to be zero at $y = 0$ and $y = b$ while at $y = b/2$ the potential equals 200 V. The potential rises linearly from $y = 0$ to $y = b/2$ and declines linearly from $y = b/2$ to $y = b$. (a) Find $V(x, y)$ within the trough. (b) Evaluate the potential at $x = a/2$, $y = b/2$ when $a = b$.

6.6 SOLUTION OF LAPLACE'S EQUATION IN CYLINDRICAL COORDINATES

In Example 5, we considered the cylindrical coordinate case where r was the only variable. The solution for $V(r)$ was given by (6.4-25). In this section we shall look briefly at two- and three-dimensional solutions. From the inside back cover of the text, we can write Laplace's equation as

$$\nabla^2 V = \frac{1}{r}\frac{\partial}{\partial r}\left(r\frac{\partial V}{\partial r}\right) + \frac{1}{r^2}\frac{\partial^2 V}{\partial \phi^2} + \frac{\partial^2 V}{\partial z^2} = 0 \tag{1}$$

and following the procedure of Sec. 6.5, we assume a product solution of the form

$$V = R(r)\Phi(\phi)Z(z) \tag{2}$$

Substituting (2) into (1) and dividing through by $R\Phi Z$, we get

$$\frac{1}{R}\frac{d^2R}{dr^2} + \frac{1}{rR}\frac{dR}{dr} + \frac{1}{\Phi r}\frac{d^2\Phi}{d\phi^2} + \frac{1}{Z}\frac{d^2Z}{dz^2} = 0 \tag{3}$$

Using the same procedure as in Sec. 6.5, we write

$$\frac{1}{R}\frac{d^2R}{dr^2} + \frac{1}{rR}\frac{dR}{dr} + \frac{1}{\Phi r}\frac{d^2\Phi}{d\phi^2} = -\frac{1}{Z}\frac{d^2Z}{dz^2}$$

or

$$\frac{1}{Z}\frac{d^2Z}{dz^2} = \mp A^2 \tag{4}$$

where A^2 is a positive real number. The solution to (4) is

$$Z = D_1 \cos(Az) + D_2 \sin(Az) \tag{5a}$$

or

$$Z = D_1' \cosh(Az) + D_2' \sinh(Az) \tag{5b}$$

depending on whether the minus or plus sign is used in (4).

The remainder of the Laplacian is now

$$\frac{1}{R}\frac{d^2R}{dr^2} + \frac{1}{rR}\frac{dR}{dr} + \frac{1}{\Phi r}\frac{d^2\Phi}{d\phi^2} = \pm A^2 \tag{6}$$

and the solution to (6) is much more complex than the solution to (4). However, when we look at the physical structures that we are primarily interested in, we realize that periodicity in ϕ is required, i.e.,

$$\Phi(0) = \Phi(2n\pi) \tag{7}$$

and thus Φ should probably be of the form

$$\Phi = C_1 \cos(n\phi) + C_2 \sin(n\phi) \tag{8}$$

and thus

$$\frac{1}{\Phi}\frac{d^2\Phi}{d\phi^2} = -n^2 \tag{9}$$

and (6) becomes

$$\frac{1}{R}\frac{d^2R}{dr^2} + \frac{1}{rR}\frac{dR}{dr} + \frac{-n^2}{r^2} = \pm A^2$$

or after multiplying through by R,

$$\frac{d^2R}{dr^2} + \frac{1}{R}\frac{dR}{dr} + \left(-A^2 - \frac{n^2}{r^2}\right)R = 0 \tag{10a}$$

or

$$\frac{d^2R}{dr^2} + \frac{1}{R}\frac{dR}{dr} + \left(A^2 - \frac{n^2}{r^2}\right)R = 0 \tag{10b}$$

where (10a) results from $(1/z)(d^2Z/dz^2) = -A^2$, and (10b) results from $(1/z) \cdot (d^2Z/dz^2) = A^2$.

The solution to (10a) is

$$R = B_1 I_n(\mathrm{Ar}) + B_2 K_n(\mathrm{Ar}) \tag{11a}$$

and the solution to (10b) is

$$R = B_3 J_n(\mathrm{Ar}) + B_4 N_n(\mathrm{Ar}) \tag{11b}$$

where $I_n(\mathrm{Ar})$ = modified Bessel function of first kind of order n with argument Ar
$K_n(\mathrm{Ar})$ = modified Bessel function of second kind of order n with argument Ar
$J_n(\mathrm{Ar})$ = Bessel function of first kind of order n with argument Ar
$N_n(\mathrm{Ar})$ = Bessel function of second kind of order n with argument Ar

Plots of these Bessel functions and modified Bessel functions are shown in Figs. 6-6 and 6-7. Other references should be consulted for an in-depth discussion of these functions.[4]

[4]S. Ramo, J. R. Whinnery, and T. Van Duzer, *Fields and Waves in Communication Electronics*, 2nd ed., John Wiley & Sons, Inc., New York, 1984.

It is clear from Figs. 6-6 and 6-7 that for V to remain finite as $r \to 0$, either $N_n(\mathrm{Ar})$ or $K_n(\mathrm{Ar})$ must equal zero, depending on whether the Bessel function or modified Bessel function applies. We now have for the complete solutions in three dimensions,

$$V = [B_1 I_n(\mathrm{Ar})][C_1 \cos(n\phi) + C_2 \sin(n\phi)][D_1 \cos(Az) + D_2 \sin(Az)] \tag{12a}$$

or

$$V = [B_3 J_n(\mathrm{Ar})][C_1 \cos(n\phi) + C_2 \sin(n\phi)][D_1' \cosh(Az) + D_2' \sinh(Az)] \tag{12b}$$

depending on whether the minus sign (12a) or the plus sign (12b) is used in (4).

If $n = 0$, (9), degenerates to

$$\frac{1}{\Phi}\frac{d^2\Phi}{d\phi^2} = 0 \tag{13}$$

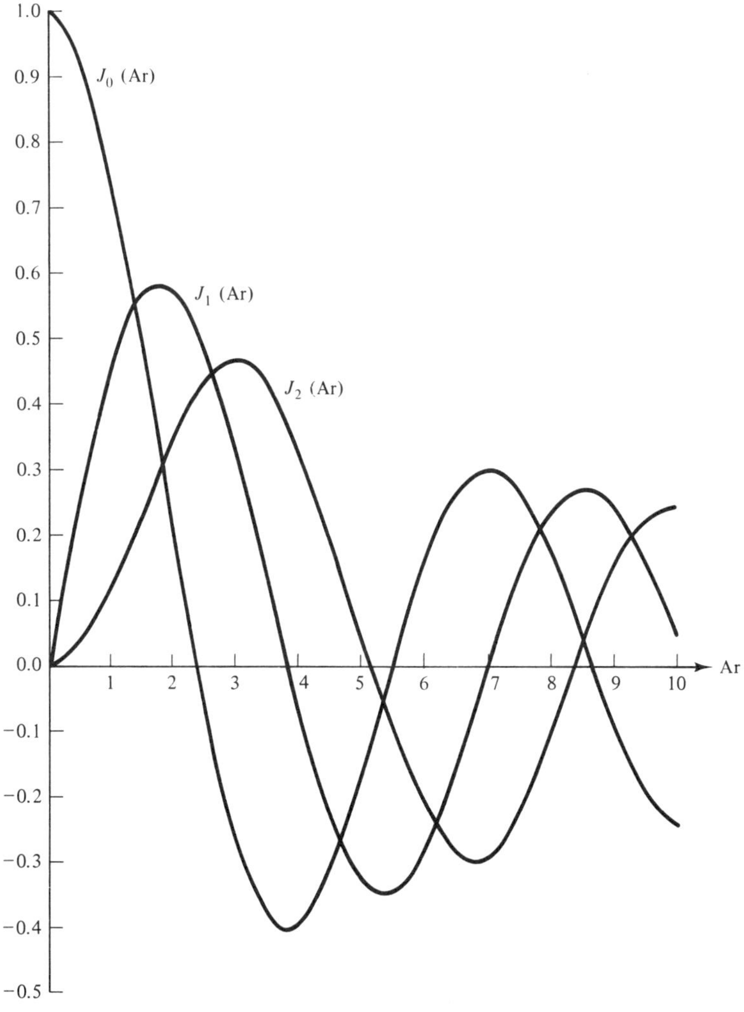

(a)

Figure 6-6 (a) Bessel functions of the first kind.

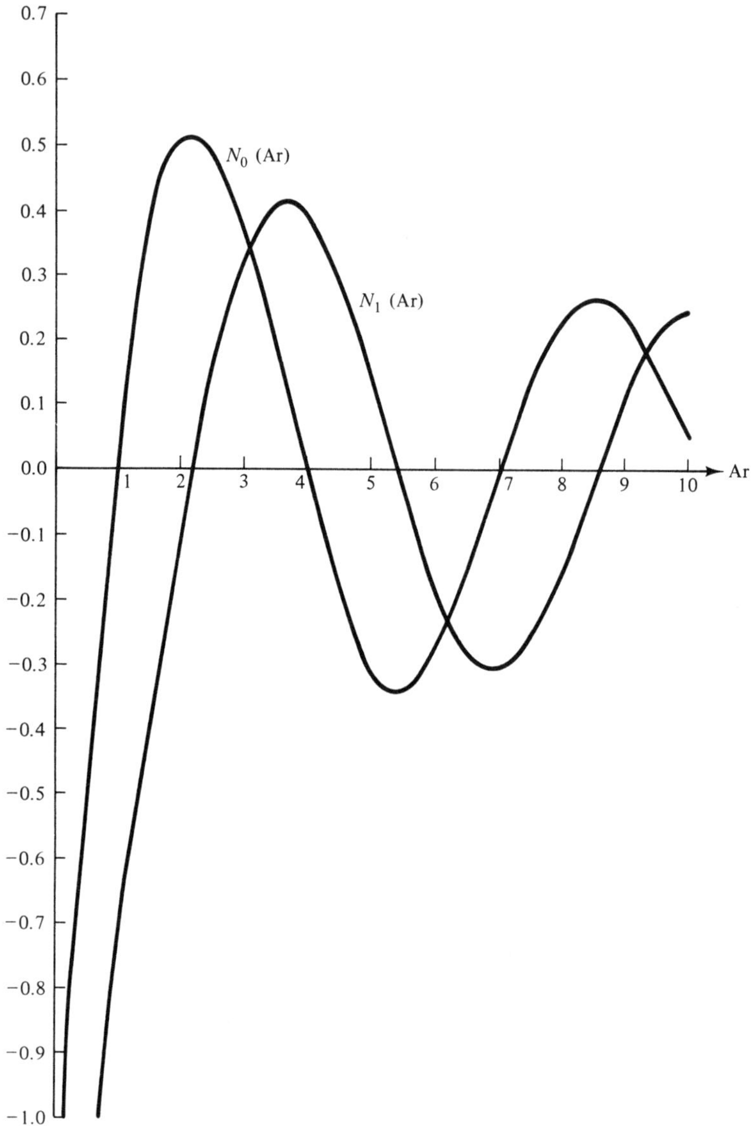

(b)

Figure 6-6 (b) Bessel functions of the second kind.

for which

$$\Phi = C_3\phi + C_4 \tag{14}$$

This is the kind of solution demonstrated in the one-dimensional case of Example 6.

For no variation with ϕ,

$$V = [B_1 I_0(\text{Ar})][D_1 \cos(Az) + D_2 \sin(Az)] \tag{15a}$$

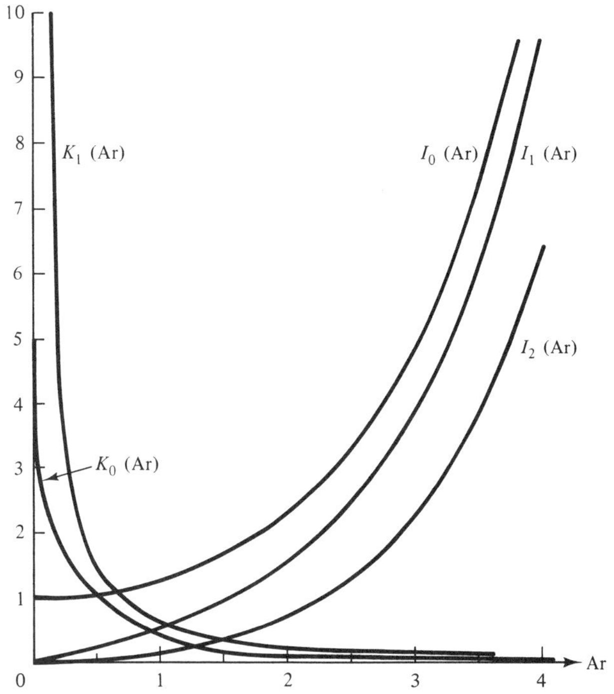

Figure 6-7 Modified Bessel functions.

or

$$V = [B_3 J_0(\text{Ar})][D_1' \cosh(Az) + D_2' \sinh(Az)] \tag{15b}$$

If $A = 0$, then (4) becomes

$$\frac{1}{Z}\frac{d^2Z}{dz^2} = 0 \tag{16}$$

and

$$Z = D_3 z + D_4 \tag{17}$$

The solution for R is the same as the one-dimensional problem of Example 5, and the complete solution, for no variation with ϕ *and* for $A = 0$, is

$$V = [B_7 \ln(r) + B_8](D_3 z + D_4) \tag{18}$$

The potential solution (18) is the potential for a coaxial cable with a linear voltage drop in the z-direction. A summary of product solutions $V = R(r)\Phi(\phi)Z(z)$ to Laplace's equation in cylindrical coordinates is given in Table 6-1 for various combinations of $n \neq 0$, $n = 0$, $A \neq 0$, $A = 0$. It should be kept in mind that these are forms of solutions only. Boundary conditions for a particular case may require a Fourier series of such solutions as was the case in Sec. 6.5 for rectangular harmonics.

TABLE 6-1 SOLUTIONS $V = R(r)\Phi(\phi)Z(z)$ TO LAPLACE'S EQUATION IN CYLINDRICAL COORDINATES

$\frac{1}{Z^2}\frac{d^2Z}{dz^2}$	$n \neq 0$	$n = 0$
$-A^2$	$[B_1 I_n(Ar)][C_1 \cos(n\phi) + C_2 \sin(n\phi)][D_1 \cos(Az) + D_2 \sin(Az)]$	$[B_1 I_0(Ar)][C_3\phi + C_4][D_1 \cos(Az) + D_2 \sin(Az)]$
A^2	$[B_3 J_n(Ar)][C_1 \cos(n\phi) + C_2 \sin(n\phi)][D_1' \cosh(Az) + D_2' \sinh(Az)]$	$[B_3 J_0(Ar)][C_3\phi + C_4][D_1' \cosh(Az) + D_2' \sinh(Az)]$
0	$[B_5 r^n + B_6 r^{-n}][C_1 \cos(n\phi) + C_2 \sin(n\phi)][D_3 z + D_4]$	$[B_7 \ln(r) + B_8][C_3\phi + C_4][D_3 z + D_4]$

6.7 SOLUTION OF POISSON'S EQUATION

Several examples of the application of Poisson's equation (6.2-4) can be found in the analysis of a *pn* junction within a semiconductor. In most of these examples the charge distribution is assumed, and the solution becomes a relatively straightforward mathematical solution. For this reason, these examples will not be discussed and the student is referred to texts on physical electronics.[5]

In this section we will take on the task of solving Poisson's equation for the case when the charge distribution is an unknown distribution of free electrons in vacuum. Our solution will not only give us a solution for the potential field, but also for the charge distribution.

Example 11

Find the potential field between the cathode and anode of a parallel-plate thermionic diode, as shown in Fig. 6-8(a). The cathode is heated so that electrons are emitted thermionically and then accelerated toward the positively charged anode.

Solution. We shall assume that a space charge, produced by the electrons between the parallel plates, exists such that the electrons are emitted with zero initial velocity and that the current is not limited by the cathode temperature but can be controlled by the anode potential V_0. This operating condition is commonly called a *space-charge-limited* condition. As electrons move from the cathode to the anode with a velocity $\overline{U}$, a space charge density ρ_v is produced and is given by

$$\rho_v = \frac{J}{U} \tag{5.2-4}$$

where J is the current density in (A m^{-2}) and U is the electron velocity in (m s^{-1}). Since ρ_v is negative for electrons, we find that J must be negative for positive U. Since the current through any plane parallel to the cathode will be the same, we deduce that J is uniform throughout the region between the cathode and anode. Our problem boils down to solving Poisson's equation, in $V = V(y)$,

$$\nabla^2 V = \frac{\partial^2 V}{\partial y^2} = \frac{d^2 V}{dy^2} = -\frac{\rho_v}{\epsilon_0} = -\frac{J}{U\epsilon_0} \tag{1}$$

with the following boundary conditions on V:

$$V = 0 \quad \text{at } y = 0 \tag{2}$$

$$V = V_0 \quad \text{at } y = y_0 \tag{3}$$

$$-\left.\frac{\partial V}{\partial y}\right|_{y=0} = 0 \tag{4}$$

Boundary condition (4) comes about from the equilibrium condition that is reached at the cathode, $\overline{E} = 0$, under space-charge-limited conditions. Since the initial velocity of the electron at $y = 0$ is zero, we have, from the conservation of energy,

$$\boxed{\tfrac{1}{2} m_e U^2 = -q_e V} \tag{5}$$

[5]Myron F. Uman, *Introduction to the Physics of Electronics*, Prentice-Hall, Inc., Englewood Cliffs, N.J., 1974, pp. 261–269.

where m_e is the mass of the electron and q_e is the charge on the electron (negative quantity). Solving for U in (5) and substituting into (1), we have

$$\frac{d^2V}{dy^2} = -\frac{J}{\epsilon_0}\left(-\frac{m_e}{2q_e V}\right)^{1/2} \tag{6}$$

Let $\dfrac{-J}{\epsilon_0}\left(-\dfrac{m_e}{2q_e}\right)^{1/2} = K$; thus, (6) becomes

$$\frac{d^2V}{dy^2} = KV^{-1/2} \tag{7}$$

Equation (7) can be integrated readily by first multiplying both sides by $2(dV/dy)$ to obtain

$$2\frac{dV}{dy}\left(\frac{d^2V}{dy^2}\right) = 2\frac{dV}{dy}KV^{-1/2} \tag{8}$$

The left side of (8) can be found equal to

$$2\frac{dV}{dy}\left(\frac{d^2V}{dy^2}\right) = \frac{d}{dy}\left(\frac{dV}{dy}\right)^2 \tag{9}$$

Now, substitute (9) into (8) and integrate with respect to y, to obtain

$$\int \frac{d}{dy}\left(\frac{dV}{dy}\right)^2 dy = \int 2KV^{-1/2}\frac{dV}{dy}dy \tag{10}$$

The right side of (10) can be written as

$$\int 2KV^{-1/2}\,dV \tag{11}$$

Thus, (10) becomes

$$\int \frac{d}{dy}\left(\frac{dV}{dy}\right)^2 dy = \int 2KV^{-1/2}\,dV \tag{12}$$

Integrating (12), we have

$$\left(\frac{dV}{dy}\right)^2 = 4KV^{1/2} + A \tag{13}$$

The constant of integration A can be evaluated by applying boundary conditions (2) and (4) to give $A = 0$; thus, (13) becomes

$$\left(\frac{dV}{dy}\right)^2 = 4KV^{1/2} \tag{14}$$

Taking the square root of (14), we have

$$\frac{dV}{dy} = 2\sqrt{K}\,V^{1/4} \tag{15}$$

Rearranging (15), we obtain

$$V^{-1/4}\,dV = 2\sqrt{K}\,dy \tag{16}$$

Integrating (16), we obtain

$$V^{3/4} = \tfrac{3}{2}\sqrt{K}\,y + B \tag{17}$$

Applying boundary condition (2), we find $B = 0$. Thus, (17) yields

$$V = (\tfrac{3}{2}\sqrt{K}\,y)^{4/3} \tag{18}$$

Now, from boundary condition (3), eq. (18) becomes

$$V_0 = (\tfrac{3}{2}\sqrt{K}\,y_0)^{4/3} \tag{19}$$

Dividing (18) by (19) and solving for V, we have

$$V = V_0\left(\frac{y}{y_0}\right)^{4/3} \quad \text{(V)} \tag{20}$$

Equation (20) is our desired solution for V between the cathode and anode of our thermionic diode. This problem is unique in that it involves the dynamics of a charged particle with a mass m_e (kg).

From (20), we find that

$$E = -\frac{dV}{dy} = -\frac{4V_0}{3y_0}\left(\frac{y}{y_0}\right)^{1/3} \qquad (\text{V m}^{-1}) \tag{21}$$

Through the use of (19) and the expression for K, we find that

$$J = -\frac{4\epsilon_0}{9y_0^2}\left(\frac{-2q_e}{m_e}\right)^{1/2} V_0^{3/2} \qquad (\text{A m}^{-2}) \tag{22}$$

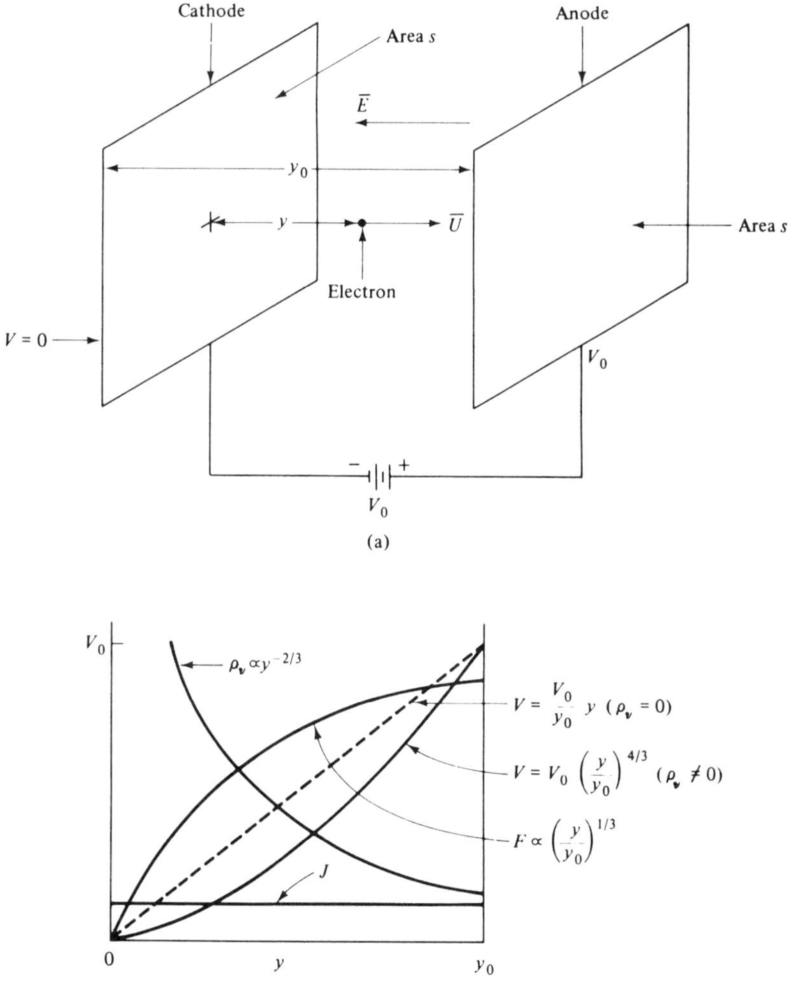

Figure 6-8 (a) Graphical display of a parallel-plate vacuum diode with a thermionic cathode (emitter of electrons) on the left and a positively charged anode (collector of electrons) on the right. A single electron is shown a distance y from the cathode, traveling toward the anode with a velocity U. (b) Plots of ρ_v, V, E, and J versus y.

If we substitute (20) and (22) into the right side of (6), which equals $-\rho_v/\epsilon_0$, and solve for ρ_v, we obtain

$$\rho_v = -\frac{4\epsilon_0}{9y_0^2}\left(\frac{y}{y_0}\right)^{-2/3} V_0 \qquad (\text{C m}^{-3}) \tag{23}$$

If we multiply (22) by the area s of the anode and neglect fringing effects, we will obtain the current I flowing between the cathode and the anode; thus,

$$I = -\frac{4\epsilon_0 s}{9y_0^2}\left(\frac{-2q_e}{m_e}\right)^{1/2} V_0^{3/2} \tag{24}$$

where the minus sign indicates that the conventional current flows from the anode to the cathode. From (24), we can say that

$$I \propto V_0^{3/2} \tag{25}$$

Equation (22) is commonly referred to as the Child-Langmuir law for thermionic diodes operating under "space-charge conditions." A plot of ρ_v, V, E, and J versus y is shown in Fig. 6-8(b). The dashed curve represents the potential solution, where $\rho_v = 0$, $\nabla^2 V = 0$.

Problem 6.7-1 Through the use of (19) and the expression for K, derive (22).

Problem 6.7-2 Find the current I that flows in a thermionic diode when $V_0 = 150$ V, $s = 10^{-4}$ m^2, and $y_0 = 10^{-3}$ m.

Problem 6.7-3 Through the use of (20), (22), and the right-hand side of (6), which equals $-\rho_v/\epsilon_0$, derive (23).

REVIEW QUESTIONS

1. In all electrostatic solutions, up to Chapter 6, was it necessary to know the charge distribution? *Sec. 6.1*
2. What is the mathematical form of Poisson's equation for a homogeneous medium? *Eq. (6.2-4)*
3. If $\nabla^2 V = 0$ at a point, what is ρ_v equal to? *Eq. (6.2-4)*
4. What is the mathematical form of Laplace's equation for a homogeneous medium? *Eq. (6.2-5)*
5. What is $\nabla\epsilon$ equal to in a homogeneous media? *Example 1*
6. A potential solution to an electrostatic problem satisfies Laplace's equation and all boundary potentials. What can we say about that solution? *Sec. 6.3*
7. What is the uniqueness theorem? *Sec. 6.3*
8. In the proof of the uniqueness theorem, can we impose that $(\partial V/\partial n)|_b$ be satisfied on part or all of the boundary? *Sec. 6.3*
9. In the solution of Laplace's equation, how are the constants of integration evaluated? *Example 4*
10. In an electrostatic problem, where two concentric spheres are involved, the potential will be a function of what spherical variable? *Example 7*
11. Can Laplace's equation be used to solve for the potential field in a two-dielectric capacitor? *Example 9*
12. In the application of Laplace's equation to a two-dielectric region, what two boundary conditions are applied at the dielectric-dielectric boundary? *Example 9*
13. What is the separation of variables solution? *Sec. 6.5*
14. In the solution of Laplace's equation in two variables, the function X is a function of what variables? *Example 10*
15. Why was $(1/X)(d^2X/dx^2)$ set equal to A_1^2 (constant)? *Example 10*
16. What is the separation equation? *Eq. (6.5-7)*
17. In the solution of the trough problem, what was the general solution for X? *Eq. (6.5-10)*
18. In the solution of the trough problem, why were we required to use an infinite number of solutions? *Example 10*

19. How was C_m evaluated in our trough problem? *Eq. (6.5-25)*
20. What is meant by a space-charge-limited condition? *Example 11*
21. How did J (current density) vary with y in the thermionic diode example? *Example 11*
22. Under a space-charge condition, what is E at the cathode? *Eq. (6.7-4)*
23. Under a space-charge condition, where is ρ_v the greatest? *Eq. (6.7-23), Fig. 6-8(b)*
24. In the Child-Langmuir law, the current is proportional to the anode potential raised to what power? *Eq. (6.7-25)*

PROBLEMS

6-1. A potential field is described by $V = 5x^2$ (V). (a) Is Laplace's equation satisfied? (b) If not, find the charge density in the region.

6-2. A potential field in a region between two concentric cylinders is found to be $V = -10^{-6}r$ (V). Find the total charge per meter contained between the cylinders if $r_a = 1$ cm, and $r_b = 3$ cm.

6-3. The potential field in a region about the origin is found to be $V = -10^{-6}r_s^2$ V. How much charge exists in a sphere centered at the origin and of 2-m radius?

6-4. Two potential functions that satisfy Laplace's equation and the potential at a boundary at $x = 5$ are: $V_1 = 6x$ (V), and $V_2 = 40x - 170$ V. Plot the functions for $0 \leqq x \leqq 5$, and explain how the uniqueness theorem would show that these potentials cannot both be valid solutions for the region.

6-5. Two potential solutions in cylindrical coordinates satisfy Laplace's equation and one boundary. The solutions are $V_1 = 100 \ln r_a$ and $V_2 = 50 \ln(r_c/r_a) + 100 \ln r_a$. Use the uniqueness theorem to show that $V_1 \neq V_2$.

6-6. For a particular problem, two potential solutions are found: $V_1 = 90/r_b$ and $V_2 = (100r_b - 10r_s)/r_s r_b$. (a) Show that both solutions satisfy Laplace's equation. (b) Show that both solutions yield the same potential at $r_s = r_b$. (c) Explain how the uniqueness theorem should be applied to show that V_1 and V_2 cannot both be valid solutions.

6-7. For the potential solutions of Prob. 6-6, find: (a) the zero potential reference radius; (b) the charge at the origin.

6-8. Two conducting planes are oriented in a charge-free region as shown in Fig. 6-2 with the reference plane at $\phi_1 = 0°$ with a potential of 0 V, and with a potential of 100 V on the plane at $\phi_2 = 45°$. Two potential functions, $V_1 = 400\phi/\pi$ and $V_2 = 141.4 \sin \phi$ (V), both satisfy the boundary conditions at $\phi = 0°$ and at $\phi = 45°$. Which potential function is valid, and why?

6-9. Which of the following potential functions satisfy Laplace's equation in two-dimensional rectangular coordinates?

(a) $V_1 = \sinh\left(\dfrac{m\pi x}{b}\right) \sin\left(\dfrac{m\pi y}{b}\right)$ (V)

(b) $V_2 = \sinh\left(\dfrac{m\pi x}{b}\right) \cos\left(\dfrac{m\pi y}{b}\right)$ (V)

(c) $V_3 = \cos\left(\dfrac{m\pi x}{b}\right) \sin\left(\dfrac{m\pi y}{b}\right)$ (V)

(d) $V_4 = \cosh\left(\dfrac{m\pi x}{b}\right) \cos\left(\dfrac{m\pi y}{b}\right)$ (V)

6-10. In Example 4, the potentials were specified at the conductor boundaries. In a Neumann formulation of the problem, deduce: (a) the $(\partial V/\partial n)|_b$ specifications at the two conductor boundaries; (b) the relationship of this specification to the surface charge densities on the two surfaces. [*Hint:* See (6.4-9), (6.4-11), (6.4-12), (6.4-13), and (6.4-14).]

6-11. Two concentric and conducting cylinders of radii r_a and r_b are located on the z axis, as shown in Fig. 5-21. If $\epsilon = 4\epsilon_0$, $\rho_v = 0$ between the cylinders, $V = 0$ at r_a, and $\partial V/\partial r = -400\ \text{V m}^{-1}$ at r_b, solve Laplace's equation to find: (a) V; (b) $\overline{E}$; (c) ρ_s on both conductors.

6-12. A coaxial capacitor, such as that pictured in Fig. 5-21, has the space between conductors filled with a dielectric whose ϵ_r is given by $\epsilon_r = 4r_b/r$. (This can be approximated in practice by wrapping the center conductor with tape, with each successive layer having a slightly smaller dielectric constant.) If $V = 100\ \text{V}$ at $r = r_a$ and 0 V at $r = r_b$, find $\overline{E}$ between conductors and the expression for V as a function of r. Does Laplace's equation apply?

6-13. A spherical capacitor is formed by two concentric spheres of radii $r_b = 4$ cm and $r_a = 1.5$ cm when the region between the spheres is filled with a dielectric whose $\epsilon = 6\epsilon_0$. When $V = 0$ at r_a and $V = 0.1$ V at r_b, find through the solution of Laplace's equation: (a) the potential field V; (b) $\overline{E}$; (c) the current between the two spheres, if the medium is a homogeneous conducting material whose $\sigma = 6\ \text{Sm}^{-1}$.

6-14. Two finite conducting planes are oriented as shown in Fig. 6-2 with the reference of 0 V at $\phi = 0°$ and the second plane of V_0 (V) at 30°. The "gap" between plates is described by a gap between the z axis and $r = r_a$ for each plate. Each conducting plate then extends from $r_a \le r_c \le r_b$ in the radial direction, and is 1 m long in the z direction. Find the capacitance of this wedge-shaped capacitor as a function of r_a and r_b. Assume air between plates.

6-15. A finite cone above a finite ground plane is formed by allowing $\theta_1 = \pi/6$ and $\theta_2 = \pi/2$ in Fig. 6-3. When the space between the cone and ground plane is filled with a homogeneous dielectric $\epsilon = \epsilon_0$ and $0 \leqq r_s \leqq 2$, find approximations for: (a) V; (b) $\overline{E}$; (c) ρ_s on both surfaces; (d) the capacitance between the cone and the ground plane.

6-16. Place two cylindrical and concentric regions of dielectric in the two-conductor cylindrical capacitor of Fig. 5-21. Let region 1 dielectric have an $\epsilon_1 = 1.5\epsilon_0$ within $r_a < r_c < r'$ and region 2 dielectric have an $\epsilon_2 = 4\epsilon_0$ within $r' < r_c < r_b$. If $V = 0$ at r_a and $V = V_0$ at r_b, find through the use of Laplace's equation: (a) the potential fields V_1 and V_2; (b) $\overline{E}_1$ and $\overline{E}_2$; (c) the capacitance of L-meter length.

6-17. For the two-dimensional electrostatic problem shown in Fig. 6-9, find: (a) V field; (b) V at $y = 0$ and $z = b/4$ when $a = b$.

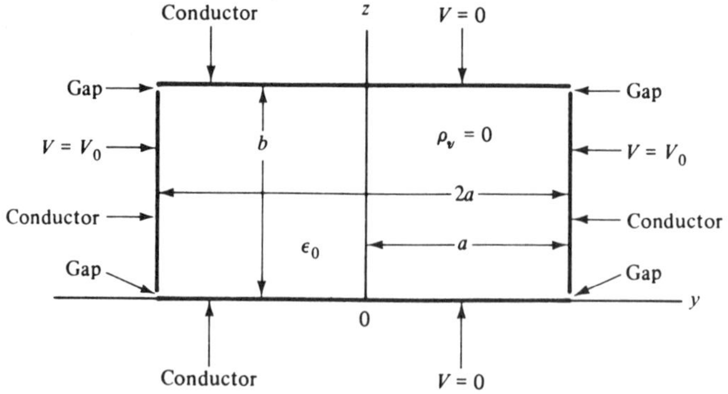

Figure 6-9 Graphical formulation of a two-variable electrostatic problem formed by four conductors of infinite extent in the x direction for Prob. 6-17.

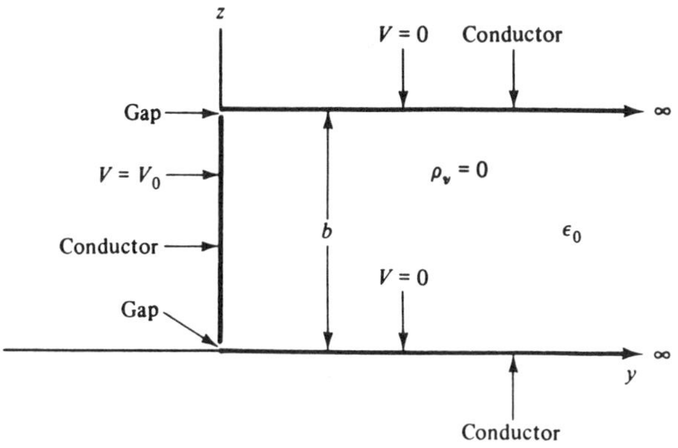

Figure 6-10 Graphical formulation of a two-variable electrostatic problem formed by three conductors for Prob. 6-18.

6-18. For the two-dimensional electrostatic problem shown in Fig. 6-10, find: (a) the potential field V; (b) V at $y = b$, $z = b$.

6-19. From the results of Prob. 6-17, calculate V at $y = a/2$, $z = b/2$; $y = 0$, $z = b/2$. Let $a = b$.

6-20. In all our solutions of Laplace's equation for the potential field, explain why the V solutions are independent of ϵ. Does this mean that Laplace's equation also holds for the case where a homogeneous conducting medium with conductivity σ fills the region of interest?

6-21. In the solution of Poisson's equation in Example 11 for a thermionic diode, prove that Poisson's equation and the boundary conditions (6.7-2), (6.7-3), and (6.7-4) are satisfied.

6-22. From (5.2-4) and the results of Example 11, obtain an expression for the electron velocity U in Example 11.

7

Graphical Solution (Curvilinear Squares), Numerical Solution (Iteration), Image Solution, Analog Solution, and Experimental Solution

7.1 INTRODUCTION

In Chapter 6 we solved Laplace's and Poisson's equations to obtain solutions for the potential field V in electrostatic problems. These solutions are referred to as *formal* or *exact solutions*. In this chapter we study graphical, numerical, image, analog, and experimental solution methods, most of which are based on Laplace's equation and do not require the knowledge of the distribution of charge. The accuracy of the solutions obtained will depend on the solution method.

7.2 GRAPHICAL SOLUTION (CURVILINEAR SQUARES)

The graphical solution method that we shall study in this section is commonly called the *curvilinear squares solution*. This solution is an approximate solution whose accuracy depends on the graphical dexterity and fortitude of an individual in drawing the equipotential surfaces and flux lines that form curvilinear squares. As we develop the rules for this curvilinear squares solution method, we shall find that it is based on Laplace's equation. This method will be applied to electrostatic problems where $\rho_v = 0$ and the boundaries are equipotentials. Due to graphics limitations, we shall limit ourselves to electrostatic problems of constant cross section, such as a long coaxial cable, eccentric cable, and the rectangular trough of Chapter 6. Thus, the fields cannot vary in a direction normal to the cross section. We shall divide this solution method into two parts; the first part will focus on a dielectric medium while the second part will focus on a conducting medium.

7.2-1 Curvilinear Squares Solution in a Dielectric Medium

The end results of the curvilinear squares method of solution for an eccentric cable on constant cross section is shown in the upper half of Fig. 7-1(a). Due to symmetry, only half of the cross section is plotted. From this plot, we will be able to find V, $\overline{E}$, $\overline{D}$ at any point between the conductors, ρ_s at any point on the conductor surfaces, and the capacitance per meter length (depth).

Let us now backtrack a bit and set up the rules that will govern the construction of these curvilinear squares. These rules are based on our previously attained knowledge of electrostatic fields and they are:

I. Electric flux lines and equipotential surfaces intersect at right angles. This follows from $\overline{E} = -\nabla V$ (see Example 4-10).
II. The surface of a conductor is an equipotential surface.
III. Electric flux lines begin on a positive charge and end on a negative charge. Since charge is present only on the conductors, flux lines will begin and end on conductor surfaces; thus, between the conductors $\nabla \cdot \overline{D} = 0$ since $\rho_v = 0$.
IV. The $\overline{E}$ and $\overline{D}$ are perpendicular to the conductor surfaces.

From rules I and III, we note that the graphical construction is governed by $\overline{E} = -\nabla V$ and $\nabla \cdot \overline{D} = 0$. Thus, Laplace's equation is satisfied in this solution method. All these rules have been adhered to in the construction of the curvilinear squares in the upper half of the eccentric cable of constant cross section shown in Fig. 7-1(a). Before we embark on the actual graphical construction of curvilinear squares, we shall define a curvilinear square as a planar geometric figure formed by four curved lines that intersect at right angles and, through successive subdivision, will yield squares as shown in Fig. 7-1(b). In drawing curvilinear squares, the distances $\Delta\ell_n$ and $\Delta\ell_t$ are made approximately equal.

Let us now, through the previously stated rules, draw curvilinear squares as shown in the lower portion of Fig. 7-1(a). From point A, we start a flux line perpendicular to the inner conductor and head toward the outer conductor to make a perpendicular intersection at point A'. Now, let us draw the line B-B' in a similar manner. Curvilinear square 1 is now formed by drawing an equipotential surface, a curved line C-C' in a two-dimensional plot, perpendicular to the two flux lines. Additional equipotential surfaces are drawn to form curvilinear squares 2, 3, 4, etc. Our first attempt ended in a failure since we have two equal potential lines intersecting at points G and G'. To prevent this from happening, we must make the curvilinear squares smaller as we travel toward the region where the conductors are closest. This means that we must start our graphical construction again and employ the benefits of our previous failure. After repeating the construction several times, the curvilinear squares plot found in the upper half of Fig. 7-1(a) is obtained.

For convenience in the following development, we shall define a flux tube as a three-dimensional volume whose ends are formed by the area bounded by the two conductors and flux lines A-A' and B-B', and of d (m) in depth. All flux lines that start

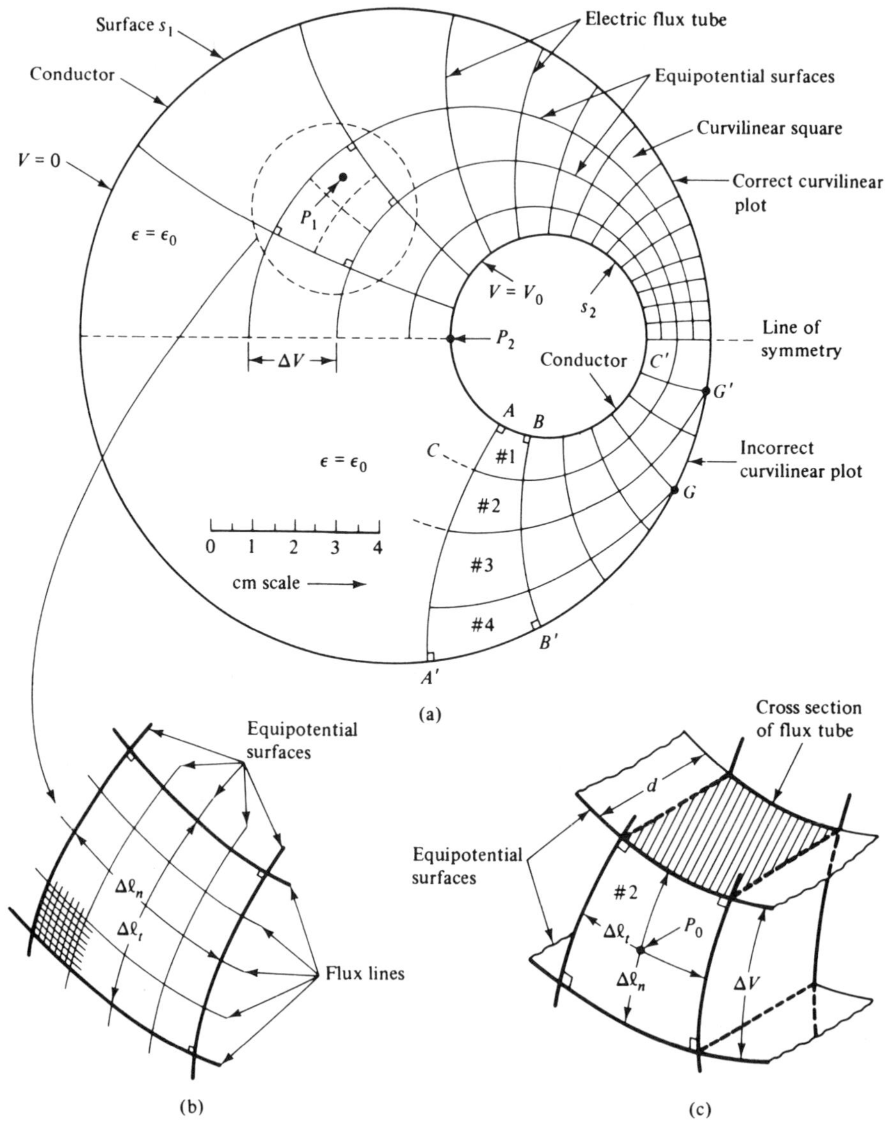

Figure 7-1 (a) Curvilinear squares plot for an eccentric cable of constant cross section. The upper half is a correct plot; the lower half is incorrect. (b) General curvilinear square and its subdivisions. (c) Expanded view of curvilinear square 2 that forms the ends of a curvilinear cell of depth d (m). The crosshatched region is the cross section of a flux tube that extends between the two conductors.

between the points A-B are thus guided to the outer conductor and intersect the outer conductor between the points A'-B', as indicated in Fig. 7-1(a).

Figure 7-1(c) shows part of the flux tube bound by the flux lines A-A' and B-B'. This is, in fact, our curvilinear square 2. The crosshatched area is the cross section of the flux tube of d (m) depth. If $\Delta\Psi_E$ lines of electric flux flow in the flux tube, we can approximate the $|\overline{E}|$ at point P_0, at the center of the curvilinear square, to be

$$|\overline{E}| \cong \frac{|\Delta\Psi_E|}{\epsilon d\, \Delta\ell_t} \tag{1}$$

where $d\,\Delta\ell_t$ is the cross-sectional area of the flux tube that passes through P_0. The $|\overline{E}|$ at P_0 can also be approximated as

$$|\overline{E}| \cong \frac{|\Delta V|}{\Delta\ell_n} \tag{2}$$

Equating (1) and (2) and rearranging, we have

$$\frac{|\Delta\Psi_E|}{|\Delta V|} \cong \epsilon d\left(\frac{\Delta\ell_t}{\Delta\ell_n}\right) \tag{3}$$

Now, since we have drawn curvilinear squares $\Delta\ell_t/\Delta\ell_n = 1$, thus (3) reduces to

$$\frac{|\Delta\Psi_E|}{|\Delta V|} \cong \epsilon d \quad \text{(constant)} \tag{4}$$

Since $|\Delta\Psi_E|$ is constant throughout a flux tube, we see that ΔV volts exist between any two adjacent equipotential surfaces (lines) when we construct curvilinear squares.

The capacitance for d (m) depth can be found from

$$C = \frac{|Q|}{|V|} \cong \frac{N_t|\Delta\Psi_E|}{N_n|\Delta V|} \quad \text{(F)} \tag{5}$$

where N_t equals the number of tubes, and N_n equals the number of ΔV divisions between conductors or the number of curvilinear squares between conductors. Stated in another way, N_t equals the number of curvilinear squares tangent to one of the conductors while N_n equals the number of curvilinear squares normal to the two conductors. Substituting (4) into (5), we obtain

$$\boxed{C \cong \frac{N_t}{N_n}\epsilon d \qquad \text{(F)}} \tag{6}$$

Thus, we see that from a graphical construction of curvilinear squares we are able to obtain an approximate expression for the capacitance of d (m) depth of our electrostatic problem. Accuracy of about 5% can be obtained in capacitance calculations by a beginner who follows a few simple rules. It should be mentioned that N_t and N_n need not be integer values.

For more detailed development on drawing of curvilinear squares, the student is referred to the two references cited.[1,2]

The magnitude of ΔV can be found by dividing the potential between the conductors by N_n; thus,

$$\boxed{|\Delta V| \cong \frac{V_0}{N_n} \qquad \text{(V)}} \tag{7}$$

[1] A. D. Moore, *Fundamentals of Electrical Design,* McGraw-Hill Book Company, New York, 1927.

[2] S. Ramo, J. R. Whinnery, and T. Van Duzer, *Fields and Waves in Communication Electronics,* John Wiley & Sons, Inc., New York, 1965.

Knowing $|\Delta V|$, between adjacent equipotential surfaces (lines), we can find the potential at any point between the two conductors. The curvilinear square in which the potential is to be found may have to be subdivided into smaller squares to obtain greater accuracy. Now, from (7) and (2) we determine $|\overline{E}|$ to be

$$|\overline{E}| \cong \frac{|\Delta V|}{\Delta \ell_n} \cong \frac{V_0}{N_n\,\Delta \ell_n} \qquad (\mathrm{V\ m^{-1}}) \tag{8}$$

Through the use of (8), we find $|\overline{E}|$ and thus $|\overline{D}|$ at the conductor surfaces to obtain ρ_s on the conductors,

$$\rho_s = |\overline{D}|\bigg|_{\text{cond.}} = \epsilon|\overline{E}|\bigg|_{\text{cond.}} \cong \frac{V_0\epsilon}{N_n\,\Delta \ell_n}\bigg|_{\text{cond.}} \qquad (\mathrm{C\ m^{-2}}) \tag{9}$$

where $|\overline{D}|\,|_{\text{cond.}}$ and $|\overline{E}|\,|_{\text{cond.}}$ are normal to the conductor surface.

It should be noted that the electric field intensity becomes greater when the squares of a curvilinear squares plot become smaller. The accuracy of a curvilinear squares solution increases with the number of curvilinear squares and neatness of construction.

From (6), when $N_t = N_n = 1$, we have

$$C_0 \cong \epsilon d \qquad (\mathrm{F}) \tag{10}$$

where C_0 is the capacitance of a curvilinear cell, i.e., between two adjacent equipotential surfaces of a flux tube of depth d, as shown in Fig. 7-1(c). Now, since there are N_t of these identical cells in parallel and N_n in series, we find that the total capacitance of depth d becomes

$$C \cong \frac{N_t}{N_n} C_0 \qquad (\mathrm{F}) \tag{11}$$

The concept of curvilinear cell capacitance will become useful when we apply the curvilinear squares solution to find resistance.

Example 1

For the curvilinear squares plot of Fig. 7-1(a): (a) find the capacitance of 1 m length (depth); (b) check part (a) through the use of the exact expression

$$C = 2\pi\epsilon/\cosh^{-1}\left[\frac{(r_a^2 + r_b^2 - c^2)}{2r_a r_b}\right]$$

where r_a and r_b are the conductor radii, and c is the separation between the conductor axes; (c) find potential at P_1; (d) $|\overline{E}|$ at P_1; (e) ρ_s at P_2. Assume that $V_0 = 100$ V, $\epsilon = 2\epsilon_0$, radius of inner conductor $r_a = 2.25$ cm, radius of outer conductor $r_b = 7.5$ cm, and the distance between axes $c = 3.75$ cm.

Solution. (a) From Fig. 7-1(a), we find that $N_t = 13.9 \times 2$ and $N_n = 4$. Through the use of (6), we have

$$C \cong \frac{N_t}{N_n}\epsilon d = \frac{13.9 \times 2}{4}(2\epsilon_0)(1) = 120 \text{ pF m}^{-1}$$

(b) Using the exact expression,

$$C = \frac{2\pi\epsilon}{\cosh^{-1}\left[\dfrac{r_a^2 + r_b^2 - c^2}{2r_a r_b}\right]} = \frac{2\pi(2\epsilon_0)}{\cosh^{-1}\left[\dfrac{(2.25)^2 + (7.5)^2 - (3.75)^2}{2(2.25)(7.5)}\right]}$$

$$= 128.27 \text{ pF m}^{-1}$$

Thus, our graphical solution is approximately 6.5% in error.

(c) From (7),

$$|\Delta V| \cong \frac{V_0}{N_n} = \frac{100}{4} = 25 \text{ V}$$

If we subdivide the curvilinear square in which P_1 is found, we find that there are 1.25 ΔV's as measured from the outer conductor. Thus,

$$V_{P_1} = 1.25\,\Delta V = 31.25 \text{ V}$$

(d) From (8), $|\overline{E}|$ can be found. To obtain greater accuracy, let us subdivide the large curvilinear square into four smaller squares to obtain $\Delta V' = \Delta V/2 = 12.5$ V and $\Delta \ell_n' = 1.0$ cm, as found on the smaller squares. Thus,

$$|\overline{E}| = \frac{\Delta V'}{\Delta \ell_n'} = \frac{\frac{25}{2}}{10^{-2}} = 1.25 \text{ kV m}^{-1}$$

(e) Let us subdivide the square in which we find P_2 into four squares to obtain $\Delta V' = \Delta V/2 = 12.5$ V and $\Delta \ell_n' = 0.4$ cm as found on the smaller squares. Thus, from (9),

$$\rho_s = \epsilon|\overline{E}|\Big|_{\text{cond.}} = 2\epsilon_0 \frac{\Delta V'}{\Delta \ell_n'}\Big|_{\text{cond.}} = 2\epsilon_0 \frac{12.5}{0.4 \times 10^{-2}} = 55.3 \text{ nC m}^{-2}$$

Problem 7.2-1 Two infinite length, concentric, and conducting cylinders of radii $r_a = 2$ cm and $r_b = 5$ cm are located on the z axis, as in Fig. 5-21. If $\epsilon = 3\epsilon_0$, $\rho_v = 0$, $V = V_0$ (V) at r_a, $V = 0$ at r_b, find through the use of curvilinear squares and check through the use of exact expressions: (a) capacitance per meter length; (b) $|D_n|$ at r_a; (c) $|D_n|$ at r_b; (d) ρ_s at r_a; (e) ρ_s at r_b. (f) See Example 6-5 for exact expressions and calculate percent errors based on exact values.

Problem 7.2-2 Two infinite and parallel circular conducting cylinders have axes in the $z = 0$ plane. Each has a radius of $r_a = 3$ cm, with their axes at $y = b = +10$ cm and $y = -10$ cm. Through the use of curvilinear squares: (a) find the capacitance per meter length; (b) check the results of part (a) through the use of the exact expression $C = \pi\epsilon_0/\cosh^{-1}(b/r_a)$ (F m^{-1}); (c) $|\overline{E}|$ at the origin; (d) find ρ_s at $y = 7$ cm. Assume that the potentials of the cylinders at $y = 10$ cm and at $y = -10$ cm are zero and 200 V, respectively, while $\epsilon = \epsilon_0$.

Problem 7.2-3 (a) Through the use of curvilinear squares, make an equipotential plot for the trough of Fig. 6-4 when $a = b$ and $\epsilon = \epsilon_0$. (b) When $V_0 = 100$ V, what is the potential at $(a/2, b/2, 0)$? Check the result of part (b) with the results of Prob. 6.5-1. (c) Which area of the cross section contributes the greatest to the capacitance between the trough and cap?

7.2-2 Curvilinear Square Solutions in a Conducting Medium

Let us now assume that the dielectric of Fig. 7-1(a) is replaced by a conducting medium of uniform σ and that a steady current flows between the two cylindrical conductor shells. Our first task is to show that under these conditions Laplace's equation holds in the conducting medium. From (5.2-11), we find that $\nabla \cdot \overline{J} = 0$ for a steady current. Now, from (5.3-3), $\overline{J} = \sigma\overline{E}$ and thus we find that $\nabla \cdot (\sigma\overline{E}) = 0$. If we set $\overline{E} = -\nabla V$, we obtain $\nabla \cdot (-\sigma\nabla V) = 0$; thus,

$$\boxed{\nabla^2 V = 0 \text{ V m}^{-2}} \tag{12}$$

which is Laplace's equation within a conducting medium. This now allows us to use the method of curvilinear squares in a manner quite parallel to that for the dielectric medium. The rules for drawing curvilinear squares in a conducting medium are:

I. Electric flux lines or current flow lines and equipotential surfaces intersect at right angles since $\overline{E} = -\nabla V$ and $\overline{E} = \overline{J}/\sigma$.
II. The two cylindrical surfaces, s_1 and s_2, are equipotential surfaces.
III. Electric flux lines or current flow lines begin on s_2 and end on s_1. The inner surface s_2 is assumed to be at a higher potential than the outer surface s_1. The current at any cross section between s_1 and s_2 is the the same since $\nabla \cdot \overline{J} = 0$.
IV. The $\overline{E}$, $\overline{D}$, and $\overline{J}$ are perpendicular to the surfaces s_1 and s_2.

A close examination of the above rules will reveal that we can replace the electric flux tubes in the dielectric medium by current flux tubes. Again, if we concentrate on curvilinear square 2 of Fig. 7-1(c), we find at point P_0 that

$$|\overline{J}| \cong \frac{\Delta I}{\Delta \ell_t d} \cong \sigma|\overline{E}| \tag{13}$$

where ΔI is the current flowing in a current flux tube instead of an electric flux tube. Also, at P_0 we have

$$|\overline{E}| \cong \frac{|\Delta V|}{\Delta \ell_n} \tag{14}$$

Equating $|\overline{E}|$ in (13) and (14) and rearranging, we have

$$\frac{|\Delta V|}{\Delta I} \cong \frac{1}{\sigma d}\left(\frac{\Delta \ell_n}{\Delta \ell_t}\right) \quad (\Omega) \tag{15}$$

For curvilinear squares, $(\Delta \ell_n/\Delta \ell_t) = 1$; thus (15) becomes

$$\frac{|\Delta V|}{\Delta I} \cong \frac{1}{\sigma d} = R_0 \quad (\Omega) \tag{16}$$

Equation (16) is the resistance R_0 of a curvilinear cell. Thus, the total resistance between surfaces s_1 and s_2 and depth d becomes

$$\boxed{R \cong R_0\frac{N_n}{N_t} \cong \frac{1}{\sigma d}\left(\frac{N_n}{N_t}\right) \quad (\Omega)} \tag{17}$$

The curvilinear cell conductance, from (16), becomes

$$\boxed{G_0 \cong \sigma d \qquad \text{(S)}} \tag{18}$$

From (17), the conductance between s_1 and s_2 and depth d (m) becomes

$$\boxed{G = \frac{1}{R} \cong \sigma d\left(\frac{N_t}{N_n}\right) \qquad \text{(S)}} \tag{19}$$

Thus, we can obtain R and G graphically through the use of (17) and (19), respectively. Note the similarity between (6) and (19). This similarity will be taken advantage of when we discuss analog solutions.

Example 2

When the dielectric between the eccentric cylinders of Fig. 7-1(a) is replaced by graphite, find for a length (depth) of 1 m: (a) the conductance between the cylinders; (b) the current flowing between the cylinders when $V_0 = 10$ V; (c) G_0; (d) R_0. Note $\sigma = 3 \times 10^4$ for graphite, from Table 5-1.

Solution. (a) Let us use the curvilinear squares plot of Fig. 7-1(a) and (19) to obtain

$$G \cong \sigma d\left(\frac{N_t}{N_d}\right) = (3 \times 10^4)(1)\left(\frac{13.9 \times 2}{4}\right) = 209 \text{ kS} \tag{20}$$

$$\text{(b) } I \cong GV_0 = (209 \times 10^3)(10) = 2090 \text{ kA} \tag{21}$$

$$\text{(c) From (18), } G_0 \cong \sigma d = \sigma = 3 \times 10^4 \text{ S}$$

$$\text{(d) } R_0 = \frac{1}{G_0} = \frac{1}{3 \times 10^4} = 33.3\ \mu\Omega$$

Problem 7.2-4 For the copper slab of Fig. 7-2, find the resistance R between the surfaces s_1 and s_2: (a) with the cooling hole shown; (b) without the cooling hole. Note that the location of the hole relative to s_1 and s_2 is not critical as long as it is not too close to the end surfaces.

Problem 7.2-5 Through the use of (5.4-5) and the dimensions of a curvilinear square, $\Delta\ell_t$, $\Delta\ell_n$, and d, show that $R_0 \cong 1/\sigma d$ for a curvilinear cell.

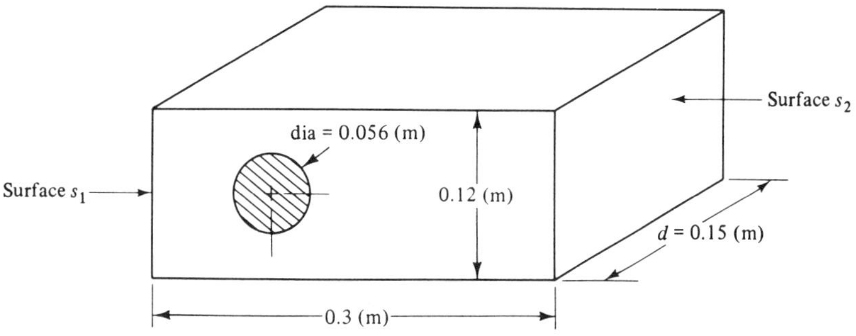

Figure 7-2 Graphical display of a copper slab with a cooling hole for Prob. 7.2-4.

7.3 NUMERICAL SOLUTION METHOD (ITERATION)

In this section we shall develop a *numerical solution* method that can be accomplished through the use of paper and pencil or a digital computer. This method can be applied to electrostatic problems, where the potential field is a function of one, two, or three variables and the boundary potentials are known. We will limit our discussion to potential fields that are functions of two variables. The accuracy of this solution method will depend on the amount of time available. High accuracy can be obtained through the use of digital computers.

The numerical solution method is based on the average-value property of Laplace's equation: the potential at a point is the average of the potentials about the point. We will develop this property by evaluating Laplace's equation at point P_0 in a potential field that is a function of two variables, x and y, as shown in Fig. 7-3. Thus, Laplace's equation becomes

$$\frac{\partial^2 V}{\partial x^2} + \frac{\partial^2 V}{\partial y^2} = 0 \tag{1}$$

Let us divide the space into squares measuring Δh by Δh, in a z = constant plane, and evaluate (1) at P_0 in terms of the known potentials V_1, V_2, V_3, and V_4. First, we will evaluate the partial derivatives at midpoints a and c to obtain

$$\left.\frac{\partial V}{\partial x}\right|_a \cong \frac{V_1 - V_0}{\Delta h} \tag{2}$$

$$\left.\frac{\partial V}{\partial x}\right|_c \cong \frac{V_0 - V_3}{\Delta h} \tag{3}$$

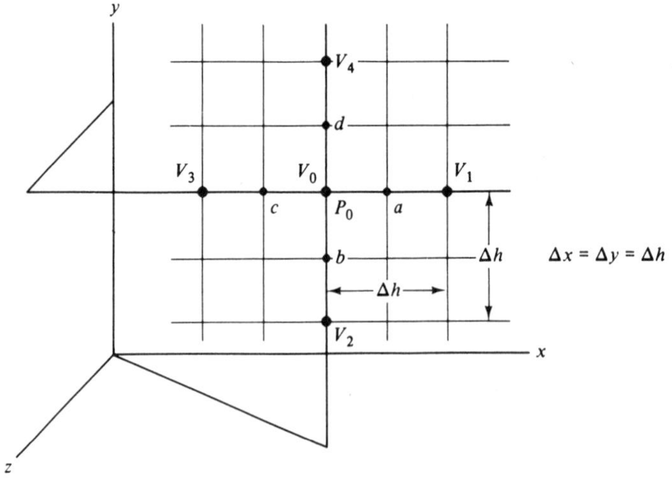

Figure 7-3 Graphical display of a region in a potential field that is divided into $\Delta x = \Delta y = \Delta h$ squares for the evaluation of the average-potential property of Laplace's equation at point P_0.

Now

$$\left.\frac{\partial^2 V}{\partial x^2}\right|_{P_0} = \frac{\partial}{\partial x}\left(\frac{\partial V}{\partial x}\right)\bigg|_{P_0} = \frac{\left.\frac{\partial V}{\partial x}\right|_a - \left.\frac{\partial V}{\partial x}\right|_c}{\Delta h} \tag{4}$$

Substituting (2) and (3) into (4), we obtain

$$\left.\frac{\partial^2 V}{\partial x^2}\right|_{P_0} = \frac{V_1 + V_3 - 2V_0}{(\Delta h)^2} \tag{5}$$

In a similar manner, we can show that

$$\left.\frac{\partial^2 V}{\partial y^2}\right|_{P_0} \cong \frac{V_4 + V_2 - 2V_0}{(\Delta h)^2} \tag{6}$$

Substituting (5) and (6) into (1), we obtain

$$\boxed{V_0 \cong \frac{V_1 + V_2 + V_3 + V_4}{4} \quad \text{(V)}} \tag{7}$$

Thus, we have proven that V_0 is the average of the potentials about the point P_0. This concept can be extended to Laplace's equation in three variables. The accuracy of (7) will depend on Δh (size of squares), and the exact value will be obtained when $\Delta h \to 0$.

We shall now apply this average value of potential concept to an infinite length conducting trough, as shown in Fig. 7-4, where the width is equal to 1.5 times the height. The cross section is divided into equal squares, and an initial approximation of potentials is made at each interior corner. The potentials at the corners on the boundary are known. When a computer is used, the initial potential values are set equal to zero. If a calculator or hand calculations are used, we should begin with reasonably good estimated values for our first set of interior corner potentials. Since the final solution is obtained by successive recalculation of the potential at each interior corner, better initial potential approximations will yield a final solution in fewer iterations. The numerical solution method is also called the *iteration method*. A set of potential approximations for interior corners is given in Table 7-1. Using row-column position designation found in a matrix, the approximate potential V_{33} is found by using the large diagonal of Fig. 7-4, with the following results:

$$V_{33} = \frac{V_{15} + V_{55} + V_{51} + V_{11}}{4} = \frac{100 + 0 + 0 + 50}{4} = 37.5 \text{ V}$$

where V_{11} is the gap voltage. Now, V_{22} and V_{42} are obtained through the use of smaller diagonals, yielding

$$V_{22} = \frac{V_{13} + V_{33} + V_{31} + V_{11}}{4} = \frac{100 + 37.5 + 0 + 50}{4} = 46.9 \text{ V}$$

and

$$V_{42} = \frac{V_{33} + V_{53} + V_{51} + V_{31}}{4} = \frac{37.5 + 0 + 0 + 0}{4} = 9.4 \text{ V}$$

TABLE 7-1 SET OF INITIAL APPROXIMATIONS FOR POTENTIALS AT INTERIOR CORNERS OF THE ELECTROSTATIC PROBLEM AS SHOWN IN FIG. 7-4

50 (V)	100 (V)	100 (V)	100 (V)	100 (V)
V_{11}	V_{12}	V_{13}	V_{14}	V_{15}
0 V_{21}	V_{22} 46.9	V_{23} 63.3	V_{24} 68.8	V_{25} 63.3
0 V_{31}	V_{32} 23.4	V_{33} 37.5	V_{34} 40.6	V_{35} 37.5
0 V_{41}	V_{42} 9.4	V_{43} 16.4	V_{44} 18.8	V_{45} 16.4
V_{51}	V_{52}	V_{53}	V_{54}	V_{55}
0	0	0	0	0

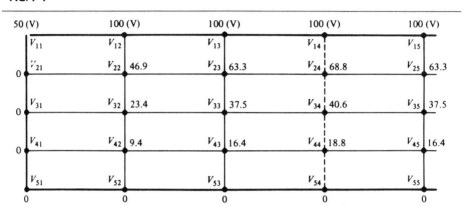

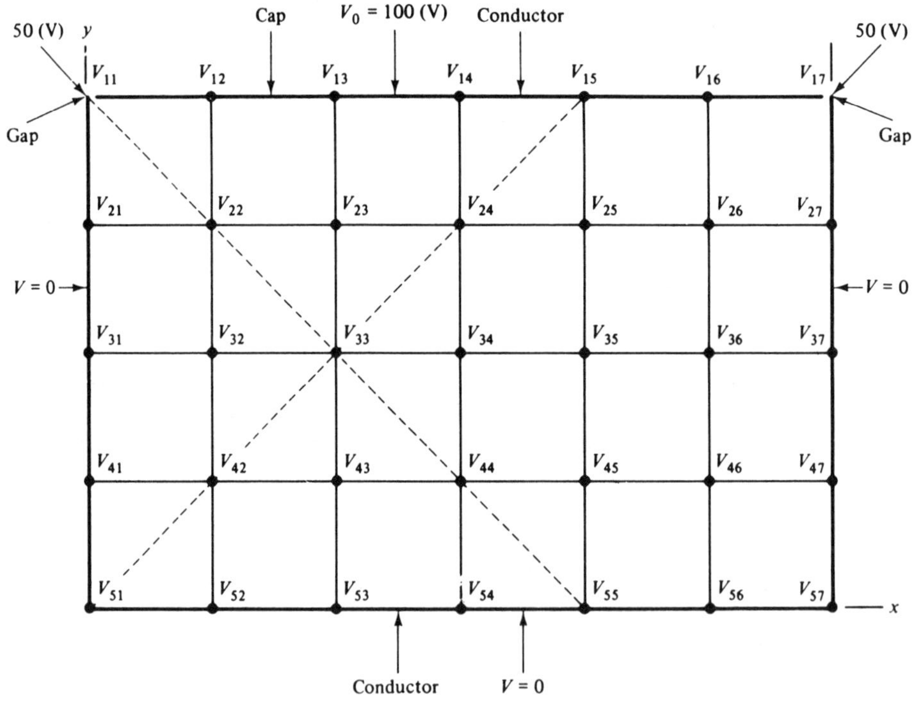

Figure 7-4 Rectangular trough for the evaluation of potential through the use of the numerical solution method (iteration). The potentials have the row-column subnotations as used in a matrix.

Next, V_{24} and V_{44} are obtained through the use of small diagonals to yield

$$V_{24} = \frac{V_{15} + V_{35} + V_{33} + V_{13}}{4} = \frac{100 + 37.5 + 37.5 + 100}{4} = 68.8 \text{ V}$$

and

$$V_{44} = \frac{V_{35} + V_{55} + V_{53} + V_{33}}{4} = \frac{37.5 + 0 + 0 + 37.5}{4} = 18.8 \text{ V}$$

where $V_{35} = V_{33}$ due to symmetry. Also due to symmetry, we find $V_{26} = V_{22}$ and $V_{46} = V_{42}$. The remaining potentials V_{32}, V_{23}, V_{34}, and V_{43} can be approximated through the use of (7) without using diagonals to yield the values shown in Table 7-1. The diagonal-average is used only in finding the initial approximations. Using these approximations, all potentials are now improved by applying (7) to find V_{22}, V_{23}, V_{24}, V_{32}, V_{33}, V_{34}, V_{42}, V_{43}, and V_{44}. The newest values of potential are used. Note that we have taken advantage of symmetry. Upon completion of this round of recalculation, the process is repeated

TABLE 7-2 COMPLETE SET OF POTENTIAL SOLUTIONS FOR THE TROUGH OF FIG. 7-4

V_{11}: 50 (V)	V_{12}: 100 (V)	V_{13}: 100 (V)	V_{14}: 100 (V)	V_{15}: 100 (V)
V_{21}: 0	V_{22}: 46.9 46.7 46.6 46.4 46.2 46.0 46.0 46.0 45.9 45.9	V_{23}: 63.3 63.2 62.3 61.9 61.6 61.5 61.4 61.3 61.2 61.2	V_{24}: 68.8 66.8 66.0 65.6 65.4 65.2 65.1 65.0 65.0 65.0	V_{25}: 63.3 63.2 62.3 61.9 61.6 61.5 61.4 61.3 61.2 61.2
V_{31}: 0	V_{32}: 23.4 23.4 23.1 22.8 22.6 22.5 22.4 22.4 22.4 22.4	V_{33}: 37.5 35.9 35.2 34.8 34.5 34.3 34.2 34.1 34.1 34.1	V_{34}: 40.6 39.4 38.6 38.2 37.9 37.7 37.6 37.6 37.6 37.6	V_{35}: 37.5 35.9 35.2 34.8 34.5 34.3 34.2 34.1 34.1 34.1
V_{41}: 0	V_{42}: 9.4 10.0 9.8 9.6 9.5 9.4 9.4 9.4 9.4 9.4	V_{43}: 16.4 16.2 15.8 15.5 15.3 15.2 15.2 15.1 15.1 15.1	V_{44}: 18.8 18.0 17.6 17.3 17.1 17.0 17.0 17.0 17.0 17.0	V_{45}: 16.4 16.2 15.8 15.5 15.3 15.2 15.2 15.1 15.1 15.1
V_{51}: 0	V_{52}: 0	V_{53}: 0	V_{54}: 0	V_{55}: 0

Line of symmetry (V_{14}–V_{54} column)

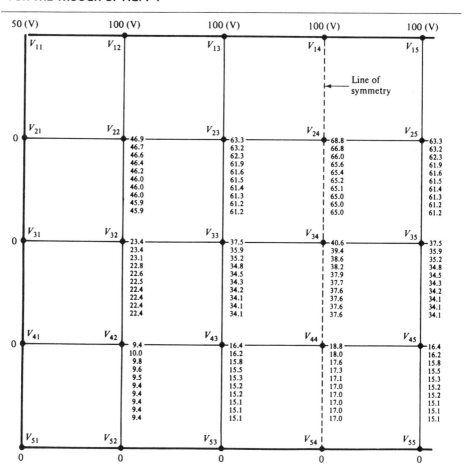

starting with V_{22} and using the same sequence as above. This iteration process is continued until there is no change in all of the recalculated values after they are rounded off at the desired digit position. A complete set of potential solutions is shown in Table 7-2, where the potentials were rounded off to the tenth's digit position.[3] Rounding off to a lower digit position does not necessarily increase the accuracy of the final potential values. Increased accuracy can only be obtained by increasing the number of squares. This leads to a greater number of calculations since the number of potential points and iterations will increase.

Equipotential lines can be drawn from the final potentials by approximating locations of equal potential and then connecting them by a smooth curve. From this plot we can obtain $\overline{E}$ and $\overline{D}$ everywhere and ρ_s on the conductors. In parts of a problem where the cross section cannot be divided into squares, we are not allowed to use (7) since $\Delta x \neq \Delta y = \Delta h$. A problem at the end of the chapter will pursue this condition. For a more complete discussion of the numerical solution method, see J. B. Scarborough.[4]

Problem 7.3-1 Divide the trough of Fig. 7-4 into six squares, and through the numerical solution method find the potentials at the two interior points. Assume the same boundary potentials and round off to the tenth's digit position. Compare the results with those in Table 7-2.

Problem 7.3-2 Let $a = b = 8$ (cm) in the trough of Fig. 6-4, and divide the interior space into 16 squares. (a) Now, through the use of the numerical solution method, find the potentials at the nine interior points when $V_0 = 100$ V. (b) Check the result found at the point $x = a/4$ and $y = b/4$ with that found in Prob. 6.5-1, or through the use of (6.5-28).

7.4 IMAGE SOLUTION METHOD

There are many mathematically formidable electrostatic problems that can be solved exactly and with great ease through the use of *image theory*. A good example of image theory exists in a problem of an infinite length line charge parallel to a grounded conductor of infinite extent, as suggested in Fig. 7-5(a). The charge on the conductor plane per unit length in the x direction will be $-\rho_\ell$ (C m^{-1}) when the line charge is $+\rho_\ell$ (C m^{-1}). Image theory allows us to replace the infinite conductor by an infinite length line charge of uniform $-\rho_\ell$ located at a (m) to the left of the origin, as shown in Fig. 7-5(b). The potential $V(y, z)$ at any point in the right half-space in Fig. 7-5(a) and (b) will be identical.

From Fig. 7-5(b), it can be seen that $\overline{E}$ is perpendicular to the $y = 0$ plane, and thus the absolute potential is equal to zero on this plane. Since the potential on the $y = 0$ plane is the same in Fig. 7-5(a) and (b), the uniqueness theorem assures us that $V(y, z)$ is the same in both right half-spaces. The right half-space includes the $y = 0$ plane.

Through the use of the image theory, we can find $V(y, z)$, $\overline{E}$, and $\overline{D}$ everywhere, and ρ_s on the conductor of the electrostatic problem of Fig. 7-5(a), by replacing the infinite conductor with an infinite length line charge and solving the problem of Fig. 7-5(b) for the desired quantities. The ρ_s variation in the z direction on the infinite conductor

[3] For a hundredth's position of five, the tenth's position was made even.

[4] J. B. Scarborough, *Numerical Mathematical Analysis*, 6th ed., The Johns Hopkins University Press, Baltimore, Md., 1966.

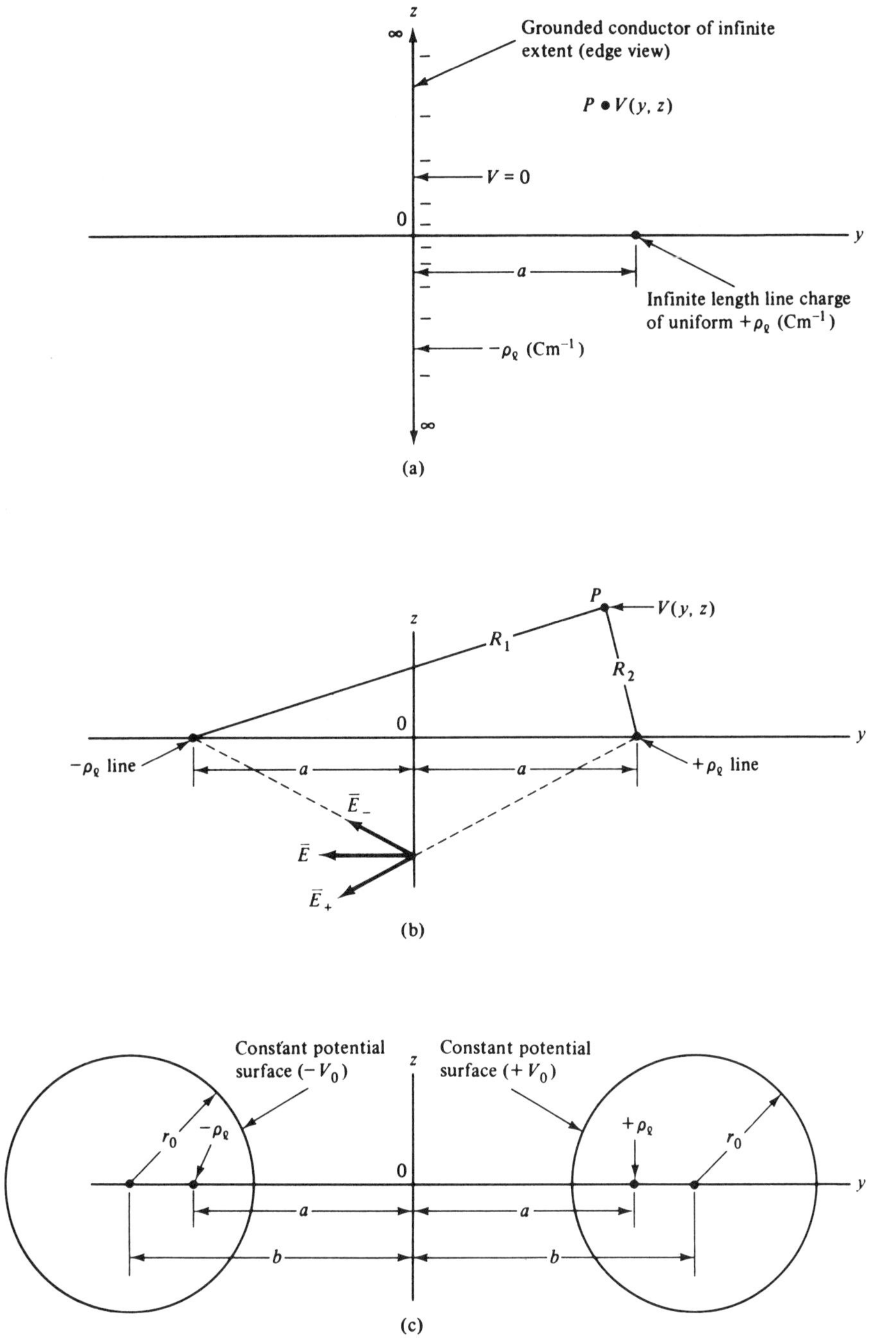

Figure 7-5 (a) Infinite line of charge parallel to a grounded conductor plane of infinite extent. (b) The conductor plane has been replaced by an image line charge a meters to the left of the plane. The $\bar{E}$ at the $y = 0$ plane is perpendicular to the $y = 0$ plane. (c) Constant $+V_0$ and $-V_0$ potential surfaces (circles) as found in Example 3.

can be found from the knowledge of $\overline{D}$ in the $y = 0$ plane. Thus, a formidable problem has been solved exactly and with great ease.

Example 3

Find the potential $V(y, z)$ in both half-spaces of the electrostatic problem shown in Fig. 7-5(b), and prove that the constant potential surfaces are circular cylinders.

Solution. The potential $V(y, z)$, due to two line charges, becomes

$$V(y, z) = V_+ + V_- \tag{1}$$

where V_+ and V_- are the potentials at point P due to the $+\rho_\ell$ line and the $-\rho_\ell$ line acting alone, respectively. With the origin as the potential reference point, we have

$$V_+ = V_{PO+} = \frac{+\rho_\ell}{2\pi\epsilon} \ln\left(\frac{a}{R_2}\right) \tag{2}$$

and

$$V_- = V_{PO-} = \frac{-\rho_\ell}{2\pi\epsilon} \ln\left(\frac{a}{R_1}\right) \tag{3}$$

as found in (5.10-11). Thus, (1) becomes

$$V(y, z) = V_+ + V_- = \frac{+\rho_\ell}{2\pi\epsilon} \ln\left(\frac{a}{R_2}\right) + \frac{-\rho_\ell}{2\pi\epsilon} \ln\left(\frac{a}{R_1}\right)$$

$$= \frac{\rho_\ell}{2\pi\epsilon} \ln\left(\frac{R_1}{R_2}\right) \tag{4}$$

From Fig. 7-5(b), we find $R_1 = [(a + y)^2 + z^2]^{1/2}$ and $R_2 = [(a - y)^2 + z^2]^{1/2}$. Substituting these expressions into (4), we obtain

$$\boxed{V(y, z) = \frac{\rho_\ell}{2\pi\epsilon} \ln\left[\frac{(a + y)^2 + z^2}{(a - y)^2 + z^2}\right]^{1/2} \quad \text{(V)}} \tag{5}$$

Let us now set $V(y,z)$ equal to a constant potential V_0 and find the equation of the equal potential surface. Thus, (5) becomes

$$\boxed{V_0 = \frac{\rho_\ell}{2\pi\epsilon} \ln\left[\frac{(a + y)^2 + z^2}{(a - y)^2 + z^2}\right]^{1/2} \quad \text{(V)}} \tag{6}$$

Rearranging (6), we obtain

$$\frac{2\pi\epsilon V_0}{\rho_\ell} = \ln\left[\frac{(a + y)^2 + z^2}{(a - y)^2 + z^2}\right]^{1/2} \tag{7}$$

and

$$e^{2\pi\epsilon V_0/\rho_\ell} = K_0 = \left[\frac{(a + y)^2 + z^2}{(a - y)^2 + z^2}\right]^{1/2} \tag{8}$$

where K_0 is a constant for a particular equipotential surface, V_0 in this case. The last two terms of (8) will expand into

$$y^2 + z^2 + a^2 - 2ay\frac{(K_0^2 + 1)}{(K_0^2 - 1)} = 0 \tag{9}$$

Adding $a^2(K_0^2 + 1)^2/(K_0^2 - 1)^2$ to both sides of (9) to complete the square, we obtain

$$\left(y - a\frac{K_0^2 + 1}{K_0^2 - 1}\right)^2 + z^2 = \left(\frac{2K_0 a}{K_0^2 - 1}\right)^2 \tag{10}$$

Equation (10) is an equation of a circle of the form

$$(y - b)^2 + z^2 = r_0^2 \tag{11}$$

where y and z are the coordinates of a point on the constant V_0 potential surface, b is the y coordinate of the center of the circle, and r_0 is the radius of the circle shown in Fig. 7-5(c). From (10) and (11), we have

$$\boxed{b = a\frac{K_0^2 + 1}{K_0^2 - 1} \quad \text{(m)}} \tag{12}$$

and

$$\boxed{r_0 = \frac{2K_0 a}{K_0^2 - 1} \quad \text{(m)}} \tag{13}$$

A complete set of constant potential surfaces (circles) is shown in Fig. 7-6. The electric flux lines have been added to complete the plot.

From the results of Example 3, the image theory can be extended to even more formidable electrostatic problems as we shall discover in Examples 4 and 5.

Example 4

Find the potential field $V(y, z)$ about two infinite-length cylindrical conductors separated $2b$ (m) and of r_0 (m) radii, as shown in Fig. 7-7(a).

Solution. Let us return to Example 3 and, in particular, Fig. 7-5(c). If we place two extremely thin conducting cylindrical shells of radii r_0 at the $+V_0$ and $-V_0$ constant potential locations, the potentials on the two shells will be $+V_0$ and $-V_0$, respectively, as found in Fig. 7-7(b). The charges on the surfaces of the thin shell conductors will be as shown in Fig. 7-7(b). The charge on the right shell will be equal to $+\rho_\ell$ on the outer surface and $-\rho_\ell$ on the inner surface. The reverse is true for the left shell. Note that the ρ_s is not uniform over the circumference of the shell. Now, let us discharge the charge on the $+\rho_\ell$ and $-\rho_\ell$ lines by shorting them to the $-\rho_\ell$ on the inner surface of the right and $+\rho_\ell$ on the inner surface of the left shell, respectively. Thus, we find that the charge is present only on the outer surfaces of the thin shells, as shown in Fig. 7-7(c). From the uniqueness theorem, we find that $V(y, z)$ external to the two cylinders will be identical to that found in the same region due to two line charges that are located as in Fig. 7-6 or 7-7(b).

The $\overline{E}$ field for $r < r_0$, inside the cylinder, is found to be zero. Thus, we can replace the thin conducting shell with a solid conducting shell, as shown in Fig. 7-7(a). Our solution for $V(y,z)$ external to the cylindrical conductors can be found through the use of image line charges located at appropriate locations. A relationship between a, b, and r_0 can be found by forming the difference of the squares of (12) and (13) to eliminate K_0 that yields

$$\boxed{a = (b^2 - r_0^2)^{1/2} \quad \text{(m)}} \tag{14}$$

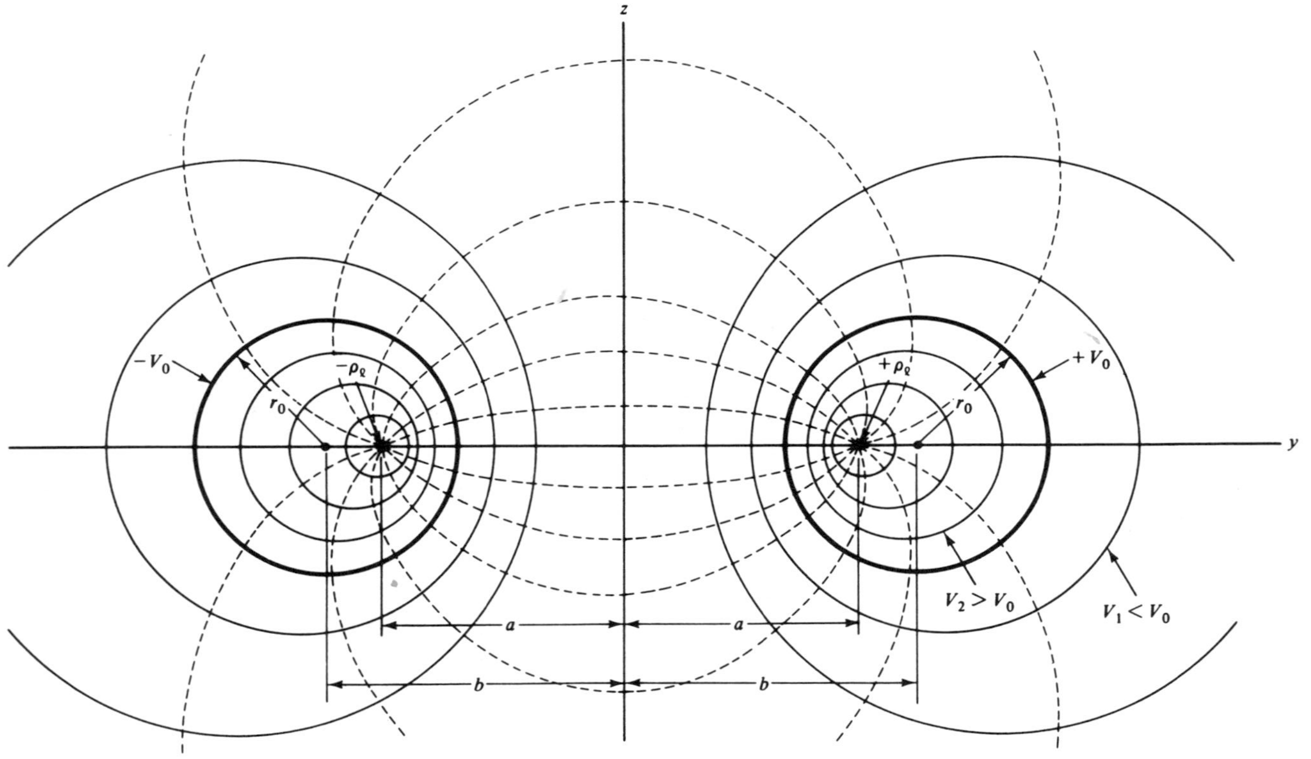

Figure 7-6 Complete plot of constant potential surfaces (circles) and electric flux lines about two infinite line charges for Example 3.

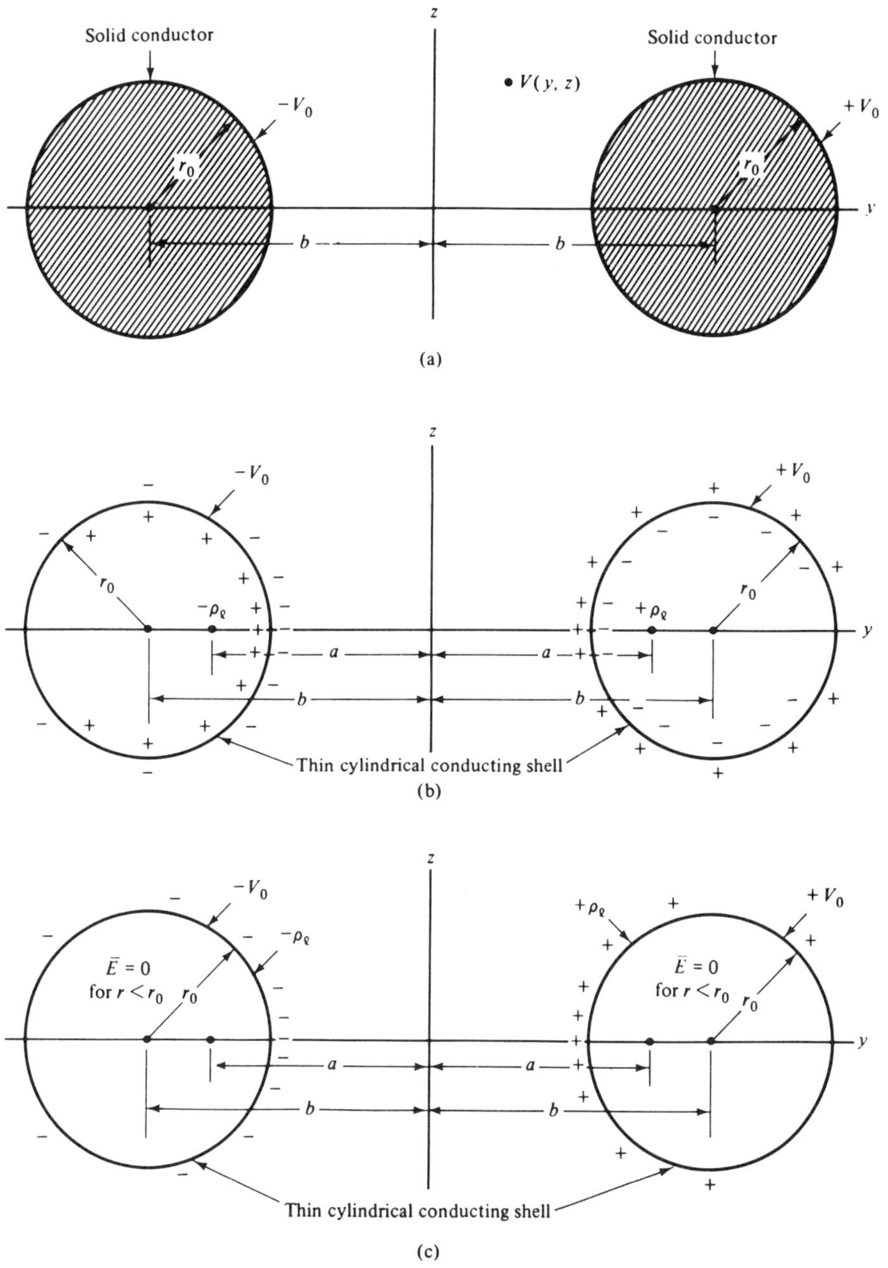

Figure 7-7 Graphical display for Example 4: (a) two infinite-length cylindrical conductors; (b) charges on two extremely thin conducting cylindrical shells of radii r_0 at the $+V_0$ and $-V_0$ constant potential surfaces as produced by two line charges; (c) the line charges in part (b) cancel the charges on the inner surfaces of the thin conductor.

Note that a is independent of K_0 and thus V_0. Therefore, the potential $V(y, z)$, exterior to the two conductors, can be found through the use of (5) when a is obtained from (14). The $\overline{E}$ and $\overline{D}$ anywhere external to the conductors and on the conductor surface can be obtained through the use of $\overline{E} = -\nabla V(y, z)$ or from the sum of the $\overline{E}$'s from the two image line charges acting alone.

Example 5

Find the capacitance per meter length of the two cylindrical conductors shown in Fig. 7-7(a).

Solution. Let us solve for a in (12) and (13), equate, and then solve for K_0 to obtain

$$K_0 = \frac{b \pm (b^2 - r_0^2)^{1/2}}{r_0} \tag{15}$$

where the plus and minus signs go with $+V_0$ and $-V_0$ constant potential surfaces, respectively. Now, from (6), (8), and (15), we obtain

$$\boxed{V_0 = \frac{\rho_\ell}{2\pi\epsilon} \ln (K_0) = \frac{\rho_\ell}{2\pi\epsilon} \ln \left[\frac{b + (b^2 - r_0^2)^{1/2}}{r_0}\right]} \tag{16}$$

for the potential of the right conductor with respect to the origin or the $y = 0$ plane. The potential between conductors is thus equal to $2V_0$ (V). The capacitance per meter length for two cylinders each of radius r_0 (m) separated by $2b$ (m) between centers is

$$\boxed{C = \frac{\rho_\ell}{2V_0} = \frac{\rho_\ell}{\dfrac{\rho_\ell}{\pi\epsilon} \ln \left[\dfrac{b + (b^2 - r_0^2)^{1/2}}{r_0}\right]} = \frac{\pi\epsilon}{\ln \left[\dfrac{b + (b^2 - r_0^2)^{1/2}}{r_0}\right]} \quad (\text{F m}^{-1})} \tag{17}$$

One can easily show that (17) reduces to

$$C \approx \frac{\epsilon\pi}{\ln (2b/r_0)} \quad (\text{F m}^{-1}) \tag{18}$$

when the separation between wires $2b$ is much greater than the radius r_0.

Example 6

Obtain the expression for the capacitance per meter between a circular cylindrical conductor and a conductor plane of infinite extent, as shown in Fig. 7-8.

Solution. Assume the infinite conductor plane to be at zero potential and the circular cylindrical conductor to be at V_0 potential. For the given b and r_0, through the use of (14) we can find the location of the equivalent $+\rho_\ell$ line charge that replaces the circular cylinder. The infinite conductor plane is replaced by the $-\rho_\ell$ line charge, as shown in Fig. 7-8. Now, the $V(y, z)$ between the infinite plane and the circular conductor is the same as that found using the two line charges. From (16), the expression for the constant potential V_0 becomes

$$\boxed{V_0 = \frac{\rho_\ell}{2\pi\epsilon} \ln \left[\frac{b + (b^2 - r_0^2)^{1/2}}{r_0}\right]} \tag{16}$$

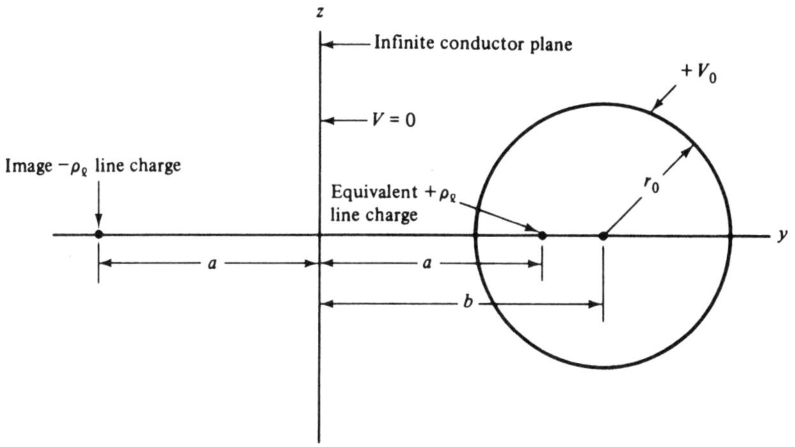

Figure 7-8 Graphical display of a circular cylindrical conductor parallel to an infinite conductor plane for Example 6.

where ρ_ℓ is the charge per unit length on either the circular cylinder conductor or the equivalent $+\rho_\ell$ line charge. The capacitance per meter between cylinder and plane is

$$C = \frac{\rho_\ell}{V_0} = \frac{2\pi\epsilon}{\ln\left[\dfrac{b + (b^2 - r_0^2)^{1/2}}{r_0}\right]} \quad (\text{F m}^{-1}) \tag{19}$$

This problem could have been generated by placing a thin infinite conductor plane in the $y = 0$ plane of Fig. 7-7(c) and then eliminating the left-hand circular cylindrical conductor.

The more general problem of the capacitance between parallel cylinders of different radii, both for external cylinders or for one cylinder inside the other, is included as an exercise for the student at the end of the chapter (see Prob. 7-17).

The image theory can be extended to a point charge above an infinite conductor plane, conductor sphere above an infinite conductor plane, and two conductor spheres (see Corson and Lorrain[5]).

A current-carrying conductor above an infinite conductor plane can be imaged in the same manner as charges since current is due to moving charges. Let us find the image of a filamentary conductor carrying a current I parallel to and a distance of b (m) above an infinite conductor plane, as shown in Fig. 7-9(a). A $+dq$ charge, traveling to the right with a velocity $\overline{U}$, has been singled out in the real conductor above the infinite conductor plane. The image of this charge is $-dq$ located b (m) below the infinite conductor plane and travels with velocity $\overline{U}$ to the right. Thus, the conventional current in the image conductor is $-I$ and flows in the opposite direction, as shown in Fig. 7-9(a). Figure 7-9(b) and (c) show the images of perpendicular and slanted conductors, respectively. In all cases, the current in the image conductor is opposite to the current in the

[5] Dale Corson and Paul Lorrain, *Introduction to Electromagnetic Fields and Waves,* W. H. Freeman and Company, Publishers, San Francisco, 1962, pp. 134–143.

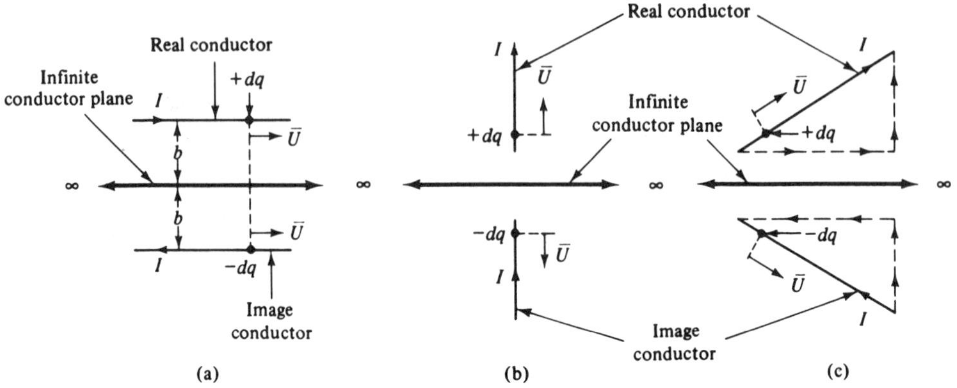

Figure 7-9 Image of a conductor above an infinite conductor plane: (a) parallel; (b) perpendicular; (c) slanted.

real conductor when we assume the conductor plane as reference. These image concepts are used extensively in finding the electromagnetic fields from an antenna above a ground plane.

Problem 7.4-1 For the infinite grounded conductor plane separated 5 m from a line charge of $\rho_\ell = 4 \times 10^{-9}$ C m^{-1}, shown in Fig. 7-5(a), find: (a) the potential at the point $(0, 6, 6)$; (b) the location of the center of the constant potential cylinder, and its radius, that passes through $(0, 6, 6)$. Assume that $\epsilon = \epsilon_0$.

Problem 7.4-2 Find the ρ_s distribution on the grounded plane in Prob. 7.4-1.

Problem 7.4-3 Show that the approximate capacitance per meter between a long conductor parallel to a ground plane and that plane is

$$C \approx \frac{2\pi\epsilon}{\ln(4H/d)} \qquad (\text{F m}^{-1})$$

where H is the height of the wire above the ground plane and d is the diameter of the wire.

7.5 ANALOG SOLUTION METHOD

In the curvilinear squares solution for a dielectric medium, we found the capacitance, d (m) in depth, of a two-variable electrostatic configuration to be

$$\boxed{C \cong \frac{N_t}{N_n}\epsilon d \qquad (\text{F})} \qquad (7.2\text{-}6)$$

Now, in the curvilinear squares solution for a conducting medium, we found the conductance, d (m) in depth, of a two-variable electrostatic configuration to be

$$\boxed{G \cong \frac{N_t}{N_n}\sigma d \qquad (\text{S})} \qquad (7.2\text{-}19)$$

If the above two-variable configurations are identical, we will find N_t/N_n to be the same in (7.2-6) and (7.2-19). If we divide (7.2-6) by (7.2-19), we obtain

$$\boxed{\frac{C}{\epsilon} = \frac{G}{\sigma}} \tag{1}$$

This equation indicates that if we have a solution for the capacitance C, no matter how obtained, we can solve for the conductance G of the same two-variable configuration through the use of (1) and the knowledge of the ϵ of the dielectric medium and the σ of the conducting medium. We can say that (1) is due to the analog between the electric flux in the dielectric medium and the current in the conductor medium. This analog is developed in Table 7-3 for the steady current case, $\partial\rho_v/\partial t = 0$; thus, $\nabla \cdot \overline{J} = 0$.

From this it can be seen that the current I is the analog of the electric flux Ψ_E, which is why in the curvilinear squares solution of a conductor medium we used current tubes while in a dielectric medium we used electric flux tubes. Note again that Laplace's equation is satisfied in both cases. It should be pointed out that the analog between $\overline{J}$ and $\overline{D}$ applies to one-, two-, and three-variable electrostatic problems. Thus, we can see that this is a powerful solution technique.

The curvilinear squares solution in a conductor medium, as found in Sec. 7.2-2, is the analog of the curvilinear squares solution in a dielectric medium, as found in Sec. 7.2-1.

TABLE 7-3 ANALOG BETWEEN QUANTITIES IN A DIELECTRIC MEDIUM AND A CONDUCTING MEDIUM

Dielectric medium ($\rho_v = 0$)	Conductor medium (steady current) ($\partial\rho_v/\partial t = 0$)
$\nabla \cdot \overline{D} = 0$	$\nabla \cdot \overline{J} = 0$ (steady current)
$\overline{D} = \epsilon\overline{E}$	$\overline{J} = \sigma\overline{E}$
$\overline{E} = -\nabla V$	$\overline{E} = -\nabla V$
$\nabla \cdot (\epsilon\overline{E}) = 0$	$\nabla \cdot (\sigma\overline{E}) = 0$
$\nabla \cdot (-\epsilon\nabla V) = 0$	$\nabla \cdot (-\sigma\nabla V) = 0$
$\nabla^2 V = 0$ (Laplace's equation)	$\nabla^2 V = 0$ (Laplace's equation)
$\Psi_E = \int_s \overline{D} \cdot \overline{ds}$ (flux of $\overline{D}$)	$I = \int_s \overline{J} \cdot \overline{ds}$ (flux of $\overline{J}$)
$C = \dfrac{\lvert Q\rvert}{\lvert V_{ab}\rvert} = \dfrac{\left\lvert\int_s \overline{D} \cdot \overline{ds}\right\rvert}{\left\lvert\int_b^a -\overline{E} \cdot \overline{d\ell}\right\rvert} \cong \dfrac{N_t}{N_n}\epsilon d$	$G = \dfrac{\lvert I\rvert}{\lvert V_{ab}\rvert} = \dfrac{\left\lvert\int_s \overline{J} \cdot \overline{ds}\right\rvert}{\left\lvert\int_b^a -\overline{E} \cdot \overline{d\ell}\right\rvert} \cong \dfrac{N_t}{N_n}\sigma d$

$$\overline{D} \longleftrightarrow \overline{J}$$
$$\overline{E} \longleftrightarrow \overline{E}$$
$$\epsilon \longleftrightarrow \sigma$$
$$V \longleftrightarrow V$$
$$\Psi_E \longleftrightarrow I$$
$$C \longleftrightarrow G$$

Problem 7.5-1 The capacitance of a two-variable electrostatic problem was found by curvilinear squares solution to equal 10^{-6} F m^{-1} when the region between the conductor boundaries was filled with a dielectric whose $\epsilon_r = 4$. Find the conductance of the same two-variable configuration when the dielectric medium is replaced by a conductor medium whose $\sigma = 3 \times 10^4$ S m^{-1}.

Problem 7.5-2 Draw the equivalent circuit of a lossy capacitor whose dielectric has the properties $\epsilon_r = 5$ and $\sigma = 10^{-4}$ S m^{-1}, if the capacitor has a capacitance of 300 pF when the dielectric is air.

7.6 EXPERIMENTAL SOLUTION METHODS

When the current-electric flux analog is used in a laboratory measurement of a conductance G (or resistance R) to find capacitance C through the use of (7.5-1), or measurement of potential to obtain a potential field plot, the procedure is commonly referred to as an *experimental solution method.* Two commonly used laboratory setups are the uniform resistance paper, such as Teledeltos paper,[6] and the electrolytic tank, as illustrated in Fig. 7-10(a) and (b), respectively.

The outline of a two-variable electrostatic problem, trough in this case, is painted on the Teledeltos paper with a conducting silver paint, as shown in Fig. 7-10(a). In this setup, a potential V_0 is applied between the cap and walls of the trough and a high resistance voltmeter is used to probe constant potential contours or to measure potential at

[6] Supplied by the Pasco Scientific Co., San Leandro, Calif.

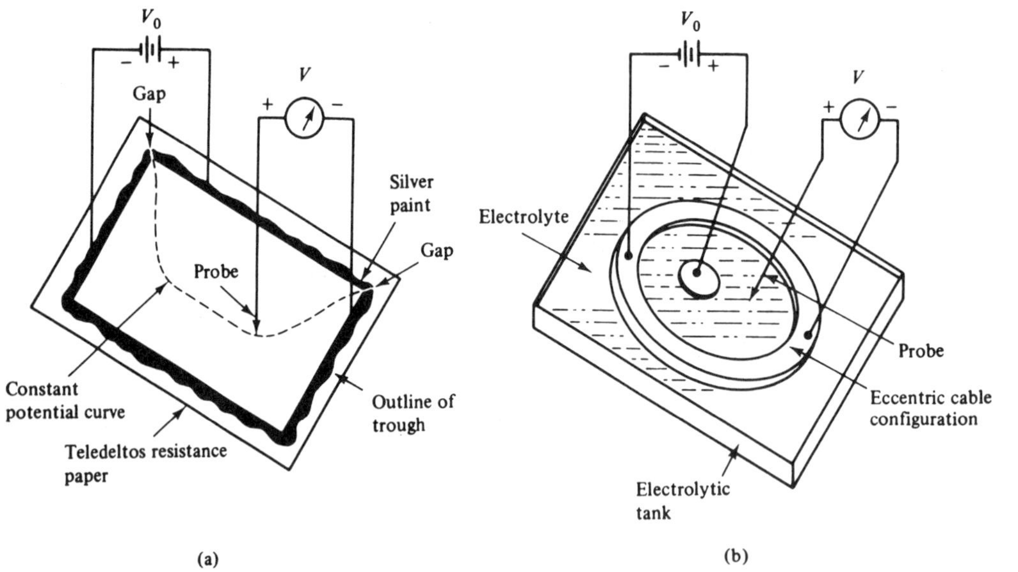

Figure 7-10 (a) Teledeltos resistance paper with the outline of a trough, painted with a conducting silver paint, used to obtain an equipotential field plot. (b) Electrolytic tank, with partially submerged metal configuration of an eccentric cable cross section, used to obtain an equipotential field plot. Both laboratory setups can also be used to obtain capacitance through conductance measurements.

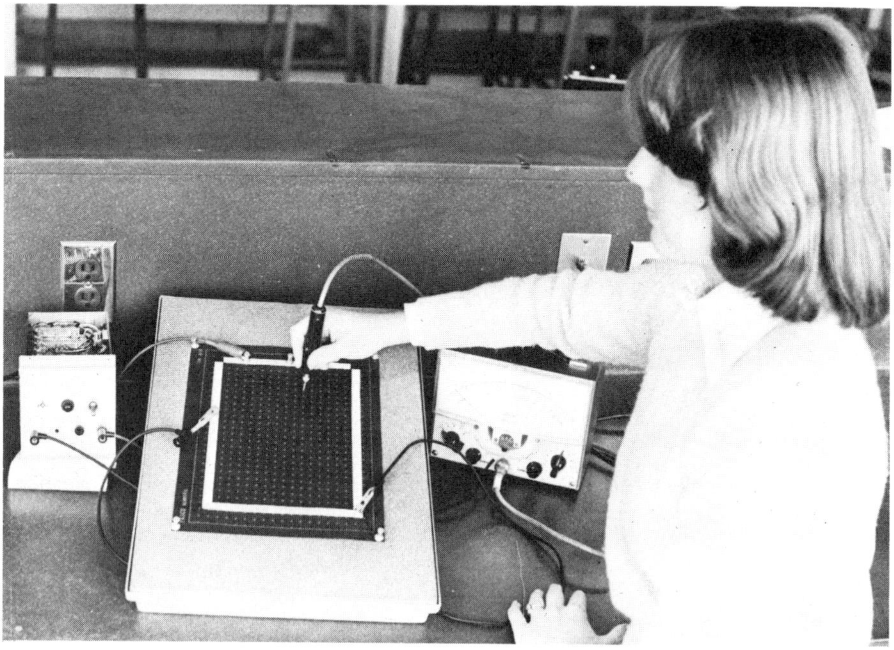

Figure 7-11 Actual laboratory setup, using resistance paper to obtain field measurements in a two-variable electrostatic trough problem.

predetermined square locations. The capacitance per meter of a dielectric filled trough can be obtained through the conductance (or resistance) measurement and the use of (7.5-1). The resistance paper must have uniform resistance, and the resistance per square must be known.[7] An actual laboratory setup is shown in Fig. 7-11.

Figure 7-10(b) shows a partially submerged metal configuration of a two-variable eccentric cable cross section in a conducting electrolytic solution. Measurements identical to those made on the resistance paper can also be made in an electrolytic tank. The depth of electrolyte must be uniform to obtain accurate results.

Example 7

A resistance paper is used to obtain the capacitance of a two-variable electrostatic problem. The cross-sections of the two conductors were drawn with conducting silver paint on the resistance paper and the resistance between the two electrodes was found to be 6660 Ω. If the resistance of the resistance paper is 10^5 Ω per square and the ϵ_r of an identical dielectric cross section is equal to 6.5, find the capacitance per meter of this two-variable electrostatic problem.

Solution. The expression for the conductance of the resistance paper, as measured between the two electrodes, from (7.2-18) and (7.2-19) is

$$G_p = \sigma_p d_p \left(\frac{N_t}{N_n}\right) = G_{0p}\left(\frac{N_t}{N_n}\right) \tag{7.2-19}$$

[7] Resistance per square is the resistance between opposite edges of any size square and equals R_0 (curvilinear cell resistance).

where the subscript p has been used to indicate resistance paper. The curvilinear cell conductance G_{0p} of the resistance paper is

$$G_{0p} = \sigma_p d_p = \frac{1}{R_{0p}} = 10^{-5}\ \text{S}$$

Now, $G_p = \frac{1}{6660}$ (S), and from (7.2-19) we can solve for N_t/N_n:

$$\frac{N_t}{N_n} = \frac{G_p}{G_{0p}} = \frac{\frac{1}{6660}}{10^{-5}} = 15$$

Now, from (7.2-6) we have

$$C = \frac{N_t}{N_n}\epsilon d \qquad (7.2\text{-}6)$$

Thus, the capacitance per meter becomes

$$\frac{C}{d} = 15(\epsilon_0\epsilon_r) = 15(8.85 \times 10^{-12})(6.5) = 863\ \text{pF m}^{-1}$$

Problem 7.6-1 A resistor is to be formed from the electrode configuration of Example 7. If the resistance is to be 0.5 Ω for $d = 1$ m, find the conductivity required for the material.

7.7 SUMMARY OF SOLUTION METHODS

In this chapter the following solution methods have been discussed:

1. Graphical (curvilinear squares)
   (a) In dielectric medium
   (b) In conductor medium (analog of solution in dielectric medium)
2. Numerical (iteration)
3. Image
4. Analog
5. Experimental

Table 7-4 shows how the above solution methods related to:

1. Number of coordinate variables involved
2. Quantities solvable
3. Solution accuracy (percent error)
4. Computer adaptability
5. Fundamental underlying concept

REVIEW QUESTIONS

1. What are the four basic rules that govern the construction of curvilinear squares? *Sec. 7.2-1*
2. Is the curvilinear squares solution method based on Laplace's equation? *Sec. 7.2-1*
3. What is the definition of N_t and N_n? *Sec. 7.2-1*
4. What is a curvilinear square? *Sec. 7.2-1*
5. Is ΔV between any adjacent equipotential surfaces in a curvilinear squares solution equal to a constant potential difference? *Sec. 7.2-1*
6. What is $(N_t/N_n)\epsilon d$ equal to? *Eq. (7.2-6)*
7. What is ϵd equal to? *(Eq. 7.2-10)*
8. Is greater accuracy obtained when smaller curvi-

TABLE 7-4 SUMMARY OF SOLUTION METHODS

Solution method	Number of coordinate variables involved	Quantities solvable	Solution accuracy (% error)	Computer adaptability	Fundamental underlying concept
Graphical (curvilinear squares					
Dielectric media	1 or 2	$V, \overline{E}, \overline{D}, \rho_s, C$	5–10%; depends on number of squares and graphical technique	No	Laplace's equation
Conductor media	1 or 2	$V, \overline{E}, \overline{D}, \overline{J}, G$			Laplace's equation[a]
Numerical (iteration)	2; can be extended to 3	$V, \overline{E}, \overline{D}, \rho_s$	Depends on number of squares	Yes	Laplace's equation
Image	2 (line charges) 3 (point charges)	$V, \overline{E}, \overline{D}, \rho_s, C$	Exact	No	Uniqueness theorem
Analog	1, 2, or 3	C, G, R	Depends on original solution	No	Identical equation forms
Experimental					
Resistance paper	1 or 2	V, R, G, C	Depends on equipment and technique used	Automatic probing and display through computer use	(Analog) identical equation forms
Electrolytic tank	1, 2, or 3				

[a] This solution is also an analog of the curvilinear squares solution in a dielectric media.

linear squares are used in the curvilinear squares solution method? *Sec. 7.2-1*

9. In applying the curvilinear squares solution method to a conductor medium, what are the tubes of flux replaced by? *Sec. 7.2-2*
10. What is σd equal to? *(Eq. 7.2-18)*
11. What is $(N_t/N_n)\sigma d$ equal to? *(Eq. 7.2-19)*
12. What is the average-value property of Laplace's equation? *Sec. 7.3, Eq. (7.3-7)*
13. Is the accuracy of the numerical solution method increased with an increase in the number of squares? *Sec. 7.3*
14. Is the numerical solution method based on Laplace's equation? *(Eq. 7.3)*
15. Is the numerical solution method adaptable to digital computer use? *Sec. 7.3*
16. Will the number of iterations needed to complete a numerical solution be decreased if good initial potential approximations are made? *Sec. 7.3*
17. In the image solution method, what is the infinite conductor plane replaced by? *Sec. 7.4*
18. Equipotential surfaces about two infinite line charges or circular cylindrical conductors are what type of geometric figures? *Sec. 7.4*
19. Is the solution for capacitance, (7.4-17), obtained through the use of image theory, an exact solution? *Sec. 7.4*
20. Can the image theory be extended to a point charge above an infinite conductor plane? *Sec. 7.4*
21. Can the image theory be extended to a current-carrying conductor above an infinite conductor plane? *Sec. 7.4*
22. What is the analog of $\nabla \cdot \overline{D} = 0$ in a dielectric medium equal to in a conductor medium? *Sec. 7.5*
23. What is the physical interpretation of $C/\epsilon = G/\sigma$, (7.5-1)? *Sec. 7.5*
24. In the analog solution method, was the ρ_v a time-varying quantity? *Sec. 7.5*
25. What is the name of the resistance paper used in the experimental solution method? *Sec. 7.6*
26. How is an electrolytic tank used in an experimental solution method? *Sec. 7.6*

PROBLEMS

7-1. Work Example 5-5 by means of a curvilinear squares solution, and check results with those found in the Example 5-5.

7-2. For the capacitor of Fig. 7-12 find, through the curvilinear squares solution: (a) the capacitance between surfaces s_1 and s_2 with the notch as shown; (b) the capacitance without the notch.

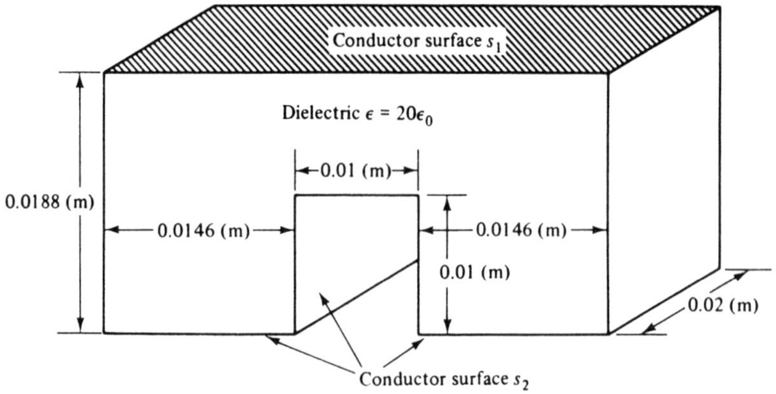

Figure 7-12 Graphical display of a notched capacitor for Prob. 7-2.

7-3. The edges of two infinite conductor planes are welded together to form an inverted letter V whose vertex is at the origin and whose axis is along the z axis. Another infinite conductor plane is located in the $z = 2$ m plane, above and parallel to the welded edge. If the interior angle of the inverted V structure is $\pi/4$ and the potential between the inverted V and the infinite conductor plane is $+V_0$ (V), find through a curvilinear squares solution: (a) the potential at the point $(0, 5, 0)$; (b) ρ_s at the point $(0, 0, 2)$ on the infinite plane; (c) the location of $|\rho_s|_{max}$; (d) the capacitance per meter between the plane and inverted V for a width of 2 m on each side of the vertex.

7-4. For the structure of Prob. 7-3: (a) Which portion contributes the greatest to the capacitance? (b) What happens to the $|\bar{E}|_{max}$ and $|\rho_s|_{max}$ as the interior angle of the inverted V becomes smaller? As the angle approaches zero, we approximate a lightning rod below a charged cloud.

7-5. When the dielectric between the two concentric and conducting cylinders of Prob. 7.2-1 is replaced by a conducting material whose $\sigma = 5$ S m^{-1}, find: (a) the conductance per meter; (b) the total current flowing between the cylinders when $V_0 = 15$ V.

7-6. (a) Through the use of the curvilinear squares solution method, find the resistance between surfaces s_1 and s_2 of the aluminum conductor configuration shown in Fig. 7-13. Assume all dimensions are in meters. (b) Repeat part (a) if all the dimensions are in centimeters.

7-7. A coaxial transmission line has a square cross section outer conductor whose inside dimensions are 2 cm on each edge, and a circular inner conductor with a diameter of 0.5 cm. If the space between conductors is filled with polystyrene ($\epsilon_r = 2.5$), find the capacitance per meter of the line, using curvilinear squares.

7-8. (a) Find the capacitance per meter for a coaxial transmission line with a cross section as shown in Fig. 5-21 filled with polystyrene, if $r_a = 0.25$ cm and $r_b = 1$ cm. Use exact equation from Sec. 5-10. (b) Repeat for $r_b = \sqrt{2}$ and compare the results of parts (a) and (b) with Prob. 7-7.

7-9. Find the capacitance per meter for the transmission-line cross section shown in Fig. 7-14. Use curvilinear squares.

7-10. Through the use of the numerical solution method, find the potential at point P of the two-variable electrostatic problem whose cross section is shown in Fig. 7-15. Round off potential values to tenth's position.

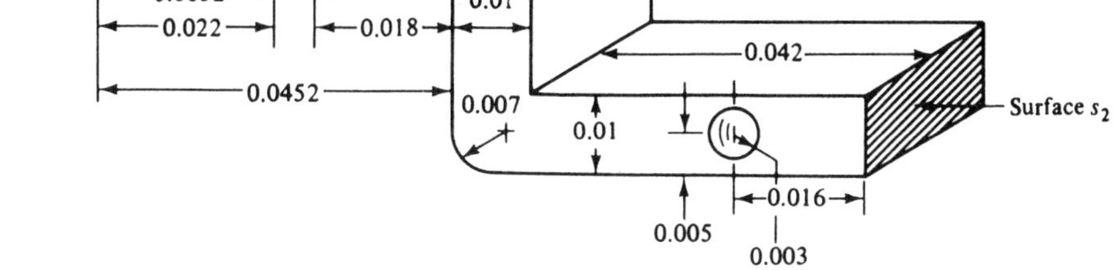

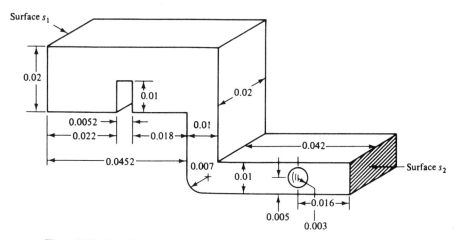

Figure 7-13 Graphical display of a conductor for use in Prob. 7-6. All dimensions are in meters.

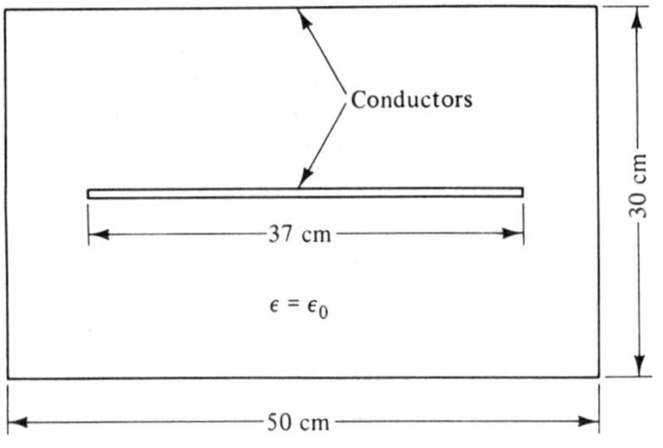

Figure 7-14 Cross section of a transmission line for use in Prob. 7-9.

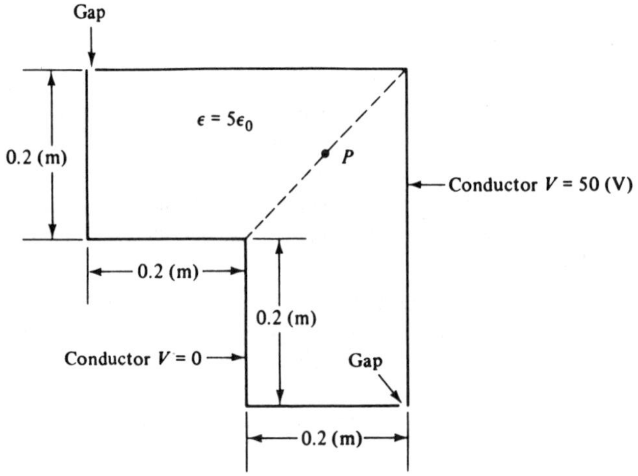

Figure 7-15 Graphical display of a two-variable electrostatic problem for Prob. 7-10.

7-11. (a) For the point locations shown in Fig. 7-16(a), show that the average-value property of Laplace's equation yields

$$V_0 = \frac{V_1}{(1 + \Delta h_1/\Delta h_3)(1 + \Delta h_1 \Delta h_3/\Delta h_2 \Delta h_4)} + \frac{V_2}{(1 + \Delta h_2/\Delta h_4)(1 + \Delta h_2 \Delta h_4/\Delta h_3 \Delta h_1)} + \frac{V_3}{(1 + \Delta h_3/\Delta h_1)(1 + \Delta h_3 \Delta h_1/\Delta h_4 \Delta h_2)} + \frac{V_4}{(1 + \Delta h_4/\Delta h_2)(1 + \Delta h_4 \Delta h_2/\Delta h_1 \Delta h_3)}$$

(b) Through the use of the numerical solution method, find the potential V_0 from the given adjacent point potentials, as found in Fig. 7-16(b).

7-12. (a) From the results of the image solution method, show that the capacitance between two parallel wires of the same diameter is approximately $C = \pi\epsilon/\ln(d/r_0)$ (F m^{-1}) where d is the spacing between wire centers and r_0 is the radius of each wire, for the case where $d \gg r_0$. (b) Now calculate the capacitance between two No. 14 wires separated by 1 cm, using the exact solution (7.4-17) and the approximate solution derived in part (a). The diameter of a No. 14 wire is 0.0641 in. Assume that $\epsilon = \epsilon_0$.

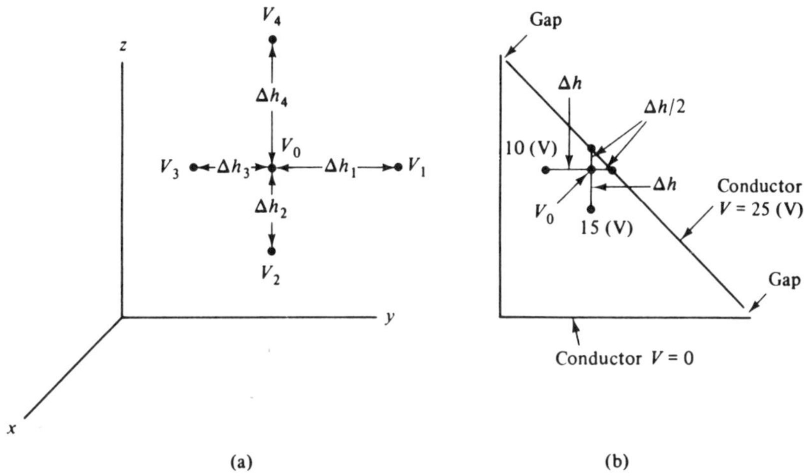

Figure 7-16 Graphical display for Prob. 7-11.

7-13. For the two circular cylindrical conductors of Fig. 7-7(a), with $b = 6$ m, $r_0 = 3$ m, $\epsilon = \epsilon_0$, and $\rho_\ell = 4 \times 10^{-9}$ C m^{-1}, find: (a) the locations of the $+\rho_\ell$ and $-\rho_\ell$ equivalent line charges; (b) the potential from conductor to conductor; (c) the capacitance per meter length of the two circular cylinders; (d) the location and values of $\rho_{s(\max)}$ on the right circular cylinder.

7-14. Find the capacitance per meter length of the two circular cylindrical conductors shown in Fig. 7-7(a), by the image solution method, when $b = 0.035$ m and $r_0 = 0.005$ m.

7-15. Find the capacitance per meter length between a circular conductor and a conductor plane of infinite extent, as shown in Fig. 7-8. Use the image solution method; assume that $b =$ 0.035 m and $r_0 = 0.004$ m.

7-16. For the two circular cylindrical conductors of Fig. 7-7(a), find the $\overline{E}$ at the rectangular point (0, 8, 0) when $b = 0.06$ m, $r_0 = 0.02$ m, and $V_0 = 100$ V.

7-17. From the image solution method, and referring to Fig. 7-17, show that the capacitance per meter between two parallel conducting cylinders of radii r_1 and r_2 is given by

$$C = \frac{2\pi\epsilon}{\cosh^{-1}\left[\pm\dfrac{D^2 - r_1^2 - r_2^2}{2r_1r_2}\right]} \qquad (\text{F m}^{-1})$$

where the plus sign applies to Fig. 7-17(a), and the minus sign applies to Fig. 7-17(b). [*Hint:* Use the identity $\cosh^{-1} x = \ln(x + \sqrt{x^2 - 1})$, and note from (7.4-14) that $(b_1^2 - r_1^2)^{1/2} = (b_2^2 - r_2^2)^{1/2} = a$, where a is the distance of the equivalent line source from the origin.]

7-18. For a steady current (non-time varying), show that $\oint_s \overline{J} \cdot \overline{ds} = 0$, starting with $\nabla \cdot \overline{J} = 0$ within a conductor.

7-19. The capacitance of a three-variable electrostatic problem was found by laboratory measurement to equal 5×10^{-4} F when the region between conductor boundaries was filled with a dielectric whose $\epsilon_r = 12$. Find the conductance of the same three-variable configuration when the dielectric medium is replaced by a conductor medium whose $\sigma = 3 \times 10^4$ S m^{-1}.

7-20. A quarter infinite space is formed by two infinite conductor planes that are welded together to form a $\pi/2$ angle between the planes. One plane is the $z = 0$ plane and extending over $-\infty \leqq x \leqq \infty$ and $0 \leqq y \leqq \infty$, while the other plane is in the $y = 0$ plane and extending over $-\infty \leqq x \leqq \infty$ and $0 \leqq z \leqq \infty$. An infinite line charge of $\rho_\ell = 10^{-6}$ C m^{-1} is found

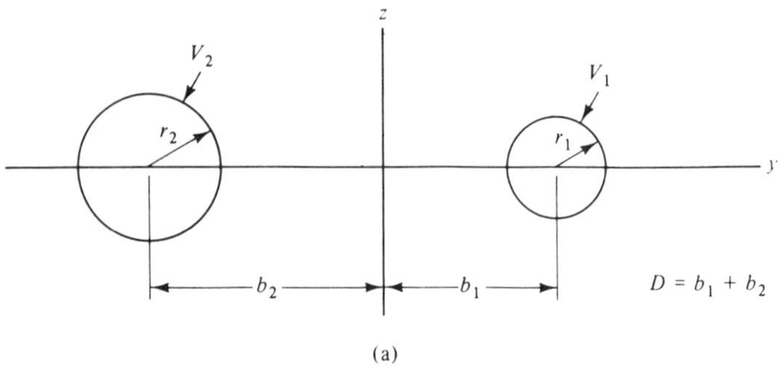

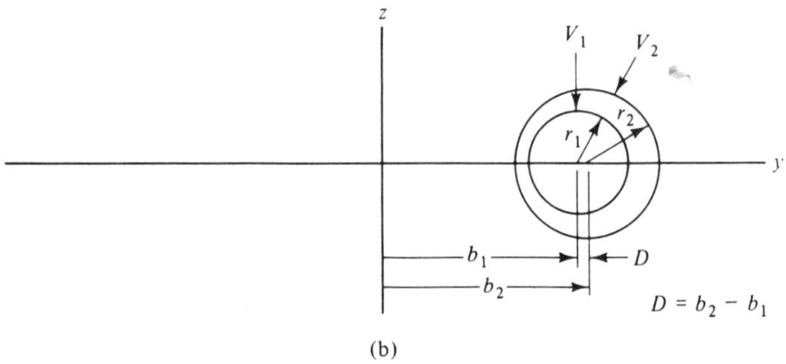

Figure 7-17 Illustration for Problem 7-17: (a) parallel cylinders of different radii; (b) parallel cylinders, one inside the other.

parallel to the x axis at $(x, 2, 2)$. Through the use of image theory, find: (a) the locations of the image line charges that will replace the two infinite conductor sheets; (b) the potential at the point $(0, 4, 3)$.

8

Biot-Savart Law, Ampère's Circuital Law, Curl, Stokes' Theorem, Magnetic Flux Density Vector, Vector Magnetic Potential, and Maxwell's Equations

8.1 INTRODUCTION

In Chapters 2 through 7, we have studied electric charges and the fields they produce, which are $\overline{E}$, $\overline{D}$, and V. We have also shown that moving charges give rise to the electric current concept. In this chapter, we shall find that a current will produce a magnetic field that will in turn produce a force on magnetic material, magnets, and other currents. We shall also find that magnet poles cannot be isolated in the same manner as electric charges, and thus our magnetic field theory will not totally parallel electric field theory.

In this chapter, we will be restricted to steady currents that are defined by $\nabla \bullet \overline{J} = -\partial \rho_v / \partial t = 0$. These steady currents will produce a steady magnetic field that we shall call a *magnetostatic field*. A magnetostatic field can also be produced by a permanent magnet and an electric field changing linearly with time. Our study of magnetostatic fields shall begin with types of current configurations and the experimental law of Biot and Savart, proposed in 1820.

8.2 ELECTRIC CURRENT CONFIGURATIONS

In our study of magnetostatic fields, we will encounter three basic current configurations or distributions: (1) filamentary, (2) surface, (3) volume. These current configurations, which form closed loops of current, are shown in Fig. 8-1. The filamentary or line current configuration shown in Fig. 8-1(a) has a differential cross section ds. The product $I\,\overline{d\ell}$ will be called the *current element* and can be visualized as the point source that will give rise to a magnetostatic field, as we shall find in Sec. 8.3. When a current

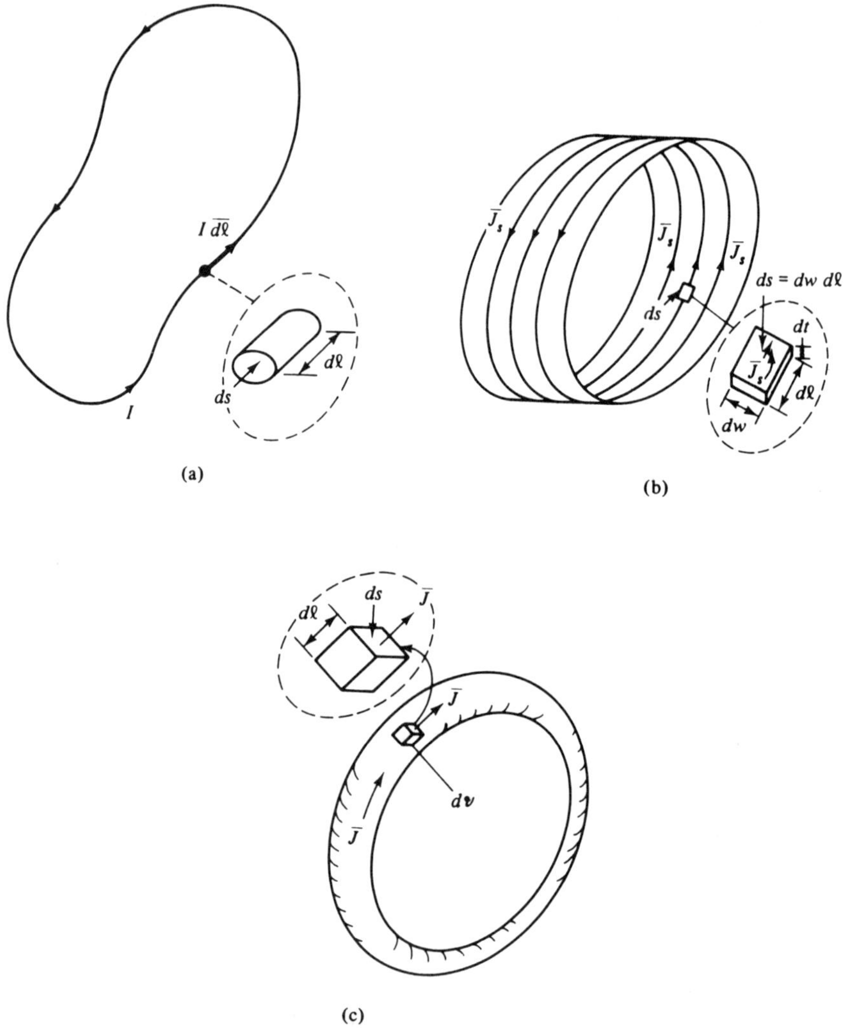

Figure 8-1 Closed loops of current configurations: (a) filamentary or line current of differential cross section ds and a point source current element of $I\,\overline{d\ell}$; (b) surface current of differential thickness dt and a point source current element of $\overline{J}_s\,ds$; (c) volume current within a finite cross section and a point source current element of $\overline{J}\,dv$.

flows only on the surface of a conductor, we have the surface current configuration of Fig. 8-1(b). In this case, the current flows within the differential thickness dt, and $\overline{J}_s$ is the surface current density (sheet current density) in amperes per width of sheet in meters (A m^{-1}). The current element for this configuration is $\overline{J}_s\,ds$, where ds is the differential surface area over which $\overline{J}_s$ flows. Figure 8-1(c) illustrates the volume current configuration that exists in a closed loop of finite cross section. The current element for this configuration is $\overline{J}\,dv$. The current density $\overline{J}$ flows perpendicular to the surface ds and has the units of (A m^{-2}). Note that the ds in Fig. 8-1(b) is defined differently in the case of the surface current configuration. The current elements for these current configurations are summarized below.

$$\begin{aligned} &\text{filamentary current element} = I\,\overline{d\ell} \quad (\text{A} \cdot \text{m}) \quad (\text{aap}) \\ &\text{surface current element} = \overline{J}_s\,ds \quad (\text{A} \cdot \text{m}) \quad (\text{aap}) \\ &\text{volume current element} = \overline{J}\,dv \quad (\text{A} \cdot \text{m}) \quad (\text{aap}) \end{aligned} \tag{1}$$

Through the use of enlarged views of the current elements found in Fig. 8-1, we can readily show that

$$I\,\overline{d\ell} = \overline{J}_s\,ds = \overline{J}\,dv \qquad (\text{A} \cdot \text{m}) \tag{2}$$

8.3 MAGNETOSTATIC FIELD INTENSITY FROM THE BIOT-SAVART LAW

It is an experimental fact that the magnetic field from a long line of current is as indicated by Fig. 8-2, where the $\overline{H}$ field is given by

$$\overline{H} = \frac{\hat{\phi} I}{2\pi r_c} \qquad (\text{A m}^{-1}) \tag{1}$$

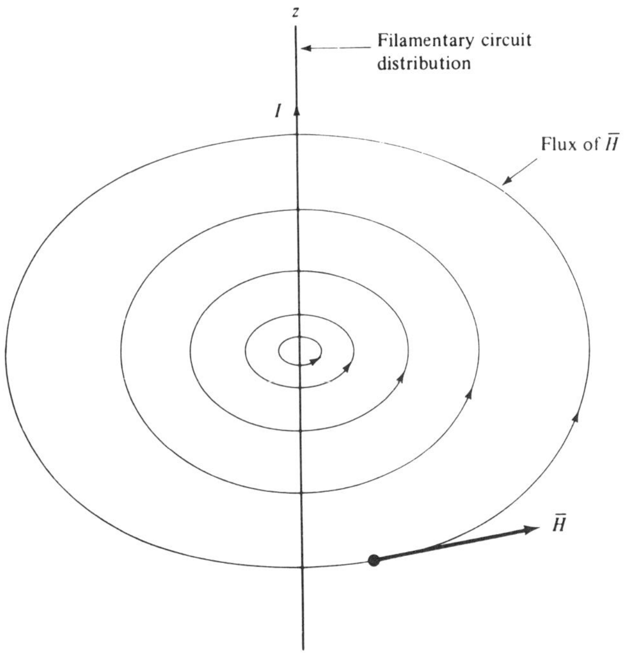

Figure 8-2 Flux of the magnetostatic vector $\overline{H}$ shown in a plane perpendicular to a filamentary current distribution of infinite length.

where r_c is the radius of a particular circle on which $\overline{H}$ is measured. Thus the magnetic field from a long line of current varies inversely as the distance from the current path. This is the principle by which one type of clamp-on ammeter operates. The strength of the magnetic field from the current is measured by the force exerted on a small magnet, which is connected to a pointer operating over a meter scale calibrated in amperes. Equation (1) is derived from the Biot-Savart law in Example 1 below. The Biot-Savart law gives the differential magnetic field intensity $\overline{dH}$ at a point P_2, produced by a current element $I_1\overline{d\ell_1}$ at point P_1, which is filamentary and differential in length, as shown in Fig. 8-3. This law can best be stated in vector form as

$$\boxed{\overline{dH_2} = \frac{I_1\overline{d\ell_1} \times \hat{a}_{R_{12}}}{4\pi R_{12}^2} \quad (\text{A m}^{-1})} \tag{2}$$

where the subscripts indicate the point to which the quantities refer, and

I_1 = filamentary current at P_1 (A)

$\overline{d\ell_1}$ = vector length of current path (vector direction same as conventional current) at P_1 (m)

$\hat{a}_{R_{12}}$ = unit vector directed from the current element $I_1\overline{d\ell_1}$ to the location of $\overline{dH_2}$, from P_1 to P_2

R_{12} = scalar distance between the current element $I_1\overline{d\ell_1}$ to the location of $\overline{dH_2}$, the distance between P_1 and P_2 (m)

$\overline{dH_2}$ = vector magnetostatic field intensity at P_2 (A m^{-1})

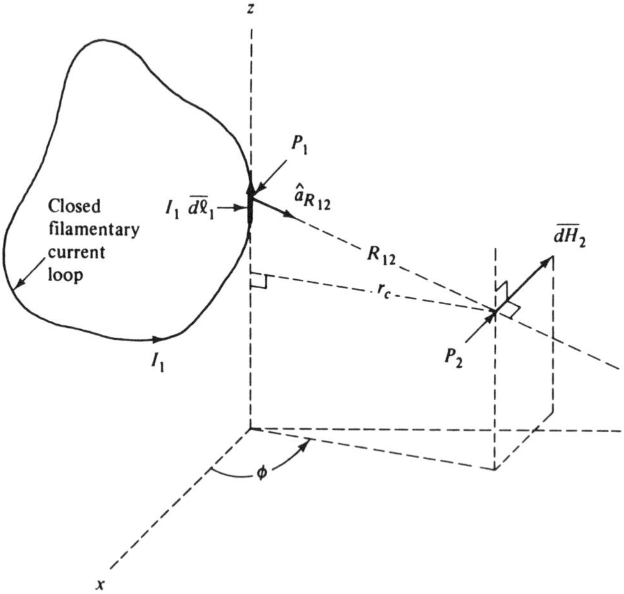

Figure 8-3 Graphical display of the vector magnetostatic field intensity $\overline{dH_2}$ at P_2 produced by a current element $I_1\overline{d\ell_1}$ at P_1. The $+z$ directed $I_1\overline{d\ell_1}$ was selected, and a z = constant plane was constructed to show more clearly the direction of $\overline{dH_2}$ as given by the Biot-Savart law. Note that $\overline{dH_2}$ is perpendicular to a plane containing $\overline{d\ell_1}$ and $\hat{a}_{R_{12}}$.

Figure 8-3 shows graphically the relationship between the quantities found in the Biot-Savart law (2) when a current element $I_1 \overline{d\ell}_1$ is singled out from a closed loop of filamentary current I_1. The direction of $\overline{dH}_2$ comes from $\overline{d\ell}_1 \times \hat{a}_{R_{12}}$ and thus is perpendicular to $\overline{d\ell}_1$ and $\hat{a}_{R_{12}}$. It should be noted that the direction of $\overline{dH}_2$ is also governed by the right-hand rule and is in the direction of the fingers of the right hand when we grasp the current element so that the thumb points in the direction of $\overline{d\ell}_1$.

The Biot-Savart law can be looked upon as the Coulomb's law of magnetostatics. In electrostatics, we found that Coulomb's law gave the differential electric field intensity $\overline{dE}_2$ at P_2 due to a differential charge dQ_1, located at P_1, as

$$\overline{dE}_2 = \frac{dQ_1 \hat{a}_{R12}}{4\pi\epsilon_0 R_{12}^2} \qquad (\text{V m}^{-1}) \tag{3}$$

Equation (2) is a *building block* equation in magnetostatics as (3) is in electrostatics.

The differential form of the Biot-Savart law (1) cannot be proven experimentally since it is impossible to produce in space an isolated steady current element $I\,\overline{d\ell}$ because we must use conductors to feed the current in and out of the $I\,\overline{d\ell}$ element. The total $\overline{H}$ field is obtained if we integrate (2) over a closed path of current to give

$$\overline{H}_2 = \oint_\ell \frac{I\,\overline{d\ell} \times \hat{a}_R}{4\pi R^2} \qquad (\text{A m}^{-1}) \tag{4}$$

where the subscripts on $\hat{a}$ and R have been dropped. Equation (4), the integral form of the Biot-Savart law, can be proven experimentally.

Through the use of (8.2-2) and (2) we find the building block equations for surface current and volume current elements to be

$$\overline{dH}_2 = \frac{\overline{J}_{s1} \times \hat{a}_{R12}\,ds_1}{4\pi R_{12}^2} \qquad (\text{A m}^{-1}) \tag{5}$$

and

$$\overline{dH}_2 = \frac{\overline{J}_1 \times \hat{a}_{R12}\,dv_1}{4\pi R_{12}^2} \qquad (\text{A m}^{-1}) \tag{6}$$

The integral forms become

$$\overline{H}_2 = \int_s \frac{\overline{J}_s \times \hat{a}_R\,ds}{4\pi R^2} \qquad (\text{A m}^{-1}) \tag{7}$$

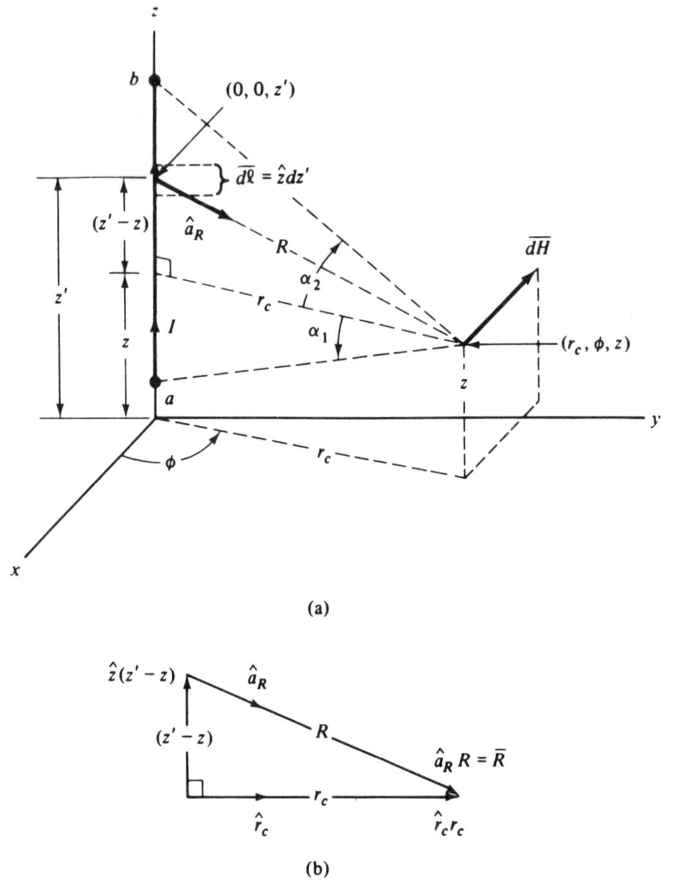

Figure 8-4 (a) Graphical display for finding the $\overline{H}$ field about a finite length of filamentary current distribution along the z axis for Example 1. (b) Auxiliary sketch for finding $\hat{a}_R$.

and

$$\overline{H}_2 = \int_v \frac{\overline{J} \times \hat{a}_R \, dv}{4\pi R^2} \qquad (\text{A m}^{-1}) \tag{8}$$

Example 1

For a filamentary current distribution of finite length and along the z axis, as shown in Fig. 8-4, find: (a) $\overline{H}$; (b) $\overline{H}$ when the current extends from $-\infty$ to $+\infty$.

Solution. (a) Let us specify the locations of the current element (source of $\overline{H}$ field) in terms of primed variables, and the field location in terms of unprimed variables. From (4), the expression for $\overline{H}$ in cylindrical functional form becomes

$$\overline{H}(r_c, \phi, z) = \int_a^b \frac{I\,\overline{d\ell}' \times \hat{a}_R(r_c', \phi', z', r_c, \phi, z)}{4\pi R^2(r_c', \phi', z', r_c, \phi, z)} \qquad (\text{A m}^{-1}) \tag{9}$$

Equation (9) parallels (2.5-3) for the $\overline{E}$ field from a line charge.

Through the use of (2), Fig. 8-4, and the results of Example 2-4, the expression for $\overline{dH}$ becomes

$$\overline{dH} = \frac{I(\hat{z}\,dz') \times [\hat{r}_c r_c - \hat{z}(z' - z)]}{4\pi[r_c^2 + (z' - z)^2]^{3/2}} \tag{10}$$

Using $\hat{z} \times \hat{r}_c = \hat{\phi}$ and $\hat{z} \times \hat{z} = 0$, (10) becomes

$$\overline{dH} = \frac{I\hat{\phi} r_c\,dz'}{4\pi[r_c^2 + (z' - z)^2]^{3/2}} \tag{11}$$

The $\overline{H}$ due to a finite current length becomes

$$\overline{H} = \frac{\hat{\phi} I r_c}{4\pi}\int_a^b \frac{dz'}{[r_c^2 + (z' - z)^2]^{3/2}} \tag{12}$$

Through the use of (2.5-4), (12) becomes

$$\overline{H} = \frac{\hat{\phi} I}{4\pi r_c}\, \frac{z' - z}{[r_c^2 + (z' - z)^2]^{1/2}}\Bigg|_a^b \tag{13}$$

and

$$\boxed{\overline{H} = \frac{\hat{\phi} I}{4\pi r_c}\left\{\frac{b - z}{[r_c^2 + (b - z)^2]^{1/2}} - \frac{a - z}{[r_c^2 + (a - z)^2]^{1/2}}\right\} \quad (\text{A m}^{-1})} \tag{14}$$

In terms of α_1 and α_2 [see Fig. 8-4(a)], (14) becomes

$$\boxed{\overline{H} = \frac{\hat{\phi} I}{4\pi r_c}(\sin\alpha_2 + \sin\alpha_1) \qquad (\text{A m}^{-1})} \tag{15}$$

(b) For a filamentary current of infinite length, $a = -\infty$ and $b = \infty$, we see that $\alpha_2 = \pi/2$ and $\alpha_1 = \pi/2$. Thus, (15) becomes

$$\boxed{\overline{H} = \frac{\hat{\phi} I}{2\pi r_c} \qquad (\text{A m}^{-1})} \tag{16}$$

The flux of $\overline{H}$ from (16) will be in the $\hat{\phi}$ direction and its density will decrease with r_c, as already indicated in Fig. 8-2. Later in this chapter we shall define a flux density vector $\overline{B}$.

Example 2

Find the expression for the $\overline{H}$ field along the axis of the circular current loop carrying a current I, as shown in Fig. 8-5.

Solution. Through the use of (1), Fig. 8-5, and the results of Example 2-6, the expression for $\overline{dH}$ on the axis of the loop becomes

$$\overline{dH} = \frac{I(\hat{\phi}' r_c'\,d\phi') \times (\hat{z}z - \hat{r}_c' r_c')}{4\pi(r_c'^2 + z^2)^{3/2}} \tag{17}$$

Using $\hat{\phi}' \times \hat{z} = \hat{r}_c' = \hat{x}\cos\phi' + \hat{y}\sin\phi'$ and $\hat{\phi}' \times (-\hat{r}_c') = \hat{z}$, (17) becomes

$$\overline{dH} = \frac{I\hat{z} r_c'^2\,d\phi'}{4\pi(r_c'^2 + z^2)^{3/2}} \tag{18}$$

where the $\hat{r}_c'$ component was omitted due to symmetry. The total $\overline{H}$ becomes

$$\overline{H} = \int_0^{2\pi} \overline{dH} = \frac{\hat{z} I r_c'^2}{4\pi(r_c'^2 + z^2)^{3/2}}\int_0^{2\pi} d\phi' \qquad (\text{A m}^{-1}) \tag{19}$$

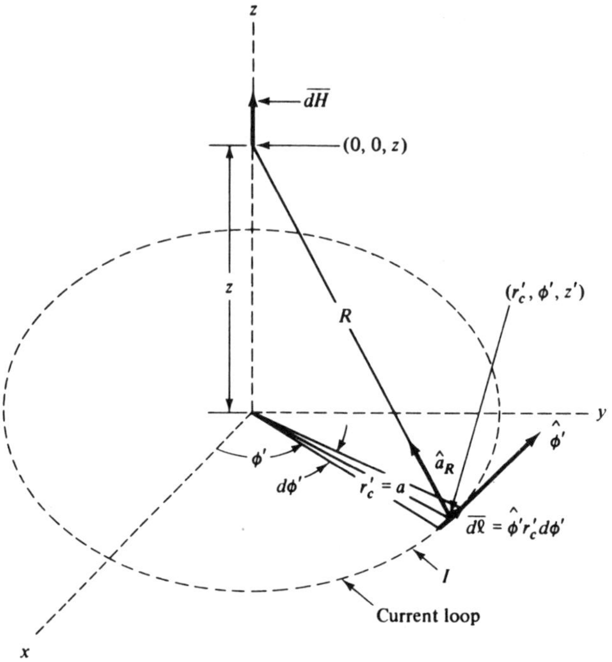

Figure 8-5 Graphical display for finding the $\overline{H}$ along the axis of a circular current loop for Example 2.

$$\overline{H} = \frac{\hat{z} I r_c'^2}{2(r_c'^2 + z^2)^{3/2}} = \hat{z}\frac{Ia^2}{2(a^2 + z^2)^{3/2}} \qquad (\text{A m}^{-1}) \tag{20}$$

It should be noted that $\overline{H}$ is in the $+\hat{z}$ direction above and below the loop (along the z axis), as dictated by the right-hand rule and the Biot-Savart law.

Example 3

Find the $\overline{H}$ field along the axis of a solenoid closely wound with a filamentary current-carrying conductor, as suggested in Fig. 8-6.

Solution. Figure 8-6(a) shows a solenoid containing a length ℓ of conductor winding and of N turns. Figure 8-6(b) is a cross-sectional view while Fig. 8-6(c) replaces the N turns with a flat surface (sheet) current. A three-dimensional sketch of the equivalent surface current is shown in Fig. 8-6(d), where the total surface current is NI (A). Since the surface current flows in a width ℓ, the surface current density $J_s = NI/\ell$ (A m^{-1}). The current in the dz length of Fig. 8-6(c) thus becomes $J_s\,dz = (NI/\ell)\,dz$ (A).

Let us now concentrate on Fig. 8-6(c) and view the dz length as a thin current loop that carries a current of $(NI/\ell)\,dz$ (A). From (20), we find that this current loop will produce, at the center of the solenoid, a $\overline{dH}$ field equal to

$$\overline{dH} = \hat{z}\frac{\left(\frac{NI}{\ell}\,dz\right)a^2}{2(a^2 + z^2)^{3/2}} \qquad (\text{A m}^{-1}) \tag{21}$$

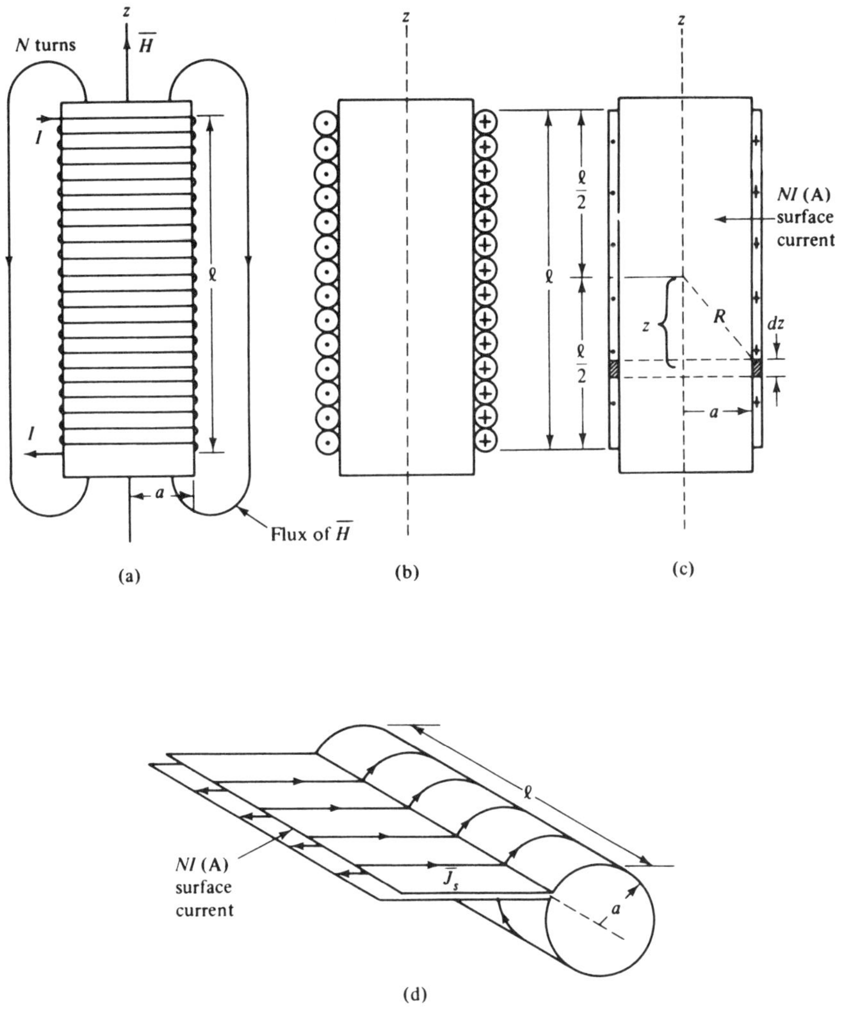

Figure 8-6 (a) Closely wound solenoid. (b) Cross section of the solenoid. (c) Turns are replaced by an equivalent flat surface current. (d) Three-dimensional view of equivalent surface of NI (A) and one turn.

where I has been replaced by $(NI/\ell)\,dz$. Integrating (21), we obtain

$$\overline{H} = \hat{z}\frac{NIa^2}{2\ell}\int_{-\ell/2}^{+\ell/2}\frac{dz}{(a^2+z^2)^{3/2}} = \hat{z}\frac{NI}{(4a^2+\ell^2)^{1/2}} \qquad (\text{A m}^{-1}) \tag{22}$$

for the magnetostatic field intensity at the center of the solenoid. If the length of the solenoid is much greater than its radius, $\ell \gg a$, (22) reduces to

$$\boxed{\overline{H} \cong \hat{z}\frac{NI}{\ell} = \hat{z}J_s \qquad (\text{A m}^{-1})} \tag{23}$$

The $\overline{H}$ at the end of the solenoid is obtained by integrating (21) from 0 to ℓ to obtain

$$\overline{H} = \hat{z}\frac{NI}{2(a^2 + \ell^2)^{1/2}} \quad (\mathrm{A\ m^{-1}}) \tag{24}$$

Now, for $\ell \gg a$, (24) reduces to

$$\boxed{\overline{H} \cong \hat{z}\frac{NI}{2\ell} = \hat{z}\frac{J_s}{2} \quad (\mathrm{A\ m^{-1}})} \tag{25}$$

which is one-half that value at the center of the solenoid as obtained in (23).

Problem 8.3-1 Through the use of (2), Fig. 8-4, and the results of Example 2-4, obtain (11) and (14).

Problem 8.3-2 A current of 1 (A) flows in the configuration shown in Fig. 8-4. Find $\overline{H}$ at the rectangular point $(4, 5, 0)$ when: (a) $a = -\infty$, $b = +\infty$; (b) $a = 0$, $b = \infty$; (c) $a = -2$, $b = +2$.

Problem 8.3-3 If the circular current loop of Fig. 8-5 is shaped into a square current loop whose legs are a meters long and parallel to the axes, find: (a) the $\overline{H}$ field at the origin; (b) the $\overline{H}$ field along the $+z$ axis through the use of (15).

Problem 8.3-4 Through the use of (2), Fig. 8-5, and the results of Example 2-6, obtain (17) and (20).

Problem 8.3-5 A solenoid is 0.4 m long and has a radius of 0.005 m. If it is closely wound and contains 1000 turns, find: (a) H at the center through the use of (22) and (23) and calculate the percent error; (b) H at the ends through the use of (24) and (25) and calculate the percent error.

Problem 8.3-6 By modifying the limits in (22), obtain the H field at any point on the solenoid axis at a distance b (m) from the center of the coil.

8.4 AMPERE'S CIRCUITAL LAW

In our study of electrostatic fields, we were fortunate to have had Gauss's law to simplify our solutions in cases of symmetrical charge distribution. In magnetostatics, a parallel law exists in the form of Ampère's circuital law. Through Ampere's circuital law, we will be able to solve quite formidable magnetostatic problems in cases of symmetrical current distributions.

Let us evaluate the integral $\oint_\ell \overline{H} \cdot \overline{d\ell}$ about a concentric closed loop that encloses the filamentary current I of infinite length, as suggested in Fig. 8-7(a). Through the use of (8.3-16), we obtain

$$\oint_\ell \overline{H} \cdot \overline{d\ell} = \oint_\ell \left(\frac{\hat{\phi} I}{2\pi r_c}\right) \cdot (\hat{\phi} r_c\, d\phi) = \frac{I}{2\pi}\int_0^{2\pi} d\phi = I = I_{\mathrm{en}} \quad (\mathrm{A}) \tag{1}$$

where I_{en} is the current enclosed by the closed loop. The formulation obtained by equating the first and last terms of (1),

$$\boxed{\oint_\ell \overline{H} \cdot \overline{d\ell} = I_{\mathrm{en}} \quad (\mathrm{A})} \tag{2}$$

is called *Ampère's circuital law.*

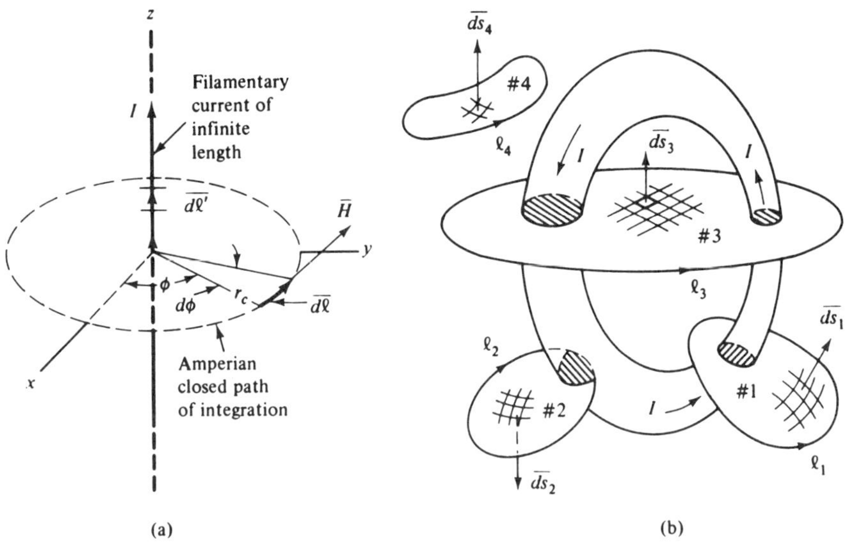

Figure 8-7 (a) Graphical display for Ampère's circuital law. (b) Several examples of closed loops about a volume current distribution for the interpretation of I_{en}.

The positive-sense relationship between the direction of the closed path of integration and I_{en} is formed by the right-hand rule. Thus, the direction of the positive I_{en} is in the direction of the thumb of the right hand as the fingers point in the direction of the path of integration, as shown in Fig. 8-7(a). Figure 8-7(b) shows four closed paths about a volume charge distribution. To more clearly show the I_{en}, we shall draw a surface whose periphery is the path of integration. The current enclosed by the path is thus the current that flows through this surface. Loop 1 of Fig. 8-7(b) thus encloses the total current I while loop 2 encloses only part of the current I. Loop 3 encloses zero current while loop 4 encloses zero current also. Thus, in (2) the $\overline{H}$ field must vary along a closed path to yield a closed loop integral of $\overline{H} \cdot \overline{d\ell}$ equal to the I_{en}. The positive sense of $\overline{ds}$ on the surface enclosed by the loops is also found by the right-hand rule. The surface direction will become important when we later relate the closed loop integral of $\overline{H} \cdot \overline{d\ell}$ to surface integrals over the surface whose periphery is the closed loop.

Example 4

Through the use of Ampère's circuital law, find the $\overline{H}$ field about a filamentary current I of infinite length along the z axis.

Solution. Let us construct a closed concentric loop perpendicular to the infinite length current, as suggested in Fig. 8-7(a). Through the use of the Biot-Savart law, we can argue that $\overline{dH}$ from any current element $I\,\overline{d\ell}'$ will be in the $\hat{\phi}$ direction at any point on the closed amperian path. Thus, $\overline{H}$ can be expressed as

$$\overline{H} = \hat{\phi} H_\phi$$

Along the amperian closed path, $\overline{d\ell} = \hat{\phi} r_c\, d\phi$. Substituting these expressions and $I_{en} = I$ into Ampère's circuital law, we obtain

$$\oint_\ell \overline{H} \cdot \overline{d\ell} = \oint_\ell (\hat{\phi} H_\phi) \cdot (\hat{\phi} r_c\, d\phi) = H_\phi r_c \oint_0^{2\pi} d\phi = H_\phi 2\pi r_c = I \qquad \text{(A)} \qquad (3)$$

where H_ϕ and r_c were taken out from under the integral since they are constants over the amperian path of integration selected. Solving for H_ϕ in the last two terms of (3), we obtain for $\overline{H}$:

$$\boxed{\overline{H} = \hat{\phi}H_\phi = \frac{\hat{\phi}I}{2\pi r_c} \qquad (\text{A m}^{-1})} \tag{4}$$

which agrees with (8.3-16).

Example 5

Through the use of Ampère's circuital law, find the $\overline{H}$ field inside and outside an infinite length conductor of finite cross section that carries a current I(A) uniformly distributed over its cross section, as shown in Fig. 8-8.

Solution. Draw an amperian closed loop of integration that is concentric with the axis of the conductor, as shown in Fig. 8-8(a). Now, let us argue that $\overline{H}$ along the amperian path is of the form $\overline{H} = \hat{\phi}H_\phi$. We can establish this fact by considering the conductor to be built up from an infinite number of filamentary currents of infinite length. Two of these filamentary currents are shown in the cross-sectional view found in Fig. 8-8(b). At point P, the resultant $\overline{dH}_r$ from a pair of filamentary currents, equally spaced about the axis of the conductor, is shown to be in the $\hat{\phi}$ direction. This is true for any pair in the plane A or in the kth plane. Also, we will find that this applies when P is inside the conductor. The contributions from all of the filamentary pairs thus allows us to state that $\overline{H} = \hat{\phi}H_\phi$ about our

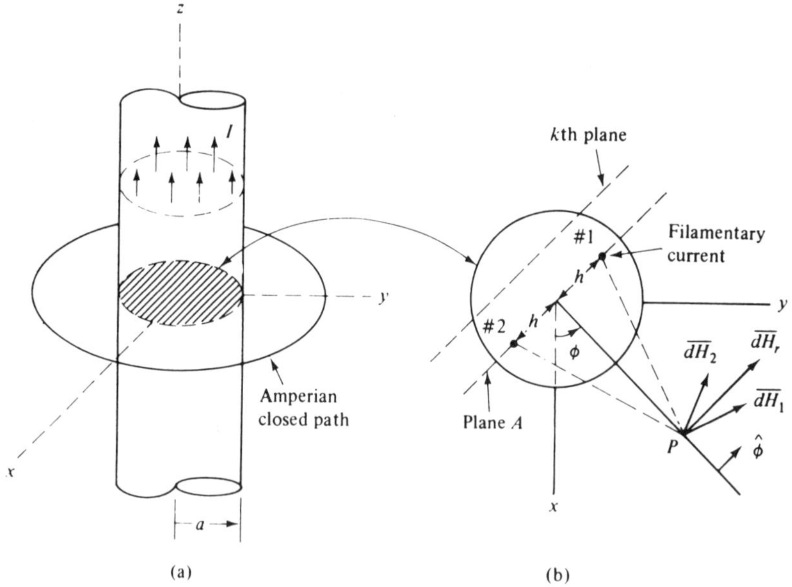

Figure 8-8 (a) Graphical display for finding the $\overline{H}$ field inside and outside a conductor of finite cross section. (b) Symmetrical pairs of filamentary currents produce a resultant field in the $\hat{\phi}$ direction.

amperian path of integration. Along the amperian path, $\overline{d\ell} = \hat{\phi} r_c \, d\phi$. Substituting these expressions into Ampère's circuital law, we obtain

$$\oint_\ell (\hat{\phi} H_\phi) \bullet (\hat{\phi} r_c \, d\phi) = I_{\text{en}} = I$$

where I is the current enclosed by the closed amperian path outside the conductor. Now, H_ϕ and r_c are constant over the amperian path and thus yield

$$H_\phi r_c \int_0^{2\pi} d\phi = H_\phi 2\pi r_c = I \tag{5}$$

Solving for H_ϕ in the last two terms of (5), we obtain for $\overline{H}$:

$$\overline{H} = \frac{\hat{\phi} I}{2\pi r_c} \quad (a < r_c) \qquad (\text{A m}^{-1}) \tag{6}$$

Note that (6) and (4) are identical for $a < r_c$.

Let us now find $\overline{H}$ for $r_c < a$. In this case, $\overline{H}$ is still in the $\hat{\phi}$ direction and can be expressed as $\overline{H} = \hat{\phi} H_\phi$. The current enclosed, I_{en}, by the amperian closed loop, with $r_c < a$, is equal to $I r_c^2 / a^2$ (A). Substituting these facts into Ampère's circuital law, we obtain

$$\oint_\ell (\hat{\phi} H_\phi) \bullet (\hat{\phi} r_c \, d\phi) = I_{\text{en}} = \frac{I r_c^2}{a^2} \tag{7}$$

Again, H_ϕ and r_c are constant over a concentric amperian loop, and (7) can be expressed as

$$H_\phi r_c \int_0^{2\pi} d\phi = H_\phi 2\pi r_c = \frac{I r_c^2}{a^2} \tag{8}$$

Solving for H_ϕ in the last two terms of (8), $\overline{H}$ becomes

$$\boxed{\overline{H} = \hat{\phi} \frac{I r_c}{2\pi a^2} \quad (r_c < a) \qquad (\text{A m}^{-1})} \tag{9}$$

Example 6

Through the use of Ampère's circuital law, find the $\overline{H}$ field above and below a surface current distribution of infinite extent with a surface current density $\overline{J}_s = \hat{y} J_{sy}$ (A m^{-1}).

Solution. Construct an amperian closed path about the surface current as shown in Fig. 8-9. Applying Ampère's circuital law to this path, we obtain

$$\oint_\ell \overline{H} \bullet \overline{d\ell} = \int_1^2 + \int_2^3 + \int_3^4 + \int_4^1 = I_{\text{en}} = J_{sy} \ell \tag{10}$$

From the construction found in Fig. 8-9(b), we can see that $\overline{H}$ above and below the surface current will be in the $\hat{x}$ and $-\hat{x}$ directions, respectively. Thus, (10) reduces to

$$\int_1^2 (\hat{x} H_x) \bullet (\hat{x} \, dx) + \int_3^4 (-\hat{x} H_x) \bullet (\hat{x} \, dx) = J_{sy} \ell \tag{11}$$

where $\int_2^3$ and $\int_4^1$ are zero since $\overline{H}$ is perpendicular to $\overline{d\ell}$ (path direction). For symmetrical location of line segments (1-2) and (3-4), we deduce that $|\overline{H}|$ above and below the surface current are equal and (11) becomes

$$H_x \ell + H_x \ell = J_{sy} \ell \tag{12}$$

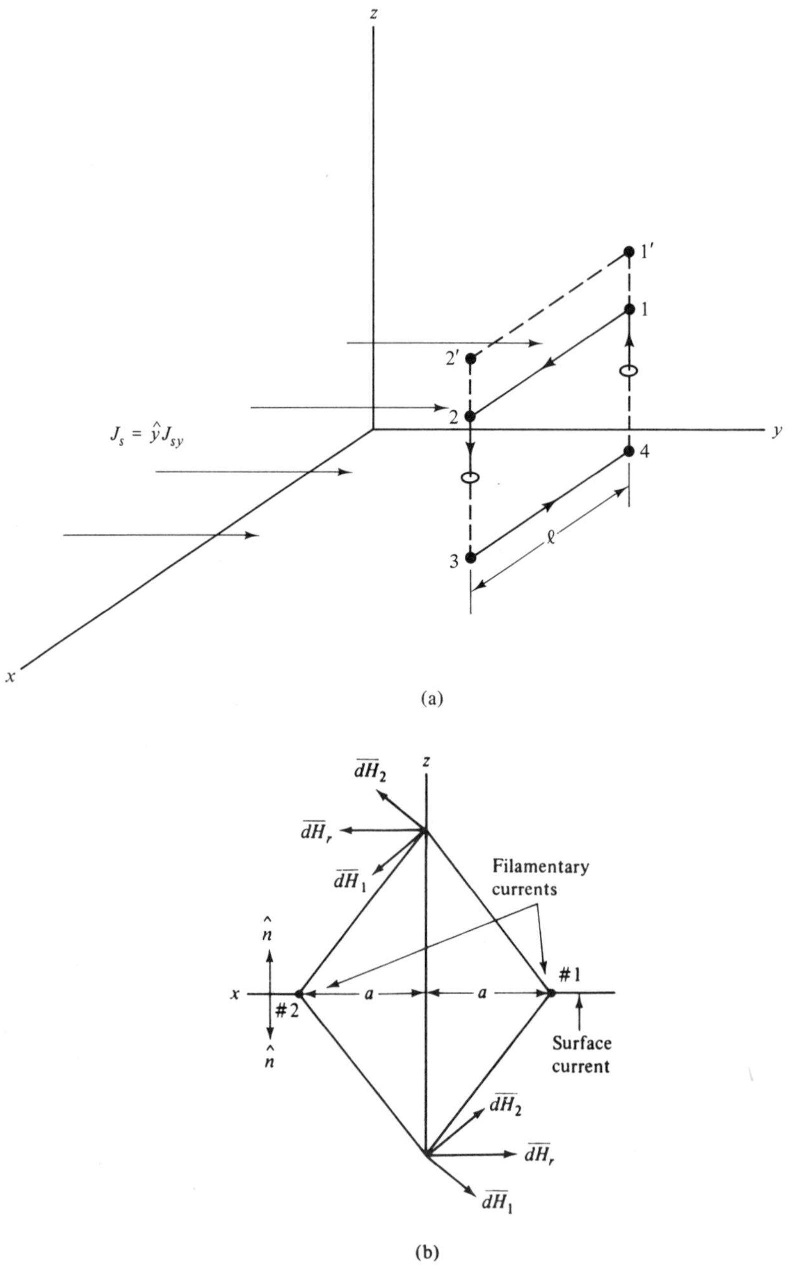

Figure 8-9 (a) Graphical display for finding the $\overline{H}$ field above and below a surface current of infinite extent. (b) Symmetrical pairs of filamentary currents produce $\overline{H}$ fields that are parallel to the plane of the surface current.

Thus, we obtain

$$\boxed{H_x = \frac{J_{sy}}{2} \quad (z > 0) \qquad (\mathrm{A\ m^{-1}})} \tag{13}$$

and

$$\boxed{H_x = -\frac{J_{sy}}{2} \quad (z < 0) \qquad (\mathrm{A\ m^{-1}})} \tag{14}$$

Equations (13) and (14) can be written in vector form:

$$\boxed{\overline{H} = \tfrac{1}{2}\overline{J}_s \times \hat{n} \qquad (\mathrm{A\ m^{-1}})} \tag{15}$$

where $\hat{n}$ is a unit vector at the surface current and is directed into the region where $\overline{H}$ is being found. If a new amperian loop takes on the points (1′-2′-3-4) as shown in Fig. 8-9, we will find that the $\overline{H}$ does not change from the line segment location (1-2) to (1′-2′) since the current enclosed does not change over this new amperian loop. Thus, the $\overline{H}$ fields do not vary in magnitude with z above and below the surface current, but they are in reverse directions.

The $\overline{H}$ field between two parallel surface currents of infinite extent and of equal current density flowing in opposite directions can readily be shown through the use of (15) to be

$$\boxed{\overline{H} = \overline{J}_s \times \hat{n} \qquad (\mathrm{A\ m^{-1}})} \tag{16}$$

Example 7

Through the use of Ampère's circuital law, find the $\overline{H}$ field in all regions of an infinite-length coaxial cable carrying a uniform and equal current I in opposite directions in the inner and outer conductors. Assume the inner conductor to have a radius of a (m) and the outer conductor to have an inner radius of b (m) and an outer radius of c (m). Assume that the cable's axis is along the z axis.

Solution. Through the use of symmetrical pairs of filamentary currents as in Example 5 and shown in Fig. 8-8(b), we can argue that the $\overline{H}$ field in all regions will be in the $\hat{\phi}$ direction and thus $\overline{H} = \hat{\phi}H_\phi$.

For a concentric amperian closed loop drawn in the region ($r_c < a$) of the inner conductor, Ampère's circuital law becomes

$$\oint_\ell \overline{H} \bullet \overline{d\ell} = \oint_\ell (\hat{\phi}H_\phi) \bullet (\hat{\phi}r_c\,d\phi) = H_\phi r_c \int_0^{2\pi} d\phi = H_\phi 2\pi r_c = I_{\mathrm{en}} \tag{17}$$

Now, $I_{\mathrm{en}} = Ir_c^2/a^2$ for the amperian closed loop inside the inner conductor. Thus, the last two terms in (17) become

$$H_\phi 2\pi r_c = I_{\mathrm{en}} = \frac{Ir_c^2}{a^2}$$

yielding

$$H_\phi = \frac{Ir_c}{2\pi a^2}$$

and

$$\boxed{\overline{H} = \hat{\phi}\frac{Ir_c}{2\pi a^2} \quad (r_c < a) \qquad (\text{A m}^{-1})} \tag{18}$$

Note that (18) is the same as (9) in Example 5, inside an isolated conductor of radius a and carrying a uniform current I.

For a concentric amperian closed loop drawn in the region $(a < r_c < b)$, Ampère's circuital law becomes

$$\oint_\ell \overline{H} \cdot \overline{d\ell} = \oint_\ell (\hat{\phi}H_\phi) \cdot (\hat{\phi}r_c\, d\phi) = H_\phi r_c \int_0^{2\pi} d\phi = H_\phi 2\pi r_c = I \tag{19}$$

where $I_{\text{en}} = I$ (A). Solving for H_ϕ from the last two terms in (19), we obtain for $\overline{H}$:

$$\boxed{\overline{H} = \frac{\hat{\phi}I}{2\pi r_c} \quad (a < r_c < b) \qquad (\text{A m}^{-1})} \tag{20}$$

Note that (20) is the same as (6) in Example 5, outside an isolated conductor of radius a.

For a concentric amperian closed loop in the region $(b < r_c < c)$, Ampère's circuital law becomes

$$\oint_\ell \overline{H} \cdot \overline{d\ell} = \oint_\ell (\hat{\phi}H_\phi) \cdot (\hat{\phi}r_c\, d\phi) = H_\phi r_c \int_0^{2\pi} d\phi = H_\phi 2\pi r_c = I_{\text{en}} \tag{21}$$

Now, $I_{\text{en}} = I - I[(r_c^2 - b^2)/(c^2 - b^2)]$ for the amperian closed loop inside the outer conductor. It should be noted that some of the enclosed current flows in the reverse direction. Substituting for I_{en} into (21) and solving for H_ϕ from the last two terms, we obtain for $\overline{H}$:

$$\boxed{\overline{H} = \hat{\phi}\frac{I}{2\pi r_c}\left(\frac{c^2 - r_c^2}{c^2 - b^2}\right) \quad (b < r_c < c) \qquad (\text{A m}^{-1})} \tag{22}$$

For a concentric amperian closed loop drawn in the region $(c < r_c)$, we find the enclosed current to be zero, and thus $\overline{H}$ is zero.

Let us now return to the single-turn surface-current solenoid of Fig. 8-6(d). If we let ℓ go to infinity, we can prove through Biot and Savart's law and symmetrically positioned thin current loops that the $\overline{H}$ field for $r_c > a$ is zero. Now, if we construct a rectangular amperian closed loop half inside and half outside the solenoid, we obtain

$$\boxed{\overline{H} = \hat{z}J_s \quad (r_c < a) \qquad (\text{A m}^{-1})} \tag{23}$$

when the axis of the solenoid is along the z axis.

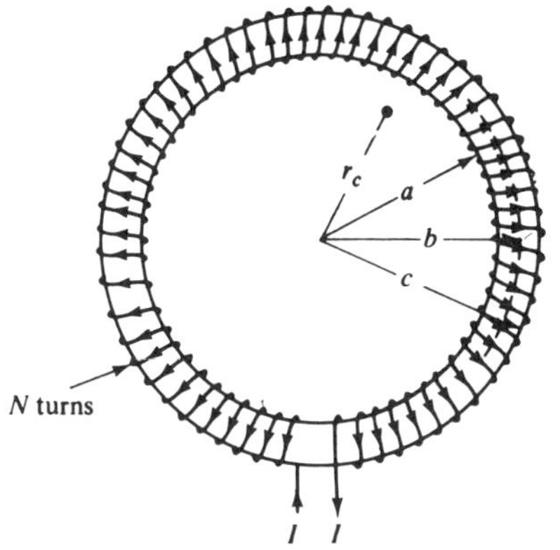

Figure 8-10 Toroid for Prob. 8.4-5.

For a solenoid of finite length ℓ and wound with N closely spaced turns of thin conductor that carries a current I, we find through the use of Ampère's circuital law that

$$\overline{H} \cong \hat{z}\frac{NI}{\ell} \quad \text{(near center of solenoid)} \qquad (\text{A m}^{-1}) \tag{24}$$

This result is based on $\overline{H} = 0$ outside the solenoid and that $\ell \gg a$ [see (8.3-23)].

Problem 8.4-1 Through the use of Ampère's circuital law, find the $\overline{H}$ field inside and outside an infinite length conductor whose radius is 0.002 (m) that carries a current, $\overline{J} = \hat{z}6r_c$ (A m^{-2}).

Problem 8.4-2 Through the use of Ampère's circuital law, find the $\overline{H}$ field inside and outside an infinite-length hollow conducting tube whose radius is 0.002 m that carries a current, $I = 10^{-7}$ A, directed in the positive z direction. Consider the thickness of the tube very small.

Problem 8.4-3 A parallel-plate transmission line has one conducting plate in the $z = 0$ plane carrying a surface current density, $\overline{J}_s = \hat{y}4$ (μA m^{-1}) and the other plate at $z = 0.02$ (m) carrying $\overline{J}_s = -\hat{y}4$ (μA m^{-1}). Find the $\overline{H}$ field: (a) between the two plates; (b) directly below the plate at $z = 0$; and (c) directly above the plate at $z = 0.02$ m.

Problem 8.4-4 (a) Work Example 7 when the current $I = 10$ A is a surface current at $r_c = a$, and the current $-I = 10$ A is a surface current at $r_c = b$. (b) Plot the results of this problem and Example 7 to the same scale when $a = 0.01$ m, $b = 0.04$ m, and $c = 0.045$ m for a range $(0 \leqq r_c \leqq 0.05)$.

Problem 8.4-5 Through the use of Ampère's circuital law, find in the midplane of the toroid of Fig. 8-10: (a) $\overline{H}$ for $(a < r_c < c)$; (b) $\overline{H}$ for $(0 \leqq r_c < a)$; (c) $\overline{H}$ for $(r_c > c)$.

8.5 CURL

In our study of electrostatic fields, we applied Gauss's law to a point in space to obtain the divergence concept. Here we shall apply Ampère's circuital law to a point in space to obtain the *curl concept*.

Figure 8-11(a) shows three amperian closed loops, each parallel to one of the rectangular planes over which we will apply Ampère's circuital law. The result will be di-

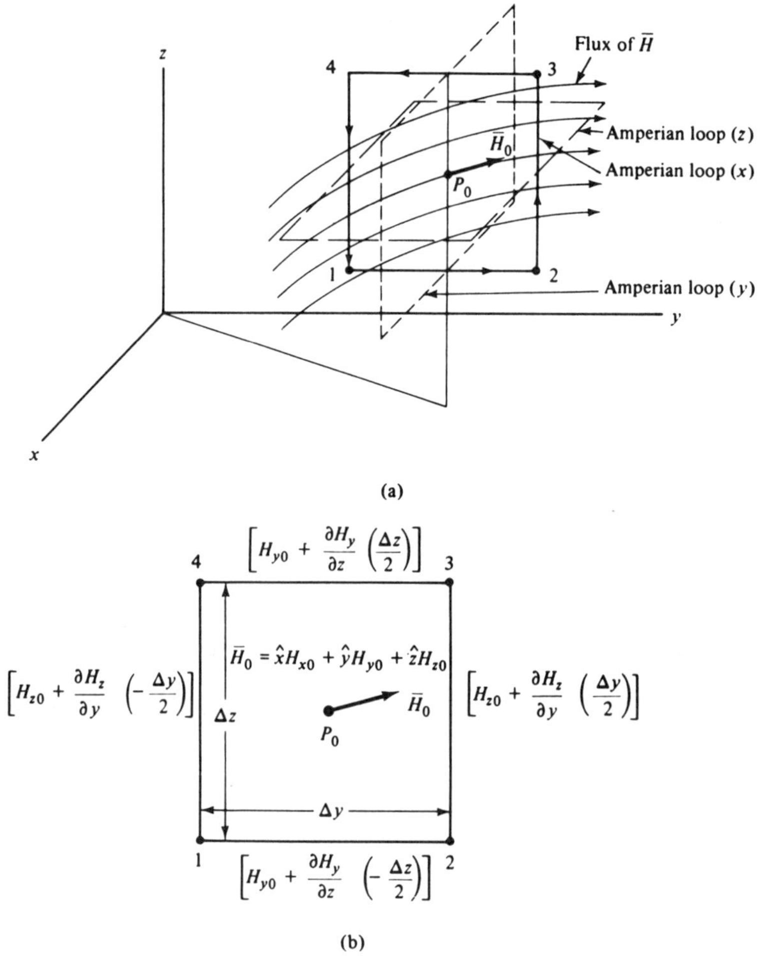

Figure 8-11 (a) Graphical display for the evaluation of $\nabla \times \overline{H}$ concept. (b) Expanded view of the amperian loop parallel to x = constant plane.

vided by the area Δs of the loop, and the limit as $\Delta s \to 0$ will be taken to obtain a scalar component of the curl of $\overline{H}$ at P_0. Let us assume that these amperian closed loops are embedded in an $\overline{H}$ field that we shall designate as $\overline{H}_0 = \hat{x}H_{x0} + \hat{y}H_{y0} + \hat{z}H_{z0}$ at P_0. We will single out the amperian closed loop (x), shown in Fig. 8-11(b), over which we shall apply Ampère's circuital law. All $\overline{H}$ components perpendicular to the direction of travel about the loop will not contribute to the closed path integral of $\overline{H} \cdot \overline{d\ell}$. If we use the first two terms of Taylor's expansion for $\overline{H}$ and only the field components that are directed along the path of integration, we will obtain approximate $\overline{H}$ field values along each path segment, as shown in Fig. 8-11(b). Using these results, we obtain

$$\oint_{\ell(x)} \overline{H} \cdot \overline{d\ell} = \int_1^2 + \int_2^3 + \int_3^4 + \int_4^1 = I_{en} = J_x \,\Delta y\, \Delta z \qquad \text{(A)} \qquad (1)$$

where

$$\int_1^2 = \left[H_{y0} + \frac{\partial H_y}{\partial z}\left(-\frac{\Delta z}{2}\right)\right]\Delta y, \qquad \int_2^3 = \left[H_{z0} + \frac{\partial H_z}{\partial y}\left(\frac{\Delta y}{2}\right)\right]\Delta z$$

$$\int_3^4 = -\left[H_{z0} + \frac{\partial H_y}{\partial z}\left(\frac{\Delta z}{2}\right)\right]\Delta y, \qquad \int_4^1 = -\left[H_{z0} + \frac{\partial H_z}{\partial y}\left(-\frac{\Delta y}{2}\right)\right]\Delta z$$

Thus,

$$\int_1^2 + \int_2^3 + \int_3^4 + \int_4^1 = \left(\frac{\partial H_z}{\partial y} - \frac{\partial H_y}{\partial z}\right)\Delta y\,\Delta z \tag{2}$$

Through the use of (1) and (2), let us form

$$\lim_{\Delta y\,\Delta z \to 0}\left[\frac{\oint_{\ell(x)} \overline{H} \bullet \overline{d\ell}}{\Delta y\,\Delta z}\right] = \left(\frac{\partial H_z}{\partial y} - \frac{\partial H_y}{\partial z}\right) = \lim_{\Delta y\,\Delta z \to 0}\left[\frac{I_{\text{en}}}{\Delta y\,\Delta z}\right] = J_x \tag{3}$$

where the first term is the definition for the x component of the curl of $\overline{H}$, and J_x is the x component of the current density vector at P_0.

If we now duplicate this development for the amperian loop (y) and amperian loop (z), shown in Fig. 8-11(a), we obtain

$$\lim_{\Delta x\,\Delta z \to 0}\left[\frac{\oint_{\ell(y)} \overline{H} \bullet \overline{d\ell}}{\Delta x\,\Delta z}\right] = \left(\frac{\partial H_x}{\partial z} - \frac{\partial H_z}{\partial x}\right) = \lim_{\Delta x\,\Delta z \to 0}\left[\frac{I_{\text{en}}}{\Delta x\,\Delta z}\right] = J_y \tag{4}$$

and

$$\lim_{\Delta x\,\Delta y \to 0}\left[\frac{\oint_{\ell(z)} \overline{H} \bullet \overline{d\ell}}{\Delta x\,\Delta y}\right] = \left(\frac{\partial H_y}{\partial x} - \frac{\partial H_x}{\partial y}\right) = \lim_{\Delta x\,\Delta y \to 0}\left[\frac{I_{\text{en}}}{\Delta x\,\Delta y}\right] = J_z \tag{5}$$

From (3), (4), and (5) it should be noted that the scalar curl components of $\overline{H}$ are equal to scalar components of a current density $\overline{J}$ at P_0. Thus, we define the curl of $\overline{H}$ as

$$\boxed{\text{curl } \overline{H} \triangleq \sum_{k=x,y,z} \hat{k} \lim_{\Delta s_{(k)} \to 0}\left[\frac{\oint_{\ell(k)} \overline{H} \bullet \overline{d\ell}}{\Delta s_{(k)}}\right] \qquad (\text{A m}^{-2}) \quad (\text{aap})} \tag{6}$$

where $\Delta s_{(k)}$ is the area bounded by the kth amperian closed loop. If we combine in vector form the second and fourth terms of (3), (4), and (5), we obtain

$$\boxed{\text{curl } \overline{H} = \left[\hat{x}\left(\frac{\partial H_z}{\partial y} - \frac{\partial H_y}{\partial z}\right) + \hat{y}\left(\frac{\partial H_x}{\partial z} - \frac{\partial H_z}{\partial x}\right) + \hat{z}\left(\frac{\partial H_y}{\partial x} - \frac{\partial H_x}{\partial y}\right)\right] = \overline{J}} \tag{7}$$

where $\overline{J} = \hat{x}J_x + \hat{y}J_y + \hat{z}J_z$. From (1.12-4), we note that $\nabla \times \overline{H}$ will give the left-hand side of (7); thus, (7) can be written in shorthand vector operator form as

$$\boxed{\nabla \times \overline{H} = \overline{J} \qquad (\text{A m}^{-2}) \quad (\text{aap})} \tag{8}$$

Equation (8) is commonly referred to as the *point form* of Ampère's circuital law as well as Maxwell's second of four equations for static fields. Maxwell's first equation, $\nabla \cdot \overline{D} = \rho_v$, was introduced in Chapter 3. Equation (7) can be expressed in determinant form as

$$\text{curl } \overline{H} = \begin{vmatrix} \hat{x} & \hat{y} & \hat{z} \\ \dfrac{\partial}{\partial x} & \dfrac{\partial}{\partial y} & \dfrac{\partial}{\partial z} \\ H_x & H_y & H_z \end{vmatrix} \quad (\text{A m}^{-2}) \quad (\text{aap}) \tag{9}$$

In cylindrical and spherical coordinates, (6) will take on the following forms:

$$\nabla \times \overline{H} = \hat{r}_c \left[\frac{1}{r_c}\left(\frac{\partial H_z}{\partial \phi}\right) - \frac{\partial H_\phi}{\partial z} \right] + \hat{\phi}\left[\frac{\partial H_{r_c}}{\partial z} - \frac{\partial H_z}{\partial r_c} \right] + \frac{\hat{z}}{r_c}\left[\frac{\partial}{\partial r_c}(r_c H_\phi) - \frac{\partial H_{r_c}}{\partial \phi} \right]$$

Cyl. $\tag{10}$

$$\nabla \times \overline{H} = \frac{\hat{r}_s}{r_s \sin\theta}\left[\frac{\partial(H_\phi \sin\theta)}{\partial\theta} - \frac{\partial H_\theta}{\partial\phi} \right] + \frac{\hat{\theta}}{r_s}\left[\frac{1}{\sin\theta}\left(\frac{\partial H_{r_s}}{\partial\phi}\right) - \frac{\partial(r_s H_\phi)}{\partial r_s} \right] + \frac{\hat{\phi}}{r_s}\left[\frac{\partial(r_s H_\theta)}{\partial r_s} - \frac{\partial H_{r_s}}{\partial\theta} \right]$$

Spher.

$$\tag{11}$$

The concepts of this section can be summarized by the following equation:

$$\underbrace{\text{curl } \overline{H}}_{\text{Concept}} = \underbrace{\nabla \times \overline{H}}_{\substack{\text{Vector} \\ \text{analysis} \\ \text{compact} \\ \text{symboli-} \\ \text{zation}}} \triangleq \underbrace{\sum_{k=x,y,z} \hat{k} \lim_{\Delta s_{(k)} \to 0} \left[\frac{\oint_{\ell_{(k)}} \overline{H} \cdot \overline{d\ell}}{\Delta s_{(k)}} \right]}_{\substack{\text{Defining equation} \\ \text{(vector form)}}} = \underbrace{\begin{vmatrix} \hat{x} & \hat{y} & \hat{z} \\ \dfrac{\partial}{\partial x} & \dfrac{\partial}{\partial y} & \dfrac{\partial}{\partial z} \\ H_x & H_y & H_z \end{vmatrix}}_{\substack{\text{Mathematical} \\ \text{relationship} \\ \text{resulting from} \\ \text{application of the} \\ \text{defining equation} \\ \text{in the rectangular} \\ \text{coordinate system}}} = \underbrace{\overline{J}}_{\substack{\text{Physical} \\ \text{quantity at} \\ \text{point } P_0}} \quad (\text{A m}^{-2}) \quad (\text{aap})$$

$$\tag{12}$$

The defining equation for the scalar component of $\nabla \times \overline{H}$ in the kth direction from (6) is

$$\begin{array}{l} \text{Component of curl of } \overline{H} \\ \text{(in } k\text{th direction)} \end{array} \triangleq \lim_{\Delta s_{(k)} \to 0}\left[\frac{\oint_{\ell_{(k)}} \overline{H} \cdot \overline{d\ell}}{\Delta s_{(k)}} \right] \quad (\text{A m}^{-2}) \quad (\text{aap}) \tag{13}$$

In word form, the definition for the kth scalar component of $\overline{H}$ is

> The kth scalar component of curl of $\overline{H}$ (in the kth direction) is equal to the quotient of the integral about a closed amperian path in a plane normal to the kth direction and the area enclosed, as the amperian path and the area enclosed shrinks to zero.

A handy "curl meter" in the form of a pinwheel, that works in some cases as suggested in Fig. 8-12(b), may be used to indicate curl of a vector field. In Fig. 8-12(a), we find a flux plot of the $\overline{H}$ field above and below a uniform surface current of infinite extent. Now, if the pinwheel is placed above or below the surface current there will be no motion, and thus the $\nabla \times \overline{H}$ is equal to zero. For this concept to work, we must visualize the $\overline{H}$ as some force that can act on the pinwheel. Since $\overline{H}$ does not vary with z, we thus deduce that the pinwheel will not rotate. If the $\overline{H}$ field above the surface current increased with z, the pinwheel would rotate clockwise and a $-\hat{x}$-directed curl would exist. This can also be indicated by placing a small amperian closed loop in the plane of the paper and noting the result of integrating $\overline{H} \bullet \overline{d\ell}$ about this small loop. In Fig. 8-12(c), we find the flux of $\overline{H}$ plotted inside and outside an infinite length conductor of radius a (m) and carrying a uniform current. If we now place the pinwheel inside or outside the conductor, we find it quite hard to decide if the pinwheel will rotate. The only sure way to find if a curl of a vector field exists is to use (7), (10), or (11). If we take the curl of the $\overline{H}$ fields inside and outside the conductor in Fig. 8-12(c), we obtain $\nabla \times \overline{H}_{(\text{inside})} = \overline{J}$ and $\nabla \times \overline{H}_{(\text{outside})} = 0$. This is consistent with (8), $\nabla \times \overline{H} = \overline{J}$.

In Chapter 4, we found that $\oint_\ell \overline{E} \bullet \overline{d\ell} = 0$ for an electrostatic field. Thus, from (6) we find that $\nabla \times \overline{E} = 0$ for an electrostatic field. The expression $\nabla \times \overline{E} = 0$ is Maxwell's third equation for static fields.

Example 8

Find the curl of $\overline{H}$ for the following $\overline{H}$ fields: (a) $\overline{H} = \hat{\phi}(I/2\pi r_c)$ about a filamentary current; (b) $\overline{H} = \hat{\phi}(Ir_c/2\pi a^2)$ inside an infinite length conductor of radius a (m); (c) $\overline{H} = \hat{x}J_s/2$ about a surface current of infinite extent and of uniform J_s (A m^{-1}); (d) $\overline{H} = \hat{\phi}I/2\pi r_c[(c^2 - r_c^2)/(c^2 - b^2)]$ inside the outer conductor of a coaxial cable of Example 7.

Solution

$$\overline{H} = \hat{\phi}H_\phi = \hat{\phi}\frac{I}{2\pi r_c} \quad \text{and} \quad H_{r_c} = H_z = 0$$

(a) From (10), we have

$$\nabla \times \overline{H} = \frac{\hat{z}}{r_c}\left[\frac{\partial}{\partial r_c}(r_c H_\phi)\right] = \frac{\hat{z}}{r_c}\left[\frac{\partial}{\partial r_c}\left(\frac{I}{2\pi}\right)\right] = 0$$

(b) From (10), we have

$$\nabla \times \overline{H} = \frac{\hat{z}}{r_c}\left[\frac{\partial}{\partial r_c}(r_c)\left(\frac{Ir_c}{2\pi a^2}\right)\right] = \frac{\hat{z}}{r_c}\left(\frac{2r_c I}{2\pi a^2}\right) = \hat{z}\frac{I}{\pi a^2} = \hat{z}J \qquad (\text{A m}^{-2})$$

(c) $\nabla \times \overline{H} = 0$ since $H_x =$ constant, and $H_y = H_z = 0$.

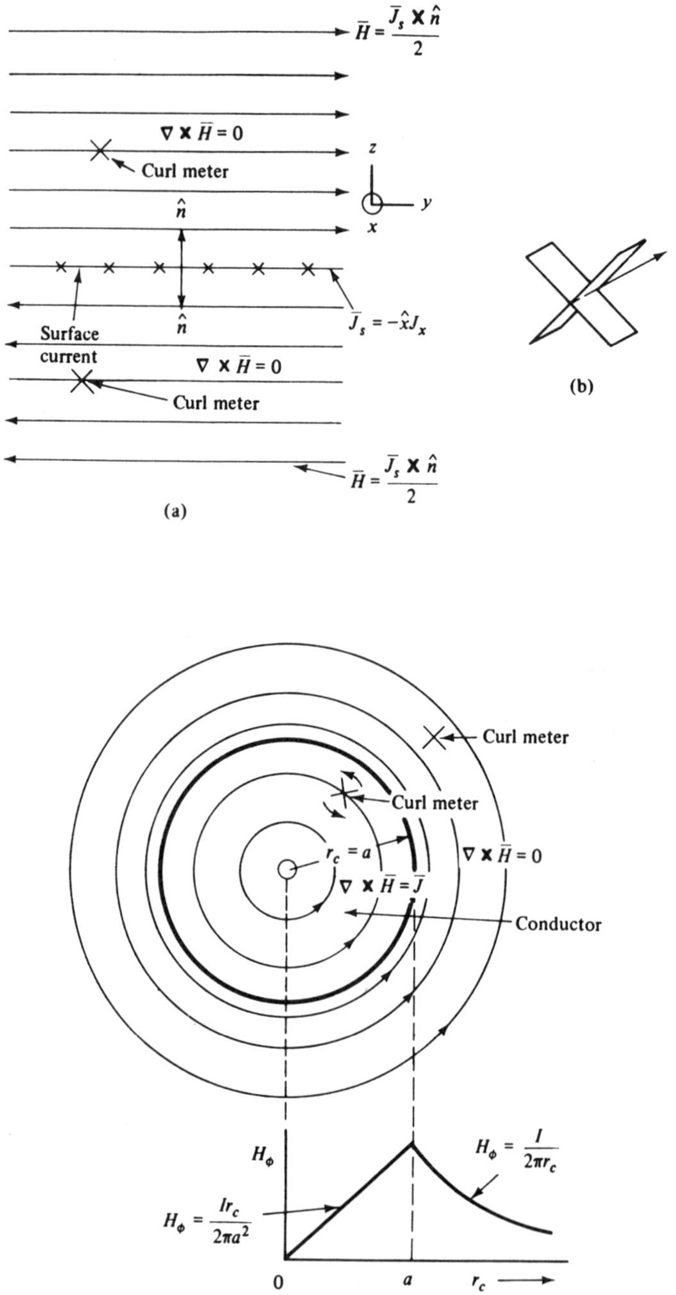

(c)

Figure 8-12 Graphical display to illustrate how a pinwheel "curl meter" works: (a) flux plot of $\overline{H}$ about a surface current of infinite extent; (b) pinwheel curl meter, (c) flux plot inside and outside an infinite length conductor of finite cross section and carrying a uniform current I (A).

(d) $$\nabla \times \overline{H} = \frac{\hat{z}}{r_c}\left[\frac{\partial}{\partial r_c}(r_c)\left(\frac{I}{2\pi r_c}\right)\left(\frac{c^2 - r_c^2}{c^2 - b^2}\right)\right]$$
$$= \frac{\hat{z}}{r_c}\left[\frac{-I}{2\pi}\frac{2r_c}{(c^2 - b^2)}\right] = -\hat{z}\frac{I}{\pi(c^2 - b^2)} = \overline{J} \qquad (\text{A m}^{-2})$$

In each of the cases it is clearly seen that $\nabla \times \overline{H} = \overline{J}$, as required by (8).

Problem 8.5-1 Find the curl of the following fields: (a) $\overline{E} = \hat{r}_s(10^{-3}/r_s^2)$ (V m^{-1}); (b) $\overline{E} = \hat{x}10^{-2}x$ (V m^{-1}); (c) $\overline{H} = \hat{\phi}Kr_s$ (A m^{-1}); (d) $\overline{F} = \hat{x}10y^2 + \hat{y}5x^2$.

Problem 8.5-2 For $\overline{H} = \hat{\phi}Kr_c$ (A m^{-1}) find the closed loop integral of $\overline{H} \cdot \overline{d\ell}$ about the following amperian closed paths: (a) from P_{cyl}. $(a, 0, 0)$ along a circular path in the $+\hat{\phi}$ direction with $r_c = a$ and $0 \leqq \phi \leqq 2\pi$, all in the $z = 0$ plane; (b) from P_{cyl} $(a, -\pi/2, 0)$ along circular path with $r_c = a$ and $-\pi/2 \leq \phi \leq 3\pi/2$ to $(a, 3\pi/2, 0)$; (c) from P_{cyl} $(a, -\pi/2, 0)$, along a circular path in the $+\hat{\phi}$ direction with $r_c = a$ to P_{cyl}. $(a, \pi/2, 0)$ and back to the $P_{\text{cyl}}(a, -\pi/2, 0)$ along a straight line, all in the $z = 0$ plane. Note that the closed loop integrals depend on the paths since $\nabla \times \overline{H} \neq 0$.

Problem 8.5-3 Repeat problem 8.5-2 for $\overline{H} = \hat{\phi}K/r_c$.

Problem 8.5-4 In a certain region, $\overline{H} = \hat{\phi}10^{-3}r_c^2$ (A m^{-1}). Find the current that flows through a fictitious loop of $r_c = 0.03$ (m) in the $z = 2$ m plane and centered at $(0, 0, 2)$.

Problem 8.5-5 A flat conducting slab has a width w (m) in the y direction, has one surface at $z = 0$ and the other surface at $z = d$, and carries a current density $\overline{J} = \hat{x}J_{x0}e^{-\alpha z}$ (A m^{-2}). Find: (a) the total current flowing in the slab, (b) under the assumption that $\overline{H}(z = 0) = -\overline{H}(z = d)$, find $\overline{H}(z)$ for $0 \leq z \leq d$.

8.6 STOKES' THEOREM

In our study of electrostatic fields we were able to relate a closed surface integral of $\overline{D} \cdot \overline{ds}$ to a volume integral of $\nabla \cdot \overline{D}\, dv$ through divergence theorem. In this section, we shall derive a parallel theorem, called *Stokes' theorem*, that will relate the closed loop integral of $\overline{H} \cdot \overline{d\ell}$ to a surface integral of $\nabla \times \overline{H} \cdot \overline{ds}$.

Let us consider a surface s of Fig. 8-13 that is subdivided into incremental surfaces Δs. From the defining equation of the scalar component of the curl of $\overline{H}$, (8.5-13), we can express Ampère's circuital law about the closed loop $\ell_{\Delta s_k}$ as

$$\oint_{\ell_k} \overline{H} \cdot \overline{d\ell_k} \cong (\nabla \times \overline{H}) \cdot \overline{\Delta s_k} \tag{1}$$

Let us now sum both sides of (1) over the surface s to obtain

$$\sum_{k=1}^{m} \oint_{\ell_k} \overline{H} \cdot \overline{d\ell_k} \cong \sum_{k=1}^{m} (\nabla \times \overline{H}) \cdot \overline{\Delta s_k} \tag{2}$$

From Fig. 8-13, we note that the integral values over interior line segments add up to zero since adjacent loop integrals traverse common interior line segments in reverse directions. Thus, the left side of (2) will be the sum of integral values over the periphery of the large surface s. In (2), let $\Delta s_k \to 0$; thus, $m = \infty$ and we obtain

$$\boxed{\oint_{\ell} \overline{H} \cdot \overline{d\ell} = \int_{s} (\nabla \times \overline{H}) \cdot \overline{ds}} \tag{3}$$

where the loop ℓ encloses the large surface s. Equation (3) is called *Stokes' theorem*.

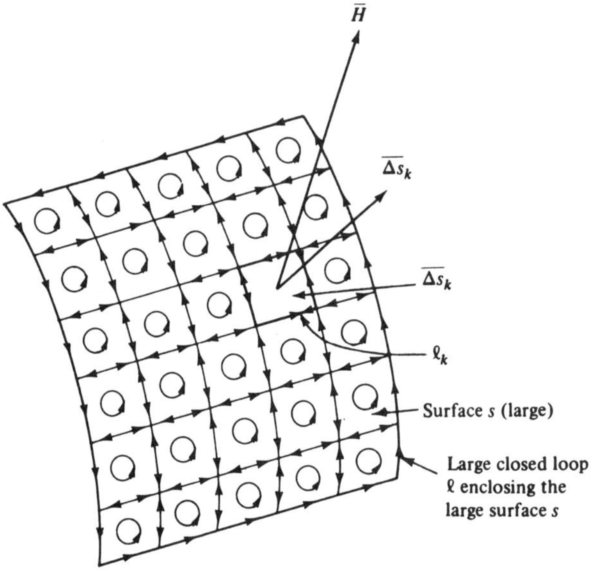

Figure 8-13 Graphical display for the development of Stokes' theorem.

Example 9

In Example 5, we found that $\overline{H} = \hat{\phi} I r_c / 2\pi a^2$ (A m^{-1}) inside an infinite length conductor of radius a (m). Evaluate both sides of Stokes' theorem to find the total current flowing in the conductor.

Solution. Let us evaluate the left-hand side over the periphery of the conductor; thus, the right-hand side must be integrated over the cross section of the conductor. Thus, we have

$$\oint_\ell \left(\hat{\phi}\frac{Ir_c}{2\pi a^2}\right)\bigg|_{r_c=a} \bullet (\hat{\phi} a\, d\phi) = \int_s \left[\nabla \times \left(\hat{\phi}\frac{Ir_c}{2\pi a^2}\right)\right] \bullet (\hat{z} r_c\, d\phi\, dr_c)$$

$$\int_0^{2\pi} \frac{I}{2\pi}\, d\phi = \int_0^a \int_0^{2\pi} \left(\hat{z}\frac{I}{\pi a^2}\right) \bullet (\hat{z} r_c\, d\phi\, dr_c)$$

$$I = I \quad \text{(A)} \tag{4}$$

Problem 8.6-1 Problem 8.5-2 was solved by evaluating the left-hand side of Stokes' theorem. Now, evaluate the right-hand side of Stokes' theorem for the $\overline{H}$ fields of Prob. 8.5-2 over the surfaces bounded by the given paths.

8.7 MAGNETIC FLUX DENSITY VECTOR

Up to this point in this chapter we have discussed only the magnetic field intensity $\overline{H}$ (A m^{-1}) in free space. We now introduce magnetic flux density vector $\overline{B}$ that is related to $\overline{H}$ in free space through

$$\boxed{\overline{B} = \mu_0 \overline{H} \qquad \text{(T)}} \tag{1}$$

where the unit of $\overline{B}$ is webers per square meter (Wb m^{-2}) or, in SI units, teslas (T). The term μ_0 is a constant called the *permeability of free space* and has the value of

$$\boxed{\mu_0 = 4\pi \times 10^{-7} \qquad (\text{H m}^{-1})} \tag{2}$$

where the unit is henrys per meter (H m^{-1}).

In Chapter 9, where we consider magnetic material, we shall find that the relationship between $\overline{B}$ and $\overline{H}$ will not be as simple as indicated by (1).

The vector $\overline{B}$ is a member of the flux-density fields, as we have found $\overline{D}$ to be in the electrostatic area. In the case of $\overline{D}$, we found Ψ_E to be the associated flux, as indicated by (3.3-1). Now, for the $\overline{B}$, we find a magnetic flux Ψ_m that is related to $\overline{B}$ through

$$\boxed{\Psi_m = \int_s \overline{B} \bullet \overline{ds} \qquad (\text{Wb})} \tag{3}$$

where the unit is the weber. The flux Ψ_m in (3) is the magnetic flux that passes through the surface s.

In electrostatic field work, we found that

$$\boxed{\Psi_E = \oint_s \overline{D} \bullet \overline{ds} = Q_{\text{en}}} \tag{4}$$

where the last two terms constitute Gauss's law. A net electric flux will emanate from a closed surface since we were able to isolate (separate) positive and negative charges. Lines of electric flux begin on positive charge and terminate on negative charge.

In magnetics, magnet poles have not been isolated; thus, we do not have a source, as an isolated charge, where magnetic lines of flux begin or terminate. Now, if we integrate (3) over a closed surface, we obtain

$$\boxed{\Psi_m = \oint_s \overline{B} \bullet \overline{ds} = 0} \tag{5}$$

Equation (5) thus dictates that magnetic flux lines are closed and do not terminate on isolated magnet poles (magnetic charge). From (5) and the defining equation for divergence (3.5-10), we obtain

$$\boxed{\nabla \bullet \overline{B} = 0(\text{T m}^{-1})} \tag{6}$$

Equation (6) is the fourth Maxwell's equation for static fields.

Example 10

For $\overline{H} = \hat{\phi}10^3 r_c$ (A m^{-1}), find the Ψ_m that passes through a plane surface defined by, ($\phi = \pi/2$), ($2 \le r_c \le 4$), and ($0 \le z \le 2$).

Solution. From (3), we have

$$\Psi_m = \int_s \overline{B} \cdot \overline{ds} = \int_0^2 \int_2^4 (\mu_0 \hat{\phi} 10^3 r_c) \cdot \hat{\phi}(dr_c\, dz)$$

$$= \mu_0 10^3 \int_0^2 \int_2^4 r_c\, dr_c\, dz = \mu_0 10^3 (12)$$

$$= 150.8 \times 10^{-4} \text{ Wb}$$

Problem 8.7-1 Through the use of the results of Example 7, find the Ψ_m between the radii a and b and a 1-m length of the cable of Example 7.

Problem 8.7-2 A toroid, such as that pictured in Fig. 8-10, has a rectangular cross section, $h(c - a)$ (m^2). If the toroid has N turns carrying current I, find the total flux for $a < r_c < c$.

8.8 MAXWELL'S EQUATIONS

With the introduction of $\nabla \cdot \overline{B} = 0$ in the preceding section, we now have four basic equations that are commonly called *Maxwell's equations in point form*. The point forms and integral forms of Maxwell's equations are given below as found to date in our study of electrostatic and magnetostatic fields.

$$\nabla \cdot \overline{D} = \rho_v$$
$$\nabla \times \overline{E} = 0$$
$$\nabla \times \overline{H} = \overline{J} \qquad (1)$$
$$\nabla \cdot \overline{B} = 0$$

Point form

$$\int_v \nabla \cdot \overline{D}\, dv = \oint_s \overline{D} \cdot \overline{ds} = \int_v \rho_v\, dv = Q_{en}$$
$$\int_s \nabla \times \overline{E} \cdot \overline{ds} = \oint_\ell \overline{E} \cdot \overline{d\ell} = 0$$
$$\int_s \nabla \times \overline{H} \cdot \overline{ds} = \oint_\ell \overline{H} \cdot \overline{d\ell} = \int_s \overline{J} \cdot \overline{ds} = I_{en} \qquad (2)$$
$$\int_v \nabla \cdot \overline{B}\, dv = \oint_s \overline{B} \cdot \overline{ds} = 0$$

Integral form

In electrostatic fields we have already considered dielectrics, and thus $\overline{D} = \epsilon \overline{E}$ is the expression that relates $\overline{D}$ and $\overline{E}$. We have studied magnetic fields only in free space, and thus $\overline{B} = \mu_0 \overline{H}$. In Chapter 9, we shall study magnetic materials and will find that $\overline{B} = \mu \overline{H}$.

Problem 8.8-1 Starting with $\nabla \cdot \overline{D} = \rho_v$ and the divergence theorem, obtain Gauss's law.

Problem 8.8-2 Starting with $\nabla \times \overline{H} = \overline{J}$ and Stokes' theorem, obtain Ampère's circuital law.

8.9 VECTOR MAGNETIC POTENTIAL

In our study of electrostatic fields, we found a potential function V such that $-\nabla V = \overline{E}$. The gradient of V, ∇V, a concept that was developed in Sec. 4.5 could have been found from the identity $\nabla \times \nabla f = 0$, where f is any scalar, and $\nabla \times \overline{E} = 0$ in the electrostatic case. From these two equations we note that there must exist a scalar f whose gradient, ∇f, can be made equal to $\overline{E}$. If we make $f = V$ and add a minus sign, we obtain the relationship $\overline{E} = -\nabla V$ that we obtained in Sec. 4.5.

In a parallel manner, we find an identity (see Table B-4),

$$\boxed{\nabla \bullet \nabla \times \overline{A} = 0} \tag{1}$$

where $\overline{A}$ is any vector. Now, in magnetostatics we find that $\nabla \bullet \overline{B} = 0$, which in conjunction with (1) leads to

$$\boxed{\overline{B} = \nabla \times \overline{A}} \tag{2}$$

where $\overline{A}$ will be called a *vector magnetic potential*. Thus, if we are able to find a vector magnetic potential $\overline{A}$ whose curl is equal to $\overline{B}$, then $\nabla \bullet \overline{B}$ is assured to be zero through (1), as required by (8.7-6). In many cases, it will be much easier to find $\overline{A}$ first and then form $\nabla \times \overline{A}$ to obtain $\overline{B}$.

Let us now find a differential equation in terms of $\overline{A}$ parallel to Poisson's equation $\nabla^2 V = -\rho_v/\epsilon$ for the scalar potential V in electrostatics. From (8.5-8) and $\overline{B} = \mu_0 \overline{H}$, we obtain

$$\boxed{\nabla \times \frac{\overline{B}}{\mu_0} = \overline{J}} \tag{3}$$

where $\overline{J}$ is due to free charges in motion. Let us now use (2) in (3) to obtain

$$\nabla \times (\nabla \times \overline{A}) = \mu_0 \overline{J} \tag{4}$$

Through the use of a vector identity from Table B-4,

$$\nabla \times (\nabla \times \overline{A}) = \nabla(\nabla \bullet \overline{A}) - \nabla^2 \overline{A} \tag{5}$$

and (4), we obtain

$$\boxed{\nabla(\nabla \bullet \overline{A}) - \nabla^2 \overline{A} = \mu_0 \overline{J}} \tag{6}$$

Equation (6) is thus a vector differential equation involving $\overline{A}$ and $\overline{J}$, the source of $\overline{A}$. The vector magnetic potential $\overline{A}$ in (2) can be shown to have many functional forms and still its curl will be equal to $\overline{B}$. For example, if

$$A_x = f(y,z) + g(x), \quad A_y = f'(x,z) + g'(y), \quad \text{and} \quad A_z = f''(x,y) + g''(z)$$

we will find that the $\nabla \times \overline{A}$ will not depend on $g(x)$, $g'(y)$, and $g''(z)$ since the derivatives of these functions yield zero in the curl operation. Thus, we can have an infinite number of $\overline{A}$'s whose curls give us the desired $\overline{B}$. Vector analysis tells us that a vector is unique if its curl, divergence, and value at one point are specified. Through $\nabla \times \overline{A} = \overline{B}$ we have specified its curl. Let $\nabla \bullet \overline{A} = 0$ to specify its divergence. This choice will be self-evident when we cover time-varying ideas in a later chapter. The last condition on $\overline{A}$ will be satisfied by allowing $\overline{A}$ to equal zero as $r_s \to \infty$. Thus, (6) becomes

$$\boxed{\nabla^2\overline{A} = -\mu_0\overline{J} \quad \text{(aap)}} \tag{7}$$

Equation (7) can be called the *vector Poisson equation* because it has a form similar to Poisson's equation $\nabla^2 V = -\rho_v/\epsilon$.

We shall now obtain a solution for $\overline{A}$ based on the solution found for V from a ρ_v distribution in electrostatics. Let us first expand the left side of (7) through the use of (1.12-6) to obtain

$$\nabla^2\overline{A} = \hat{x}\nabla^2 A_x + \hat{y}\nabla^2 A_y + \hat{z}\nabla^2 A_z \tag{8}$$

Thus, (7) will yield three scalar differential equations:

$$\nabla^2 A_x = -\mu_0 J_x, \qquad \nabla^2 A_y = -\mu_0 J_y, \qquad \nabla^2 A_z = -\mu_0 J_z \tag{9}$$

where each scalar equation is similar in form to Poisson's equation in free space.

$$\boxed{\nabla^2 V = -\frac{\rho_v}{\epsilon_0}} \tag{10}$$

In Chapter 4, we found the solution for V from a ρ_v distribution in free space to be

$$\boxed{V = \int_{v'} \frac{\rho_v' \, dv'}{4\pi\epsilon_0 R} \quad \text{(V)}} \tag{4.4-10}$$

where the integral is carried over the ρ_v distribution. Equation (4.4-10) must be a solution of (10).

Using a mathematical analogy between (9) and (10), the solutions for the components of $\overline{A}$ become

$$\boxed{A_x = \int_{v'} \frac{\mu_0 J_x' \, dv'}{4\pi R}, \qquad A_y = \int_{v'} \frac{\mu_0 J_y' \, dv'}{4\pi R}, \qquad A_z = \int_{v'} \frac{\mu_0 J_z' \, dv'}{4\pi R}} \tag{11}$$

Let us combine the components of $\overline{A}$ in (11) to form

$$\boxed{\overline{A} = \int_{v'} \frac{\mu_0 \overline{J}' \, dv'}{4\pi R} \quad (\text{Wb m}^{-1})} \tag{12}$$

which is the inferred vector solution of (7).

For a surface current and a filamentary current, we have

$$\boxed{\overline{A} = \int_{s'} \frac{\mu_0 \overline{J}_s' \, ds'}{4\pi R} \quad (\text{Wb m}^{-1})} \tag{13}$$

and

$$\overline{A} = \int_{\ell'} \frac{\mu_0 I' \overline{d\ell'}}{4\pi R} \quad (\text{Wb m}^{-1}) \tag{14}$$

Example 11

Find the vector magnetic potential $\overline{A}$ from a differential current element $I\overline{d\ell} = I\overline{dz}$ at the origin.

Solution. From (12) and Fig. 8-14, wc have

$$\overline{A} = \int_{v'} \frac{\mu_0 \overline{J}' \, dv'}{4\pi R} = \frac{\mu_0}{4\pi r_s} \int_{dz'} \int_{ds'} \overline{J}'(ds' \, dz')$$
$$= \frac{\mu_0}{4\pi r_s} \int_{dz'} \int_{ds'} \hat{z}(J_z' \, ds') \, dz' = \frac{\mu_0}{4\pi r_s} \int_{dz'} \hat{z} I \, dz' \tag{15}$$

where $\int_{ds} J_z \, ds = I$. Integrating (15), we obtain

$$\overline{A} = \hat{z} \frac{\mu_0 I \, dz'}{4\pi r_s} \quad (\text{Wb m}^{-1}) \tag{16}$$

This result could have been obtained directly from (14).

Example 12

From the vector magnetic potential $\overline{A}$ of the current element of Example 11, find the $\overline{B}$ through the use of $\overline{B} = \nabla \times \overline{A}$.

Solution. In Fig. 8-14, we have projected $\overline{A}$ onto the $\hat{\theta}$ and $\hat{r}_s$ directions to obtain the $\overline{A}$ in spherical coordinates

$$\overline{A} = \hat{r}_s |\overline{A}| \cos\theta - \hat{\theta} |\overline{A}| \sin\theta \tag{17}$$

Substituting from (16) into (17), we obtain

$$\overline{A} = \hat{r}_s \left(\frac{\mu_0 I \, dz'}{4\pi r_s} \right) \cos\theta - \hat{\theta} \left(\frac{\mu_0 I \, dz'}{4\pi r_s} \right) \sin\theta \tag{18}$$

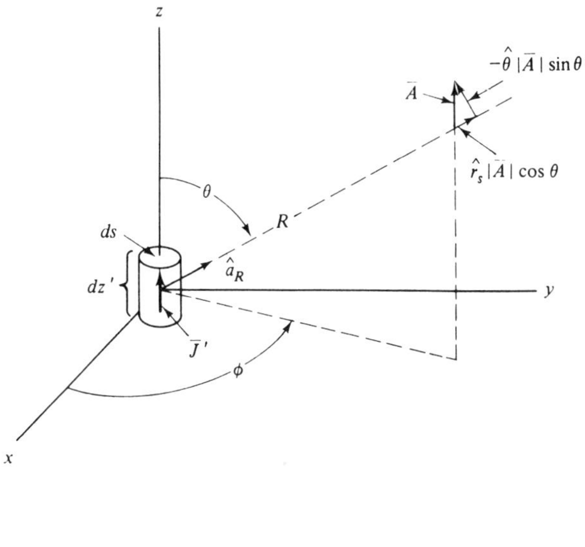

Figure 8-14 Graphical display for finding the vector magnetic potential $\overline{A}$ due to a current element $I\overline{dz}$ for Example 11.

Now, form $\nabla \times \overline{A}$ in spherical coordinates to obtain

$$\overline{B} = \nabla \times \overline{A} = \frac{\hat{\phi}}{r_s}\left[\frac{\partial}{\partial r_s}(r_s A_\theta) - \frac{\partial A_{r_s}}{\partial \theta}\right]$$

$$= \frac{\hat{\phi}}{r_s}\left[\frac{\partial}{\partial r_s}\left(\frac{-r_s \mu_0 I\, dz'}{4\pi r_s}\sin\theta\right) - \frac{\partial}{\partial \theta}\left(\frac{\mu_0 I\, dz'}{4\pi r_s}\cos\theta\right)\right]$$

$$= \hat{\phi}\frac{\mu_0 I\, dz'}{4\pi r_s^2}\sin\theta \qquad \text{(T)} \tag{19}$$

Thus, we now have two methods for finding $\overline{B}$: $\overline{B} = \nabla \times \overline{A}$ and through the Biot-Savart law.

Problem 8.9-1 Starting with (16), find $\overline{A}$ for an infinite line of current, and then using (2) and (8.7-1), derive the $\overline{H}$ field from that current. Compare with (8.3-16).

8.10 DERIVATION OF THE BIOT-SAVART LAW

From (8.9-2) and (8.9-12), we obtain for $\overline{B}$:

$$\overline{B} = \nabla \times \overline{A} = \nabla \times \int_{v'} \frac{\mu_0 \overline{J}'\, dv'}{4\pi R} \tag{1}$$

At this point it might be wise to express (1) in functional form so that we keep clear the functional dependence of each term. Thus,

$$\overline{B}(x, y, z) = \nabla \times \int_{v'} \frac{\mu_0 \overline{J}'(x', y', z')\, dv'(x', y', z')}{4\pi R(x, y, z, x', y', z')} \tag{2}$$

where ∇ operates on the unprimed variables only. The prime variables denote the source point while the unprimed variables denote the field point. Note that

$$R = [(x - x')^2 + (y - y')^2 + (z - z')^2]^{1/2}$$

Thus, (2) becomes

$$\overline{B} = \int_{v'} \frac{\mu_0}{4\pi} \nabla \times \left(\frac{\overline{J}'}{R}\right) dv' \tag{3}$$

From an identity in vector analysis,

$$\nabla \times \left(\frac{\overline{J}'}{R}\right) = \nabla\left(\frac{1}{R}\right) \times \overline{J}' + \frac{1}{R}\nabla \times \overline{J}' \tag{4}$$

A problem at the end of this chapter calls for the proof, by the student, of this identity. The last term of (4) is zero since $\overline{J}'$ is a function of the prime variables and ∇ performs operations only on unprimed variables. The term $\nabla(1/R)$ can readily be expressed as

$$\nabla\left(\frac{1}{R}\right) = -\frac{\hat{a}_R}{R^2} \tag{5}$$

where $\hat{a}_R$ is the unit vector from the source point to the field point. Thus, (4) becomes

$$\nabla \times \left(\frac{\overline{J}'}{R}\right) = \left(-\frac{\hat{a}_R}{R^2}\right) \times \overline{J}' = \overline{J}' \times \frac{\hat{a}_R}{R^2} \tag{6}$$

Substituting (6) into (3), we obtain

$$\boxed{\overline{B} = \int_{v'} \frac{\mu_0}{4\pi}\left(\frac{\overline{J}' \times \hat{a}_R}{R^2}\right) dv'} \tag{7}$$

which is the integral form of the Biot-Savart law (8.3-7)

Problem 8.10-1 Show that $\nabla(1/R) = -\hat{a}_R/R^2$.

8.11 SUMMARY

The introduction of the Biot-Savart law gave us an expression for $\overline{dH}$, the magnetic field intensity, from a current element $I\,\overline{d\ell}$. The total $\overline{H}$ from a closed path of current distribution was then found through the integration of $\overline{dH}$ over the entire current distribution. For symmetrical current distribution, the $\overline{H}$ field was found through the usc of Ampère's circuital law. The curl of $\overline{H}$ concept is well summarized by (8.5-12). The relationship between curl of $\overline{H}$, Stokes' theorem, and Ampère's circuital law are displayed in (1).

$$\oint_\ell \overline{H} \cdot \overline{d\ell} = \int_s \overline{J} \cdot \overline{ds} = \int_s (\nabla \times \overline{H}) \cdot \overline{ds} = I_{\text{en}} \tag{1}$$

Maxwell's equations; Curl of $\overline{H}$; Stokes' theorem; Ampere's circuital law

A display of $\overline{B}$ relationships parallel to that of $\overline{D}$ in (3.7-1) is shown in (2).

$$\Psi_m = \oint_s \overline{B}_s \cdot \overline{ds} = 0 = \int_v \nabla \cdot \overline{B}\, dv = 0 \tag{2}$$

Maxwell's equations; Divergence of $\overline{B}$; Divergence theorem; Gauss's law in $\overline{B}$; Magnetic flux lines close on themselves; Magnet poles cannot be isolated

It should be noted that $\Psi_m = 0$ in (2) since $\overline{B}_s \cdot \overline{ds}$ was integrated over a closed surface.

REVIEW QUESTIONS

1. What is a filamentary current element? *Sec. 8.2*
2. Does $I\,\overline{d\ell} = \overline{J}_s ds = \overline{J}\,dv$? *Eq. (8.2-2)*
3. Is $\overline{J}_s$ defined as a sheet current density in amperes per meter width? *Sec. 8.2*
4. What is the Biot-Savart law? *Eq. (8.3-2)*
5. Can the Biot-Savart law be proven experimentally in differential form? *Sec. 8.3*
6. Is the direction of $\overline{dH}$ the same as $\overline{d\ell} \times \hat{a}_R$? *Eq. (8.3-2)*

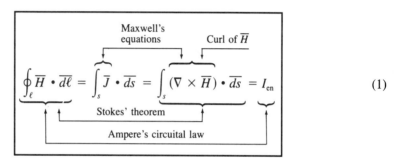

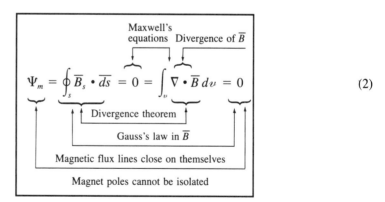

7. What is the expression for $\overline{H}$ about an infinite filamentary current along the z axis? *Eq. (8.3-16)*
8. What is the direction of $\overline{H}$ along the axis of a filamentary circular loop of current? *Eq. (8.3-20)*
9. Where is the $\overline{H}$ field the greatest along the axis of a solenoid? *Example 3*
10. What is the mathematical form of Ampere's circuital law? *Eq. (8.4-2)*
11. What law in electrostatics is parallel to Ampère's circuital law? *Sec. 8.4*
12. Is the $\overline{H}$ field at $r_c > a$ for a long conductor that carries a uniform current I the same as the $\overline{H}$ field about an infinite length filamentary current? *Example 5, Eq. (8.4-6)*
13. Is the $\overline{H}$ above a surface current $\overline{J}_s$ in the same direction as $\overline{J}_s$? *Example 6, Eq. (8.4-15)*
14. Is $\overline{H} = \overline{J}_s \times \hat{n}$ a boundary condition? *Eq. (8.4-16)*
15. What is the $\overline{H}$ field outside a coaxial cable? *Example 7*
16. What is the definition for $\nabla \times \overline{H}$? *Eq. (8.5-6)*
17. What is $\nabla \times \overline{H}$ equal to in terms of a physical quantity? *Eq. (8.5-7)*
18. What is the determinant form for $\nabla \times \overline{H}$? *Eq. (8.5-9)*
19. What is the mathematical expression for Stokes' theorem? *Eq. (8.6-3)*
20. In Stokes' theorem, how is the surface s related to the amperian closed loop? *Sec. 8.6*
21. What is the relationship between $\overline{H}$ and $\overline{B}$ in free space? *Eq. (8.7-1)*
22. What is the value for μ_0? *Eq. (8.7-2)*
23. Why is $\oint_s \overline{B} \cdot \overline{ds} = 0$? *Sec. 8.7*
24. Are magnetic flux lines closed? *Sec. 8.7*
25. What is $\nabla \cdot \overline{B}$ equal to? *Eq. (8.7-6)*
26. How many Maxwell's equations are there? *Eq. (8.8-1)*
27. What is $\overline{A}$? *Sec. 8.9, Eq. (8.9-2)*
28. The ∇ operator operates on what variables? *Sec. 8.10*
29. Can the Biot-Savart law be derived from $\overline{B} = \nabla \times \overline{A}$? *Sec. 8.10*

PROBLEMS

8-1. Calculate the current required to produce an H field of 79.58 A m^{-1} (1 oersted) at the center of a circular loop having a radius of 1 cm.

8-2. Assume that a uniformly wound solenoidal coil, $\ell = 10$ cm, $a = 1$ cm, is used to produce a field of 40 A m^{-1} at the center of the solenoid. If $N = 5000$ turns, find the current required.

8-3. Find the number of turns required to produce a field of 40 A m^{-1} at the center of a uniformly wound solenoid whose length is 5 cm, and whose radius is 1 cm, for a current of 5 mA.

8-4. Two identical circular filamentary current loops with axes along the z axis are located at $z = \pm 0.5$ m. If a current of 10 A flows in the same direction in each loop and their radii equals 1.0 m: (a) plot the magnitude of $\overline{H}$ along the z axis for a range of $-0.5 \leq z \leq 0.5$; (b) over what range of z, about $z = 0$, is the $|\overline{H}|$ within 5%? Use the results found for a single loop. This coil configuration is called a *Helmholtz coil* and provides a large region of extremely uniform $|\overline{H}|$ for experimental work.

8-5. For Prob. 8-4, find the first and second derivatives of H_z with respect to z, for $z = 0$. What is the significance of these derivatives?

8-6. A surface current distribution of infinite extent, located in the $z = 0$ plane, carries a uniform current density $\overline{J}_s = \hat{y}10$ (A m^{-1}). Using the results for a filamentary current distribution of infinite extent, find the $\overline{H}$ above and below the surface current.

8-7. A lightning discharge from a cloud to earth carries a current of 100,000 A for 10 μs. If the cloud height above the ground is 2 km, find the H field generated by the strike at a distance of 10 km from the site of the strike. Assume an image current beneath the earth's surface as per Fig. 7-9.

8-8. Using the results from Prob. 8.3-6 and a programmable calculator, plot the normalized ($NI = 1$) $\overline{H}$ field as a function of distance from the center of the solenoid, for a solenoid whose ℓ is: (a) $2a$; (b) $10a$ where a is the radius. Plot the field out to $\ell/4$ beyond the ends of the solenoid.

8-9. Through the use of the Biot-Savart law and Ampère's circuital law give arguments that, outside a single-turn surface-current solenoid of infinite length-along the z axis: (a) $H_\phi = 0$; (b) $H_r = 0$; (c) $H_z = 0$ [see Fig. 8-6(d)].

8-10. (a) Calculate the approximate axial field, in the plane of the loop, for a flat spiral of N turns, where the initial radius is a and the final radius is $a + N\delta$. Assume a current I, and start with the Biot-Savart law. (b) Compare the field obtained in part (a) with that from an N-turn loop whose radius is $[a + (a + N\delta)]/2$. Try several numerical examples.

8-11. Find $\nabla \times \overline{H}$ inside and outside a long circular cross-section conductor carrying a uniform current density $\overline{J} = \hat{z}K$ A m^{-2}, $0 \le r_c \le a$, where a is the radius of the conductor.

8-12. Find the $\nabla \times \overline{H}$ in all regions of the coaxial cable of Example 7.

8-13. Given a vector $\overline{G} = \hat{\phi} r_s \sin\theta \cos\phi$, evaluate both sides of Stokes' theorem when the amperian loop is defined by $z = 0$, $r_s = 5$, and the surface is the hemisphere defined by $r_s = 5$ and $z > 0$.

8-14. Find the magnetic flux, due to a z-directed filamentary current I_0 of infinite extent, that flows through a rectangular cross section defined by $\phi = \pi/2$, $2 < r_c < 4$ m, and $3 < z < 6$ m.

8-15. For the filamentary circular current loop of Fig. 8-5, find: (a) the vector magnetic potential $\overline{A}$; (b) $\overline{B}$ through the use of $\overline{A}$.

8-16. Assume that in a given location, the earth's magnetic flux density is 5×10^{-5} T. To cancel the effect of the earth's field on a stationary magnetic field sensor, the sensor is encircled by a single loop of current with the axis of the loop parallel to the earth's field. If the radius of the loop is 1 cm, find the current required to cancel the earth's field.

8-17. Find the total magnetic flux in the space between two parallel wires of radius a separated by distance d (between centers) if each wire is carrying current I but in opposite directions.

8-18. Prove (8.10-4) through expansion of both sides.

8-19. Prove (8.10-5) by expansion of $\nabla(1/R)$ when

$$R = [(x - x')^2 + (y - y')^2 + (z - z')^2]^{1/2}$$

8-20. (a) Referring to Fig. 8-4, find the vector magnetic potential $\overline{A}$ at the point $P(r_c, \phi, z)$ for a line length L, if a is at $z' = -L/2$ and b is at $z' = L/2$. (b) Find A for a very short L, i.e., as $L \to 0$, but for the line still centered at $z = 0$.

9

Magnetic Forces, Magnetic Polarization, Magnetic Material, Reluctance, Magnetic Circuits, and Inductance

9.1 INTRODUCTION

The introduction to Chapter 2 should be reread by the student since several of the concepts mentioned there are relevant to the introduction of this chapter. In this chapter, we shall introduce the experimental results of Ampère; i.e., $\overline{dF} = I\,\overline{d\ell} \times \overline{B}$, the force on a current element immersed in a magnetic field $\overline{B}$. The total force on a closed loop of current distribution will be found through integration. Thus, using the results from the Biot-Savart law and those of Ampère, we will be able to find the force between two current elements.

The magnetic dipole will be introduced. This concept will lead us to magnetic material and thus to a more general relationship between $\overline{H}$ and $\overline{B}$. The concepts of magnetic circuits, inductance, and magnetic energy density will also be developed. We start this chapter with an extended study of the field concept as a solution method.

9.2 THE CHARGE-FIELD-CHARGE SOLUTION CONCEPT

Before we embark on finding the force between current elements (moving electric charges), we will attempt to obtain a deeper insight of and respect for the field concept. Faraday developed the field concept during his work with electromagnetism. The field concept can be viewed as a solution concept in finding the force between static or moving charges. Before Faraday's time, the force between two point charges was considered to be an instantaneous and direct interaction between the charges. Coulomb's force law, with vector notation added,

$$\overline{F}_2 = \frac{Q_1 Q_2}{4\pi\epsilon_0 R_{12}^2}\hat{a}_{R12} \qquad \text{(N)} \tag{1}$$

is a good example of this *action-at-a-distance* reasoning in finding the force $\overline{F}_2$ that Q_1 produces on Q_2.

Through the introduction of the field concept, the solution for finding $\overline{F}_2$ is separated into two parts: first, we find the field produced by Q_1; second, we find the force that this field produces on Q_2 when immersed in it. Thus, the field plays an intermediate part in finding forces between charges. This interplay between charge Q_1 and the field, and the field and Q_2, in the solution process for finding $\overline{F}_2$ can be viewed as the *charge-field-charge solution* concept. A display of the action-at-a-distance solution concepts for two point charges is shown in Fig. 9-1. In Fig. 9-1(a), the charge Q_1 produces a field $\overline{E}_t$, force on a test charge Q_t at a distance R_{1t} from Q_1, cqual to

$$\overline{E}_t = \frac{Q_1 \hat{a}_{R_{1t}}}{4\pi\epsilon_0 R_{1t}^2} \qquad (\text{V m}^{-1}) \tag{2}$$

Now, the force produced by the field on the charge Q_2 is

$$\overline{F}_2 = \overline{E}_t \big|_{\text{at } Q_2} Q_2 \qquad (\text{N}) \tag{3}$$

where $\overline{E}_t$ is evaluated at the location of Q_2. It should be noted that the action-at-a-distance solution (1) comes about through the elimination of the $\overline{E}_t$ field in (2) and (3), as displayed in Fig. 9-1(b). Thus, it might seem that our solution for the force on Q_2 due to Q_1 could be solved with greater ease by using the action-at-a-distance concept. This is true if we were only interested in forces between point static charges or between differential current elements carrying a steady current. When the charge or current distributions become more complex, much is gained by separating the solution into two parts through the use of the field concept. Also, in the case of accelerated charges (non-steady current), the action-at-a-distance concept does not take into account the finite time it takes a distance charge Q_2 to sense the motion of Q_1 so that it can react to the new position of Q_1 as indicated by experiments. The action-at-a-distance concept assumes that any change in position of Q_1 produces an instantaneous change in the force on Q_2. In the

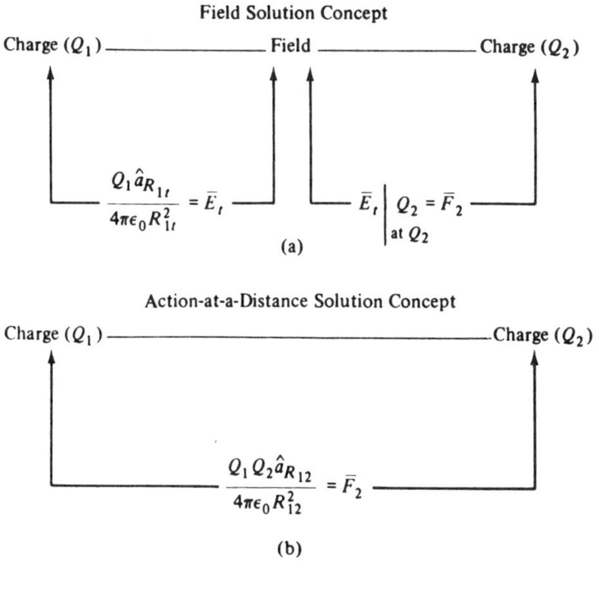

Figure 9-1 Solution concept display: (a) field concept (charge-field-charge concept); (b) action-at-a-distance concept.

field solution concept, charge Q_2 learns of the motion of Q_1 through a field disturbance that is propagated from Q_1 to Q_2 with the speed of light in free space. Accelerated charges in a transmitting antenna will produce forces on charges in a distant receiving antenna after a time delay of ℓ/U (s), where ℓ is the distance between the antennas and U is the propagation speed of the field disturbance. Thus, the field concept, charge-field-charge solution concept is indispensable in the study of interactions between moving charges and is basic to all scientific theory involving forces between particles. Our present gravitational concepts are based on a gravitational field that is defined as the force per unit mass. The force on another mass in this field is then obtained through the product of the gravitational field intensity and the mass.

In magnetostatics, the Biot-Savart law allows us to go from a conductor current-element $I_1\overline{d\ell}_1$ to magnetic field $\overline{dH}$. In Sec. 9.3, we shall introduce Ampère's experimental results, which will allow us to go from a magnetic field $\overline{dH}$ to the force on a conductor current-element $I_2\overline{d\ell}_2$. This last step will complete our charge (moving)-field-charge (moving) solution concept.

Example 1

For two finite and parallel line charges, derive the expression for the force on one of the lines: (a) in terms of the field concept, charge-field-charge; (b) in terms of the action-at-a-distance concept.

Solution. (a) From Fig. 9-2, the $\overline{E}_2$ field at the location of line 2 due to line 1 is

$$\overline{E}_2 = \int_a^b \frac{(\rho_{\ell_1}\, d\ell_1)\hat{a}_{R12}}{4\pi\epsilon_0 R_{12}^2} \qquad (\text{V m}^{-1}) \tag{4}$$

Now, the force on line 2 immersed in the $\overline{E}_2$ field is

$$\overline{F}_2 = \int_c^d (\rho_{\ell_2}\, d\ell_2)\overline{E}_2 \qquad (\text{N}) \tag{5}$$

Thus, our solution has been separated into two parts, (4) and (5).

(b) The force on $(\rho_{\ell_2}\, d\ell_2)$ due to $(\rho_{\ell_1}\, d\ell_1)$ is, from Coulomb's force law,

$$d(\overline{dF}_2) = \frac{(\rho_{\ell_2}\, d\ell_2)\,(\rho_{\ell_1}\, d\ell_1)}{4\pi\epsilon_0 R_{12}^2}\hat{a}_{R12} \tag{6}$$

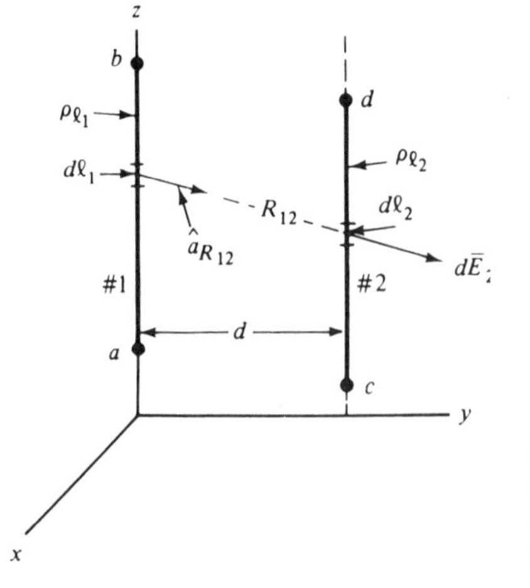

Figure 9-2 Graphical display for finding the force between two line charges through the use of the field concept and the action-at-a-distance concept for Example 1.

Now, $\overline{dF}_2$ on $(\rho_{\ell_2} d\ell_2)$ is obtained by integrating over line 1 to give

$$\overline{dF}_2 = (\rho_{\ell_2} d\ell_2) \int_a^b \frac{(\rho_{\ell_1} d\ell_1)}{4\pi\epsilon_0 R_{12}^2} \hat{a}_{R_{12}} \tag{7}$$

The force $\overline{F}_2$ is obtained by integrating over line 2 to give

$$\overline{F}_2 = \int_c^d (\rho_{\ell_2} d\ell_2) \int_a^b \frac{(\rho_{\ell_1} d\ell_1)}{4\pi\epsilon_0 R_{12}^2} \hat{a}_{R_{12}} \quad \text{(N)} \tag{8}$$

Equation (8) has been obtained without the use of the field concept. It should be noted that (8) can be obtained by substituting (4) into (5).

Problem 9.2-1   Neglecting fringing effects at the edges, show: (a) that the force between two parallel charged plates, of area s (m²) each, is equal to the electric field from one plate times the charge on the other plate; (b) that this force can also be obtained from the expression for energy density (4.6-18) by assuming that the energy stored in the volume is equal to the force times the separation of plates.

9.3 FORCE BETWEEN CONDUCTOR CURRENT-ELEMENTS

From Ampère's experimental results, the force of translation on a conductor current-element $I\,\overline{d\ell}$ immersed in a magnetic field $\overline{B}$ is

$$\boxed{\overline{dF} = I\,\overline{d\ell} \times \overline{B} \quad \text{(N)}} \tag{1}$$

Let us now find the force of translation between the two conductor current-elements, $I_1\,\overline{d\ell}_1$ and $I_2\,\overline{d\ell}_2$, as found in Fig. 9-3. From the Biot-Savart law (8.3-2), $\overline{dH}_2$ at $P_2(x_2, y_2, z_2)$ due to $I_1\,\overline{d\ell}_1$ is found to be

$$\boxed{\overline{dH}_2 = \frac{I_1\,\overline{d\ell}_1 \times \hat{a}_{R_{12}}}{4\pi R_{12}^2} \quad \text{(A m}^{-1}\text{)}} \tag{2}$$

Thus, our problem has been divided into two parts: first, finding the field due to a conduction current-element; second, finding the force on another conductor current-element.

From $\overline{dB}_2 = \mu_0\,\overline{dH}_2$ and (2), eq. (1) becomes[1]

$$\boxed{d(\overline{dF}_2) = I_2\,\overline{d\ell}_2 \times \frac{\mu_0 I_1\,\overline{d\ell}_1 \times \hat{a}_{R_{12}}}{4\pi R_{12}^2} \quad \text{(N)}} \tag{3}$$

where $d(\overline{dF}_2)$ is the force on $I_2\,\overline{d\ell}_2$ immersed in the $\overline{dB}_2$ field of $I_1\,\overline{d\ell}_1$, or the force between the two conductor current-elements. Note that in (3) we have eliminated the $\overline{B}$ field, and the expression is now in terms of the conductor current-elements, quite parallel to Coulomb's force expression between two charges. Thus, (3) represents the action-

---

[1] For parallel side-by-side current-elements, equation (3) may be written as $d|\overline{dF}| = \mu_0\, dQ_{m1}\, dQ_{m2}/4\pi R_{12}^2$, where dQ_{m1} and dQ_{m2} are magnetic poles in (A · m).

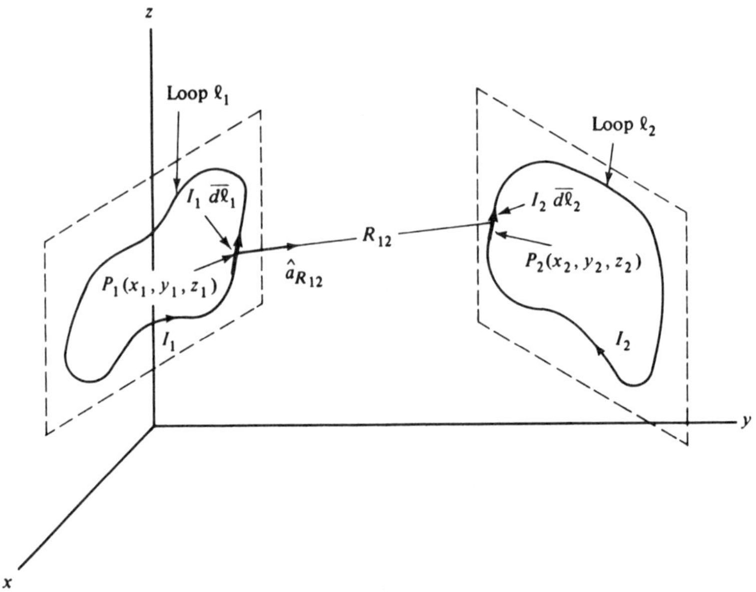

Figure 9-3 Graphical display for finding the force between two filamentary conductor current loops carrying currents I_1 and I_2.

at-a-distance solution concept of our problem. If we integrate the last part of (3) over the loop ℓ_1, we obtain

$$\boxed{\overline{dF}_2 = I_2\overline{d\ell}_2 \times \oint_{\ell_1} \frac{\mu_0 I_1 \overline{d\ell}_1 \times \hat{a}_{R_{12}}}{4\pi R_{12}^2} = I_2\overline{d\ell}_2 \times \overline{B}_2 \qquad \text{(N)}} \tag{4}$$

where $\overline{dF}_2$ is the force on $I_2\overline{d\ell}_2$ immersed in the $\overline{B}_2$ field of loop ℓ_1, or the force on $I_2 d\ell_2$ due to loop ℓ_1. Integrating (4) over loop ℓ_2 yields

$$\boxed{\overline{F}_2 = \oint_{\ell_2} I_2\overline{d\ell}_2 \times \oint_{\ell_1} \left[\frac{\mu_0 I_1 \overline{d\ell}_1 \times \hat{a}_{R_{12}}}{4\pi R_{12}^2}\right] \qquad \text{(N)}} \tag{5}$$

where $\overline{F}_2$ is the total force of translation on conductor current loop ℓ_2 due to conductor current loop ℓ_1. Equation (5) can be simplified to read

$$\boxed{\overline{F}_2 = \frac{\mu_0 I_1 I_2}{4\pi} \oint_{\ell_2} \left[\oint_{\ell_1} \frac{(\hat{a}_{R_{12}} \times \overline{d\ell}_1)}{R_{12}^2}\right] \times \overline{d\ell}_2 \qquad \text{(N)}} \tag{6}$$

Some authors choose to start the magnetostatic theory with (5) since it can readily be verified experimentally and the integral over loop ℓ_1 can be identified as the integral form of the Biot-Savart law in terms of $\overline{B}$. Equations (5) and (6) can be extended to surface and volume conductor currents by replacing $I\,\overline{d\ell}$ by $\overline{J}_s\,ds$ and $\overline{J}\,dv$, respectively, and integrating over the surface distribution of $\overline{J}_s$ or the volume distribution of $\overline{J}$.

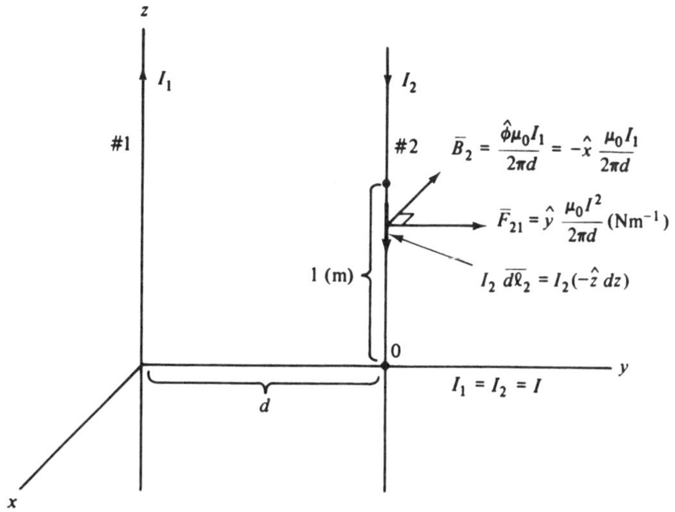

Figure 9-4 Graphical display for finding the force per meter between two infinite and parallel filamentary current-carrying conductors in Example 2.

Example 2

Find the force per meter between two infinite and parallel filamentary current-carrying conductors that are separated d (m) and carry a current I (A) in opposite directions.

Solution. Let us assume that conductor 1 is along the z axis and that conductor 2 is at $y = d$, as shown in Fig. 9-4. The $\overline{B}_2$ field at the location of conductor 2, from (8.3-16), is

$$\overline{B}_2 = \mu_0 \overline{H}_2 = \mu_0 \frac{\hat{\phi} I_1}{2\pi r_c}\bigg|_{\substack{r_c=d \\ \phi=\pi/2}} = \frac{-\hat{x}\mu_0 I_1}{2\pi d} \qquad \text{(T)} \tag{7}$$

Substituting (7) into (1) and integrating over 1 meter, we obtain for $\overline{F}_2$:

$$\overline{F}_2 = \int_0^1 I_2 \overline{d\ell}_2 \times \left(\frac{-\hat{x}\mu_0 I_1}{2\pi d}\right) = \int_0^1 I_2(-\hat{z}\, dz_2) \times \left(\frac{-\hat{x}\mu_0 I_1}{2\pi d}\right)$$

$$= \hat{y}\mu_0 \frac{I_1 I_2}{2\pi d} = \hat{y}\frac{\mu_0 I^2}{2\pi d} \qquad (\text{N m}^{-1}) \tag{8}$$

The same result could have been obtained from (5) or (6).

Example 3

Find the force of translation on an arbitrary current-carrying conductor loop in a uniform $\overline{B}$ field.

Solution. From (1), the force of translation on a closed loop of conduction current becomes

$$\overline{F} = \oint_\ell I\, \overline{d\ell} \times \overline{B} \tag{9}$$

For a uniform $\overline{B}$ and a steady current I, (9) becomes

$$\overline{F} = -I\overline{B} \times \oint_\ell \overline{d\ell} = 0 \tag{10}$$

since $\oint_\ell \overline{d\ell} = 0$. It should be noted that the torque $\overline{T}$ on the loop may not be zero even though the force of translation is zero.

Problem 9.3-1 Show that $\overline{F}_2 = -\overline{F}_1$ in the problem of Example 2.

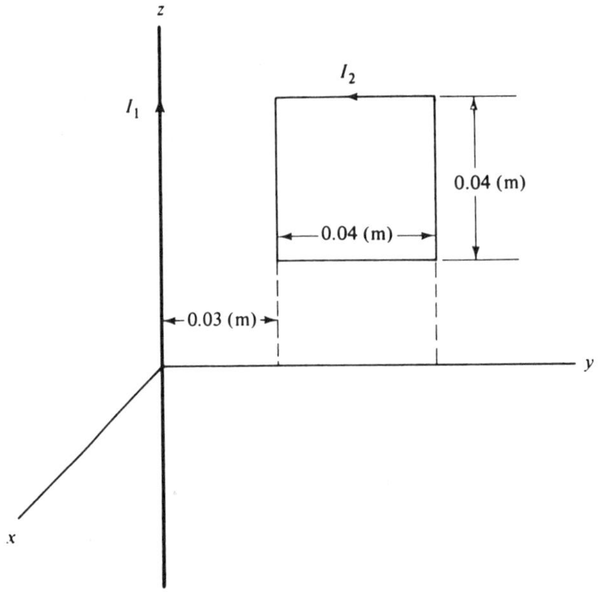

Figure 9-5 Graphical display for finding the force of translation on a square current-carrying loop near an infinite filamentary current-carrying conductor of Prob. 9.3-2.

Problem 9.3-2 Find the force of translation on a square conductor current loop when immersed in the $\overline{B}$ field of a current-carrying filamentary conductor of infinite length along the z axis, as shown in Fig. 9-5. Assume that $I_1 = 15$ A and $I_2 = 50$ A.

9.4 FORCE ON MOVING POINT CHARGES

In the previous section, all the forces were either on a conductor current-element or between two conductor current-elements. All of the expressions thus were in terms of current-elements $I\,\overline{d\ell}$, $\overline{J}_s\,ds$, or $\overline{J}\,dv$. In this section, we shall extend the concept of force on conductor current-elements to force on moving point charges. This concept will give us a better insight into the mechanics of force on conductor current-elements and charges moving in free space.

In Chapter 5, we found that the current density $\overline{J} = \rho_v \overline{U}$, (5.2-4), applied to convection currents as well as conduction currents. Let us first express (9.3-1) in terms of the conductor current-element $\overline{J}\,dv$ to obtain

$$\overline{dF} = (\overline{J}\,dv) \times \overline{B} \tag{1}$$

Substituting (5.2-4) into (1), we obtain

$$\boxed{\overline{dF} = (\rho_v \overline{U}\,dv) \times \overline{B} = (\rho_v\,dv)\overline{U} \times \overline{B} = dQ\,\overline{U} \times \overline{B} \qquad \text{(N)}} \tag{2}$$

where $\overline{dF}$ can be interpreted as the force on a point charge $dQ = \rho_v\,dv$ that moves with a velocity $\overline{U}$ when immersed in a magnetic field $\overline{B}$. Dividing (2) by dv, we obtain

$$\boxed{\frac{\overline{dF}}{dv} = \rho_v \overline{U} \times \overline{B} \qquad (\text{N m}^{-3})} \tag{3}$$

as the force per unit volume of ρ_v distribution. Thus, we have extended our concept of force on a conductor current-element to force on a moving point charge in either a conductor or free space. From (2) or (3), we find that the force is perpendicular to $\overline{U}$ and $\overline{B}$. Thus, the charge $dQ = \rho_v \, dv$ can neither gain nor lose energy from the magnetic field $\overline{B}$.

In a conductor, ρ_v becomes the electron charge density ρ_{ve} while $\overline{U}$ is the drift velocity $\overline{U}_d$ of the electrons. From (2), we see that the force is on the moving electrons. This force on the electrons causes a slight displacement between the free moving electrons and the stationary positive ions that form the lattice structure of the conductor. The Coulomb force between the electrons and the lattice ions transfers the magnetic force on the moving electrons to the lattice structure and thus to the conductor. The slight displacement between the electrons and the positive lattice ions also gives rise to a Hall potential, which shall be discussed in the next section. A simple electric motor, using these concepts, is shown in Fig. 9-6(a). The force on the moving electron and thus on the conductor is shown in Fig. 9-6(b).

Another extremely important effect exists when a conductor is moved by some applied force $\overline{F}_a$ in a magnetic field, as suggested in Fig. 9-7(a). The velocity due to $\overline{F}_a$ is imparted to both the free electrons and the positive lattice ions. From (2), we note that a force $\overline{dF}$ due to motion will exist on both the electrons and the positive lattice ions, but only the free electrons will experience translation. Figure 9-7(b) shows the direction of the force $\overline{dF}_e$ on the electron and the direction of an apparent electric field intensity

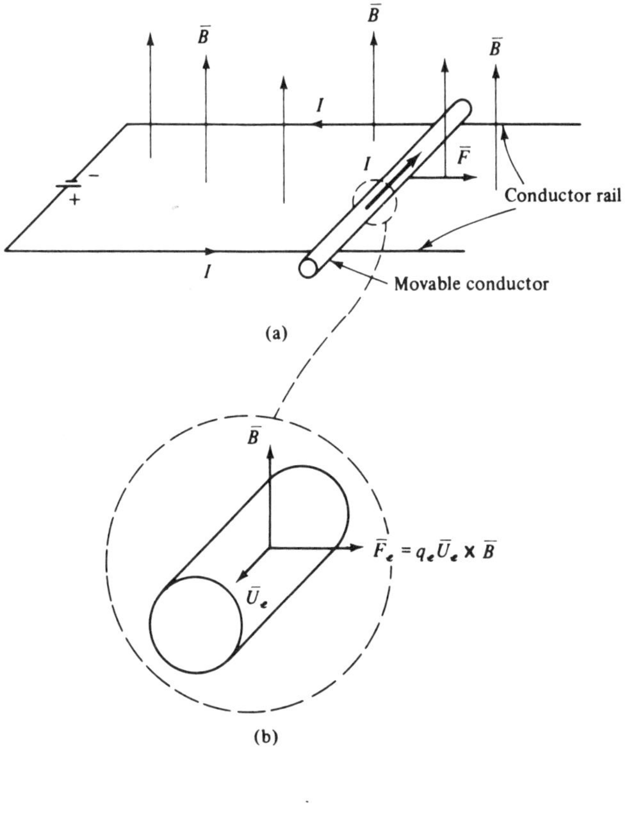

Figure 9-6 (a) Graphical display of a simple electric motor effect produced by moving charges (current) in a conductor immersed in a $\overline{B}$ field. (b) Vector force and velocity experienced by an electron.

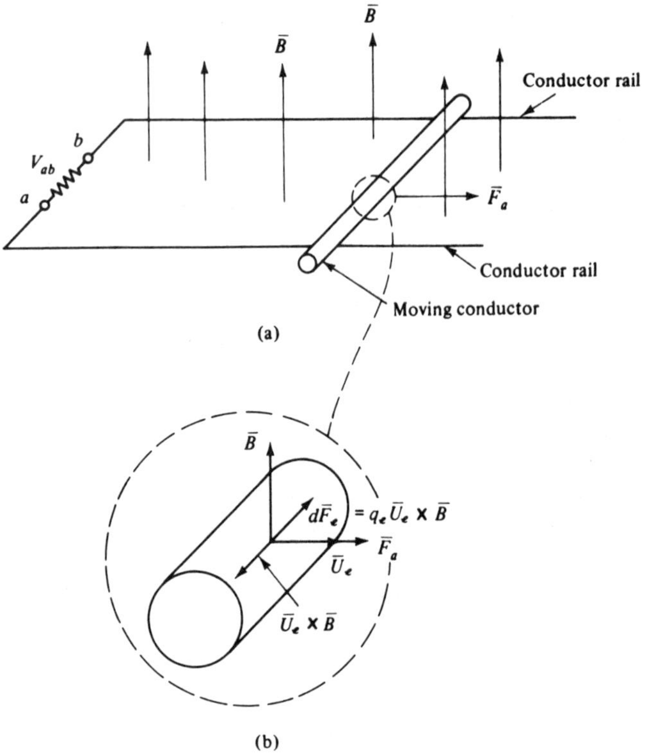

Figure 9-7 (a) Graphical display of force on an electron in a conductor that is moved by an applied force $\overline{F}_a$. This display can be viewed as a simple electric generator. (b) Expanded view of vector forces and fields.

$\overline{U} \times \overline{B}$, that acts on the electron. This apparent electric field intensity is obtained by dividing (2) by dQ to obtain

$$\boxed{\frac{\overline{dF}}{dQ} = \overline{U} \times \overline{B} \qquad (\mathrm{N\ C^{-1}})} \tag{4}$$

The movement of the electrons in the conductor will give rise to a current and a build-up of potential between points a and b of Fig. 9-7(a). This effect is fundamental to generation of electricity and is basic to the operation of an electric generator. More will be said about this effect in Chapter 10.

The electron beam in a cathode-ray tube is a good example of convection current. It can be deflected by placing a $\overline{B}$ field perpendicular to the direction of the beam. In this case, the force $\overline{dF}$ can be viewed as either on each electron or on a volume dv of electrons.

In previous chapters, we found that a charge dQ immersed in an electric field experiences a force.

$$\boxed{\overline{dF} = \overline{E}\, dQ} \tag{5}$$

Equation (5) not only applies to stationary charges but also to moving charges. For a moving charge, a component of $\overline{dF}$ can exist along its trajectory, and thus an exchange of energy can take place between the $\overline{E}$ field and the charge, as given in Sec. 4.2. When dQ is immersed in combined $\overline{E}$ and $\overline{B}$ fields, the combined force from (2) and (5) becomes

$$\boxed{\overline{dF} = (dQ\overline{E} + dQ\overline{U} \times \overline{B}) \qquad \text{(N)}} \tag{6}$$

Equation (6) is known as the *Lorentz force equation,* whose solution is required in determining the motion of a point charge in combined $\overline{E}$ and $\overline{B}$ fields. The first and second terms of (6) are the electric and magnetic forces, respectively, on the charge dQ.

Example 4

Derive the general scalar differential equations that must be solved to obtain the trajectory of a point charge immersed in combined $\overline{E}$ and $\overline{B}$ fields.

Solution. Let us assume the charge to be Q (C) and its mass m (kg). From Newton's force law and the Lorentz force equation, we have

$$Q(\overline{E} + \overline{U} \times \overline{B}) = m\overline{a} = m\frac{d\overline{U}}{dt} \tag{7}$$

Expanding $\overline{E}$, $\overline{B}$, and $\overline{U}$ in rectangular component form and equating like component terms in (7), we obtain

$$Q(E_x + U_yB_z - U_zB_y) = m\frac{dU_x}{dt} \tag{8}$$

$$Q(E_y + U_zB_x - U_xB_z) = m\frac{dU_y}{dt} \tag{9}$$

$$Q(E_z + U_xB_y - U_yB_x) = m\frac{dU_z}{dt} \tag{10}$$

With the aid of initial conditions on $\overline{U}$, the solutions of (8), (9), and (10) can be obtained. If we express $U_x = dx/dt$, $U_y = dy/dt$, and $U_z = dz/dt$, we can obtain the trajectory of the charge with the aid of added initial conditions on x, y, and z.

Problem 9.4-1 A point charge of mass m and charge Q is injected with an initial velocity of $\overline{U}_0 = \hat{y}5 + \hat{x}3 + \hat{z}10$ (m s^{-1}) at the origin, into a region where $\overline{E} = -\hat{z}5$ (V m^{-1}) and $\overline{B} = 0$. Find the velocity at time t.

Problem 9.4-2 Repeat Prob. 9.4-1 with $\overline{B} = \hat{z}2$ (T). Show that the trajectory, projected onto the x-y plane, is a circle with a period of $T = 2\pi m/QB_z$.

9.5 HALL EFFECT

As mentioned in Sec. 9.4, the magnetic force on moving electrons in a conductor will produce a slight displacement between the electrons and the positive ions of the stationary lattice structure. This displacement of electrons will give rise to surface charges, as suggested in Fig. 9-8. The buildup of the surface charge will stop when the transversal electric field $\overline{E}_T$, due to surface charges, is just cancelled by the equivalent electric field $\overline{U}_- \times \overline{B}$, due to the magnetic force on the electron, or

$$\overline{E}_T + \overline{U}_- \times \overline{B} = 0 \tag{1}$$

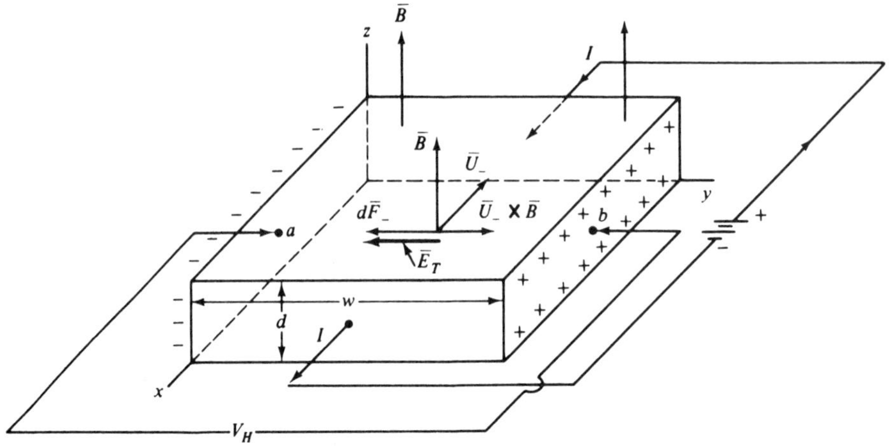

Figure 9-8 Graphical display for finding the Hall potential V_H.

A potential V_H, the *Hall potential*, will be found between directly opposite points on the side walls. The Hall potential, due to surface charges, is equal to

$$V_H = V_{ab} = -\int_b^a \overline{E}_T \bullet \overline{d\ell} = -\int_b^a (-\overline{U}_- \times \overline{B}) \bullet \overline{d\ell} = -U_x B_z w \qquad \text{(V)} \qquad (2)$$

where w is the width of the slab as found in Fig. 9-8. Through the use of $\overline{U}_- = \overline{J}/\rho_{ve}$, (2) will become

$$\boxed{V_H = -\frac{J_x}{|\rho_{ve}|} B_z w = -\frac{I/wd}{|\rho_{ve}|} B_z w = \frac{IB_z}{n_e q_e d} \qquad \text{(V)}} \qquad (3)$$

where n_e is the free electron density. If the carrier of current I is a positive charge, we will find $\overline{U}_+$ to be opposite to $\overline{U}_-$ but $\overline{dF}_+ = \overline{dF}_-$. In this case, the positive carriers will be forced to the left, and V_H will be the negative of that found for negative carriers.

The Hall voltage polarity can be used as an indicator of the polarity of the carrier of I in a semiconductor material. Thus, we can distinguish between N-type and P-type semiconductors. The magnitude of the Hall voltage can also be used as an indicator of the magnitude of $\overline{B}$ field. Hall effect sensors are widely used for measurement of magnetic fields, for clamp-on current probes, and for ignition systems for automobiles. Figure 9-9 illustrates application of the Hall generator to the measurement of ac power. Since the Hall voltage V_H is proportional to the product of I and B, V_H in Fig. 9-9 is proportional to the product of line voltage and line current. The average value of V_H is proportional to true average power, taking into account the phase angle between voltage and current. Clamp-on power probes are commercially available. Other applications of the Hall effect are found in an article by Epstein.[2]

Problem 9.5-1 Make two sketches like Fig. 9-8, one for positive carriers (holes) and one for negative carriers (electrons). Let the polarity of $\overline{B}$ and the direction of current be the same in each and show: (a) that the polarity of V_H is reversed for the different carriers, but (b) that $\overline{dF} = q\overline{U} \times \overline{B}$ is the same for holes or electrons.

[2]M. Epstein, "Hall-Effect Devices," *IEEE Trans. Magn.*, September 1967, pp. 352–359.

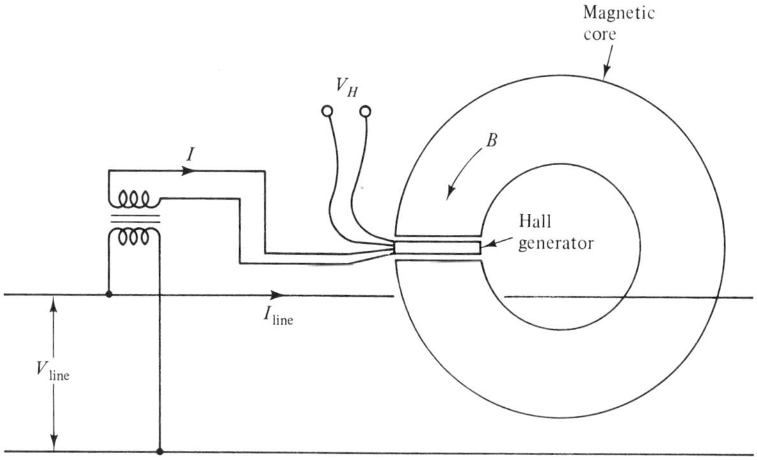

Figure 9-9 Application of Hall generator to measurement of ac power. I is proportional to V_{line}, B is proportional to I_{line}, and V_H is thus proportional to the product of V_{line} and I_{line}. The magnetic core is used to achieve flux concentration.

Problem 9.5-2 Find the Hall potential for a copper strip 0.04 m wide and 1.0 mm thick, carrying a current of 100 A, and immersed in a $\overline{B} = \hat{z}2$ (T), as in Fig. 9-8. Assume that $n_e = 8.4 \times 10^{28}\ \text{m}^{-3}$.

9.6 TORQUE ON CURRENT LOOPS

In Example 3, we found that the force of translation on a conductor current loop in a uniform $\overline{B}$ field was equal to zero. Even though the force of translation equals zero, a torque may exist on the conductor current loop.

Let us find the torque on a differential conductor current loop in a $\overline{B}$ field, as shown in Fig. 9-10(a). The loop lies in a constant z plane with its sides parallel to the coordinate axes. The magnetic field along the differential sides of the loop will be assumed to be constant since any variations in $\overline{B}$ over the differential loop will give rise to higher-order differential effects when the total torque is found. The torque arms $\overline{\ell}_{T1}$, $\overline{\ell}_{T2}$, $\overline{\ell}_{T3}$, and $\overline{\ell}_{T4}$ will be measured from the axes of rotation that bisect the loop along the x and y directions. The differential force on side 1 is

$$\overline{dF}_1 = I\,\overline{d\ell}_1 \times \overline{B} = I(\hat{x}\,dx) \times (\hat{x}B_x + \hat{y}B_y + \hat{z}B_z) = I\,dx(\hat{z}B_y - \hat{y}B_z) \tag{1}$$

The torque on side 1 is

$$\overline{dT}_1 = \overline{\ell}_{T1} \times \overline{dF}_1 = \left(-\hat{y}\frac{dy}{2}\right) \times [I\,dx(\hat{z}B_y - \hat{y}B_z)]$$

$$= \left(-\frac{I}{2}\,dx\,dy\right)\hat{x}B_y \tag{2}$$

Following the same procedure, the torque on sides 2, 3, and 4 becomes

$$\overline{dT}_2 = \left(\frac{I}{2}\,dx\,dy\right)\hat{y}B_x \tag{3}$$

$$\overline{dT}_3 = \left(-\frac{I}{2}\,dx\,dy\right)\hat{x}B_y = \overline{dT}_1 \tag{4}$$

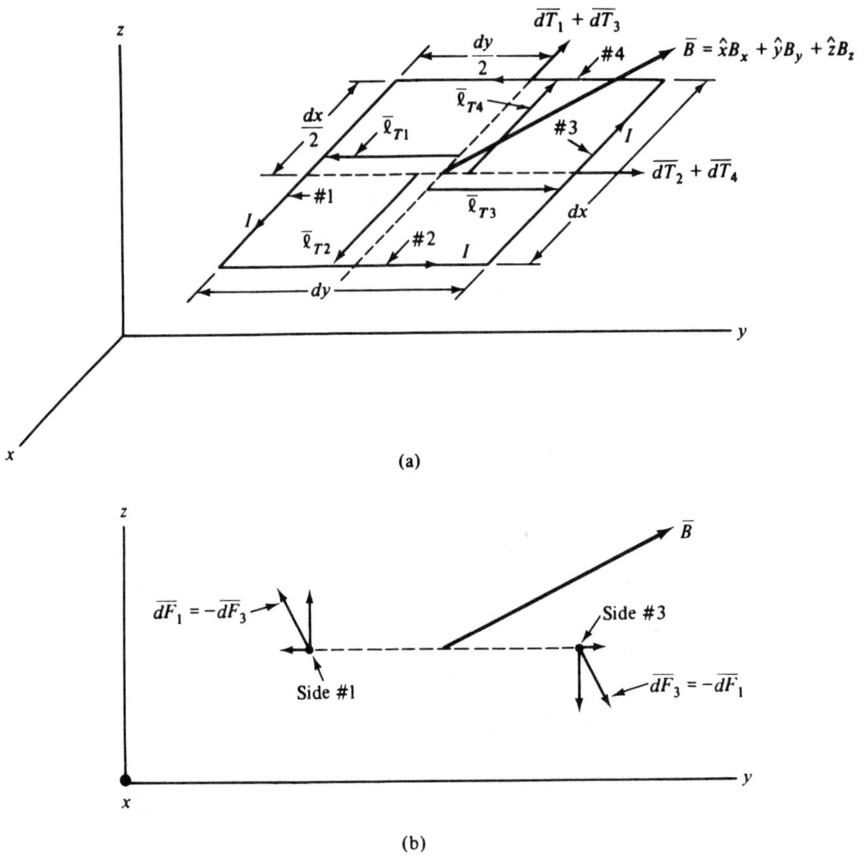

Figure 9-10 (a) Graphical display for finding the torque on a differential conductor current loop. (b) Vector diagram of forces acting on sides 1 and 3.

$$\overline{dT}_4 = \left(\frac{I}{2}\,dx\,dy\right)\hat{y}B_x = \overline{dT}_2 \tag{5}$$

The total torque is obtained by adding (2), (3), (4) and (5); thus,

$$\overline{dT} = I\,dx\,dy(-\hat{x}B_y + \hat{y}B_x) = I\,dx\,dy(\hat{z} \times \overline{B}) \tag{6}$$

A vector diagram of forces on sides 1 and 3 is found in Fig. 9-10(b). Note that the forces produce a torque even though the sum of forces is equal to zero.

If we associate $\hat{z}$ with $dx\,dy$, (6) can be expressed as

$$\boxed{\overline{dT} \triangleq I\,\overline{ds} \times \overline{B} \qquad (\text{N} \cdot \text{m})} \tag{7}$$

where $\overline{ds}$ is the vector area of the loop.

Let us now define the magnetic dipole moment $\overline{dm}$ as

$$\boxed{\overline{dm} = I\,\overline{ds} \qquad (\text{A} \cdot \text{m}^2)} \tag{8}$$

Thus, (7) can be expressed as

$$\boxed{\overline{dT} = \overline{dm} \times \overline{B} \qquad (\text{N} \cdot \text{m})} \tag{9}$$

Note that (9) is similar in form to (5.8-4),

$$\boxed{\overline{dT} = \overline{p} \times \overline{E}_a \qquad (\text{N} \cdot \text{m})} \tag{5.8-4}$$

where $\overline{p}$ is the electric dipole moment and $\overline{E}_a$ the applied electric field. Through (5.8-4), we found that electric dipoles formed in a dielectric material aligned themselves with the applied electric field. Through (9), we shall find that magnetic dipoles in a magnetic field will align themselves with the applied $\overline{B}_a$ field. This will thus lead us into the study of magnetic material. It can be shown that (9) will also apply to differential circular current loops as well as square ones.

The torque on a non-differential current loop must in general be obtained by summation of forces on the conductor at all points, using the $\overline{B}$ field at the actual conductor location. The integral

$$\boxed{\overline{T} = I \int_s \overline{ds} \times \overline{B} \qquad (\text{N} \cdot \text{m})} \tag{10}$$

applies only when B is uniform within the area s and we may then express (10) as

$$\overline{T} = -I\overline{B} \times \int_s \overline{ds} = -I\overline{B} \times \overline{s} = I\overline{s} \times \overline{B} = \overline{m} \times \overline{B} \qquad (\text{N} \cdot \text{m}) \tag{11}$$

where $\overline{m}$ is the magnetic dipole moment of a planar conductor current loop whose vector area is $\overline{s}$.

Before embarking on the study of magnetic material, we find the $\overline{B}$ field of a differential current loop to equal[3]

$$\boxed{\overline{B} \cong \frac{\mu_0(\pi a^2)I}{4\pi r_s^3}(\hat{r}_s 2 \cos\theta + \hat{\theta} \sin\theta)} \tag{12}$$

where a is the radius of the differential loop whose axis is along the z axis. Equation (12) is also true for the field from a bar magnet, where the magnetic moment is $Q_m \ell_{\text{eff}}$ instead of $I\pi a^2$, where Q_m is the pole strength at the end of the magnet and ℓ_{eff} is the effective length of the magnet between poles. The north pole of the magnet corresponds to the positive end of the electric dipole. Then, with $r_s \gg \ell_{\text{eff}}$,

$$\overline{B} = \frac{\mu_0 Q_m \ell_{\text{eff}}}{4\pi r_s^3}(\hat{r}_s 2 \cos\theta + \hat{\theta} \sin\theta) \qquad (\text{T}) \tag{13}$$

[3] D. T. Paris and F. K. Hurd, *Basic Electromagnetic Theory,* McGraw-Hill Book Company, New York, 1969, pp. 206–207.

compared to

$$\boxed{\overline{E} \cong \frac{Qd}{4\pi\epsilon_0 r_s^3}(\hat{r}_s 2\cos\theta + \hat{\theta}\sin\theta)} \qquad (4.5\text{-}26)$$

for the electric dipole. Thus, we see that the $\overline{B}$ field from a differential current loop and the $\overline{E}$ field from an electric dipole have the same form. For this reason, the differential current loop is commonly referred to as a *magnetic dipole*. The $\overline{E}$ field plot of Fig. 4-9(b) can thus be used for the $\overline{B}$ field plot of a magnetic dipole.

Example 5

A magnetic dipole whose $\overline{dm} = \hat{r}_c 10^{-5}$ (A · m²) is located at the center of a long solenoid along the z axis. If the solenoid is 0.02 m long and carries a current of 50 A in 2000 turns, find: (a) the initial torque on the dipole; (b) the final direction of $\overline{dm}$ if the dipole is allowed to turn.

Solution. (a) From (8.4-24), the $\overline{B}$ along the axis of a long solenoid is

$$\overline{B} = \mu_0\left(\hat{z}\frac{NI}{\ell}\right) = \hat{z}\mu_0\frac{(2000)(50)}{0.02} = \hat{z}6.28 \qquad \text{(T)}$$

From (9), we have

$$\overline{dT} = \overline{dm} \times \overline{B} = (\hat{r}_c 10^{-5}) \times (\hat{z}6.28) = -\hat{\phi}62.8 \qquad (\mu\text{N} \cdot \text{m})$$

(b) The $\overline{B}$ field of the solenoid will cause the magnetic dipole to rotate until $\overline{dm}$ is in the $\overline{B}$ direction and the torque is equal to zero. With this alignment of $\overline{dm}$ and the $\overline{B}$ of the solenoid, the resultant $\overline{B}$ field is now the sum of the $\overline{B}$ fields of the solenoid and that of the magnetic dipole.

Problem 9.6-1 A differential magnetic dipole has an initial $\overline{dm} = \hat{r}_c 5$ (μA · m²) when placed 3 m from an infinite length conductor along the z axis. If the current in the conductor is 10 A in the z direction, find the initial and final torques on the magnetic dipole that is free to rotate in the $\overline{B}$ field.

Problem 9.6-2 A differential square loop of current is found at the origin in the $z = 0$ plane. Find: (a) the expression for the vector magnetic potential $\overline{A}$ at a general distant point; (b) the expression for $\overline{B}$ through the use of $\overline{B} = \nabla \times \overline{A}$. Note that $\overline{B}$ will become (9.6-12) with (πa^2) replaced by $(d\ell)^2$.

9.7 MAGNETIC MATERIAL

The prominent characteristic of magnetic material is magnetic polarization—the alignment of its atomic magnetic dipoles when a magnetic field is applied. Through this alignment, we shall find that the magnetic field will be altered since the magnetic fields of the dipoles will combine with the applied magnetic field. In most cases, the resultant magnetic field will be increased.

9.7-1 Magnetic Polarization (Magnetization)

In our study of magnetic material, we shall use the classical atomic model of electrons orbiting about a positive nucleus, to obtain quantitative results and a qualitative theory. Through this approach, our background of quantum mechanics need not be very deep to obtain satisfactory conceptional results.

An atomic magnetic dipole with its associated magnetic moment is the result of three sources of moments on the atomic scale: the orbiting electron, the electron spin, and the nuclear spin. An electron in orbit about the nucleus possesses an angular momentum, and thus an orbital magnetic moment that is equivalent to a magnetic current loop of equal moment. The orbiting electron gives rise to a current I that encircles a surface ds. In atoms having many orbital electrons, the total orbital magnetic moment is the vector sum of the individual orbital magnetic moments. From quantum mechanics, we find that each electron also has its own spin angular momentum and thus an electron spin magnetic moment. Spin magnetic moment of only those electrons in unfilled shells will contribute to the resultant magnetic moment of the atom. In a similar manner, the nucleus possesses a spin angular momentum and thus a nuclear spin magnetic moment. In general, the nuclear spin magnetic moment has minimal effect on magnetic material properties since its moment is smaller by a factor of 10^{-3} than that due to an orbital electron or electron spin.

The net effect of these sources of magnetic moment on the atomic or microscopic base can be represented by a magnetic current loop whose magnetic dipole moment $\overline{dm}$ is equal to that of the atom. From (9.6-8), we have

$$\overline{dm} = I\,\overline{ds} \qquad (\text{A} \cdot \text{m}^2) \tag{9.6-8}$$

where I is now a bound current, bound to the atom through Coulomb forces. Figure 9-11(a) represents an assemblage of such atomic moments within a magnetic mate-

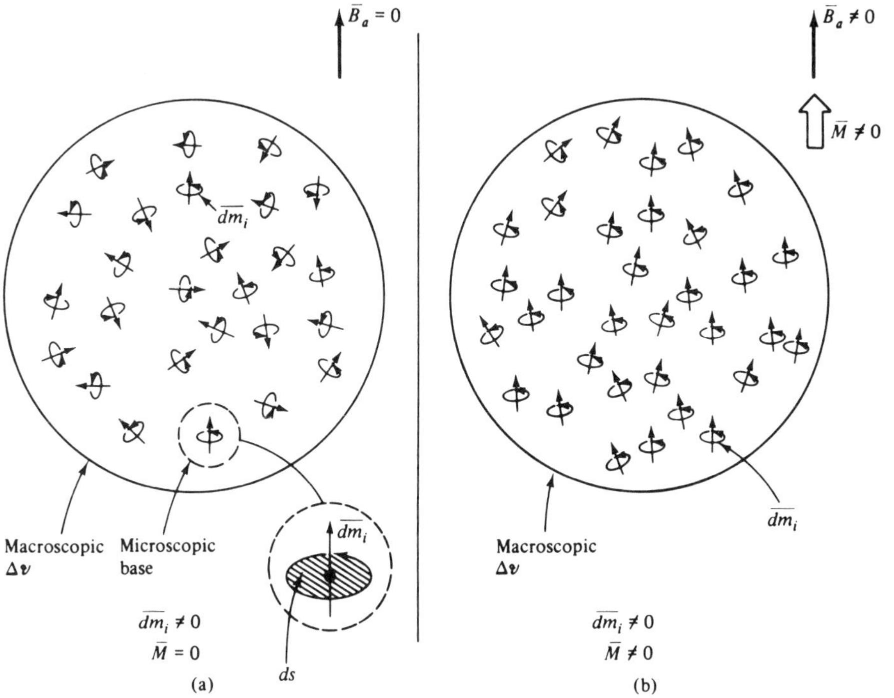

Figure 9-11 Magnetic dipole moments in a magnetic material: (a) the applied field $\overline{B}_a = 0$, thus $\overline{M} = 0$ on the macroscopic base; (b) the applied field $\overline{B}_a \neq 0$ and the $\overline{dm}$'s tend to align themselves in the direction of $\overline{B}_a$, thus, $\overline{M} \neq 0$.

rial when the applied field $\overline{B}_a$ is zero. Due to random orientation, the sum of the $\overline{dm}$'s on a macroscopic base is zero. Now, if the applied field is made non-zero the $\overline{dm}$'s will tend to align more or less in the direction of $\overline{B}_a$ since a torque $\overline{dT} = \overline{dm} \times \overline{B}_a$ now exists on the $\overline{dm}$'s. This is illustrated in Fig. 9-11(b). Under this condition, a net moment will now exist in the $\overline{B}$ direction that will affect the $\overline{B}$ field inside and outside the magnetic material. We shall advance from the microscopic based $\overline{dm}$'s to a macroscopic based $\overline{M}$, magnetic dipole moment per unit volume, through the defining equation

$$\overline{M} \triangleq \lim_{\Delta v \to 0} \left[\frac{1}{\Delta v} \sum_{i=1}^{n\Delta v} \overline{dm}_i \right] \qquad (\text{A m}^{-1}) \tag{1}$$

where n is the volume dipole density and $\Delta v \to 0$ on the macroscopic base. The magnetic dipole per unit volume $\overline{M}$ is called the *magnetization vector* within the magnetic material. Note that (1) is similar to the definition of $\overline{P}$ in a dielectric [see (5.8-2)]. Through the use of (1), we find that $\overline{M} = 0$ in Fig. 9-11(a) and $\overline{M} \neq 0$ in Fig. 9-11(b). If the dipole moments in Fig. 9-11(b) become totally aligned and $\overline{B}_a$ is uniform, (1) becomes

$$\overline{M} = n\,\overline{dm} = nI\,\overline{ds} \qquad (\text{A m}^{-1}) \tag{2}$$

In order to simplify our graphic display of magnetization within a magnetic material, we shall assume total alignment of $\overline{dm}$'s and thus $\overline{M}$'s.

9.7-2 Bound Magnetization Current Densities $\overline{J}_{sm}$ and $\overline{J}_m$

Let us now consider a slab of magnetic material with a uniform $\overline{B}_a$, as suggested in Fig. 9-12. The $\overline{M}$ will thus be uniform and give rise to a non-zero bound magnetization surface current density $\overline{J}_{sm}$ (A m^{-1}) on the surfaces of the material. In this figure, the microscopic current loops of moment $\overline{dm}$ are used to illustrate the formation of the surface currents. Within the material, the current due to adjacent current loops will cancel to produce a bound magnetization current density $\overline{J}_m$ (A m^{-2}) equal to zero. A non-zero $\overline{J}_m$ will occur within the material when $\overline{M}$ varies, as suggested in Fig. 9-13. The bound currents between adjacent loops do not cancel, and thus there exists a $\overline{J}_m$ within the material. These magnetization currents are commonly called *amperian currents*. We should note that the magnetization vector $\overline{M}$ has a non-zero $\partial M_y/\partial x$. This term can be recognized as the z component of $\nabla \times \overline{M}$.

To find expressions for $\overline{J}_{sm}$ and $\overline{J}_m$, let us consider a loop ℓ' within a slab of magnetic material, as suggested in Fig. 9-14(a). A volume $dv' = \overline{ds}' \cdot \overline{d\ell}'$ is constructed about a $\overline{d\ell}'$ portion of this loop such that ds' is equal to the area associated with $\overline{dm}$ [Fig. 9-14(b)]. Any magnetic dipole that is found within dv' will direct a current I upward through the surface s. The bound magnetization current flowing upward through s, due to $n\,dv'$ magnetic dipoles within dv', becomes

$$dI_m = I(n\,\overline{ds}' \cdot \overline{d\ell}') = (nI\,\overline{ds}') \cdot (\overline{d\ell}') \qquad (\text{A}) \tag{3}$$

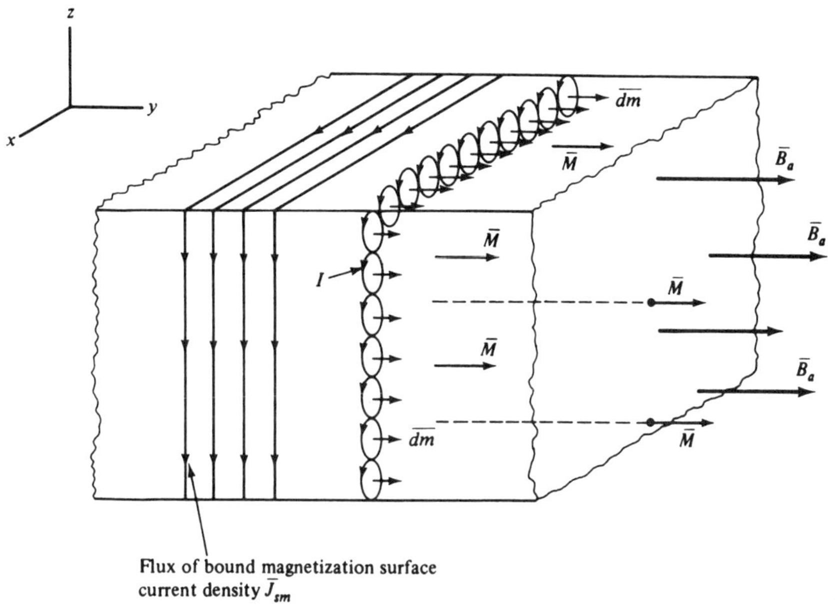

Figure 9-12 Alignment of $\overline{dm}$'s within a magnetic material, under uniform $\overline{B}_a$ conditions, to form a non-zero $\overline{J}_{sm}$, bound magnetization surface current density on the slab surfaces, and a $\overline{J}_m = 0$, bound magnetization current density within the material.

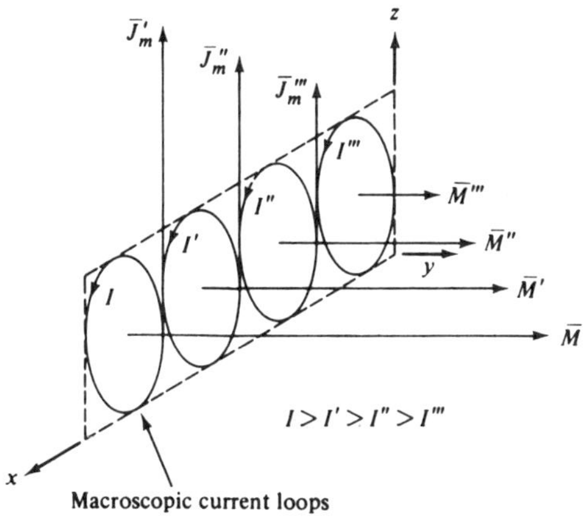

Figure 9-13 Graphical illustration of the magnetization mechanism that gives rise to a non-zero $\overline{J}_m$ within a magnetic material. The magnetization vector $\overline{M}$ has a non-zero $\partial M_y / \partial x$.

Substituting (2) into (3), we have

$$\boxed{dI_m = \overline{M} \cdot \overline{d\ell'} \qquad \text{(A)}} \tag{4}$$

The total I_m through the ℓ' loop is

$$I_m = \oint_{\ell'} \overline{M} \cdot \overline{d\ell'} \tag{5}$$

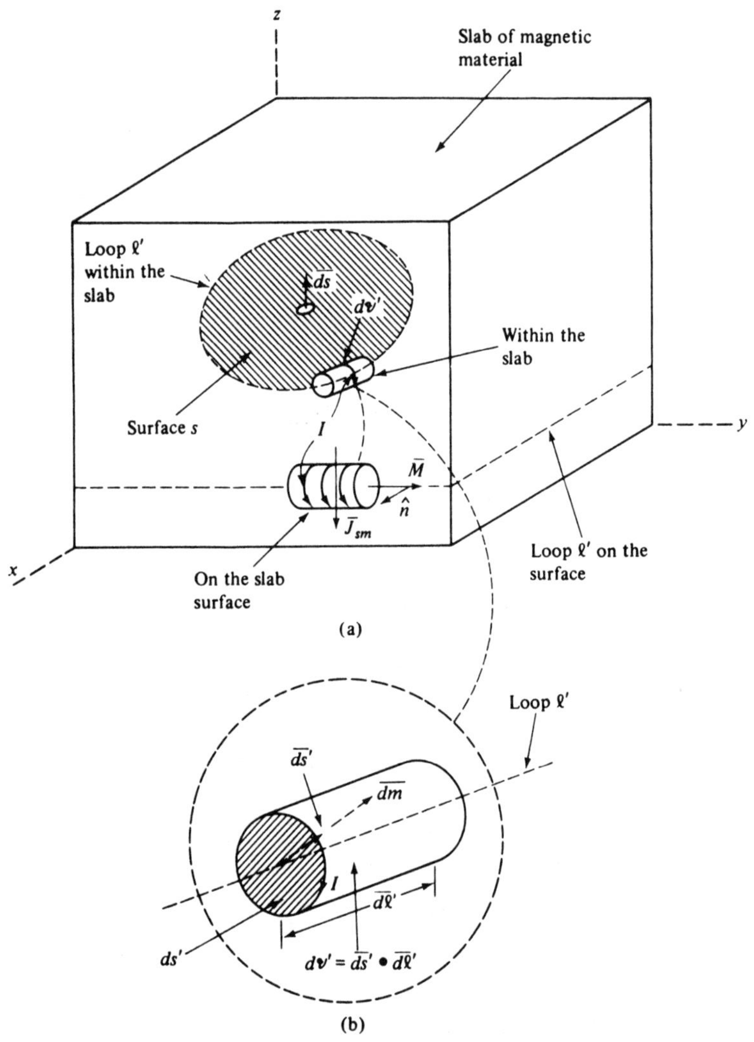

Figure 9-14 Graphical display for finding expressions for $\bar{J}_{sm}$ (A m^{-1}) and $\bar{J}_m$ (A m^{-2}): (a) slab of magnetic material with closed loop ℓ' within the material and on the slab surface; (b) expanded view of dv' about loop ℓ'.

The current I_m can also be equated to

$$I_m = \int_s \bar{J}_m \cdot \overline{ds} \tag{6}$$

where s is the surface bound by the loop ℓ', and $\bar{J}_m$ is the current density within the loop. Equating the right-hand sides (5) and (6) and using Stokes' theorem, we have

$$I_m = \int_s \bar{J}_m \cdot \overline{ds} = \oint_\ell \bar{M} \cdot \overline{d\ell'} = \int_s \nabla \times \bar{M} \cdot \overline{ds} \tag{7}$$

From (7), we obtain

$$\overline{J}_m = \nabla \times \overline{M} \qquad (\text{A m}^{-2}) \quad (\text{aap}) \tag{8}$$

where the surface s was shrunk to a point. Equation (8) is the desired expression for the bound magnetization current density within the magnetic material.

Let us now take loop ℓ' to the slab surface, as shown in Fig. 9-14(a). At the surface of the slab, the current dI_m becomes a bound surface current flowing downward. In (4), the path length $\overline{d\ell'}$ will be tangent to the surface of the slab and (4) can be written

$$dI_m = M_{\tan}\, d\ell' \tag{9}$$

where $M_{\tan}$ is the tangent component of $\overline{M}$ at the slab surface. From (9),

$$M_{\tan} = \frac{dI_m}{d\ell'} \triangleq J_{sm} \qquad (\text{A m}^{-1}) \tag{10}$$

where J_{sm} is our desired bound magnetization surface current density. Equation (10) can be expressed in vector form as

$$\overline{J}_{sm} = \overline{M} \times \hat{n} \quad (\text{on the surface}) \qquad (\text{A m}^{-1}) \quad (\text{aap}) \tag{11}$$

where $\hat{n}$ is our unit vector normal to the surface of the slab, and $\overline{M}$ is the magnetization vector at the surface.

Example 6

A long cylinder of magnetic material of radius a (m) is found along the z axis. When the magnetization $\overline{M} = \hat{z}10$ (A m^{-1}) within the cylinder, find: (a) $\overline{J}_m$; (b) $\overline{J}_{sm}$.

Solution. (a) From (9.7-8), $\overline{J}_m = \nabla \times \overline{M} = 0$ within the magnetic material.

(b) From (9.7-11),

$$\overline{J}_{sm} = \overline{M} \times \hat{n}\big|_a = (\hat{z}10) \times (\hat{r}_c)\big|_a = \hat{\phi}10 \qquad (\text{A m}^{-1})$$

at $r_c = a$ on the magnetic material surface.

Problem 9.7-1 A long bar of magnetic material is found parallel to the y axis with its cross section defined by $0.1 \geq x \geq 0$ and $0.2 \geq z \geq 0$. If $\overline{M} = \hat{y}3x$ (A m^{-1}), find: (a) $\overline{J}_m$ within the material; (b) $\overline{J}_{sm}$ on the four surfaces; (c) the total bound current that flows through a cross section of the bar at $z = 0.1$ plane for a length of 1 meter in the y direction.

9.7-3 Effect of Magnetization on Magnetic Fields

Due to magnetization in a material, we have seen the formation of bound magnetization surface current density $\overline{J}_{sm}$ and bound magnetization current density $\overline{J}_m$. These bound currents produce a $\overline{B}$ field as if they were currents due to movement of free charges. Thus, we must alter the free space equation (8.9-3) to read

$$\nabla \times \frac{\overline{B}}{\mu_0} = (\overline{J} + \overline{J}_m) \quad (\text{aap}) \tag{12}$$

where $\overline{B}$ is the resultant field due to currents associated with free charges and currents associated with magnetization. Substituting (8) into (12), we have

$$\nabla \times \frac{\overline{B}}{\mu_0} = \overline{J} + \nabla \times \overline{M} \tag{13}$$

Rearranging (13), we obtain

$$\nabla \times \left(\frac{\overline{B}}{\mu_0} - \overline{M}\right) = \overline{J} \tag{14}$$

Let us now define

$$\boxed{\left(\frac{\overline{B}}{\mu_0} - \overline{M}\right) \triangleq \overline{H}} \tag{15}$$

Thus, (14) becomes

$$\boxed{\nabla \times \overline{H} = \overline{J}} \tag{8.5-8}$$

which is our former (8.5-8) and now is also true in a magnetic material. Equation (15) is the general expression for $\overline{H}$ that reduces to $\overline{H} = \overline{B}/\mu_0$ for free space since $\overline{M} = 0$. From (15), we obtain

$$\boxed{\overline{B} = \mu_0(\overline{H} + \overline{M})} \tag{16}$$

The magnetization vector $\overline{M}$ can be eliminated through the relationship between $\overline{M}$ and $\overline{H}$. In an isotropic magnetic material, we find

$$\boxed{\overline{M} = \chi_m \overline{H}} \tag{17}$$

where χ_m is the magnetic susceptibility of the material. Substituting (17) into (16), we obtain

$$\overline{B} = \mu_0 \overline{H}(1 + \chi_m) \tag{18}$$

Now, define

$$\boxed{\mu_r \triangleq 1 + \chi_m} \tag{19}$$

to obtain

$$\overline{B} = \mu_0 \mu_r \overline{H} \tag{20}$$

Let us define

$$\boxed{\mu \triangleq \mu_0 \mu_r} \tag{21}$$

where μ is the permeability of the material. Thus, (20) becomes

$$\boxed{\overline{B} = \mu\overline{H}} \tag{22}$$

where μ takes into account the magnetization of the material. Representative values of several materials are found in Table 9-1.

TABLE 9-1 REPRESENTATIVE VALUES FOR PERMEABILITY μ_r FOR SEVERAL MATERIALS

Material	Type	μ_r
Bismuth	Diamagnetic	0.9999834
Silver	Diamagnetic	0.99998
Copper	Diamagnetic	0.999991
Vacuum	Nonmagnetic	1.00
Aluminum	Paramagnetic	1.00002
Nickel chloride	Paramagnetic	1.00004
Cobalt	Ferromagnetic	250
Nickel	Ferromagnetic	600
Mild steel	Ferromagnetic	2,000
Iron	Ferromagnetic	5,000
Mumetal	Ferromagnetic	100,000
Supermalloy	Ferromagnetic	800,000

The concepts of linearity, homogeneity, and isotropy also apply to magnetic materials. For a non-isotropic magnetic material, χ_m becomes a matrix (tensor) susceptibility and thus has nine terms in general. In equation (22), μ becomes a matrix (tensor) permeability while (16) remains unaltered with $\overline{B}$, $\overline{H}$, and $\overline{M}$ not in the same direction in general.

In (16), we have an expression that relates the resultant $\overline{B}$, $\overline{H}$, and the magnetization vector $\overline{M}$ at a point. This expression was developed through the addition of $\overline{J}_m$ to (12). The induced field $\overline{B}_i$, due to magnetization $\overline{M}$,[4] can be found by starting with the induced vector magnetic potential $d\overline{A}_i$, due only to the magnetic dipoles $\overline{M}\,dv$ within a volume dv. The expression for $d\overline{A}_i$ is similar to (5.8-26),

$$\boxed{d\overline{A}_i = \frac{\mu_0}{4\pi}\left(\frac{\overline{M} \times \hat{a}_R}{R^2}\right) dv} \tag{23}$$

Through mathematical manipulation, $\overline{A}_i$ is found to be

$$\boxed{\overline{A}_i = \frac{\mu_0}{4\pi}\oint_s \left(\frac{\overline{M} \times \hat{n}}{R}\right) ds + \frac{\mu_0}{4\pi}\int_v \frac{\nabla \times \overline{M}}{R}\, dv} \tag{24}$$

[4] See Dale R. Corson and Paul Lorrain, *Introduction to Electromagnetic Fields and Waves*, W. H. Freeman and Company, Publishers, San Francisco, 1962, pp. 262–265.

where the first integral is taken over all surfaces containing a non-zero $\overline{M}$ and the second integral is taken over all volume excluding the surfaces. In (24), the terms $(\overline{M} \times \hat{n})$ and $(\nabla \times \overline{M})$ can be recognized as $\overline{J}_{sm}$ and $\overline{J}_m$, respectively. From $\overline{B}_i = \nabla \times \overline{A}_i$ and (24), the expression for the induced magnetic field $\overline{B}_i$ becomes

$$\boxed{\overline{B}_i = \frac{\mu_0}{4\pi} \oint_s \frac{\overline{J}_{sm} \times \hat{a}_R}{R^2}\, ds + \frac{\mu_0}{4\pi} \int_v \frac{\overline{J}_m \times \hat{a}_R}{R^2}\, dv} \tag{25}$$

Thus, the induced field $\overline{B}_i$, due to magnetization only, can be obtained by treating $\overline{J}_{sm}$ and $\overline{J}_m$ as if they were conduction densities. Note that (25) is the Biot-Savart law in terms of a bound surface current element $\overline{J}_{sm}\, ds$ and a bound volume current element $\overline{J}_m\, dv$. The resultant $\overline{B}$ is

$$\boxed{\overline{B} = \overline{B}_a + \overline{B}_i} \tag{26}$$

where $\overline{B}_a$ is the applied field.

Example 7

A closely wound long solenoid has a concentric magnetic rod inserted as shown in Fig. 9-15(a). In the center region, find: (a) $\overline{H}$, $\overline{B}$, and $\overline{M}$ in both air and the magnetic rod; (b) the ratio of the $\overline{B}$ in the rod to the $\overline{B}$ in the air; (c) $\overline{J}_{sm}$ on the surface of the rod and $\overline{J}_m$ within the rod. Assume that the permeability of the rod equals $5\mu_0$.

Solution. (a) From $\nabla \times \overline{H} = \overline{J}$, now true for steady currents in free space and magnetic material, we find that Ampère's circuital law takes on the same form as found in Chapter 8, i.e.,

$$\boxed{\oint_\ell \overline{H} \cdot \overline{d\ell} = I_{\text{en}}}$$

where I_{en} is the current due to motion of free charges and does not include any bound magnetization current I_m. Let us apply Ampère's circuital law to the closed path P_1-P_2-P_3-P_4 to obtain

$$\oint_\ell \overline{H} \cdot \overline{d\ell} = \int_{P2}^{P3} (\hat{z}H_z) \cdot (\hat{z}\, dz) = H_z d = I_{\text{en}} = \frac{NI}{\ell} d \tag{27}$$

where H_z, inside the solenoid, is the only non-zero $\overline{H}$ field over the path. Solving for H_z, we have in the air region

$$H_z = \frac{NI}{\ell} = J_s \tag{28}$$

If we apply Ampère's circuital law to the closed path P_1-P_2'-P_3'-P_4-P_1, we will find that H_z in the rod will be the same as in the air since Ampère's circuital law does not include any I_m in its I_{en} term.

From (17), the $\overline{M}$ in the rod is

$$\overline{M} = \hat{z}M_z = \chi_m(\hat{z}H_z) \tag{29}$$

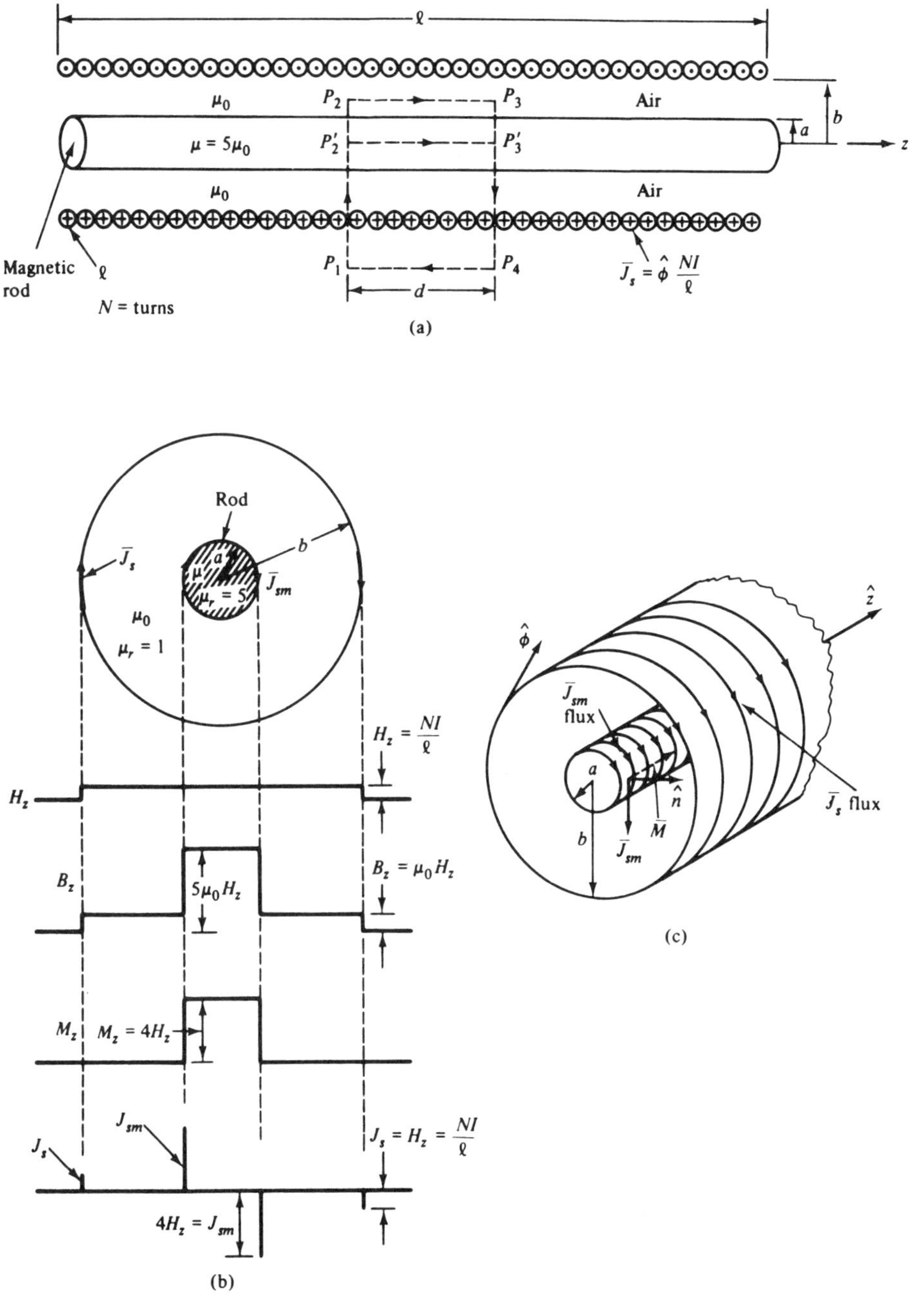

Figure 9-15 (a) Long solenoid with a concentric magnetic rod. (b) Plots of H, B, M, J_s, and J_{sm} along the cross section. (c) $\overline{J}_s$ on the solenoid and $\overline{J}_{sm}$ on the magnetic rod for Example 7.

The $\overline{M}$ in the air is zero since χ_m of air is zero. From (22), the $\overline{B}$ in air becomes $\mu_0(\hat{z}H_z)$ while $\overline{B}$ in the rod becomes $\mu_0\mu_r(\hat{z}H_z)$.

(b) From the above we see that the ratio of $\overline{B}$ in the rod to that in the air is μ_r. Thus, the $\overline{B}$ in the rod has been increased by a factor of μ_r due to magnetization.

(c) From (9.7-11), we have

$$\overline{J}_{sm} = \overline{M} \times \hat{n} = (\hat{z}M_z) \times (\hat{r}_c) = \hat{\phi}M_z \tag{30}$$

From (9.7-8), we have

$$\overline{J}_m = \nabla \times \overline{M} = \nabla \times (\hat{z}M_z) = 0 \tag{31}$$

within the magnetic rod. The results are plotted in Fig. 9-15(b) for $\mu_r = 5$. Figure 9-15(c) shows the $\overline{J}_s$ on the surface of the solenoid due to free charges flowing in the N turns and $\overline{J}_{sm}$ on the surface of the magnetic rod due to magnetization. The $\overline{J}_m$ within the rod, from (31), is zero.

Problem 9.7-2 The toroid of Fig. 8-10 is filled with a magnetic material whose $\chi_m = 40$. Assume $2\pi b > (c - a)$, so that $\overline{H} \cong \hat{\phi}NI/2\pi b$ for $(a < r_c < c)$, and find: (a) $\overline{B}$; (b) M; (c) $\overline{J}_{sm}$; (d) $\overline{J}_m$. Also assume that $a = 0.2$ m, $b = 0.22$ m, $c = 0.24$ m, $N = 1000$ turns, and $I = 5$ A.

9.7-4 Magnetic Material Classification

In Sec. 9.7-1, we found that an atomic dipole moment was basically due to the orbital electron moment and the electron spin moment. A macroscopic viewpoint was obtained through the magnetization vector $\overline{M}$ and the concepts of $\overline{J}_{sm}$ and $\overline{J}_m$ were developed. These two current densities proved to be the sources of $\overline{M}$ and led us to the effect of magnetization on magnetic fields.

In this section we shall study, on a limited scale, the nature and classification of magnetic material. Magnetic material can be classified into two main groups on the microscopic scale in the absence of an applied magnetic field. Group A has a zero dipole moment while group B has a non-zero dipole moment. Group A material is called diamagnetic while group B is subdivided into paramagnetic, ferromagnetic, anti-ferromagnetic, and ferrimagnetic.

In a diamagnetic atom, the electron spin moments oppose each other as well as the orbiting electron moments to produce a zero net $\overline{dm}$ in the absence of an applied magnetic field. When a magnetic field is applied, a Lorentz force $q_e\overline{U}_e \times \overline{B}_a$ will be exerted on the orbiting electrons. This force will add to or subtract from the centrifugal force due to the electron's angular velocity ω_0 about the nucleus, depending on the direction of $\overline{B}_a$ and the moment of the orbital electron. From quantum mechanics, a fixed orbit must be maintained; thus, the angular velocity must be changed by $\pm\Delta\omega$. This causes a change of electron orbital current, thereby reflecting a small change in the net moment of the electron. For the case where $\overline{B}_a$ has a positive projection onto the direction of the moment of the orbiting electron, the moment of the orbiting electron will be reduced by a small amount. For the case where $\overline{B}_a$ has a negative projection onto the direction of the moment of the orbiting electron, the moment of the orbiting electron will be increased by a small amount. Now, if we start with two orbiting electrons whose moments cancel, we shall find that a small net moment will exist opposite to the applied $\overline{B}_a$. Thus, the atom now has a net non-zero moment in the presence of an applied magnetic field. On the macroscopic base, there will exist a magnetization vector $\overline{M}$ opposite to the applied magnetic field. This negative $\overline{M}$ leads to a slightly negative susceptibility χ_m and a relative permeability that is slightly less than one. Bismuth is a good example of a diamagnetic material whose $\chi_m = -1.66 \times 10^{-5}$ and $\mu_r = 0.9999834$. Diamagnetism is found in all material but is obscured by other stronger forms of magnetization that we shall soon discuss.

In a paramagnetic atom, a small net non-zero moment exists in the absence of an applied $\overline{B}_a$ field. Due to random orientation of the atoms, the net $\overline{M} = 0$ on the macroscopic base in the absence of an applied $\overline{B}_a$ field. In an applied magnetic field, however, there will be a slight alignment of the atomic moments with the applied field to produce an $\overline{M} \neq 0$, thus slightly increasing the magnetic field within the material. From Table 9-1, we see that aluminum is paramagnetic with $\chi_m = 2 \times 10^{-5}$ and $\mu_r = 1.00002$.

The ferromagnetic atom has a strong magnetic moment in the absence of an applied $\overline{B}_a$ field. Metals such as nickel, cobalt, and iron exhibit ferromagnetism. In diamagnetic and paramagnetic atoms, the electron spin moments usually are cancelled by equal and opposite spin moments. However, in ferromagnetic atoms, quantum mechanics shows that as many as five electrons have aligned spin moments. In addition, strong interaction forces are found between ferromagnetic atoms that cause spin moments of many atoms to align in parallel and thus form regions of strong magnetization called *domains*. Domains have many shapes and sizes, ranging from a few micrometers to a few millimeters, depending on the material and its past history. On a macroscopic base, the domains are randomly oriented and thus give rise to $\overline{M} = 0$ in the absence of an applied field. In the presence of an applied magnetic field, extremely large values $\overline{M}$, $\overline{B}$, χ_m, and μ_r will result. The magnetization curve, a plot of B versus H, will be found to be non-linear with multivalued behavior when H is varied over positive and negative values. Figure 9-16 shows a B-H magnetization curve for a virgin ferromagnetic material as the applied H varies from zero to H_3. The curve begins at the origin and goes through P_1, P_2, and P_3. As the applied H increases from zero, the resulting B increases very rapidly until P_2 is reached. Up to this point, all of the easy domain alignment has taken place and saturation is beginning to set in. At P_3, most of the domains have been aligned with the applied H field. Now, if the applied H field is reduced to zero, we will find that the magnetization curve is not retraced. The B-H plot takes the dashed path from P_3 to P_4, at which point we find that a residual B_r exists, even though the applied H equals zero. Thus, we have created a permanent magnet. If we reverse the applied H to a value H_c, we find that $B = 0$ and H_c is called the *coercive force*. A variation of applied H, as suggested in Fig. 9-16, will trace the dashed curve that is called a *hysteresis loop*. Energy must be expended in tracing a hysteresis loop since losses are incurred in the non-reversible domain movements. The loss is proportional to the area of the hysteresis loop and is called the *hysteresis loss* of the material. Thus, ferromagnetic material with thin hysteresis loops are desirable in the design of transformer and induction cores.

Magnetic recording tape consists of ferromagnetic material (iron oxide) coated on a plastic base. The magnetic material is magnetized by the applied field of a recording head, while the playback head converts the magnetized areas into electric signals.

The μ at P_2 of the magnetization curve, based on $\mu \triangleq B/H$, is found to have maximum value. If we define the differential permeability as dB/dH, its maximum value will

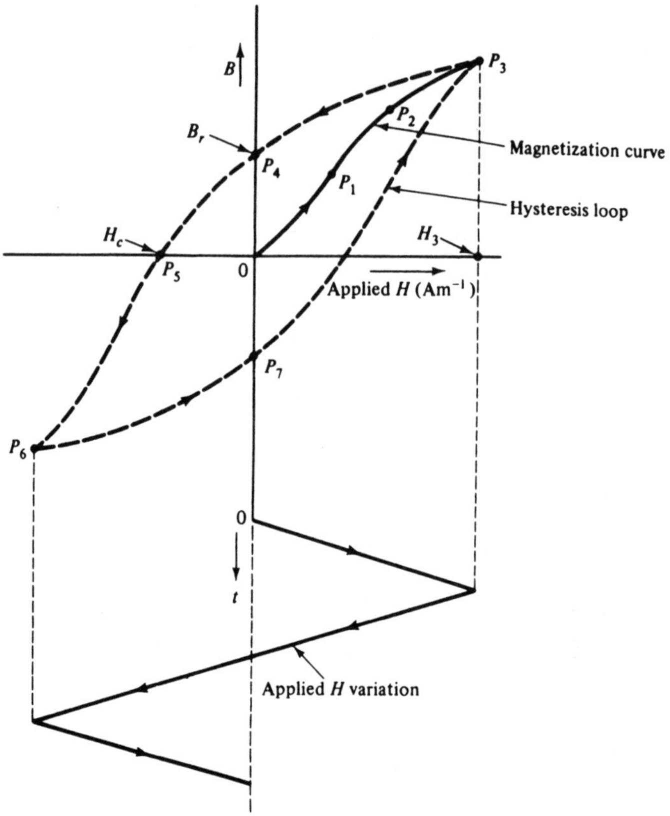

Figure 9-16 Magnetization curve for a virgin ferromagnetic material is shown from the origin to P_3. The dashed plot is a B-H hysteresis loop when the applied H is varied over one cycle.

be found at the point P_1. The permeability concept is always applied to the magnetization curve and not to the hysteresis loop.

In some materials, the interaction forces between atoms produce an antiparallel alignment of the electron spin moments of adjacent atoms. If the opposing moments of adjacent atoms are approximately equal, the material is called *antiferromagnetic*, and its relative permeability is slightly greater than one. If the opposing moments of adjacent atoms are greatly unequal, the material is called *ferrimagnetic*, and its relative permeability is relatively large but not as large as ferromagnetic materials. A special group of ferrimagnetic materials are the ferrites, which possess large resistivity as well as large permeability. The large resistivity (low conductivity) property makes ferrites very well-suited as magnetic core material when high frequency fields are applied. The large resistivity reduces the magnitude of the induced eddy currents, and thus the ohmic core losses are relatively low.

Example 8

Show that the orbiting electron has its angular velocity reduced by $-q_e B_a/2m_e$ when immersed in a B_a field that is in the same direction as the orbiting electron moment.

Solution. Figure 9-17(a) shows an electron orbiting with an angular velocity ω_0 when placed in a $\overline{B}_a = 0$ field. The force $\overline{F}_E$ is due to Coulomb forces between charges while $\overline{F}'_{\text{cent}} = \hat{r}_c m_e \omega_0^2 r_c$ is the balancing centrifugal force. When $\overline{B}_a \neq 0$, an additional force $\overline{F}_m = q_e \overline{U}_e \times \overline{B}_a$ acts on the electron. This Lorentz force will be found in the radial $\hat{r}_c$ direction for our electron, as shown in Fig. 9-17(b). When $\overline{F}_m$ is much smaller than $\overline{F}_E$, quantum mechanics requires that the electron remain in the same orbit but with a change in its angular velocity. This new angular velocity we will denote as ω; thus, $\omega_0 - \omega = \Delta\omega$. From Fig. 9-17(b), the sum of forces on the orbiting electron becomes

$$\overline{F}_E = \overline{F}_m + \overline{F}_{\text{cent}} \tag{32}$$

The force $F_E = m_e \omega_0^2 r_c$ will not change with a change in ω, and (32) becomes

$$m_e \omega_0^2 r_c = -q_e U_e B_a + m_e \omega^2 r_c \tag{33}$$

where $-q_e$ is a positive quantity. Substituting $U_e = \omega r_c$ and canceling out r_c, (33) becomes

$$m_e \omega_0^2 = -q_e \omega B_a + m_e \omega^2 \tag{34}$$

Rearranging, we have

$$-q_e \omega B_a = m_e(\omega_0^2 - \omega^2) = m_e(\omega_0 - \omega)(\omega_0 + \omega) \tag{35}$$

When $\omega_0 \cong \omega$, we can express $(\omega_0 + \omega) \cong 2\omega$ and (35) becomes

$$-q_e \omega B_a = m_e(\Delta\omega)(2\omega) \tag{36}$$

Solving for $\Delta\omega$, we have

$$\Delta\omega = \frac{-q_e B_a}{2m_e} = \frac{|q_e| B_a}{2m_e} \tag{37}$$

From (37), we see that $\Delta\omega$ is a positive quantity for our example and from $\Delta\omega = (\omega_0 - \omega)$, ω will be less than ω_0 and thus $m' < m$. This change in angular velocity of an orbiting electron takes place in a diamagnetic material where two orbital moments cancel each other when $\overline{B}_a = 0$. The orbiting electron of opposite moment (opposite to that assumed in this example) will experience an increase in ω when $\overline{B}_a \neq 0$ and thus $m' > m$. The associated changes in moments of a pair of opposite orbiting electrons will thus produce a small net moment in a direction opposite to the applied $\overline{B}_a$ and thus gives rise to diamagnetism.

Problem 9.7-3 Review Prob. 9.4-2 and show that even in the absence of a central charge in Fig. 9-17, that if $\overline{B}_a \neq 0$, the electron will orbit a central point with an angular velocity $\omega = |q_e|\,|B_a|/m_e$.

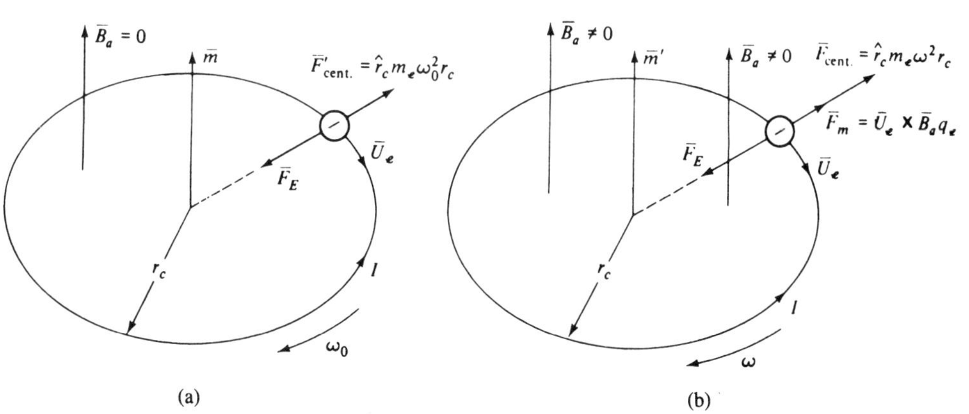

Figure 9-17 (a) Electron orbiting in a stable orbit with $\overline{B}_a = 0$. (b) Orbiting electron immersed in a $\overline{B}_a \neq 0$ field.

Problem 9.7-4 Repeat Example 8 for an electron orbiting in the reverse direction and show that $\Delta\omega = q_e B_a/2m_e$.

Problem 9.7-5 For an orbital electron whose angular velocity is ω in a circular orbit of r_c radius, show that: (a) the moment equals $-q_e[(\omega/2)r_c^2]$; (b) the change in moment $\Delta m = -q_e^2(B_a/4m_e)r_c^2 = -3.7 \times 10^{-29}$ A m^2 when $r_c = 5.1 \times 10^{-11}$ m and $B_a = 2$ T. The orbital electron moment for hydrogen is 9.1×10^{-24} A · m^2; thus Δm is extremely small even though a $B_a = 2$ T is a strong field.

9.8 MAGNETIC BOUNDARY CONDITIONS

The procedure to find the relationship between $\overline{B}$'s $\overline{H}$'s and $\overline{M}$'s at a boundary involving magnetic material is identical to that used in Secs. 5.6 and 5.9. From $\nabla \cdot \overline{B} = 0$ and $\nabla \times \overline{H} = \overline{J}_s$ (current due to free charges), we can obtain the integral forms

$$\oint_s \overline{B} \cdot \overline{ds} = 0 \tag{1}$$

and

$$\oint_\ell \overline{H} \cdot \overline{d\ell} = I_{\text{en}} \quad \text{(due to free charges)} \tag{2}$$

as found in Sec. 8.8.

Applying (1) to the small cylinder of Fig. 9-18, as $\Delta h \to 0$, yields

$$\oint_s \overline{B} \cdot \overline{ds} = B_{1n}\,\Delta s - B_{2n}\,\Delta s = 0 \tag{3}$$

From (3), we have

$$\boxed{B_{1n} = B_{2n}} \tag{4}$$

The boundary condition relating tangent field components can be obtained by applying (2) to the closed loop of Fig. 9-18, as $\Delta h \to 0$, to yield

$$\boxed{H_{1t} - H_{2t} = J_s \quad \text{(due to free charges)}} \tag{5}$$

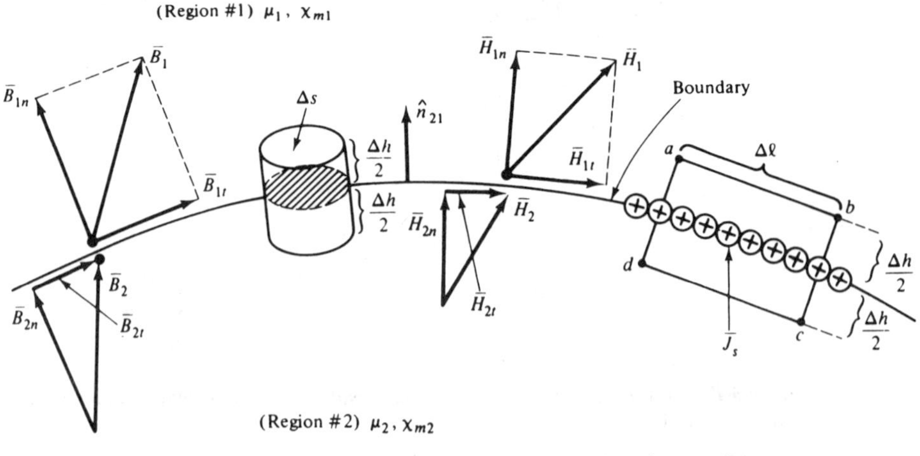

Figure 9-18 Graphical display for obtaining magnetic boundary conditions.

where J_s is perpendicular to the directions of H_{1t} and H_{2t}. The vector form of (5) becomes

$$\boxed{\hat{n}_{21} \times (\overline{H}_1 - \overline{H}_2) = \overline{J}_s} \tag{6}$$

where $\hat{n}_{21}$ is a normal unit vector from region 2 to region 1.

Through the use of $\overline{B} = \mu\overline{H}$ and equations (4) and (5), we obtain

$$\boxed{\mu_1 H_{1n} = \mu_2 H_{2n}} \tag{7}$$

and

$$\boxed{\frac{B_{1t}}{\mu_1} - \frac{B_{2t}}{\mu_2} = J_s} \tag{8}$$

From $\nabla \times \overline{M} = \overline{J}_m$, we can obtain the integral form

$$\oint_\ell \overline{M} \bullet \overline{d\ell} = I_m \quad \text{(enclosed magnetic current)} \tag{9}$$

Applying (9) at the boundary leads to

$$\boxed{M_{1t} - M_{2t} = J_{sm} \quad \text{(bound surface current density)}} \tag{10}$$

Through the use of (5) and (9.7-17), we obtain

$$\boxed{\frac{M_{1t}}{\chi_{m1}} - \frac{M_{2t}}{\chi_{m2}} = J_s \quad \text{(due to free charges)}} \tag{11}$$

Also from (7) and (9.7-17), we obtain

$$\boxed{\mu_1 \frac{M_{1n}}{\chi_{m1}} = \mu_2 \frac{M_{2n}}{\chi_{m2}}} \tag{12}$$

Equations (4), (5), (7), (8), (10), (11), and (12) make up the set of boundary conditions at a magnetic boundary.

Example 9

At a boundary between two magnetic regions, show that $(\tan \alpha_1/\tan \alpha_2) = \mu_{r1}/\mu_{r2}$ when α_1 and α_2 are defined as in Fig. 9-19. Assume that $\overline{J}_s = 0$ at the boundary and that both regions are isotropic.

Solution. From (4) and (5), we have

$$B_{1n} = B_{2n} \tag{4}$$

and

$$H_{1t} = H_{2t} \tag{13}$$

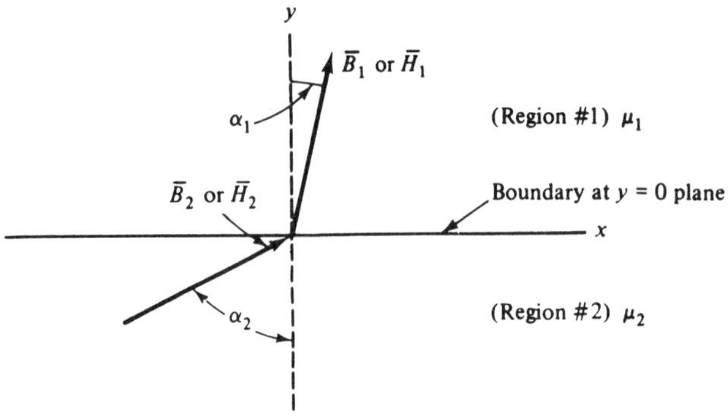

Figure 9-19 Graphical display for finding a relationship between α_1 and α_2 at a magnetic boundary.

Now, from Fig. 9-19 we obtain

$$B_{1n} = B_1 \cos \alpha_1 \tag{14}$$

$$B_{2n} = B_2 \cos \alpha_2 \tag{15}$$

$$H_{1t} = H_1 \sin \alpha_1 \tag{16}$$

and

$$H_{2t} = H_2 \sin \alpha_2 \tag{17}$$

Substituting (14) and (15) into (4), and (16) and (17) into (13), we obtain

$$\boxed{\frac{\tan \alpha_1}{\tan \alpha_2} = \frac{\mu_1}{\mu_2} = \frac{\mu_{r1}}{\mu_{r2}}} \tag{18}$$

Problem 9.8-1 A toroidal current-carrying winding, such as that illustrated in Fig. 8-10, is wound about an iron toroid with a small air gap. The length of the iron path is 50 cm, and the air gap is 1 mm. If the flux density in the iron is 1.8 T, and neglecting fringing at the gap edges, find: (a) B_{gap}; (b) H_{iron}; (c) H_{gap}. $\mu_{r_{iron}} = 5000$.

Problem 9.8-2 Assume that the toroidal current-carrying winding referred to in Prob. 9.8-1 is very loosely wound around a complete iron toroid (no air gap) and that B_{iron} is 1.8 T. Find: (a) B in the air space between the iron toroid and the coil; (b) H_{air}; (c) H_{iron}. μ_r for the iron core is 5000.

Problem 9.8-3 For the boundary shown in Fig. 9-19, let region 1 be air ($\mu_{r1} = 1$) and region 2 be soft iron whose $\mu_{r2} = 6000$. Find: (a) α_1 when $\overline{B}_2$ is normal to the boundary ($\alpha_2 = 0$); (b) α_1 when $\overline{B}_2$ is almost tangent ($\alpha_2 = 88°$). This problem clearly indicates that at the boundary between a low μ_{r1} and a high μ_{r2}, the angle α_2 can vary from zero to say 88° (B_2 almost tangent) with $\alpha_1 \cong 0°$ over this range.

9.9 MAGNETIC CIRCUITS

The toroid of Fig. 9-20(a) is a *simple magnetic circuit* that yields a large magnetic flux Ψ_m within the ferromagnetic core. Here the turns are closely wound and distributed over the entire core to ensure that most of the flux exists within the core. Now, if the permeability of the core is much greater than μ_0, we shall find that the flux will restrict itself largely to the magnetic core, even though the turns are not wound uniformly and the cross section is not constant, as suggested in Fig. 9-20(b). The ease of flux flow

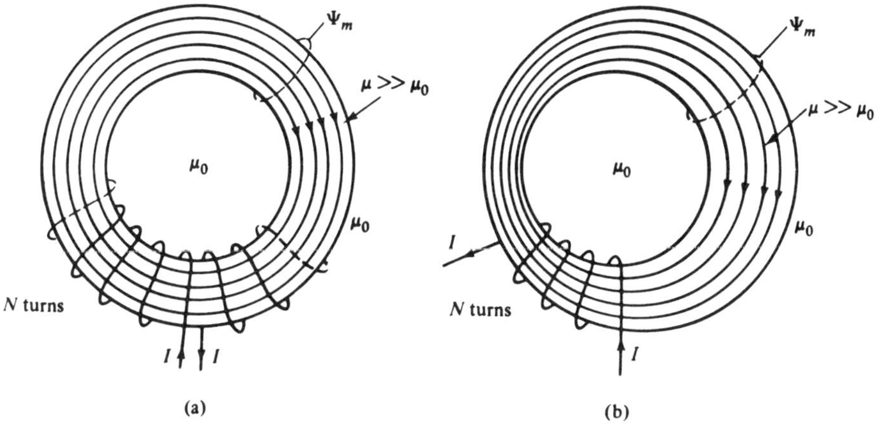

Figure 9-20 (a) Uniform toroid closely wound with N turns. (b) Non-uniform toroid wound with N turns non-uniformly spaced with $\mu \gg \mu_0$ in both magnetic cores.

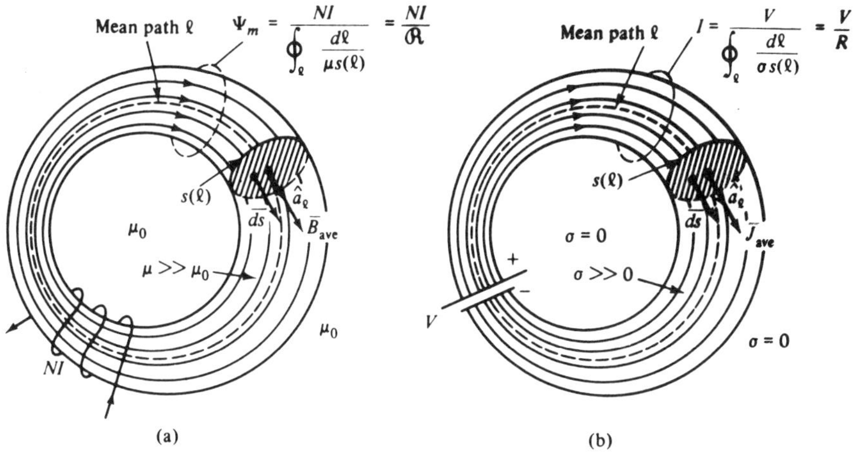

Figure 9-21 (a) Non-uniform magnetic circuit. (b) Non-uniform electric circuit.

in a high permeability material path parallels the ease of flow of electric current in a high conductivity material path. This parallelism seems to analog magnetic flux to electric current.

Let us form two dimensionally similar circuits, as suggested in Fig. 9-21. Figure 9-21(a) is a high permeability magnetic circuit excited by NI while Fig. 9-21(b) is a high conductivity electric circuit excited by voltage source V. If all the flux Ψ_m is found in the magnetic core (zero leakage flux), the flux in any cross section of Fig. 9-21(a) will be

$$\Psi_m = \int_{s(\ell)} \overline{B} \cdot \overline{ds} \tag{1}$$

where the cross section $s(\ell)$ is a function of the location along the mean path ℓ. When $\overline{B}$ is relatively uniform over the cross section $s(\ell)$, (1) can be written

$$\Psi_m = B_{ave} s(\ell) \tag{2}$$

where B_{ave} is the average value over the cross section $s(\ell)$. Let us now apply Ampère's circuital law over the mean path ℓ to obtain

$$\oint_{\ell} \overline{H}_{\text{ave}} \bullet \overline{d\ell} = NI \tag{3}$$

where $\overline{H}_{\text{ave}}$ can be viewed as an average $\overline{H}$ over the cross section $s(\ell)$. Through the use of $\overline{d\ell} = \hat{a}_\ell\, d\ell$, $\overline{H}_{\text{ave}} = \overline{B}_{\text{ave}}/\mu$, and (2), eq. (3) becomes

$$\oint_{\ell} \left(\frac{\hat{a}_\ell \Psi_m}{\mu s(\ell)}\right) \bullet \hat{a}_\ell\, d\ell = NI \tag{4}$$

The flux Ψ_m was assumed constant through any cross section along the path ℓ and can be removed from under the integral in (4) to obtain

$$\Psi_m \oint_{\ell} \frac{d\ell}{\mu s(\ell)} = NI \tag{5}$$

The integral in (5) will be a function of μ and the geometry of the magnetic circuit and is called the *reluctance* $\mathcal{R}$:

$$\boxed{\mathcal{R} \triangleq \oint_{\ell} \frac{d\ell}{\mu s(\ell)} \qquad (\text{A Wb}^{-1} \text{ or } \text{H}^{-1})} \tag{6}$$

Solving (5) for Ψ_m and using (6), we obtain

$$\boxed{\Psi_m = \frac{NI}{\mathcal{R}}} \tag{7}$$

Equations (6) and (7) indicate a strong parallelism to the resistance and current equations in a conductive circuit. Table 9-2 lists the analog terms and equations between the magnetic and electric circuits of Fig. 9-21 that substantiate this parallelism.

TABLE 9-2 ANALOG TERMS AND EQUATIONS BETWEEN MAGNETIC AND ELECTRIC CIRCUIT TERMS UNDER STEADY FLUX AND STEADY CURRENT CONDITIONS

Magnetic circuit	Electric circuit
$\overline{B} = \mu\overline{H}$	$\overline{J} = \sigma\overline{E}$
$\overline{H}$	$\overline{E}$
μ	σ
$\Psi_m = \int_s \overline{B} \bullet \overline{ds}$	$I = \int_s \overline{J} \bullet \overline{ds}$
NI	V
$\mathcal{R} = \oint_{\ell} \frac{d\ell}{\mu s(\ell)}$	$R = \oint_{\ell} \frac{d\ell}{\sigma s(\ell)}$
$\frac{1}{\mathcal{R}} = \mathcal{P}$	$\frac{1}{R} = G$

For a magnetic circuit of length ℓ, constant cross sections s, and constant permeability, (6) will yield

$$\boxed{\mathcal{R} = \frac{\ell}{\mu s} \qquad (\mathrm{H}^{-1})} \tag{8}$$

The reciprocal of reluctance (analogous to conductance G) is termed the *permeance* $\mathcal{P}$ of the magnetic circuit. The term NI in (5) and (7) is called the *magnetomotive force* (mmf) that "forces" flux through a magnetic circuit.

A series magnetic circuit made up of four legs, each with constant cross-sectional areas, is shown in Fig. 9-22(a). If we apply (5) to this circuit, we obtain

$$\Psi_m\left(\frac{\ell_1}{\mu_1 s_1} + \frac{\ell_2}{\mu_2 s_2} + \frac{\ell_3}{\mu_3 s_3} + \frac{\ell_4}{\mu_4 s_4}\right) = NI \tag{9}$$

Solving (9) for Ψ_m and using reluctances, we obtain

$$\boxed{\Psi_m = \frac{NI}{\mathcal{R}_1 + \mathcal{R}_2 + \mathcal{R}_3 + \mathcal{R}_4} = \frac{NI}{\mathcal{R}}} \tag{10}$$

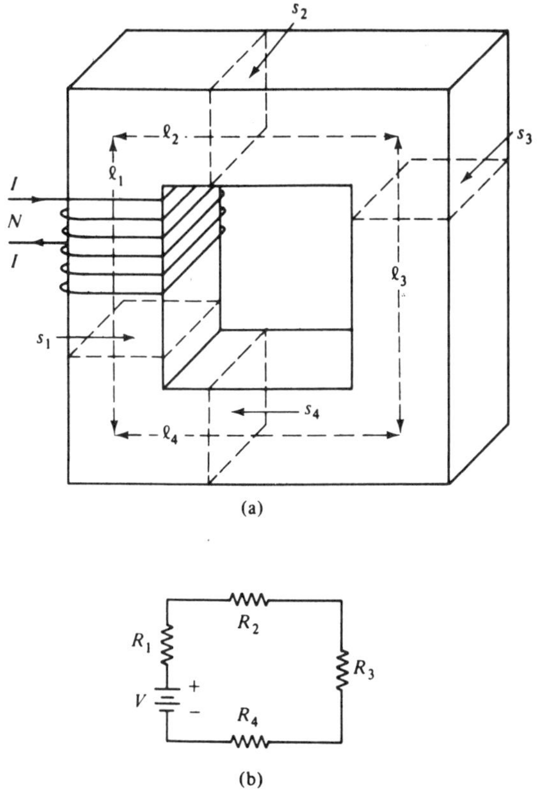

Figure 9-22 (a) Series magnetic circuit containing four legs, each of constant cross section. (b) Analogous electrical circuit.

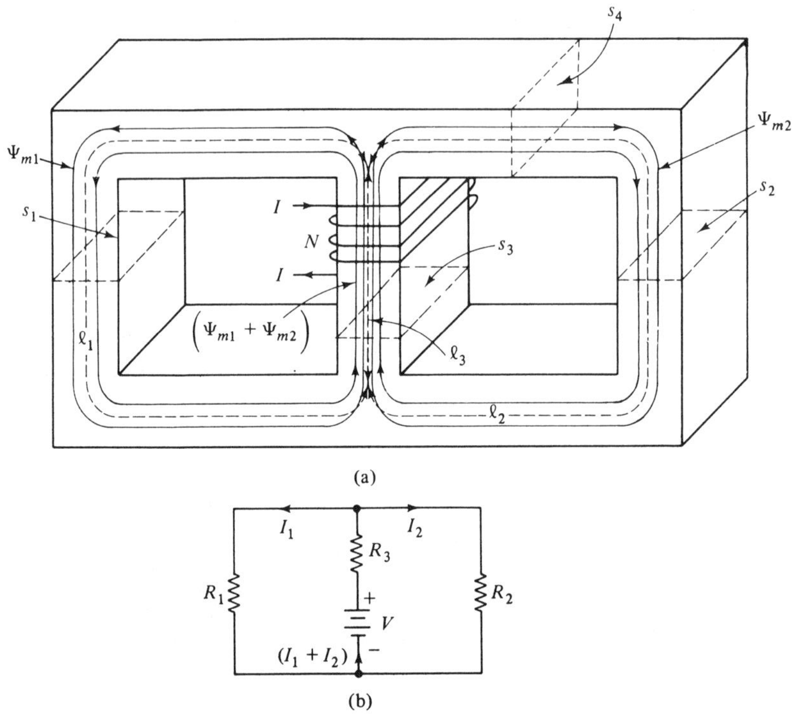

Figure 9-23 (a) Parallel magnetic circuit excited by a magnetomotive force on the center leg. (b) Analogous electric circuit.

where

$$\mathcal{R} = \mathcal{R}_1 + \mathcal{R}_2 + \mathcal{R}_3 + \mathcal{R}_4 = \frac{\ell_1}{\mu_1 s_1} + \frac{\ell_2}{\mu_2 s_2} + \frac{\ell_3}{\mu_3 s_3} + \frac{\ell_4}{\mu_4 s_4} \tag{11}$$

The analogous electrical circuit is shown in Fig. 9-22(b).

The magnetic circuit concept can be extended to more complicated magnetic circuits, as shown in Fig. 9-23(a), with its analogous electric circuit shown in Fig. 9-23(b). The system of equations involving Ψ_{m1}, Ψ_{m2}, NI, and the reluctances will be identical in form to those involving I_1, I_2, V, and the resistances in the electric circuit.

Example 10

The magnetic circuit of Fig. 9-22(a) is made from steel with $\ell_1 = \ell_3 = 0.3$ m, $\ell_2 = \ell_4 = 0.2$ m, $s_1 = s_2 = s_3 = s_4 = 6 \times 10^{-4}$ m^2, and $I = 0.05$ A. Assume a linear magnetization curve such that, in the steel, $H_s = 200B_s$ and find: (a) $\mathcal{R}$; (b) N to produce $B_s = 1$ T in the ℓ_1 leg.

Solution. (a) From $B_s = \mu_s H_s$ and $H_s = 200B_s$, $\mu_s = \frac{1}{200}$. Now, from (11),

$$\mathcal{R} = \frac{\ell_1}{\mu_s s_1} + \frac{\ell_2}{\mu_s s_2} + \frac{\ell_3}{\mu_s s_3} + \frac{\ell_4}{\mu_s s_4}$$

$$= \frac{0.3}{(\frac{1}{200})(6 \times 10^{-4})} + \frac{0.2}{(\frac{1}{200})(6 \times 10^{-4})} + \frac{0.3}{(\frac{1}{200})(6 \times 10^{-4})} + \frac{0.2}{(\frac{1}{200})6 \times 10^{-4})}$$

$$= 10^5 + 0.667 \times 10^5 + 10^5 + 0.667 \times 10^5 = 3.33 \times 10^5 \text{ H}^{-1}$$

(b) For $B_s = 1$ (T) in leg ℓ_1, we find

$$\Psi_m = B_s(s_1) = 1 \times 6 \times 10^{-4} \text{ Wb}$$

Thus, from (10)

$$N = \frac{\Psi_m \mathcal{R}}{I} = \frac{(6 \times 10^{-4})(3.33 \times 10^5)}{0.05} = 4000 \text{ turns}$$

or

$$NI = (4000)(0.05) = 200 \text{ A Turns}$$

From (9) or (10), we can write

$$\Psi_m(\mathcal{R}_1 + \mathcal{R}_2 + \mathcal{R}_3 + \mathcal{R}_4) = NI \qquad \text{(A turns)} \quad \text{(mmf)} \tag{12}$$

where the term such as $\Psi_m\mathcal{R}_1$ represents the mmf across the reluctance $\mathcal{R}_1$ to produce a flux of Ψ_m. The sum of these mmf's thus must equal NI and is analogous to the sum of potentials across all series R's must equal the applied V.

From Ampère's circuital law, for a single-loop magnetic circuit such as in Fig. 9-22,

$$\oint_\ell \overline{H} \cdot \overline{d\ell} = I_{\text{en}} \tag{8.4-2}$$

Equation (12) can also be written as

$$\Psi_m(\mathcal{R}_1 + \mathcal{R}_2 + \mathcal{R}_3 + \mathcal{R}_4) = NI = H_1\ell_1 + H_2\ell_2 + H_3\ell_3 + H_4\ell_4 \tag{13}$$

where H_1 is the H field for the portion of the path ℓ_1, H_2 is the H field for the portion ℓ_2, etc. For example, consider a toroidal core with air gap as shown in Fig. 9-24. The total ampere-turns are equal to the sum of $H_s\ell_s$ and $H_g\ell_g$, i.e.,

$$NI = H_s\ell_s + H_g\ell_g \tag{14}$$

where ℓ_s is the mean length in the steel core and ℓ_g is the length of air gap.

A graphical solution for H_s and B_s within the material is permissible if the magnetization curve for the core material is known. This magnetization curve is illustrated in Fig. 9-25. Superimposed on the magnetization curve plot is an "air gap line," similar to a

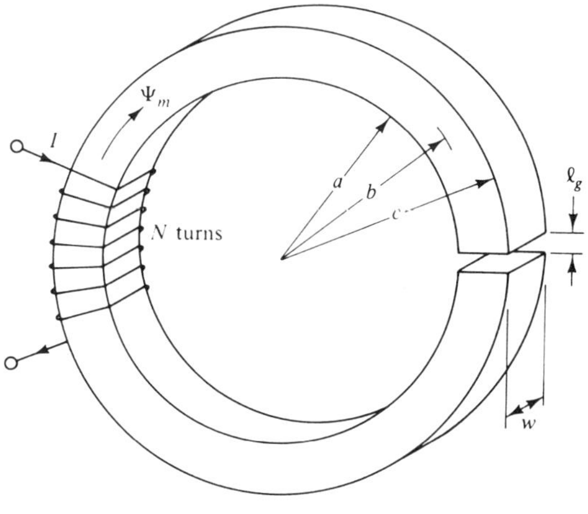

Figure 9-24 Toroidal core with air gap. The rectangular cross section is typical of ferrite or tape-wound cores used for inductors and transformers.

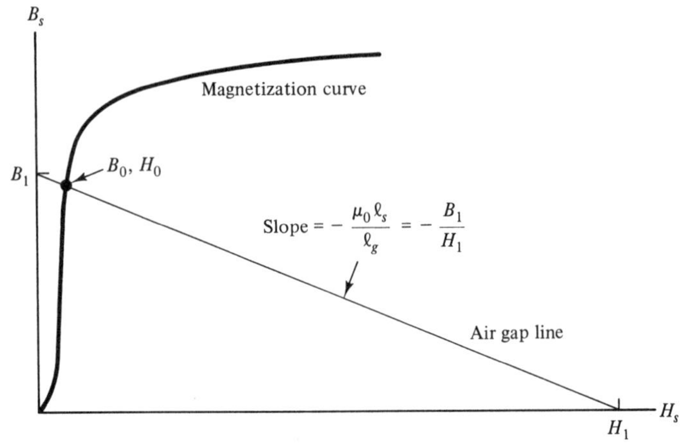

Figure 9-25 Graphical solution of magnetic circuit with air gap. $B_1 = \mu_0 NI/l_g$, $H_1 = NI/l_s$.

"load line" used in non-linear electronic circuit analysis. The intercept point B_1, is calculated by assuming that the μ_r of the core is infinite and the ampere-turns are entirely across the air gap. The H_1 intercept is obtained by assuming that μ_r is zero and all the ampere-turns appear along the core material. The intersection of the air gap line with the magnetization curve represents a simultaneous solution that satisfies both the air gap and the B_s and H_s in the core. Under the assumption that the cross-sectional areas of the core and that of the air gap are equal, the equation for the air gap line can be shown to be (see Prob. 9.9-2)

$$B_s = \frac{\mu_0 NI}{\ell_g} - \frac{\mu_0 \ell_s H_s}{\ell_g} \tag{15}$$

which gives the B_1 intercept as $\mu_0 NI/\ell_g$ and the slope of the air gap line as $-\mu_0 \ell s/\ell_g$. The H_1 intercept is obtained by setting $B_g = 0$, and is NI/ℓ_s.

A gapped core is used to linearize the reluctance of a magnetic circuit. This comes about from the fact that the reluctance of the air gap, which is linear, is usually much greater than that of the non-linear reluctance of the core. Inductors designed to carry large currents usually use gapped cores. Magnetization curves for silicon steel and cast steel are shown in Fig. 9-26. A gapped core is frequently used to prevent saturation for an inductor that carries a large dc current. We will discuss inductors in the next section.

Problem 9.9-1 Assume that for the toroidal core of Fig. 9-24, $b = 2.5$ cm, $w = 2$ cm, $c - a = 1.0$ cm, $\ell_g = 0.1$ mm, and $\mu_r = 5000$. Neglecting fringing, find: (a) $\mathcal{R}$; and (b) N to produce $B_g = 1$ T across the air gap if $I = 10$ A.

Problem 9.9-2 A toroidal magnetic circuit contains an air gap of length $\ell_g = 1.5$ mm. If the mean steel length $\ell_s = 0.3$ m, the constant cross section $s = 6 \times 10^{-4}$ m^2, and $NI = 1000$ A turns, find the B_s in the steel. Assume silicon sheet steel (see Fig. 9-26) and zero fringing at the gap. [*Hint:* Construct an air gap line where the ordinate at $H_s = 0$ and the abscissa for $B_s = 0$ are used to construct the line by the point-slope method after rearranging (14) to solve for B_s. Show that the slope is $-\ell_s \mu_0/\ell_g$.]

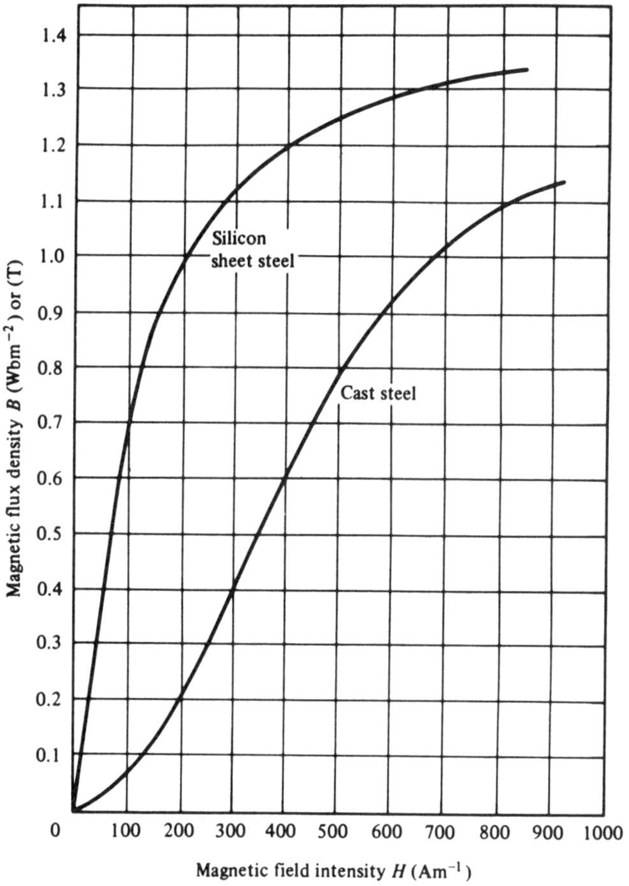

Figure 9-26 Typical magnetization curves for silicon sheet steel and cast steel.

9.10 SELF-INDUCTANCE AND MUTUAL INDUCTANCE

The *inductance* is a circuit concept that becomes quite important when dealing with time-varying current and thus time-varying fields. From circuit theory, the student has found that the induced potential across a wire-wound coil, such as a solenoid or a toroid, is equal to

$$\boxed{V_L = L\frac{dI}{dt} \qquad \text{(V)}} \tag{1}$$

where L is the inductance of the coil, and I is the time-varying current flowing through the coil. From this point on, we shall call the coil an *inductor*.

In a capacitor, the energy is stored in the electric field $\overline{E}$ and is found to equal $\frac{1}{2}CV^2$. In an inductor, the energy is stored in the magnetic field, as suggested in

Fig. 9-27. The presence of this stored energy is sensed by the persistence of energy delivered to the lamp for a short time after the switch in Fig. 9-27 is opened. The energy stored in the inductor can be found through the use of (1), and

$$W_m = \int_{t=0}^{t=t_0} V_L I\,dt = \int_{t=0}^{t=t_0} \left(L\frac{dI}{dt}\right) I\,dt = \int_0^I LI\,dI = \frac{1}{2}LI^2 \quad \text{(J)} \tag{2}$$

where $t = 0$ is the time the switch is closed. Equations (1) and (2) are circuit equations that can be used to define the inductance L. We shall return to (2) after we have discovered how to calculate W_m. Equation (1) will be useful in Chapter 10 when we begin our study of induced potential and time-varying fields.

At this point, we shall define the inductance of an inductor as

$$L \triangleq \frac{\Lambda}{I} \quad \text{(flux linkage A}^{-1}\text{)} \quad \text{(H)} \tag{3}$$

where the unit of inductance is the henry, Λ (lambda) is the total flux linkage of the inductor, and I is the current flowing in the inductor. A flux linkage of one exists when 1 Wb of flux links one turn of the inductor. If all the flux link all the turns, the total flux linkage is equal to

$$\Lambda = \Psi_m N \quad \text{(Wb turns)} \tag{4}$$

where Ψ_m is the total flux produced by the inductor. When all the flux does not link all turns, as suggested in Fig. 9-27, the total flux linkage will be less than found by (4). In a coaxial cable or a two-wire line, the inductance produced by the flux internal to the

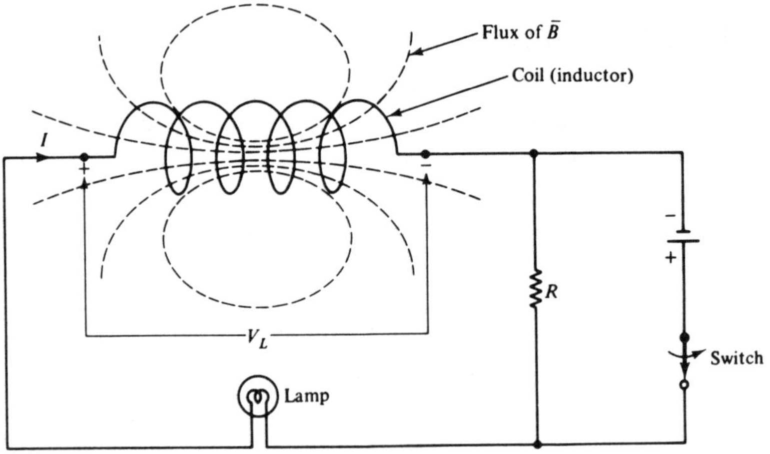

Figure 9-27 Simple electric circuit that shows the effect of energy stored in a magnetic field of an inductor.

conductor, is called *internal inductance*. The internal flux links only a portion of the total current I. The inductance defined by (3) is commonly called *self-inductance* since the linkages are self-produced by the inductor.

Mutual inductance exists between two magnetic circuits that share a common flux linkage, as suggested in Fig. 9-28. The mutual inductance M_{12} is defined as

$$M_{12} = \frac{\Lambda_{12}}{I_1} \quad \text{(H)} \tag{5}$$

where Λ_{12} is the linkage of circuit 2 produced by I_1 in circuit 1. For a linear magnetic medium, it can be shown that $M_{12} = M_{21}$.

Example 11

Obtain the expression for the self-inductance of the long solenoid shown in Fig. 8-6.

Solution. Assume that all the flux Ψ_m links all N turns and that B does not vary over the cross-section area of the solenoid; thus, from (4)

$$\Lambda = \Psi_m N = B(\pi a^2)N$$

Through the use of $\overline{B} = \mu\overline{H}$, we obtain

$$\Lambda = (\mu H)(\pi a^2)N = \left(\frac{\mu NI}{\ell}\right)(\pi a^2)N = \frac{\mu N^2 I}{\ell}(\pi a^2) \quad \text{(Wb turns)} \tag{6}$$

From (3) and (6)

$$\boxed{L = \frac{\Lambda}{I} = \frac{\mu N^2 \pi a^2}{\ell} \quad \text{(H)}} \tag{7}$$

Example 12

Obtain the expression for the self-inductance of the toroid shown in Fig. 8-10.

Solution. Assume that we have many closely spaced turns of filamentary wire and that the mean magnetic path length $2\pi b$ is much greater than $(c - a)$, the diameter of the cross section of the toroid. Thus, all the flux will be within the toroid, and the B can be assumed to be constant throughout the cross section of the toroid. From (3) and the results of

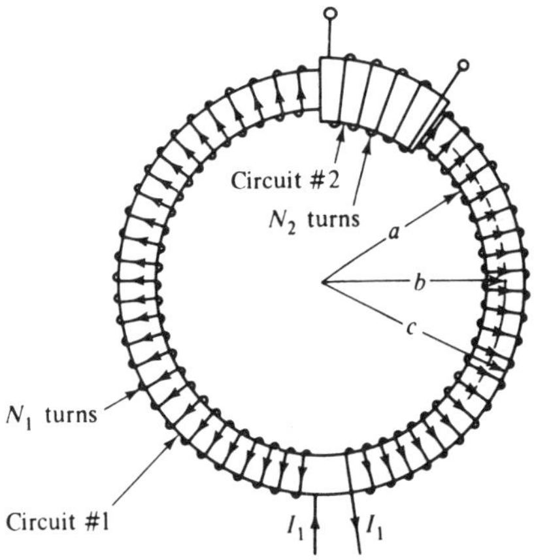

Figure 9-28 Two circuits coupled by a common magnetic flux that leads to mutual inductance between the two circuits.

Prob. 8.4-5(a), we have

$$L = \frac{\Lambda}{I} = \frac{\Psi_m N}{I} = \frac{B\left(\frac{\pi(c-a)^2}{4}\right)N}{I} = \frac{\mu \frac{NI}{2\pi b} sN}{I}$$

$$\boxed{L = \frac{\mu N^2 s}{2\pi b} \qquad \text{(H)}} \tag{8}$$

where b is the mean radius and s is the cross-sectional area of the toroid.

Example 13

Obtain the expression for the self-inductance per meter of the coaxial cable of Fig. 5-21 when the current flow is restricted to the surface of the inner conductor and the inner surface of the outer conductor.

Solution. The Ψ_m will exist only between r_a and r_b, and it will link all the current I. From (3) and (8.4-20), we have

$$L = \frac{\Lambda}{I} = \frac{\Psi_m}{I} = \int_0^1 \int_{ra}^{rb} \frac{(\mu H)(dr_c\, dz)}{I} = \int_0^1 \int_{ra}^{rb} \left(\frac{\mu I}{2\pi r_c}\right)\left(\frac{dr_c\, dz}{I}\right)$$

$$\boxed{L = \frac{\mu}{2\pi} \ln\left(\frac{r_b}{r_a}\right) \qquad (\text{H m}^{-1})} \tag{9}$$

Example 14

Find the expression for the mutual inductance between circuit 1 and circuit 2, as shown in Fig. 9-28.

Solution. Let us assume mean path $2\pi b$ is much greater than $(c - a)$, as in Example 12. From (5) and Example 12, we have

$$M_{12} = \frac{\Lambda_{12}}{I_1} = \frac{\Psi_{m(12)} N_2}{I_1} = \frac{B_{12}\left(\frac{\pi(c-a)^2}{4}\right)N_2}{I_1}$$

$$= \frac{\mu \frac{N_1 I_1}{2\pi b} sN_2}{I_1} = \frac{\mu N_1 N_2 s}{2\pi b} \qquad \text{(H)} \tag{10}$$

Problem 9.10-1 Find the self-inductance of a long coil, as shown in Fig. 8-6, when $\mu = \mu_0$, $N = 2000$ turns, $a = 1$ cm, and $\ell = 50$ cm.

Problem 9.10-2 Find the self-inductance of a toroid, as shown in Fig. 9-24, when $\mu = \mu_0$, and $N = 1500$ turns. Assume a rectangular cross section with $a = 0.2$ m, $c = 0.21$ m, and with a core width of 2 cm. Assume zero air gap.

Problem 9.10-3 Find the self-inductance for the toroid of Fig. 9-24 for the core described in Prob. 9.9-1 with $N = 6$ turns.

Problem 9.10-4 If the current I in Example 13 is distributed uniformly in the center conductor, show that the internal inductance, due to the linkage produced by the flux inside the inner conductor, is equal to $\mu/8\pi$ (H m^{-1}).

Problem 9.10-5 The voltage V_2 induced across the terminals of circuit 2 in Fig. 9-28 is equal to $M_{12}(dI_1/dt)$. Find: (a) M_{12} if $N_1 = 1000$ turns, $N_2 = 200$ turns, $\mu = 1000\mu_0$, $b = 0.2$ m, $s = 2 \times 10^{-4}$ m^2, and $I_1 = 5 \cos(120\pi t)$; (b) $V_{2_{\text{rms}}}$.

9.11 MAGNETIC ENERGY DENSITY AND FORCE ON MAGNETIC MATERIAL

The energy density in an electric field $\overline{E}$ has been found to be equal to $\frac{1}{2}\overline{D} \cdot \overline{E}$. This expression was developed in Chapter 4 when we found the energy to build up a system of charges. The expression for the energy density in a magnetic field is not as easily obtained. We will derive the magnetic energy expression in a round-about manner through the use of (9.10-2) and (9.10-3). Let us substitute (9.10-3) into (9.10-2) to obtain

$$W_m = \frac{1}{2}LI^2 = \frac{1}{2}\left(\frac{\Lambda}{I}\right)I^2 = \frac{1}{2}\Lambda I \tag{1}$$

The energy in the magnetic field of a toroid of cross-sectional area s becomes

$$W_m = \tfrac{1}{2}BsNI \tag{2}$$

Rearranging (2) and multiplying the numerator and denominator by $2\pi b$, we obtain

$$W_m = \tfrac{1}{2}B\frac{NI}{2\pi b}(s2\pi b) \qquad \text{(J)} \tag{3}$$

where $s2\pi b$ is the volume υ in which Ψ_m exists (volume of the toroid) and $NI/2\pi b$ equals H. Let us now divide (3) by the volume υ to obtain the magnetic energy density w_m:

$$w_m = \frac{W_m}{\upsilon} = \tfrac{1}{2}BH = \tfrac{1}{2}\mu H^2 \qquad (\text{J m}^{-3}) \tag{4}$$

The more general expression for magnetic energy density is expressed as

$$\boxed{w_m = \tfrac{1}{2}\overline{B} \cdot \overline{H} \qquad (\text{J m}^{-3})} \tag{5}$$

Note the similarity in form to $\frac{1}{2}\overline{D} \cdot \overline{E}$. Equation (5) will be accepted, without additional proof, to be the magnetic energy density in any magnetic field.

Now that we have an expression for the magnetic energy density, we can solve for the inductance in (9.10-2) to obtain

$$\boxed{L = \frac{2W_m}{I^2} \qquad \text{(H)}} \tag{6}$$

where W_m is obtained by a volume integral of (5).

We are now in a position to find the force of attraction between two poles on either side of an air gap, as shown in Fig. 9-24. Let us assume zero flux fringing in the air gap; thus, the B in the ferromagnetic material and in the air gap are equal. Now, let us assume that the gap length is changed by dx when we apply a force F. Under the condition that B is not changed, we find that the energy in the gap has been increased by the amount $F\,dx$. This energy increase in the gap can also be found through the use of (4); thus,

$$dW_m = \frac{1}{2}\left(\frac{B^2}{\mu_0}\right)(s\,dx) \tag{7}$$

where $s\,dx$ represents the change in the volume of the gap. Equating (7) to $F\,dx$ and solving for F, we obtain

$$\boxed{F = \frac{B^2 s}{2\mu_0} \quad \text{(N)}} \tag{8}$$

where F can be viewed as the force of attraction due to two magnetic poles. Electromagnets and their force of attraction are found in many electromechanical devices such as valves, relays, etc. Equation (8) is analogous to equation (4.6-21) for the force of attraction between two charged plates, where the pole faces correspond to the surfaces of the plates.

Example 15

The air gap of Fig. 9-24 is found to have $B = 1.4$ T when $s = 10^{-4}$ m^2. Find the force of attraction F across the air gap.

Solution. From (8), we have

$$F = \frac{B^2 s}{2\mu_0} = \frac{(1.4)^2(10^{-4})}{2(4\pi \times 10^{-7})} = 78 \text{ N}$$

Since 1 N = 0.224 lb force, $F = 17.47$ lb force.

Problem 9.11-1 Obtain the expression for the self-inductance of the coaxial cable of Example 13 through the use of (5) and (6).

Problem 9.11-2 Neglecting the reluctance of the iron or steel core, find the alternating and average force across the air gap of Fig. 9-24 when $N = 1000$ turns, $I = 3.0 \cos 377t$ (A), $s = 8 \times 10^{-4}$ m^2, and $\ell_g = 2$ mm.

9.12 DEMAGNETIZING FACTOR—EFFECTIVE PERMEABILITY

It is tempting at this point to assume that the inductance of an air core solenoid is increased by the μ_r of the core when a ferromagnetic core is inserted into the coil. Although this approaches the case for very long solenoids, it is far from true for short coils. There is an *effective permeability* defined as

$$\mu_{\text{eff}} = \frac{B}{H_a} \tag{1}$$

where B is the flux density induced into the core, and H_a is the applied H field, i.e., the H field that would exist if the core were not present to distort the field. Equation (1) takes into account the demagnetizing effect of the poles formed at the ends of a magnetized rod. For a given material and shape,

$$\mu_{\text{eff}} = \frac{\mu_0\mu_r}{1 + (\mu_r - 1)N} \tag{2}$$

where N is the *demagnetizing factor,* a function of the shape of the core relative to the direction of the applied field.[5] A similar relation is true for dielectrics, where the factor is called the *depolarizing factor.* In the SI system of units, the depolarizing factor and the demagnetizing factor are the same. It is strictly a function of shape. The demagnetizing factor for a sphere is $\frac{1}{3}$, which means that even for infinite μ_r, the effective permeability can be at most equal to $3\mu_0$. For very long slender rods, with the applied field parallel to the rod axis, N approaches zero. The derivation of (2) is left for Prob. 9-22.

REVIEW QUESTIONS

1. Does the action-at-a-distance solution involve the field concept? *Sec. 9.2*
2. What is the expression for the force of translation on a current element $I\overline{d\ell}$ in a magnetic field $\overline{B}$? *Eq. (9.3-1)*
3. What is the expression for the force per unit volume of a ρ_v distribution? *Eq. (9.4-3)*
4. What are the units of $\overline{U} \times \overline{B}$? *Eq. (9.4-4)*
5. What is the Lorentz force equation? *Eq. (9.4-6)*
6. What is the Hall effect? *Sec. 9.5*
7. What is the expression for the torque on a differential current loop in a magnetic field $\overline{B}$? *Eq. (9.6-7)*
8. What is the expression for the dipole moment $\overline{dm}$? *Eq. (9.6-8)*
9. What is the defining equation for the magnetization vector $\overline{M}$? *Eq. (9.7-1)*
10. What is the expression for the bound magnetization current density $\overline{J}_m$? *Eq. (9.7-8)*
11. What is the expression for the bound magnetization current density $\overline{J}_{sm}$? *Eq. (9.7-11)*
12. Is $\nabla \times \overline{H} = \overline{J}$ also true in a magnetic material? ($\overline{J}$ is due to free charges.) *Eq. (9.7-14), (9.7-15)*
13. What is χ_m? *Eq. (9.7-17)*
14. Is the μ_r of a diamagnetic material less than one? *Sec. 9.7-4*
15. Is the energy loss in a magnetic material proportional to the area of the hysteresis loop? *Sec. 9.7-4*
16. Do ferromagnetic materials have a large μ_r? *Sec. 9.7-4*
17. At a magnetic-magnetic boundary, what is the relationship between the normal B components? *Eq. (9.8-4)*
18. What is the expression for the reluctance of a magnetic circuit? *Eq. (9.8-6)*
19. What is the relationship between NI, $\mathcal{R}$, and Ψ_m in a magnetic circuit? *(Eq. 9.9-7)*
20. What is the defining equation for self-inductance in terms of flux linkages? *Eq. (9.10-3)*
21. What is a flux linkage? *Sec. 9.10*
22. What is the expression for the energy density in a magnetic field? *Eq. (9.11-4)*

PROBLEMS

9-1. Calculate the force between two equal and opposite magnetic poles having pole strengths of 10 A · m each and which are separated by a distance of 1 mm. These could be poles of two long bar magnets, where the other pole pairs are separated by a large distance such that the force between them is negligible. [*Hint:* Refer to footnote in Sec. 9.3.]

9-2. If pole strengths were given in webers instead of ampere-meters, the Coulomb force law would be

$$F = \frac{Q_{m1} Q_{m2}}{4\pi\mu_0 R^2} \qquad \text{(N)}$$

[5] J. K. Watson, *Applications of Magnetism,* John Wiley & Sons, Inc., New York, 1980.

For a force of attraction of 10 N, find the pole strengths Q_{m1} and Q_{m2}, assuming that they are of the same magnitude, and that R is 1 mm. If you have worked Prob. 9-1, what is the conversion between pole strength in ampere-meters and pole strength in webers? [*Note:* In SI units, pole strength is in ampere-meters.]

9-3. Two infinite and parallel conductor sheets carry surface currents $\overline{J}_s$ in opposite directions. If $J_s = 5$ A m^{-1} and the separation $d = 1$ mm, find the force of translation (repulsion) per square meter.

9-4. To the filamentary current-carrying conductor of infinite length and the square current-carrying loop of Fig. 9-5, another current-carrying conductor of infinite length is added at $y = 0.1$ m, $x = 0$, $-\infty \le z \le \infty$. Let the current I_3 in the added conductor be in the reverse direction to I_1, and find: (a) the force of translation on the square loop when $I_1 =$ 10 A, $I_2 = 50$ A, and $I_3 = 10$ A; (b) the same force when I_3 is reversed.

9-5. Show that the energy exchange between a moving charged particle and the magnetic field is zero by showing, through expansion in component form, that $\overline{F} \bullet \overline{U} = 0$. The expression $\overline{F} \bullet \overline{U} = (Q\overline{U} \times \overline{B}) \bullet \overline{U}$ and is equal to the rate of change of energy in a magnetic field.

9-6. A charged particle is injected into a uniform magnetic field $\overline{B} = \hat{z}2$ (T) with an initial velocity of $\hat{y}$ ($m\ s^{-1}$) at the origin when $t = 0$. If the charge on the particle is equal to 1 C and its mass is 1 kg, show that: (a) $(d^2U_y/dt^2) + 4U_y = 0$; (b) $(d^2U_x/dt^2) + 4U_x = 0$; (c) the parametric equations of the particle path are $y = \frac{1}{2}\sin 2t$ and $x = \frac{1}{2}(1 - \cos 2t)$; (d) the kinetic energy is equal to 0.5 J and is independent of time.

9-7. A bar magnet whose pole strength is 1 A m and whose length is 10 cm is suspended by a string. If the magnet is initially positioned such that its axis is perpendicular to the earth's field of 40 A m^{-1}, find the magnitude of the torque acting on the bar magnet.

9-8. A square filamentary conductor loop carrying a current of 25 A is centered in the $z = 0$ plane, with legs parallel to the coordinate axes. If the legs are 10 cm long and the current flows CCW as viewed along the $+z$ axis toward the origin, find the vector torque $\overline{T}$ on the loop when: (a) $\overline{B} = (0.5\hat{x} + 1.0\hat{y})$ (T); (b) $\overline{B} = |y|\hat{y}$ (T).

9-9. Find the torque $\overline{T}$ that a $\overline{dm}_1$ at the origin produces on a $\overline{dm}_2$ at the spherical point $(3, \pi/4, \pi/2)$. Let $\overline{dm}_1 = \hat{z}10^{-3}$ (A · m^2) and $\overline{dm}_2 = \hat{r}_s 10^{-3}$ (A · m^2).

9-10. An infinite slab of magnetic material fills the space between two infinite and parallel conductor current sheets, with $\overline{J}_s = 2\hat{x}$ (A m^{-1}) at $z = 0.01$ and $\overline{J}_s = -2\hat{x}$ (A m^{-1}) at $z = -0.01$. Find $\overline{H}, \overline{B}, \overline{M}$, and χ_m everywhere when $\mu_r = 3$ for the magnetic material. Assume that $\mu_r = 1$ for $|z| > 0.01$.

9-11. For the gapped core in Fig. 9-24, find $\overline{H}$, $\overline{B}$, and $\overline{M}$ both in the core and in the air gap. Assume N turns and current I.

9-12. Solve Prob. 9-10 when the magnetic material is found only between the $z = 0$ plane and the $z = -0.01$ plane.

9-13. For the magnetic circuit example (Example 10), find B_s when $N = 250$ turns, $I = 0.05$ A, and $\mu_r = 5000$.

9-14. If leg ℓ_3 of Fig. 9-22(a) and Example 10 has an air gap of length $\ell_g = 1.5$ mm, find; (a) $\mathcal{R}$; and (b) N to produce $B_g = 1$ T in the gap. Assume $\mu_r = 5000$ and zero fringing at the gap.

9-15. Find the flux in each of the legs of the magnetic circuit of Fig. 9-23(a) when $N = 500$, $I = 10$ A, $\ell_1 = 0.3$ m, $\ell_2 = 0.4$ m, $\ell_3 = 0.1$ m, $s_1 = s_2 = s_4 = 10^{-3}$ m^2, $s_3 = 1.5 \times 10^{-3}$ m^2, and $\mu_r = 3000$.

9-16. For the magnetic circuit of Example 10, find: (a) the self-inductance when $N = 3000$ turns (use the results of Example 10); (b) the inductance for a 0.1-mm air gap cut in leg ℓ_3.

9-17. Show that when a gap is cut in a core like that in Fig. 9-22, the inductance is $L = L_m L_g / (L_m + L_g)$, where L_m is the inductance of an ungapped core, and $L_g = N^2/\mathcal{R}_g$.

9-18. A single layer of $N_2 = 200$ turns is wound directly on a long solenoid of $N_1 = 1000$ turns. Obtain the expression for the mutual inductance M_{12} when $\mu = \mu_0$, the radius of the inner solenoid is a (m), and its length is ℓ (m). Assume that all the flux of the inner solenoid links all the turns of the outer solenoid.

9-19. Repeat Prob. 9.11-2 and consider the reluctance of the material. For the magnetic material, assume an effective length of 30 cm, cross-sectional area $s = 8 \times 10^{-4}$ m^2, and $H_s = 300B_s$.

9-20. Show that the lifting force of a magnet pole on a magnetized slab of ferromagnetic material is $F = BQ_m/2$ (N), where Q_m is the pole strength of the magnet and B is the total flux density between the magnet pole and the slab of magnetized material. Assume that the pole face and surface of the object being lifted are smooth and parallel. Neglect fringing. Note that one must define Q_m such that the result is compatible with (9.11-8).

9-21. Show that

$$\overline{T} = \overline{m} \times \overline{B} \qquad \text{and} \qquad F = \frac{\mu_0 Q_{m1} Q_{m2}}{4\pi R_{12}^2} \quad \text{(N)}$$

for a Coulomb's law between "isolated" magnetic poles are compatible with the unit of magnetic pole strength being the ampere-meter, with $Q_m = \Psi_m/\mu_0$.

9-22. Derive (9.12-2) by starting with the following:

1. The demagnetizing field $\overline{H}_d$ is opposite to the applied $\overline{H}$ field and is given by $\overline{H}_d = -N\overline{M}$.
2. $\overline{B} = \mu_0(\overline{M} + \overline{H})$.
3. $\overline{H} = \overline{H}_a + \overline{H}_d$.
4. $\mu_r = 1 + \chi_m$.
5. $\mu_{\text{eff}} = \overline{B}/\overline{H}_a$.
6. $\overline{M} = \chi_m \overline{H}$.

9-23. A long magnetized bar (permanent magnet) carries a flux density of 0.1 T, and the magnetic moment is 200 A · m^2. If the volume of the bar is 0.002 m^3, illustrate the direction and magnitude of the H field in the bar. Assume uniform magnetization and assume that B is uniform throughout the length of the bar. No external ampere-turns are applied.

9-24. The magnetic moment of a magnet can be measured by measuring the axial H field at a distance from the magnet that is large compared to ℓ_{eff}. If the magnet is approximately 2 cm in length and the axial H field at 30 cm from the magnet is 0.8 A m^{-1}, find the magnetic moment (A · m^2) of the magnet.

10

Faraday's Law, Time-Varying Fields, Potential Functions, and Boundary Relations

10.1 INTRODUCTION

In the first nine chapters of this book, we have considered mostly non-time-varying fields, fields that are derived either from static charges or charges in unaccelerated motion. The moving charges of Chapters 5 and 8 give rise to electrostatic and magnetostatic fields. In this chapter we shall develop the concept of an electromotive force (emf) voltage source which is generated by a time-changing magnetic field, a concept basic to the understanding of the electric generator, transformer, inductor, etc., and to the understanding of eddy current, an important source of loss in transformer cores. This concept is known as *Faraday's law.*

The concept of displacement current is also introduced to complete Maxwell's equations for time-varying fields. The Maxwell equations given in (8.8-2) turn out to be a special case where there is no time variation of the fields. For time-varying fields, we will find that the $\overline{E}$ field is not just the negative gradient of a scalar potential function, but it is also related to the time-changing magnetic vector potential as well.

An *RLC* circuit is used to demonstrate the relation between field and circuit equations. Finally, we shall extend the boundary relations already developed for static fields to time-varying fields.

10.2 EMF AND FARADAY'S LAW

We found in Sec. 4.3 that for static fields

$$\oint_{\ell} \overline{E} \cdot \overline{d\ell} = 0 \tag{1}$$

because of the conservative property of the static $\overline{E}$ field. However, the static $\overline{E}$ field is due to a charge distribution; i.e., it is the negative gradient of the scalar electric potential. In this section, we need to consider also the type of electric field that exists within a battery that forces charge from the negative terminal of the battery to the positive terminal. An electric field may be defined by

$$\boxed{\overline{E} = \frac{\overline{F}}{Q} \qquad (\text{N C}^{-1}) \quad (\text{V m}^{-1})} \tag{2}$$

then the direction of the $\overline{E}$ field, whether that field is due to a charge distribution or due to an emf source such as a battery, is the direction that a positive charge would tend to move. The two types of electric field are illustrated in Fig. 10-1(a) and (b).

Within the battery, the chemical emf field $\overline{E}_e$ is directed in the direction of positive current through the battery, but, external to the battery, the field due to the charge distri-

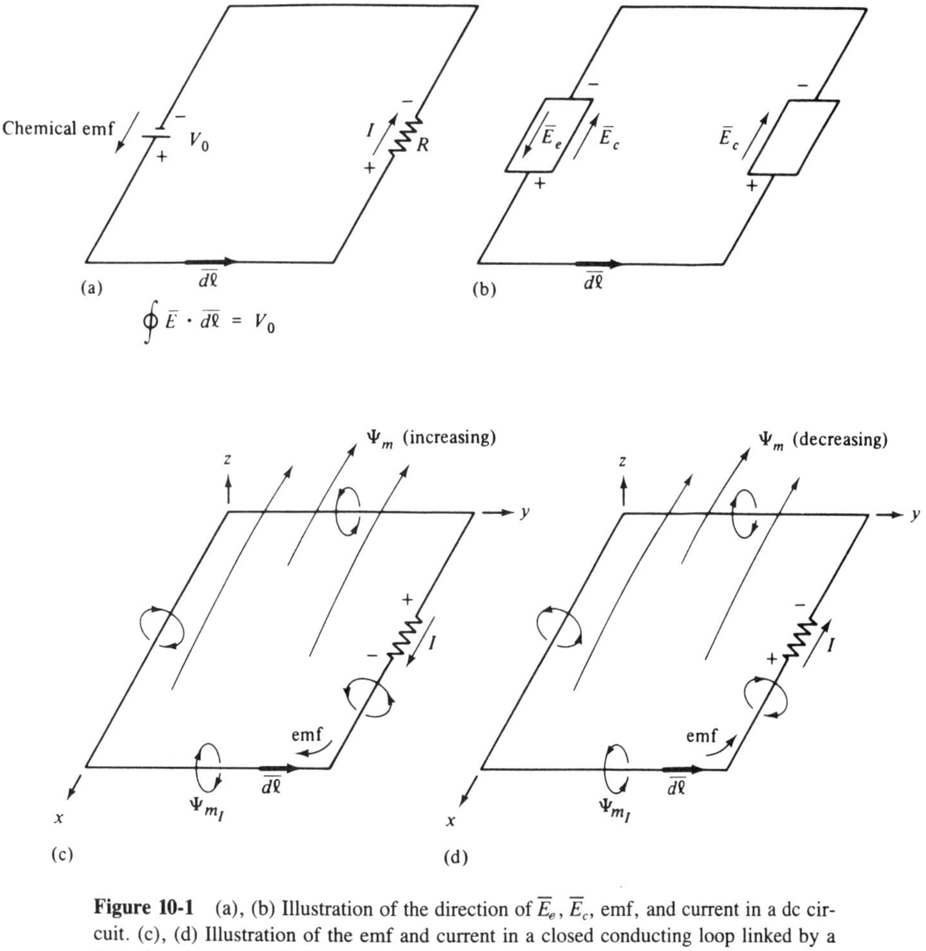

Figure 10-1 (a), (b) Illustration of the direction of $\overline{E}_e$, $\overline{E}_c$, emf, and current in a dc circuit. (c), (d) Illustration of the emf and current in a closed conducting loop linked by a time-changing magnetic field.

bution, $\overline{E}_c$, is in just the opposite direction. That is, if a conducting path is connected across the battery, positive charge will move in the direction indicated by $\overline{E}_c$.

Now, just as is true for the fields discussed in Sec. 4.3,

$$\boxed{\oint_{\ell} \overline{E}_c \bullet \overline{d\ell} = 0 \text{ V}} \tag{3}$$

but for the total field $\overline{E}_e + \overline{E}_c$,

$$\boxed{\oint_{\ell} \overline{E} \bullet \overline{d\ell} = \oint_{\ell} \overline{E}_e \bullet \overline{d\ell} = \text{emf} \qquad \text{(V)}} \tag{4}$$

We generally credit Faraday with the discovery that a time-changing magnetic field induces an emf in a closed loop linked by that field. This experimental fact is called *Faraday's law* and is written as

$$\boxed{\text{emf} = -N\frac{d\Psi_m}{dt} \qquad \text{(V)}} \tag{5}$$

where N is the number of turns of the loop or number of times the time-changing flux is encircled by the loop. The loop referred to is a closed path. If the closed path is also a conducting path, then current will flow in the direction of the induced emf. For a filamentary conducting loop, the current is given by[1]

$$I = \frac{-N\,d\Psi_m/dt}{R} \qquad \text{(A)} \tag{6}$$

where R is the resistance of the loop, and Ψ_m is the net flux linking the loop. The net flux is the sum of the applied flux and the flux resulting from the current flow, thus

$$\Psi_m = \Psi_{m\text{net}} = \Psi_{m_a} + \Psi_{m_I} \qquad \text{(Wb)} \tag{7}$$

where Ψ_{m_a} is the flux linking the loop in the absence of any current flow, and Ψ_{m_I} is the flux due to the induced current.

In Fig. 10-1(c), a time-changing magnetic flux is linking the closed conducting loop and is increasing in the positive z direction. The induced emf is in a clockwise direction when viewed from the positive z axis looking toward the origin. The flux due to the induced current is in a direction to oppose the increasing flux that induced the current.[2]

In Fig. 10-1(d), the flux is in the same direction as in part (c) but is decreasing. The induced emf is consequently in the counterclockwise direction as indicated. The flux Ψ_{m_I} in Fig. 10-1(c) and (d) is in the direction to oppose the changing applied flux. Note that the minus sign in (5) and (6) is in accordance with the positive sense of circu-

[1] See also (10.9-7).

[2] Lenz's law.

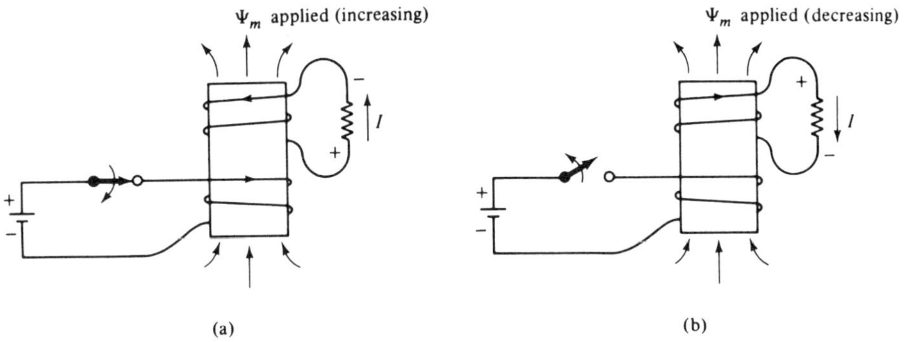

Figure 10-2 Illustration of momentarily induced current in a coupled circuit: (a) closing the switch to the battery; (b) opening the battery switch.

lation in the x-y plane for an increasing Ψ_m in the positive z direction; i.e., a positive response would be a counterclockwise emf in Fig. 10-1(c), whereas it is experimentally observed that the emf is in the clockwise direction.

Figure 10-2 indicates how a momentary applied flux might be generated and indicates the polarity of the induced current for (a) when the switch is closed and (b) when the switch is opened.

In general, Faraday's law manifests itself in either or both a stationary circuit linked by a time-changing flux, such as a transformer or the coupled circuit depicted in Fig. 10-2, or the flux may be stationary, but the circuit is moving relative to the flux in such a way as to produce a time-changing flux enclosed by the circuit. A rotating machine generates an emf by the latter mechanism.

10.3 STATIONARY CIRCUIT—TIME-VARYING FIELD

In Fig. 10-3(a), a path ℓ is linked by a time-changing magnetic flux. The positive senses of $\overline{B}$, $\overline{d\ell}$, $\overline{ds}$, and of the emf are shown by arrows. The time relationship between a sinusoidal $\overline{B}$ and the emf is shown in Fig. 10-3(b). When the rate of change of $\overline{B}$ is maximum, as found at $t = 0$, the generated emf has its maximum negative value. This means that the actual direction of the emf at $t = 0$ is clockwise as viewed from the positive z axis toward the origin. Applying (10.2-5) and finding Ψ_m by integrating over the area enclosed by the loop,

$$\boxed{\text{emf} = -\frac{d}{dt}\left(\int_s \overline{B} \bullet \overline{ds}\right) = -\int_s \frac{\partial \overline{B}}{\partial t} \bullet \overline{ds} \qquad \text{(V)}} \qquad (1)$$

where the partial derivative is required under the integral sign because, in general, $\overline{B} = \overline{B}(x, y, z, t)$ and the circuit is stationary.

For example, if $\overline{B} = \hat{z}B_0 \sin \omega t$ and the loop is in the x-y plane, then

$$\text{emf} = -\omega \cos \omega t \int_y \int_x B_0 \, dx\, dy = -V_0 \cos \omega t \qquad \text{(V)}$$

where $V_0 = \omega B_0 s$.

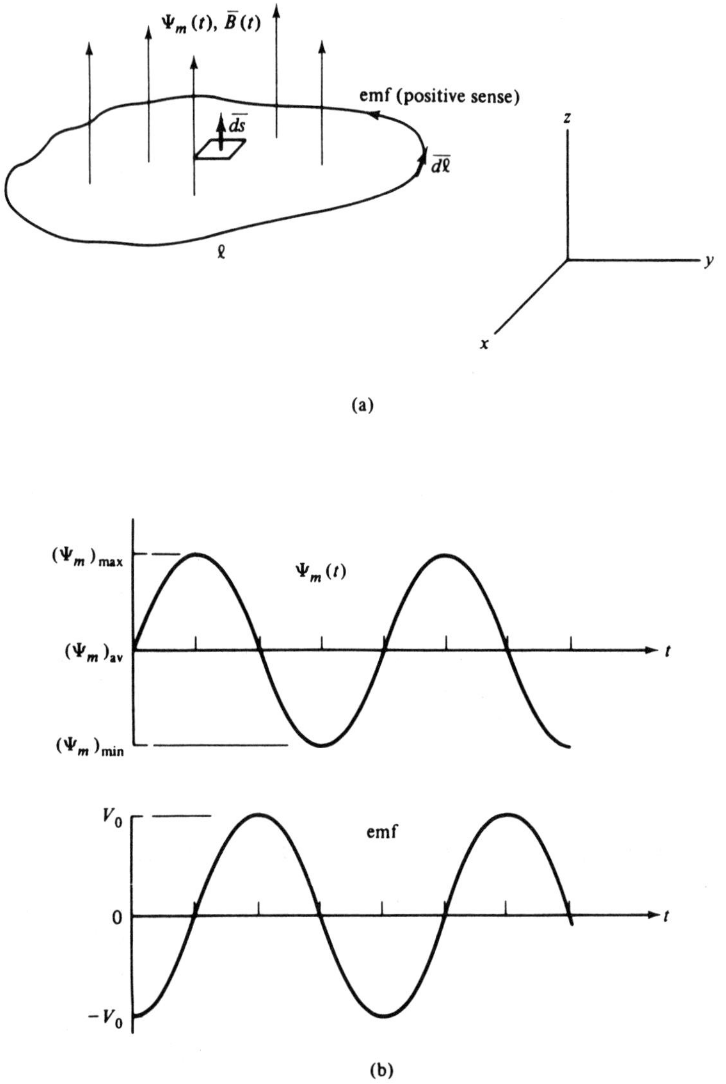

Figure 10-3 (a) Time-changing magnetic field linking a closed loop. (b) Time relationship between the time-changing magnetic field and the emf induced in the loop.

The emf is equal to the line integral of the emf producing $\overline{E}$ field around the closed loop,

$$\boxed{\text{emf} = \oint_\ell \overline{E}_e \cdot \overline{d\ell} = \oint_\ell \overline{E} \cdot \overline{d\ell} \qquad \text{(V)}} \tag{2}$$

Equating (1) and (2) yields

$$\oint_\ell \overline{E} \cdot \overline{d\ell} = -\int_s \frac{\partial \overline{B}}{\partial t} \cdot \overline{ds} \tag{3}$$

Applying Stokes' theorem to (3),

$$\int_s \nabla \times \overline{E} \bullet \overline{ds} = \oint_\ell \overline{E} \bullet \overline{d\ell} = -\int_s \frac{\partial \overline{B}}{\partial t} \bullet \overline{ds} \qquad (4)$$

and equating the integrands of the first and last terms of (4),

$$\boxed{\nabla \times \overline{E} = -\frac{\partial \overline{B}}{\partial t} \qquad (\text{V m}^{-2})} \qquad (5)$$

Equation (5) indicates that the curl of $\overline{E}$ is not, in general, equal to zero for time-varying fields, but (5) reduces to $\nabla \times \overline{E} = 0$ for the static field.

Example 1

Referring to Fig. 10-3(a), a circular loop described by the equation $x^2 + y^2 = 16$ is located in the x-y plane centered at the origin. The $\overline{B}$ field is described by

$$\overline{B} = \hat{z}2\sqrt{x^2 + y^2} \cos \omega t \qquad (\text{T})$$

Find the total emf induced in the loop.

Solution. From (1),

$$\text{emf} = -\int_y \int_x -2\omega \sin \omega t \sqrt{x^2 + y^2}\, \hat{z} \bullet \hat{z}\, dx\, dy$$

$$= 2\omega \sin \omega t \iint \sqrt{x^2 + y^2}\, dx\, dy$$

Changing to cylindrical coordinates,

$$\text{emf} = 2\omega \sin \omega t \int_0^{2\pi} \int_0^4 r_c\, dr_c r_c\, d\phi$$

$$= 2 \sin \omega t (2\pi) \frac{4^3}{3}$$

$$= \frac{4\omega\pi \times 64}{3} \sin \omega t$$

$$= 268\omega \sin \omega t \qquad (\text{V})$$

Problem 10.3-1 In Fig. 10-4 assume that the magnetic flux in the core is $3.75 \times 10^{-3} \sin 377t$ (Wb). If $R_1 = 4$ kΩ and $R_2 = 1$ kΩ, find the rms voltages V_1 and V_2 read by ideal voltmeters.

Problem 10.3-2 If the number of turns on the V_p side of the core in Fig. 10-4 is 500 and $V_p = 100 \cos 377t$ (V), find the peak-to-peak amplitude of Ψ_m.

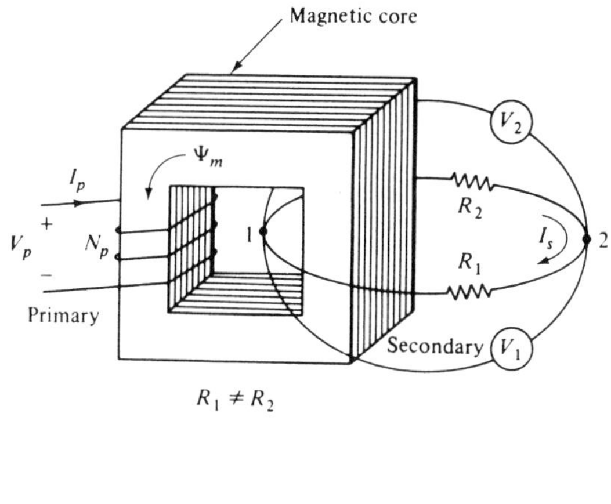

Figure 10-4 Illustration for Probs. 10.3-1, 10.3-2, and 10.3-3.

Problem 10.3-3 If in Fig. 10-4, the flux density B in the core is given by $B = 1.5 \cos 2513t$ (T), what core cross-sectional area would be required to develop 120 V_{rms} across a 1000-turn winding N_p?

10.4 MOVING CIRCUIT—STATIC FIELD

A basic alternator consists of a single rotating loop in a static magnetic field. In Fig. 10-5, the $\overline{B}$ field is in the positive z direction, and the loop is rotating about the y axis at ω rad/sec. At the position shown, the unit vector $\hat{n}$ normal to the plane of the loop is at an angle ωt with respect to the $\overline{B}$ field. If positive flux is for $\overline{B} \cdot \hat{n}$ positive and if positive emf is as indicated, the emf is increasing in the positive direction at the position of the loop shown. We can show this by applying Faraday's law,

$$\text{emf} = -\frac{d\Psi_m}{dt} = -\frac{d}{dt}\int_s \overline{B} \cdot \overline{ds} \quad \text{(V)} \tag{1}$$

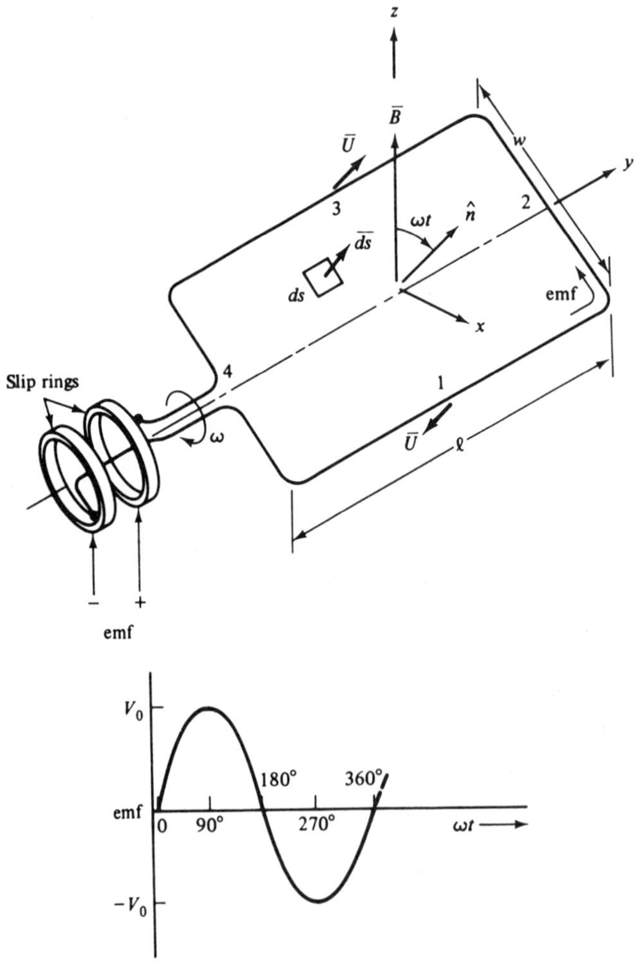

Figure 10-5 Generation of an emf by a loop rotating in a static magnetic field.

and if $\overline{B}$ is a uniform field,

$$\text{emf} = -\frac{d}{dt}\overline{B} \bullet \hat{n}s = -\frac{d}{dt}|\overline{B}|s \cos \omega t$$

or

$$V_0 \sin \omega t = \omega B s \sin \omega t \tag{2}$$

where s is the total area enclosed by the loop.

If $\overline{B}$ is not a uniform field, then $\overline{B} \bullet \overline{ds}$ must be integrated over the area enclosed by the loop.

Another example of an emf being produced by motion of a conductor has already been illustrated in Fig. 9-7. An apparent electric field $\overline{U} \times \overline{B}$ is produced along the length of a conductor moving through a magnetic field $\overline{B}$ with a velocity $\overline{U}$. This is referred to as a *motional emf field* $\overline{E}_m$*:*

$$\boxed{\overline{E}_m = \overline{U} \times \overline{B} \qquad (\text{V m}^{-1})} \tag{3}$$

The total emf in the circuit of Fig. 9-7 is

$$IR = V_{ba} = \text{emf} = \int_{\substack{\text{length of bar}\\ \text{between rails}}} \overline{U} \times \overline{B} \bullet \overline{d\ell} \tag{4}$$

where $\overline{U} \times \overline{B}$ is directed out of the paper, making point a positive with respect to point b.

It is also possible to apply (4) to each side of the rotating loop and add the emf's to get the same result given by (2) (see Example 2).

If the $\overline{B}$ field is also time-varying, then we have emf due to both transformer coupling $\partial\overline{B}/\partial t$ as well as due to the motional $\overline{E}_m$ field, $\overline{U} \times \overline{B}$. Then the total emf is given by

$$(\text{emf})_{\text{total}} = -\int_s \frac{\partial \overline{B}}{\partial t} \bullet \overline{ds} + \int_\ell \overline{U} \times \overline{B} \bullet \overline{d\ell} \qquad (\text{V}) \tag{5}$$

Example 2

Referring to Fig. 10-5, let the $\overline{B}$ field be directed along the positive z axis, and let the axis of rotation be the y axis. Sides 1 and 3 are ℓ in length, and sides 2 and 4 are of length w. Use (4) to show that the emf is the same as given in (2).

Solution

$$\begin{aligned}(\text{emf})_1 &= \int_0^\ell \overline{U} \times \overline{B} \bullet \overline{d\ell} = [(-\hat{z}U \cos \omega t - \hat{x}U \sin \omega t) \times B\hat{z}] \bullet \hat{y}\ell \\ &= [(-\hat{x}U \sin \omega t) \times \hat{z}B] \bullet \hat{y}\ell \quad \text{since } \hat{z} \times \hat{z} = 0 \\ &= \hat{y} \bullet \hat{y}UB\ell \sin \omega t = UB\ell \sin \omega t \\ &= \omega \frac{w}{2} B\ell \sin \omega t\end{aligned}$$

$$(\text{emf})_2 = \int_0^w \overline{U} \times \overline{B} \bullet \overline{d\ell} = 0 \quad \text{since } \overline{U} \times \overline{B} \text{ is perpendicular to } \overline{d\ell} \text{ along side 2}$$

$$(\text{emf})_3 = \int_\ell^0 \overline{U} \times \overline{B} \cdot \overline{d\ell} = [(\hat{z}U \cos \omega t + \hat{x}U \sin \omega t) \times B\hat{z}] \cdot (-\hat{y}\ell)$$
$$= [\hat{x}U \sin \omega t \times \hat{z}B] \cdot (-\hat{y}\ell)$$
$$= \hat{y} \cdot \hat{y}UB\ell \sin \omega t = UB\ell \sin \omega t$$
$$= \omega \frac{w}{2} B\ell \sin \omega t$$

$(\text{emf})_4 = 0$ for the same reason as side 2

Then the total emf around the rotating loop is obtained by summing the emf's from all the sides:

$$\text{total emf} = 2\left(\omega \frac{w}{2} B\ell \sin \omega t\right)$$

or $V_0 \sin \omega t = \omega Bs \sin \omega t$, as indicated by (2).

Problem 10.4-1 Referring to Fig. 9-7(a), determine: (a) how the force required to move the bar varies with the resistance of the total loop; (b) how the power dissipated in R varies with the value of R for a given velocity of the bar. Assume that B is not changed significantly by the current in the loop.

Problem 10.4-2 Referring to Fig. 9-7(a), let the length of the bar between the conducting rails be ℓ, and let the velocity be U_0. Find the emf if $|\overline{B}| = B_0 \sin \omega t$. Assume that R is a very high resistance such that the flux from the current is negligible.

Problem 10.4-3 A circular loop of 10 cm diameter is spinning in air in a static magnetic field. When the axis of rotation is adjusted for maximum voltage from the loop's slip rings, the rms value of the voltage is 1 mV. (a) Find the flux density B of the magnetic field if the loop is spinning at 10,000 rpm. (b) Find H.

10.5 BASIC TRANSFORMERS

A typical transformer illustrated in Fig. 10-6 consists of a pair of windings wound about a common core. If a current I_1 is supplied to N_1 and if the N_2 terminals are open-circuited, then the alternating core flux is given by the Ohm's law of magnetic circuits,[3]

$$\Psi_m = \frac{N_1 I_1}{\mathcal{R}} = \frac{N_1 I_1}{\ell/\mu_0\mu_\Delta s} = \frac{\mu_0\mu_\Delta N_1 I_1 s}{\ell} \quad \text{(Wb)} \tag{1}$$

where $\mathcal{R}$ is the total reluctance of the core of mean length ℓ as defined in Sec. 9.9, and s is the cross-sectional area of the core. Applying Faraday's law, but ignoring the minus sign,[4]

$$V_2 = N_2 \frac{d\Psi_m}{dt} \quad \text{(V)} \tag{2}$$

Substituting (1) into (2) gives

$$V_2 = \frac{\mu_0\mu_\Delta N_1 N_2 s}{\ell}\left(\frac{dI_1}{dt}\right) = M\frac{dI_1}{dt} \quad \text{(V)} \tag{3}$$

[3] Actually, $\mathcal{R}$ is $N\Delta I/\Delta\Psi$ and $\mu_0\mu_\Delta = \Delta B/\Delta H$ as compared to $\mu_0\mu_r = B/H$ for non-time-varying fields. Because of the non-linear nature of ferromagnetic materials, $\mu_\Delta \neq \mu_r$ in general.

[4] Changing the polarity of the output voltage is accomplished by merely reversing the leads.

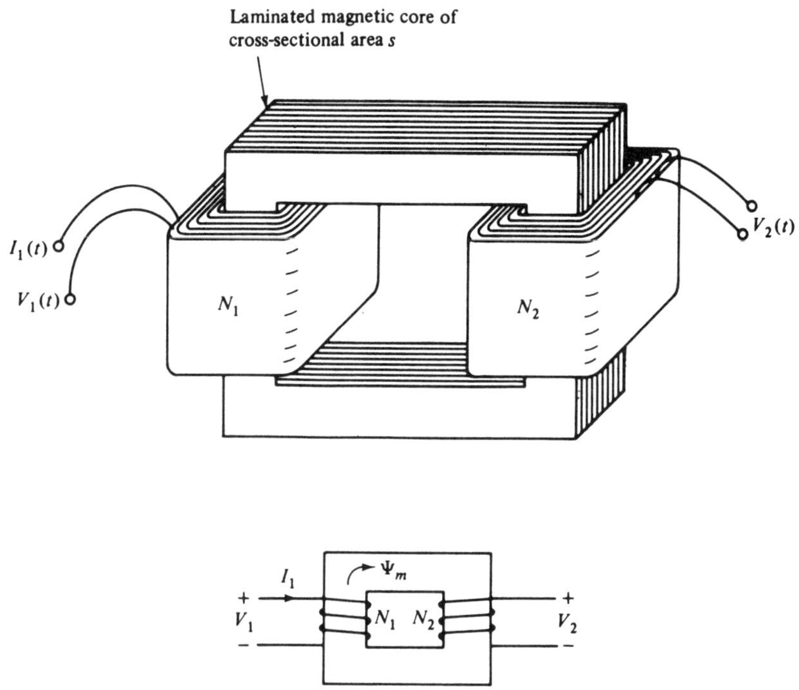

Figure 10-6 Basic transformer.

where M is the mutual inductance as defined in Sec. 9.10. When we are dealing with sinusoidal functions only, then

$$\boxed{|V_2|_{\text{open circuit}} = \omega M |I_1| \qquad \text{(V)}} \tag{4}$$

The current I_1 in (4) is the *magnetizing current* and is that current necessary to produce the required flux Ψ_m. When a load is placed on the output or secondary side of the transformer, I_1 then consists of both magnetizing current *and* transformed load current, as explained below. If we neglect the winding resistance and assume that all flux that links N_1 also links N_2, and that this flux is the magnetizing flux Ψ_m,

$$|V_1| = \omega N_1 |\Psi_m| = \omega N_1 s |B|$$

and

$$|V_2| = \omega N_2 |\Psi_m| = \omega N_2 s |B|$$

or

$$\boxed{\frac{|V_1|}{|V_2|} = \frac{N_1}{N_2}} \tag{5a}$$

In addition, since the flux Ψ_m must remain at the same magnitude for a constant magnitude of sinusoidal applied voltage, the load ampere-turns $N_2 I_2$ must be canceled by additional primary ampere-turns $N_1 I_1$. Thus, the total primary current is

$$I_1 = I_{1_{mag}} + \left(\frac{N_2}{N_1}\right) I_2 \tag{5b}$$

Equations (5a) and (5b) differ from those for an ideal transformer only by the the $I_{1_{mag}}$ term. The magnitude of this magnetizing current is related to the width of the hysteresis loop (see Fig. 9-16). The wider the loop, the more magnetizing current required to "magnetize" the core. It should be emphasized that Ψ_m in the core is dependent only on the voltage across the winding and is not a function of load current I_2. For a given transformer and given voltage, $I_{1_{mag}}$ and Ψ_m are constant amplitude even though the load is variable.

A practical power transformer has a core of some variety of steel with a saturation flux density B_{sat}. Since μ_r decreases very rapidly as saturation is approached, it is not practical to operate at B_{sat} or greater, and the maximum voltage developed across a winding is

$$|V|_{max} = \omega N s B_{max} \qquad \text{(V)} \tag{6}$$

where B_{max} is somewhat less than B_{sat}.

When voltage is given as an rms value,

$$\boxed{|V_{rms}|_{max} = \frac{2\pi f N s B_{max}}{\sqrt{2}} = 4.44 f N s B_{max} \qquad \text{(V)}} \tag{7}$$

Equation (7) shows the desirability of using as high a frequency as possible to reduce the required core cross section or number of turns (400 Hz is widely used instead of 60 Hz for military applications to reduce the weight of transformers and motors).

While Faraday's law is the mechanism by which voltage is developed in a winding of a transformer, it is also the cause of eddy current loss in the transformer core. As illustrated in Fig. 10-7, the time-changing magnetic field through the core produces a circulating or eddy current around the magnetic flux lines. The eddy current produces a

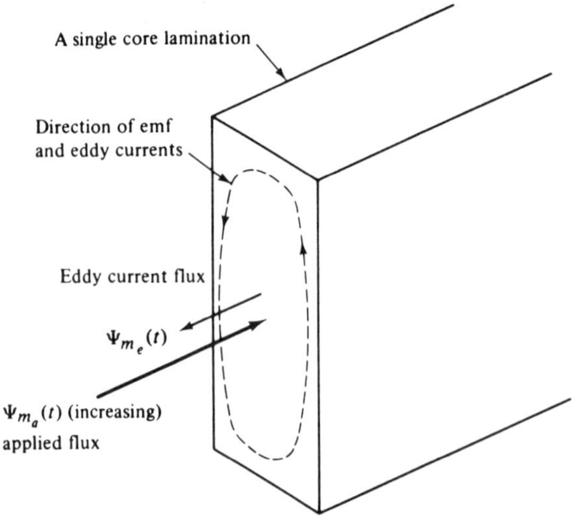

Figure 10-7 Graphical display of the effect of eddy currents in reducing the total flux through a core lamination.

time-changing flux in opposition to the applied flux, thereby reducing the total time-changing flux through the core and thus reducing the effective cross-sectional area of the core. There is also an I^2R loss from eddy current that is part of the total core loss.

Eddy current is reduced in practical transformer design by using a high resistivity core material, or by laminating the material and insulating the laminations from each other, or by both techniques. Since eddy current is induced by the time-changing field, it is proportional to frequency. For tens of kilohertz and higher, ferrite cores are used because of the extremely high resistivity of the material. From 60 Hz to 20 kHz or so, various ferromagnetic metal alloys are used in laminated or tape-wound cores. The lamination thickness used is closely related to skin depth, a topic covered in Chapter 11. Ferrite is used at lower frequencies, but it is not attractive for low frequency, high voltage applications because of its relatively low B_{sat}, thus tending to make the transformer rather bulky to get the necessary turns and cross-sectional area.

Problem 10.5-1 An available transformer core has a cross-sectional area of 5 cm^2 and a saturation flux density of 1.8 T. Find the number of turns required for a 120-V rms winding at 60 Hz.

Problem 10.5-2 A given transformer has been tested and its mutual inductance was found to be 5 H. Assuming no leakage flux, i.e., Ψ_m links both N_1 and N_2, find the total primary current if $V_1 = 120$ V, 60 Hz; $N_1 = 600$ turns; $N_2 = 50$ turns; and the load across N_2 is 8 Ω. Assume sinusoidal voltages and currents.

Problem 10.5-3 A particular transformer has a primary voltage rating of 120 V, 400 Hz. Assuming that the transformer was designed specifically for 400 Hz, what would be the maximum safe primary voltage at 60 Hz? What would happen if this 400-Hz transformer were connected to 120 V, 60 Hz?

10.6 DISPLACEMENT CURRENT

By definition, the divergence of current density at a point is the net outward flow of current density per unit volume. This must be numerically equal to the rate of decrease of charge density within the region under consideration. That is, from (5.2-11),

$$\boxed{\nabla \bullet \overline{J} = -\frac{\partial \rho_v}{\partial t} \qquad (\text{A m}^{-3})} \tag{1}$$

and from Maxwell's equations (Sec. 8.8), $\overline{J} = \nabla \times \overline{H}$ or

$$\nabla \bullet \nabla \times \overline{H} = -\frac{\partial \rho_v}{\partial t} = -\frac{\partial}{\partial t}\nabla \bullet \overline{D} = -\nabla \bullet \frac{\partial \overline{D}}{\partial t} \qquad (\text{A m}^{-3}) \tag{2}$$

But the divergence of the curl is identically zero; therefore, there must be a term missing from the right-hand side of (2) for $\partial \rho_v / \partial t = 0$ cannot be accepted as generally true. Then adding the missing term to (2),

$$\nabla \bullet \nabla \times \overline{H} = 0 = -\frac{\partial \rho_v}{\partial t} + \frac{\partial \rho_v}{\partial t}$$

and substituting $\nabla \bullet \overline{J}$ for $-\partial \rho_v / \partial t$, and $\nabla \bullet \overline{D}$ for ρ_v in the added term,

$$\nabla \bullet \nabla \times \overline{H} = \nabla \bullet \overline{J} + \frac{\partial}{\partial t}\nabla \bullet \overline{D}$$

or

$$\nabla \times \overline{H} = \overline{J} + \frac{\partial \overline{D}}{\partial t} \qquad (\text{A m}^{-2}) \tag{3}$$

Admittedly, the derivation of (3) is somewhat of a slight-of-hand trick, but the result justifies the means, for without the $\partial \overline{D}/\partial t$ term there would be no electromagnetic wave propagation. But let us examine (3) carefully. First of all, we are already familiar with current density $\overline{J}$. Let us now refer to it as conduction current density $\overline{J}_c$ and refer to $\partial \overline{D}/\partial t$ as displacement current density whose existence was postulated by Maxwell. Then,

$$\nabla \times \overline{H} = \overline{J}_c + \frac{\partial \overline{D}}{\partial t} = \overline{J}_c + \overline{J}_d \qquad (\text{A m}^{-2}) \tag{4}$$

A simple example of displacement current is the "current" through a capacitor. We have become accustomed to thinking of the current in a series *RC* circuit as being the same through the resistor or capacitor. But suppose the dielectric material between the plates of the capacitor is perfectly non-conducting; suppose it is air, for example. How can current flow through such a capacitor? The answer is that current in the conventional sense does not flow through a perfect insulator. But if an ammeter is placed in series with an *RC* circuit to which an alternating source of voltage is applied, current is measured. To resolve this dilemma, refer to Fig. 10-8.

Let I_c be the conduction current in the wires and I_d the current through the capacitor. If Q is the magnitude of the charge on each plate of area s, then in the wire leads attached to the capacitor the current is given by

$$I_c = \frac{dQ}{dt} = s\frac{d\rho_s}{dt} = s\frac{dD}{dt}$$

$$= \epsilon s\frac{dE}{dt} = \frac{\epsilon s}{d}\left(\frac{dV}{dt}\right)$$

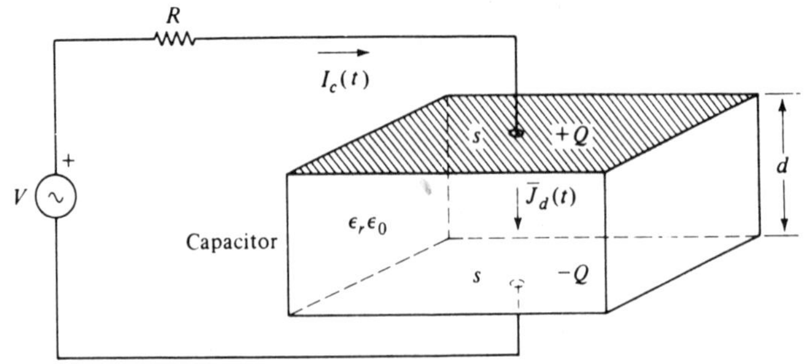

Figure 10-8 Graphical display of displacement current density through a capacitor.

or

$$\boxed{I_{c_{\text{wires}}} = C\frac{dV}{dt} \qquad \text{(A)}} \tag{5}$$

since $C = \epsilon s/d$. But from circuit theory, the capacitor current is given by

$$\boxed{I_d = C\frac{dV}{dt} \qquad \text{(A)}} \tag{6}$$

We see then that

$$I_c = I_d$$

or

$$(J_c)\,(\text{conductor cross section}) = (J_d)\,(\text{area of plates})$$

$$J_c s_{\text{conductor}} = s_{\text{plates}}\frac{dD}{dt}$$

Thus, the conduction current I_c through the wires and the displacement current I_d through the capacitor are numerically equal in a series circuit, but they are conceptually quite different. Between the plates of a perfect capacitor there is no flow of charges but merely an oscillation of the electric field. Even the concept of bound charges is not necessary to explain the existence of displacement current, for a capacitor could have a vacuum between the plates and still have a displacement current.

Example 3

A coaxial cable has a dielectric with an ϵ_r of 4. The inner conductor has a radius of 1.0 mm, and the inside radius of the outer conductor is 5.0 mm. Find the displacement current between the two conductors per meter length of the cable for an applied voltage, $V = 100\cos(12\pi \times 10^6 t)$ (V).

Solution. From (5.10-12), for a cable length L,

$$C = \frac{0.2\pi\epsilon \text{L}}{\ln(r_b/r_a)}$$

or

$$\frac{C}{\text{L}} = \frac{(2\pi)(4)(8.854 \times 10^{-12})}{\ln 5}$$

$$= 1.3826 \times 10^{-10}\ \text{F m}^{-1}$$

$$\frac{I_d}{\text{L}} = \frac{C}{\text{L}}\left(\frac{dV}{dt}\right) = -(1.38 \times 10^{-10})(12\pi \times 10^6)100\sin(12\pi \times 10^6 t)$$

$$= -0.520\sin(12\pi \times 10^6 t) \qquad (\text{A m}^{-1})$$

From Faraday's law and from the concept of displacement current density, we have modified the two curl equations of the Maxwell equation set. Letting (10.3-5) and

(3) replace the static curl equations, Maxwell's equations at a point become

$$
\begin{aligned}
\nabla \times \overline{E} &= -\frac{\partial \overline{B}}{\partial t} \\
\nabla \times \overline{H} &= \overline{J}_c + \frac{\partial \overline{D}}{\partial t} \\
\nabla \bullet \overline{D} &= \rho_v \\
\nabla \bullet \overline{B} &= 0
\end{aligned}
\qquad (7)
$$

We shall frequently refer to these equations.

Problem 10.6-1 Find the rms value of the displacement current density in Example 3, at $r_c = 1$ mm and at $r_c = 5$ mm.

Problem 10.6-2 (a) Using the free space version of (7), i.e., letting $\epsilon = \epsilon_0$, $\mu = \mu_0$, $J_c = 0$, and $\rho_v = 0$, show that if $\overline{E}$ is given by $\overline{E} = \hat{x}E_0 \cos(\omega t - \beta z)$ (V m^{-1}), where $\beta = \omega\sqrt{\mu_0\epsilon_0}$, then $|\overline{E}|/|\overline{H}| = E_x/H_y = \sqrt{\mu_0/\epsilon_0}$. (b) Now find the numerical value of $\sqrt{\mu_0/\epsilon_0}$ and its units. (c) Give the expression for $\overline{H}$ that is derived from the above. [*Hint:* Of the set of equations in (7), use only the curl $\overline{E}$ equation.]

Problem 10.6-3 Using the E_x function given in Prob. 10.6-2, and the expression for β also given in that problem, find the value and units of ω/β. Does this number look familiar?

10.7 LOSSY DIELECTRICS

We need to pause briefly at this point to emphasize the phasor notation that shall be used throughout much of the remainder of this text when dealing with the steady alternating state. Phasors will be written as script letters to distinguish them from time functions. For example, the phasor $\mathcal{V}$ is defined by

$$
V = V(t) = V_0 \cos(\omega t + \theta) = R_e[\mathcal{V} e^{j\omega t}] \qquad \text{(V)} \qquad (1)
$$

where V_0 is the amplitude of the time function, and where $\mathcal{V}$ is, in general, complex; i.e.,

$$
\mathcal{V} = |\mathcal{V}| \angle\theta = |\mathcal{V}| e^{j\theta}
$$

and, from (1),

$$
|\mathcal{V}| = V_0 \qquad (2)
$$

We recall from earlier courses that a j operator in front of a phasor advances the phase by 90°, that differentiation with respect to time is accomplished by simply multiplying by $j\omega$ (integration by dividing by $j\omega$), and that the $e^{j\omega t}$ term need not be explicitly shown, for it will be common to terms on both sides of any such equation. Using phasor notation, we express Maxwell's curl equations as

$$\boxed{\nabla \times \overline{\mathscr{E}} = -j\omega\overline{\mathscr{B}} = -j\omega\mu\overline{\mathscr{H}} \qquad (\text{V m}^{-2})} \tag{3}$$

and

$$\boxed{\nabla \times \overline{\mathscr{H}} = \overline{\mathscr{J}}_c + j\omega\overline{\mathscr{D}} = \sigma\overline{\mathscr{E}} + j\omega\epsilon'\overline{\mathscr{E}} \qquad (\text{A m}^{-2})} \tag{4}$$

where $\overline{\mathscr{E}}$, $\overline{\mathscr{H}}$, $\overline{\mathscr{B}}$, etc. are vector phasors.

The two terms of the right-hand side of (4) are the conduction current density and displacement current density. If we factor out $\mathscr{E}$, we have

$$\boxed{\nabla \times \overline{\mathscr{H}} = (\sigma + j\omega\epsilon')\overline{\mathscr{E}} = j\omega\epsilon\overline{\mathscr{E}}} \tag{5}$$

where we have now introduced a complex permittivity ϵ defined by

$$\boxed{\epsilon = \epsilon'\left(1 - j\frac{\sigma}{\omega\epsilon'}\right) = \epsilon' - j\epsilon''} \tag{6}$$

where $\sigma/\omega = \epsilon''$, or

$$\epsilon_0\epsilon_r = \epsilon_0(\epsilon_r' - j\epsilon_r'') \qquad (\text{F m}^{-1})$$

From (4), the ratio $\sigma/\omega\epsilon'$ is the ratio of the magnitude of the conduction current density to the magnitude of the displacement current density and is called the *loss tangent*. A lossless capacitor has a loss tangent of zero. When the lossy capacitor is replaced by its equivalent resistance in parallel with a perfect capacitor, the loss tangent is the ratio of the current through the resistor to that through the capacitor. Since the resistance and capacitance have common boundaries, i.e., the plates of the lossy capacitor,

$$I_{\text{cond}} = GV = \frac{V}{R}, \qquad I_{\text{displ}} = j\omega CV$$

and

$$\boxed{\frac{|I_{\text{cond}}|}{|I_{\text{displ}}|} = \frac{1/R}{\omega C} = \frac{\sigma}{\omega\epsilon'} = \frac{\epsilon''}{\epsilon'} = \text{loss tangent}} \tag{7}$$

When the loss tangent is very small, it is essentially the same as the power factor of the lossy capacitor.

$$\underset{(1/R \ll \omega C)}{\text{PF}} = \frac{1/R}{\sqrt{(1/R)^2 + \omega^2 C^2}} \approx \frac{1}{\omega RC} = \frac{\epsilon''}{\epsilon'} \tag{8}$$

The effective conductivity of the material may also have a frequency dependence because of dielectric hysteresis loss, but a discussion of that phenomenon is appropriate to a more advanced course.

Example 4

Consider a parallel-plate capacitor having a plate area of 1.0 cm^2 each and where the plates are separated by a distance of 0.1 mm by a dielectric having the following properties at 1 GHz:

$$\epsilon_r' = 2, \qquad \sigma = 10^{-7}\ \text{S m}^{-1}$$

Find the equivalent circuit for this capacitor and calculate the conduction current, displacement current, and the loss tangent if 1 V at 1 GHz is applied across the capacitor.

Solution. The capacitance is

$$C = \frac{\epsilon_0 \epsilon_r' s}{d} = \frac{(8.854 \times 10^{-12})(2)(10^{-4})}{10^{-4}} = 17.7\ \text{pF}$$

The resistance between plates is

$$R = \frac{d}{\sigma s} = \frac{10^{-4}}{(10^{-7})(10^{-4})} = 10^7\ \Omega$$

Thus, the equivalent circuit is a 17.7-pF capacitor in parallel with a 10-MΩ resistor. For 1 V at 1 GHz,

$$I_d = C\frac{dV}{dt}$$

or

$$\mathcal{I}_d = j\omega C \mathcal{V} = j2\pi(10^9)(17.7 \times 10^{-12})(1) = j0.11\ \text{A}$$

and

$$\mathcal{I}_c = \frac{\mathcal{V}}{R} = G\mathcal{V} = \frac{1}{10^7} = 0.1\ \mu\text{A}$$

The loss tangent is

$$\frac{0.1 \times 10^{-6}}{0.11} \approx 10^{-6}$$

Problem 10.7-1 Draw the equivalent circuit for a lossy capacitor and find the expression for power loss per unit volume in terms of σ and E. For convenience, assume parallel plates of area s and separation d.

Problem 10.7-2 Find the average power loss per unit volume for a dielectric having a dielectric constant of 2.5 and a loss tangent of 0.0005 if the field strength is 1 kV m^{-1} at 500 MHz.

Problem 10.7-3 Find the average power dissipated by a capacitor whose separation between plates is 0.05 mm, with a plate area of 3 cm^2 each, a dielectric constant of 5, and a loss tangent of 0.0004, for an applied voltage of 120 V_{rms} at 20 kHz.

10.8 SCALAR ELECTRIC AND VECTOR MAGNETIC POTENTIALS

It was pointed out in Sec. 8.9 that since $\nabla \cdot \overline{B} = 0$, $\overline{B}$ can be represented as the curl of some vector $\overline{A}$,

$$\boxed{\nabla \times \overline{A} = \overline{B} \qquad (\text{T})} \tag{1}$$

where $\overline{A}$ is called the *vector magnetic potential* for reasons that were discussed in that same section. Substituting (1) into the curl equation for $\overline{E}$, we have

$$\nabla \times \overline{E} = -\frac{\partial}{\partial t}\nabla \times \overline{A} = -\nabla \times \frac{\partial \overline{A}}{\partial t} \qquad (\text{V m}^{-2}) \tag{2}$$

since the order of partial differentiation can be interchanged. We then write (2) as

$$\nabla \times \left(\overline{E} + \frac{\partial \overline{A}}{\partial t}\right) = 0 \tag{3}$$

and since the curl of a gradient is identically zero, we can write

$$\overline{E} + \frac{\partial A}{\partial t} = -\nabla V \tag{4}$$

or

$$\boxed{\overline{E} = -\nabla V - \frac{\partial \overline{A}}{\partial t} \qquad (\text{V m}^{-1})} \tag{5}$$

where V is a scalar potential. If $\partial\overline{A}/\partial t = 0$, then $\overline{E} = -\nabla V$ as for static fields.

Substituting (1) and (5) into the middle two equations of (10.6-7) gives

$$\frac{1}{\mu}\nabla \times (\nabla \times \overline{A}) = \overline{J}_c + \epsilon\frac{\partial}{\partial t}\left(-\nabla V - \frac{\partial \overline{A}}{\partial t}\right)$$

or

$$\frac{1}{\mu}[\nabla(\nabla \bullet \overline{A}) - \nabla^2\overline{A}] = \overline{J}_c - \epsilon\nabla\frac{\partial V}{\partial t} - \epsilon\frac{\partial^2 \overline{A}}{\partial t^2} \tag{6}$$

and

$$\epsilon\nabla \bullet \left(-\nabla V - \frac{\partial \overline{A}}{\partial t}\right) = \rho_v \tag{7}$$

If we let

$$\boxed{\nabla \bullet \overline{A} = -\mu\epsilon\frac{\partial V}{\partial t} \qquad (\text{T})} \tag{8}$$

then (6) becomes

$$\boxed{\nabla^2\overline{A} = -\mu\overline{J}_c + \mu\epsilon\frac{\partial^2 \overline{A}}{\partial t^2} \qquad (\text{Wb m}^{-3})} \tag{9}$$

and (7) becomes

$$\boxed{\nabla^2 V = -\frac{\rho_v}{\epsilon} + \mu\epsilon\frac{\partial^2 V}{\partial t^2} \qquad (\text{V m}^{-2})} \tag{10}$$

by direct substitution of (8) into (6) and (7).

We are able to make the choice (8) for $\nabla \cdot \overline{A}$ simply because up to this point it is required only that $\nabla \times \overline{A} = \overline{B}$, which leaves the choice of $\nabla \cdot \overline{A}$ wide open. Once $\nabla \cdot \overline{A}$ is chosen, the vector $\overline{A}$ is then uniquely determined; i.e., a vector may be specified by its curl and divergence.[5] The question then naturally arises, why this particular choice? The answer is that (8) is a choice that produces symmetry between (9) and (10) and that predicts wave propagation for the case where $\overline{J}_c$ and ρ_v are zero, e.g., free space. *For free space,*

$$\boxed{\nabla^2\overline{A} = \mu_0\epsilon_0\frac{\partial^2\overline{A}}{\partial t^2} \qquad (\text{Wb m}^{-3})} \tag{11}$$

and

$$\boxed{\nabla^2 V = \mu_0\epsilon_0\frac{\partial^2 V}{\partial t^2} \qquad (\text{V m}^{-2})} \tag{12}$$

are waves in $\overline{A}$ and V that propagate at the velocity $1/\sqrt{\mu_0\epsilon_0}$. We shall study plane waves in the next chapter, where we will be more concerned with $\overline{E}$ and $\overline{H}$ rather than $\overline{A}$ and V. However, as we have just seen, $\overline{H}$ can be derived from $\overline{A}$, and $\overline{E}$ from V and $\overline{A}$. If we are to have plane waves with $\overline{E}$ and $\overline{H}$ fields propagating at the same velocity, then the V and $\overline{A}$ potentials must propagate at the same velocity as $\overline{E}$ and $\overline{H}$; thus, the required symmetry between (9) and (10) or (11) and (12). We would not be able to make a good choice for $\nabla \cdot \overline{A}$ at this point without being able to look ahead, unless we were just lucky.

Problem 10.8-1 Find the numerical value of the velocity $1/\sqrt{(\mu_0\epsilon_0)}$. Is this number familiar if the units are in meters per second?

10.9 FIELD EQUATIONS AND CIRCUITS (RLC CIRCUIT)

It is assumed that when we begin a study of time-varying fields we have already become well acquainted with *RLC* circuits. It is our purpose in this section to tie the field concepts we have been discussing to the more familiar circuit concepts. In Fig. 10-9(a) we have a conducting loop with a gap, the loop being linked by a time-changing magnetic field. The gap forms a capacitor. The emf generated by the time-changing field causes a conduction current to flow through the conducting portion of the loop and a displacement current to flow across the gap. Let

$$\overline{B}_a = \text{the applied } \overline{B} \text{ field} = \nabla \times \overline{A}_a$$

and

$$\overline{B}_i = \text{the induced } \overline{B} \text{ field resulting from current flow through the loop} = \nabla \times \overline{A}_i$$

[5] See W. K. H. Panofsky and M. Phillips, *Classical Electricity and Magnetism,* 2nd ed., Addison-Wesley Publishing Company, Inc., Reading, Mass., 1962, pp. 2–5.

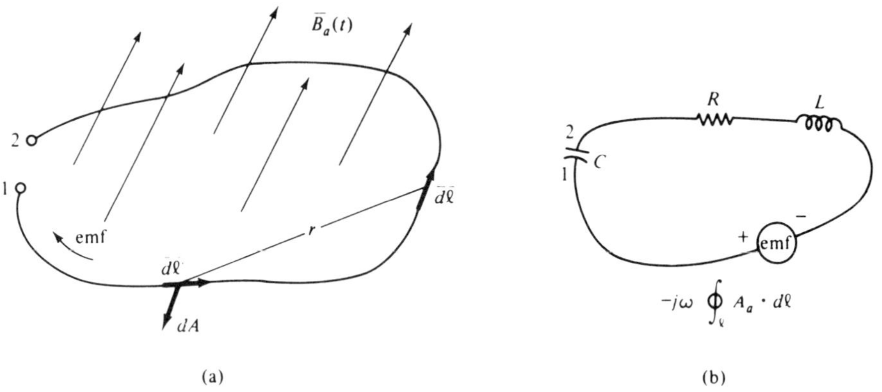

Figure 10-9 Illustration of the equivalence between (a) a loop with a gap being linked by a time-changing magnetic field, and (b) an emf source in series with an *RLC* circuit.

Then for

$$\boxed{\overline{A}_i = \frac{\mu}{4\pi}\int_\ell \frac{I\,\overline{d\ell}}{r} \qquad (\text{Wb m}^{-1})} \tag{1}$$

and, using (10.8-5), the integral of $\overline{E} \cdot \overline{d\ell}$ around the loop is

$$\boxed{\oint_\ell \overline{E} \cdot \overline{d\ell} = -\oint_\ell \nabla V \cdot \overline{d\ell} - \oint_\ell \frac{\partial \overline{A}}{\partial t} \cdot \overline{d\ell} \qquad (\text{V})} \tag{2}$$

$$\begin{aligned} IR + \frac{Q}{C} &= 0 - \oint_\ell \frac{\partial \overline{A}_a}{\partial t} \cdot \overline{d\ell} - \oint_\ell \frac{\partial \overline{A}_i}{\partial t} \cdot \overline{d\ell} \\ &= (\text{emf})_{\text{applied}} - \frac{\mu}{4\pi}\left(\frac{d}{dt}\right)\oint_\ell \left(\oint_{\ell'} \frac{I\,\overline{d\ell'}}{r}\right) \cdot \overline{d\ell} \end{aligned} \tag{3}$$

Let us examine these results, beginning with (2). The left-hand side of the equation is just a straightforward summation of the product of $\overline{E} \cdot \overline{d\ell}$ around the loop. This is simply *IR* for the conducting portion and the voltage across the capacitor for the gap portion of the loop. The first term on the right-hand side of (2) is zero because ∇V is a conservative vector field (the same property as for the ∇V in static fields); therefore, the closed line integral is zero. The second term on the right-hand side of (2) is the net emf,

$$\text{emf} = -\frac{d\Psi_m}{dt} = -\int_s \frac{\partial \overline{B}}{\partial t} \cdot \overline{ds} = -\int_s \frac{\partial}{\partial t}(\nabla \times \overline{A}) \cdot ds$$

and by Stokes' theorem

$$\text{emf} = -\oint_\ell \frac{\partial}{\partial t}\overline{A} \cdot \overline{d\ell} = -\frac{d}{dt}\oint_\ell \overline{A} \cdot \overline{d\ell} \qquad (\text{V})$$

But the net emf consists of the applied emf and the counter emf developed by the current in the loop,

$$\frac{d}{dt}\oint_{\ell}\overline{A}\bullet\overline{d\ell} = \frac{d}{dt}\oint_{\ell}\overline{A}_a\bullet\overline{d\ell} + \frac{d}{dt}\oint_{\ell}\overline{A}_i\bullet\overline{d\ell} \qquad \text{(V)}$$

The vector magnetic potential from the current loop is already given by (1), where I is the induced current. We must then integrate around the same loop twice; therefore, a $\overline{d\ell}'$ is used for one of the integrations. Now, equating (3) to the *RLC* circuit of Fig. 10-9(b), we have, upon rearranging (3),

$$(\text{emf})_{\text{applied}} = IR + \frac{Q}{C} + \frac{\mu}{4\pi}\left(\frac{d}{dt}\right)\oint_{\ell}\left(\oint_{\ell'}\frac{I\,\overline{d\ell}'}{r}\right)\bullet\overline{d\ell} \tag{4}$$

$$= IR + \frac{Q}{C} + L\frac{dI}{dt} \qquad \text{(V)} \tag{5}$$

In the process of equating (4) and (5), we have derived a formula for inductance. This is called *Neumann's formula;* i.e.,

$$L = \frac{\mu}{4\pi}\oint_{\ell}\left(\oint_{\ell'}\frac{\overline{d\ell}'}{r}\right)\bullet\overline{d\ell} \qquad \text{(H)} \tag{6}$$

The use of (6) in finding the inductance of even a single circular loop is a formidable problem, however, and we refer the reader to other sources.[6]

Referring back to eq. (10.2-6), we would consider a *closed* conducting loop as a series *RL* circuit, and

$$I = \frac{-N\dfrac{d\Psi_m}{dt}}{R} = \frac{(\text{emf})_a - L\dfrac{dI}{dt}}{R} \qquad \text{(A)} \tag{7}$$

where the net emf is given by

$$-N\frac{d\Psi_{m_a}}{dt} - L\frac{dI}{dt}$$

which is the same as

$$-N\frac{d\Psi_{m_a}}{dt} - N\frac{d\Psi_{m_I}}{dt}$$

In phasor notation, (7) can be written as

$$\mathcal{I} = \frac{(\text{emf})_a - j\omega L\mathcal{I}}{R}$$

[6] For example, see C. T. A. Johnk, *Engineering Electromagnetic Fields and Waves,* John Wiley & Sons, Inc., New York, 1975, p. 331.

or

$$\mathcal{I} = \frac{(\text{emf})_a}{R + j\omega L} \tag{8}$$

for the current in a closed conducting loop exposed to an ambient time-changing magnetic field. The emf in the loop would be simply $(\text{emf})_a$ if the current were zero, i.e., $(\text{emf})_a$ is the open-circuit voltage.

Problem 10.9-1 The open-circuit voltage induced in a conducting loop was found to be 10 V, and the resistance of the loop was measured to be 1.15 Ω. If the magnetic field is assumed to uniform, what is the inductance of the loop for a frequency of 10 kHz if the closed-loop current is $5\angle -55^\circ$ A?

10.10 BOUNDARY RELATIONS

In Secs. 5.6, 5.9, and 9.8, boundary relations at an interface between two materials were derived for normal and tangential components of $\overline{E}$ and $\overline{D}$ or $\overline{H}$ and $\overline{B}$. In summary, they were

$$\boxed{E_{1t} = E_{2t} \qquad (\text{V m}^{-1})} \tag{1}$$

$$\boxed{D_{1n} - D_{2n} = \rho_s \text{ on the interface} \qquad (\text{C m}^{-2})} \tag{2}$$

$$\boxed{H_{1t} - H_{2t} = J_s, \text{ a surface current density} \qquad (\text{A m}^{-1})} \tag{3}$$

and

$$\boxed{B_{1n} = B_{2n} \qquad (\text{T})} \tag{4}$$

We shall see that all of the above are still true for time-varying fields and, in addition, $B_n = 0$ at a perfectly conducting surface.

Referring to Fig. 10-10, we will integrate $\overline{E} \bullet \overline{d\ell}$ around the loop shown. From (10.6-7),

$$\oint_\ell \overline{E} \bullet \overline{d\ell} = \int_\ell \nabla \times \overline{E} \bullet \overline{ds} = -\int_s \frac{\partial \overline{B}}{\partial t} \bullet \overline{ds}$$

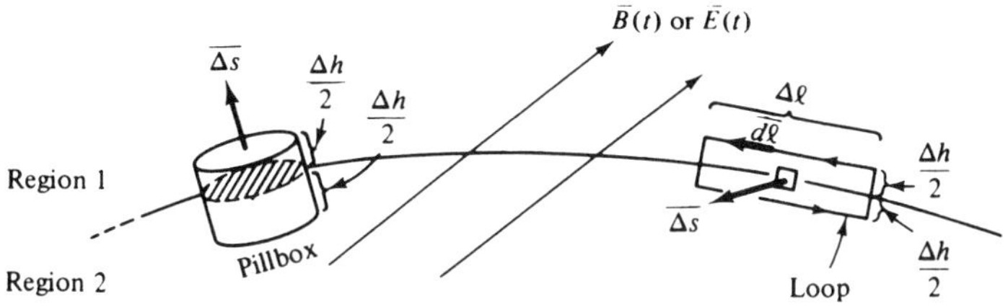

Figure 10-10 Graphical display for obtaining boundary conditions (see also Fig. 9-18).

As $\Delta h \rightarrow 0$, we have

$$E_{2t}\,\Delta\ell - E_{1t}\,\Delta\ell = -\frac{\partial \overline{B}}{\partial t} \bullet \overline{\Delta s} = 0$$

because $\overline{\Delta s} \rightarrow 0$. Therefore,

$$E_{1t} = E_{2t}$$

We next apply Gauss's law for $\overline{D}$ to the pillbox,

$$\oint_s \overline{D} \bullet \overline{ds} = \int_v \nabla \bullet \overline{D}\,dv = \int_v \rho_v\,dv = Q_{\text{en}}$$

and as $\Delta h \rightarrow 0$,

$$D_{1n}\,\Delta s - D_{2n}\,\Delta s = \rho_s\,\Delta s$$

or

$$D_{1n} - D_{2n} = \rho_s$$

Integrating around the loop for $\overline{H} \bullet \overline{d\ell}$, we have

$$\oint_\ell \overline{H} \bullet \overline{d\ell} = \int_s \nabla \times \overline{H} \bullet \overline{ds} = \int_s \overline{J}_c \bullet \overline{ds} + \int_s \frac{\partial \overline{D}}{\partial t} \bullet \overline{ds}$$

and as $\Delta h \rightarrow 0$, $\overline{\Delta s} \rightarrow 0$, and

$$\int_s \frac{\partial \overline{D}}{\partial t} \bullet \overline{ds} \longrightarrow 0,$$

then

$$\oint_{\substack{\ell \\ \Delta h \to 0}} \overline{H} \bullet \overline{d\ell} = (J_{s_{\text{normal to loop}}})\,\Delta\ell$$

Therefore,

$$H_{1t} - H_{2t} = J_s$$

where J_s is a surface current density normal to the loop. This, of course, assumes that a surface current has zero thickness such that the dimension Δh of the loop may go to zero without affecting the magnitude of surface current enclosed. We will consider the imperfect conducting boundary in the next chapter, where the so-called surface current has a thickness equal to the skin depth.

For the normal $\overline{B}$ field, we have

$$\oint_s \overline{B} \bullet \overline{ds} = \int_v \nabla \bullet B\,dv = \text{magnetic charge enclosed} = 0$$

and as $\Delta h \rightarrow 0$ for the pillbox of Fig. 10-10,

$$B_{1n}\,\Delta s - B_{2n}\,\Delta s = 0$$

or

$$B_{1n} = B_{2n}$$

Now suppose region 2 is a perfect conductor. Then within region 2 there will be eddy current generated by dB_n/dt and, since the resistance of the material is zero, infinite eddy current would flow for a non-zero emf. But since eddy current produces a magnetic flux that opposes the time-changing flux that induces it, the inducing flux may be cancelled completely by only a finite value of eddy current. For a perfect conductor, this finite value of eddy current will be just sufficient to cause cancellation of the time-varying normal component of $\overline{B}$; therefore, the boundary condition attained is that the normal component of $\overline{B}$ is zero, the emf induced in the conducting material is zero, and the eddy current is a finite value. Then we add to the previous boundary conditions,

$$\boxed{B_n = 0} \tag{5}$$

at a perfect conducting surface for time-varying fields.

Problem 10.10-1 To illustrate the boundary condition given by (5), let $\overline{B}_a(t)$ be an applied uniform sinusoidal magnetic field normal to a single conducting loop with zero resistance. Use phasor diagrams to show that the magnetic flux from the current in the loop cancels the emf that produces it and that the net time-varying $\overline{B}$ field linking the loop is zero.

Problem 10.10-2 Referring to Fig. 9-23, assume that an alternating voltage is applied to the winding N and that a shorted zero resistance turn encircles the leg ℓ_2 of the core. Describe, using a sketch of the flux in each leg of the core, how the shorted turn affects the flux distribution. Is this an application of (5)?

REVIEW QUESTIONS

1. How can one define an electric field unambiguously? *Sec. 10.2*
2. What is the meaning of the minus sign in *(10.2-5)*?
3. Is a closed conducing path necessary for an emf to be induced in a wire? *Sec. 10.2*
4. For a loop rotating in a magnetic field, what are the important factors determining the magnitude of the emf induced in the loop? *Sec. 10.4*
5. What is meant by "motional emf"? *Sec. 10.4*
6. What is an ideal transformer? *Sec. 10.5*
7. What is eddy current? *Sec. 10.5*
8. What is displacement current? *Sec. 10.6*
9. Define what is meant by a phasor. *Sec. 10.7*
10. What is complex permittivity? Loss tangent? *Sec. 10.7*
11. How does the vector magnetic potential enter into the determination of the electric field? *Sec. 10.8*
12. What is the numerical value of $1/\sqrt{\mu_0\epsilon_0}$?
13. Are there any boundary conditions for time-varying fields that are different from those for static fields? *Sec. 10.10*

PROBLEMS

10-1. For the circuit of Fig. 10-11, if the polarity on the microammeter is as marked, which way will the needle deflect when the switch is closed? When opened?

10-2. A small circular search coil is used to search for stray magnetic fields around a transformer. The coil diameter is 1.5 cm, its length is essentially zero, and its output is connected to an infinite input impedance voltmeter. For a stray field of 1 mT at 60 Hz, how many turns would be required for the search coil to produce a 1-mV input to the voltmeter?

10-3. A step change, occurring over a period of 5 ms, of 20 nT (nanoteslas) in the earth's magnetic field results from a sunspot activity. Assuming that the earth's field is directed 20° from a vertical to the earth's surface, how many volts would be induced in a conducting

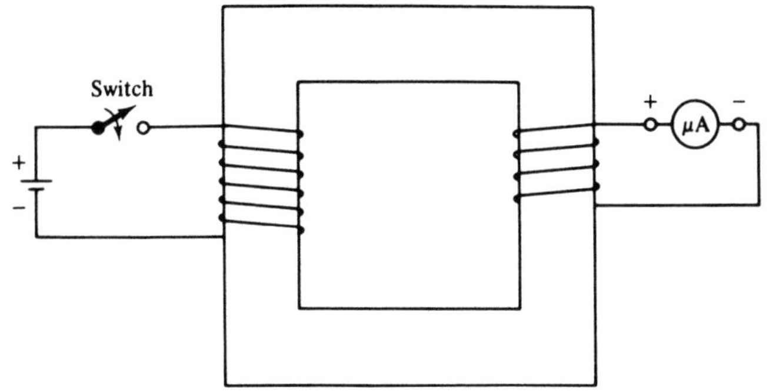

Figure 10-11 Illustration for use with Prob. 10-1.

loop around a surface area that is 300 ft by 50 ft? Neglect the flux produced by current in the loop.

10-4. A filamentary conducting loop whose resistance is 2.0 Ω is situated in a region where a time-changing magnetic field threads through the loop. When the loop is opened with a small gap, a voltage of 10 V is measured across the open ends of the loop. When the loop is closed, the measured current is $3.53 \angle -45°$ A. If an additional 2 Ω is placed in series with the loop, what would be the new value of current? Draw the equivalent circuit.

10-5. Referring to Fig. 10-12, an aluminum sheet has a velocity normal to a magnetic field $\overline{B}$. Find the voltage induced between the edges of the sheet.

10-6. An open-circuited rectangular conducting loop, 5 cm × 10 cm, is spinning in air at 8000 rpm. If the H field normal to the loop axis is 40 A m^{-1}, what is the rms voltage induced in the loop?

10-7. A single-turn rotating loop having an area of 30 cm^2 and a resistance of 20 Ω has its axis normal to a magnetic flux density of 0.5×10^{-4} T. Find the average torque on the loop if the speed of rotation is 5000 rpm.

10-8. A 10-cm-diameter single-turn circular conducting loop is spinning about an axis perpendicular to a magnetic field at a rate of 10,000 rpm. The short circuit current induced in the loop is 100 A rms. If the resistance of the loop is 0.1 Ω, find: (a) the horsepower required to spin the loop; (b) the average torque on the loop. (1 hp = 746 W.)

10-9. A transformer core is constructed of a permalloy having a saturation flux density of 0.75 T. The primary is to be connected to $V = 48 \cos 2000\pi t$ (V). If the primary turns are not to exceed 750, what is the minimum core cross-sectional area required?

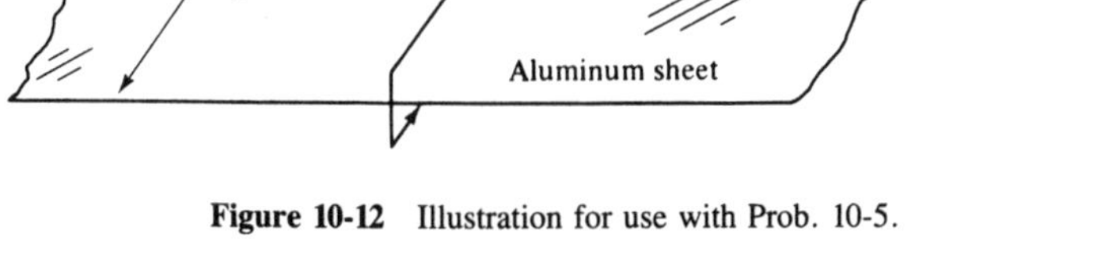

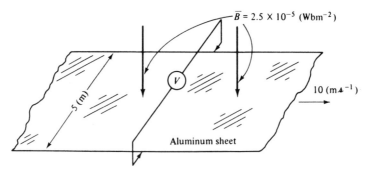

Figure 10-12 Illustration for use with Prob. 10-5.

10-10. A tape-wound toroidal core is to be used as a power transformer at 60 Hz. It is to be operated at a primary voltage of 117 V rms. Two possible choices of cores have cross-sectional areas of 5.7 cm^2 and 7.6 cm^2. Find the number of primary turns required for each core for a B_{max} of 1.7 T.

10-11. For the circuit of Fig. 10-4, find the ideal voltmeter readings and the rms current I_s if $\Psi_m = 4 \times 10^{-3} \sin(2513t)$, $R_1 = 1$ kΩ, and $R_2 = 5$ kΩ.

10-12. A basic transformer is being used as a current transformer; i.e., it is desired that $I_2 = (N_1/N_2)I_1$ to a close approximation. If R_2 = resistance of the secondary circuit, including the resistance of the winding N_2, $\mathcal{R}$ = reluctance of the core, show that

$$\mathcal{I}_2 = \frac{-j\omega N_1 \mathcal{I}_1 N_2}{R_2\mathcal{R}\left(1 + j\dfrac{N_2^2\omega}{R_2\mathcal{R}}\right)}$$

and that

$$|\mathcal{I}_2| \longrightarrow \left(\frac{N_1}{N_2}\right)|\mathcal{I}_1|$$

as $\omega N_2^2/R_2\mathcal{R}$ becomes much larger than unity; i.e., $\omega L_2/R_2 \gg 1$.

10-13. Referring to Fig. 10-13, assume the following:

$$(\mu_\Delta)_{core} = 4000$$
$$I_1 = 5 \sin 377t$$
$$N_1 = 300$$
$$\ell_m = \text{mean core length} = 28 \text{ cm}$$
$$\ell_g = \text{gap} = 1 \text{ mm}$$
$$s_c = \text{cross-sectional area of core} = 4 \text{ cm}^2$$
$$s_\ell = \text{area enclosed by square loop} = 36 \text{ cm}^2$$

Find the rms voltage read by an infinite impedance voltmeter.

10-14. For the data in Prob. 10-13, find the length of air gap required such that the reluctance of the core is only 1% of the total reluctance. Now repeat Prob. 10-13 using this new air gap.

10-15. The open-circuit secondary voltage V_2 for a transformer is 30 V. If $I_1 = 0.004$ A, $f = 60$ Hz, $N_1 = 395$ turns, $N_2 = 215$ turns, find: (a) V_1; (b) M, the mutual inductance; (c) $\mathcal{R}$, the reluctance; (d) L_1.

10-16. A lossy capacitor has a resistance of 20 MΩ. At $f = 20$ MHz, the loss tangent is 0.003. (a) Find the displacement current for an applied voltage of 100 V. (b) Find the power dissipated in the capacitor.

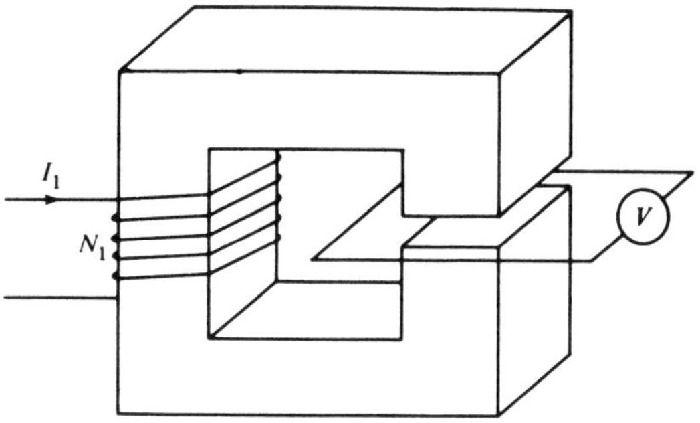

Figure 10-13 Illustration for use with Prob. 10-13.

10-17. A capacitor has a capacitance of 100 pF. If the conductivity of the dielectric is 10^{-6} S m^{-1}, find the resistance of the capacitor if the dielectric constant is 4.

10-18. A resistor, constructed of material having a conductivity of 10^{-6} S m^{-1} and a dielectric constant of 3, has a resistance of 265 kΩ. Find the capacitance of the equivalent capacitor in parallel with the resistor.

10-19. An air dielectric capacitor has a plate area of 1 cm^2 for each plate and a plate separation of 0.1 mm. Find the displacement current for an applied voltage of $100 \sin 3.14 \times 10^4 t$ (V).

10-20. The dielectric constant of a certain capacitor is 6.5, and the loss tangent is 0.001 for operating at 5 MHz. Find the power loss in watts per cubic meter for operating at 1000 V rms across the capacitor if the plates are separated by 0.005 m.

10-21. A transformer with a single flux path such as shown in Fig. 10-4 has a shorted turn on the secondary. Assuming no other load on the secondary, show that the effective inductance as seen from the primary side is now

$$L = \frac{N_p \Psi_m}{\mathcal{I}_p} = \frac{N_p^2}{\mathcal{R}\left(1 + j\dfrac{\omega}{\mathcal{R}R}\right)}$$

where $\mathcal{R}$ is the core reluctance and R the resistance of the shorted turn.

10-22. A zero-resistance closed wire ring with a radius of 0.5 cm is placed in a region where the alternating B field has a amplitude of 10^{-4} T. Find the approximate current induced in the ring.

11

Propagation and Reflection of Plane Waves

11.1 INTRODUCTION

An illustration of a traveling wave that is familiar to all of us is that of dropping a small rock or pebble near the center of a quiet pond. The waves set up by the water oscillation travel outward from the disturbance toward the bank at a velocity determined by the parameters of the water. If a fisherman has a line in the water with a cork on it, the cork bobs up and down as the wave passes by. This traveling water wave exhibits an oscillatory motion to an observer in a fixed position, such as the cork bobbing; it also presents an oscillatory pattern with distance: for example, if a snapshot is taken of the waves at some instant of time. If a miniature surfer were riding near the crest of one of these waves as it travels toward the bank, he would see just a fixed pattern since he would be traveling with the wave at the wave velocity.

Whether one is talking about the $\overline{E}$ and $\overline{H}$ fields of an electromagnetic wave or voltage and current waves in transmission lines to be studied in Chapter 12, the water wave analogy can be helpful in understanding the concepts of a traveling wave.

The experimental laws of electromagnetics may be summed up by Maxwell's equations, which we have already been exposed to briefly, and in this chapter we shall use these equations to derive the propagation phenomena for a type of traveling wave called a *plane wave* in free space, lossless dielectrics, and conducting media. We shall also examine the phenomena of reflections at interfaces between different materials.

11.2 MAXWELL'S EQUATIONS

Maxwell's equations, in both the time-dependent and phasor form, are summarized at this point for convenience in the various derivations in this chapter. The differential forms are

$$\nabla \times \overline{E} = -\mu \frac{\partial \overline{H}}{\partial t} \quad (1) \qquad \nabla \times \overline{\mathcal{E}} = -j\omega\mu\overline{\mathcal{H}} \quad (1a)$$

$$\nabla \times \overline{H} = \overline{J}_c + \epsilon' \frac{\partial \overline{E}}{\partial t} \quad (2) \qquad \nabla \times \overline{\mathcal{H}} = (\sigma + j\omega\epsilon')\overline{\mathcal{E}} \quad (2a)$$

$$\nabla \cdot \overline{D} = \rho_v \quad (3) \qquad \nabla \cdot \overline{\mathcal{D}} = \rho_v \quad (3a)$$

$$\nabla \cdot \overline{B} = 0 \quad (4) \qquad \nabla \cdot \overline{\mathcal{B}} = 0 \quad (4a)$$

or, in integral form,

$$\oint_\ell \overline{E} \cdot \overline{d\ell} = \text{emf} = -\int_s \frac{\partial \overline{B}}{\partial t} \cdot \overline{ds} \quad (5) \qquad \oint_\ell \overline{\mathcal{E}} \cdot \overline{d\ell} = -\int_s j\omega\mu\overline{\mathcal{H}} \cdot \overline{ds} \quad (5a)$$

$$\oint_\ell \overline{H} \cdot \overline{d\ell} = I_{\text{en}} = \int_s \left(\overline{J}_c + \frac{\partial \overline{D}}{\partial t}\right) \cdot \overline{ds} \quad (6) \qquad \oint_\ell \overline{\mathcal{H}} \cdot \overline{d\ell} = \int_s (\sigma + j\omega\epsilon')\overline{\mathcal{E}} \cdot \overline{ds} \quad (6a)$$

$$\oint_s \overline{D} \cdot \overline{ds} = Q_{\text{en}} = \int_v \rho_v \, dv \quad (7) \qquad \oint_s \overline{\mathcal{D}} \cdot \overline{ds} = \int_v \rho_v \, dv \quad (7a)$$

$$\oint_s \overline{B} \cdot \overline{ds} = 0 \quad (8) \qquad \oint_s \overline{\mathcal{B}} \cdot \overline{ds} = 0 \quad (8a)$$

In free space or a non-conducting dielectric,

$$\left.\begin{aligned} \overline{J}_c &= 0 \\ \text{or} \quad \sigma &= 0 \\ \text{and} \quad \nabla \cdot \overline{D} &= 0 \end{aligned}\right\} \quad (9)$$

and the symmetry thereby created between the curl equations in $\overline{E}$ and $\overline{H}$ leads directly to the derivation of a wave equation in either $\overline{E}$ or $\overline{H}$.

11.3 PLANE WAVES IN FREE SPACE OR LOSSLESS DIELECTRIC

In free space or a lossless dielectric, (11.2-2) and (11.2-3) become

$$\nabla \times \overline{H} = \epsilon \frac{\partial \overline{E}}{\partial t} \quad (1)$$

$$\nabla \cdot \overline{D} = \nabla \cdot \overline{E} = 0 \quad (2)$$

Then, by taking the curl of (11.2-1) and substituting (1),

$$\nabla \times (\nabla \times \overline{E}) = -\mu\left(\nabla \times \frac{\partial \overline{H}}{\partial t}\right) = -\mu\epsilon \frac{\partial^2 \overline{E}}{\partial t^2}$$

From vector identity (19) in Table B-4,

$$\nabla \times (\nabla \times \overline{E}) = \nabla(\nabla \bullet \overline{E}) - \nabla^2 \overline{E} = -\mu\epsilon \frac{\partial^2 \overline{E}}{\partial t^2} \tag{3}$$

or

$$\boxed{\nabla^2 \overline{E} = \mu\epsilon \frac{\partial^2 \overline{E}}{\partial t^2} \qquad (\text{V m}^{-3})} \tag{4}$$

when (2) is substituted into (3).

When expanded in cartesian coordinates, eq. (4) is a set of wave equations, one of which is

$$\boxed{\frac{\partial^2 E_x}{\partial x^2} + \frac{\partial^2 E_x}{\partial y^2} + \frac{\partial^2 E_x}{\partial z^2} = \mu\epsilon \frac{\partial^2 E_x}{\partial t^2} \qquad (\text{V m}^{-3})} \tag{5}$$

for the x component in cartesian coordinates.

Taking the curl of (1), then substituting (11.2-1) and (11.2-4) leads to

$$\boxed{\nabla^2 \overline{H} = \mu\epsilon \frac{\partial^2 \overline{H}}{\partial t^2} \qquad (\text{A m}^{-3})} \tag{6}$$

and a set of wave equations in $\overline{H}$. Solutions of (4) and (6) are of the form

$$f = f\left(t \pm \frac{x}{U}\right) \tag{7}$$

as can be demonstrated by taking partial derivatives with respect to x and t, giving

$$\frac{\partial^2 f}{\partial x^2} = \frac{1}{U^2}\left(\frac{\partial^2 f}{\partial t^2}\right) \tag{8}$$

That (7) is a wave function can be seen by setting the argument $(t \pm x/U)$ equal to a constant. If $(t - x/U)$ is constant, then x must be increasing at the rate U as t increases. Whatever value of the function f one may choose, that value travels in the x direction. A surfer rides at a more or less constant amplitude point on a wave that is traveling toward the shore. The surfer must travel with the wave to stay at the constant amplitude point. If a snapshot is taken of the wave, it will reveal a spatial distribution, but to an observer at a fixed location the wave is a function of time. It is only to an observer traveling with the wave that it appears to be constant. If the argument $(t + x/U)$ is constant, the x must be decreasing at the rate U as t increases. Whenever a function has an argument of the form $(t \pm x/U)$, it is a wave function, and the second partial derivatives with respect to time and with respect to distance will differ only by a constant. For example,

$$E_x = E_x^+ \cos \omega\left(t - \frac{\beta}{\omega} z\right) + E_x^- \cos \omega\left(t + \frac{\beta}{\omega} z\right) \qquad (\text{V m}^{-1}) \tag{9}$$

is a sinusoidal function that consists of a wave of amplitude E_x^+ traveling in the positive z direction with a velocity ω/β, and a wave of amplitude E_x^- traveling in the minus z direction, also with a velocity ω/β. Two partial differentiations of (9) with respect to z and t give

$$\boxed{\frac{\partial^2 E_x}{\partial z^2} = \frac{\beta^2}{\omega^2}\left(\frac{\partial^2 E_x}{\partial t^2}\right) \qquad (\text{V m}^{-3})} \tag{10}$$

from which it is clear that ω/β is velocity, corresponding to the U in (8) and to $1/\sqrt{\mu\epsilon}$ in (4) and (6).

Applying (11.2-1) to (9) and equating components on both sides of the equation, we have

$$\hat{y}\frac{\partial E_x}{\partial z} = -\mu\frac{\partial \overline{H}}{\partial t}$$

or

$$\beta E_x^+ \sin\omega\left(t - \frac{\beta}{\omega}z\right) - \beta E_x^- \sin\omega\left(t + \frac{\beta}{\omega}z\right) = -\mu\frac{\partial H_y}{\partial t} \tag{11}$$

from which one obtains, by integrating with respect to time,

$$\begin{aligned} H_y &= \frac{\beta}{\omega\mu}E_x^+ \cos\omega\left(t - \frac{\beta}{\omega}z\right) - \frac{\beta}{\omega\mu}E_x^- \cos\omega\left(t + \frac{\beta}{\omega}z\right) \\ &= H_y^+ \cos\omega\left(t - \frac{\beta}{\omega}z\right) - H_y^- \cos\omega\left(t + \frac{\beta}{\omega}z\right) \qquad (\text{A m}^{-1}) \end{aligned} \tag{12}$$

Thus, a y component for $\overline{H}$ results when an x component is assumed for $\overline{E}$. Both $\overline{E}$ and $\overline{H}$ are in time phase and together form a TEM (transverse electromagnetic) forward traveling wave in the positive z direction with a velocity ω/β and a reverse traveling wave in the negative z direction with the same velocity. Each of these waves is called a *uniform plane wave* because the transverse fields $\overline{E}$ and $\overline{H}$ lie in a plane that is normal to the velocity and are functions only of the direction of travel and of time. The amplitude ratio of $\overline{E}$ to $\overline{H}$ for the waves in either direction is called the *intrinsic impedance* of the material in which the wave is traveling and is given by

$$\eta = \frac{E_x^+}{H_y^+} = -\frac{E_x^-}{H_y^-} = \frac{\omega\mu}{\beta} = U\mu \qquad (\Omega) \tag{13}$$

By comparing (6) and (8), it is clear that

$$\boxed{U = \frac{1}{\sqrt{\mu\epsilon}} \qquad (\text{m s}^{-1})} \tag{14}$$

and

$$\eta = U\mu = \frac{\mu}{\sqrt{\mu\epsilon}} = \sqrt{\frac{\mu}{\epsilon}} \quad (\Omega) \tag{15}$$

For free space,

$$\eta_0 = \sqrt{\frac{\mu_0}{\epsilon_0}} \approx 377\Omega = 120\pi \quad (\Omega) \tag{16}$$

and

$$U = c = \frac{1}{\sqrt{\mu_0\epsilon_0}} \approx 3 \times 10^8 \text{ m s}^{-1} \tag{17}$$

Note the βz has the same units as ωt, i.e., radians. β is called a *phase constant* and has units of radians per meter. If λ is the wavelength in meters,

$$\beta = \frac{2\pi}{\lambda} = \frac{\omega}{U} = \frac{2\pi f}{U}$$

or, by substituting (14) for U,

$$\beta = \omega\sqrt{\mu\epsilon} \quad (\text{m}^{-1}) \tag{18}$$

Multiplication by β converts a length into a phase angle.

The E_x^+ and H_y^+ fields of (9) and (12) comprise a uniform plane wave traveling in the positive z direction, and the E_x^- and H_y^- fields a plane wave traveling in the negative z direction. Either wave is unbounded and does not diminish either with time or distance. The significance of the two directions of traveling waves will be made clear when reflections are considered. In this text, the amplitudes are E_x^+ and H_y^+ for the incident wave and E_x^- and H_y^- for the reflected wave. The incident or forward traveling wave will be shown traveling in the positive z direction, as represented in Fig. 11-1. As we shall see when we study the radiation fields from antennas, the actual $\overline{E}$ and $\overline{H}$ fields of the radiated wave vary inversely with distance from the source, but the idealized plane wave is a good model with which to begin the study of propagation and reflection characteristics of waves. At a great distance from the source, a radiated wave looks like a section of an idealized plane wave.

Problem 11.3-1 In a certain lossless dielectric, $\mu = \mu_0$, $\epsilon = 3.4\epsilon_0$, an $\overline{E}$ field is represented by $\overline{E} = \hat{x}10 \sin(3\pi \times 10^8 t - \beta z)$ (V m^{-1}). Find: (a) E_x^+; (b) f; (c) β; (d) $\overline{U}$; (e) η; (f) H_y^+; (g) λ.

Problem 11.3-2 Completely describe the E and the H field components for a plane wave whose E field is described in sinusoidal form as $\overline{E} = \hat{x}10 \sin(3\pi \times 10^8 t + \beta z)$ (V m^{-1}). Indicate polarity and whether forward or reverse wave. Assume that $\mu = \mu_0$ and $\epsilon_r = 3.4$.

Problem 11.3-3 The intrinsic impedance of a lossless material is known to be 300 Ω. If $\mu = \mu_0$, find the velocity of propagation and the wavelength for a 1 GHz wave.

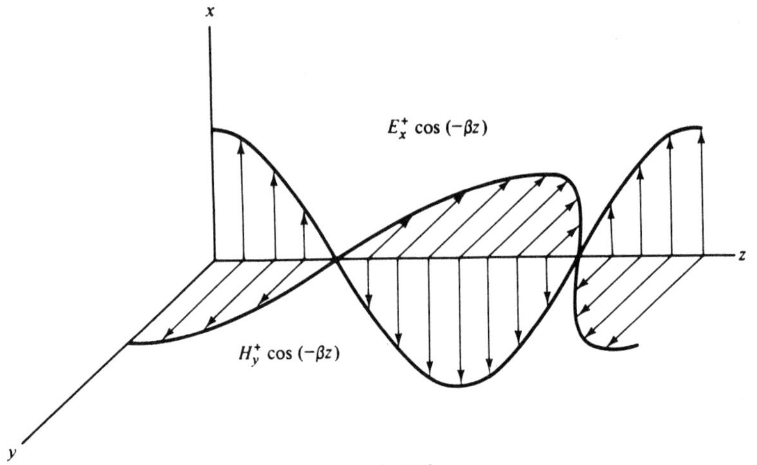

(a) $t = 0$

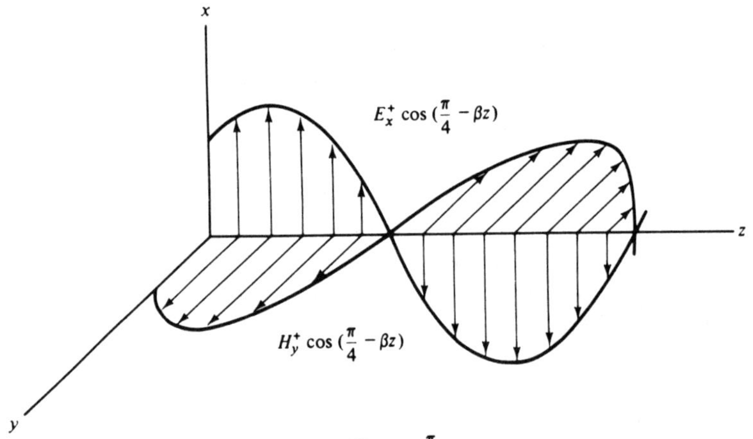

(b) $t = \frac{\pi}{4\omega}$

Figure 11-1 Graphical representation of the forward $\overline{E}$ and $\overline{H}$ waves of (11.3-9) and (11.3-12): (a) $t = 0$; (b) $t = \pi/4\omega$.

11.4 POWER FLOW IN LOSSLESS MEDIUM—THE POYNTING VECTOR

Power flow can perhaps best be illustrated by the planar dc transmission line of Fig. 11-2. From the study of static fields, the $\overline{E}$ and $\overline{H}$ fields between the planar conductors are

$$\left.\begin{aligned} |\overline{E}| &= \frac{V}{d} \qquad (\text{V m}^{-1}) \\ \text{and} \qquad |\overline{H}| &\approx \frac{I}{w} \qquad (\text{A m}^{-1}) \end{aligned}\right\} \tag{1}$$

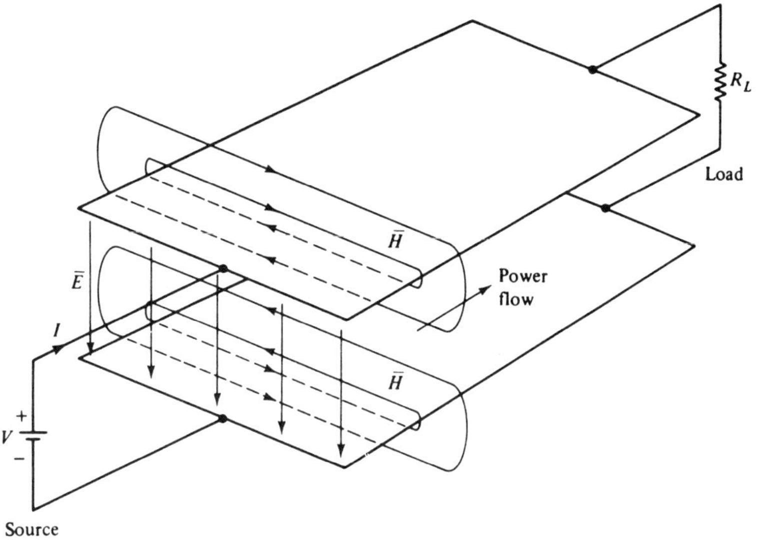

(a)

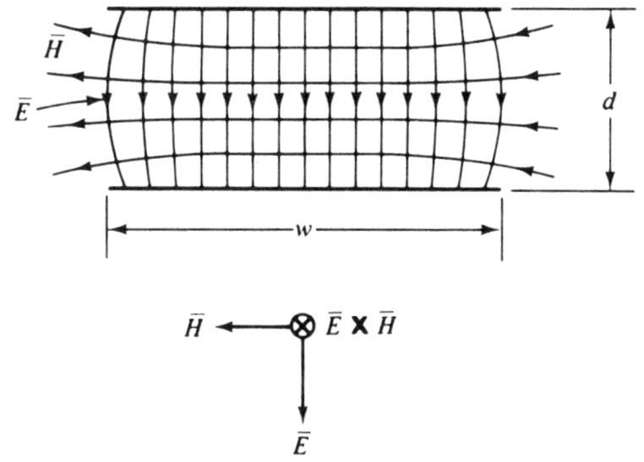

(b)

Figure 11-2 Graphical representation of power flow.

where edge effects are neglected. The direction of power flow can be seen to be the direction given by $\overline{E} \times \overline{H}$, and from circuit theory the magnitude of the power flow is given by

$$P = VI \qquad \text{(W)}$$

but the power density flow is given by

$$P_{\text{density}} = \frac{VI}{wd} = |\overline{E}|\,|\overline{H}|$$

or

$$\boxed{\overline{P}_{\text{density}} = \overline{\mathcal{P}} = \overline{E} \times \overline{H} \qquad (\text{W m}^{-2})} \tag{2}$$

where $\overline{\mathcal{P}}$, which gives the direction as well as the magnitude of the power density flow, is called the *Poynting vector.* A more rigorous derivation of $\overline{\mathcal{P}}$ follows.

The total energy density in a region where there are both electric and magnetic fields is

$$\boxed{w = w_E + w_M = \tfrac{1}{2}\epsilon E^2 + \tfrac{1}{2}\mu H^2 \qquad (\text{J m}^{-3})} \tag{3}$$

Now let $\overline{\mathcal{P}}$ represent a positive outward flow of power density from a closed surface. Then

$$-\frac{\partial}{\partial t}\int_v \left(\frac{1}{2}\epsilon E^2 + \frac{1}{2}\mu H^2\right) dv = \oint_s \overline{\mathcal{P}} \bullet \overline{ds} \qquad (\text{W}) \tag{4}$$

relating the net outward flow of power to the rate of change of energy within the closed surface. By the divergence theorem,

$$\oint_s \overline{\mathcal{P}} \bullet \overline{ds} = \int_v \nabla \bullet \overline{\mathcal{P}}\, dv$$

or

$$\begin{aligned} \nabla \bullet \overline{\mathcal{P}} &= -\frac{\partial}{\partial t}\left(\frac{1}{2}\epsilon E^2 + \frac{1}{2}\mu H^2\right) \\ &= -\left(\epsilon \overline{E} \bullet \frac{\partial \overline{E}}{\partial t} + \mu \overline{H} \bullet \frac{\partial \overline{H}}{\partial t}\right) \qquad (\text{W m}^{-3}) \end{aligned} \tag{5}$$

By substituting the curl relations for $\epsilon\, \partial \overline{E}/\partial t$ and $\mu\, \partial \overline{H}/\partial t$ from Maxwell's equations (11.3-1) and (11.2-1) and then applying the vector identity (12) in Table B-4,

$$\begin{aligned} \nabla \bullet \overline{\mathcal{P}} &= -\left[\epsilon \overline{E} \bullet \frac{1}{\epsilon}(\nabla \times \overline{H}) - \mu \overline{H} \bullet \left(\frac{\nabla \times \overline{E}}{\mu}\right)\right] \\ &= -[\nabla \bullet \overline{H} \times \overline{E}] = \nabla \bullet \overline{E} \times \overline{H} \end{aligned}$$

or

$$\boxed{\overline{\mathcal{P}} = \overline{E} \times \overline{H} \qquad (\text{W m}^{-2})} \tag{6}$$

and

$$\boxed{\overline{\mathcal{P}}_{\text{av}} = \tfrac{1}{2}\overline{E}_{\text{peak}} \times \overline{H}_{\text{peak}} = \overline{E}_{\text{rms}} \times \overline{H}_{\text{rms}} \qquad (\text{W m}^{-2})} \tag{7}$$

The power flow through a given cross-sectional area in a lossless medium is

$$P = \iint_{\text{cross-section}} \overline{E}_{\text{rms}} \times \overline{H}_{\text{rms}} \cdot \overline{ds} \qquad (\text{W}) \tag{8}$$

Example 1

Consider a planar transmission line as illustrated in Fig. 11-2 with $w = 10$ cm and $d = 1$ mm. Let the plates be in the planes $x = 0$ and $x = 1$ mm, with the positive z direction being the direction of power flow toward the load. Assume that $V = 100$ (V) and $I = 30$ (mA) dc load current. Neglecting the resistance of conductors and fringing effects at the edges, calculate the load power using voltage and current, and then using (8).

Solution. Using voltage and current, we have

$$P = (100)(30 \times 10^{-3}) = 3 \qquad \text{(W)}$$

The $\overline{E}$ field is

$$\overline{E} = \frac{-\hat{x}100}{0.001} = -\hat{x} \times 10^5 \qquad (\text{V m}^{-1})$$

and the $\overline{H}$ field is

$$\overline{H} = \frac{-\hat{y}(30 \times 10^{-3})}{0.1} = -\hat{y}(0.3) \qquad (\text{A m}^{-1})$$

Then

$$P = \int_{x=0}^{0.001} \int_{y=0}^{0.1} -\hat{x}10^5 \times -\hat{y}(0.3) \cdot \hat{z}\, dx\, dy$$

$$= 3 \times 10^4 x \Big|_{x=0}^{0.001} y \Big|_{y=0}^{0.1} = 3 \times 10^4 \times 10^{-4} = 3 \qquad \text{(W)}$$

which is the same as calculated using V and I.

11.5 PLANE WAVES IN LOSSY DIELECTRICS

It will be convenient in this section to use phasor notation, thus implying sinusoidal steady state. The required set of Maxwell's equations have already been given by (11.2-1a) through (11.2-8a). The material being considered here will, in general, have a conductivity σ, a dielectric constant ϵ_r, and a permeability μ. Unless specified otherwise, $\mu = \mu_0$. The complex permittivity, as discussed in Sec. 10.7, is

$$\boxed{\epsilon = \epsilon' - j\epsilon'' = \left(\epsilon_r'\epsilon_0 - j\frac{\sigma}{\omega}\right) \qquad (\text{F m}^{-1})} \tag{1}$$

The ϵ in equations for a lossless material may be replaced by (1) to give the corresponding equation for lossy or conducting material. For example, the intrinsic impedance of a lossy material is

$$\boxed{\eta = \sqrt{\frac{\mu}{\epsilon}} = \sqrt{\frac{\mu}{\epsilon'\left(1 - j\dfrac{\sigma}{\omega\epsilon'}\right)}} = \frac{\sqrt{\mu/\epsilon'}}{\sqrt{1 - j\dfrac{\sigma}{\omega\epsilon'}}} \qquad (\Omega)} \tag{2}$$

Consider a wave traveling in the positive z direction having an x component of $\overline{E}$ and a y component of $\overline{H}$. From the Maxwell curl equations, we get

$$\frac{\partial \mathscr{E}_x}{\partial z} = -j\omega\mu\mathscr{H}_y \qquad (\text{V m}^{-2}) \tag{3}$$

$$-\frac{\partial \mathscr{H}_y}{\partial z} = j\omega\epsilon'\left(1 - j\frac{\sigma}{\omega\epsilon'}\right)\mathscr{E}_x \qquad (\text{A m}^{-2}) \tag{4}$$

Now let us define γ such that differentiation with respect to z is performed by simply multiplying by $-\gamma$, just as differentiation with respect to time is by multiplying by $j\omega$. Then (3) and (4) become

$$\gamma \mathscr{E}_x = j\omega\mu \mathscr{H}_y \qquad (\text{V m}^{-2}) \tag{5}$$

and

$$\gamma \mathscr{H}_y = j\omega\epsilon'\left(1 - j\frac{\sigma}{\omega\epsilon'}\right)\mathscr{E}_x \qquad (\text{A m}^{-2}) \tag{6}$$

Dividing (5) by (6) gives the intrinsic impedance

$$\eta = \frac{\gamma \mathscr{E}_x}{\gamma \mathscr{H}_y} = \frac{\mathscr{E}_x}{\mathscr{H}_y} = \frac{\mathscr{H}_y \mu}{\mathscr{E}_x \epsilon'\left(1 - j\dfrac{\sigma}{\omega\epsilon'}\right)} \qquad (\Omega)$$

Then

$$\frac{\mathscr{E}_x^2}{\mathscr{H}_y^2} = \frac{\mu}{\epsilon'\left(1 - j\dfrac{\sigma}{\omega\epsilon'}\right)}$$

or

$$\eta = \frac{\sqrt{\mu/\epsilon'}}{\sqrt{1 - j\dfrac{\sigma}{\omega\epsilon'}}} \qquad (\Omega)$$

the result already given by (2).

If we now equate $\mathscr{E}_x/\mathscr{H}_y$ from (5) with $\mathscr{E}_x/\mathscr{H}_y$ from (6),

$$j\frac{\omega\mu}{\gamma} = \frac{\gamma}{j\omega\epsilon'\left(1 - j\dfrac{\sigma}{\omega\epsilon'}\right)}$$

and

$$\gamma^2 = (j\omega)^2\mu\epsilon'\left(1 - j\frac{\sigma}{\omega\epsilon'}\right)$$

or

$$\boxed{\gamma = j\omega\sqrt{\mu\epsilon'}\sqrt{1 - j\frac{\sigma}{\omega\epsilon'}} \qquad (\text{m}^{-1})} \tag{7}$$

where only the positive square root is used. Note that as $\sigma \to 0$, $\gamma \to j\omega\sqrt{\mu\epsilon'} = j\beta$. Then γ has the value of $j\beta$ for the lossless case but is, in general, a complex number having both a real and imaginary part,

$$\gamma = \alpha + j\beta \tag{8}$$

Example 2

Consider the function, $E_x = E_{x0}^+ e^{-\alpha z} \cos(\omega t - \beta z)$. Show that $\partial E_x/\partial z$ gives the same result in the time domain as does multiplication by $-\gamma$, using phasors.

Solution.

$$\frac{\partial E_x}{\partial z} = -\alpha E_{x0}^{+} e^{-\alpha z} \cos(\omega t - \beta z) + \beta E_{x0}^{+} e^{-\alpha z} \sin(\omega t - \beta z)$$
$$= E_{x0}^{+} e^{-\alpha z}\left[-\alpha \cos(\omega t - \beta z) - \beta \cos\left(\omega t - \beta z + \frac{\pi}{2}\right)\right]$$

and, by transforming to phasor notation, $\partial E_x/\partial z$ would be replaced by

$$\mathscr{E}_{x0}^{+} e^{-\alpha z}[-\alpha e^{-j\beta z} - \beta e^{-j\beta z} e^{j(\pi/2)}] = \mathscr{E}_{x0}^{+} e^{-(\alpha+j\beta)z}(-\alpha - j\beta)$$
$$= \mathscr{E}_{x0}^{+} e^{-\gamma z}(-\gamma) = -\gamma \mathscr{E}_x \qquad (\text{V m}^{-2})$$

The term α is known as the *attenuation constant* and has units of nepers per meter. The term γ is known as the *propagation constant*, and since a neper is a dimensionless quantity as is the radian we rewrite (8) with dimensions,

$$\boxed{\gamma = \alpha + j\beta \qquad (\text{m}^{-1})} \tag{8}$$

Example 3

An $\mathscr{E}$ field given by $\hat{x}100e^{-\gamma z}$ (V m^{-1}) is traveling through a material ($\epsilon_r' = 4, \sigma = 0.1$ S m^{-1}, $\mu = \mu_0$) and the frequency is 2.45 GHz. Find α and β and the decibels per meter attenuation in the material.

Solution. From (7),

$$\gamma = j2\pi(2.45)(10^9)\sqrt{4\mu_0\epsilon_0}\sqrt{1 - j\frac{0.1}{4\omega\epsilon_0}}$$
$$= j1.539(10^{10})\frac{\sqrt{4}}{3 \times 10^8}(1.008 \angle -5.197^\circ)$$
$$= 103.44 \angle 84.8^\circ$$
$$= 9.37 + j103 = \alpha + j\beta \qquad (\text{m}^{-1})$$
$$\frac{|\mathscr{E}_x|_{z=1}}{|\mathscr{E}_x|_{z=0}} = e^{-\alpha(1)} = e^{-9.37}$$

The dB attenuation is equal to

$$10 \log_{10} \frac{|\mathscr{E}_x|_0^2}{|\mathscr{E}_x|_1^2} = 20 \log_{10} e^{9.37}$$
$$= (20 \log_{10} e)(9.37)$$
$$= 81.39 \text{ dB m}^{-1}$$

Then

$$\alpha = 9.37 \text{ Np m}^{-1}$$
$$\beta = 103 \text{ rad m}^{-1}$$
$$\text{Atten.} = 81.39 \text{ dB m}^{-1}$$

At a depth equal to $1/\alpha$, the wave is attenuated to e^{-1} times its initial value. This depth is referred to as the *skin depth*. We will discuss this further in Sec. 11.6.

It would be well at this point to summarize the characteristics of electromagnetic wave travel in lossy material compared to lossless material. A comparison of velocity, intrinsic impedance, and propagation constant is best accomplished by means of Table 11-1.

TABLE 11-1 COMPARISONS OF VELOCITY, IMPEDANCE, AND THE PROPAGATION CONSTANT FOR FREE SPACE, LOSSLESS MATERIAL, AND CONDUCTING OR LOSSY MATERIAL

	Velocity	Intrinsic impedance	Propagation constant
Free space	$c = \dfrac{1}{\sqrt{\mu_0\epsilon_0}} = \dfrac{\omega}{\beta}$ $\approx 3 \times 10^8$ (m s^{-1})	$\eta = \sqrt{\dfrac{\mu_o}{\epsilon_0}} \approx 120\pi\ (\Omega)$ $\approx 377\Omega$	$\gamma = j\beta$ $= j\omega\sqrt{\mu_0\epsilon_0}\ (\text{m}^{-1})$
Lossless dielectric	$U = \dfrac{1}{\sqrt{\mu\epsilon}} = \dfrac{c}{\sqrt{\mu_r\epsilon_r}}$ $= \dfrac{\omega}{\beta}$ (m s^{-1})	$\eta = \sqrt{\dfrac{\mu}{\epsilon}} = \sqrt{\dfrac{\mu_0\mu_r}{\epsilon_0\epsilon_r}}$ $\approx 120\pi\sqrt{\dfrac{\mu_r}{\epsilon_r}}\ (\Omega)$	$\gamma = j\beta$ $= j\omega\sqrt{\mu_0\epsilon_0\mu_r\epsilon_r}\ (\text{m}^{-1})$
Lossy dielectric, $\epsilon = \epsilon' - j\dfrac{\sigma}{\omega}$	$U = \dfrac{\omega}{\beta}$ (m s^{-1})	$\eta = \dfrac{\sqrt{\mu/\epsilon'}}{\sqrt{1 - j\dfrac{\sigma}{\omega\epsilon'}}}\ (\Omega)$	$\gamma = \alpha + j\beta$ $= j\omega\sqrt{\mu\epsilon'}\left(\sqrt{1 - j\dfrac{\sigma}{\omega\epsilon'}}\right)\ (\text{m}^{-1})$
Slightly lossy dielectric, $\dfrac{\sigma}{\omega\epsilon'} \ll 1$	$U = \dfrac{\omega}{\beta}$ (m s^{-1})	$\eta \approx \sqrt{\dfrac{\mu}{\epsilon'}}\left(1 + j\dfrac{\sigma}{2\omega\epsilon'}\right)\ (\Omega)$	$\gamma = \alpha + j\beta$ $\approx j\omega\sqrt{\mu\epsilon'}\left(1 - j\dfrac{\sigma}{2\omega\epsilon'}\right)\ (\text{m}^{-1})$
Good conductor $\dfrac{\sigma}{\omega\epsilon'} \gg 1$	$U = \dfrac{\omega}{\beta}$ (m s^{-1})	$\eta \approx \sqrt{\dfrac{\omega\mu}{\sigma}}\ \underline{/45°}\ (\Omega)$	$\gamma = \alpha + j\beta$ $\approx \sqrt{\pi f\mu\sigma}\,(1 + j)\ (\text{m}^{-1})$

When a dielectric material is only slightly conducting, i.e., $\sigma/\omega\epsilon' \ll 1$, then the impedance and propagation constant may be approximated by taking only the first two terms of the binomial expansion of η and γ. Thus,

$$\eta_{\sigma/\omega\epsilon' \ll 1} \approx \sqrt{\frac{\mu}{\epsilon'}}\left(1 + \frac{1}{2}j\frac{\sigma}{\omega\epsilon'}\right) \qquad (\Omega) \tag{9}$$

and

$$\begin{aligned}\gamma_{\sigma/\omega\epsilon' \ll 1} &\approx j\omega\sqrt{\mu\epsilon'}\left(1 - \frac{1}{2}j\frac{\sigma}{\omega\epsilon'}\right)\\ &= \frac{1}{2}\sigma\sqrt{\frac{\mu}{\epsilon'}} + j\omega\sqrt{\mu\epsilon'}\\ &= \alpha + j\beta \qquad (\text{m}^{-1})\end{aligned} \tag{10}$$

The phase constant β for a slightly lossy material is approximately the same as for a lossless material having the same permittivity. Calculations for plane waves in slightly lossy materials are handled in the same way as calculations for lossless materials except that the amplitude of the $\overline{E}$ or $\overline{H}$ field varies exponentially with distance.

$$E_{x\,\text{low loss}} \approx E_{x0}^{+} e^{[(-1/2)\sigma]\sqrt{(\mu/\epsilon')}\,z} \cos\left(\omega t - \omega\sqrt{\mu\epsilon'}\,z\right)$$

or

$$\mathscr{E}_{x\,\text{low loss}} \approx \mathscr{E}_{x0}^{+} e^{[(-1/2)\sigma]\sqrt{(\mu/\epsilon')}\,z} e^{-j\omega\sqrt{\mu\epsilon'}\,z} \qquad (\text{V m}^{-1}) \tag{11}$$

The terms "low loss" and "high loss" are, of course, relative. As a rule of thumb, we establish the following criteria:

$$\left.\begin{aligned}&\text{A material is considered low loss if } \frac{\sigma}{\omega\epsilon'} < 0.1\,.\\ &\text{A material is considered high loss if } \frac{\sigma}{\omega\epsilon'} > 10\,.\end{aligned}\right\} \tag{12}$$

The following example is for the "in between" region where the material cannot be considered either low or high loss.

Example 4

Calculate the intrinsic impedance for the material of Example 3, and then calculate both the impedance and propagation constant using the low loss approximations of (9) and (10). Compare the results.

Solution. First calculate $\sigma/\omega\epsilon'$.

$$\frac{\sigma}{\omega\epsilon'} = \frac{0.1}{2\pi(2.45 \times 10^{9})4(8.854 \times 10^{-12})} = 0.1834$$

Then

$$\sqrt{1 - j\frac{\sigma}{\omega\epsilon'}} = \sqrt{1 - j0.1834} = 1.0083 \underline{/-5.197^\circ}$$

$$\eta = \frac{\sqrt{\mu/\epsilon'}}{\sqrt{1 - j\dfrac{\sigma}{\omega\epsilon'}}} = \frac{377/\sqrt{4}}{1.0083 \underline{/-5.197^\circ}} = 186.9 \underline{/5.197^\circ}$$

$$= 186.2 + j16.93 \qquad (\Omega)$$

By low loss approximation,

$$\gamma \approx \frac{1}{2}\sigma\sqrt{\frac{\mu}{\epsilon'}} + j\omega\sqrt{\mu\epsilon'}$$

$$= \frac{1}{2}(0.1)\frac{377}{\sqrt{4}} + j2\pi(2.45 \times 10^9)\frac{\sqrt{4}}{3 \times 10^8}$$

$$= 9.43 + j102.6 = \alpha + j\beta \qquad (\text{m}^{-1})$$

compared to

$$9.37 + j103$$

from Example 3.

$$\eta_{\text{low loss}} \approx \sqrt{\frac{\mu}{\epsilon'}} + \frac{1}{2}j\frac{\sigma}{\omega\epsilon'}\sqrt{\frac{\mu}{\epsilon'}}$$

$$= \frac{377}{\sqrt{4}} + \frac{1}{2}j0.1834\frac{377}{\sqrt{4}}$$

$$= 188.5 + j17.29 \qquad (\Omega)$$

compared to

$$186.2 + j16.93 \qquad (\Omega)$$

calculated by the exact method.

Problem 11.5-1 Find the intrinsic impedance and propagation constant for seawater at 10 kHz, 100 kHz, 10 MHz, and 1 GHz under the assumption that $\sigma = 5$ S m^{-1}, $\epsilon_r' = 80$, and $\mu = \mu_0$.

Problem 11.5-2 A 9-GHz wave is propagating through a material that has a dielectric constant of 2.4 and a loss tangent of 0.005. Find α, β, and the wavelength within the material.

Problem 11.5-3 An electromagnetic wave traveling in the positive z direction has an $\overline{E}$ field amplitude of 75 V m^{-1} at $z = 0$ within the material of Examples 3 and 4. Find the amplitudes of $\overline{E}$ and $\overline{H}$ at $z = 0.2$ m if the frequency is 2.45 GHz.

11.6 PLANE WAVES IN GOOD CONDUCTORS

A good conductor is a high loss material and is defined here as a material for which $\sigma/\omega\epsilon' > 100$. This covers a broad range of materials, for it puts seawater and copper in the same classification at frequencies of 10 kHz or less even though their conductivities have a ratio of about 10^7. Using the foregoing definition of a conductor, (11.5-2) and (11.5-7) become

$$\boxed{\eta_{\text{conductor}} \approx \frac{\sqrt{\mu/\epsilon'}}{\sqrt{-j\dfrac{\sigma}{\omega\epsilon'}}} = \sqrt{\frac{\omega\mu}{\sigma}}\,\angle 45^\circ \qquad (\Omega)} \tag{1}$$

and

$$\gamma_{\text{conductor}} \approx j\omega\sqrt{\mu\epsilon'}\sqrt{-j\frac{\sigma}{\omega\epsilon'}} = \sqrt{\omega\mu\sigma}\,\angle 45^\circ \qquad (\text{m}^{-1})$$

$$= \sqrt{\pi f\mu\sigma} + j\sqrt{\pi f\mu\sigma}$$
$$= \alpha + j\beta \tag{2}$$

It is seen from (1) and (2) that the impedance phase angle is approximately 45°, and that the α and β are equal for good conductors. This impedance angle is interesting. As can be readily calculated from (11.5-2), a perfect dielectric (zero conductivity) has an impedance angle of 0°, and the best conductor has an impedance angle of 45°. The impedance angle for any material lies between 0° and 45°. This is in sharp contrast to circuits where the impedance angle can vary from −90° to +90°. Note also that the impedance seen by a plane wave has only a positive angle. This means that for a plane wave in a conducting material the phase of the $\mathscr{H}$ field may lag the phase of the $\mathscr{E}$ field by as much as 45°.

From (2), we see that α and β vary as the square root of frequency. This means that at high frequencies attenuation is large and phase shift is large, meaning short wavelengths. Consider a wave traveling in the positive z direction through a conducting material, and let $z_2 > z_1$.

$$\frac{(\mathscr{E}_x)_{z_2}}{(\mathscr{E}_x)_{z_1}} = \frac{\mathscr{E}_{x0}^{+}e^{-\alpha z_2}e^{-j\beta z_2}}{\mathscr{E}_{x0}^{+}e^{-\alpha z_1}e^{-j\beta z_1}} = e^{-\alpha(z_2-z_1)}e^{-j\beta(z_2-z_1)} \tag{3}$$

When the distance $(z_2 - z_1) = 1/\alpha$, then

$$\frac{(\mathscr{E}_x)_{z_2}}{(\mathscr{E}_x)_{z_1}} = e^{-\alpha(1/\alpha)}e^{-j\beta(1/\beta)} = e^{-1}e^{-j} \tag{4}$$

In a distance of $1/\alpha$, the amplitude of either $\mathscr{E}$ or $\mathscr{H}$ is attenuated to e^{-1} times its initial value, and the phase differs by 1 radian from its initial value. This distance of $1/\alpha$ has already been referred to as the *skin depth,* where it refers to the fact that a plane wave entering a conducting material from air, e.g., is reduced to e^{-1} times its value at the surface in one skin depth (see Fig. 11-3). In a good conductor,

$$\underset{(\sigma/\omega\epsilon' \gg 1)}{\text{skin depth, } \delta} = \frac{1}{\alpha} = \frac{1}{\sqrt{\pi f\mu\sigma}} \qquad (\text{m}) \tag{5}$$

Example 5

Find the phase and amplitude of the $\mathscr{E}$ field at a depth of 0.1 mm into a copper sheet relative to that entering the surface for a 1 GHz wave directed normal to the sheet.

Solution. First,

$$\frac{\sigma}{\omega\epsilon'} = \frac{5.8 \times 10^7}{2\pi(10^9)(8.854 \times 10^{-12})} = 1.04 \times 10^9$$

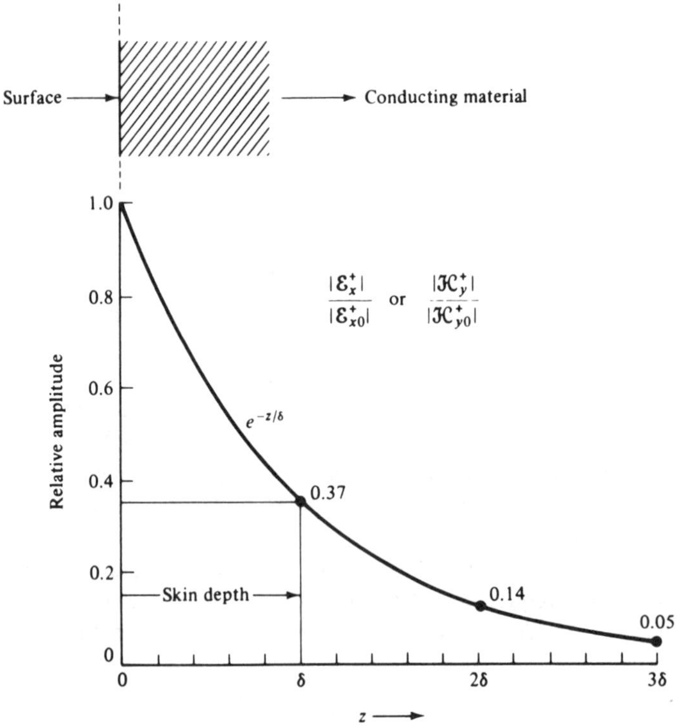

Figure 11-3 Illustration of the decay in amplitude of the $\overline{E}$ or $\overline{H}$ field for an electromagnetic wave entering a conducting material, with the depth of penetration into the material expressed in skin depth units.

which indicates that copper is a very good conductor at this frequency. Next, applying (2),

$$\alpha = \beta = \sqrt{\pi(10^9)(4\pi \times 10^{-7})(5.8 \times 10^7)}$$
$$= 4.79 \times 10^5 \text{ m}^{-1}$$

At a depth of 10^{-4} m,

$$\frac{\mathscr{E}_x^+}{\mathscr{E}_{x0}^+} = e^{-(4.79\times 10^5)(10^{-4})} = e^{-47.9} = 1.58 \times 10^{-21}$$

and

$$\beta z = (4.79 \times 10^5)(10^{-4}) = 47.9 \text{ rad}$$

There is a conduction current density, $\mathcal{J}_x = \sigma\mathscr{E}_x$ induced in a conducting sheet on which a plane wave is incident. Referring to Fig. 11-4, $\mathscr{H}_{y0}$ is the boundary value of the $\overline{H}$ field. The magnitude H_y^+ of the $\overline{H}$ field within the material varies exponentially in the same way as the $\overline{E}$ field discussed above. At a depth d, where d is several skin depths, the $\overline{H}$ field is near zero and does approach zero as d approaches infinity. Then for a width $\Delta\ell$ of the material, the total current in the sheet is given by

$$\begin{aligned} \mathcal{J} &= \oint_\ell \overline{\mathscr{H}} \bullet \overline{d\ell} = \mathscr{H}_{y0}^+ \Delta\ell - \mathscr{H}_y^+(d)\,\Delta\ell \\ &\underset{d\to\infty}{=} \mathscr{H}_{y0}^+ \Delta\ell, \quad \text{since } \mathscr{H}_y^+(d) \to 0 \end{aligned} \tag{6}$$

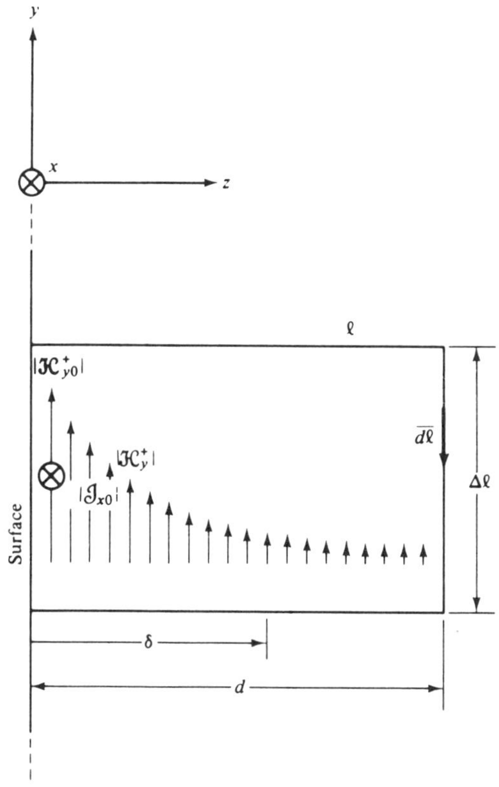

Figure 11-4 Illustration of the path used in $\oint_\ell \overline{H} \cdot \overline{d\ell}$ used to investigate the boundary condition for the tangential $\overline{H}$ field at a good conducting surface.

But also,

$$\mathcal{J} = \Delta\ell \int_0^\infty \mathcal{J}_x \, dz = \Delta\ell \int_0^\infty \mathcal{J}_{x0} e^{-\gamma z} \, dz$$
$$= \frac{\Delta\ell \mathcal{J}_{x0} e^{-\gamma z}}{-\gamma}\bigg|_0^\infty = \frac{\Delta\ell \mathcal{J}_{x0}}{\gamma} = \mathcal{J}_s \, \Delta\ell \tag{7}$$

Equating (6) and (7),

$$\boxed{\begin{aligned}\mathcal{H}^+_{y0} &= \frac{\mathcal{J}_{x0}}{\gamma} = \frac{\mathcal{J}_{x0}(\alpha - j\beta)}{\alpha^2 + \beta_2} = \frac{1}{2}\mathcal{J}_{x0}\,\delta(1 - j) \\ &= \mathcal{J}_s \qquad (\text{A m}^{-1})\end{aligned}} \tag{8}$$

which says that the tangential $\overline{H}$ field at a conducting boundary of at least several skin depths in thickness is equal to $\mathcal{J}_s$, the effective "sheet current" density. The real and imaginary parts of this sheet current density are the same, each being equal to the product of one-half the current density (A m^{-2}) at the surface times the skin depth.

Now, since the $\overline{E}$ field and the $\overline{H}$ field are both attenuated with depth of penetration into the material and since $\overline{E} \times \overline{H}$ represents power density flow, then the differ-

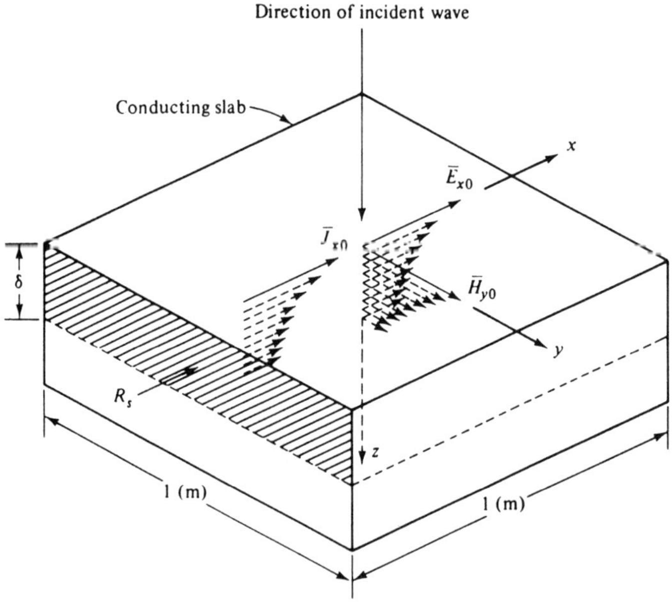

Figure 11-5 Graphical illustration of exponential decay of the amplitudes of the $\overline{E}$ and $\overline{H}$ fields and the conduction current density for a wave entering a conducting slab.

ence between $(\overline{E} \times \overline{H})_{\text{in}}$ and $(\overline{E} \times \overline{H})_{\text{out}}$ must be equal to the (W m^{-2}) absorbed in the material. Referring to Fig. 11-5,

$$\mathcal{J}_x = \sigma \mathcal{E}_x \qquad (\text{A m}^{-2})$$

and since we are dealing with a good conductor, the displacement current density is negligible in comparison with the conduction current density, and

$$\begin{aligned} \nabla \times \overline{\mathcal{H}} &= \overline{\mathcal{J}}_c + \epsilon \frac{\partial \overline{\mathcal{E}}}{\partial t} = (\sigma + j\omega\epsilon')\overline{\mathcal{E}} \\ &\approx \sigma \overline{\mathcal{E}} = \overline{\mathcal{J}}_c \qquad (\text{A m}^{-2}) \end{aligned} \tag{9}$$

Then, since there is only a y component of $\overline{H}$ and an x component of $\overline{E}$, (9) becomes

$$-\frac{\partial \mathcal{H}_y^+}{\partial z} = -\frac{\partial (\mathcal{H}_{y0}^+ e^{-\gamma z})}{\partial z} = \mathcal{J}_x = \mathcal{J}_{x0} e^{-\gamma z} \tag{10}$$

or

$$\gamma \mathcal{H}_{y0}^+ = \mathcal{J}_{x0} = \sigma \mathcal{E}_{x0}^+ \qquad (\text{A m}^{-2}) \tag{11}$$

which agrees with (8), where $\mathcal{J}_{x0}$ is the current density at $z = 0$. The power loss per unit surface area due to the current density $\mathcal{J}_x$ is, for a thickness dz,

$$\frac{1}{2}|I_{dz}|^2 R_{dz} = |\mathcal{J}_x(1)\,dz|^2 \frac{1}{2\sigma(1)\,dz} = \frac{|\mathcal{J}_x|^2\,dz}{2\sigma} \qquad (\text{W})$$

and for the total loss per unit surface area for a material thick enough to absorb all the incident power,

$$\begin{aligned}\mathcal{P}_{\text{loss}} &= \int_0^\infty \frac{|\mathcal{J}_x|^2}{2\sigma}\,dz = \int_0^\infty \frac{|\mathcal{J}_{x0}|^2 e^{-2(z/\delta)}}{2\sigma}\,dz \qquad \text{(W)}\\ &= \frac{|\mathcal{J}_{x0}|^2\,\delta}{4\sigma}\\ &= \tfrac{1}{2}|\mathcal{J}_s|^2 R_s\end{aligned} \tag{12}$$

where $\mathcal{J}_s$ is defined by (8), and where

$$\boxed{R_s = \frac{1}{\sigma\delta} = \frac{1}{\sigma(1)\delta} \qquad (\Omega)} \tag{13}$$

is the skin resistance or surface resistivity, which is the resistance of the material with a cross-sectional area of 1 m by δ and a length of 1 m. Equations (12) and (13) are particularly useful when computing wall losses in waveguides.

The power density entering the material is

$$\mathcal{P}_{\substack{\text{ave}\\ \text{entering}}} = \frac{1}{2}|\mathcal{H}_{y0}^+|^2 \mathcal{R}_e(\eta) \qquad (\text{W m}^{-2}) \tag{14}$$

and from (1),

$$\begin{aligned} &= \frac{1}{2}|\mathcal{H}_{y0}^+|^2 \mathcal{R}_e\left(\sqrt{\frac{\omega\mu}{\sigma}}\,\angle 45^\circ\right)\\ &= \frac{1}{2}|\mathcal{H}_{y0}^+|^2 \sqrt{\frac{\pi f \mu}{\sigma}} = \frac{1}{2}|\mathcal{H}_{y0}^+|^2 \frac{1}{\sigma\delta}\\ &= \frac{1}{2}|\mathcal{H}_{y0}^+|^2 R_s\end{aligned} \tag{15}$$

Therefore, from (12),

$$\boxed{|\mathcal{H}_{y0}^+| = |\mathcal{J}_s| \qquad (\text{A m}^{-1})} \tag{16}$$

where $\mathcal{J}_s$ is a sheet current density. We see that (16) is in agreement with (8), which was obtained from a different approach. We have equated the power density entering a thick (several skin thicknesses) material to the power absorbed in the material. This led to (16), which verifies (8).

Problem 11.6-1 Calculate the skin depth of copper at 60 Hz and at 6 GHz.

Problem 11.6-2 Repeat Prob. 11.6-1 for iron, assuming a μ_r of 1500.

Problem 11.6-3 Calculate the power loss per square meter of wall surface for a thick copper wall if the entering $\overline{E}$ field at the surface is 10^{-6} V m^{-1} at 1 GHz.

11.7 REFLECTION OF PLANE WAVES—NORMAL INCIDENCE

Referring to Fig. 11-6, where the incident wave is traveling from left to right in the positive z direction, the boundary relations as discussed in Sec. 10.10 require that

$$\mathscr{E}_{t_1} = \mathscr{E}_{t_2} = \mathscr{E}^+_{x20} \qquad (\text{V m}^{-1}) \tag{1}$$

$$\mathscr{H}_{t_1} - \mathscr{H}_{t_2} = \mathscr{J}_s \qquad (\text{A m}^{-1}) \tag{2}$$

where $\mathscr{E}_{t_1}$ and $\mathscr{H}_{t_1}$ are the tangential fields on the incident side of the boundary, or region 1, and $\mathscr{E}_{t_2}$ and $\mathscr{H}_{t_2}$ are the tangential fields on the transmitting side of the boundary, or region 2.

The intrinsic impedances on each side of the boundary are

$$\eta_1 = \frac{\mathscr{E}^+_{x10}}{\mathscr{H}^+_{y10}} \qquad (\Omega) \tag{3}$$

$$\eta_2 = \frac{\mathscr{E}^+_{x20}}{\mathscr{H}^+_{y20}} \qquad (\Omega) \tag{4}$$

where $\mathscr{E}^+_{x10}$ and $\mathscr{H}^+_{y10}$ are the complex amplitudes of the incident or forward traveling fields at the boundaries. $\mathscr{E}^+_{x20}$ and $\mathscr{H}^+_{y20}$ are the boundary field complex amplitudes of the forward traveling wave to the right of the boundary, i.e., the transmitted wave.

From (11.6-8), we see that $\mathscr{J}_s = 0$ for a zero thickness boundary, and (2) may be rewritten as

$$\mathscr{H}_{t_1} = \mathscr{H}_{t_2} = \mathscr{H}^+_{y20} \qquad (\text{A m}^{-1}) \tag{5}$$

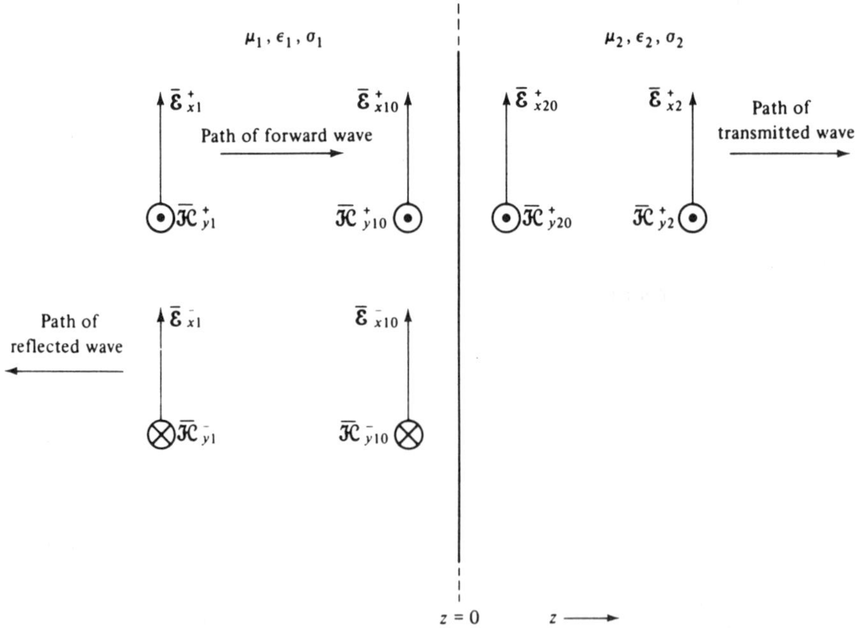

Figure 11-6 Graphical display of the $\overline{\mathscr{E}}$ and $\overline{\mathscr{H}}$ fields for a traveling wave crossing a boundary between two different materials. Also shown is the reflected wave resulting from satisfaction of the boundary conditions for tangential fields.

It is not possible in general for $\mathscr{E}^+_{x10}/\mathscr{H}^+_{y10}$ to be equal to $\mathscr{E}^+_{x20}/\mathscr{H}^+_{y20}$ unless $\eta_1 = \eta_2$. Therefore, additional components of $\overline{E}$ and $\overline{H}$ are required on the incident side of the boundary to make $\mathscr{E}_{t_1} = \mathscr{E}_{t_2}$ and $\mathscr{H}_{t_1} = \mathscr{H}_{t_2}$. But recall from Sec. 11.3 that solutions to the wave equations may include a reverse traveling wave as well as a forward traveling wave. In this case, let us refer to the reverse wave as the reflected wave. As we shall see, the concept of a reflected wave satisfies all the boundary conditions as well as (3) and (4).

Since $\overline{E} \times \overline{H}$ gives the direction of power flow, i.e., the direction of wave travel, a reflected wave must have an $\overline{E} \times \overline{H}$ direction in the $-z$ direction if the forward wave is in the z direction. Let the boundary values of the complex amplitudes of the reflected wave be $\mathscr{E}^-_{x10}$ and $\mathscr{H}^-_{y10}$. Now the reflected $\overline{E}$ field may have the same polarity as the incident $\overline{E}$ field, in which case the reflected $\overline{H}$ field will be reversed in polarity from the $\overline{H}$ field, or the $\overline{E}$ field is reversed if the reflected $\overline{H}$ field has the same polarity as the incident $\overline{H}$ field. We have:

$$\left.\begin{aligned} \mathscr{E}_{t_1} &= (\mathscr{E}^+_{x10} + \mathscr{E}^-_{x10}) \qquad (\text{V m}^{-1}) \\ &\text{and} \\ \mathscr{H}_{t_1} &= \mathscr{H}^+_{y10} + \mathscr{H}^-_{y10} \end{aligned}\right\} \tag{6}$$

but

$$\left.\begin{aligned} \hat{x}\mathscr{E}^+_{x10} \times \hat{y}\mathscr{H}^{+*}_{y10} &= \hat{z}\mathscr{E}^+_{x10}\mathscr{H}^{+*}_{y10} \\ &\text{and} \\ \hat{x}\mathscr{E}^-_{x10} \times \hat{y}\mathscr{H}^{-*}_{y10} &= \hat{z}\mathscr{E}^-_{x10}\mathscr{H}^{-*}_{y10} \end{aligned}\right\} \tag{7}$$

differ by 180° in polarity. The expressions in (7) are the total complex Poynting vectors for the forward and the reverse traveling waves, respectively.

The impedance for the reflected wave is, of course, the same as for the incident wave, but since η_1 has been defined for the forward wave, and the sense of $\overline{E} \times \overline{H}$ is reversed for the reflected wave,

$$\boxed{\frac{\mathscr{E}^+_{x10}}{\mathscr{H}^+_{y10}} = \eta_1 = -\frac{\mathscr{E}^-_{x10}}{\mathscr{H}^-_{y10}} \qquad (\Omega)} \tag{8}$$

Let us define a complex reflection coefficient ρ by

$$\boxed{\rho \triangleq \frac{\mathscr{E}^-_{x10}}{\mathscr{E}^+_{x10}}} \tag{9}$$

and from (8), we also see that

$$\rho = -\frac{\mathscr{H}^-_{y10}}{\mathscr{H}^+_{y10}} \tag{10}$$

We can now write (6) as

$$\begin{aligned} \mathscr{E}_{t_1} &= \mathscr{E}^+_{x10}(1 + \rho) \\ \mathscr{H}_{t_1} &= \mathscr{H}^+_{y10}(1 - \rho) = \frac{\mathscr{E}^+_{x10}}{\eta_1}(1 - \rho) \end{aligned} \tag{11}$$

From (1), (2), (3), and (5), and substituting for $\mathscr{E}_{t_1}$ and $\mathscr{H}_{t_1}$ from (11), we have

$$\frac{\mathscr{E}_{t_1}}{\mathscr{H}_{t_1}} = \frac{\mathscr{E}_{t_2}}{\mathscr{H}_{t_2}} = \frac{\mathscr{E}_{x10}^{+}(1+\rho)}{\mathscr{E}_{x10}^{+}(1-\rho)}\eta_1 = \eta_2 \tag{12}$$

and solving for ρ, we get

$$\boxed{\rho = \frac{\eta_2 - \eta_1}{\eta_2 + \eta_1}} \tag{13}$$

Now to the left of the boundary, we have the total fields,

$$\begin{aligned} \mathscr{E}_{x1} &= \mathscr{E}_{x10}^{+}e^{-\gamma_1 z} + \mathscr{E}_{x10}^{-}e^{\gamma_1 z} \\ &= \mathscr{E}_{x10}^{+}(e^{-\gamma_1 z} + \rho e^{\gamma_1 z}) \qquad (\text{V m}^{-1}) \end{aligned} \tag{14}$$

$$\begin{aligned} \mathscr{H}_{y1} &= \mathscr{H}_{y10}^{+}e^{-\gamma_1 z} + \mathscr{H}_{y10}^{-}e^{\gamma_1 z} \\ &= \frac{\mathscr{E}_{x10}^{+}}{\eta_1}(e^{-\gamma_1 z} - \rho e^{\gamma_1 z}) \qquad (\text{A m}^{-1}) \end{aligned} \tag{15}$$

and to the right of the boundary,

$$\begin{aligned} \mathscr{E}_{x2} &= (\mathscr{E}_{x10}^{+} + \mathscr{E}_{x10}^{-})e^{-\gamma_2 z} = \mathscr{E}_{x20}^{+}e^{-\gamma_2 z} \\ &= \mathscr{E}_{x10}^{+}(1+\rho)^{-\gamma_2 z} \qquad (\text{V m}^{-1}) \end{aligned} \tag{16}$$

$$\begin{aligned} \mathscr{H}_{y2} &= (\mathscr{H}_{y10}^{+} + \mathscr{H}_{y10}^{-})e^{-\gamma_2 z} = \mathscr{H}_{y20}^{+}e^{-\gamma_2 z} \\ &= \frac{\mathscr{E}_{x10}^{+}}{\eta_1}(1-\rho)e^{-\gamma_2 z} \qquad (\text{A m}^{-1}) \end{aligned} \tag{17}$$

A transmission coefficient τ is defined as

$$\boxed{\tau \triangleq \frac{\mathscr{E}_{x20}^{+}}{\mathscr{E}_{x10}^{+}} = \frac{\mathscr{E}_{x10}^{+} + \mathscr{E}_{x10}^{-}}{\mathscr{E}_{x10}^{+}} = 1 + \rho}$$

$$= 1 + \frac{\eta_2 - \eta_1}{\eta_2 + \eta_1}$$

$$\boxed{\tau = \frac{2\eta_2}{\eta_2 + \eta_1}} \tag{18}$$

All the $\overline{\mathscr{E}}$ and $\overline{\mathscr{H}}$ fields can now be defined in terms of the incident field at the boundary. We have

$$\mathscr{E}_{x1}^{+} = \mathscr{E}_{x10}^{+}e^{-\gamma_1 z} \tag{19}$$

$$\mathscr{H}_{y1}^{+} = \frac{\mathscr{E}_{x10}^{+}}{\eta_1}e^{-\gamma_1 z} \tag{20}$$

$$\mathscr{E}_{x1}^{-} = \rho\mathscr{E}_{x10}^{+}e^{\gamma_1 z} \tag{21}$$

$$\mathscr{H}_{y1}^{-} = -\frac{\rho}{\eta_1}\mathscr{E}_{x10}^{+}e^{\gamma_1 z} \tag{22}$$

$$\mathscr{E}_{x2}^{+} = \tau\mathscr{E}_{x10}^{+}e^{-\gamma_2 z} \tag{23}$$

$$\mathscr{H}_{y2}^{+} = \frac{\tau}{\eta_2}\mathscr{E}_{x10}^{+}e^{-\gamma_2 z} \tag{24}$$

In sinusoidal notation,

$$E_{x1}^{+} = E_{x10}^{+} e^{-\alpha_1 z} \cos(\omega t - \beta_1 z) \tag{25}$$

$$H_{y1}^{+} = \frac{E_{x10}^{+}}{|\eta_1|} e^{-\alpha_1 z} \cos(\omega t - \beta_1 z - \theta_{\eta_1}) \tag{26}$$

$$E_{x1}^{-} = |\rho| E_{x10}^{+} e^{\alpha_1 z} \cos(\omega t + \beta_1 z + \theta_\rho) \tag{27}$$

$$H_{y1}^{-} = \frac{|\rho|}{|\eta_1|} E_{x10}^{+} e^{\alpha_1 z} \cos(\omega t + \beta_1 z - \theta_{\eta_1} + \theta_\rho + \pi) \tag{28}$$

$$E_{x2}^{+} = |\tau| E_{x10}^{+} e^{-\alpha_2 z} \cos(\omega t - \beta_2 z + \theta_\tau) \tag{29}$$

$$H_{y2}^{+} = \frac{|\tau|}{|\eta_2|} E_{x10}^{+} e^{-\alpha_2 z} \cos(\omega t - \beta_2 z - \theta_{\eta_2} + \theta_\tau) \tag{30}$$

where $\eta_1 = |\eta_1| \angle\theta_{\eta_1}$, $\eta_2 = |\eta_2| \angle\theta_{\eta_2}$, $\rho = |\rho| \angle\theta_\rho$, and $\tau = |\tau| \angle\theta_\tau$. Note also that, from the notation established in Sec. 10.7, $E_{x10}^{+} = |\mathscr{E}_{x10}^{+}|$, etc.

The amplitudes of the $\overline{E}$ field waves are plotted in Fig. 11-7 as a function of distance along the z axis. The amplitude of the reflected wave, $E_{x10}^{-} e^{\alpha_1 z}$, can at most be equal to the incident wave amplitude at $z = 0$ and that only if $|\rho| = 1$. The amplitude of the transmitted wave at $z = 0$ can be less or greater than that of the incident wave, depending on whether $|1 + \rho|$ or $|\tau|$ is less than or greater than 1.

Example 6

A plane wave is normally incident from medium 1 on an interface with medium 2, where $\epsilon_{r_1}' = 2$, $\sigma_1 = 0.2$ S m^{-1}, $\epsilon_{r_2}' = 4$, and $\sigma_2 = 0.1$ S m^{-1}. Find γ_1, γ_2, η_1, η_2, and the amplitudes of E_{x10}^{-}, H_{y10}^{+}, H_{y10}^{-}, E_{x20}^{+}, H_{y20}^{+}, in terms of E_{x10}^{+} at $f = 10$ GHz.

Solution. Find $\sigma_1/\omega\epsilon_1'$ and $\sigma_2/\omega\epsilon_2'$.

$$\frac{\sigma_1}{\omega\epsilon_1'} = \frac{0.2}{20\pi(10^9)(2)(10^{-9}/36\pi)} = 0.18$$

$$\frac{\sigma_2}{\omega\epsilon_2'} = \frac{0.1}{20\pi(10^9)(4)(10^{-9}/36\pi)} = 0.05$$

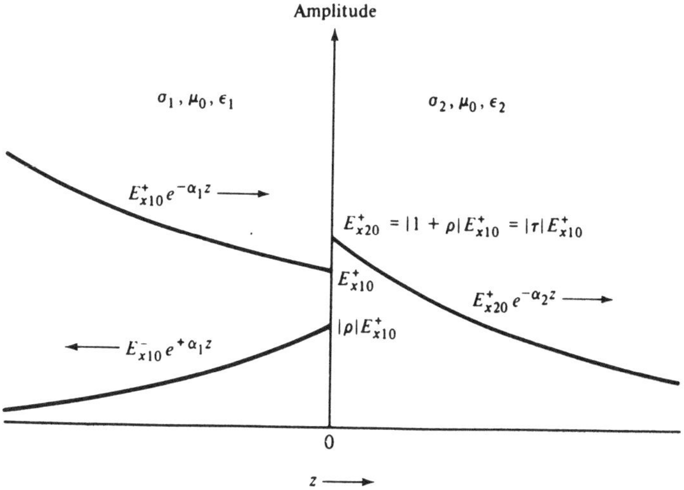

Figure 11-7 Illustration of how the incident, reflected, and transmitted wave amplitudes vary with respect to the boundary between medium 1 and medium 2.

From Table 11-1,

$$\eta_1 = \frac{\sqrt{\mu_0/\epsilon_1'}}{\sqrt{1 - j0.18}} = \frac{377/\sqrt{2}}{\sqrt{1 - j0.18}} = 264.46 \underline{/5.10^\circ} \quad (\Omega)$$

$$= 263.41 + j23.52 \quad (\Omega)$$

$$\eta_2 = \frac{\sqrt{\mu_0/\epsilon_2'}}{\sqrt{1 - j0.05}} = \frac{377/\sqrt{4}}{\sqrt{1 - j0.05}} = 188.26 \underline{/1.43^\circ} \quad (\Omega)$$

$$= 188.20 + j4.69 \quad (\Omega)$$

Also from Table 11-1,

$$\begin{aligned}\gamma_1 &= j\omega\sqrt{\mu\epsilon_1'}\,\sqrt{1 - j0.18}\\ &= j20\pi(10^9)\frac{\sqrt{2}}{3 \times 10^8}1.01 \underline{/-5.10^\circ}\\ &= 299 \underline{/84.9^\circ}\\ &= 26.6 + j297.97 \quad (\text{m}^{-1})\\ &= \alpha_1 + j\beta_1\end{aligned}$$

and

$$\begin{aligned}\gamma_2 &= j20\pi(10^9)\frac{\sqrt{4}}{3 \times 10^8}1.0012 \underline{/-1.43^\circ}\\ &= 419.4 \underline{/88.57^\circ}\\ &= 10.46 + j419.25 \quad (\text{m}^{-1})\\ &= \alpha_2 + j\beta_2\end{aligned}$$

We must find ρ and τ to compute amplitudes. From (13),

$$\begin{aligned}\rho &= \frac{\eta_2 - \eta_1}{\eta_2 + \eta_1} = \frac{188.20 + j4.69 - 263.41 - j23.52}{188.20 + j4.69 + 263.41 + j23.52}\\ &= 0.17 \underline{/-169.52^\circ}\\ &= -0.1685 - j0.0312\end{aligned}$$

From (18),

$$\begin{aligned}\tau &= 1 + \rho = 1 - 0.1685 - j0.0312\\ &= 0.8315 - j0.0312 = 0.8321 \underline{/-2.15^\circ}\end{aligned}$$

Then

$$\begin{aligned}E_{x10}^- &= |\rho|E_{x10}^+ = 0.17E_{x10}^+ \quad (\text{V m}^{-1})\\ H_{y10}^+ &= \frac{E_{x10}^+}{|\eta_1|} = \frac{E_{x10}^+}{|264.46|} = 0.0038E_{x10}^+ \quad (\text{A m}^{-1})\\ H_{y10}^- &= |\rho|H_{y10}^+ = (0.17)(0.0038E_{x10}^+) = 0.00064E_{x10}^+ \quad (\text{A m}^{-1})\\ E_{x20}^+ &= |\tau|E_{x10}^+ = 0.8321E_{x10}^+ \quad (\text{V m}^{-1})\\ H_{y20}^+ &= \frac{E_{x20}^+}{|\eta_2|} = \frac{0.8321E_{x10}^+}{188.26} = 0.00442E_{x10}^+ \quad (\text{A m}^{-1})\end{aligned}$$

Example 7

Find the values of the phasors $\mathscr{E}_{x10}^-$, $\mathscr{H}_{y10}^+$, $\mathscr{H}_{y10}^-$, $\mathscr{E}_{x20}^+$, and $\mathscr{H}_{y20}^+$ relative to the phasor $\mathscr{E}_{x10}^+$.

Solution. The magnitudes of the phasors have already been determined from Example 6, and the angles can be found using (25) through (30).

From Example 6 and (26),

$$\begin{aligned}\frac{\mathscr{H}_{y10}^+}{\mathscr{E}_{x10}^+} &= 0.0038 \underline{/-\theta_{\eta 1}} \quad (\text{S})\\ &= 0.0038 \underline{/-5.10^\circ} \quad (\text{S})\end{aligned}$$

From Example 6 and (27),

$$\frac{\mathscr{E}_{x10}^{-}}{\mathscr{E}_{x10}^{+}} = 0.17 \underline{/\theta_\rho}$$
$$= 0.17 \underline{/-169.52^\circ}$$

From Example 6 and (28),

$$\frac{\mathscr{H}_{y10}^{-}}{\mathscr{E}_{x10}^{+}} = 0.00064 \underline{/-\theta_{\eta 1} + \theta_\rho + 180^\circ}$$
$$= 0.00064 \underline{/5.38^\circ} \qquad \text{(S)}$$

From Example 6 and (29),

$$\frac{\mathscr{E}_{x20}^{+}}{\mathscr{E}_{x10}^{+}} = 0.8321 \underline{/\theta_\tau}$$
$$= 0.8321 \underline{/-2.15^\circ}$$

From Example 6 and (30),

$$\frac{\mathscr{H}_{y20}^{+}}{\mathscr{E}_{x10}^{+}} = 0.00442 \underline{/-\theta_{\eta 2} + \theta_\tau}$$
$$= 0.00442 \underline{/-3.58^\circ} \qquad \text{(S)}$$

The complex values of these phasors are plotted in Fig. 11-8, where the angle of $\mathscr{E}_{x10}^{+}$ is used as a reference for all the other phasors.

The total amplitude of the combined incident and reflected waves has an oscillatory-shaped pattern as a function of distance from the interface. This is because the phase of the incident wave advances as the source is approached, whereas the phase

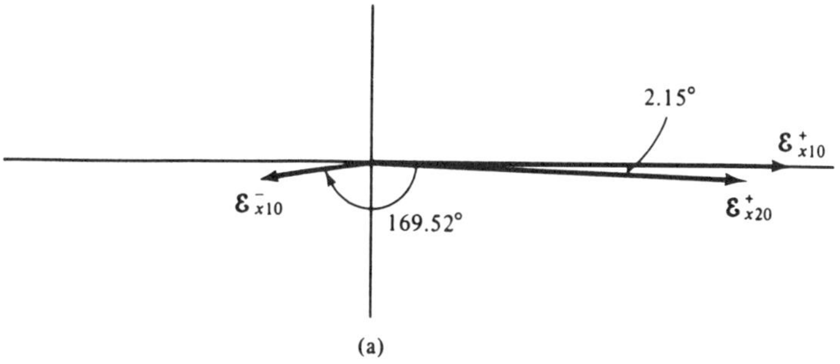

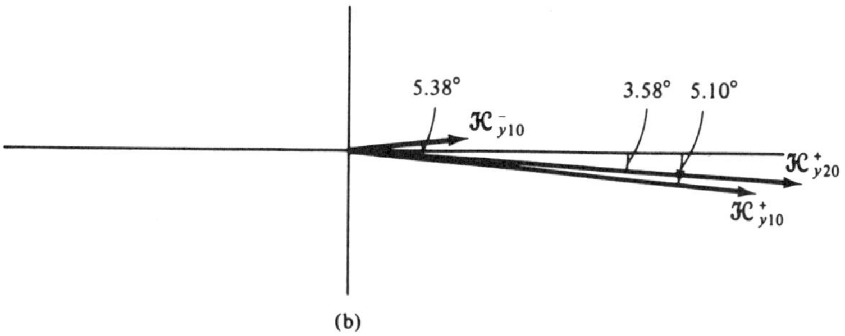

Figure 11-8 Complex values of the phasors calculated in Example 7. Parts (a) and (b) are not the same amplitude scale, but the angle reference is the same.

of the reflected wave is retarded. The "source" for the reflected wave is the point of reflection, i.e., the interface between medium 1 and medium 2. This oscillatory pattern referred to is known as a *standing wave pattern*. We will discuss this further when we consider the reflections of plane waves at the interface between two lossless dielectrics and at the interface between a lossless dielectric and a good conductor. In lossy media, the standing wave pattern diminishes in amplitude with distance from the point of reflection because the relative amplitude of the reflected wave to the incident wave decreases in that direction, and at a great distance from the point of reflection, the reflected wave has negligible effect because of attenuation.

Problem 11.7-1 Repeat the calculations of Example 6, using $f = 1$ GHz.

Problem 11.7-2 Repeat the calculations of Example 7, using the results of Prob. 11.7-1, and make a plot similar to that of Fig. 11-8.

11.8 PLANE WAVE NORMALLY INCIDENT ON DIELECTRIC TO DIELECTRIC INTERFACE

Consider the case where $\sigma_1 = \sigma_2 = 0$, and $\mu_1 = \mu_2 = \mu_0$; then ϵ_1 and ϵ_2 are both positive real numbers. In dealing with such lossless dielectrics, the attenuation constants α_1 and α_2 are both zero ($\gamma_1 = j\beta_1, \gamma_2 = j\beta_2$), the intrinsic impedances η_1 and η_2 are real, and the reflection coefficient is real. From (11.7-13) we have, then,

$$\left.\begin{aligned} \rho &= \text{positive real if } \eta_2 > \eta_1 \\ &= \text{negative real if } \eta_2 < \eta_1 \end{aligned}\right\} \tag{1}$$

Then from (11.7-13), (11.7-21), and (11.7-22) we see that the reflected $\overline{E}$ field at the interface is either in phase (if ρ is positive) or 180° out of phase (if ρ is negative) with the incident $\overline{E}$ field, and that the reflected $\overline{H}$ field is either 180° out of phase (ρ positive) or in phase (ρ negative) with the incident $\overline{H}$ field. By equating tangential fields on both sides of the boundary, we obtain

$$\mathscr{E}^+_{x20} = \mathscr{E}^+_{x10} + \mathscr{E}^-_{x10} = \mathscr{E}^+_{x10}(1 + \rho) \qquad (\text{V m}^{-1}) \tag{2}$$

$$\begin{aligned} \mathscr{H}^+_{y20} = \mathscr{H}^+_{y10} + \mathscr{H}^-_{y10} &= \frac{\mathscr{E}^+_{x10}}{\eta_1}(1 - \rho) \\ &= \mathscr{H}^+_{y10}(1 - \rho) \qquad (\text{A m}^{-1}) \end{aligned} \tag{3}$$

or, since ρ is positive or negative real,

$$|\mathscr{E}^+_{x20}| = |\mathscr{E}^+_{x10}|\,(1 \pm |\rho|) \tag{4}$$

$$|\mathscr{H}^+_{y20}| = |\mathscr{H}^+_{y10}|\,(1 \mp |\rho|) \tag{5}$$

where the upper algebraic signs in (4) and (5) apply when $\eta_2 > \eta_1$, and the lower signs apply when $\eta_2 < \eta_1$.

The relationship between the incident, reflected, and transmitted vectors is displayed in Fig. 11-9 for $\eta_2 < \eta_1$ and for $\eta_2 > \eta_1$. In phasor form, the fields are

$$\mathscr{E}^+_{x1} = \mathscr{E}^+_{x10}e^{-j\beta_1 z} \tag{6}$$

$$\mathscr{H}^+_{y1} = \frac{\mathscr{E}^+_{x10}}{\eta_1}e^{-j\beta_1 z} \tag{7}$$

$$\mathscr{E}^-_{x1} = \rho\mathscr{E}^+_{x10}e^{j\beta_1 z} \tag{8}$$

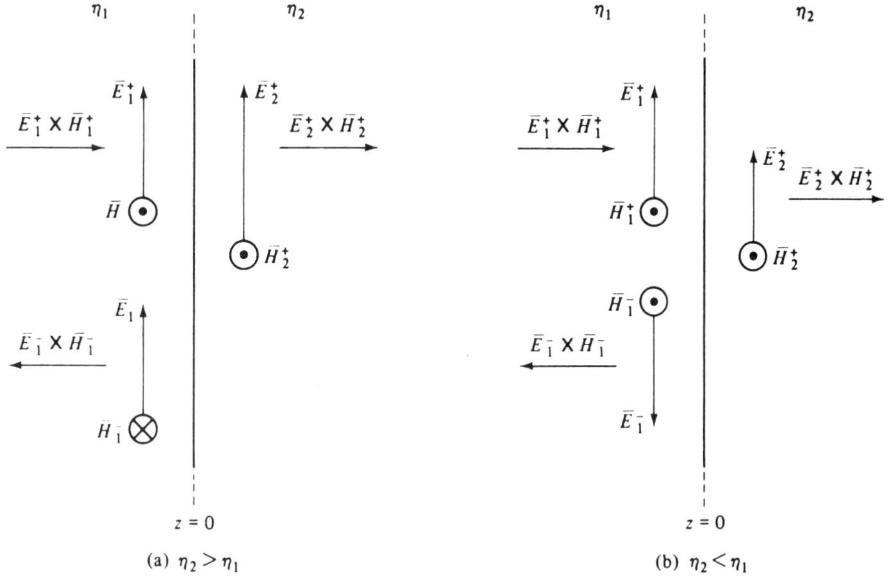

Figure 11-9 Graphical display of the satisfying of boundary conditions for two cases.

$$\mathcal{H}_{y1}^{-} = \frac{-\rho}{\eta_1}\mathcal{E}_{x10}^{+}e^{j\beta_1 z} \tag{9}$$

$$\mathcal{E}_{x2}^{+} = \tau\mathcal{E}_{x10}^{+}e^{-j\beta_2 z} \tag{10}$$

$$\mathcal{H}_{y2}^{+} = \frac{\tau}{\eta_2}\mathcal{E}_{x10}^{+}e^{-j\beta_2 z} \tag{11}$$

If we add (6) and (8) we get the total $\overline{E}$ field to the left of the boundary.

$$\mathcal{E}_{x1} = \mathcal{E}_{x10}^{+}(e^{-j\beta_1 z} + \rho e^{j\beta_1 z}) \tag{12}$$

Factoring out an $e^{-j\beta_1 z}$, then dividing both sides of the equation by $\mathcal{E}_{x10}^{+}$, and then taking the magnitude of both sides, we have

$$\begin{aligned}\left|\frac{\mathcal{E}_{x1}}{\mathcal{E}_{x10}^{+}}\right| &= |e^{-j\beta_1 z}(1 + \rho e^{2j\beta_1 z})| \\ &= |1 + \rho e^{2j\beta_1 z}|\end{aligned} \tag{13}$$

since $|e^{-j\beta_1 z}| = 1$. The magnitude of the total $\overline{E}$ field is then proportional to $|1 + \rho e^{2j\beta_1 z}|$ and can be solved graphically by what is referred to as a *crank diagram*. In Fig. 11-10 the length of the resultant arm is proportional to the total $\overline{E}$ field. For ρ positive, $|1 + \rho e^{2j\beta_1 z}|$ is a *maximum* at $z = 0, -\lambda_1/2, -\lambda_1, -3\lambda_1/2, -2\lambda$, etc.

$$2\beta_1 z = 2\left(\frac{2\pi}{\lambda}\right)z = \frac{4\pi}{\lambda_1}z$$

$$e^{2j\beta_1 z} = 1 \quad \text{for } 2\beta_1 z = 0, -2\pi, -4\pi, \ldots$$

or

$$z = 0, -\frac{\lambda_1}{2}, -\lambda_1, -\frac{3\lambda_1}{2}, \ldots \tag{14}$$

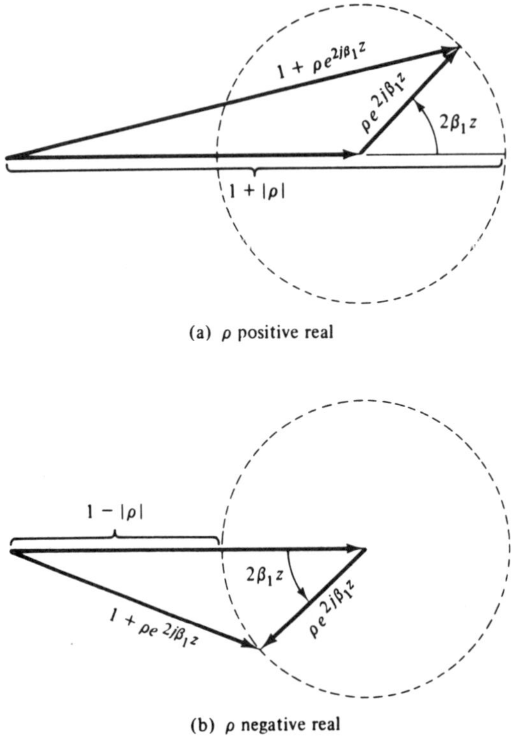

Figure 11-10 Crank diagram used to illustrate how maxima and minima occur as a function of z.

Also, for ρ positive, we find that $|1 + \rho e^{2j\beta_1 z}|$ is a *minimum* at $z = -\lambda_1/4, -3\lambda_1/4, -5\lambda_1/4$, etc.

$$e^{2j\beta_1 z} = -1 \quad \text{for } 2\beta_1 z = -\pi, -3\pi, -5\pi, \ldots$$

or

$$z = -\frac{\lambda_1}{4}, -\frac{3\lambda_1}{4}, -\frac{5\lambda_1}{4}, \ldots \tag{15}$$

And for the $\overline{\mathcal{H}}$ field, by adding (7) and (9) we get

$$\mathcal{H}_{y1} = \frac{\mathcal{E}_{x10}^{+}}{\eta_1}(e^{-j\beta_1 z} - \rho e^{j\beta_1 z}) \tag{16}$$

or

$$\left|\frac{\mathcal{H}_{y1}}{\mathcal{E}_{x10}^{+}}\right| = \frac{1}{\eta_1}|1 - \rho e^{2j\beta_1 z}| \tag{17}$$

We see that $|1 - \rho e^{2j\beta_1 z}|$ has a maximum where $|1 + \rho e^{2j\beta_1 z}|$ has a minimum and vice versa. The ratio of maximum to minimum for either the $\overline{E}$ field or the $\overline{H}$ field is

$$\boxed{S \triangleq \frac{|1 \pm \rho e^{2j\beta_1 z}|_{\max}}{|1 \pm \rho e^{2j\beta_1 z}|_{\min}} = \frac{1 + |\rho|}{1 - |\rho|}} \tag{18}$$

where S is referred to as the *standing wave ratio*. We see from the foregoing that

$$S = \frac{E_{\text{max}}}{E_{\text{min}}} = \frac{H_{\text{max}}}{H_{\text{min}}} = \frac{1 + |\rho|}{1 - |\rho|} \tag{19}$$

but that an H_{max} is $\lambda/4$ from an E_{max}, and E_{max} and E_{min} are $\lambda/4$ apart, etc. Also, when ρ is negative, the positions of maxima and minima are interchanged from the case where ρ is positive. The standing wave patterns for ρ positive ($\eta_2 > \eta_1$) and ρ negative ($\eta_2 < \eta_1$) are displayed in Fig. 11-11.

Note from Fig. 11-11 that the amplitudes of $\mathscr{E}_{x2}$ and $\mathscr{H}_{y2}$ for all $z > 0$ are the amplitudes at the interface. Also, $|\mathscr{E}_{x2}|/|\mathscr{H}_{y2}| = \eta_2$ for all $z > 0$, but $|\mathscr{E}_{x1}|/|\mathscr{H}_{y1}| \neq \eta_1$ for all $z < 0$. In region 1, the ratio of the total $\overline{\mathscr{E}}$ field to the total $\overline{\mathscr{H}}$ field may be thought of as an input impedance viewed from the source, much like that for transmission lines that are discussed in Chapter 12. From (12) and (16),

$$\frac{\mathscr{E}_{x1}}{\mathscr{H}_{y1}} = \eta_{\text{in}} = \frac{\mathscr{E}_{x1}^{+} + \mathscr{E}_{x1}^{-}}{\mathscr{H}_{y1}^{+} + \mathscr{H}_{y1}^{-}} = \frac{\mathscr{E}_{x10}^{+}(e^{-j\beta_1 z} + \rho e^{j\beta_1 z})}{\dfrac{\mathscr{E}_{x10}^{+}}{\eta_1}(e^{-j\beta_1 z} - \rho e^{j\beta_1 z})} \tag{20}$$

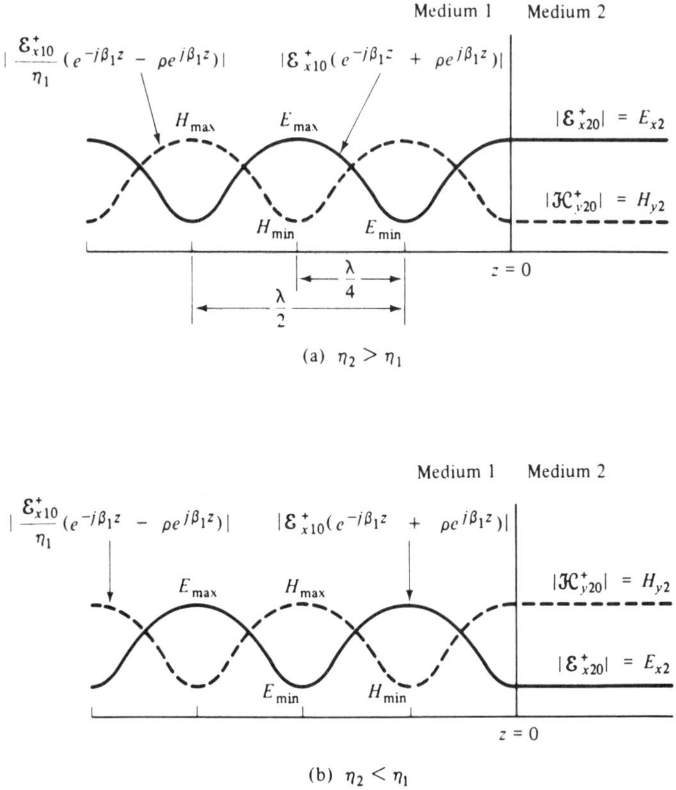

Figure 11-11 Graphical illustration of the standing wave pattern caused by reflection from a boundary located at $z = 0$.

and substituting (11.7-13) for ρ,

$$\eta_{in} = \eta_1 \frac{e^{-j\beta_1 z} + \dfrac{\eta_2 - \eta_1}{\eta_2 + \eta_1} e^{j\beta_1 z}}{e^{-j\beta_1 z} - \dfrac{\eta_2 - \eta_1}{\eta_2 + \eta_1} e^{j\beta_1 z}} \tag{21}$$

$$= \eta_1 \left[\frac{\eta_2(e^{-j\beta_1 z} + e^{j\beta_1 z}) + \eta_1(e^{-j\beta_1 z} - e^{j\beta_1 z})}{\eta_1(e^{-j\beta_1 z} + e^{j\beta_1 z}) + \eta_2(e^{-j\beta_1 z} - e^{j\beta_1 z})} \right]$$

and after dividing numerator and denominator by $(e^{-j\beta_1 z} + e^{j\beta_1 z})$, we have

$$\boxed{\eta_{in} = \eta_1 \left[\frac{\eta_2 - j\eta_1 \tan \beta_1 z}{\eta_1 - j\eta_2 \tan \beta_1 z} \right] \qquad (\Omega)} \tag{22}$$

We see that $\eta_{in} = \eta_1$ if $\eta_1 = \eta_2$, in which case there is no reflection from the interface and hence no standing wave pattern. We shall refer to (22) a little later when we discuss two interfaces, and the equation also has a counterpart in the study of transmission lines.

An incident plane wave carries a power density described by the Poynting vector $\overline{\mathcal{P}}_1^+$ directed in the positive z direction.

$$\overline{\mathcal{P}}_1^+ = \overline{E}_1^+ \times \overline{H}_1^+ \qquad (\text{W m}^{-2}) \tag{23}$$

Likewise, the reflected wave carries a power density

$$\overline{\mathcal{P}}_1^- = \overline{E}_1^- \times \overline{H}_1^- \qquad (\text{W m}^{-2}) \tag{24}$$

directed in the negative z direction. On side 2 of the boundary, the transmitted Poynting vector is

$$\overline{\mathcal{P}}_2^+ = \overline{E}_2^+ \times \overline{H}_2^+ \qquad (\text{W m}^{-2}) \tag{25}$$

Since we are dealing with lossless materials where the $\overline{E}$ and $\overline{H}$ fields are in time phase and since $\overline{E}$ and $\overline{H}$ are orthogonal in space, we can write the equation

$$\mathcal{P}_1^+ = |\mathscr{E}_{x1}^+ \mathscr{H}_{y1}^+| = \frac{|\mathscr{E}_{x1}^+|^2}{\eta_1} \qquad (\text{W m}^{-2}) \tag{26}$$

for the magnitude of the instantaneous Poynting vector for the incident wave.[1] For the average power density in the incident or forward wave,

$$\mathcal{P}_{1\,av}^+ = \frac{1}{2} \frac{|\mathscr{E}_{x1}^+|^2}{\eta_1} = \frac{1}{2} \frac{|\mathscr{E}_{x10}^+|^2}{\eta_1} \qquad (\text{W m}^{-2}) \tag{27}$$

and for the reflected wave,

$$\mathcal{P}_{1\,av}^- = \frac{1}{2} \frac{|\mathscr{E}_{x10}^-|^2}{\eta_1} \qquad (\text{W m}^{-2}) \tag{28}$$

and

$$\mathcal{P}_{2\,av}^+ = \frac{1}{2} \frac{|\mathscr{E}_{x20}^+|^2}{\eta_2} \qquad (\text{W m}^{-2}) \tag{29}$$

for the transmitted wave. But since $\mathscr{E}_{x10}^- = \rho \mathscr{E}_{x10}^+$ and $\mathscr{E}_{x20}^+ = \tau \mathscr{E}_{x10}^+$, we have

$$\mathcal{P}_{1\,av}^- = \frac{1}{2} \frac{|\rho \mathscr{E}_{x10}^+|^2}{\eta_1} \tag{30}$$

[1] If the material is not lossless, then (26) is written as

$$\mathcal{P}_1^+ = \mathscr{E}_{x1}^+ \mathscr{H}_{y1}^{+*} = \text{complex power density}$$

and

$$\mathcal{P}^+_{2\,\text{av}} = \frac{1}{2}\frac{|\tau\mathcal{E}^+_{x10}|^2}{\eta_2} \tag{31}$$

With some algebraic manipulation, we can show that

$$\boxed{\mathcal{P}^+_{2\,\text{av}} = (\mathcal{P}^+_{1\,\text{av}} - \mathcal{P}^-_{1\,\text{av}}) \qquad (\text{W m}^{-2})}$$

$$\frac{1}{2}\frac{|\mathcal{E}^+_{x20}|^2}{\eta_2} = \frac{1}{2}\frac{|\mathcal{E}^+_{x10}|^2}{\eta_1}(1 - |\rho|^2) \tag{32}$$

or

$$\frac{1}{2}|\mathcal{E}^+_{x20}|\,|\mathcal{H}^+_{y20}| = \frac{1}{2}|\mathcal{E}^+_{x10}|\,|\mathcal{H}^+_{y10}|\,(1 - |\rho|^2)$$

for a dielectric to dielectric interface.

Example 8

Referring to Fig. 11-9, the incident $\overline{E}$ field is given by $\mathcal{E}^+_{x1} = 100e^{-j\beta_1 z}$ (V m^{-1}) and the wave is incident from air. The material of region 2 is characterized by $\mu_r = 1$, $\epsilon'_r = 3.4$. Find η_1, η_2, λ_2/λ_1, ρ, $\mathcal{P}^+_{1\,\text{ave}}$, $\mathcal{P}^-_{1\text{ave}}$ and $\mathcal{P}^+_{2\,\text{ave}}$.

Solution

$$\eta_1 = \sqrt{\frac{\mu_1}{\epsilon_1}} = \sqrt{\frac{\mu_0}{\epsilon_0}} = 377\ \Omega$$

$$\eta_2 = \sqrt{\frac{\mu_0}{3.4\epsilon_0}} = \frac{377}{\sqrt{3.4}} = 204\ \Omega$$

$$\mathcal{H}^+_{y1} = \frac{100}{377}e^{-j\beta_1 z} = 0.265e^{-j\beta_1 z} \qquad (\text{A m}^{-1})$$

$$\rho = \frac{\eta_2 - \eta_1}{\eta_2 + \eta_1} = \frac{204 - 377}{204 + 377} = -0.298$$

$$\mathcal{E}^-_{x1} = -0.298(100)e^{j\beta_1 z} = -29.8e^{j\beta_1 z} \qquad (\text{V m}^{-1})$$

$$\mathcal{H}^-_{y1} = \frac{29.8}{377}e^{j\beta_1 z} = 0.079e^{j\beta_1 z} \qquad (\text{A m}^{-1})$$

$$\mathcal{E}^+_{x2} = (100 - 29.8)e^{-j\beta_2 z} = 70.2e^{-j\beta_2 z} \qquad (\text{V m}^{-1})$$

$$\mathcal{H}^+_{y2} = \frac{70.2}{204}e^{-j\beta_2 z} = 0.344e^{-j\beta_2 z} \qquad (\text{A m}^{-1})$$

$$\frac{\lambda_2}{\lambda_1} = \frac{\beta_1}{\beta_2} = \frac{\omega\sqrt{\mu_1\epsilon_1}}{\omega\sqrt{\mu_2\epsilon_2}} = \frac{1}{\sqrt{3.4}}$$

$$\mathcal{P}^+_{1\,\text{ave}} = \frac{1}{2}(100)\,(0.265) = 13.25\ \text{W m}^{-2}$$

$$\mathcal{P}^-_{1\,\text{ave}} = \frac{1}{2}(29.8)\,(0.079) = 1.17\ \text{W m}^{-2}$$

$$\mathcal{P}^+_{2\,\text{ave}} = \frac{1}{2}(70.2)\,(0.344) = 12.05\ \text{W m}^{-2}$$

Note that

$$\mathcal{P}^+_{2\,\text{ave}} = \mathcal{P}^+_{1\,\text{ave}} - \mathcal{P}^-_{1\,\text{ave}}$$

Problem 11.8-1 A plane wave is normally incident from air on a semi-infinite slab of dielectric material, $\epsilon_r' = 2.4$. If the frequency is 2.45 GHz, find: (a) the reflection coefficient; (b) the standing wave ratio in front of the dielectric slab; (c) the wavelengths in air and in the slab; (d) the percentage of the incident power that is reflected from the interface.

Problem 11.8-2 Sketch the standing wave patterns for the E and H fields for Prob. 11.8-1, indicating the distances of the nearest maximum and minimum from the air-dielectric interface.

Problem 11.8-3 Rework Example 8 with the 100-V m^{-1} $\overline{E}$ field incident from the dielectric, with region 2 being air.

11.8-1 The Two-Interface Problem

For many applications, a standing wave ratio much greater than unity is undesirable. A large value for S indicates that power is being reflected from some interface and that the transmission path between the source and some distant receiving point is not as efficient as it would be if $S = 1$. We see from (11.7-13) that when two media of different intrinsic impedance are joined together, the interface will not, in general, pass plane waves at normal incidence without reflection. At optical frequencies, the air-glass interface for lenses is a good example. Lenses are coated to reduce reflection, and we shall examine here the principles by which this coating is effective.

Returning to (22), if we substitute a distance ℓ from the interface for $-z$, we have

$$\eta_{\text{in}} = \eta_1 \frac{\eta_2 + j\eta_1 \tan \beta_1 \ell}{\eta_1 + j\eta_2 \tan \beta_1 \ell} \quad (\Omega) \tag{33}$$

which is independent of positive or negative direction. Now, for the two-interface arrangement shown in Fig. 11-12, on side 2 of the 1-2 interface, the input impedance looking toward region 3 is

$$\eta^{+}_{\text{in}_{1-2}} = \eta_2 \frac{\eta_3 + j\eta_2 \tan \beta_2 d}{\eta_2 + j\eta_3 \tan \beta_2 d} \quad (\Omega) \tag{34}$$

Now, if it should happen that $\eta^{+}_{\text{in}_{1-2}} = \eta_1$, then there would be no reflection from the 1-2 interface. It would appear as though medium 1 is connected to another medium of

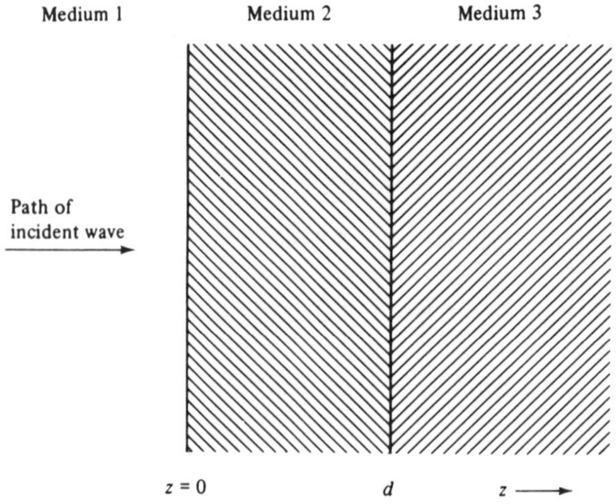

Figure 11-12 Illustration of the case where an incident wave encounters two successive dielectric-dielectric interfaces.

the same intrinsic impedance. Let us examine (34) to see how this condition could be brought about. Setting $\eta_1 = \eta^+_{\text{in}_{1-2}}$,

$$\begin{aligned}\eta_1 &= \eta_2 \frac{\eta_3 + j\eta_2 \tan \beta_2 d}{\eta_2 + j\eta_3 \tan \beta_2 d} \quad (\Omega)\\ &= |\eta_1| + j0\end{aligned} \tag{35}$$

since η_1 is real (we are still assuming lossless dielectrics).

There are two ways in which (35) could be true. If $\tan \beta_2 d = 0$ and $\eta_3 = \eta_1$, then (35) becomes $\eta_1 = \eta_1$, but that restricts the situation to where region 1 and region 3 are the same impedances (probably the same material), and $\tan \beta_2 d$ being zero means that d must be some integral number of half-wavelengths. This case has an application for radome covers for radar antennas, where the dielectric on either side of the cover is air. Equation (35) tells us that if the thickness is $\lambda/2$, then $\beta_2 d = \pi$ and $\eta_1 = \eta_3 = \eta_{\text{air}}$, regardless of the impedance η_2.

The other way that (35) can be true is for $\tan \beta_2 d \to \infty$, in which case (35) is written as

$$\begin{aligned}\eta_1\Big|_{\tan \beta_2 d \to \infty} &= \eta_2 \frac{\eta_3 + j\eta_2(\infty)}{\eta_2 + j\eta_3(\infty)}\\ &= \frac{\eta_2^2}{\eta_3} \quad (\Omega)\end{aligned} \tag{36}$$

where $\beta_2 d = \pi/2, 3\pi/2, 5\pi/2$, etc., or $d = \lambda_2/4, 3\lambda_2/4, 5\lambda_2/4$, etc. The requirement for no reflection at the 1-2 interface is

$$\left.\begin{aligned} d &= \text{odd multiple of } \lambda_2/4 \\ &\text{and} \\ \eta_2 &= \sqrt{\eta_1 \eta_3} \end{aligned}\right\} \tag{37}$$

For the lens coating application, a $\lambda/4$ thickness of coating material, $\eta_{\text{coating}} = \sqrt{\eta_{\text{air}}\eta_{\text{glass}}}$ applied to each side of the lens will eliminate (theoretically) reflections from the air-glass interface. We say that the $\lambda/4$ coating "matches" the impedance of glass to that of air. Note that this technique applies whether going from air to glass or from glass to air.

An additional comment should be made, however, to the effect that we are talking about sinusoidal steady state when we apply $\lambda/2$ or $\lambda/4$ matching techniques. There are reflections at all of the interfaces and multiple reflections between interfaces, but these reflections can be caused to cancel each other by appropriate spacing of interfaces and choice of ϵ'_r. The steady-state condition allows for no *net* reflection if the $\lambda/2$ or $\lambda/4$ plus $\eta_2 = \sqrt{\eta_1\eta_3}$ condition is met. A similar technique is used in transmission lines to match lines having different characteristic impedances (see Sec. 12.10).

Problem 11.8-4 Find the thickness and dielectric constant for an effective lens coating for $\lambda_{\text{air}} = 500$ nm. Assume that $\epsilon'_{r_{\text{glass}}} = 2.34$.

Problem 11.8-5 Calculate the dielectric constant and minimum thickness of material required to match air to plexiglass, $\epsilon'_r = 3.45$, at 9 GHz.

Problem 11.8-6 Calculate the thickness for a "transparent" plexiglass ($\epsilon'_r = 3.45$) radome for a radar operating at 6 GHz.

11.9 A PLANE WAVE NORMALLY INCIDENT ON A GOOD CONDUCTOR

When the intrinsic impedance of a good conductor as given by (11.6-1) is substituted into (11.7-13), the reflection coefficient for a wave incident from a dielectric onto a conducting surface is given by

$$\rho = \frac{\eta_2 - \eta_1}{\eta_2 + \eta_1} = \frac{\sqrt{\dfrac{\omega\mu_2}{\sigma_2}}\underline{/45^\circ} - \sqrt{\dfrac{\mu_1}{\epsilon_1}}}{\sqrt{\dfrac{\omega\mu_2}{\sigma_2}}\underline{/45^\circ} + \sqrt{\dfrac{\mu_1}{\epsilon_1}}} \tag{1}$$

Multiplying numerator and denominator by $\sqrt{\epsilon_2/\mu_2}$, we have

$$\rho = \frac{\sqrt{\dfrac{\omega\epsilon_2}{\sigma_2}}\underline{/45^\circ} - \sqrt{\dfrac{\mu_1\epsilon_2}{\mu_2\epsilon_1}}}{\sqrt{\dfrac{\omega\epsilon_2}{\sigma_2}}\underline{/45^\circ} + \sqrt{\dfrac{\mu_1\epsilon_2}{\mu_2\epsilon_1}}} \tag{2}$$

Since by definition $\sigma/\omega\epsilon > 100$ for a good conductor and since the second term in the numerator and denominator is on the order of unity,

$$\underset{(\eta_2 \ll \eta_1)}{\rho} \approx \frac{\sqrt{\dfrac{1}{100}}\underline{/45^\circ} - 1}{\sqrt{\dfrac{1}{100}}\underline{/45^\circ} + 1} \approx -1 \tag{3}$$

The transmission coefficient is given by (11.7-18) and is

$$\underset{(\eta_2 \ll \eta_1)}{\tau} = \frac{2\eta_2}{\eta_2 + \eta_1} \approx \frac{2\eta_2}{\eta_1} \tag{4}$$

For a perfect conductor,

$$\rho = -1, \qquad \tau = 0 \tag{5}$$

and the incident E and H fields are totally reflected.

Example 9

A 1-MHz plane wave is incident from air on an infinite copper sheet. If the amplitude of the incident E field is 100-V m^{-1}, find ρ, τ, δ, and the $\mathscr{E}$ and $\mathscr{H}$ fields at a distance of one skin depth into the copper.

Solution

$$\eta_2 = \sqrt{\frac{2\pi(10^6)4\pi(10^{-7})}{5.8(10^7)}}\underline{/45^\circ}$$

$$= 3.68 \times 10^{-4}\underline{/45^\circ} = (2.61 + j2.61) \times 10^{-4} \qquad (\Omega)$$

$$\rho = \frac{(2.61 + j2.61)10^{-4} - 377}{(2.61 + j2.61)10^{-4} + 377}$$

$$= \frac{377\underline{/179.99^\circ}}{377\underline{/0.00004^\circ}} \approx -1$$

$$\tau = \frac{2\eta_2}{\eta_2 + \eta_1} = \frac{2(3.68 \times 10^{-4}) \angle 45^\circ}{377 \angle 0.00004^\circ}$$
$$= 1.95 \times 10^{-6} \angle 45^\circ$$
$$\delta = \frac{1}{\sqrt{\pi f \mu \sigma}} = \frac{1}{\sqrt{\pi(10^6)4\pi(10^{-7})5.8(10^7)}}$$
$$= 6.61 \times 10^{-5} \text{ m}$$

At a depth δ,

$$|\mathscr{E}_{x2}^+| = (1.95 \times 10^{-6})(100)e^{-1} = 7.17 \times 10^{-5} \text{ V m}^{-1}$$
$$|\mathscr{H}_{y2}^+| = \frac{7.17 \times 10^{-5}}{3.68 \times 10^{-4}} = 0.195 \text{ A m}^{-1}$$

Referring back to (11.8-12) and for $\rho = -1$, the total $\mathscr{E}$ field on the incident side of the boundary is

$$\boxed{\begin{aligned}\mathscr{E}_{x1} &= \mathscr{E}_{x10}^+(e^{-j\beta_1 z} - e^{j\beta_1 z})\\ &= -2j\mathscr{E}_{x10}^+ \sin \beta_1 z \qquad (\text{V m}^{-1})\end{aligned}} \tag{6}$$

and the total $\mathscr{H}$ field is given by (11.8-16) with $\rho = -1$,

$$\boxed{\begin{aligned}\mathscr{H}_{y1} &= \frac{\mathscr{E}_{x10}^+}{\eta_1}(e^{-j\beta_1 z} + e^{j\beta_1 z})\\ &= 2\frac{\mathscr{E}_{x10}^+}{\eta_1} \cos \beta_1 z = 2\mathscr{H}_{y10}^+ \cos \beta_1 z \qquad (\text{A m}^{-1})\end{aligned}} \tag{7}$$

The magnitudes $|\mathscr{E}_{x1}|$ and $|\mathscr{H}_{y1}|$ are plotted versus distance from the dielectric-conducting interface of Fig. 11-13. Also, as partially shown in Fig. 11-13, at the boundary,

$$\mathscr{E}_{\text{bound}} = \mathscr{E}_{x10}^+ + \mathscr{E}_{x10}^- \approx \mathscr{E}_{x10}^+(1 - 1) = 0 \tag{8}$$
$$\mathscr{H}_{\text{bound}} = \mathscr{H}_{y10}^+ + \mathscr{H}_{y10}^- \approx \mathscr{H}_{y10}^+(1 + 1) = 2\mathscr{H}_{y10}^+ \tag{9}$$

Example 10

Using the approximate formulas (8) and (9), find the values of $\mathscr{E}$ and $\mathscr{H}$ at the boundary and at a depth δ into the copper sheet for Example 9.

Solution. From (8),

$$\mathscr{E}_{\text{bound}} = 0$$

From (9),

$$\mathscr{H}_{\text{bound}} = 2\mathscr{H}_{\text{incident}} = 2\mathscr{H}_{y10}^+$$
$$= \frac{2(100)}{377} = 0.531 \text{ A m}^{-1}$$

At a depth δ into the copper,

$$(\mathscr{E}_{x2}^+)_\delta \approx 0$$

and

$$(\mathscr{H}_{y2}^+)_\delta = 0.531e^{-1} = 0.195 \text{ A m}^{-1}$$

which is the same answer obtained in Example 9 for the $\overline{H}$ field.

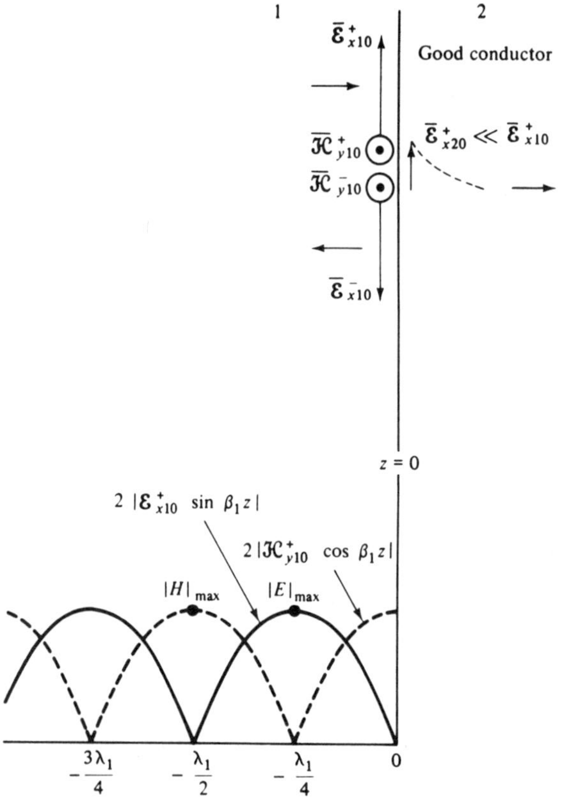

Figure 11-13 Illustration of a plane wave normally incident on a good conductor.

Problem 11.9-1 A 100-V m^{-1}, 2.45-GHz wave is normally incident from air on a perfectly conducting surface. At distances of $\lambda/4$ and $\lambda/2$ from the reflecting surface, determine the magnitudes of the $\overline{E}$ and $\overline{H}$ fields. Express the distances in centimeters.

11.10 REFLECTION OF PLANE WAVES—OBLIQUE INCIDENCE

Only one case will be considered, that of a vertically polarized (E field in the plane of incidence, H field parallel to the boundary)[2] wave incident on a dielectric-to-dielectric boundary as shown in Fig. 11-14, where $\mathscr{E}_i$, $\mathscr{E}_r$, $\mathscr{E}_t$ and $\mathscr{H}_i$, $\mathscr{H}_r$, $\mathscr{H}_t$ refer to incident, reflected, and transmitted fields at the boundary. The angles of incidence, reflection, and transmission are represented by θ_i, θ_r, and θ_t.

The boundary relations for tangential fields are

$$\mathscr{E}_i \cos \theta_i + \mathscr{E}_r \cos \theta_r = \mathscr{E}_t \cos \theta_t \tag{1}$$

and

$$\mathscr{H}_i + \mathscr{H}_r = \mathscr{H}_t \tag{2}$$

It is evident from the geometry of Fig. 11-14(a) that the line formed by the intersection of the wavefront with the boundary must progress to the right with a velocity v_{p1},

[2]The plane of incidence is defined by the plane that is parallel to the direction of wave travel and perpendicular to the boundary.

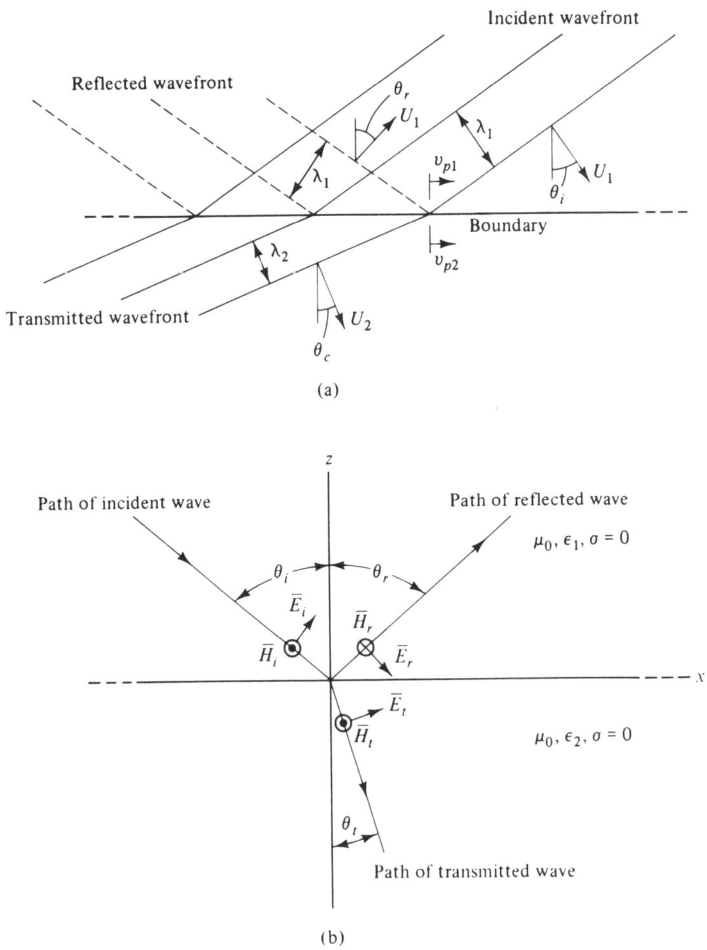

Figure 11-14 (a) Geometry for oblique reflection. (b) Illustration for oblique reflection where the E field is in the plane of incidence.

and that if tangential components are to remain matched on both sides of the boundary, the intersections of the reflected and transmitted wavefronts must also move along the boundary to the right with the same velocity, i.e., $v_{p1} = v_{p2}$. We refer to this velocity as *phase velocity,* and will encounter it again in the discussion of waveguides. It is also evident from Fig. 11-14(a) that the relation between phase velocity and velocity of the wavefront in a particular medium is

$$v_{p1} = \frac{U_1}{\sin \theta_i} = \frac{U_1}{\sin \theta_r} = \frac{U_2}{\sin \theta_t} = v_{p2} \tag{3}$$

from which we get

$$\frac{U_1}{U_2} = \frac{\sin \theta_i}{\sin \theta_t} = \frac{\sqrt{1/\mu_0\epsilon_1}}{\sqrt{1/\mu_0\epsilon_2}} = \sqrt{\frac{\epsilon_2}{\epsilon_1}} \tag{4}$$

and

$$\theta_r = \theta_i \tag{5}$$

since U_1, the wave velocity in region 1, is the velocity of the reflected wave as well as the incident wave.

Substituting (5) and $\mathcal{E} = \sqrt{\mu/\epsilon}\,\mathcal{H}$ into (1), we have[3]

$$\sqrt{\frac{\mu_0}{\epsilon_1}}\cos\theta_i(\mathcal{H}_i - \mathcal{H}_r) = \sqrt{\frac{\mu_0}{\epsilon_2}}\mathcal{H}_t\cos\theta_t \tag{6}$$

Equations (2) and (6) now represent two equations in $\mathcal{H}_i$ and $\mathcal{H}_r$ from which we can solve for the reflection coefficient, defined as

$$\rho_{vert} = \frac{\mathcal{E}_r\cos\theta_i}{\mathcal{E}_i\cos\theta_i} = -\frac{\mathcal{H}_r}{\mathcal{H}_i} \tag{7}$$

From (4) we obtain

$$\sin^2\theta_t = \frac{\epsilon_1}{\epsilon_2}\sin^2\theta_i$$

and by an identity, (8)

$$\cos^2\theta_t = 1 - \frac{\epsilon_1}{\epsilon_2}\sin^2\theta_i$$

Substituting equation (8) into (6), dividing both sides of the equation by $\sqrt{(\mu_0/\epsilon_2)(1 - (\epsilon_1/\epsilon_2)\sin^2\theta_i)}$, and subtracting from (2) eliminates $\mathcal{H}_t$ and gives

$$\mathcal{H}_i\left[1 - \frac{\sqrt{\dfrac{\mu_0}{\epsilon_1}}\cos\theta_i}{\sqrt{\dfrac{\mu_0}{\epsilon_2}\left(1 - \dfrac{\epsilon_1}{\epsilon_2}\sin^2\theta_i\right)}}\right] + \mathcal{H}_r\left[1 + \frac{\sqrt{\dfrac{\mu_0}{\epsilon_1}}\cos\theta_i}{\sqrt{\dfrac{\mu_0}{\epsilon_2}\left(1 - \dfrac{\epsilon_1}{\epsilon_2}\sin^2\theta_i\right)}}\right] = 0 \tag{9}$$

Thus, the reflection coefficient is given by

$$\rho_{vert} = -\frac{\mathcal{H}_r}{\mathcal{H}_i} = \frac{\sqrt{\dfrac{\mu_0}{\epsilon_2}\left(1 - \dfrac{\epsilon_1}{\epsilon_2}\sin^2\theta_i\right)} - \sqrt{\dfrac{\mu_0}{\epsilon_1}}\cos\theta_i}{\sqrt{\dfrac{\mu_0}{\epsilon_1}\left(1 - \dfrac{\epsilon_1}{\epsilon_2}\sin^2\theta_i\right)} + \sqrt{\dfrac{\mu_0}{\epsilon_1}}\cos\theta_i}$$

$$= \frac{-\dfrac{\epsilon_2}{\epsilon_1}\cos\theta_i + \sqrt{\dfrac{\epsilon_2}{\epsilon_1} - \sin^2\theta_i}}{\dfrac{\epsilon_2}{\epsilon_1}\cos\theta_i + \sqrt{\dfrac{\epsilon_2}{\epsilon_1} - \sin^2\theta_i}} \tag{10}$$

One can easily show that for $\theta_i = 0°$, i.e., for normal incidence, (10) reduces to the reflection coefficient for normal incidence given by (11.7-13). However, when one considers the polarity of the vertical component of the electric field, it becomes apparent that what had been called a negative reflection coefficient becomes a positive reflection coefficient, and vice versa. When applied to problems dealing with vertical antennas on a ground plane, the phase relation between the direct path from the antenna to the distant field point and that of the reflected path is correctly computed only if the algebraic sign of (10) is reversed from that given. This is illustrated in Figs. 11-15 and 11-16. Of course,

[3]Note that $\mathcal{E}_i = \sqrt{\mu_0/\epsilon_1}\,\mathcal{H}_i$, $\mathcal{E}_r = -\sqrt{\mu_0/\epsilon_1}\,\mathcal{H}_r$, and $\mathcal{E}_t = \sqrt{\mu_0/\epsilon_2}\,\mathcal{H}_t$.

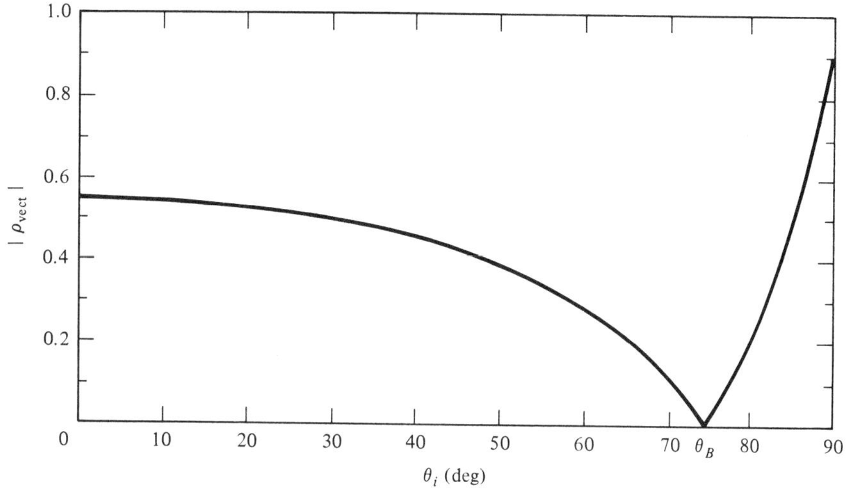

Figure 11-15 Magnitude of ρ_{vert} versus angle of incidence for $\epsilon_2/\epsilon_1 = 12$. Note that ρ_{vert} is negative for $0 < \theta_i < \theta_B$ and is positive for $\theta_B < \theta_i < 90°$. The Brewster angle is computed in Prob. 11.10-5.

the phase reverses as the incidence angle exceeds the Brewster angle. The Brewster angle θ_B is the incidence angle for which the numerator of (10) is zero, and we find

$$\sin^2 \theta_B = \frac{\epsilon_2}{\epsilon_1 + \epsilon_2} \tag{11}$$

When $\sin^2 \theta_i$ in (10) is greater than ϵ_2/ϵ_1,

$$|\rho_{vert}| = \left|\frac{-a + jb}{a + jb}\right| = 1 \tag{12}$$

where $a = \epsilon_2/\epsilon_1 \cos \theta_i$, and $b = \sqrt{\sin^2 \theta_i - \epsilon_2/\epsilon_1}$, and we have total reflection. Note, however, that total reflection occurs only when $\epsilon_1 > \epsilon_2$. The critical angle for total reflection is

$$\theta_c = \sin^{-1} \sqrt{\frac{\epsilon_2}{\epsilon_1}} = \sin^{-1}\left(\frac{n_2}{n_1}\right) \tag{13}$$

where the index of refraction $n = \sqrt{\epsilon_r}$. For incidence angles greater than θ_c, we have total reflection. We shall have occasion to use (13) in Sec. 13.7. The reflection coefficient for a horizontally polarized wave is derived in Prob. 11-24.

Problem 11.10-1 Show that when $\theta_i = 0°$, (10) reduces to (11.7-13).

Problem 11.10-2 Derive (11) from (10).

Problem 11.10-3 Find the Brewster angle for a vertically polarized plane wave incident from air into glass, $\epsilon_r = 2.34$. Also find the transmitted angle θ_t.

Problem 11.10-4 Find the critical angle θ_c for a wave traveling from glass toward air. Assume that $\epsilon_{r\text{glass}} = 2.34$.

Problem 11.10-5 Find the Brewster angle for the interface depicted in Fig. 11-15.

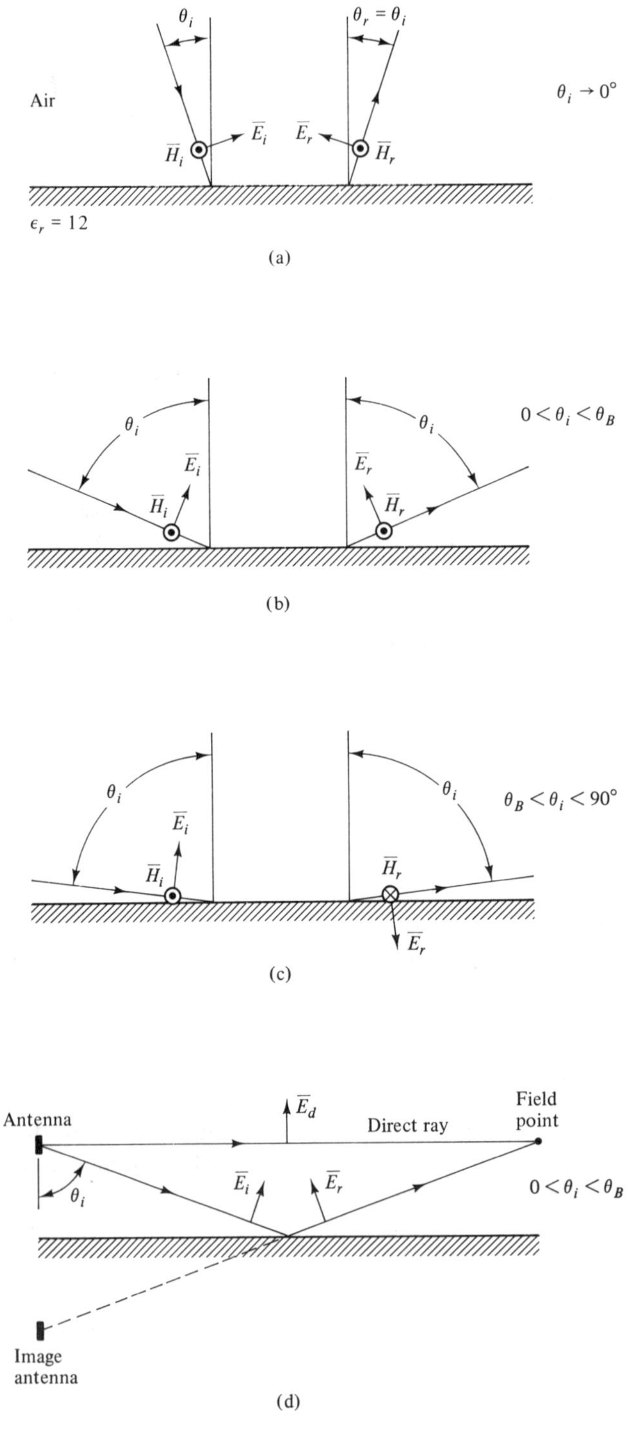

Figure 11-16 (a) The reflection coefficient, as calculated by (10), approaches -0.55 as θ_i approaches 0°. This is the reflection coefficient for normal incidence. (b) The reflected E field is becoming more parallel with the incident E field as θ_i approaches θ_B. (c) The reflected field E reverses when $\theta_i > \theta_B$. (d) When θ_i is in the range $0 < \theta_i < \theta_B$, the reflected E field appears to come from an image antenna. ρ_{vert} should be considered positive for this example when applied to antenna patterns and the algebraic sign of (10) should be reversed.

REVIEW QUESTIONS

1. What set of equations sum up the laws of electromagnetics? *Sec. 11.2*
2. Write a general form of a function which represents a wave phenomenon. *Sec. 11.3*
3. What is intrinsic impedance? *Sec. 11.3*
4. How practical is the study of uniform plane waves? *Sec. 11.3*
5. How does the velocity of propagation vary with the dielectric constant? With relative permeability? *Sec. 11.3*
6. What is the direction of power flow in a dc circuit? How is this related to $\overline{E}$ and $\overline{H}$? Does this apply to time-varying voltages and current? *Sec. 11.4*
7. Define propagation constant, attenuation constant, and phase constant. *Sec. 11.5*
8. How would one distinguish between a "high loss" and a "low loss" material? *Sec. 11.5*
9. Are both seawater and copper considered good conductors? Under what circumstances? *Sec. 11.6*
10. What is skin depth? *Sec. 11.6*
11. What causes reflections? *Sec. 11.7*
12. Define reflection coefficient and transmission coefficient. What can cause these coefficients to be complex? *Sec. 11.7*
13. What is a standing wave pattern? *Sec. 11.7*
14. At a dielectric-to-dielectric interface, what determines whether the reflected $\overline{E}$ field or the reflected $\overline{H}$ field is reversed in polarity from that of the incident wave? *Sec. 11.8*
15. How is a crank diagram related to a standing wave pattern? *Sec. 11.8*
16. Is there any power absorbed at a dielectric-to-dielectric interface? *Sec. 11.8*
17. What is the function of the coating on a lens? *Sec. 11.8*
18. In a good conductor, what is the phase angle between the $\overline{E}$ and the $\overline{H}$ fields? *Sec. 11.9*
19. For a perfectly conducting surface, how much energy is lost in the reflection of an EM wave arriving from air? *Sec. 11.9*
20. What is the standing wave ratio in front of a perfectly reflecting surface for a wave normally incident from a perfect dielectric? *Sec. 11.9*
21. Is it possible to have no reflection from a dielectric-to-dielectric interface even though $\eta_1 \neq \eta_2$? *Sec. 11.10*

PROBLEMS

11-1. A magnetic field propagating through a certain material may be described by $\overline{H} = \hat{z}(0.03)\cos(10^9 t - 4x)$ A m^{-1}. Find: (a) the direction of wave travel; (b) the velocity of wave; (c) the wavelength; (d) a description of the electric field in the same material.

11-2. An $\overline{E}$ field is given by $\overline{E} = \hat{z}50\cos(10^9 t - 5x)$ (V m^{-1}). Find: (a) direction of wave travel; (b) velocity of the wave; (c) wavelength; (d) complete description of the $\overline{H}$ field.

11-3. Write the expression for a sinusoidal plane wave having the $\overline{E}$ field polarized in the $+z$ direction, the $\overline{H}$ field polarized in the $+y$ direction, $f = 1$ GHz, and $U = 10^8$ m s^{-1}.

11-4. For the field described in Prob. 11-1, find: (a) the average power density in the wavefront; (b) the dielectric constant of the medium, assuming that $\mu_r = 1$; (c) the wavelength.

11-5. The average power density in a wavefront in air is 1 mW cm^{-2}. Find the rms values of the $\overline{E}$ and $\overline{H}$ fields.

11-6. A plane wave in a certain medium has a power density of 5 W m^{-2} and the rms electric field intensity is 31.62 V m^{-1}. Find: (a) the intrinsic impedance; (b) the rms value of the $\overline{H}$ field; (c) the velocity of propagation of the wave.

11-7. Assume that at 2.45 GHz, a plane wave average power density of 1 mW cm^{-2} is considered safe for human exposure. A rms field strength meter measures 50 V m^{-1} for the electric field in the region. Is the region safe?

11-8. Find the skin depth and velocity of propagation of an EM wave in seawater at 10 kHz, 100 kHz, 10 MHz, and 1 GHz (see Prob. 11.5-1).

11-9. A plane wave at 100 MHz, traveling through a lossy material, has a phase shift of 1 rad m^{-1} and its amplitude is reduced 50% for every meter traveled. Find α, β, U, and the skin depth. Also find the attenuation in dB ft^{-1}.

11-10. A 30-MHz plane wave is normally incident from air onto a semi-infinite dielectric slab, $\epsilon_r = 4$. If the amplitude of the incident electric field is 80 V m^{-1}, find: (a) ρ; (b) $\mathcal{P}_{inc}$ (W m^{-2}); (c) $\mathcal{P}_{trans}$ (W m^{-2}); (d) distance from boundary to nearest maximum of the electric field.

11-11. A 100-MHz plane wave is normally incident from air onto a semi-infinite slab of dielectric (region 2). If the standing wave ratio in front of the slab is 1.5, and if E_{min} is at the boundary, find: (a) η_2; (b) ϵ_{r2}; (c) ρ; (d) distance d from the boundary to the nearest E_{max}.

11-12. A plane wave is incident from air onto a dielectric, $\epsilon_r = 9$ at normal incidence. For a frequency of 1 GHz, find: (a) ρ; (b) S; (c) λ in air and in the dielectric; (d) percent power density reflected from the interface; (e) distance of first E_{max} from the interface.

11-13. A plane wave having an E field amplitude of 100 V m^{-1} is normally incident from air onto a semi-infinite slab of dielectric material $\epsilon_r = 2$. Find: (a) amplitude of the E and H fields in the dielectric; (b) velocity of propagation in the dielectric; (c) average power density propagating in the dielectric.

11-14. At a boundary between two materials, the E field in region 1 is given by $E_{x10}^{+} = 100 \cos \omega t$ (V m^{-1}) and $E_{x10}^{-} = -35 \cos(\omega t + 30°)$ (V m^{-1}). (a) Assuming that the wave is incident from region 1, find ρ, the complex reflection coefficient. (b) Find the magnitude of the total E field at the boundary.

11-15. The reflection coefficient is $0.3 \angle 20°$ for a plane wave incident from air onto a lossy dielectric material. Find: (a) the impedance of the material; (b) the standing wave ratio S in air in front of the material; (c) for an incident electric field of $\mathcal{E}^+ = \hat{x}100e^{j\beta\ell}$ (V m^{-1}), where ℓ is the distance from the boundary, find the total E field at the boundary.

11-16. The skin depth of a given conductor has been determined to be 0.8 mm at a frequency of 1000 Hz. If the tangential electric field at the boundary between air and the conductor is 100 V m^{-1}, determine the amplitude and phase shift with respect to the boundary for a wave incident from air at a depth of 3 mm into the conductor.

11-17. A plane wave is normally incident on a conducting sheet. The frequency is 10 GHz and the skin depth is 0.001 mm. What is the velocity of the wave in the conducting material?

11-18. An $\overline{E}$ field is given by

$$\overline{E} = \hat{y}50e^{-0.001x} \sin(377 \times 10^9 t - 3770x) + \hat{y}25e^{0.001x} \sin(377 \times 10^9 t + 3770x) \qquad (\text{V m}^{-1})$$

Find: (a) the direction of polarization of the $\overline{E}$ vector; (b) the direction of travel of the wave components; (c) frequency; (d) |velocity| of wave travel; (e) S; (f) skin depth in the material in which the waves are traveling.

11-19. A plane wave is normally incident from air onto an aluminum slab ($\sigma = 3.82 \times 10^7$ S m^{-1}). Find the percentage of the incident power density that is absorbed by the aluminum slab at a frequency of 500 MHz.

11-20. A vertically polarized (E field in the plane of incidence) wave is incident from air onto a dielectric slab whose dielectric constant is 10. Find: (a) the Brewster angle; (b) the angle of transmission into the dielectric when $\theta_i = \theta_B$.

11-21. When the incident wave is incident from medium 1 onto medium 2: (a) What are the limiting values of the Brewster angle if $\eta_1 < \eta_2$? (b) What are the limiting values if $\eta_1 > \eta_2$? Assume that $\mu_1 = \mu_2 = \mu_0$.

11-22. Referring to Fig. 13-10, find the critical angle of incidence for total reflection for a wave traveling from a dielectric, $\epsilon_r = 2.0$, to a dielectric whose ϵ_r is 1.5.

11-23. Show that, in general, for a lossy dielectric the skin depth is given by

$$\delta = \frac{1}{\alpha} = \frac{67.52}{f}\left[\sqrt{(\epsilon_r')^2 + \frac{\sigma^2}{\omega^2\epsilon_0^2}} - \epsilon_r'\right]^{-1/2}$$

where $\epsilon' = \epsilon_0\epsilon_r'$, and f is the frequency in megahertz.

11-24. Using boundary conditions similar to eqs. (11.10-1) and (11.10-2), except with $\overline{E}$ and $\overline{H}$ interchanged, derive the equation for the reflection coefficient for oblique incidence for a horizontally polarized wave,

$$\rho_{\text{hor}} = \frac{\cos\theta_i - \sqrt{\dfrac{\epsilon_2}{\epsilon_1} - \sin^2\theta_i}}{\cos\theta_i + \sqrt{\dfrac{\epsilon_2}{\epsilon_1} - \sin^2\theta_i}}$$

12

Transmission Lines

12.1 INTRODUCTION

A transmission line consists of two or more parallel conductors used to connect a source to a load or to connect one circuit to another. A two-conductor line may carry megawatts of power from a transmitter to an antenna, or it may be used to carry picowatts of received signal power from an antenna to a receiver. Three-conductor transmission lines are used for 60-Hz, three-phase transmission. There is a certain amount of redundancy between Chapter 11 on plane waves, and the present chapter on transmission lines. The mathematics of wave phenomena, propagation constants, reflection coefficients, and the concept of standing wave ratio are common to both chapters. However, skin effect can best be understood by considering a plane wave propagating in a bulk material. Although Chapter 12 is designed to follow Chapter 11, there would be no great difficulty in treating them in reverse order. As a general rule, voltage and current are easier to work with than electric and magnetic fields.

In this text, we use the term "transmission line" to mean a pair of conductors operating in the TEM (transverse electromagnetic) mode, which means that no component of either the electric or magnetic field is in the direction of transmission. The transverse $\overline{E}$ and $\overline{H}$ fields have the same configuration for the highest frequency transmitted as for dc fields, but here we will be concerned primarily with voltage and current instead of the fields. In contrast to ordinary circuit theory, where resistance, capacitance, conductance, and inductance are represented as lumped constant elements, the R, C, G, and L of transmission lines are considered as distributed parameters. For example, R may be given in ohms per meter or ohms per mile, L is in henrys per meter (or mile), etc. As we shall see, this distributed characteristic means that the transmission line is not a low pass filter as it would appear from a lumped constant schematic representation.

We begin the chapter with the derivation of the transmission line equations in the time domain. For the special case of lossless lines, we derive the expressions for characteristic impedance (resistance) and reflection coefficient. We take a brief look at transients on transmission lines, examining the case where either a step function of voltage is applied to the line input, or a step change occurs in the load impedance. The remainder of the chapter is devoted to sinusoidal steady state, where we treat the general lossy line as well as the lossless line, and introduce the Smith chart for use in single-stub and double-stub matching problems. The concepts of input impedance, standing wave ratio, and quarter-wave matching sections are developed in these sections.

12.2 GENERAL EQUATIONS FOR LINE VOLTAGE AND CURRENT—TIME DOMAIN

Consider the electrical circuit shown in Fig. 12-1. If the two conductors connecting the source to the load are of uniform cross section and spacing throughout their length, and if the dielectric medium in which they are in contact is also uniform throughout their length, we refer to the pair as a uniform transmission line. The parameters of the transmission line are its resistance, inductance, conductance, and capacitance per unit length of the line pair. These parameters are typically given as

$$\begin{array}{ll} R & (\Omega\ \mathrm{m}^{-1}) \\ L & (\mathrm{H\ m}^{-1}) \\ G & (\mathrm{S\ m}^{-1}) \\ C & (\mathrm{F\ m}^{-1}) \end{array} \qquad (1)$$

respectively. As indicated in Fig. 12-2, uniform transmission lines may assume a variety of different cross-sectional geometries.

The uniform transmission line is modeled schematically in Fig. 12-3, but it must be kept in mind that the R, L, G, and C shown as lumped elements on the schematic are actually distributed parameters. Using partial derivatives to distinguish between derivatives with respect to time and with respect to distance, the voltage changes with z because of voltage drop across the series distributed R and L.

$$\frac{\partial V}{\partial z} = -RI - L\frac{\partial I}{\partial t} \qquad (2)$$

Likewise, the current is shunted by both the shunt conductance and shunt capacitance as it moves toward the load.

$$\frac{\partial I}{\partial z} = -GV - C\frac{\partial V}{\partial t} \qquad (3)$$

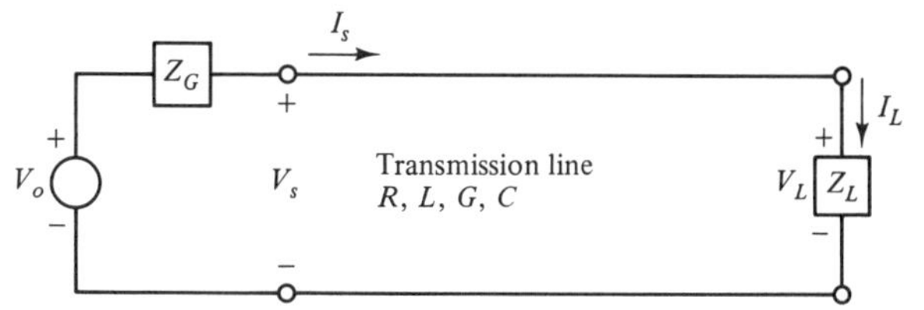

Figure 12-1 Schematic representation of a circuit employing a transmission line to connect a source V_0 to a load Z_L.

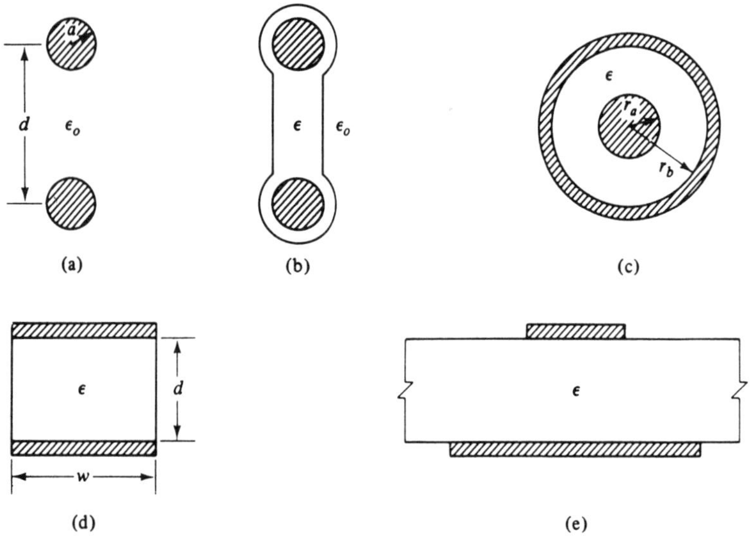

Figure 12-2 Cross sections of various types of transmission lines: (a) parallel line; (b) parallel wires imbedded in dielectric material; (c) coaxial conductors; (d) planar; (e) strip line.

Taking the derivatives of (2) with respect to z and (3) with respect to t, and making appropriate substitutions, yields

$$\frac{\partial^2 V}{\partial z^2} = RGV + (RC + LG)\frac{\partial V}{\partial t} + LC\frac{\partial^2 V}{\partial t^2} \tag{4}$$

Likewise, taking derivatives of (3) with respect to z and (2) with respect to t, and making appropriate substitutions, yields

$$\frac{\partial^2 I}{\partial z^2} = RGI + (RC + LG)\frac{\partial I}{\partial t} + LC\frac{\partial^2 I}{\partial t^2} \tag{5}$$

Equations (4) and (5) represent the general time-domain differential equations for a uniform transmission line. However, we shall delay working with the general lossy line until we consider sinusoidal steady state.

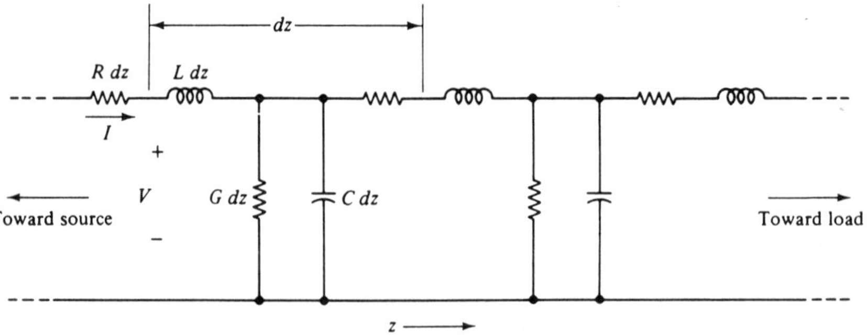

Figure 12-3 Lumped constant representation of a transmission line.

For the lossless line, $R = G = 0$, and (4) and (5) become

$$\frac{\partial^2 V}{\partial z^2} = LC\frac{\partial^2 V}{\partial t^2} \tag{6}$$

and

$$\frac{\partial^2 I}{\partial z^2} = LC\frac{\partial^2 I}{\partial t^2} \tag{7}$$

Equations (6) and (7) are clearly wave equations having solutions of the form

$$V(z,t) = V_1(t - \sqrt{LC}\,z) + V_2(t + \sqrt{LC}\,z) \tag{8}$$

and

$$I(z,t) = I_1(t - \sqrt{LC}\,z) + I_2(t + \sqrt{LC}\,z) \tag{9}$$

as can be demonstrated by substituting (8) and (9) into (6) and (7), respectively. The velocity of the wave is

$$U = \frac{1}{\sqrt{LC}} \tag{10}$$

The function V_1 is a forward traveling wave, for to keep the argument $t - z/U$ constant as t increases, z must increase at a velocity U. Likewise, the argument for V_2 is constant with increasing t only if z is becoming more negative with a velocity U.

A simple sinusoidal example of (8) is the expression

$$V(z,t) = A\cos\omega\left(t - \frac{\beta z}{\omega}\right) + B\cos\omega\left(t + \frac{\beta z}{\omega}\right) \tag{11}$$

where we see

$$U = \frac{\omega}{\beta} = \frac{2\pi f}{\beta} \tag{12}$$

or

$$\beta = \frac{2\pi f}{U} = \frac{2\pi}{\lambda} \tag{13}$$

where we recall the fundamental relation between wavelength, frequency, and velocity as

$$\lambda f = U \tag{14}$$

Substituting (11) into (6), we have

$$-\beta^2 A\cos\omega\left(t - \frac{\beta z}{\omega}\right) - \beta^2 B\cos\omega\left(t + \frac{\beta z}{\omega}\right)$$
$$= -\omega^2 LCA\cos\omega\left(t - \frac{\beta z}{\omega}\right) - \omega^2 LCB\cos\omega\left(t + \frac{\beta z}{\omega}\right)$$

or

$$\frac{\omega^2}{\beta^2} = \frac{1}{LC} = U^2 \tag{15}$$

Of course, (10) is the general case, and (12) and (15) apply only for sinusoidal excitation.

It is convenient to define V^+ and V^- as voltage functions at the load where we also set $z = 0$. Thus, $V^+ = V_1(t - 0)$ or simply $V_1(t)$, and V^- is $V_2(t)$. Likewise, we define I^+ as $I_1(t)$ and I^- as $I_2(t)$. Then, at the load,

$$V_L = V^+ + V^- \tag{16}$$

and

$$I_L = I^+ + I^- \tag{17}$$

We shall now determine the functions I^+ and I^- in terms of V^+ and V^-. We use the lossless form of (2), namely,

$$\frac{\partial V}{\partial z} = -L\frac{\partial I}{\partial t} \tag{18}$$

and perform the indicated differentiations on the solutions (8) and (9). We get

$$\frac{\partial I}{\partial t} = -\frac{1}{L}[-\sqrt{LC}\,V_1'(t - \sqrt{LC}\,z) + \sqrt{LC}\,V_2'(t + \sqrt{LC}\,z) \tag{19}$$

where the primes indicate derivatives of the functions. For example, if V_1 is a cosine function, V_1' is a minus sine function, etc. Integrating with respect to t,

$$I = \frac{1}{\sqrt{L/C}}[V_1(t - \sqrt{LC}\,z) - V_2(t + \sqrt{LC}\,z)] + \text{const.} \tag{20}$$

We set the constant of (20) equal to zero (this means that initial condition of current on the line is zero), and finally, we have an expression for current that replaces (9),

$$I(z,t) = \frac{1}{R_0}[V_1(t - \sqrt{LC}\,z) - V_2(t + \sqrt{LC}\,z)] \tag{21}$$

where we define the characteristic resistance R_0 by

$$R_0 = \sqrt{\frac{L}{C}} \qquad (\Omega) \tag{22}$$

It is clear now that

$$I^+ = \frac{V^+}{R_0} \tag{23}$$

and

$$I^- = \frac{-V^-}{R_0} \tag{24}$$

and the load resistance R_L may be expressed as

$$R_L = \frac{V_L}{I_L} = \frac{V^+ + V^-}{(1/R_0)(V^+ - V^-)} \tag{25}$$

and if we define a reflection coefficient ρ_L as

$$\rho_L = \frac{V^-}{V^+} \tag{26}$$

we have, after dividing top and bottom of (25) by V^+,

$$R_L = \frac{R_0(1 + \rho_L)}{1 - \rho_L}$$

and solving for ρ_L,

$$\rho_L = \frac{R_L - R_0}{R_L + R_0} \tag{27}$$

for the load reflection coefficient.

We will look at the reflection coefficient again in the frequency domain, where we will find that in general ρ_L is a complex number. For the present case, we will consider only resistive loads on lossless lines for which case ρ_L is a positive or negative real number. Propagation of voltage and current on a lossless line is analogous to the propagation of electric and magnetic fields through a lossless medium, and reflection of voltage at a resistive load is analogous to reflection of an electric field from an interface between two lossless dielectrics. It should also be pointed out that the velocity of propagation on the transmission line, operating in the TEM mode, is the same as that for a plane wave propagating through the same bulk material that serves as the dielectric between conductors of the transmission line. Thus,

$$\mu\epsilon = LC \tag{28}$$

but

$$\frac{\mu}{\epsilon} \neq \frac{L}{C} \tag{29}$$

It will be left as an exercise for the student to demonstrate (28) and (29) using formulas for L and C for a lossless coaxial line (Prob. 12.2-1).

Example 1

A lossless transmission line has an inductance per unit length of 0.5 μH m^{-1} and a capacitance per unit length of 200 pF m^{-1}. The source is sinusoidal with a frequency of 1000 Hz, and the amplitude of the voltage across a 35-Ω load is 100 V. Find: (a) R_0; (b) ρ_L; (c) U; (d) λ; (e) V^+; (f) V^-; (g) β.

Solution

(a) $R_0 = \sqrt{\dfrac{L}{C}} = \sqrt{\dfrac{5 \times 10^{-7}}{2 \times 10^{-10}}} = 50\ \Omega$

(b) $\rho_L = \dfrac{35 - 50}{35 + 50} = -0.176$

(c) $U = \dfrac{1}{\sqrt{LC}} = \dfrac{1}{\sqrt{(5 \times 10^{-7})(2 \times 10^{-10})}} = 10^8$ m s^{-1}

(d) $\lambda = \dfrac{U}{f} = \dfrac{10^8}{10^3} = 10^5$ m

(e) Assume that $V_L = 100 \cos 6280t$. Then $|V_L| = 100\ V = |V^+ + V^-| = |V^+||1 - 0.176| = 0.824|V^+|$. Then $|V^+| = 100/0.824 = 121.36\ V$, or $V^+ = 121.36 \cos 6280t\ V$.

(f) Then $|V^-| = 0.176(121.36) = 21.36\ V$, or $V^- = -21.36 \cos 6280t\ V$.

(g) $\beta = \dfrac{2\pi}{\lambda} = \dfrac{2\pi}{10^5} = 6.28 \times 10^{-5}$ rad/m.

Problem 12.2-1 Using the formulas for capacitance and inductance for a coaxial cable derived in Chapters 5 and 9 (Examples 5-14 and 9-13), demonstrate that (28) and (29) given above are true.

Problem 12.2-2 Given an air dielectric transmission line that has a capacitance per unit length of 300 pF m^{-1}, find the resistance of the load R_L for which the reflection coefficient ρ_L is zero.

12.3 TRANSIENTS ON LOSSLESS TRANSMISSION LINES WITH RESISTIVE LOADS

If the transmission line is lossless, it is clear from (12.2-8) and (12.2-21) that, for a change in forward voltage applied to the input to a transmission line, the ratio of change in forward voltage to change in forward current is a constant and is equal to the characteristic resistance R_0. If a step of voltage is applied, the resulting change in current is also a step change. At the first instant after a change is applied, the voltage and current waves that propagate from that point are not at all influenced by the load connected to the other end of the line. It is not until reflected waves from the load arrive back at the sending end that the load enters into boundary conditions at the sending end. A step of voltage applied to either an open or a shorted line will cause the same initial voltage and current waves to propagate toward the load.

An in-depth treatment of transients on transmission lines involves the use of Laplace transforms. However, when only step functions are considered with resistive loads, it is convenient to work entirely in the time domain, and that will be the approach taken in this treatment. This simplified approach serves quite well for step functions or rectangular pulses encountered in digital circuits and for switching action that instantaneously changes the output impedance of a source or the input impedance of a load.

We shall consider first an ideal battery of voltage V_0, connected through an internal resistance R_G to a lossless line connected to a resistive load R_L. In Fig. 12-4, distance

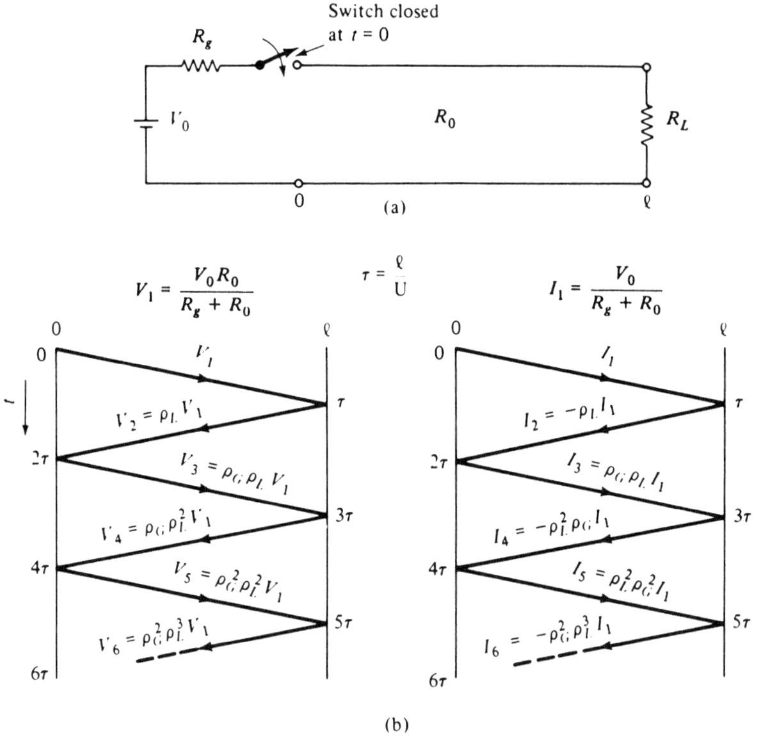

Figure 12-4 (a) Introduction of a step function into a transmission line circuit. (b) Use of the distance-time plot to tabulate voltage and current as a function of time.

is measured from the sending end, and the switch is closed at $t = 0$, giving at the input terminals,

$$V_{\text{in}}(0+) = V(0, 0+) = \frac{V_0 R_0}{R_G + R_0} = V_1 \qquad \text{(V)} \tag{1}$$

and

$$I_{\text{in}}(0+) = I(0, 0+) = \frac{V_0}{R_G + R_0} = I_1 \qquad \text{(A)} \tag{2}$$

where $V(0, 0+)$ is the voltage at zero distance and at $t = 0+$, etc. The waves, of amplitude V_1 and I_1, propagate toward the load end at the velocity

$$U = \frac{1}{\sqrt{LC}} \qquad (\text{m } s^{-1}) \tag{3}$$

It will be convenient to let the transit time from one end of the line to the other be represented by τ, where

$$\tau = \frac{\ell}{U} \qquad (s) \tag{4}$$

Then after τ seconds, the voltage and current waves reach the load end, and

$$V_L(\tau) = V(\ell, \tau) = \frac{V_0 R_0}{R_G + R_0}(1 + \rho_L) \tag{5}$$

and

$$I_L(\tau) = I(\ell, \tau) = \frac{V_0}{R_G + R_0}(1 - \rho_L) \tag{6}$$

where

$$\rho_L = \frac{R_L - R_0}{R_L + R_0} \tag{7}$$

to distinguish the load reflection coefficient from the sending-end or generator-end reflection coefficient,

$$\rho_G = \frac{R_G - R_0}{R_G + R_0} \tag{8}$$

which is seen by the load-reflected wave as it arrives back at the sending end. The reflected waves at time τ (or $\tau+$) are

$$V_2 = V_1 \rho_L = \frac{V_0 R_0}{R_G + R_0} \rho_L \tag{9}$$

and

$$I_2 = -I_1 \rho_L = -\frac{V_0}{R_G + R_0} \rho_L \tag{10}$$

and at $t = 2\tau+$, where the waves have been reflected from the sending end,

$$V_3 = \rho_L \rho_G V_1 = \rho_G V_2 \tag{11}$$

and

$$I_3 = \rho_L \rho_G I_1 = -\rho_G I_2 \tag{12}$$

These, and subsequent reflections from both the load end and the sending end, are illustrated on a distance-time plot in Fig. 12-4.

Example 2

In Fig. 12-4, let $R_G = 25\ \Omega$, $R_0 = 50\ \Omega$, and $R_L = 75\ \Omega$. The battery voltage is 10 V, and the switch is closed at $t = 0$. Find and plot V_L, I_L as a function of time, and find the steady-state values.

Solution

$$\rho_L = \frac{75 - 50}{75 + 50} = 0.2$$

$$\rho_G = \frac{25 - 50}{25 + 50} = -0.333$$

$$V_1 = \frac{10(50)}{25 + 50} = 6.67\ \text{V}$$

$$I_1 = \frac{10}{25 + 50} = 0.133\ \text{A}$$

Next, a distance-time plot is used as a bookkeeping scheme to keep track of the values of successive reflective waves (see Fig. 12-5). The voltages V_1, V_2, etc., and currents I_1, I_2, etc., in Fig. 12-5 are the amplitudes of the voltage and current waves traveling in the direction indicated during the time intervals labeled on either end of the directed line. Then

$$V_1 = 6.67\ \text{V}$$
$$V_2 = 6.67(0.2) = 1.33\ \text{V}$$
$$V_3 = 6.67(0.2)(-0.333) = -0.445\ \text{V}$$
$$V_4 = 6.67(0.2)(-0.333)(0.2) = -0.089\ \text{V}$$
$$V_5 = 6.67(0.2)(-0.333)(0.2)(-0.333) = 0.0297\ \text{V}$$
$$V_6 = 6.67(0.2)(-0.333)(0.2)(-0.333)(0.2) = 0.0059\ \text{V}$$

and

$$I_1 = 0.133\ \text{A}$$
$$I_2 = 0.133(-0.2) = -0.0267\ \text{A}$$
$$I_3 = 0.133(-0.2)(0.333) = -0.0089\ \text{A}$$
$$I_4 = 0.133(-0.2)(0.333)(-0.2) = 0.00178\ \text{A}$$
$$I_5 = 0.133(-0.2)(0.333)(-0.2)(0.333) = 0.00059\ \text{A}$$
$$I_6 = 0.133(-0.2)(0.333)(-0.2)(0.333)(-0.2) = -0.000118\ \text{A}$$

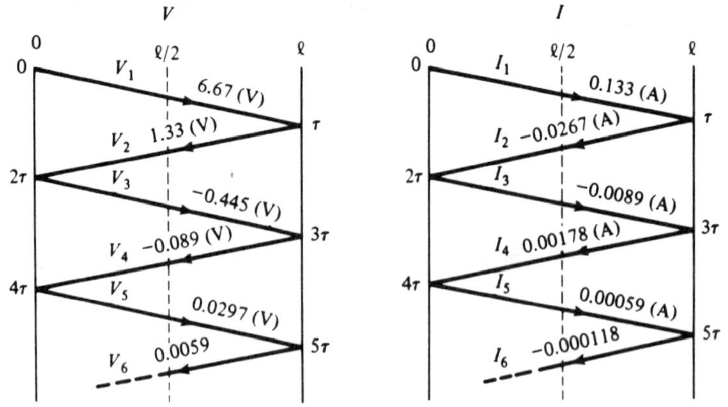

Figure 12-5 Distance-time plot for Example 2.

Using the amplitudes calculated above, the load voltage may be expressed as

$$V_L = 6.67u(t - \tau) + 1.33u(t - \tau) - 0.445u(t - 3\tau) - 0.089u(t - 3\tau) + 0.0297u(t - 5\tau) + \ldots$$

Likewise, the load current is

$$I_L = 0.133u(t - \tau) - 0.0267u(t - \tau) - 0.0089u(t - 3\tau) + 0.00178u(t - 3\tau) + 0.00059u(t - 5\tau) + \ldots$$

The values of V_L and I_L are plotted in Fig. 12-6, out to $t = 5\tau+$. Because of roundoff error, the ratio V_L/I_L as plotted is slightly greater than 75 Ω.

The steady-state values for voltage and current can be calculated by inspection,

$$V_{L_{\text{steady state}}} = V_L(t \to \infty) = \frac{10(75)}{25 + 75} = 7.5 \text{ V}$$

and

$$I_L(t \to \infty) = \frac{10}{25 + 75} = 0.1 \text{ A}$$

The Laplace transformed voltage at the load could have been written as

$$\mathcal{V}(\ell, s) = \frac{6.67}{s}[e^{-s\tau} + \rho_L e^{-s\tau} + \rho_L \rho_G e^{-3s\tau} + \rho_L^2 \rho_G e^{-3s\tau} + \rho_L^2 \rho_G^2 e^{-5s\tau} + \ldots]$$

$$= \frac{6.67}{s}\left(\frac{e^{-s\tau}(1 + \rho_L)}{1 - \rho_L \rho_G e^{-2s\tau}}\right)$$

$$= \frac{6.67}{s}\left(\frac{e^{-s\tau}(1.2)}{1 + 0.0667e^{-2s\tau}}\right)$$

and

$$V_L(\infty) = \lim_{s \to 0} s\mathcal{V}(\ell, s) = \frac{6.67(1.2)}{1.0667} = 7.5 \text{ V}$$

etc, where s is the complex frequency.

Example 3

The given conditions are the same as for Example 2, except that the voltage and current plots are required at the midpoint of the line.

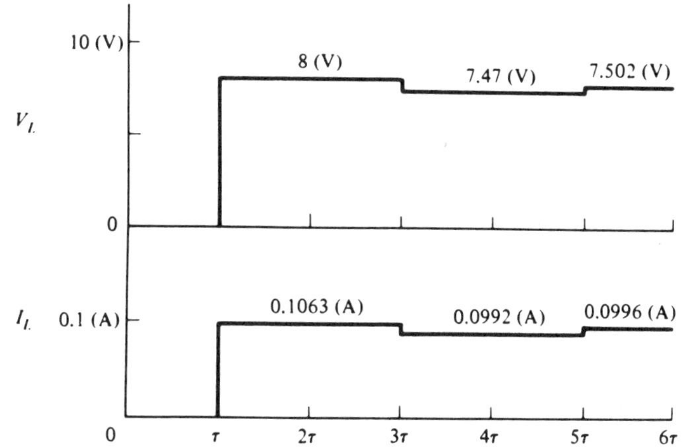

Figure 12-6 Plots of load voltage and current for Example 2.

Solution. Using the results from Example 2.

$$V\left(\frac{\ell}{2}\right) = 6.67u\left(t - \frac{\tau}{2}\right) + 1.33u\left(t - \frac{3\tau}{2}\right) - 0.445u\left(t - \frac{5\tau}{2}\right) - 0.0889u\left(t - \frac{7\tau}{2}\right) + 0.0297u\left(t - \frac{9\tau}{2}\right) + \ldots$$

Also

$$I\left(\frac{\ell}{2}\right) = 0.133u\left(t - \frac{\tau}{2}\right) - 0.0267u\left(t - \frac{3\tau}{2}\right) - 0.00889u\left(t - \frac{5\tau}{2}\right) + 0.00178u\left(t - \frac{7\tau}{2}\right) + 0.00059u\left(t - \frac{9\tau}{2}\right) + \ldots$$

The results are plotted in Fig. 12-7. The steady-state values are the same as for Example 2.

The preceding examples were for a step of voltage being applied at the input end at $t = 0$. Another interesting possibility is to introduce a step change in the load resistance at $t = 0$ in what has been up to that time a dc circuit in steady-state condition.

Example 4

The circuit of Fig. 12-8 is in steady-state condition just before the switch is closed, shorting out part of the load resistance. Draw the distance-time plots for voltage and current, and plot V_L, I_L versus time after $t = 0$.

Solution. At $t = 0-$,

$$V_L = 5\frac{40}{50} = 4 \text{ V}$$

and

$$I_L = \frac{4}{40} = 0.1 \text{ A}$$

Prior to the closing of the switch, the load voltage is in steady state at 4 V. Then at $t = 0+$, a change in V_L and I_L is forced by the closing of the switch. These changes, ΔV_L and ΔI_L, are propagated toward the generator end. Because of the way we define polarity of voltage and current (a positive current flows toward the load on the wire having a positive polarity of voltage) then for the changes ΔV_L and ΔI_L propagating toward the generator,

$$\frac{\Delta V_L}{\Delta I_L} = -R_0 = -50$$

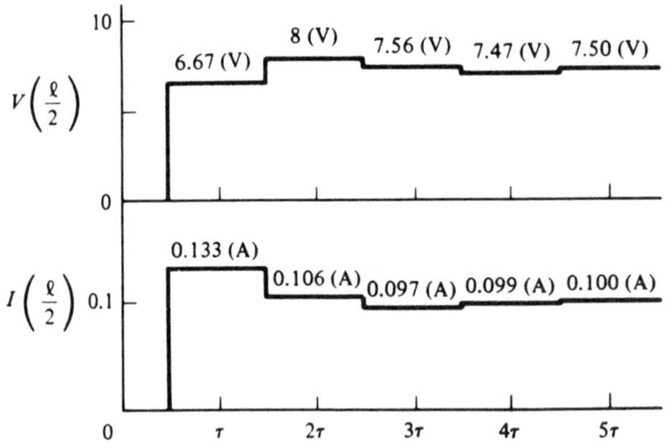

Figure 12-7 Midline voltage and current for Example 3.

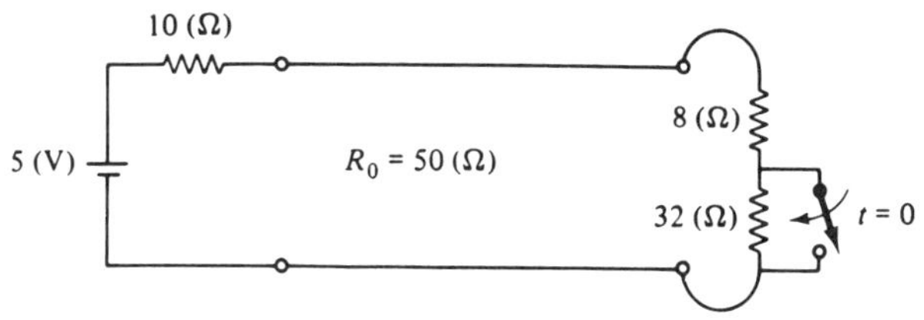

(a)

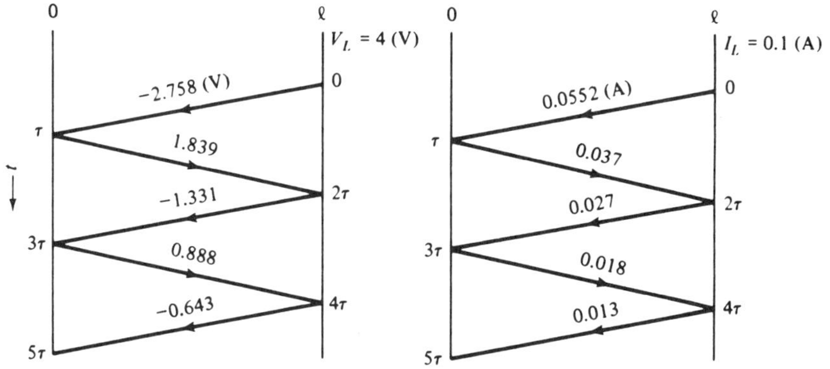

(b)

Figure 12-8 (a) Introducing a step change in the load impedance of a transmission line circuit. (b) Distance-time plot for the above circuit change. The numerical values are for use with Example 4.

and at the load, at $t = 0+$,

$$Z_L = \frac{V_L}{I_L} = 8 = \frac{V_L + \Delta V_L}{I_L + \Delta I_L}$$

or

$$8 = \frac{4 + \Delta V_L}{0.1 - \Delta V_L/50}$$

Solving for ΔV_L, we find that $\Delta V_L = -2.759$ V, and $\Delta I_L = -2.759/-50 = 0.055$ A. The new load voltage at $t = 0+ = 4 - 2.759 = 1.242$ V, and the new load current is $0.1 + 0.055 = 0.155$ A.

To calculate load voltage and current as a function of time, we must know ρ_G and ρ_L.

$$\rho_G = \frac{10 - 50}{10 + 50} = -0.667 = -\frac{2}{3}, \qquad \text{and} \quad \rho_L = \frac{8 - 50}{8 + 50} = -0.724.$$

We can now write V_L and I_L as a function of time.

$$\begin{aligned} V_L &= \{4.0 + 2.758[-u(t) + \tfrac{2}{3}u(t - 2\tau) - 0.724(\tfrac{2}{3})u(t - 2\tau) \\ &\quad + 0.724(\tfrac{2}{3})^2u(t - 4\tau) - (0.724)^2(\tfrac{2}{3})^2u(t - 4\tau) + \ldots]\} \quad \text{(V)} \\ &= [4.0 - 2.758u(t) + 0.507u(t - 2\tau) + 0.245u(t - 4\tau) + \ldots] \quad \text{(V)} \end{aligned}$$

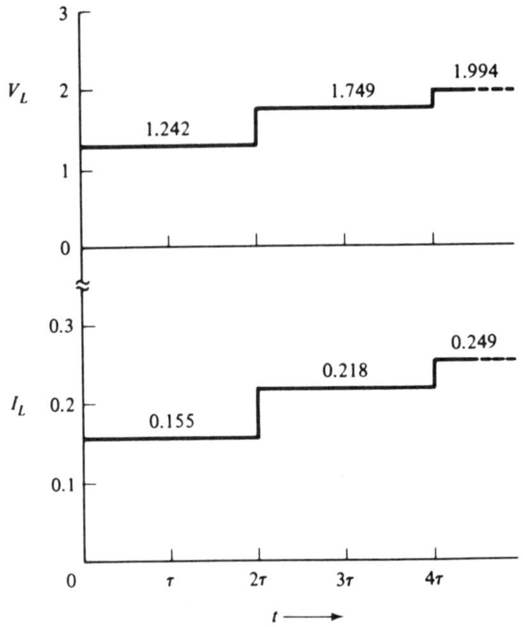

Figure 12-9 Plots of load voltage and current for Example 4.

and

$$
\begin{aligned}
I_L &= \{0.1 + 0.0552[u(t) + \tfrac{2}{3}u(t - 2\tau) + 0.724(\tfrac{2}{3})u(t - 2\tau) \\
&\quad + 0.724(\tfrac{2}{3})^2u(t - 4\tau) + (0.724)^2(\tfrac{2}{3})^2u(t - 4\tau) + \ldots]\} \qquad \text{(A)} \\
&= [0.1 + 0.0552u(t) + 0.063u(t - 2\tau) + 0.031u(t - 4\tau) + \ldots] \qquad \text{(A)}
\end{aligned}
$$

Both V_L and I_L are plotted in Fig. 12-9.

Problem 12.3-1 In Fig. 12-4, let $V_0 = 10$ V, $R_G = 10\ \Omega$, $R_0 = 50\ \Omega$, and $R_L = 40\ \Omega$. Draw the distance-time plots for voltage and current and plot $V(\ell/4)$ and $I(\ell/4)$ versus time for $0 < t < 4\tau$.

Problem 12.3-2 Use the data from Prob. 12.3-1 to plot V and I as a function of distance d from the input end of the line at $t = 7(\tau/4)$.

Problem 12.3-3 Repeat Prob. 12.3-1 with (a) $R_G = 50\ \Omega$, $R_L = 40\ \Omega$, and (b) $R_G = 10\ \Omega$, and $R_L = 50\ \Omega$. What can you say about reaching steady-state condition when either R_G or R_L is equal to R_0?

Problem 12.3-4 Referring to Fig. 12-8 and Example 4, rework this example, starting with the switch closed until $t = 0$ and then opened. Draw the distance-time plots, and plot V_L, I_L versus time out to $4\tau+$.

12.4 GENERAL EQUATIONS FOR LINE VOLTAGE AND CURRENT-FREQUENCY DOMAIN

Referring back to Figs. 12-1 and 12-3 and eqs. (12.2-2) and (12.2-3), the transformed differential equations are

$$\frac{d\mathcal{V}}{dz} = -(R + j\omega L)\mathcal{I} = -Z\mathcal{I} \qquad (\text{V m}^{-1}) \tag{1}$$

and

$$\frac{d\mathcal{I}}{dz} = -(G + j\omega C)\mathcal{V} = -Y\mathcal{V} \qquad (\text{A m}^{-1})$$

where $\mathcal{V}$ and $\mathcal{I}$ are phasors and Z and Y are complex numbers representing impedance and admittance. By differentiating (1) with respect to z and substituting (2) for $\partial\mathcal{I}/\partial z$, we get

$$\boxed{\frac{d^2\mathcal{V}}{dz^2} = -Z\frac{d\mathcal{I}}{dz} = ZY\mathcal{V} \qquad (\text{V m}^{-2})} \tag{3}$$

In a similar manner, differentiate (2) with respect to z and substitute $\partial\mathcal{V}/\partial z$ from (1) to get

$$\boxed{\frac{d^2\mathcal{I}}{dz^2} = -Y\frac{d\mathcal{V}}{dz} = ZY\mathcal{I} \qquad (\text{A m}^{-2})} \tag{4}$$

The differential equations (3) and (4) have exponential solutions of the form

$$\boxed{\mathcal{V} = \mathcal{V}^+ e^{-\sqrt{ZY}z} + \mathcal{V}^- e^{\sqrt{ZY}z} \qquad (\text{V})} \tag{5}$$

and

$$\boxed{\mathcal{I} = (\mathcal{I}^+ e^{-\sqrt{ZY}z} + \mathcal{I}^- e^{\sqrt{ZY}z}) \qquad (\text{A})} \tag{6}$$

as can easily be verified by differentiation. When the $e^{j\omega t}$ term is multiplied times each side of (5) and (6), it is evident that they are wavelike functions with complex arguments $(\omega t \pm \sqrt{ZY}z)$.

Two of the arbitrary constants, $\mathcal{I}^+$ and $\mathcal{I}^-$, can be eliminated from (6) by noting from (1) that

$$\mathcal{I} = -\frac{1}{Z}\left(\frac{d\mathcal{V}}{dz}\right)$$

and that (6) then becomes

$$\begin{aligned} \mathcal{I} &= -\frac{1}{Z}\left(\frac{d}{dz}\right)(\mathcal{V}^+ \epsilon^{-\sqrt{ZY}z} + \mathcal{V}^- e^{\sqrt{ZY}z}) \\ &= \frac{\sqrt{ZY}}{Z}(\mathcal{V}^+ e^{-\sqrt{ZY}z} - \mathcal{V}^- e^{\sqrt{ZY}z}) \end{aligned}$$

or

$$\boxed{\mathcal{I} = \frac{1}{\sqrt{Z/Y}}(\mathcal{V}^+ e^{-\sqrt{ZY}z} - \mathcal{V}^- e^{\sqrt{ZY}z}) \qquad (\text{A})} \tag{7}$$

The ratio $\sqrt{Z/Y}$ is called the *characteristic impedance* of the line, designated Z_0.

$$Z_0 \triangleq \sqrt{Z/Y} = \sqrt{\frac{R + j\omega L}{G + j\omega C}} \qquad (\Omega) \tag{8}$$

and, by analogy to plane wave propagation, the quantity $\sqrt{ZY}$ is referred to as the *propagation constant*.

$$\begin{aligned} \gamma &\triangleq \sqrt{ZY} = \sqrt{(R + j\omega L)(G + j\omega C)} \\ &= \alpha + j\beta \qquad (\mathrm{m}^{-1}) \end{aligned} \tag{9}$$

It is convenient to use the load location as a reference, i.e., let $z = 0$ at the load. Then

$$\mathcal{V}_L = \mathcal{V}^+ + \mathcal{V}^- \tag{10}$$

and

$$\mathcal{I}_L = \frac{1}{Z_0}(\mathcal{V}^+ - \mathcal{V}^-) \tag{11}$$

We now have

$$Z_L = \frac{\mathcal{V}_L}{\mathcal{I}_L} = \frac{\mathcal{V}^+(1 + \rho)}{\frac{\mathcal{V}^+}{Z_0}(1 - \rho)} = Z_0 \frac{1 + \rho}{1 - \rho}$$

where

$$\rho \triangleq \frac{\mathcal{V}^-}{\mathcal{V}^+} = \rho_L \tag{12}$$

Solving for ρ,

$$\rho = \frac{Z_L - Z_0}{Z_L + Z_0} \tag{13}$$

Drawing on our analogies with plane waves, it is obvious that ρ is a reflection coefficient [see (11.7-13)].

At a distance ℓ from the load, $\ell = -z$, and the equations for voltage and current may be written independently of any coordinate axis as

$$\begin{aligned} \mathcal{V}(\ell) &= \mathcal{V}^+ e^{\gamma \ell} + \mathcal{V}^- e^{-\gamma \ell} \\ &= \mathcal{V}^+ (e^{\gamma \ell} + \rho e^{-\gamma \ell}) \qquad (\mathrm{V}) \end{aligned} \tag{14}$$

and

$$\mathcal{I}(\ell) = \frac{\mathcal{V}^+}{Z_0}(e^{\gamma\ell} - \rho e^{-\gamma\ell}) \qquad \text{(A)} \tag{15}$$

With a little algebraic maneuvering, we note that (14) and (15) can be written as

$$\mathcal{V}(\ell) = \frac{\mathcal{V}^+}{2}(1 + \rho)(e^{\gamma\ell} + e^{-\gamma\ell}) + \frac{\mathcal{V}^+}{2}(1 - \rho)(e^{\gamma\ell} - e^{-\gamma\ell})$$

or

$$\mathcal{V}(\ell) = (\mathcal{V}_L \cosh \gamma\ell) + (\mathcal{I}_L Z_0 \sinh \gamma\ell) \qquad \text{(V)} \tag{16}$$

and

$$\mathcal{I}(\ell) = \left(\mathcal{I}_L \cosh \gamma\ell + \frac{\mathcal{V}_L}{Z_0} \sinh \gamma\ell\right) \qquad \text{(A)} \tag{17}$$

where $\mathcal{V}_L$ and $\mathcal{I}_L$ are taken from (10) and (11). If ℓ is the total line length, then (16) and (17) give the sending end voltage $\mathcal{V}_s$ and current $\mathcal{I}_s$ in terms of the load voltage and current.

It is also possible to express the load voltage and current in terms of the sending end voltage and current (see Prob. 12.4-1). Referring to Fig. 12-10, the input impedance to the line at a distance ℓ from the load is

$$Z_{\text{in}} \triangleq Z(\ell) \triangleq \frac{\mathcal{V}(\ell)}{\mathcal{I}(\ell)} = \frac{\mathcal{V}^+(e^{\gamma\ell} + \rho e^{-\gamma\ell})}{\dfrac{\mathcal{V}^+}{Z_0}(e^{\gamma\ell} - \rho e^{-\gamma\ell})}$$

$$\begin{aligned} Z_{\text{in}} &= Z_0 \left[\frac{e^{\gamma\ell} + \dfrac{Z_L - Z_0}{Z_L + Z_0} e^{-\gamma\ell}}{e^{\gamma\ell} - \dfrac{Z_L - Z_0}{Z_L + Z_0} e^{-\gamma\ell}} \right] \\ &= Z_0 \frac{Z_L + Z_0 \tanh \gamma\ell}{Z_0 + Z_L \tanh \gamma\ell} \qquad (\Omega) \end{aligned} \tag{18}$$

The voltage on the line consists of a forward wave of amplitude $|\mathcal{V}^+|e^{\alpha\ell}$ and a reverse wave of amplitude $|\rho_L \mathcal{V}^+|e^{-\alpha\ell}$. The current consists of a forward wave of amplitude $(1/|Z_0|)|\mathcal{V}^+|e^{\alpha\ell}$ and a reverse wave of amplitude $(1/|Z_0|)|\rho_L \mathcal{V}^+|e^{-\alpha\ell}$. The input

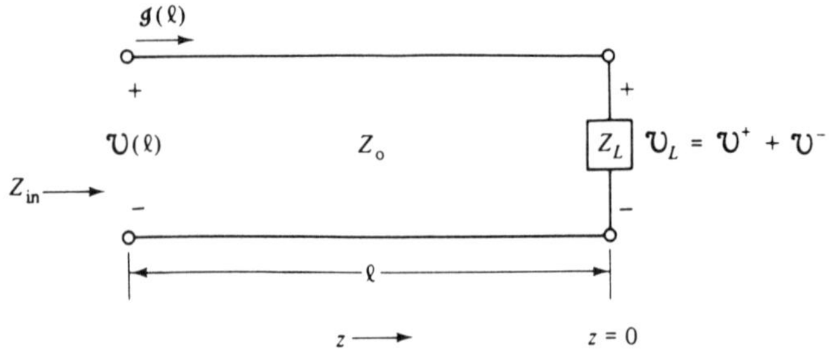

Figure 12-10 Illustration of the concept of input impedance.

impedance given by (18) is the ratio of total voltage to total current at a distance ℓ from the load.

Example 5

An open-wire (parallel wire) telephone line has the following parameters:

$$R = 4.11\ \Omega\ \text{mile}^{-1}$$
$$L = 0.00337\ \text{H mile}^{-1}$$
$$G = 0.29\ \mu\text{S mile}^{-1}$$
$$C = 0.00915\ \mu\text{F mile}^{-1}$$

Find Z_0, γ, and the input impedance 20 miles from a load of $Z_L = (50 + j50)$ (Ω) at a frequency of 1000 Hz.

Solution

$$Z_0 = \sqrt{\frac{4.11 + j2\pi(1000)(0.00337)}{0.29 + j2\pi(1000)(0.00915)}} \times 10^3 = \sqrt{\frac{21.57\,\angle 79.02^\circ}{57.49\,\angle 89.71^\circ}} \times 10^3 = \sqrt{\frac{Z}{Y}}$$
$$= 612\,\angle -5.35^\circ \qquad (\Omega)$$
$$\gamma = \sqrt{(21.57)\,\angle 79.02^\circ)(57.49\,\angle 89.71^\circ)} \times 10^{-3} = \sqrt{ZY}$$
$$= 0.00345 + j0.03504 \qquad (\text{mile}^{-1})$$
$$= \alpha + j\beta$$
$$\cosh \gamma\ell = \cosh [20(0.00345 + j0.03504)]$$
$$= \cosh (0.069 + j0.70)$$
$$= \cosh (0.069) \cos (0.70) + j \sinh (0.069) \sin (0.70)$$
$$= 0.768 + j0.045 = 0.768\,\angle 3.36^\circ$$
$$\sinh \gamma\ell = \sinh (0.069) \cos (0.7) + j \cosh (0.069) \sin (0.7)$$
$$= 0.0535 + j0.6458 = 0.648\,\angle 85.26^\circ$$
$$\tanh \gamma\ell = \frac{\sinh \gamma\ell}{\cosh \gamma\ell} = \frac{0.648\,\angle 85.26^\circ}{0.768\,\angle 3.36^\circ} = 0.844\,\angle 81.9^\circ$$
$$Z_{in} = 612\,\angle -5.35^\circ \frac{(50 + j50) + (612\,\angle -5.35^\circ)(0.844\,\angle 81.9^\circ)}{(612\,\angle -5.35^\circ) + (70.7\,\angle 45^\circ)(0.844\,\angle 81.9^\circ)}$$
$$= 616\,\angle 68.1^\circ \qquad (\Omega)$$

Example 6

Repeat Example 5, except consider $R = G = 0$, a lossless line.

Solution

$$Z_0 = \sqrt{\frac{L}{C}} = \sqrt{\frac{0.00337}{0.00915}} \times 10^3 = 607\ \Omega$$

$$\gamma = j\omega\sqrt{LC} = j2\pi(1000)\sqrt{(0.00337)(0.00915)} \times 10^{-3}$$
$$= j0.035 \text{ mile}^{-1} = j\beta$$

Since $\gamma = j\beta$, i.e., $\alpha = 0$, then

$$\tanh \gamma\ell = \tanh j\beta\ell = j \tan \beta\ell$$

and

$$Z_{\text{in}} = 607\frac{(50 + j50) + j607 \tan (0.035)(20)}{607 + j(70.7\ \angle 45^\circ) \tan (0.035)(20)}$$
$$= 601.12\ \angle 80.64^\circ \qquad (\Omega)$$

Problem 12.4-1 Derive the equations given below for load voltage and current from the sending-end voltage and current, for a line of length ℓ.

$$\mathcal{V}_L = \mathcal{V}_s \cosh \gamma\ell - \mathcal{I}_s Z_0 \sinh \gamma\ell$$

$$\mathcal{I}_L = \mathcal{I}_s \cosh \gamma\ell - \frac{\mathcal{V}_s}{Z_0} \sinh \gamma\ell$$

Problem 12.4-2 For the transmission line of Example 5, find the percent decrease in amplitude per mile for a voltage wave traveling toward the load. For a matched line, i.e., $Z_L = Z_0 = 612\ \angle -5.35^\circ\ \Omega$, find the voltage and current at the input to a 20-mile line if the load voltage is 5 mV. Assume all rms values.

Problem 12.4-3 Calculate ρ for line and load of Example 5.

Problem 12.4-4 Calculate the input impedance for the line of Example 5 for: (a) $Z_L = 0$, i.e., a short circuit; (b) Z_L = open circuit.

Problem 12.4-5 For a lossless line, where $R = G = 0$, show that (3) and (4) become

$$\frac{d^2\mathcal{V}}{dz^2} = -\omega^2 LC\mathcal{V}$$

and

$$\frac{d^2\mathcal{I}}{dz^2} = -\omega^2 LC\mathcal{I}$$

and have as solutions

$$\mathcal{V} = \mathcal{V}^+ e^{-j\omega\sqrt{LC}z} + \mathcal{V}^- e^{j\omega\sqrt{LC}z}$$

and

$$\mathcal{I} = \frac{1}{Z_0}(\mathcal{V}^+ e^{-j\omega\sqrt{LC}z} - \mathcal{V}^- e^{j\omega\sqrt{LC}z})$$

Show also that the velocity of propagation is given by

$$U = \frac{\omega}{\beta} = \frac{1}{\sqrt{LC}}$$

which corresponds to $U = 1/\sqrt{\mu\epsilon}$ for plane waves in lossless material.

12.5 POWER TRANSMISSION LINES

Although this chapter primarily concerns communications-type lines and frequencies, it is important to note that a transmission line is a transmission line, regardless of the frequency. It just happens that because of the extremely long wavelength at 60 Hz and because of the parameters of a typical power line, certain approximations may be made for short to medium-length lines that simplify the calculations.

Let a single-phase power line be represented by its pi equivalent, as shown in Fig. 12-11. The sending-end voltage and current are determined by ordinary network theory to be

$$\mathcal{V}_s = \mathcal{V}_L\left(1 + \frac{Z'Y'}{2}\right) + \mathcal{I}_L Z' \qquad \text{(V)} \tag{1}$$

$$\mathcal{I}_s = \mathcal{I}_L\left(1 + \frac{Z'Y'}{2}\right) + \mathcal{V}_L\left(Y' + \frac{Z'Y'^2}{4}\right) \qquad \text{(A)} \tag{2}$$

and when (1) and (2) are compared with (12.2-16) and (12.2-17) we see that

$$Z' = Z_0 \sinh \gamma\ell \tag{3}$$

$$\frac{Y'}{2} = Y_0 \tanh \frac{\gamma\ell}{2} \tag{4}$$

where $Y_0 = 1/Z_0$.

For a long transmission line, (12.2-16) and (12.2-17) must be used for accuracy, but for medium-length lines,

$$Z' \approx (R + j\omega L)\ell \tag{5}$$

$$Y' \approx (G + j\omega C)\ell \tag{6}$$

and for short power lines, $\cosh \gamma\ell \to 1$, $\sinh \gamma\ell \to \gamma\ell$, $Y' \to 0$, and

$$\mathcal{V}_s \approx \mathcal{V}_L + \mathcal{I}_L Z_0 \gamma\ell = \mathcal{V}_L + \mathcal{I}_L Z' \tag{7}$$

$$\mathcal{I}_s \approx \mathcal{I}_L + \frac{\mathcal{V}_L}{Z_0}\gamma\ell = \mathcal{I}_L + \mathcal{V}_L Y' \approx \mathcal{I}_L \tag{8}$$

12.6 DISTORTIONLESS LINE

From (12.4-8) and (12.4-9) we see that both the characteristic impedance and the propagation constant may be complex. A complex γ merely means that there is attenuation as well as phase shift as the voltage and current waves propagate along the line. To have distortionless transmission, the attenuation constant α must not be a function of frequency, and β must be directly proportional to frequency; i.e., a third harmonic of a

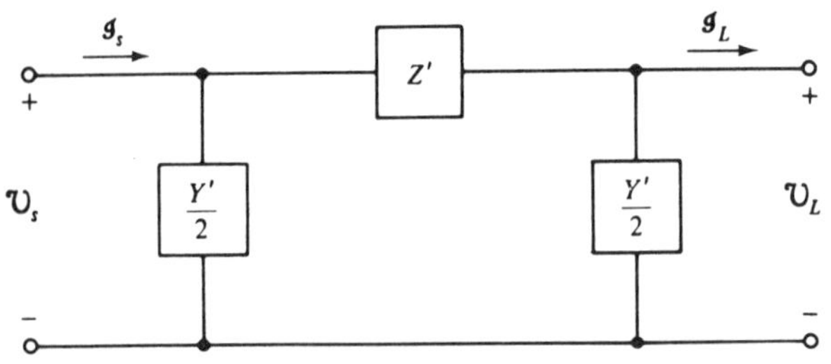

Figure 12-11 Pi representation of a power transmission line.

complex periodic waveform must be phase shifted three times as much as the fundamental in going from one end of the line to the other. When Z_0 is real; i.e., if

$$Z_0 = \sqrt{\frac{R + j\omega L}{G + j\omega C}} \text{ is real}$$

then

$$\frac{L}{R} = \frac{C}{G} \tag{1}$$

and

$$\boxed{Z_0 = \sqrt{\frac{R}{G}} = \sqrt{\frac{L}{C}} \quad (\Omega)} \tag{2}$$

and also

$$\begin{aligned}\gamma &= \sqrt{(R + j\omega L)(G + j\omega C)} \\ &= \sqrt{RG}\sqrt{\left(1 + j\omega\frac{L}{R}\right)\left(1 + j\omega\frac{C}{G}\right)} \\ &= \sqrt{RG}\left(1 + j\omega\frac{L}{R}\right) = \sqrt{RG}\left(1 + j\omega\frac{C}{G}\right)\end{aligned}$$

or

$$\boxed{\gamma = \sqrt{RG} + j\omega\sqrt{LC} = \alpha + j\beta \quad (\text{m}^{-1})} \tag{3}$$

since $L\sqrt{G}/\sqrt{R} = L\sqrt{C}/\sqrt{L} = \sqrt{LC}$.

In the early days of telephony, inductors were inserted in series with the line at frequent intervals to better approximate the condition (1). These were called *loading coils*. But since these coils were discrete elements or lumped constants, rather than distributed, the approximated distortionless line actually became a low-pass filter. The loading coils had to be removed when carrier frequencies were above the cutoff frequency of the loaded line. Thus a solution to voice frequency transmission was rendered unsatisfactory when carrier telephony was introduced.

Problem 12.6-1 Calculate the inductance of loading coils required to make the transmission line of Example 5 distortionless. Assume that the loading coils are added in series with the line at 1-mile intervals.

12.7 SKIN EFFECT, AND HIGH- AND LOW-LOSS APPROXIMATIONS

In earlier sections, equations are developed for the general transmission line with both series resistance and shunt conductance, but R and G are both assumed to be constant with frequency. Realistically, because of skin effect, the effective cross-sectional area of the conductor is reduced when the frequency is increased; thus, R must increase with frequency. What happens to G with frequency depends on the nature of the dielectric material in a more complicated way and is beyond the scope of the present discussion.

An exact analysis of skin effect in conductors of circular cross section involves the use of Bessel functions, but if the skin depth δ is small compared to the radius of the conductor, the approximate determination of R is illustrated in Fig. 12-12 for a coaxial line, where δ is calculated as for a plane surface. The total series R is, approximately,

$$R = \frac{1}{\sigma s} = \frac{1}{\sigma 2\pi r_a \delta} + \frac{1}{\sigma 2\pi r_b \delta}$$
$$= \frac{1}{2\pi\sigma\delta}\left(\frac{1}{r_a} + \frac{1}{r_b}\right) \qquad (\Omega\ \text{m}^{-1}) \tag{1}$$

where s is the current-carrying cross section of the conductors, and $\delta \approx 1/\sqrt{\pi f \mu \sigma}$, as determined in Sec. 11.6. Then

$$\boxed{\begin{aligned} R &\approx \frac{\sqrt{\pi f \mu \sigma}}{2\pi\sigma}\left(\frac{1}{r_a} + \frac{1}{r_b}\right) \\ &= \frac{1}{2}\sqrt{\frac{f\mu}{\pi\sigma}}\left(\frac{1}{r_a} + \frac{1}{r_b}\right) \qquad (\Omega\ \text{m}^{-1}) \end{aligned}} \tag{2}$$

If the transmission line losses are small in the sense that $RG \ll \omega^2 LC$, then the low-loss or high-frequency approximation may be used. This may be shown as follows. Since

$$\gamma = \sqrt{ZY} = \sqrt{(R + j\omega L)(G + j\omega C)}$$

then, if $RG \ll \omega^2 LC$,

$$\gamma = \sqrt{RG + j\omega RC + j\omega LG - \omega^2 LC} \tag{3}$$
$$= j\omega\sqrt{LC}\sqrt{1 - \frac{RG}{\omega^2 LC} + \frac{RC + LG}{j\omega LC}}$$
$$\approx j\omega\sqrt{LC}\left(1 - \frac{1}{2}j\frac{RC + LG}{\omega LC}\right)$$
$$= \frac{1}{2}\left(\frac{RC + LG}{\sqrt{LC}}\right) + j\omega\sqrt{LC} \tag{4}$$

where

$$\boxed{\underset{\text{low loss}}{\alpha} \approx \frac{1}{2}\frac{RC + LG}{\sqrt{LC}} = \frac{1}{2}\left(R\sqrt{\frac{C}{L}} + G\sqrt{\frac{L}{C}}\right) \qquad (\text{m}^{-1})} \tag{5}$$

and

$$\boxed{\underset{\text{low loss}}{\beta} \approx \omega\sqrt{LC} \qquad (\text{m}^{-1})} \tag{6}$$

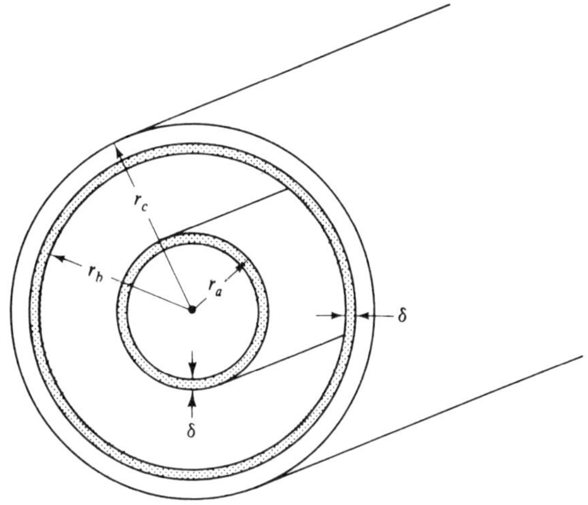

Figure 12-12 Illustration of the reduction in current-carrying cross-sectional area of coaxial conductors due to skin effect.

Note that the phase constant β is the same as for a lossless line. If $G = 0$, then (4) reduces to

$$\underset{(G=0)}{\gamma} = \left(\frac{1}{2}R\sqrt{\frac{C}{L}} + j\omega\sqrt{LC}\right) \qquad (\text{m}^{-1}) \tag{7}$$

If the series resistance and shunt conductance are appreciable such that $RG \gg \omega^2 LC$, then (3) can be written for the high-loss approximation as

$$\gamma = \sqrt{RG}\sqrt{1 + j\omega\left(\frac{C}{G} + \frac{L}{R}\right) - \omega^2\frac{LC}{RG}} \tag{8}$$

$$\approx \sqrt{RG}\sqrt{1 + j\omega\left(\frac{C}{G} + \frac{L}{R}\right)}$$

$$\approx \sqrt{RG}\left[1 + \frac{1}{2}j\omega\left(\frac{C}{G} + \frac{L}{R}\right)\right] \tag{9}$$

where

$$\boxed{\underset{\text{high loss}}{\alpha} \approx \sqrt{RG} \qquad (\text{m}^{-1})} \tag{10}$$

and

$$\boxed{\underset{\text{high loss}}{\beta} \approx \frac{1}{2}\omega\left(C\sqrt{\frac{R}{G}} + L\sqrt{\frac{G}{R}}\right) \qquad (\text{m}^{-1})} \tag{11}$$

If $R/L = G/C$, then both the low-loss and high-loss approximation formulas for α and β reduce to the exact expressions for α and β for a distortionless line; i.e.,

$$\alpha = \sqrt{RG}, \qquad \beta = \omega\sqrt{LC} \tag{12.6-3}$$

12.8 LOSSLESS LINES

A lossless line, as previously mentioned in Prob. 12.4-5, is a transmission line for which the series resistance and shunt conductance are both zero. The lossless line is also a distortionless line according to the criteria established in Sec. 12.6. While there are no truly lossless lines, most communication-type lines are low loss over short distances and may be treated as lossless to a good approximation. Even when the small loss must be considered, β is negligibly affected, and the amplitude of the voltage or current wave at any distance ℓ from the load is found by multiplying by $e^{\alpha\ell}$ for the forward traveling wave or $e^{-\alpha\ell}$ for the reflected wave.

Using the results of Prob. 12.4-5 and letting $\ell = -z$, we have

$$\boxed{\begin{aligned}\mathcal{V} &= \mathcal{V}^+ e^{j\omega\sqrt{LC}\,\ell} + \mathcal{V}^- e^{-j\omega\sqrt{LC}\,\ell} \\ &= \mathcal{V}^+(e^{j\omega\sqrt{LC}\,\ell} + \rho e^{-j\omega\sqrt{LC}\,\ell}) \qquad \text{(V)}\end{aligned}} \tag{1}$$

and

$$\boxed{\mathcal{I} = \frac{\mathcal{V}^+}{Z_0}(e^{j\omega\sqrt{LC}\,\ell} - \rho e^{-j\omega\sqrt{LC}\,\ell}) \qquad \text{(A)}} \tag{2}$$

where $\omega\sqrt{LC} = \beta$ for the voltage and current on a lossless line at a distance ℓ from the load. Although not explicitly shown, the load impedance influences the line voltage and current through the reflection coefficient,

$$\rho = \frac{Z_L - Z_0}{Z_L + Z_0} \tag{12.4-13}$$

From (12.4-8) and (12.4-9) and with $R = G = 0$,

$$\boxed{Z_0 = \sqrt{\frac{Z}{Y}} = \sqrt{\frac{0 + j\omega L}{0 + j\omega C}} = \sqrt{\frac{L}{C}} \qquad (\Omega)} \tag{3}$$

and

$$\boxed{\begin{aligned}\gamma &= \sqrt{ZY} = \sqrt{(0 + j\omega L)(0 + j\omega C)} = j\omega\sqrt{LC} \\ &= j\beta \qquad (\text{m}^{-1})\end{aligned}} \tag{4}$$

and since the velocity of the wave is ω/β,

$$U = \frac{\omega}{\beta} = \frac{1}{\sqrt{LC}} \qquad (\text{m } s^{-1}) \tag{5}$$

Since only transverse $\overline{\mathscr{E}}$ and $\overline{\mathscr{H}}$ fields are allowed, the wave equations for plane waves apply here also; namely,

$$\frac{d^2\mathscr{E}_{\text{trans}}}{dz^2} = -\omega^2\mu\epsilon\mathscr{E}_{\text{trans}} \tag{6}$$

$$\frac{d^2\mathcal{H}_{\text{trans}}}{dz^2} = -\omega^2\mu\epsilon\mathcal{H}_{\text{trans}} \tag{7}$$

where $\mathcal{E}_{\text{trans}}$ and $\mathcal{H}_{\text{trans}}$ refer to mutually transverse fields which are transverse to the z direction of propagation. Then, with

$$\frac{d^2\mathcal{V}}{dz^2} = -\omega^2 LC\mathcal{V} \tag{8}$$

and

$$\frac{d^2\mathcal{I}}{dz^2} = -\omega^2 LC\mathcal{I} \tag{9}$$

from Prob. 12.4-5, we see that

$$\boxed{U = \frac{1}{\sqrt{\mu\epsilon}} = \frac{1}{\sqrt{LC}} \qquad (\text{m } s^{-1})} \tag{10}$$

The transverse $\overline{E}$ field is directly related to the voltage at any point on the line, and the transverse $\overline{H}$ field is directly related to the current. Therefore, the velocity of the transverse fields must be the same as the velocity of the voltage and current waves, and this velocity as indicated by (10) is the velocity of a plane wave in the medium that exists between the two conductors.

Example 7

Consider a coaxial line having the radii r_a and r_b for the outside radius of the inner conductor and inside radius of the outer conductor respectively. By relating the transverse fields to a forward voltage and current wave, show that $Z_0 = \mathcal{V}^+/\mathcal{I}^+$. Also find the characteristic impedance and propagation constant from the equations for C and L developed in Chapters 5 and 9.

Solution. The capacitance per unit length, from (5.10-12), is

$$C = \frac{2\pi\epsilon}{\ln(r_b/r_a)} \qquad (\text{F m}^{-1})$$

The transverse electric field $\mathcal{E}_r$ is related to the line charge density, where r is used for r_c here for convenience,

$$\mathcal{E}_r = \frac{\rho_\ell}{2\pi\epsilon r} \qquad (\text{V m}^{-1})$$

and the transverse magnetic field $\mathcal{H}_\phi$ between conductors to the current by

$$\mathcal{H}_\phi = \frac{\mathcal{I}^+}{2\pi r} \qquad (\text{A m}^{-1})$$

But

$$\rho_\ell = C\mathcal{V}^+ = \frac{2\pi\epsilon\mathcal{V}^+}{\ln(r_b/r_a)}$$

and then

$$\frac{\mathcal{E}_r}{\mathcal{H}_\phi} = \frac{2\pi\epsilon\mathcal{V}^+}{\mathcal{I}^+\epsilon\ln(r_b/r_a)} = \sqrt{\frac{\mu}{\epsilon}} \qquad (\Omega)$$

since in cylindrical coordinates the curl equation for this example becomes

$$\hat{r}\left(-\frac{\partial \mathcal{H}_\phi}{\partial z}\right) = j\omega\epsilon\hat{r}\mathcal{E}_r$$

$$j\beta\mathcal{H}_\phi = j\omega\epsilon\mathcal{E}_r$$

or

$$\frac{\mathcal{E}_r}{\mathcal{H}_\phi} = \frac{\beta}{\omega\epsilon} = \sqrt{\frac{\mu}{\epsilon}}$$

Then

$$\frac{\mathcal{V}^+}{\mathcal{I}^+} = \frac{\mathcal{E}_r \ln (r_b/r_a)}{\mathcal{H}_\phi 2\pi} = \sqrt{\frac{\mu}{\epsilon}}\frac{\ln (r_b/r_a)}{2\pi} = Z_0$$

Using (5.10-12) for C and (9.10-9) for L,

$$\sqrt{\frac{L}{C}} = \sqrt{\frac{(\mu/2\pi)\ln (r_b/r_a)}{\dfrac{2\pi\epsilon}{\ln (r_b/r_a)}}} = \left(\sqrt{\frac{\mu}{\epsilon}}\right)\frac{\ln (r_b/r_a)}{2\pi}$$

the same as Z_0 above.

$$\sqrt{LC} = \sqrt{\left(\frac{\mu}{2\pi}\ln\frac{r_b}{r_a}\right)\frac{2\pi\epsilon}{\ln (r_b/r_a)}} = \sqrt{\mu\epsilon}$$

so that

$$\gamma = j\omega\sqrt{LC} = j\omega\sqrt{\mu\epsilon}$$

From Example 7, we observe that

$$LC = \mu\epsilon$$

but that (11)

$$\frac{L}{C} \neq \frac{\mu}{\epsilon}$$

and if we looked at a number of examples with various cross-sectional geometries, we would find that Z_0 and U for a lossless line are

$$\underset{\text{(lossless)}}{Z_0} = \sqrt{\frac{\mu}{\epsilon}} \times \text{geometrical factor}$$

$$= \sqrt{\frac{L}{C}} \quad (\Omega) \tag{12}$$

and

$$\underset{\text{(lossless)}}{U} = \sqrt{\frac{1}{LC}} = \sqrt{\frac{1}{\mu\epsilon}} \quad (\text{m } s^{-1})$$

regardless of the cross-sectional geometry. Another example illustrates this point.

Example 8

Find the characteristic impedance and propagation velocity for a lossless planar transmission line by finding L and C from static fields and applying (3) and (5).

Solution. Using the planar line shown in Fig. 11-2 and definition of L,

$$L = \frac{N\Psi_m}{\mathcal{I}} = \frac{(1)\mu\mathcal{H}\,d(1)}{\mathcal{I}}$$

per unit length of the line, i.e., $\Psi_m = \mu \mathcal{H} d(1)$. Then

$$L = \frac{\mu(\mathcal{I}/w)d}{\mathcal{I}} = \frac{\mu d}{w} \qquad (\text{H m}^{-1})$$

and

$$C = \frac{\epsilon s}{d} = \frac{\epsilon w(1)}{d} = \frac{\epsilon w}{d} \qquad (\text{F m}^{-1})$$

Then

$$Z_0 = \sqrt{\frac{L}{C}} = \sqrt{\frac{\mu d/w}{\epsilon w/d}} = \sqrt{\frac{\mu}{\epsilon}}\left(\frac{d}{w}\right) \qquad (\Omega)$$

and

$$U = \frac{1}{\sqrt{LC}} = \frac{1}{\sqrt{(\mu d/w)(\epsilon w/d)}} = \frac{1}{\sqrt{\mu\epsilon}} \qquad (\text{m s}^{-1})$$

Problem 12.8-1 The capacitance per meter of a transmission line has been determined to be 80 pF m^{-1}. Assuming it is a lossless line and for an air-dielectric medium, find Z_0 and β at $f = 500$ MHz.

Problem 12.8-2 A coaxial transmission line is filled with a material having a dielectric constant of 2.75. (a) Find the ratio r_b/r_a for $Z_0 = 50\ \Omega$. (b) Find the time delay in a 20-ft section of this line.

Problem 12.8-3 Find the Z_0 for a lossless planar transmission line for $\epsilon_r = 2.25$, spacing = 0.1 mm, conductor width = 2 mm.

12.9 INPUT IMPEDANCE AND STANDING WAVES (FOR LOSSLESS LINES)

The input impedance at a distance ℓ from the load is given by (12.4-18) for the general case, and for a lossless line $\tanh \gamma\ell$ becomes $j \tan \beta\ell$ and

$$Z_{in} = Z_0 \frac{Z_L + jZ_0 \tan \beta\ell}{Z_0 + jZ_L \tan \beta\ell} \qquad (\Omega) \tag{1}$$

is a periodic function of the distance from the load. The maxima and minima of Z_{in} occur at the maxima and minima of the voltage and current standing waves.

$$(Z_{in})_{max} = \frac{V_{max}}{I_{min}} \qquad (\Omega) \tag{2}$$

$$(Z_{in})_{min} = \frac{V_{min}}{I_{max}} \qquad (\Omega) \tag{3}$$

From (12.8-1) and (12.8-2), the maxima of voltage and minima of current occur where $e^{j\beta\ell}$ and $\rho e^{-j\beta\ell}$ are in phase. Then

$$V_{max} = |\mathcal{V}^+|(1 + |\rho|) \qquad (\text{V})$$

$$I_{min} = \frac{|\mathcal{V}^+|}{Z_0}(1 - |\rho|) \qquad (\text{A})$$

and

$$Z_{max} = \frac{V_{max}}{I_{min}} = Z_0 \frac{1 + |\rho|}{1 - |\rho|} \qquad (\Omega) \tag{4}$$

and likewise, the minima of voltage and maxima of current occur where $e^{j\beta\ell}$ and $-\rho e^{-j\beta\ell}$ are in phase, which can be seen to be $\lambda/4$ displaced from where $e^{j\beta\ell}$ and $\rho e^{-j\beta\ell}$ are in phase ($\beta = 2\pi/\lambda$). Thus,

$$V_{min} = |\mathcal{V}^+|(1 - |\rho|) \qquad (\text{V})$$

$$I_{max} = \frac{|\mathcal{V}^+|}{Z_0}(1 + |\rho|) \qquad (\text{A})$$

and

$$Z_{min} = Z_0 \frac{1 - |\rho|}{1 + |\rho|} \qquad (\Omega) \tag{5}$$

But analogous to the plane wave case and to (11.8-18), the standing wave ratio is

$$\boxed{S = \frac{V_{max}}{V_{min}} = \frac{I_{max}}{I_{min}} = \frac{1 + |\rho|}{1 - |\rho|}} \tag{6}$$

and (4) and (5) may also be written as

$$\boxed{(Z_{in})_{max} = SZ_0 \qquad (\Omega)} \tag{7}$$

$$\boxed{(Z_{in})_{min} = \frac{Z_0}{S} \qquad (\Omega)} \tag{8}$$

If Z_L is real, then ρ is either positive or negative, depending on whether $Z_L > Z_0$ or $Z_0 > Z_L$. When (6), (7), and (8) are rewritten specifically for Z_L real, then

$$S = \frac{Z_L}{Z_0} \quad \text{or} \quad \frac{Z_0}{Z_L} \tag{9}$$

whichever is greater than unity. Also for Z_L real,

$$\left.\begin{aligned} (Z_{in})_{max} &= \left(\frac{Z_L}{Z_0} Z_0\right)_{Z_L > Z_0} = Z_L = SZ_0 \\ &\text{or} \\ &= \left(\frac{Z_0}{Z_L} Z_0\right)_{Z_L < Z_0} = \frac{Z_0^2}{Z_L} = SZ_0 \end{aligned}\right\} \tag{10}$$

and

$$\left.\begin{aligned} (Z_{in})_{min} &= \left(\frac{Z_0}{Z_L} Z_0\right)_{Z_L > Z_0} = \frac{Z_0^2}{Z_L} = \frac{Z_0}{S} \\ &\text{or} \\ &= \left(\frac{Z_L}{Z_0} Z_0\right)_{Z_L < Z_0} = Z_L = \frac{Z_0}{S} \end{aligned}\right\} \tag{11}$$

The standing wave patterns for a lossless line connected to a pure resistance load are illustrated in Fig. 12-13 for $Z_L < Z_0$, $Z_L > Z_0$, and $Z_L = Z_0$.

Example 9

Referring to Fig. 12-13, let $Z_0 = 50\ \Omega$, $Z_L = 80\ \Omega$, and the load voltage = 5 V. Find ρ, S, Z_{in} at $\ell = \lambda/4, \lambda/2, 3\lambda/8$, and $V_{\max}$, $V_{\min}$, $I_{\max}$, $I_{\min}$.

Solution

$$\rho = \frac{Z_L - Z_0}{Z_L + Z_0} = \frac{80 - 50}{80 + 50} = 0.23$$

$$S = \frac{1 + |\rho|}{1 - |\rho|} = \frac{1 + 0.23}{1 - 0.23} = 1.60 = \frac{V_{\max}}{V_{\min}} = \frac{I_{\max}}{I_{\min}}$$

$$Z_{\text{in}} = 50\frac{80 + j50 \tan \beta\ell}{50 + j80 \tan \beta\ell}, \qquad \beta\ell = \frac{2\pi}{\lambda}\ell$$

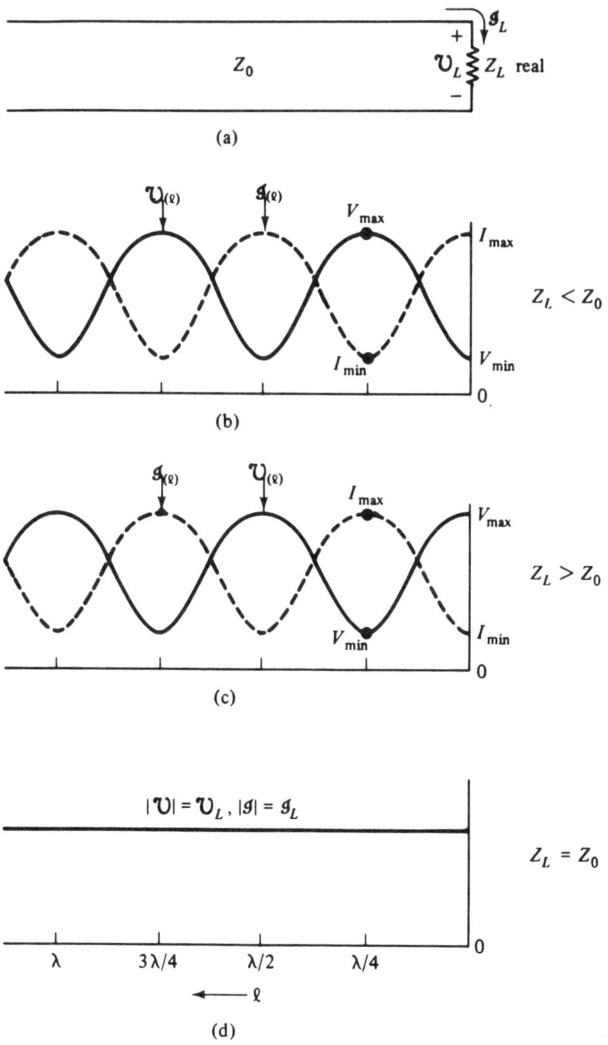

Figure 12-13 (a) Illustration of a lossless line connected to a pure resistance load. (b) Standing waves of voltage and current for a load smaller than Z_0. (c) Standing waves of voltage and current for a load larger than Z_0. (d) Absence of standing waves when $Z_L = Z_0$.

For $\ell = \lambda/4$,

$$Z_{in} = 50\frac{80 + j50 \times \infty}{50 + j80 \times \infty} = 50\frac{50}{80} = 31.25\ \Omega$$
$$= \frac{50}{S} = \frac{50}{1.60} = Z_{min}$$

For $\ell = \lambda/2$,

$$Z_{in} = 50\frac{80 + j0}{50 + j0} = 80\ \Omega$$
$$= S50 = (1.60)50 = Z_{max}$$

For $\ell = 3\lambda/8$,

$$Z_{in} = 50\frac{80 - j50}{50 - j80} = 50\frac{94.34 \angle -32.01^\circ}{94.34 \angle -57.99^\circ}$$
$$= 50 \angle 25.99^\circ \qquad (\Omega)$$
$$V_{max} = SV_{min} = \mathcal{V}_L \text{ in this case}$$

Then

$$V_{min} = \frac{\mathcal{V}_L}{S} = 3.13 \text{ V at } \lambda/4 \text{ from load}$$
$$V_{max} = \mathcal{V}_L = 5 \text{ V}$$
$$I_{min} = \frac{5}{80} = 0.063 \text{ A}$$
$$I_{max} = S\mathcal{I} = (1.6)(0.063) = 0.10 \text{ A at } \lambda/4 \text{ from load}$$

Example 10

Using the same line and load as in Example 9, find the voltage, current, and $\mathcal{V}(\ell)/\mathcal{I}(\ell)$ at $\ell = 3\lambda/8$ from the load if $\mathcal{V}_L = 5$ V.

Solution. Refer back to (12.8-1) and (12.8-2). At the load,

$$\mathcal{V}_L = \mathcal{V}^+ + \mathcal{V}^- = \mathcal{V}^+(1 + \rho) = \mathcal{V}^+(1.23)$$

Then

$$\mathcal{V}^+ = \frac{5}{1.23} = 4.065 \text{ V}$$
$$\mathcal{V}^- = \rho\mathcal{V}^+ = (0.23)(4.065) = 0.935 \text{ V}$$
$$\mathcal{V}\left(\ell = \frac{3\lambda}{8}\right) = 4.065e^{j\beta\ell} + 0.935e^{-j\beta\ell}$$
$$= 4.065e^{j3\pi/4} + 0.935e^{-j3\pi/4}$$
$$= 4.065\left(\frac{-1 + j1}{\sqrt{2}}\right) + 0.935\left(\frac{-1 - j1}{\sqrt{2}}\right)$$
$$= \frac{(-4.065 - 0.935) + j(4.065 - 0.935)}{\sqrt{2}}$$
$$= 4.17 \angle 147.95^\circ \qquad (\text{V})$$
$$\mathcal{I}\left(\ell = \frac{3\lambda}{8}\right) = \frac{4.065(-1 + j1) - 0.935(-1 - j1)}{50\sqrt{2}}$$
$$= \frac{5.899 \angle 122.047^\circ}{50\sqrt{2}} = 0.083 \angle 122.047^\circ \qquad (\text{A})$$

$$\frac{\mathcal{V}\left(\ell = \frac{3\lambda}{8}\right)}{\mathcal{I}\left(\ell = \frac{3\lambda}{8}\right)} = \frac{4.17 \angle 147.95^\circ}{0.083 \angle 122.047^\circ} = 50 \angle 25.9^\circ \quad (\Omega)$$

$$= Z_{\text{in}}\left(\ell = \frac{3\lambda}{8}\right)$$

from Example 9.

Even if the load is complex, there are two positions within each $\lambda/2$ length from the load for which Z_{in} is real, as the next example will show.

Example 11

A lossless line whose Z_0 is 50 (Ω) is connected to a load $Z_L = 50 \angle 25.99^\circ$ (Ω). Find Z_{in} at $\ell = \lambda/8$ and $3\lambda/8$.

Solution. Using (1), for $\ell = \lambda/8$,

$$Z_{\text{in}} = 50\left[\frac{(44.94 + j21.91) + j50}{50 + j(44.94 + j21.91)}\right]$$

$$= 50\frac{84.79 \angle 57.99^\circ}{52.99 \angle 57.99^\circ} = 80 \quad (\Omega)$$

For $\ell = 3\lambda/8$,

$$Z_{\text{in}} = 50\left[\frac{44.94 + j21.91 - j50}{50 - j(44.94 + j21.91)}\right]$$

$$= 50\frac{53.0 \angle -32.01}{84.80 \angle -32^\circ} = 31.25 \ (\Omega)$$

Refer to Fig. 12-14 to compare these results with those of Example 9.

The short circuit load is of major interest because of the use of shorted lines as tuning stubs, a topic we shall discuss in greater depth a little later. If $Z_L = 0$,

$$\underset{(Z_L=0)}{Z_{\text{in}}} = Z_0\left(\frac{0 + jZ_0 \tan \beta\ell}{Z_0 + j0}\right) = jZ_0 \tan \beta\ell \quad (\Omega) \tag{12}$$

and

$$\underset{(Z_L=0)}{S} = \frac{1 + |\rho|}{1 - |\rho|} = \frac{1 + 1}{1 - 1} = \infty \tag{13}$$

Some interesting special cases are considered here and in Fig. 12-15. For a shorted $\lambda/8$ line,

$$(Z_{\text{in}})_{\lambda/8\text{ shorted}} = jZ_0 \tan \frac{2\pi}{\lambda}\left(\frac{\lambda}{8}\right) = jZ_0 \quad (\Omega) \tag{14}$$

is an inductive reactance equal in magnitude to Z_0. For a shorted $\lambda/4$ line,

$$(Z_{\text{in}})_{\lambda/4\text{ shorted}} = jZ_0 \tan \frac{2\pi}{\lambda}\left(\frac{\lambda}{4}\right) = j\infty \tag{15}$$

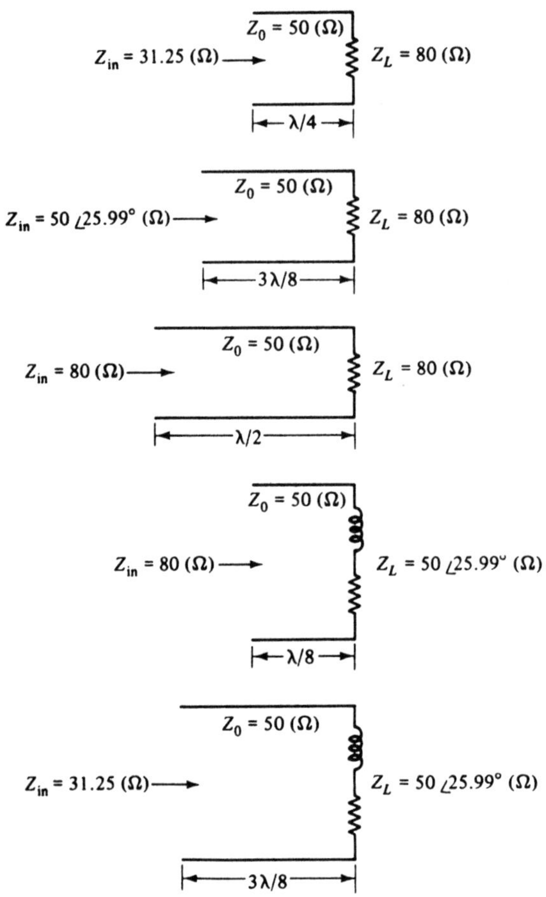

Figure 12-14 Illustration of the comparison between the results of Examples 9 and 11.

and looks like an open circuit. For a $3\lambda/8$ shorted line,

$$(Z_{\text{in}})_{3\lambda/8\,\text{shorted}} = jZ_0 \tan \frac{2\pi}{\lambda}\left(\frac{3\lambda}{8}\right) = -jZ_0 \qquad (\Omega) \tag{16}$$

and

$$(Z_{\text{in}})_{\lambda/2\,\text{shorted}} = jZ_0 \tan \frac{2\pi}{\lambda}\left(\frac{\lambda}{2}\right) = 0 \tag{17}$$

the same as at the short. It is a general rule that at a half-wavelength from any load the input impedance is the same as the load and repeats for every half-wavelength thereafter, as can be demonstrated by substituting an integral number of half-wavelengths for ℓ in (1).

The shorted line may be used to provide any value of inductive or capacitive reactance (or susceptance) from zero to infinity and thus can be used instead of an inductor or capacitor. These "stubs" are widely used for this purpose at frequencies above 100 MHz or so, where the lengths of the shorted stubs are practical for circuit design. Tuning is accomplished by moving an adjustable short to the desired position.

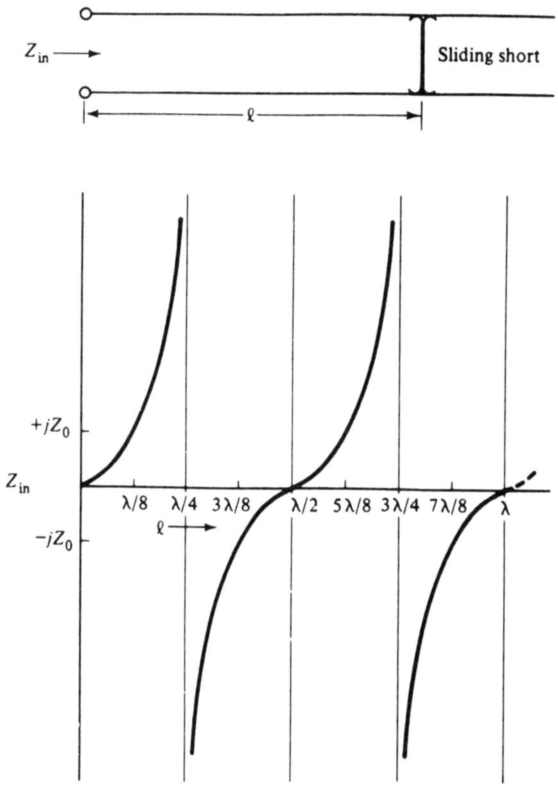

Figure 12-15 Input impedance to a lossless transmission line as a function of the distance to a sliding short circuit.

The open-circuited line theoretically has the same standing wave characteristic as the shorted line, for when $Z_L = \infty$ is substituted into (1),

$$\underset{(Z_L=\infty)}{Z_{\text{in}}} = -jZ_0 \cot \beta\ell \qquad (\Omega) \tag{18}$$

and the Z_{in} versus ℓ plot is the same as for the shorted line except that it is displaced by $\lambda/4$. The reflection coefficient for the open end is $+1$. As a practical matter, however, an open circuit can radiate some energy, especially at the higher frequencies, and this radiation appears as a resistance load; thus, the standing wave ratio is not infinity as for an ideal shorted line. Because of small losses in a shorted termination, a short can be more nearly an ideal short than an open can be an ideal open.

Problem 12.9-1 Using (1), show that $Z_0 = \sqrt{Z_{sc}Z_{oc}}$, where $Z_{sc} = Z_{\text{in}}$ for $Z_L = 0$, and $Z_{0c} = Z_{\text{in}}$ for $Z_L = \infty$.

Problem 12.9-2 Find the input impedance and standing wave ratio for a 75-Ω line connected to a 30-Ω load, if the input to the line is: (a) $\lambda/2$ from the load; (b) $3\lambda/8$ from the load.

Problem 12.9-3 A lossless coaxial line having a characteristic impedance of 75 Ω is connected to a 50-Ω load. If the dielectric has a dielectric constant of 2.3 and the frequency is

500 MHz, find the standing wave ratio on the line and the input impedance 0.5 (m) from the load.

Problem 12.9-4 Find the shortest length in wavelengths for a shorted stub that will have an input impedance of $+j30$ (Ω) if the characteristic impedance of the stub is 50 Ω.

12.10 THE $\lambda/4$ TRANSFORMER

When $\ell = \lambda/4$, the input impedance found by using (12.9-1) is

$$\underset{(\ell=\lambda/4)}{Z_{\text{in}}} = Z_0 \frac{Z_L + jZ_0 \tan(\pi/2)}{Z_0 + jZ_L \tan(\pi/2)} = \frac{Z_0^2}{Z_L} = Z_0\left(\frac{Z_0}{Z_L}\right) \quad (\Omega) \tag{1}$$

By analogy to an ordinary transformer which is used for impedance matching ($Z_{\text{in}} = (N_1^2/N_2^2)Z_{\text{out}}$), the quarter-wavelength section of transmission line is referred to as a transformer. When a transmission line of high characteristic impedance is to be connected to a low impedance load, or vice versa, the $\lambda/4$ transformer of a suitable Z_0 can act as the connecting link for impedance matching. Of course, the $\lambda/4$ transformer is inherently a narrow band device, for it can be exactly a quarter wavelength at one frequency only.

A good example of the use of the $\lambda/4$ transformer concept is found in Prob. 11.8-4, where a $\lambda/4$ lens coating is used to match the impedance of glass to that of free space. In that case, the match is required over only a small percentage bandwidth, just the visible portion of the spectrum. A poor application of the $\lambda/4$ transformer would be to connect a 300-Ω TV set to a 75-Ω cable (such as used by cable TV companies) because the transformer would be $\lambda/4$ for one channel frequency only. A broad band transformer is required for that application. The $\lambda/4$ transformer is widely used in antenna and transmission line hardware, however, because in many cases wide bandwidth is not a requirement. When a match is required at more than one frequency, but not simultaneously, a sliding adjustment may be provided for "tuning" the transformer to the desired frequency.

Example 12

Use a $\lambda/4$ transformer to match a 300-Ω line to a 75-Ω load such that there will be no standing waves on the 300-Ω line. Find Z_0 of the transformer.

Solution. For no reflection, Z_{in} at the input to the $\lambda/4$ transformer must be 300 Ω. Referring to Fig. 12-16,

$$Z_{\text{in}} = 300 = \frac{Z_{01}^2}{75}$$

or

$$Z_{01} = \sqrt{(300)(75)} = 150 \ \Omega$$

Note that $300/150 = 150/75$. Also, note that the same $\lambda/4$ transformer would match a 75-Ω line to a 300-Ω load.

In the preceding example, the 300-Ω line is connected to what looks like a 300-Ω load at a frequency for which the Z_{01} section of line is exactly $\lambda/4$. There will be no

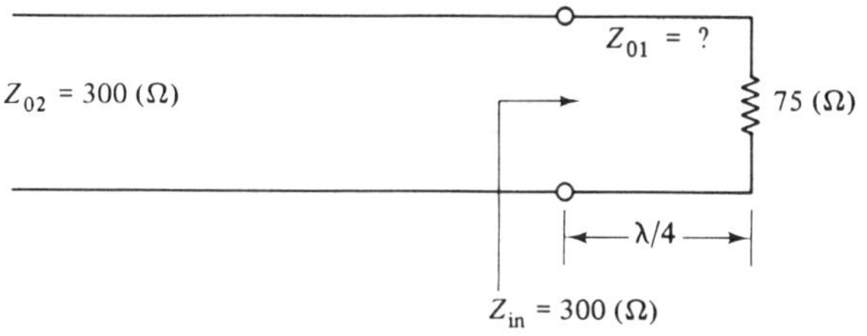

Figure 12-16 Illustration for use with Example 12.

standing waves on the Z_{02} line because of this matched condition. However, the $\lambda/4$ length of 150-Ω line is mismatched to a 75-Ω load. This means that the standing wave ratio on the $\lambda/4$ line is two. A voltage minimum will exist at the 75-Ω load, and a voltage maximum at the interface with the 300-Ω line. The overall matched condition resulting from the use of the $\lambda/4$ transformer is for a sinusoidal steady-state signal only, for on a steady-state basis the mismatch at the 300-Ω line and the 150-Ω line interface is compensated by the mismatch at the 150-Ω line and the 75-Ω load interface.

Problem 12.10-1 A 35-Ω cable and a 75-Ω cable are to be connected together, using a $\lambda/4$ transformer. If the transmission line for the transformer has an air dielectric and the frequency is 75 MHz, find the length of line and Z_0 required for matching the two lines.

Problem 12.10-2 A $\lambda/2$ section of transmission line is sometimes referred to as a half-wavelength transformer. Determine the input impedances of Z_{in1} and Z_{in2} by inspection from Fig. 12-17, using only $\lambda/4$ and $\lambda/2$ transformer characteristics.

Problem 12.10-3 A $\lambda/4$ section of 50-Ω line is connected to a load $Z_L = 50\angle{-30^\circ}$ (Ω). Find the input impedance to the $\lambda/4$ section.

12.11 POWER FLOW ON LOSSLESS LINES

If $\mathcal{V}(\ell)$ and $\mathcal{I}(\ell)$ are the total voltage and current at a distance ℓ from the load, the average input power at this point is

$$P_{\text{av in}} = \frac{1}{2}R_e[\mathcal{V}\mathcal{I}^*]$$

$$= \frac{1}{2}R_e\left[\mathcal{V}^+(e^{j\beta\ell} + \rho e^{-j\beta\ell})\frac{\mathcal{V}^{+*}}{Z_0}(e^{-j\beta\ell} - \rho^* e^{j\beta\ell})\right]$$

$$= \left(\frac{1}{2}\right)\frac{|\mathcal{V}^+|^2}{Z_0}[1 - |\rho|^2) + R_e(\rho e^{-2j\beta\ell} - \rho^* e^{2j\beta\ell})]$$

$$= \left(\frac{1}{2}\right)\frac{|\mathcal{V}^+|^2}{Z_0}(1 - |\rho|^2)$$

or

$$P_{\text{av in}} = \left[\left(\frac{1}{2}\right)\frac{|\mathcal{V}^+|^2}{Z_0} - \left(\frac{1}{2}\right)\frac{|\mathcal{V}^+|^2|\rho|^2}{Z_0}\right] \quad \text{(W)} \tag{1}$$

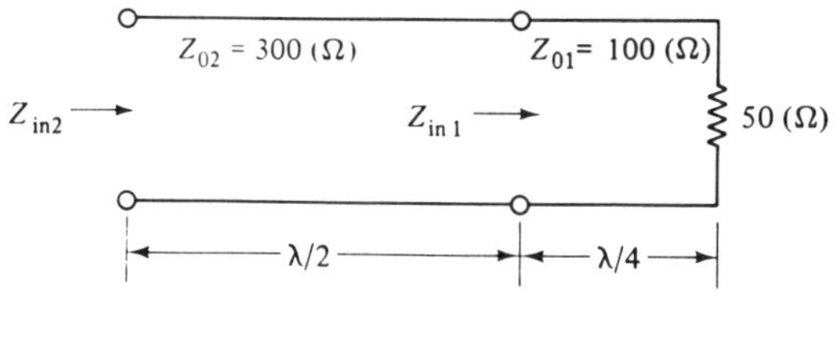

Figure 12-17 Illustration for use with Prob. 12.10-2.

The first term of (1) is the incident power, and the second term is the reflected power. The incident power is the power in the forward traveling wave, and it is the power that would be absorbed by the load if the reflection coefficient were zero, i.e., if $Z_L = Z_0$. The reflected power is returned to the source so that the net power to the load (assuming the line is lossless and therefore does not absorb any power) is that given by (1). We can show this by directly computing the power at the load.

$$
\begin{aligned}
P_{\text{load}} &= \frac{1}{2} R_e[(\mathcal{V}^+ + \mathcal{V}^-)(\mathcal{I}^+ + \mathcal{I}^-)^*] \\
&= \frac{1}{2} R_e\left[\mathcal{V}^+(1+\rho)\frac{\mathcal{V}^{+*}}{Z_0}(1-\rho^*)\right] \\
&= \left(\frac{1}{2}\right)\frac{|\mathcal{V}^+|^2}{Z_0}[(1-|\rho|^2) + R_e(\rho - \rho^*)] \\
&= \left(\frac{1}{2}\right)\frac{|\mathcal{V}^+|^2}{Z_0}(1-|\rho|^2) \quad \text{(W)}
\end{aligned}
\tag{2}
$$

Thus, we have the equivalence for lossless lines,

$$
\boxed{P_{\text{in}} = P_{\text{net}} = P_{\text{load}} = (P_{\text{incident}} - P_{\text{reflected}})} \tag{3}
$$

A suitable wattmeter placed anywhere between the source and load would measure the same power as measured at the load.

Incident and reflected power can be separately measured by using directional couplers. A directional coupler is a network that can be inserted in series with the power flow that has an output that is proportional to the power flow in one direction only (although dual directional couplers have two outputs, one for the power in each direction), and the power traveling in the opposite direction is ignored. The insertion loss of the coupler is very small, on the order of 0.1 dB. The use of directional couplers is illustrated in the following example.

Example 13

A 50-Ω line is connected to a load whose Z_L is $35 + j35$ (Ω). Find the power readings on the two power meters in Fig. 12-18 if $\mathcal{V}^+ = 50$ V, and determine the load power. The arrows on the directional couplers indicate the direction of power flow that is coupled to the power meter, and the decibel value on the coupler is the attenuation between the power on the line and the output of the directional coupler.

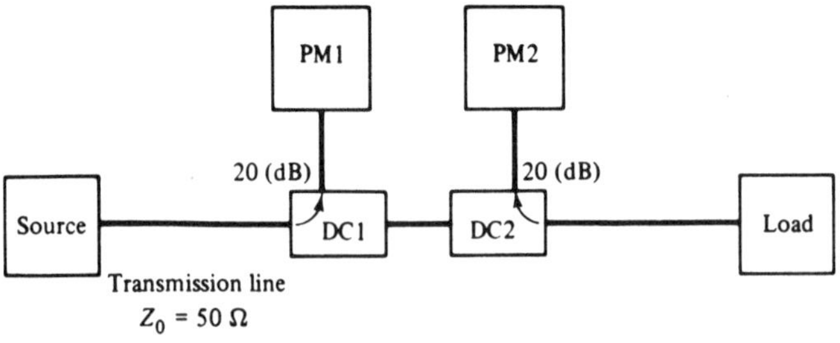

Figure 12-18 Illustration for use with Example 13.

Solution. The reflection coefficient is

$$\rho = \frac{35 + j35 - 50}{35 + j35 + 50} = \frac{-15 + j35}{85 + j35}$$

$$= \frac{38.08 \underline{/113.20^\circ}}{91.92 \underline{/22.38^\circ}} = 0.414 \underline{/90.82^\circ}$$

Then

$$P_{\text{forward}} = P_{\text{incident}} = \frac{1}{2}\left(\frac{|\mathcal{V}^+|^2}{Z_0}\right) = \frac{1}{2}\left(\frac{50^2}{50}\right) = 25 \text{ W}$$

$$P_{\text{reflected}} = \frac{1}{2}\left(\frac{|\mathcal{V}^+|^2|\rho|^2}{Z_0}\right) = \frac{50^2(0.414)^2}{2(50)}$$

$$= 4.29 \text{ W}$$

$$P_{\text{net}} = P_{\text{load}} = 25 - 4.29 = 20.71 \text{ W}$$

Then PM1 will read 20 dB less than P_{incident},

$$\text{PM1} = \tfrac{1}{100}25 = 0.25 \text{ W}$$

and PM2 will read 20 dB less than $P_{\text{reflected}}$,

$$\text{PM2} = \tfrac{1}{100}4.29 = 0.0429 \text{ W}$$

One could take the two power meter readings and quickly compute the load power,

$$P_L = 100(\text{PM1} - \text{PM2}) = 100(0.25 - 0.0429) = 20.71 \text{ W}$$

The power handling capability of a transmission line may be determined by the voltage breakdown of the line. Where this is the case, it is important to have the standing wave ratio as close to unity as possible in order to handle as much power as possible on the line. That is, since

$$V_{\text{max}} = |\mathcal{V}^+|(1 + |\rho|) \qquad \text{(V)} \tag{4}$$

and the power delivered to the load is

$$P_L = \frac{1}{2}\frac{|\mathcal{V}^+|^2}{Z_0}(1 - |\rho|^2) \tag{2}$$

$$= \frac{1}{2}\left(\frac{|\mathcal{V}^+|^2}{Z_0}\right)(1 + |\rho|)(1 - |\rho|)$$

then

$$\boxed{P_L = \frac{1}{2}\left[\frac{V_{\text{max}}^2(1 - |\rho|)}{(1 + |\rho|)Z_0}\right] = \frac{1}{2}\left(\frac{V_{\text{max}}^2}{SZ_0}\right) \qquad \text{(W)}} \tag{5}$$

which indicates the power delivered to the load is inversely proportional to the standing wave ratio if V_{max} on the transmission line is a fixed voltage.

Problem 12.11-1 A 50-Ω coaxial cable is connected to an impedance of $40 + j30$ Ω. The net input average power to the line is 10 W. Find the incident and reflected power and voltage across the load.

Problem 12.11-2 Determine which of the following two lossless transmission lines is delivering the larger power to the load: Line 1 has a Z_0 of 50 Ω, a V_{max} of 100 V, and a V_{min} of 80 V; line 2 has a Z_0 of 75 Ω, a V_{max} of 150 V, and a V_{min} of 100 V.

Problem 12.11-3 A voltage rating on a particular 75-Ω transmission line is 1000 V. In an application, the standing wave ratio on the line is 2.0 while operating at maximum safe voltage; i.e., $V_{max} = 1000$ V. Find the maximum average power delivered to the load, and determine the increase in load power possible if S were reduced to 1.2.

12.12 THE SMITH CHART

In its usual form, the Smith chart is a set of coordinates on which all values of normalized impedance or admittance can be plotted within a unit circle, a circle whose radius $|\rho| = 1$. The normalized values of impedance are obtained by dividing by Z_0. For example, if $Z = R + jX$, the normalized impedance is

$$z = r + jx = \frac{R + jX}{Z_0} \tag{1}$$

where lowercase letters are used for normalized values. Normalized admittance is

$$y = g + jb = \frac{G + jB}{Y_0} \tag{2}$$

where the characteristic admittance Y_0 is the reciprocal of Z_0.

Now, consider a load impedance Z_L. The normalized load impedance is

$$z_L = r + jx$$

and

$$\rho = \frac{z_L - 1}{z_L + 1} = \frac{r + jx - 1}{r + jx + 1} \tag{3}$$

$$= \rho_r + j\rho_i \tag{4}$$

where ρ is in general complex with real part ρ_r and imaginary part ρ_i. Solving for z_L, from (3) and (4),

$$z_L = \frac{1 + \rho}{1 - \rho} = \frac{1 + \rho_r + j\rho_i}{1 - \rho_r - j\rho_i} = r + jx \tag{5}$$

and, equating real and imaginary parts on both sides of the equation,

$$r = \frac{1 - \rho_r^2 - \rho_i^2}{(1 - \rho_r)^2 + \rho_i^2} \tag{6}$$

$$x = \frac{2\rho_i}{(1 - \rho_r)^2 + \rho_i^2} \tag{7}$$

which can be rearranged to give the following two equations,

$$\left(\rho_r - \frac{r}{1 + r}\right)^2 + \rho_i^2 = \left(\frac{1}{1 + r}\right)^2 \tag{8}$$

$$(\rho_r - 1)^2 + \left(\rho_i - \frac{1}{x}\right)^2 = \frac{1}{x^2} \tag{9}$$

Equations (8) and (9) are equations of circles, (8) being a circle centered at $[\rho_r = r/(1 + r), \rho_i = 0]$ with radius $1/(1 + r)$, and (9) being a circle centered at $(\rho_r = 1, \rho_i = 1/x)$ with radius $1/x$. Plots of constant r and constant x circles, with ρ_i as the ordinate and ρ_r as the abscissa, constitute a Smith chart used for impedance calculations. By relabling the circles, they become constant g and constant b circles, and this Smith chart is used for admittance calculations. A typical Smith chart such as obtained at a book store is shown in Fig. 12-19. The real and imaginary values of impedance or admittance are read from the chart just as one would read points on any other coordinate

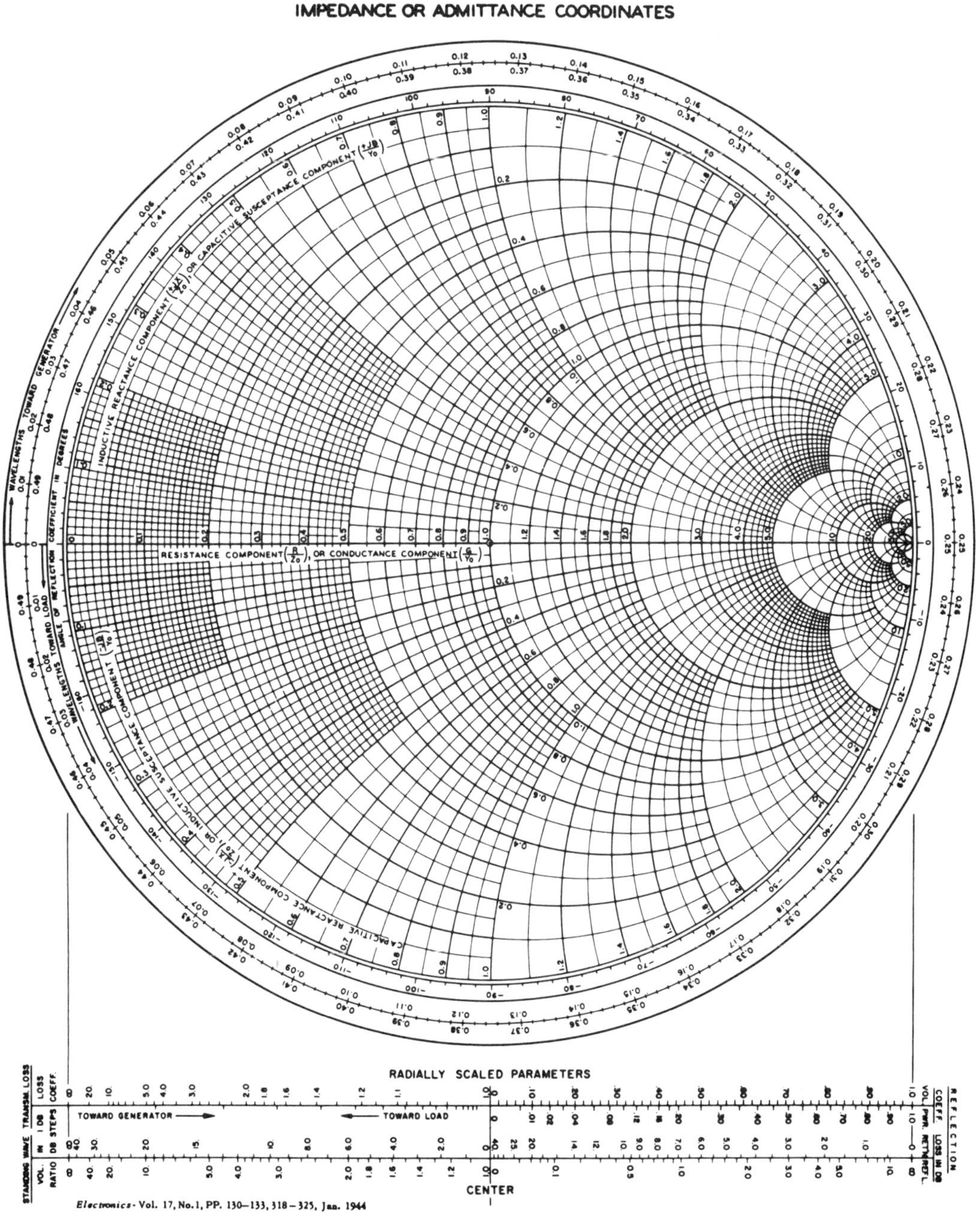

Figure 12-19 Smith chart. (Permission to reproduce Smith charts throughout this book has been granted by Phillip H. Smith, Murray Hill, N.J., under his renewal copyright issued in 1976.)

system. An example will be used to compare the cartesian coordinate system in r and x with that of the Smith chart.

Example 14

A 50-Ω transmission line is connected to a load, $Z_L = (50 + j50)$ (Ω). Using normalized values of impedance, plot the r and x values of input impedance for $\ell = 0, \lambda/16, \lambda/8, \lambda/4, 3\lambda/8$, and $\lambda/2$ from the load on a rectangular impedance chart. Using (8) and (9), construct a Smith chart and plot the same normalized impedance points.

Solution. The normalized load impedance is

$$z_L = \frac{50 + j50}{50} = 1 + j1$$

and

$$z_{in} = \frac{z_L + j \tan \beta\ell}{1 + jz_L \tan \beta\ell}$$

The values for z_{in} in the following table are plotted on the rectangular impedance chart in Fig. 12-20(a).

	a	b	c	d	e	a
ℓ	0	$\lambda/16$	$\lambda/8$	$\lambda/4$	$3\lambda/8$	$\lambda/2$
$\beta\ell$	0	$\pi/8$	$\pi/4$	$\pi/2$	$3\pi/4$	π
$\tan \beta\ell$	0	0.414	1	∞	-1	0
z_{in}	$1 + j1$	$2.275 + j0.805$	$2 - j$	$\frac{1}{2} - j\frac{1}{2}$	$0.4 + j0.2$	$1 + j1$

Now various values of r and x are substituted into (8) and (9) to construct circles that form the Smith chart of Fig. 12-20(b). The scale for $|\rho|$ is found immediately below the chart. For example, for $r = 2$, (8) gives

$$\left(\rho_r - \frac{2}{1 + 2}\right)^2 + \rho_i^2 = \left(\frac{1}{1 + 2}\right)^2$$

which is a circle centered at $\rho_r = 2/3$, $\rho_i = 0$ and with a radius of $1/3$, with the origin being the intersection of the ρ_i and ρ_r axes. Likewise, for $x = 0.5$, (9) gives

$$(\rho_r - 1)^2 + \left(\rho_i - \frac{1}{0.5}\right)^2 = \left(\frac{1}{0.5}\right)^2$$

which is a circle centered at $\rho_r = 1$, $\rho_i = 2$ with a radius of 2. A similar circle is centered at $\rho_r = 1$, $\rho_i = -2$, for $x = -0.5$. Having drawn in circles for $r = 0$, 0.4, 1.0, and 2.0, and for $x = 0.5$, 1.0, 2.0, and 5.0, the points a, b, c, d, e, from the table are plotted on the chart.

One interesting fact is immediately evident: on both the rectangular chart and the Smith chart a plot of normalized input impedance forms a circle. This is called a *standing wave circle*. On a Smith chart, all such standing wave circles, a different circle for each Z_L, are centered at the center of the chart, i.e., at the intersection of the $r = 1.0$ circle and the real axis. Also note on the Smith chart that the points a, c, d, and e are 90° apart, corresponding to $\lambda/8$ spacing. A half-wavelength spacing is 360° around the chart.

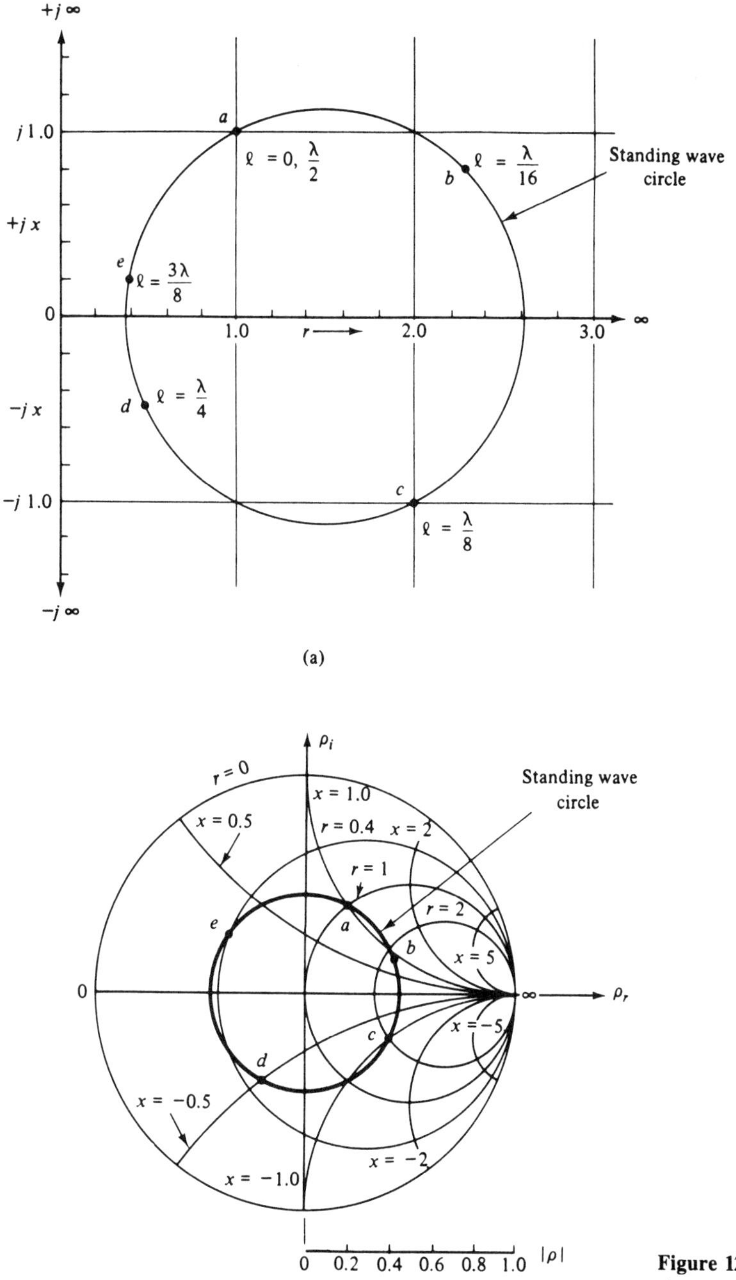

(b)

Figure 12-20 (a) Rectangular impedance chart, and (b) Smith chart, with results plotted from Example 14.

The Smith chart is a coordinate system designed to facilitate graphical calculation of transmission line problems. All values on the chart are normalized impedance or admittance. Some basic features of the chart are:

1. A circle centered at the center of the chart is a standing wave circle (constant S). The center of the chart is a circle of zero radius for a standing wave ratio of 1.0. The outer edge of the chart is for $S = \infty$. The standing wave ratio is read directly from the chart, where the S circle crosses the real axis for real values greater than 1.0.
2. Constant r or constant g circles are complete circles having their centers on the real axis and all are tangent at ∞.
3. The constant x or constant b portions of circles have their centers on a vertical axis through the ∞ point on the right-hand edge of the chart.
4. A complete trip around the chart is $\lambda/2$. Clockwise is toward generator; counterclockwise is toward load.
5. On an impedance chart, V_{min} is with z_{min}, which is on the real axis for $r < 1$. V_{max} and z_{max} are on the real axis, where $r > 1$. V_{max} and V_{min} are $\lambda/4$ (180° on chart) apart. On an admittance chart, V_{max} and y_{min} are on the real axis, where $g < 1$, and V_{min} and y_{max} are on the real axis, where $g > 1$.

If the load is a short, an open, or a pure reactance of any value, the S circle is the outer edge of the chart; i.e., $S = \infty$. Thus, if $S = \infty$, it is not possible with that information alone to determine whether the load is a short, an open, or a pure reactance. However, if another piece of information is known, such as the location of a voltage minimum with respect to the load, then one can use the Smith chart to calculate the load impedance. An example will illustrate this later in the use of the slotted line.

Problem 12.12-1 With a compass centered at the center of a Smith chart used as an impedance chart, draw a circle through the point ($r = 0.8, x = 0.9$), or $z = (0.8 + j0.9)$. What is the standing wave ratio S?

Problem 12.12-2 Let the point $0.8 + j0.9$ on the Smith chart from Prob. 12.12-1 be the normalized load impedance for a load connected to a 50-Ω line. What is the input impedance at 0.11λ from the load? Use the Smith chart.

Problem 12.12-3 Using the Smith chart as an admittance chart: (a) Find and plot the normalized load admittance for a $(65 + j35)$ (Ω) load connected to a 50-Ω line. (b) Draw an S circle through this point and find S. (c) What is the normalized input admittance at 0.122λ from the load?

12.13 MATCHING WITH SHUNT STUBS

When a transmission line is connected to a load that is not equal to its characteristic impedance, we have seen that the input impedance is a function of the distance from the load and is not equal to Z_0 anywhere. We have also seen that a quarter-wave transformer can be used to match any real impedance to Z_0 at a single frequency. Another procedure for matching the load to the line, i.e., a procedure for making the standing wave ratio unity between the source and the matching device, is to place a shorted section of line in parallel with the load at an appropriate distance from the load. This device is referred to as tuning stub, and since it is to be placed in shunt with the existing line, it will be more convenient to work with admittance instead of impedance.

The procedure for using a shunt stub is outlined in Fig. 12-21. The normalized load admittance y_L is plotted on the Smith chart, and an S circle drawn through y_L inter-

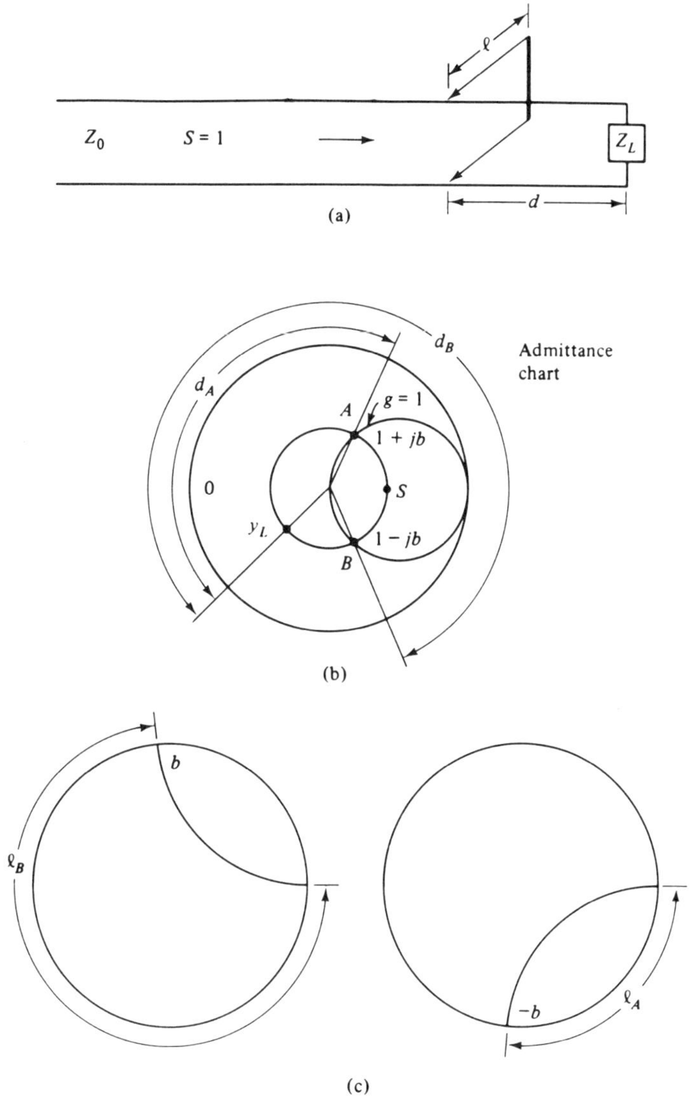

Figure 12-21 Graphical illustration of the use of a shunt shorted stub in matching a transmission line to a load: (a) the line, load, and available stub; (b) calculation of two possible locations for the stub; (c) determination of stub length required for each of the locations calculated in part (b).

sects the $g = 1$ circle at two points [Fig. 12-21(b)]. This means that at distance d_A or d_B from the load, the real part of the normalized input admittance is 1.0, and, since the shunt stub can add susceptance only, adding the correct length of the stub at either A or B will achieve a match to the line. For example, at A,

$$(y_{\text{in}})_A = 1 + jb \tag{1}$$

and, after adding the correct stub length in shunt at this point,

$$(y_{\text{in}})'_A = 1 + jb + y_{\text{stub}} = 1 + j0 \tag{2}$$

or

$$(y_{stub})_A = -jb \tag{3}$$

Likewise at B,

$$(y_{in})_B = 1 - jb \tag{4}$$

and, after adding the correct stub length,

$$(y_{in})'_B = 1 - jb + y_{stub} = 1 + j0 \tag{5}$$

or

$$(y_{stub})_B = jb \tag{6}$$

After finding the normalized admittance of the stub for either A or B from the Smith chart, the length of either stub is also found by using the Smith chart as illustrated in Fig. 12-21(c) and (d). A numerical example follows.

Example 15

A 50-Ω line is connected to a load of (30 + j60) (Ω). Find two possible locations within the first $\lambda/2$ from the load where a single shunt stub could be used for matching, and find the length of stub required in each case. Assume that Z_0 of the stub is 50 Ω.

Solution. The normalized load admittance is

$$y_L = \frac{1}{z_L} = \frac{50}{30 + j60} = 0.333 - j0.667$$

Incidentally, z_L and y_L are 180° apart on the Smith chart and are, of course, on the same S circle. When the S circle is drawn through y_L, it crosses the $g = 1$ circle at $(1 + j1.62)$ and at $(1 - j1.62)$, or points A and B in Fig. 12-22. By measuring along the edge of the chart, we see that

$$d_A = 0.10\lambda + 0.18\lambda = 0.28\lambda$$

and that

$$d_B = 0.10\lambda + 0.32\lambda = 0.42\lambda$$

The required susceptance for a match at A is

$$(y_{stub})_A = -j1.62$$

and at B,

$$(y_{stub})_B = j1.62$$

The stub lengths are found using the Smith chart as outlined in Fig. 12-21(c) and (d). We find that

$$\ell_A = 0.25\lambda - 0.162\lambda = 0.088\lambda$$

and

$$\ell_B = 0.25\lambda + 0.162\lambda = 0.412\lambda$$

Caution must be exercised when measuring distance in wavelengths around the chart because of the possibility of reading the wrong scale. One should recheck each such determination.

While it has been demonstrated by example how a single shunt stub may be used to match a load to a transmission line characteristic impedance, to do so requires that one know exactly where to place the stub. The location must correspond to where the S circle crosses the $g = 1$ circle on the Smith chart.

One does not have to be as precise in the location of a double-stub tuner. The distance from the load is more or less arbitrary (although it is possible to locate the tuner

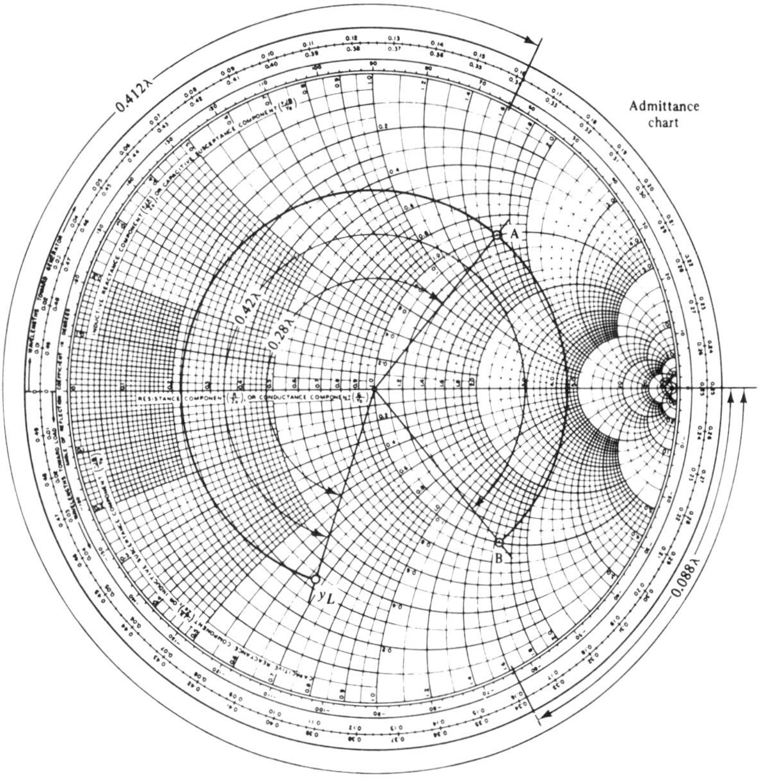

Figure 12-22 Illustration for use with Example 15.

such that a match cannot be achieved) and matching involves adjustment of ℓ_1, then ℓ_2, until a matched condition is obtained. The two stubs are a fixed distance apart, and as illustrated in Fig. 12-23, we will assume that this separation is $\lambda/8$ for this discussion.

Referring to Fig. 12-23(b) and starting at the load, moving a distance d_1 toward the generator changes the input admittance from y_L to y_1. In Fig. 12-23(c), the $g = 1$ circle is now rotated 90° (corresponding to $\lambda/8$) counterclockwise in order that all points on the $g = 1$ circle at the stub 2 location will be on the rotated circle at the stub 1 location. Then the input admittance of d_1 from the load is changed from y_1 to y_1', a point on the rotated circle, by the addition of stub 1. Next, the admittance is changed from y_1' to y_2 in going from the stub 1 location to the stub 2 location, a distance of $\lambda/8$ in this case Fig. 12-23(d). Finally, stub 2 adds the necessary susceptance to change the admittance to y_2', where $y_2' = 1 + j0$, and the match to the line has been achieved. Between the source and stub 2, the standing wave ratio is unity.

An alternative was available during this procedure. The stub 1 length could have been adjusted to give point A instead of y_1' on Fig. 12-23(c), but A is farther from the center of the chart, hence the standing wave ratio between stubs would be higher. A numerical example will now be used to analyze the double stub tuner.

Example 16

Referring to Fig. 12-23(a), a load of $(30 - j40)$ (Ω) is connected to a 50-Ω line. The stubs are $\lambda/8$ apart and stub 1 is 0.3λ from the load. Find the lengths of the stubs required for the

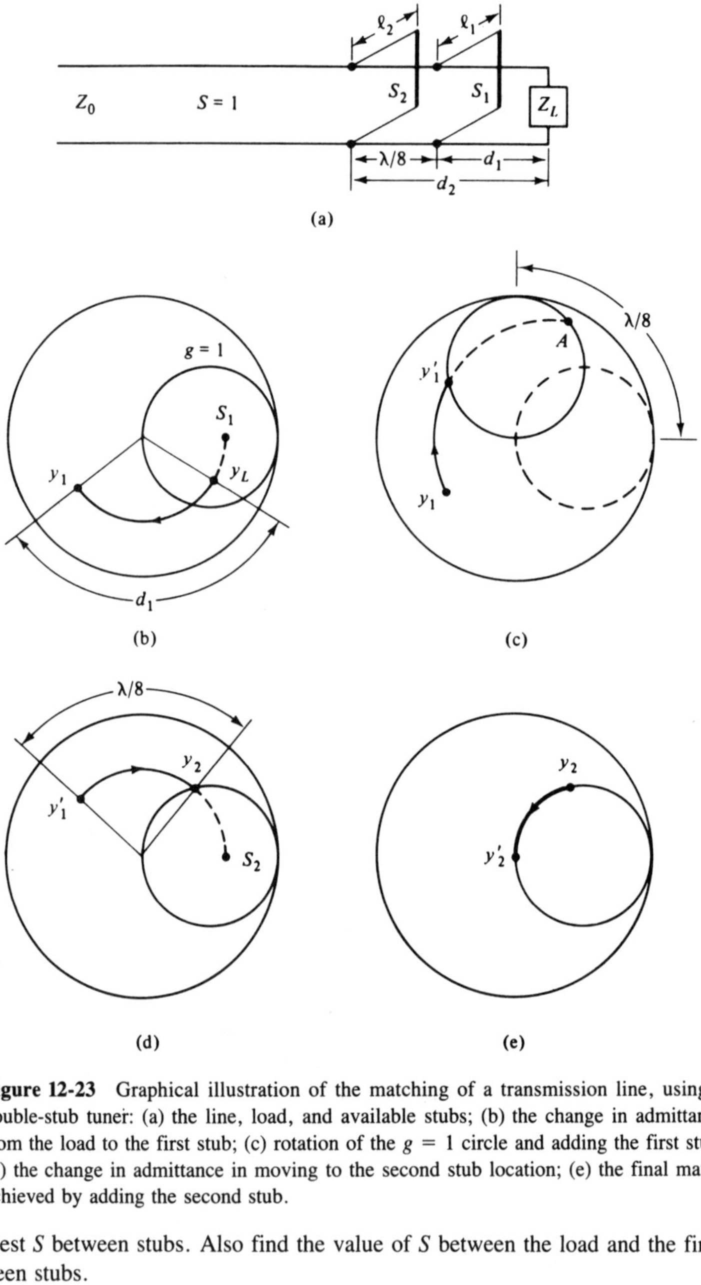

Figure 12-23 Graphical illustration of the matching of a transmission line, using a double-stub tuner: (a) the line, load, and available stubs; (b) the change in admittance from the load to the first stub; (c) rotation of the $g = 1$ circle and adding the first stub; (d) the change in admittance in moving to the second stub location; (e) the final match achieved by adding the second stub.

smallest S between stubs. Also find the value of S between the load and the first stub, and between stubs.

Solution. The normalized load admittance is

$$y_L = \frac{1}{z_L} = \frac{50}{30 - j40} = 0.6 + j0.8$$

y_L is plotted on the Smith chart in Fig. 12-24. Moving 0.3λ toward generator from y_L,

$$y_{\text{in}} = y_1 = 0.41 - j0.44$$

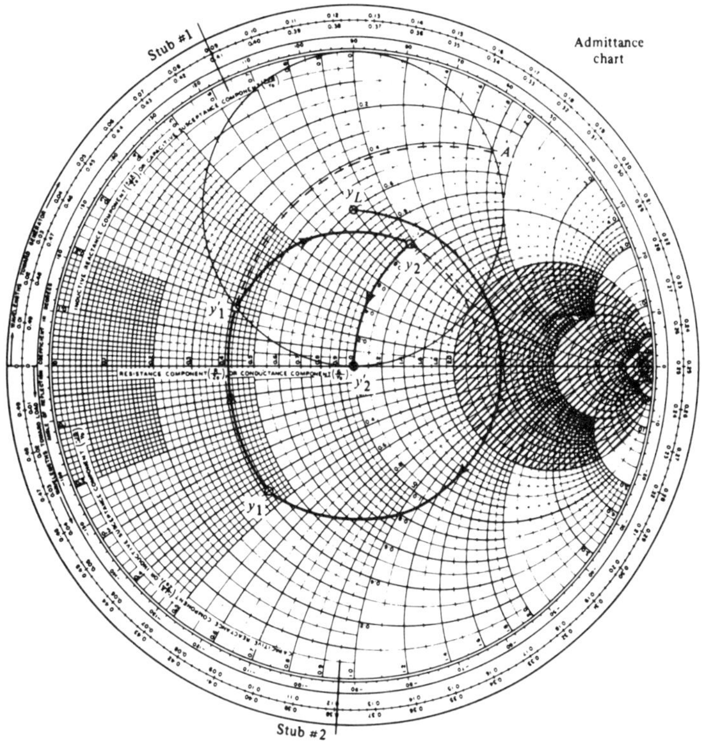

Figure 12-24 Illustration for use with Example 16.

Adding the susceptance of stub 1 to get on the rotated $g = 1$ circle by following a constant-conductance curve,

$$y_1' = 0.41 + j0.19$$

Then

$$y_{\text{stub 1}} = y_1' - y_1 = j0.63$$

Then, moving $\lambda/8$ along a constant S circle from y_1' to y_2, we find that

$$y_2 = 1 + j0.95$$

which means that $y_{\text{stub 2}} = -j0.95$. Finding stub lengths in the same manner as in Example 15, assuming that Z_0 of the stubs is 50 Ω,

$$\ell_1 = 0.25\lambda + 0.09\lambda = 0.34\lambda$$

and

$$\ell_2 = 0.25\lambda - 0.121\lambda = 0.129\lambda$$

Finally, by extending all the S circles, we find that

S between the load and stub 1 $= S_1 = 3.0$

S between stubs $= S_2 = 2.5$

S between the source and stub 2 $= 1.0$

It is possible, however, that a match cannot be achieved for a specific location of the double stub tuner. For example, if in Fig. 12-24, y_1 were inside the shaded area of the Smith chart, i.e., inside the $g = 2$ circle, there would be no value of susceptance that could be added that would put y_2' on the rotated $g = 1$ circle, and no exact match could be achieved. For a spacing between stubs other than $\lambda/8$, there would be a different unmatchable region. When a match cannot be achieved on the first trial location of the tuner, one should move the double-stub unit $\lambda/4$ farther from or closer to the load to ensure that a match is then possible.

Problem 12.13-1 Referring to Fig. 12-21(a): (a) Find two values of d, within $\lambda/2$ from the load, where a stub could be placed for matching to the line, and the stub length ℓ required in each case, where $Z_0 = 50\ \Omega$ and $Z_L = 30 + j40\ (\Omega)$. (b) Also find the standing wave ratio between the stub and load.

Problem 12.13-2 Referring to Fig. 12-23(a), Z_0 is 50 Ω, Z_L is $40 + j50\ (\Omega)$, and $d_1 = 0.05\lambda$. Find: (a) the lengths of stubs for matching to the line for the smallest S between stubs; (b) the S between stub 1 and the load; (c) the S between stubs.

12.14 SLOTTED LINE TECHNIQUE

Referring to eq. (12.9-1), the normalized input impedance is

$$z_{\text{in}} = \frac{z_L + j\tan\beta\ell}{1 + jz_L\tan\beta\ell} \tag{1}$$

and at a voltage minimum, where z_{in} is also a minimum, using (12.9-8), we have

$$(z_{\text{in}})_{\text{min}} = \frac{1}{S} = \frac{z_L + j\tan\beta\ell}{1 + jz_L\tan\beta\ell} \tag{2}$$

where ℓ is now the distance from the load to the nearest V_{min}. Solving for z_L, we have

$$1 + jz_L\tan\beta\ell = Sz_L + jS\tan\beta\ell$$

or

$$z_L = \frac{1 - jS\tan\beta\ell}{S - j\tan\beta\ell} \tag{3}$$

If $\ell < \lambda/4$, $0 < \beta\ell < \pi/2$, and $\tan\beta\ell$ is positive. If $\ell > \lambda/4$, i.e., $\lambda/4 < \ell < \lambda/2$, $\beta\ell$ is in the second quadrant and $\tan\beta\ell$ is negative. Thus, if S is known and ℓ is known, where ℓ is the location of the first V_{min}, it is possible to determine z_L. This can be accomplished conveniently using the Smith chart.

A slotted line is a section of transmission line of a shielded construction similar to a coaxial line with a longitudinal slot in the outer conductor through which a probe on a sliding carriage is able to sample the $\overline{E}$ field between the conductors. The slotted line is inserted between the source and the load, and the standing wave pattern can then be explored by moving the carriage along the line and reading the detected output (see Fig. 12-25).

The VSWR (voltage standing wave ratio) meter is designed to read directly the standing wave ratio from the detected probe output by measuring at a voltage (or $\overline{E}$ field) maximum and then at a voltage minimum. A scale is marked along the longitudinal slot to provide for measurement of the distance moved by the carriage. The primary use of the slotted line is in the measurement of S and of impedance through use of the Smith chart.

Example 17

Referring to Fig. 12-26(a), the standing wave ratio with the load connected to the slotted line is 2.3, and the distance between minima is 12 cm. With reference to the longitudinal

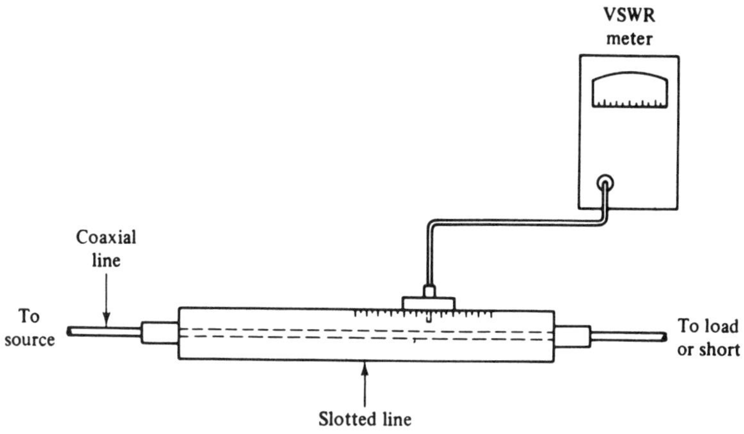

Figure 12-25 Slotted line for use with coaxial transmission lines.

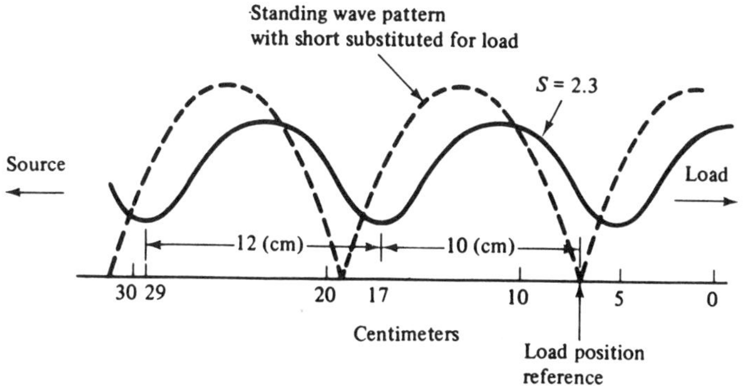

(a)

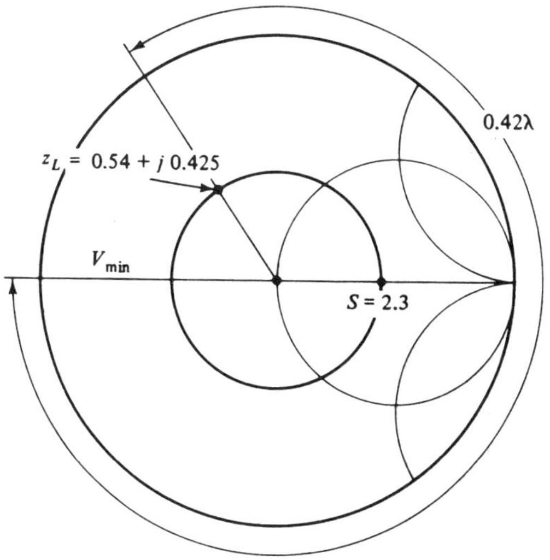

(b)

Figure 12-26 Illustrations for use with Example 17: (a) standing wave; (b) Smith chart.

scale on the line, minima are found at 5 cm, 17 cm, and 29 cm. A short circuit is now substituted for the load, and minima are found at 7 cm, 19 cm, and 31 cm. Find the frequency being transmitted and the load impedance, using the Smith chart. Assume that $Z_0 = 50\ \Omega$.

Solution. The frequency is calculated from the distance between minima. Knowing it is an air-dielectric line, 12 (cm) is $\lambda/2$ in air, and

$$f = \frac{3 \times 10^8 \text{ m s}^{-1}}{2(0.12) \text{ (m)}} = \frac{U}{\lambda}$$
$$= 1.25 \times 10^9 = 1.25 \text{ GHz}$$

A location of a minima with the short substituted for a load can be used as a load position reference. Any one of these minima may be used for such a reference, since $Z_L = 0$ is repeated every $\lambda/2$ from the short circuit, and the position of the short circuit is the position of the load when it is connected to the line. Then, referring to Fig. 12-26(a), with the load connected a minimum is found 10 cm toward the generator from the load on the $S = 2.3$ circle. Since the wavelength is 24 cm, 10 cm is 0.42λ. Then, measuring 0.42λ from V_{min}, on the impedance Smith chart we read that

$$z_L = 0.54 + j0.425$$

or

$$Z_L = Z_0 z_L = 50 z_L = 27 + j21.25 \qquad (\Omega)$$

as illustrated in Fig. 12-26(b).

Problem 12-14-1 With a load connected directly to the end of a 50-Ω slotted line, the measured S is 2.0. When a short is substituted for the load, the position of the minima is shifted 1 (cm) toward the load, and the distance between minima is 8 (cm). Find the frequency and the load impedance.

Problem 12.14-2 The same data is obtained as in Prob. 12.14-1, except that the position of the minima is shifted 1 (cm) toward the generator when the short is substituted for the load. Find the frequency and the load impedance.

12.15 LOSSY LINE ANALYSIS USING THE SMITH CHART

The load reflection coefficient was given by (12.4-13) as

$$\rho = \frac{Z_L - Z_0}{Z_L + Z_0} \tag{12.4-13}$$

or

$$\rho = \frac{(Z_L/Z_0) - 1}{(Z_L/Z_0) + 1} \tag{1}$$

in normalized form. A generalized reflection coefficient may be defined at the input to the transmission line as

$$\rho_{in} = \frac{Z_{in} - Z_0}{Z_{in} + Z_0} \tag{2}$$

where Z_{in} is defined by (12.4-18). However, it is more convenient for the present purpose to use Z_{in} in the form

$$Z_{in} = \frac{Z_0(1 + \rho e^{-2\gamma\ell})}{1 - \rho e^{-2\gamma\ell}} \tag{3}$$

which when substituted into (2), and with some algebraic simplification, reduces (2) to

$$\rho_{in} = \rho e^{-2\gamma\ell} \tag{4}$$

or

$$\rho_{in} = (|\rho|\angle\theta_{\rho L})(e^{-2\alpha\ell}\angle -2\beta\ell) \tag{5}$$

for the reflection coefficient at a distance ℓ from the load. The input impedance at the given distance from the load is treated as a load impedance at that location. Keep in mind that on the Smith chart, the radius from the center of the chart to the coordinates of the normalized impedance under consideration is $|\rho_{in}|$ for that impedance, where the measure of $|\rho_{in}|$ is scaled from the voltage reflection coefficient scale at the bottom of the Smith chart. The angle of the complex ρ_{in} is read directly from the angular position of that radius. One can easily construct a loci of z_{in} for various ℓ using (5), provided that ρ, α, and β are known. The lossy transmission line of Example 5, with a very mismatched load of $Z_L = 50 + j50\ \Omega$, will be used to illustrate.

Example 18

For a frequency of 1000 Hz as in Example 5, the numerical values for Z_0, α, and β were found to be

$$Z_0 = 612\angle -5.35^\circ\ \Omega = 609.85 - j57.09\ \Omega$$
$$\alpha = 0.00345\ \text{Np/mile} = 0.02998\ \text{dB mile}^{-1}$$
$$\beta = 0.03504\ \text{rad/mile} = 2.008^\circ\ \text{mile}^{-1}$$

We want to calculate and plot on a Smith chart values for ρ_{in} for $0 < \ell < 200$ miles.

Solution. First, the load reflection coefficient is found to be

$$\rho = \frac{50 + j50 - 609.85 + j57.09}{50 + j50 + 609.85 - j57.09}$$
$$= 0.864\angle 169.787^\circ$$

At 5 miles,

$$\rho_{in} = (0.864\angle 169.787^\circ)[e^{-(2)(5)(0.00345)}\angle -2(5)(2.008)]$$
$$= 0.835\angle 149.707^\circ$$

At 30 miles,

$$\rho_{in} = (0.864\angle 169.787^\circ)[e^{-(2)(30)(0.00345)}\angle -2(30)(2.0)]$$
$$= 0.702\angle 49.307^\circ$$

In a similar manner, ρ_{in} is calculated and plotted for a number of distances ℓ in Fig. 12-27, creating a spiral that approaches the center of the chart as ℓ approaches ∞.

To find Z_{in}, read z_{in} directly from the chart and multiply by Z_0. For example, at 20 miles, ρ_{in} was found to be $0.753\angle 89.467^\circ$. When plotted on the Smith chart of Fig. 12-27, we read

$$z_{in} = 0.28 + j0.97 = 1.01\angle 73.89^\circ$$
$$Z_{in} = (612\angle -5.35^\circ)1.01\angle 73.89^\circ$$
$$= 617.88\angle 68.55^\circ \quad (\Omega)$$

compared to $616\angle 68.1^\circ\ \Omega$ calculated in Example 5. Of course, some accuracy must be sacrificed when using the Smith chart for such calculations.

The transmission loss (1-dB steps) scale may be used to measure the transmission loss between various points on the spiral. For example, find the loss between 90 and

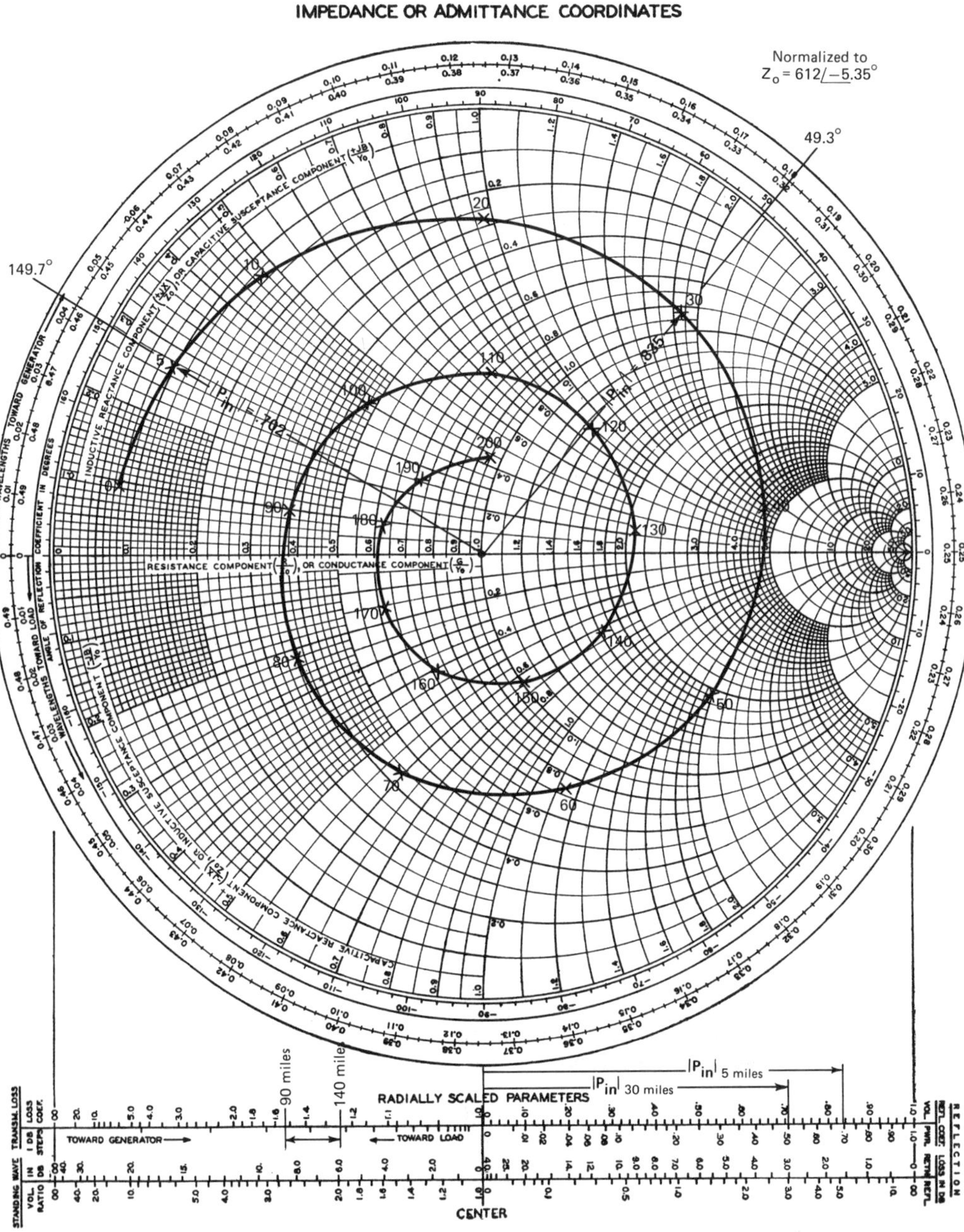

Figure 12-27 Illustration for Example 18.

140 miles. Using the transmission loss scale to measure the radius of the spiral, we read approximately 1.5 dB for the difference in radii. To check this value, we calculate

$$10 \log_{10} e^{-2\alpha\ell} = 10 \log_{10} e^{-2(0.00345)(50)} = -1.5 \text{ dB}$$

which is independent of which 50-mile section is considered.

Each time the spiral of Fig. 12-27 completes 360°, the attenuation corresponds to that for one-half wavelength of transmission line. For this particular line of Example 18, $\lambda/2 = 180°/(2.008° \text{ mile}^{-1}) = 90$ miles. Thus, for every 90 miles of travel, or for each revolution of the plot, the radius of the plot, i.e., the magnitude of the reflection coefficient, will be reduced by a factor $e^{-90(0.00345)} = 0.73$ and the attenuation in dB is $-10 \log_{10}(0.73)^2 = 2.7$ dB.

The standing wave ratio may be determined for any point on the spiral by drawing a standing wave circle through that point as was done for a lossless line. It is obvious that the reflection coefficient and standing wave ratio decrease with increasing distance from the load.

Problem 12.15-1 Using the spiral plot of Fig. 12-27 and Example 18, estimate the shortest distance from the load to where Z_{in} is purely real, and determine that value of Z_{in}. Check your estimate by calculating Z_{in} using (12.4-18).

Problem 12.15-2 Determine the standing wave ratio for Example 18 at a distance of 180 miles from the load.

Problem 12.15-3 Find Z_{in} for Example 18 as $\ell \to \infty$.

REVIEW QUESTIONS

1. What is the TEM mode? *Sec. 12.1*
2. What is meant by distributed parameters as contrasted to lumped parameters? *Secs. 12.1, 12.2*
3. What is characteristic impedance? Propagation constant? *Secs. 12.2, 12.4*
4. How does one calculate hyperbolic functions of complex numbers? *Example 5*
5. Qualitatively, what distinguishes a power transmission line from a communications-type line? *Sec. 12.5*
6. What is a distortionless line? If R and G are both zero, is the line distortionless? *Sec. 12.6*
7. How does the velocity of propagation along a lossless line compare to that of a plane wave? *Sec. 12.8*
8. How does the characteristic impedance of a lossless line relate to the intrinsic impedance of a medium? *Sec. 12.8*
9. How are S, V_{max}, V_{min}, $(Z_{in})_{max}$, $(Z_{in})_{min}$, and $|\rho|$ related for a lossless line? *Sec. 12.9*
10. For an adjustable-length lossless line with a load Z_L, how many distinct values of Z_{in} are there, in general, where Z_{in} is entirely real? *Sec. 12.9*
11. What is the range of possible input impedance values for a shorted line? *Sec. 12.9*
12. What is a quarter-wave transformer? Is it broad band? *Sec. 12.10*
13. How does the maximum power handling capability vary with the standing wave ratio? *Sec. 12.11*
14. Why are the coordinates on the Smith chart normalized?
15. Name some advantages, if any, of the Smith chart over a rectangular chart. *Sec. 12.12*
16. For a transmission line connected to a mismatched load, how many locations, within a $\lambda/2$ interval, could a single shunt stub be used to achieve a match between the source and the stub? *Sec. 12.13*
17. What advantages does the double-stub tuner have over the single-stub tuner? *Sec. 12.13*
18. Can one always achieve a match with an arbitrary location of the double stub tuner? *Sec. 12.13*
19. Describe the basic procedure for using a slotted line to measure load impedance. *Sec. 12.14*
20. How does skin effect affect the loss of a transmission line? *Sec. 12.7*
21. When a step change of voltage is applied to a transmission line, what determines the instantaneous change in input current? *Sec. 12.3*

PROBLEMS

12-1. An artificial lossless transmission line is constructed of a large number of alternate series L and shunt C sections as shown in Fig. 12-28, where $L = 1\ \mu\text{H}$ per section and $C = 400$ pF per section. Find: (a) the R_0 of the actual transmission line represented by this artificial line; (b) the velocity of transmission on the actual line being simulated if each section represents 1 m of length of the transmission line; (c) the time delay per section.

12-2. Starting with eqs. (12.2-4) and (12.2-5), derive the solutions for $V(z)$ and $I(z)$ for a dc transmission line.

12-3. (a) Find the characteristic resistance R_0 of a lossless coaxial transmission line where the ratio $r_b/r_a = 2$, and the dielectric is Teflon. (b) Find the velocity of propagation for this cable.

12-4. A 5-V dc source is connected to a section of 50-Ω transmission line ℓ meters in length. Initially, the load end of the line is open, but at $t = 0$, a 75-Ω load is connected to the line. If $Z_G = 100\ \Omega$, plot V_L, I_L versus time out to $4\tau+$ seconds.

12-5. For the circuit of Prob. 12-4, assume that the 75-Ω load has been connected long enough that steady state has been reached. The 75-Ω load is then disconnected from the line. Plot: (a) V_L, I_L versus time out to $4\tau+$; (b) the distribution of voltage on the line at $t = 3\tau/2$.

12-6. For the circuit of Fig. 12-8, let the 8-Ω resistor be R_2, the 32-Ω resistor be R_1, and the 10-Ω resistor be R_G. If V_L, I_L are the initial steady-state load voltage and current, show that the change in load voltage (ΔV_L) and the change in load current (ΔI_L) caused by shorting out R_1 is given by

$$\frac{\Delta V_L}{V_L} = \frac{-R_1}{(R_1 + R_2)(1 + R_2/R_0)} \qquad \text{where } \Delta V_L = -R_0 \Delta I_L$$

12-7. (a) Plot the load voltage and load current waveforms for the circuit of Fig. 12-29, out to $11\tau+$. (b) Repeat for the diode removed, leaving an open load. What function is served by the diode? [*Note:* This problem is lengthy and more difficult than most.]

12-8. For the ladder network in Fig. 12-30, using transmission line concepts, estimate R_L such that $Z_{in} = R_L$. Check your result by calculating the actual input impedance using your estimated value of R_L.

12-9. Find the characteristic impedance, propagation constant, and velocity of propagation for a transmission line having the following parameters:

$$R = 0.042\ \Omega\ \text{m}^{-1}$$
$$G = 5 \times 10^{-10}\ \text{S m}^{-1}$$
$$L = 5 \times 10^{-7}\ \text{H m}^{-1}$$
$$C = 30.5\ \text{pF m}^{-1}$$

Use a frequency of 1000 Hz.

12-10. Find the characteristic impedance, propagation constant, and velocity of propagation for an arbitrary frequency for a lossless transmission line having $L = 5 \times 10^{-7}\ \text{H m}^{-1}$ and $C = 150\text{pF m}^{-1}$.

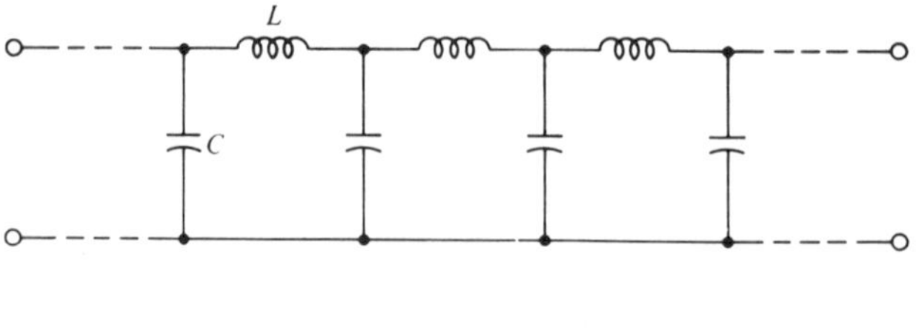

Figure 12-28 Artificial transmission line or delay line.

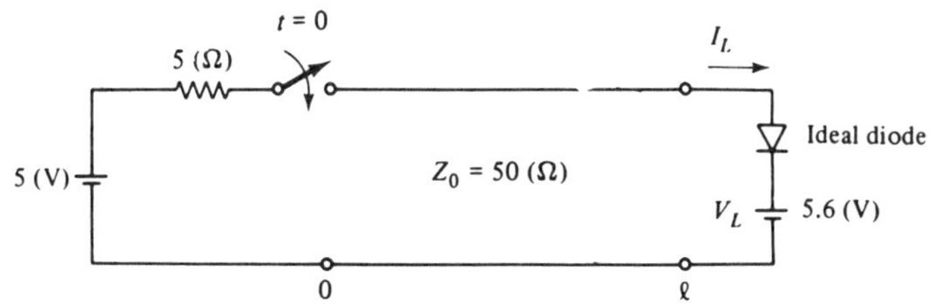

Figure 12-29 Illustration for use with Prob. 12-7.

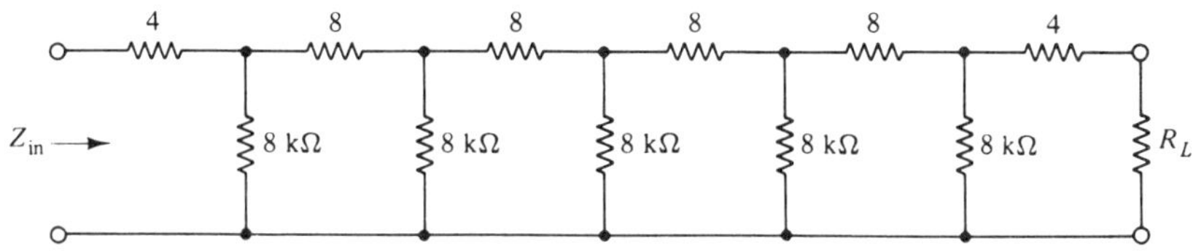

Figure 12-30 Illustration for use with Prob. 12-8.

12-11. A lossless transmission line has the specifications $L = 1.57 \times 10^{-7}$ H m^{-1} and $C =$ 15 pF m^{-1}. What is wrong with these specifications?

12-12. A coaxial line uses a lossless dielectric material having a dielectric constant of 2.3. The radii of the two conductors are 1 mm and 3.6 mm. Find Z_0 and U for this line. Find β for a frequency of 500 MHz.

12-13. A lossless line has a Z_0 of 50 Ω and a propagation velocity of 2.5×10^8 m s^{-1}. Find the series L and shunt C for the line.

12-14. A coaxial line is to be designed for a voltage rating of 1000 V and for a Z_0 of 50 Ω. If air dielectric is used and if E_{max} is 2.5×10^6 V m^{-1} to avoid breakdown, determine the minimum radii of the two conductors.

12-15. A planar transmission line has a characteristic impedance of 57 Ω. This line consists of two strips of conducting material on either side of a circuit board. By what percentage should the widths of the strips be increased or decreased to get a Z_0 of exactly 50 Ω?

12-16. The voltage on a distortionless transmission line is given by $V(\ell) = 100e^{0.0003\ell} \cos(10^5 t + 10^{-3}\ell) + 60e^{-0.0003\ell} \cos(10^5 t - 10^{-3}\ell)$ (V), where the load impedance is $Z_L = 300$ Ω, and ℓ is the distance from the load. Find: (a) α; (b) β; (c) U; (d) Z_0; (e) $|I_L|$.

12-17. A lossless transmission line is several integral wavelengths long. If Z_G is 50 Ω, Z_0 is 100 Ω, and Z_L is 50 Ω, find, for a V_L of 5 V_{rms}: (a) $V_{max_{rms}}$; (b) V^+_{rms}; (c) V^-_{rms}; (d) $V_{0_{rms}}$.

12-18. A distortionless, air-dielectric line has a capacitance of 35 pF m^{-1}. Find; (a) L (H m^{-1}); (b) Z_0 (Ω).

12-19. For the circuit of Fig. 12-10, where Z_0 is 75 Ω, and Z_L is $50 + j0$ (Ω), find: (a) ρ_L; (b) S; (c) $Z_{in_{max}}$; (d) $Z_{in_{min}}$; (e) Z_{in} at 0.75λ from the load. Assume a lossless line.

12-20. A lossless transmission line having a Z_0 of 50 Ω is connected to a 20-Ω load. Find: (a) the reflection coefficient; (b) S, the standing wave ratio; (c) Z_{in} at λ/2 from the load; (d) Z_{in} at λ/4 from the load; (e) distance in wavelengths from the load to the nearest maximum input impedance.

12-21. An ideal current source, $\mathcal{I}_G = 1\,\angle 0°$ (A), is connected to a λ/4 open-circuited 50-Ω lossless transmission line section. Find the open-circuit load-end voltage.

12-22. A voltage source having an output impedance of 30 Ω is connected to a λ/4 section of lossless 75-Ω line. What value of load R_L will absorb maximum power from the source?

12-23. A 100-Ω line, $\lambda/2$ in length, is connected to a $\lambda/4$ length of 150-Ω line, which is in turn connected to a $\lambda/4$ section of 300-Ω line terminated by a 75-Ω resistor. What is the input impedance to the first section of 100-Ω line?

12-24. A lossless transmission line is several integral wavelengths long. $Z_G = 50\ \Omega$, $Z_0 = 100\ \Omega$, $Z_L = 50\ \Omega$, and $V_{L_{rms}} = 5$ V. Find: (a) net load power; (b) incident or forward power; (c) reflected power.

12-25. For a lossless transmission line where $\mathcal{V}_s$ is given by

$$\mathcal{V}_s = \mathcal{V}^+(e^{j\beta\ell} + \rho_L e^{-j\beta\ell}) = \mathcal{V}_o - \mathcal{I}_s Z_G$$

and where $\rho_G = (Z_G - Z_0)/(Z_G + Z_0)$, show that

$$\mathcal{V}^+ = \frac{\mathcal{V}_o Z_0 e^{-j\beta\ell}}{Z_0 + Z_G}\left(\frac{1}{1 - \rho_G \rho_L e^{-2j\beta\ell}}\right)$$

[*Note:* From this result, if $Z_G = Z_0$, i.e., the generator is matched to the line, the power delivered to the load is independent of line length.]

12-26. Using the result of Prob. 12-25, find the percentage of the maximum available generator power $[(1/2)|V_o|^2/4R_G]$ that is delivered to the load for lossless line, where $Z_G = 2\ Z_0$ and $Z_L = 1.5Z_0$ if the line length is $3\lambda/8$.

12-27. For the circuit of Fig. 12-1, $Z_G = 30\ \Omega$, $Z_0 = 50\ \Omega$, and $Z_L = 70\ \Omega$. For a V_o of $100 \cos \omega t$ (V), find V_{in} and the average load power for a line length of $\lambda/4$. Assume a lossless line.

12-28. Referring to Fig. 12-31, if Z_{01} and Z_{02} are characteristic impedances for lossless lines and if Z_1 and Z_2 are both real, show that

$$\frac{V_2}{V_1} = \frac{Z_{02}}{Z_{01}}$$

where V_1 and V_2 are the voltages across Z_1 and Z_2, and consequently the power ratio is given by

$$\frac{P_2}{P_1} = \frac{V_2^2/Z_2}{V_1^2/Z_1} = \frac{V_2^2 Z_1}{V_1^2 Z_2} = \frac{Z_{02}^2 Z_1}{Z_{01}^2 Z_2}$$

Note that if $Z_1 = Z_2$, then each load has equal power if $Z_{01} = Z_{02}$.

12-29. For a certain lossless transmission line, $S = 4$, V_{max} occurs at the load, and the power dissipated in the load is 100 W. Find: (a) ρ; (b) incident power.

12-30. Using a Smith chart, find the input impedance at a distance of 0.125λ from a load of 85 Ω, connected to a 50-Ω line.

12-31. Using a Smith chart, find the input impedance of a lossless transmission line, where $L = 1\ \mu\text{H m}^{-1}$, $C = 100\ \text{pF m}^{-1}$, at a distance of 0.09 m from the load impedance of a 100-Ω resistor in series with a 0.1-μH inductor at a frequency of 500 MHz.

12-32. A 30-Ω load is connected to a 50-Ω line. A 0.1λ shunt shorted stub, whose Z_0 is 75 Ω, is connected at a distance of 0.1λ from the load. Find the total input impedance at the stub location, including the stub.

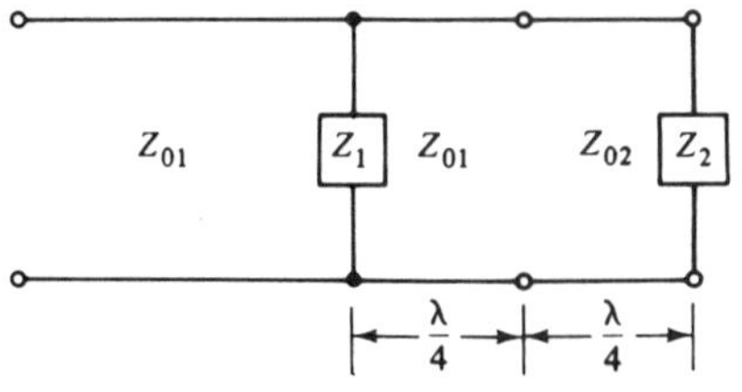

Figure 12-31 Illustration for use with Prob. 12-28.

12-33. A lossless transmission line, $Z_0 = 50\ \Omega$, has a load impedance of $(60 + j30)\ (\Omega)$. (a) Using a Smith chart, find two possible stub locations, within 0.5λ from the load, that will permit a match to be achieved, and determine the length of stub required in each case if Z_0 of the stub is also $50\ \Omega$. (b) Repeat for a stub having a Z_0 of $75\ \Omega$.

12-34. A lossless 50-Ω line is terminated by a 150-Ω load. Find a suitable stub location as close to the load as possible for matching the impedance to the line, and determine the length of a 50-Ω stub required. Is the stub inductive or capacitive?

12-35. A 50-Ω line is terminated by a $(60 + j60)\ (\Omega)$ load. An adjustable shorted shunt stub is placed 0.23λ from the load. What is the smallest standing wave ratio that can be achieved between the generator and the stub, by adjustment of this stub length? What was S before the stub was added?

12-36. A 100-Ω line is terminated by an impedance of $50 - j50(\Omega)$. Find the distance from the load to the shortest 100-Ω stub that will enable a match to the 100-Ω line to be achieved. Also find the stub length.

12-37. A 50-Ω line is terminated by a $20 - j35(\Omega)$ impedance. How far from the load should we place a fixed $\lambda/8$ shorted shunt stub for a minimum S between generator and stub?

12-38. A calculated stub location and inductive stub length have been determined for an assumed 75-Ω antenna. Assuming that a T-connector has been placed at the exact calculated location and the calculated stub length is used, where Z_0 of the line and stub is $50\ \Omega$, find the standing wave ratio between the transmitter and the stub if the actual antenna impedance is $(75 + j75)\ (\Omega)$.

12-39. Referring to Fig. 12-23(a), $Z_L = 65 + j50(\Omega)$, $d_1 = 0.115\lambda$, $d_2 = 0.115\lambda + 0.125\lambda$, and $Z_0 = 50\ \Omega$ for the line and stubs. Find the stub lengths ℓ_1 and ℓ_2 for the smallest S between stubs.

12-40. Repeat Prob. 12-39 for $d_2 = 0.115\lambda + 0.15\lambda$.

12-41. Referring to Fig. 12-23(a), $Z_L = (75 + j30)\ (\Omega)$, $d_1 = 0.375\lambda$, $d_2 = d_1 + \lambda/8$, $Z_0 = 50\ \Omega$ for the line and stubs. Find the stub lengths ℓ_1 and ℓ_2 for the smallest S between stubs.

12-42. Find the input impedance of a shorted five-wavelength-long transmission line whose characteristic impedance is $50\ \Omega$ if the attenuation is 0.2 dB per wavelength. Use the Smith chart. (Hint: Use return loss scale. The return loss is 2 dB)

13

Waveguides

13.1 INTRODUCTION

The transmission lines discussed in Chapter 12 require at least two conductors to support the TEM (transverse electromagnetic) mode, and it is common practice to restrict the term "transmission line" to a line operating in the TEM mode. But just as a coaxial or parallel-wire transmission line operating in the TEM mode may be called a *waveguide,* since the two conductors do guide the transverse electromagnetic wave from the source to the load, a waveguide may also be called a *transmission line*. However, in this text, the term "waveguide" refers to a hollow tube within which it is not possible to support the TEM mode. Two characteristics of the waveguide that distinguish it from the ordinary transmission line are the cut-off frequency at the lower end of the pass band and the fact that group velocity and phase velocity are both different from the unbounded or plane wave velocity.

In Fig. 13-1, a rectangular and circular waveguide are compared to a planar transmission line. A waveguide is basically a hollow tube (or dielectric rod) and, unlike the

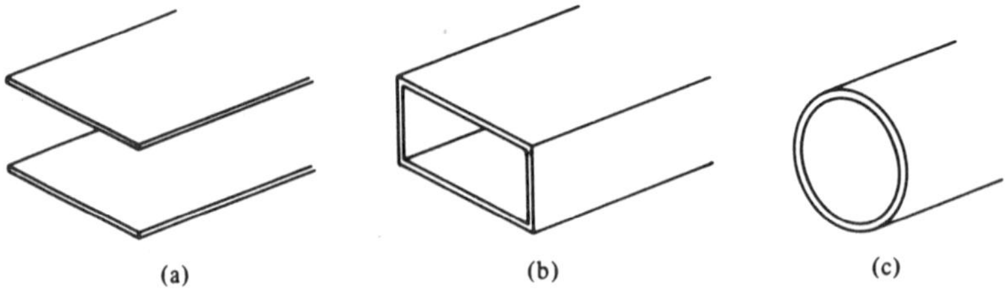

Figure 13-1 (a) Planar transmission line. (b) Rectangular waveguide. (c) Circular waveguide.

transmission line, has a cutoff frequency below which energy is not propagated through the guide. This is in contrast to the transmission line which operates from dc up to a frequency limited by skin effect and dielectric losses. A waveguide may have an arbitrary cross-sectional shape, but generally it is either rectangular or circular. In this chapter, we are going to discuss only rectangular waveguides, but first we shall examine how a waveguide mode may exist along a planar transmission line.

13.2 THE TE WAVEGUIDE MODE IN A PLANAR TRANSMISSION LINE

We have seen from Fig. 11-2 a graphical representation of power flow along a planar transmission line. In that figure, the line is operating in the TEM mode and, although dc transmission is represented, the same transverse $\overline{E}$ and $\overline{H}$ field relationship holds at any frequency. At sufficiently high frequency, another mode of transmission is possible in which the $\overline{E}$ field is transverse but the $\overline{H}$ field is not, or vice versa.

Referring to Fig. 13-2, the boundary condition that the tangential $\overline{E}$ field at a conducting boundary be zero can be met by propagating two plane waves, each with

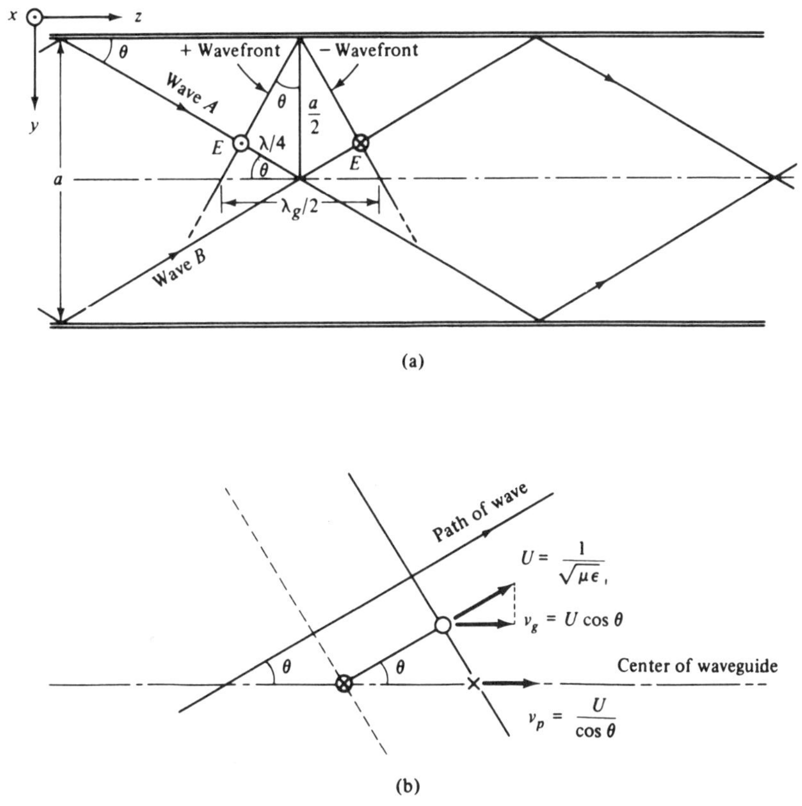

Figure 13-2 (a) Graphical illustration of two plane waves propagating generally in the z direction via a zigzag path between two conducting planes. The $\overline{E}$ field is parallel to the planes. (b) Illustrating the relation between medium velocity, group velocity, and phase velocity.

the $\overline{E}$ field parallel to the boundaries. The positive and negative wavefronts pictured in Fig. 13-2(a) must cancel at the conducting boundary; therefore, the two wavefronts must be $\lambda/2$ apart measured along the path of propagation. Since the $\overline{E}$ field is transverse to the axis of the line, this is called a *TE mode*. For the planar line, the TE as well as the TEM mode is possible as long as $a/2$ is greater than $\lambda/4$. As the wavelength gets longer, the θ approaches 90° until, when $\lambda = 2a$, a cutoff condition is reached. Then the cutoff wavelength λ_c is

$$\lambda_c = 2a \tag{1}$$

The distance between a maximum positive and a maximum negative wavefront or phase front, measured along the guide axis, is the guide half-wavelength.

$$\frac{\lambda_g}{2} = \frac{2\lambda/4}{\cos\theta}$$

or

$$\lambda_g = \frac{\lambda}{\cos\theta} \tag{2}$$

When there is a standing wave pattern due to a load mismatch, then the distance between nulls in the pattern is $\lambda_g/2$ rather than $\lambda/2$. It is the result of the zigzag paths the two waves follow down the guide that the apparent wavelength or guide wavelength is longer than the true wavelength in the medium.

Also as a result of the zigzag paths, we have to distinguish between three different velocities. First, there is the velocity in the medium given by

$$\boxed{U = \frac{1}{\sqrt{\mu\epsilon}} \qquad (\mathrm{m\ s^{-1}})} \tag{3}$$

Next, there is the group velocity v_g, which is a measure of the component of media velocity that is directed along the guide axis. The group velocity is a measure of how fast the energy (or information) is transmitted down the guide and is always slower than the true velocity in the medium. We see from Fig. 13-2(b) that

$$v_g = U\cos\theta \tag{4}$$

Finally, there is the phase velocity v_p, which is a measure of the speed at which an intersection of a phase front with the guide axis moves along the guide. The concept of phase velocity was first introduced in Sec. 11.10, where we were studying oblique reflection of plane waves.

$$v_p = \frac{U}{\cos\theta} = \frac{\omega}{\beta} \qquad (\mathrm{m\ s^{-1}}) \tag{5}$$

These three velocities are illustrated in Fig. 13-2(b). Within a certain time interval, the phase front moves from the dashed line to the solid line. The circle drawn on each phase front represents the movement of a point on the phase front, and the distance between circles, divided by the time interval, is the velocity U. The component of U along the guide axis is v_g. The distance between crosses on the two phase fronts, divided by the time interval, is the phase velocity. Then, from a study of Fig. 13-2(b) and from (4) and (5),

$$U = \sqrt{\frac{U}{\cos\theta} U \cos\theta} = \sqrt{v_p v_g} \qquad (\text{m s}^{-1}) \tag{6}$$

The variable θ can be eliminated from the various equations by noting that

$$\sin\theta = \frac{\lambda/4}{a/2} = \frac{\lambda}{2a} = \frac{\lambda}{\lambda_c} \tag{7}$$

where λ_c is obtained from (1). Then

$$\cos\theta = \sqrt{1 - \sin^2\theta} = \sqrt{1 - \frac{\lambda^2}{\lambda_c^2}} = \sqrt{1 - \frac{f_c^2}{f^2}} \tag{8}$$

where f_c is the cutoff frequency,

$$f_c = \frac{U}{\lambda_c} \tag{9}$$

and substituting (8) into (4) and (5).

$$\boxed{v_g = U\sqrt{1 - \frac{f_c^2}{f^2}} \qquad (\text{m s}^{-1})} \tag{10}$$

$$\boxed{v_p = \frac{U}{\sqrt{1 - \frac{f_c^2}{f^2}}} \qquad (\text{m s}^{-1})} \tag{11}$$

We see from (10) and (11) that as the cutoff frequency is approached from above, $v_g \to 0$ and $v_p \to \infty$. For $f < f_c$, there is no propagation in the TE mode, but propagation is still possible in the TEM mode for the planar line.

If to the parallel planes of Fig. 13-2 are added two additional sides to form a rectangular hollow pipe as in Fig. 13-3, the TE mode described above can still be supported, for the $\overline{E}$ field will be normal to the added sides and thus can meet boundary requirements. The $\overline{E}$ field is in the x direction and reaches a maximum at $y = a/2$ and is zero at $y = 0, a$. The $\overline{H}$ field, on the other hand, has only a y component at $y = a/2$ but develops only a z component at the conducting walls at $y = 0, a$. This can be qualitatively deduced from a study of Fig. 13-2 when waves A and B are treated as two separate plane waves. A more rigorous mathematical derivation of the $\overline{E}$ and $\overline{H}$ fields for the TE modes in a rectangular waveguide follows in the next section.

Problem 13.2-1 A wave is being propagated between two parallel planes as illustrated in Fig. 13-2. If $a = 10$ cm, find the cutoff frequency. Assume air between the plates.

Problem 13.2-2 If $f = 2f_c$, find v_g and v_p for the parallel plane waveguide of Prob. 13.2-1.

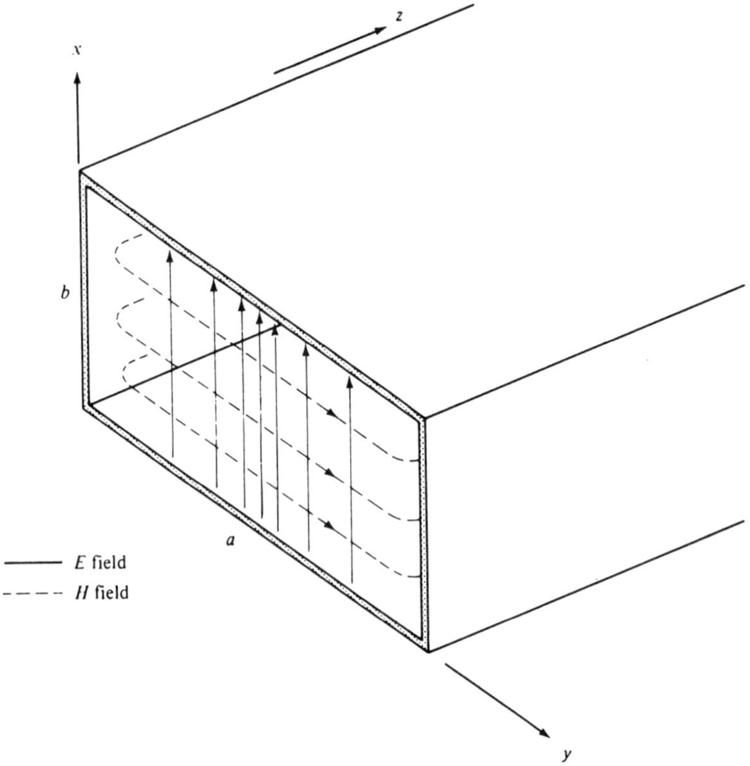

Figure 13-3 Illustration of the transverse field distribution in a waveguide cross section operating in the TE_{10} mode.

13.3 THE TE_{mn} MODES IN RECTANGULAR WAVEGUIDES

Let us assume a rectangular waveguide with the coordinate system, as shown in Fig. 13-3, filled with a lossless medium and having perfectly conducting walls. The derivation of the $\overline{E}$ and $\overline{H}$ fields for the TE_{mn} modes begins with Maxwell's equations in phasor form (for sinusoidal steady state) given by

$$\nabla \times \overline{\mathscr{E}} = -j\omega\mu\overline{\mathscr{H}} \tag{1}$$

$$\nabla \times \overline{\mathscr{H}} = j\omega\epsilon\overline{\mathscr{E}} \tag{2}$$

$$\nabla \cdot \overline{\mathscr{E}} = 0 \tag{3}$$

$$\nabla \cdot \overline{\mathscr{H}} = 0 \tag{4}$$

If we now look at these vector equations one component at a time, we have

$$\frac{\partial \mathscr{E}_z}{\partial y} - \frac{\partial \mathscr{E}_y}{\partial z} = -j\omega\mu\mathscr{H}_x \tag{5}$$

$$\frac{\partial \mathscr{E}_x}{\partial z} - \frac{\partial \mathscr{E}_z}{\partial x} = -j\omega\mu\mathscr{H}_y \tag{6}$$

$$\frac{\partial \mathscr{E}_y}{\partial x} - \frac{\partial \mathscr{E}_x}{\partial y} = -j\omega\mu\mathscr{H}_z \tag{7}$$

$$\frac{\partial \mathcal{H}_z}{\partial y} - \frac{\partial \mathcal{H}_y}{\partial z} = j\omega\epsilon \mathcal{E}_x \tag{8}$$

$$\frac{\partial \mathcal{H}_x}{\partial z} - \frac{\partial \mathcal{H}_z}{\partial x} = j\omega\epsilon \mathcal{E}_y \tag{9}$$

$$\frac{\partial \mathcal{H}_y}{\partial x} - \frac{\partial \mathcal{H}_x}{\partial y} = j\omega\epsilon \mathcal{E}_z \tag{10}$$

$$\frac{\partial \mathcal{E}_x}{\partial x} + \frac{\partial \mathcal{E}_y}{\partial y} + \frac{\partial \mathcal{E}_z}{\partial z} = 0 \tag{11}$$

$$\frac{\partial \mathcal{H}_x}{\partial x} + \frac{\partial \mathcal{H}_y}{\partial y} + \frac{\partial \mathcal{H}_z}{\partial z} = 0 \tag{12}$$

In accordance with our convention used for plane waves and transmission lines, we shall assume propagation in the $\pm z$ direction with a propagation constant γ. Thus,

$$\mathcal{E}_x^+ = \mathcal{E}_{x0}^+ e^{-\gamma z}$$

or

$$\frac{\partial}{\partial z} \rightarrow -\gamma \tag{13}$$

where $\mathcal{E}_{x0}^+$ is the complex amplitude of the forward x component of the $\overline{E}$ field at $z = 0$.

We now direct the derivation specifically toward the TE_{mn} modes by setting $\mathcal{E}_z = 0$; i.e., we allow transverse electric fields only. The procedure will now be to rewrite the curl equations and, with some algebra, obtain expressions for each of the transverse fields in terms of the single variable $\mathcal{H}_z$. The curl equations, with $\mathcal{E}_z = 0$ and applying (13), are

$$\gamma \mathcal{E}_y = -j\omega\mu \mathcal{H}_x \tag{14}$$

$$-\gamma \mathcal{E}_x = -j\omega\mu \mathcal{H}_y \tag{15}$$

$$\frac{\partial \mathcal{E}_y}{\partial x} - \frac{\partial \mathcal{E}_x}{\partial y} = -j\omega\mu \mathcal{H}_z \tag{16}$$

$$\frac{\partial \mathcal{H}_z}{\partial y} + \gamma \mathcal{H}_y = j\omega\epsilon \mathcal{E}_x \tag{17}$$

$$-\gamma \mathcal{H}_x - \frac{\partial \mathcal{H}_z}{\partial x} = j\omega\epsilon \mathcal{E}_y \tag{18}$$

$$\frac{\partial \mathcal{H}_y}{\partial x} - \frac{\partial \mathcal{H}_x}{\partial y} = 0 \tag{19}$$

and the divergence equations are

$$\frac{\partial \mathcal{E}_x}{\partial x} + \frac{\partial \mathcal{E}_y}{\partial y} = 0 \tag{20}$$

and

$$\frac{\partial \mathcal{H}_x}{\partial x} + \frac{\partial \mathcal{H}_y}{\partial y} - \gamma \mathcal{H}_z = 0 \tag{21}$$

We will now rearrange the equations such that all transverse fields are expressed in terms of $\mathcal{H}_z$ (or derivatives of $\mathcal{H}_z$).

From (14) and (15),

$$-\frac{\mathscr{E}_y}{\mathscr{H}_x} = \frac{\mathscr{E}_x}{\mathscr{H}_y} = \frac{j\omega\mu}{\gamma} = Z_{\text{TE}} \qquad (\Omega) \tag{22}$$

the wave impedance. Substituting (22) into (17), we have

$$\mathscr{H}_y = \frac{j\omega\epsilon\mathscr{E}_x - \dfrac{\partial\mathscr{H}_z}{\partial y}}{\gamma} = \frac{\dfrac{j\omega\epsilon j\omega\mu\mathscr{H}_y}{\gamma} - \dfrac{\partial\mathscr{H}_z}{\partial y}}{\gamma}$$

or

$$\mathscr{H}_y = \frac{-\gamma\dfrac{\partial\mathscr{H}_z}{\partial y}}{\gamma^2 + \omega^2\mu\epsilon} \qquad (\text{A m}^{-1}) \tag{23}$$

Substituting (22) into (18), we get

$$-\gamma\mathscr{H}_x - \frac{\partial\mathscr{H}_z}{\partial x} = j\omega\epsilon\mathscr{E}_y = j\omega\epsilon\left(\frac{-j\omega\mu\mathscr{H}_x}{\gamma}\right)$$

or

$$\mathscr{H}_x = \frac{-\gamma\dfrac{\partial\mathscr{H}_z}{\partial x}}{\gamma^2 + \omega^2\mu\epsilon} \qquad (\text{A m}^{-1}) \tag{24}$$

Now we can find $\mathscr{E}_x$ and $\mathscr{E}_y$ by simply multiplying (23) and (24) by Z_{TE} and $-Z_{\text{TE}}$, respectively. Then

$$\mathscr{E}_x = Z_{\text{TE}}\mathscr{H}_y = \frac{-j\omega\mu\dfrac{\partial\mathscr{H}_z}{\partial y}}{\gamma^2 + \omega^2\mu\epsilon} \qquad (\text{V m}^{-1}) \tag{25}$$

and

$$\mathscr{E}_y = -Z_{\text{TE}}\mathscr{H}_x = \frac{j\omega\mu\dfrac{\partial\mathscr{H}_z}{\partial x}}{\gamma^2 + \omega^2\mu\epsilon} \qquad (\text{V m}^{-1}) \tag{26}$$

The next step is to substitute (23) and (24) into the divergence equation (21) to get an equation in $\mathcal{H}_z$ only.

$$-\frac{\gamma \dfrac{\partial^2 \mathcal{H}_z}{\partial x^2}}{\gamma^2 + \omega^2 \mu \epsilon} - \frac{\gamma \dfrac{\partial^2 \mathcal{H}_z}{\partial y^2}}{\gamma^2 + \omega^2 \mu \epsilon} - \gamma \mathcal{H}_z = 0$$

or

$$\frac{\partial^2 \mathcal{H}_z}{\partial x^2} + \frac{\partial^2 \mathcal{H}_z}{\partial y^2} + (\gamma^2 + \omega^2 \mu \epsilon)\mathcal{H}_z = 0 \tag{27}$$

Using a product solution and letting

$$\mathcal{H}_z = X(x)Y(y)e^{-\gamma z} \tag{28}$$

where X is a function of x only and Y is a function of y only, then substituting (28) into (27), and dividing by $XYe^{-\gamma z}$, we get

$$\frac{1}{X}\left(\frac{d^2X}{dx^2}\right) + \frac{1}{Y}\left(\frac{d^2Y}{dy^2}\right) = -(\gamma^2 + \omega^2 \mu \epsilon) \tag{29}$$

If we let

$$\frac{1}{X}\left(\frac{d^2X}{dx^2}\right) = -k_x^2 \tag{30}$$

and

$$\frac{1}{Y}\left(\frac{d^2Y}{dy^2}\right) = -k_y^2 \tag{31}$$

then

$$\gamma^2 + \omega^2 \mu \epsilon = k_x^2 + k_y^2 = k^2$$

or

$$\gamma = \sqrt{k_x^2 + k_y^2 - \omega^2 \mu \epsilon} \tag{32}$$

Now we see the possibility that the propagation constant may be either real or imaginary, depending on whether $\omega^2 \mu \epsilon$ is less than or greater than $(k_x^2 + k_y^2)$.

By inspection, we recognize that the solutions to (30) and (31) are sinusoidal.

$$X = C_1 \sin k_x x + C_2 \cos k_x x \tag{33}$$

and

$$Y = C_3 \sin k_y y + C_4 \cos k_y y \tag{34}$$

and thus $\mathcal{H}_z = XYe^{-\gamma z}$ becomes

$$\begin{aligned}\mathcal{H}_z = [&C_1 C_3 \sin k_x x \sin k_y y + C_1 C_4 \sin k_x x \cos k_y y \\ &+ C_2 C_3 \cos k_x x \sin k_y y + C_2 C_4 \cos k_x x \cos k_y y]e^{-\gamma z}\end{aligned} \tag{35}$$

But according to boundary conditions,

$$\mathcal{E}_x = 0 \quad \text{at } y = 0, a \tag{36}$$

and

$$\mathcal{E}_y = 0 \quad \text{at } x = 0, b \tag{37}$$

Substituting (35) into (25) and satisfying (36), requires that $C_1 C_3 = 0$ and $C_2 C_3 = 0$. Similarly, substituting (35) into (26) and satisfying (37), requires that $C_1 C_3 = 0$ and

$C_1 C_4 = 0$. Therefore, we now simplify (35) to

$$\mathcal{H}_z = e^{-\gamma z}(C_2 C_4 \cos k_x x \cos k_y y) \tag{38}$$
$$= e^{-\gamma z}\mathcal{H}_{z0} \cos k_x x \cos k_y y \qquad (\text{A m}^{-1}) \tag{39}$$

where $\mathcal{H}_{z0}$ is an arbitrary amplitude.
Substituting (39) into (23), (24), (25), and (26), we have

$$\mathcal{H}_y = e^{-\gamma z}\left(\frac{\gamma \mathcal{H}_{z0} k_y \cos k_x x \sin k_y y}{\gamma^2 + \omega^2 \mu \epsilon}\right) \tag{40}$$

$$\mathcal{H}_x = e^{-\gamma z}\left(\frac{\gamma \mathcal{H}_{z0} k_x \sin k_x x \cos k_y y}{\gamma^2 + \omega^2 \mu \epsilon}\right) \tag{41}$$

$$\mathcal{E}_x = e^{-\gamma z}\left(\frac{j\omega\mu k_y \mathcal{H}_{z0} \cos k_x x \sin k_y y}{\gamma^2 + \omega^2 \mu \epsilon}\right) \tag{42}$$

$$\mathcal{E}_y = e^{-\gamma z}\left(\frac{-j\omega\mu k_x \mathcal{H}_{z0} \sin k_x x \cos k_y y}{\gamma^2 + \omega^2 \mu \epsilon}\right) \tag{43}$$

At $y = 0, a$, $\mathcal{E}_x = 0$; thus, $\sin k_y a = 0$, or

$$k_y = \frac{m\pi}{a} \qquad (\text{m}^{-1}) \tag{44}$$

Also, at $x = 0, b$, $\mathcal{E}_y = 0$; thus, $\sin k_x b = 0$, or

$$k_x = \frac{n\pi}{b} \qquad (\text{m}^{-1}) \tag{45}$$

where m and n are integers.

We can now substitute (44) and (45) into (32) to determine the conditions under which γ is real or imaginary. Making that substitution yields

$$\gamma = \sqrt{\frac{m^2\pi^2}{a^2} + \frac{n^2\pi^2}{b^2} - \omega^2\mu\epsilon} \qquad (\text{m}^{-1}) \tag{46}$$

In the pass band, γ is purely imaginary since $\omega^2\mu\epsilon > (m^2\pi^2/a^2) + (n^2\pi^2/b^2)$, and γ may be expressed as

$$\gamma = j\beta = \sqrt{\mu\epsilon}\sqrt{\omega_c^2 - \omega^2} \tag{47}$$

where

$$\omega_c = \frac{1}{\sqrt{\mu\epsilon}}\sqrt{\frac{m^2\pi^2}{a^2} + \frac{n^2\pi^2}{b^2}} = 2\pi f_c \tag{48}$$

and

$$\boxed{\begin{aligned}\beta &= \sqrt{\omega^2\mu\epsilon - \left(\frac{m^2\pi^2}{a^2} + \frac{n^2\pi^2}{b^2}\right)} \\ &= \sqrt{\beta_0^2 - k^2} \qquad (\text{m}^{-1})\end{aligned}} \tag{49}$$

where β_0 is the β for the unbounded plane wave.

From (47), we see that when $\omega > \omega_c$, γ is purely imaginary and equal to $j\beta$, but when $\omega < \omega_c$, γ is purely real, and there is no wave propagation. The apparent wavelength in the guide is expressed in terms of β by

$$\lambda_g = \frac{2\pi}{\beta} = \frac{1}{\sqrt{f^2\mu\epsilon - \frac{1}{4}\left(\frac{m^2}{a^2} + \frac{n^2}{b^2}\right)}}$$

or

$$\boxed{\lambda_g = \frac{\lambda}{\sqrt{1 - \frac{\lambda^2}{\lambda_c^2}}} \qquad (\text{m})} \tag{50}$$

and the cutoff frequency, i.e., the frequency at which

$$\omega = \frac{1}{\sqrt{\mu\epsilon}}\sqrt{\frac{m^2\pi^2}{a^2} + \frac{n^2\pi^2}{b^2}}$$

is given by

$$\boxed{f_{c\text{TE}_{mn}} = \frac{1}{2\sqrt{\mu\epsilon}}\sqrt{\frac{m^2}{a^2} + \frac{n^2}{b^2}} \qquad (\text{Hz})} \tag{51}$$

For an air-filled guide,

$$\underset{\text{(air-filled)}}{f_{c\text{TE}_{mn}}} = 1.5 \times 10^8\sqrt{\frac{m^2}{a^2} + \frac{n^2}{b^2}} \qquad (\text{Hz}) \tag{52}$$

The equations for the fields are now written in terms of the mn modes, the arbitrary constant $\mathcal{H}_{z0}$, and $\gamma^2 + \omega^2\mu\epsilon = k^2$.

$$\mathcal{H}_y = \left(\frac{\gamma\frac{m\pi}{a}\mathcal{H}_{z0}\cos\frac{n\pi x}{b}\sin\frac{m\pi y}{a}}{k^2}\right)e^{-\gamma z} \tag{53}$$

$$\mathcal{H}_x = \left(\frac{\gamma\frac{n\pi}{b}\mathcal{H}_{z0}\sin\frac{n\pi x}{b}\cos\frac{m\pi y}{a}}{k^2}\right)e^{-\gamma z} \tag{54}$$

$$\mathscr{E}_x = \left(\frac{j\omega\mu \dfrac{m\pi}{a} \mathscr{H}_{z0} \cos \dfrac{n\pi x}{b} \sin \dfrac{m\pi y}{a}}{k^2}\right) e^{-\gamma z} \tag{55}$$

$$\mathscr{E}_y = \left(\frac{-j\omega\mu \dfrac{n\pi}{b} \mathscr{H}_{z0} \sin \dfrac{n\pi x}{b} \cos \dfrac{m\pi y}{a}}{k^2}\right) e^{-\gamma z} \tag{56}$$

The most common mode, called the *dominant mode,* is the TE_{10}, where $m = 1$, $n = 0$. For the TE_{10} mode,

$$\mathscr{H}_z = \mathscr{H}_{z0} \cos \frac{\pi y}{a} e^{-\gamma z} \tag{57}$$

$$\mathscr{H}_y = \left(\frac{\gamma \dfrac{\pi}{a} \mathscr{H}_{z0} \sin \dfrac{\pi y}{a}}{k^2}\right) e^{-\gamma z} \tag{58}$$

$$\mathscr{H}_x = 0 \tag{59}$$

$$\mathscr{E}_x = \left(\frac{j\omega\mu \dfrac{\pi}{a} \mathscr{H}_{z0} \sin \dfrac{\pi y}{a}}{k^2}\right) e^{-\gamma z} \tag{60}$$

$$\mathscr{E}_y = 0 \tag{61}$$

and the cutoff frequency is

$$f_{c\mathrm{TE}_{10}} = \frac{1}{2\sqrt{\mu\epsilon}} \sqrt{\frac{1}{a^2}} = \frac{1}{2\sqrt{\mu\epsilon}\, a} \tag{62}$$

and

$$\lambda_{c\mathrm{TE}_{10}} = \frac{U}{f_c} = 2a \tag{63}$$

which agrees with (13.2-1). The TE_{10} mode is the mode partially depicted by Fig. 13-3.

From (50), we obtain the phase velocity,

$$v_p = \lambda_g f = \frac{1}{\sqrt{\mu\epsilon} \sqrt{1 - \dfrac{f_c^2}{f^2}}} \qquad (\mathrm{m\ s^{-1}}) \tag{64}$$

or

$$\boxed{v_p = \frac{U}{\sqrt{1 - \dfrac{f_c^2}{f^2}}} \qquad (\mathrm{m\ s^{-1}})} \tag{65}$$

In summarizing the relationships between various wavelengths and velocities, we have

$$\text{plane wave wavelength,} \qquad \lambda = \frac{U}{f} = \frac{2\pi}{\beta_0} \tag{66}$$

$$\text{waveguide wavelength,} \qquad \lambda_g = \frac{v_p}{f} = \frac{2\pi}{\beta} \tag{67}$$

$$\text{cutoff wavelength,} \qquad \lambda_c = \frac{U}{f_c} = \frac{2\pi}{k} \tag{68}$$

Example 1

(a) For an air-filled waveguide whose inside dimensions are 3.0 in. × 1.5 in., find the cutoff frequency and cutoff wavelength for the TE_{10} mode. (b) For the same waveguide, calculate the fields in terms of an arbitrary constant $\mathcal{H}_{z0}$ for $f = 2.45$ GHz operating in the TE_{10} mode.

Solution. $m = 1, n = 0.$

(a)
$$a = 3.0 \text{ in.} = 0.0762 \text{ m}$$
$$b = 1.5 \text{ in.} = 0.0381 \text{ m}$$
$$\lambda_c = 2a = 0.1524 \text{ m}$$
$$f_c = \frac{3 \times 10^8 \text{ m s}^{-1}}{0.1524 \text{ m}} = 1.97 \text{ GHz}$$

(b) Since $f > f_c$, and $\gamma = j\beta$,

$$\beta = \sqrt{\omega^2\mu\epsilon - \frac{\pi^2}{a^2}}$$
$$= \sqrt{\left(\frac{2\pi(2.45 \times 10^9)}{3 \times 10^8}\right)^2 - \left(\frac{\pi}{0.0762}\right)^2}$$
$$= 30.5 \text{ m}^{-1}$$
$$k^2 = k_x^2 + k_y^2 = 0 + \left(\frac{\pi}{a}\right)^2 = 1700$$
$$\frac{m\pi}{a} = \frac{(1)\pi}{0.0762} = 41.23$$
$$\frac{\gamma\pi}{a} = j\frac{\beta\pi}{a} = j(30.5)(41.23) = j1258$$
$$j\omega\mu\frac{\pi}{a} = j2\pi(2.45 \times 10^9)(4\pi \times 10^{-7})(41.23)$$
$$= j7.97 \times 10^5$$

Then

$$\mathcal{H}_z = \mathcal{H}_{z0} \cos 41.23y e^{-j30.5z}$$
$$\mathcal{H}_y = \left(\frac{j1258\mathcal{H}_{z0} \sin 41.23y}{1700}\right)e^{-j30.5z}$$
$$= j0.74\mathcal{H}_{z0} \sin 41.23y e^{-j30.5z}$$
$$\mathcal{H}_x = 0$$
$$\mathcal{E}_x = \left(\frac{j7.97 \times 10^5\mathcal{H}_{z0} \sin 41.23y}{1700}\right)e^{-j30.5z}$$
$$= j468\mathcal{H}_{z0} \sin 41.23y e^{-j30.5z}$$
$$\mathcal{E}_y = 0$$

Incidentally,

$$Z_{TE} = \frac{\mathcal{E}_x}{\mathcal{H}_y} = \frac{468}{0.74} = 633\ \Omega = j\frac{\omega\mu}{j\beta}$$

Example 2

The waveguide of Example 1 has a λ_g of 0.2 m as determined by slotted line measurements. Find the frequency, phase velocity, and group velocity.

Solution

$$\lambda_g = 0.2 = \frac{2\pi}{\beta}$$

or

$$\beta = \frac{2\pi}{0.2} = 10\pi = \sqrt{\omega^2\mu\epsilon - \frac{\pi^2}{a^2}}$$

$$\omega^2\mu\epsilon = \beta^2 + \frac{\pi^2}{a^2} = \left(\frac{2\pi}{\lambda_g}\right)^2 + \left(\frac{\pi}{a}\right)^2$$

$$f = \frac{1}{2\sqrt{\mu\epsilon}}\sqrt{\left(\frac{2}{\lambda_g}\right)^2 + \left(\frac{1}{a}\right)^2}$$

$$= 1.5 \times 10^8 \sqrt{\left(\frac{2}{0.2}\right)^2 + \left(\frac{1}{0.076}\right)^2}$$

$$= 2.47 \text{ GHz}$$

$$v_p = \frac{U}{\sqrt{1 - (f_c/f)^2}} = \frac{3 \times 10^8}{\sqrt{1 - (1.97/2.47)^2}}$$

$$= 4.97 \times 10^8 \text{ m s}^{-1}$$

$$v_g = \frac{U^2}{v_p} = \frac{(3 \times 10^8)^2}{4.97 \times 10^8} = 1.81 \times 10^8 \text{ m s}^{-1}$$

It is usually desirable to operate in one mode only in order that the signal energy propagating along the guide is not divided into components that travel at different velocities. Also, the method of exciting a particular mode or receiving a particular mode is such that it is usually preferable to work with a single mode. For this reason, some thought has to go into the choice of a waveguide size such that the operating frequency is above $f_{c\text{TE}_{10}}$ and below f_c for a higher-order mode, if the TE_{10} mode is the desired mode.

Example 3

For the 3.0 in. × 1.5 in. waveguide of Examples 1 and 2, find the frequency range over which operation would be restricted to the TE_{10} mode only.

Solution. The lowest frequency is $f_{c\text{TE}_{10}}$, or $f_{\text{lower}} = 1.97$ GHz from Example 1.

To find the upper frequency, we must look at $f_{c\text{TE}_{01}}$, $f_{c\text{TE}_{11}}$, and $f_{c\text{TE}_{20}}$. From (52) for an air-filled waveguide,

$$f_{c\text{TE}_{01}} = 1.5 \times 10^8\left(\frac{1}{0.0381}\right) = 3.94 \text{ GHz}$$

$$f_{c\text{TE}_{11}} = 1.5 \times 10^8\sqrt{\left(\frac{1}{0.0762}\right)^2 + \left(\frac{1}{0.0381}\right)^2} = 4.40 \text{ GHz}$$

$$f_{c\text{TE}_{20}} = 1.5 \times 10^8\left(\frac{2}{0.0762}\right) = 3.94 \text{ GHz}$$

Then the range for TE_{10} operation only is

$$1.97 \text{ GHz} < f < 3.94 \text{ GHz}$$

The visualization of the fields becomes difficult for the higher-order modes although they can be plotted from calculations using (39) through (43). End views and top views of the $\overline{E}$ and $\overline{H}$ field patterns are shown in Fig. 13-4 for the TE_{10}, TE_{11}, and TE_{20} modes. The patterns propagate along the waveguide at the phase velocity.

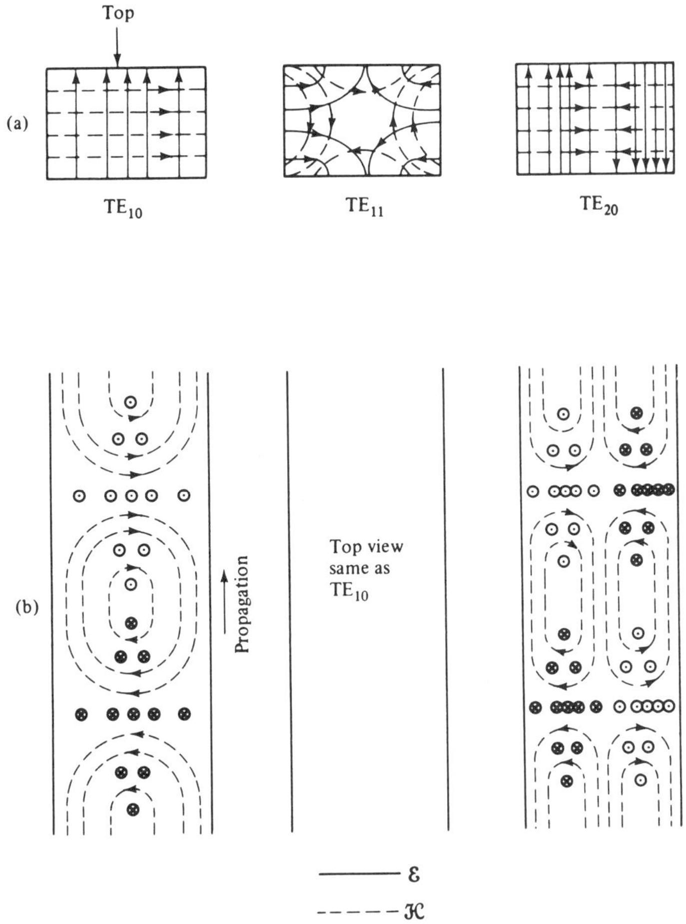

Figure 13-4 End and top views for three of the lower-order modes for rectangular waveguides.

Problem 13.3-1 Find the range of frequencies for which the following air-filled waveguides will operate in the TE_{10} mode only: (a) 7.21 cm × 3.40 cm; (b) 4.76 cm × 2.22 cm; (c) 1.91 cm × 0.95 cm.

Problem 13.3-2 A 2.29 cm × 1.02 cm air-filled waveguide is being used to transmit a signal at a frequency of 9 GHz in the TE_{10} mode. Find: (a) λ_g; (b) v_p; (c) β.

Problem 13.3-3 A 0.9 in. × 0.4 in. waveguide operating in the TE_{11} modes has a standing wave pattern for which the distance between minima is 3 cm. Find the frequency.

13.4 The TM_{mn} MODES IN RECTANGULAR WAVEGUIDES

When the field configuration allows only a transverse magnetic field, that is called the *TM mode*. The approach to the derivation of the field equations is identical to that of Sec. 13.3, except that here we let $\mathcal{H}_z = 0$ instead of $\mathcal{E}_z = 0$ and the transverse fields are all expressed as functions of $\mathscr{E}_z$ instead of $\mathscr{H}_z$. We shall start with the expanded Maxwell equations (13.3-5) through (13.3-12). With $\mathscr{H}_z = 0$, the curl equations are

$$\frac{\partial \mathscr{E}_z}{\partial y} + \gamma \mathscr{E}_y = -j\omega\mu\mathscr{H}_x \tag{1}$$

$$-\gamma\mathscr{E}_x - \frac{\partial \mathscr{E}_z}{\partial x} = -j\omega\mu\mathscr{H}_y \tag{2}$$

$$\frac{\partial \mathscr{E}_y}{\partial x} - \frac{\partial \mathscr{E}_x}{\partial y} = 0 \tag{3}$$

$$\gamma\mathscr{H}_y = j\omega\epsilon\mathscr{E}_x \tag{4}$$

$$-\gamma\mathscr{H}_x = j\omega\epsilon\mathscr{E}_y \tag{5}$$

$$\frac{\partial \mathscr{H}_y}{\partial x} - \frac{\partial \mathscr{H}_x}{\partial y} = j\omega\epsilon\mathscr{E}_z \tag{6}$$

and the divergence equations are

$$\frac{\partial \mathscr{E}_x}{\partial x} + \frac{\partial \mathscr{E}_y}{\partial y} - \gamma\mathscr{E}_z = 0 \tag{7}$$

$$\frac{\partial \mathscr{H}_x}{\partial x} + \frac{\partial \mathscr{H}_y}{\partial y} = 0 \tag{8}$$

After rearranging the equations such that all transverse fields are expressed in terms of $\mathscr{E}_z$, we have

$$\boxed{\mathscr{E}_x = \frac{-\gamma\dfrac{\partial \mathscr{E}_z}{\partial x}}{\gamma^2 + \omega^2\mu\epsilon} \qquad (\text{V m}^{-1})} \tag{9}$$

$$\boxed{\mathscr{E}_y = \frac{-\gamma\dfrac{\partial \mathscr{E}_z}{\partial y}}{\gamma^2 + \omega^2\mu\epsilon} \qquad (\text{V m}^{-1})} \tag{10}$$

$$\boxed{\mathscr{H}_y = \frac{-j\omega\epsilon\dfrac{\partial \mathscr{E}_z}{\partial x}}{\gamma^2 + \omega^2\mu\epsilon} \qquad (\text{A m}^{-1})} \tag{11}$$

$$\boxed{\mathscr{H}_x = \frac{j\omega\epsilon\dfrac{\partial \mathscr{E}_z}{\partial y}}{\gamma^2 + \omega^2\mu\epsilon} \qquad (\text{A m}^{-1})} \tag{12}$$

where the wave impedance for the TM mode waves is

$$\boxed{Z_{\text{TM}} = \frac{\mathscr{E}_x}{\mathscr{H}_y} = -\frac{\mathscr{E}_y}{\mathscr{H}_x} = \frac{\gamma}{j\omega\epsilon} \qquad (\Omega)} \tag{13}$$

If we now substitute (9) and (10) into (7), we get

$$\frac{\partial^2 \mathscr{E}_z}{\partial x^2} + \frac{\partial^2 \mathscr{E}_z}{\partial y^2} + (\gamma^2 + \omega^2\mu\epsilon)\mathscr{E}_z = 0 \tag{14}$$

which is identical in form to (13.3-27). Using a product solution, $\mathscr{E}_z = X(x)Y(y)e^{-\gamma z}$ and applying the boundary conditions at the conducting walls of the guide, we obtain

$$\begin{aligned}\mathscr{E}_z &= \mathscr{E}_{z0} \sin k_x x \sin k_y y\, e^{-\gamma z} \\ &= \mathscr{E}_{z0} \sin \frac{n\pi x}{b} \sin \frac{m\pi y}{a} e^{-\gamma z}\end{aligned} \tag{15}$$

and the transverse fields are

$$\mathscr{E}_x = \left(\frac{-\gamma \dfrac{n\pi}{b} \mathscr{E}_{z0} \cos \dfrac{n\pi x}{b} \sin \dfrac{m\pi y}{a}}{k^2} \right) e^{-\gamma z} \tag{16}$$

$$\mathscr{E}_y = \left(\frac{-\gamma \dfrac{m\pi}{a} \mathscr{E}_{z0} \sin \dfrac{n\pi x}{b} \cos \dfrac{m\pi y}{a}}{k^2} \right) e^{-\gamma z} \tag{17}$$

$$\mathscr{H}_y = \left(\frac{-j\omega\epsilon \dfrac{n\pi}{b} \mathscr{E}_{z0} \cos \dfrac{n\pi x}{b} \sin \dfrac{m\pi y}{a}}{k^2} \right) e^{-\gamma z} \tag{18}$$

$$\mathscr{H}_x = \left(\frac{j\omega\epsilon \dfrac{m\pi}{a} \mathscr{E}_{z0} \sin \dfrac{n\pi x}{b} \cos \dfrac{m\pi y}{a}}{k^2} \right) e^{-\gamma z} \tag{19}$$

The formulas for γ, β, f_c, v_g, and v_p are the same as for the TE_{mn} modes. There is one distinct departure from parallelism between the TE_{mn} and the TM_{mn} modes, however. For the TM_{mn} modes, neither m nor n can be zero, as (15) shows that $\mathscr{E}_z$ would be zero also in that case and no fields would exist at all. Therefore, the lowest-order mode for the TM_{mn} modes is the TM_{11}, but

$$f_{c\text{TM}_{11}} = f_{c\text{TE}_{11}} = \frac{1}{2\sqrt{\mu\epsilon}} \sqrt{\frac{1}{a^2} + \frac{1}{b^2}} \qquad (\text{Hz}) \tag{20}$$

The wave impedance Z_{TM} is lower than Z_{TE}. We can show this easily for a lossless guide operating in the pass band ($f > f_c$). Then $\gamma = j\beta$, and

$$\begin{aligned}\beta &= \sqrt{\omega^2\mu\epsilon - \left(\frac{m^2\pi^2}{a^2} + \frac{n^2\pi^2}{b^2}\right)} \\ &= \omega\sqrt{\mu\epsilon}\sqrt{1 - \frac{f_c^2}{f^2}}\end{aligned}$$

and

$$Z_{\text{TM}} = \frac{\gamma}{j\omega\epsilon} = \frac{\beta}{\omega\epsilon} = \frac{\omega\sqrt{\mu\epsilon}\sqrt{1 - (f_c^2/f^2)}}{\omega\epsilon}$$

$$= \sqrt{\frac{\mu}{\epsilon}}\sqrt{1 - \frac{f_c^2}{f^2}} \quad (\Omega) \tag{21}$$

whereas

$$Z_{\text{TE}} = \frac{j\omega\mu}{\gamma} = \frac{\omega\mu}{\beta} = \frac{\sqrt{\frac{\mu}{\epsilon}}}{\sqrt{1 - (f_c^2/f^2)}} \quad (\Omega) \tag{22}$$

Clearly, $Z_{\text{TE}} > Z_{\text{TM}}$.

A summary of the TM_{mn} as well as the TE_{mn} mode characteristics is provided in Table 13-1.

TABLE 13-1 FUNDAMENTAL EQUATIONS FOR TE_{mn} AND TM_{mn} MODES

TE_{mn}	TM_{mn}
$\mathscr{E}_z = 0$	$\mathscr{H}_z = 0$
$\mathscr{H}_z = (\mathscr{H}_{z0} \cos k_x x \cos k_y y)e^{-\gamma z}$	$\mathscr{E}_z = (\mathscr{E}_{z0} \sin k_x x \sin k_y y)e^{-\gamma z}$
$\mathscr{E}_x = \frac{(j\omega\mu k_y \mathscr{H}_{z0} \cos k_x x \sin k_y y)e^{-\gamma z}}{k^2}$	$\mathscr{H}_x = \frac{(j\omega\epsilon k_y \mathscr{E}_{z0} \sin k_x x \cos k_y y)e^{-\gamma z}}{k^2}$
$\mathscr{E}_y = \frac{(-j\omega\mu k_x \mathscr{H}_{z0} \sin k_x x \cos k_y y)e^{-\gamma z}}{k^2}$	$\mathscr{H}_y = \frac{(-j\omega\epsilon k_x \mathscr{E}_{z0} \cos k_x x \sin k_y y)e^{-\gamma z}}{k^2}$
$\mathscr{H}_x = \frac{(\gamma k_x \mathscr{H}_{z0} \sin k_x x \cos k_y y)e^{-\gamma z}}{k^2}$	$\mathscr{E}_x = \frac{(-\gamma k_x \mathscr{E}_{z0} \cos k_x x \sin k_y y)e^{-\gamma z}}{k^2}$
$\mathscr{H}_y = \frac{(\gamma k_y \mathscr{H}_{z0} \cos k_x x \sin k_y y)e^{-\gamma z}}{k^2}$	$\mathscr{E}_y = \frac{(-\gamma k_y \mathscr{E}_{z0} \sin k_x x \cos k_y y)e^{-\gamma z}}{k^2}$
$Z_{\text{TE}} = \frac{j\omega\mu}{\gamma} = \frac{\omega\mu}{\beta} = \frac{\sqrt{\mu/\epsilon}}{\sqrt{1 - f_c^2/f^2}}$	$Z_{\text{TM}} = \frac{\gamma}{j\omega\epsilon} = \frac{\beta}{\omega\epsilon} = \sqrt{\frac{\mu}{\epsilon}}\sqrt{1 - \frac{f_c^2}{f^2}}$

$$k^2 = k_x^2 + k_y^2 = \frac{n^2\pi^2}{b^2} + \frac{m^2\pi^2}{a^2} = \gamma^2 + \omega^2\mu\epsilon$$

$$f_c = \frac{1}{2\sqrt{\mu\epsilon}}\sqrt{\frac{m^2}{a^2} + \frac{n^2}{b^2}}$$

$$\lambda_c = \frac{U}{f_c} = \frac{2}{\sqrt{\frac{m^2}{a^2} + \frac{n^2}{b^2}}}$$

$$\gamma = j\beta = j\sqrt{\omega^2\mu\epsilon - \left[\left(\frac{m\pi}{a}\right)^2 + \left(\frac{n\pi}{b}\right)^2\right]} \quad (f > f_c)$$

$$= j\sqrt{\beta_0^2 - k^2}$$

Problem 13.4-1 For each of the three wavelengths of Prob. 13.3-1, find the cutoff frequencies for the three lowest-order TM_{mn} modes.

Problem 13.4-2 Plot the wave impedances $Z_{\text{TE}_{11}}$ and $Z_{\text{TM}_{11}}$, normalized to $\sqrt{\mu/\epsilon}$, for the frequency range $f_c < f < 100f_c$.

13.5 ATTENUATION IN AIR-FILLED RECTANGULAR WAVEGUIDES

So far we have avoided the determination of α, the real part of the propagation constant, when operating in the pass band. We found $\gamma = j\beta$ in that case. For an air-filled waveguide, the only source of loss is the I^2R losses in the walls of the guide. The tangential $\overline{H}$ fields at the walls are numerically equal to the sheet current densities in the walls (11.6-8), and knowing the surface resistivity R_s of the material (11.6-13), the power density in the material is given by (11.6-15). The total power loss in the walls is obtained by integrating the power density entering the lossy wall over all the wall area.

Referring to Fig. 13-5, P_2 is related to P_1 by

$$P_2 = P_1 e^{-2\alpha\Delta z} \qquad \text{(W)} \tag{1}$$

and the power loss in the walls of the guide in the distance Δz is

$$P_L = P_1 - P_2$$

or

$$P_2 = P_1 - P_L = P_1 e^{-2\alpha\Delta z}$$

and, dividing by P_1 and taking the natural logarithm of both sides,

$$\ln\left(1 - \frac{P_L}{P_1}\right) = -2\alpha\,\Delta z \tag{2}$$

For most practical cases $P_L/P_1 \ll 1$, and for a unit distance Δz we then have

$$\ln\left(1 - \frac{P_L}{P_1}\right) \approx -\frac{P_L}{P_1} = -2\alpha(1)$$

or

$$\alpha \approx \frac{P_L}{2P_1} \approx \frac{P_L}{2P_T} \qquad (\text{m}^{-1}) \tag{3}$$

where P_T denotes power being transmitted by the guide at a given position z (we are saying that $P_1 \approx P_T \approx P_2$, where actually $P_1 > P_T > P_2$).

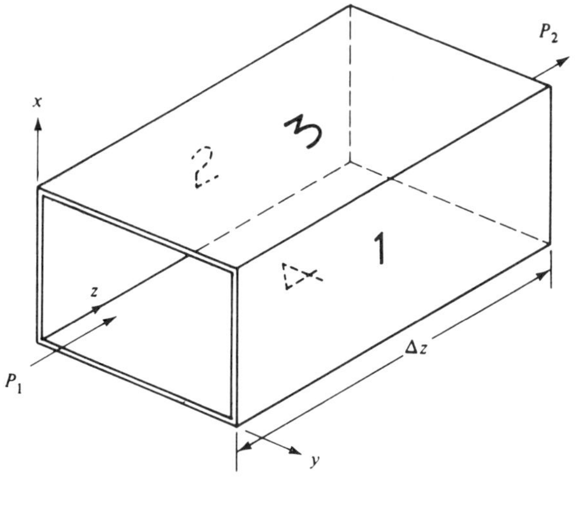

Figure 13-5 Illustration used for computation of wall losses in a rectangular waveguide.

We shall now proceed to calculate the loss in a unit (1 m) length of waveguide and use (3) to compute α. Let us consider the TE_{10} mode. Referring to Fig. 13-5, the average power loss in the four walls, taking a pair at a time, is

$$P_{L_{1+2}} = 2(\tfrac{1}{2}R_s)\int_{x=0}^{b}\int_{z=0}^{1}|\mathcal{H}_z|^2\,dz\,dx \tag{4}$$

Substituting (13.3-57) for $\mathcal{H}_z$ and letting $\gamma = j\beta$, we have

$$\begin{aligned} P_{L_{1+2}} &= R_s\int_{x=0}^{b}\int_{z=0}^{1}\mathcal{H}_{z0}^2\cos^2 k_y y\,dz\,dx \\ &= R_s\mathcal{H}_{z0}^2 b \qquad \text{(W)} \end{aligned} \tag{5}$$

and for sides 3 and 4,

$$P_{L_{3+4}} = R_s\int_{y=0}^{a}\int_{z=0}^{1}(|\mathcal{H}_y|^2 + |\mathcal{H}_z|^2)\,dz\,dy \tag{6}$$

$$\begin{aligned} &= R_s\int_0^a\int_0^1\left[\frac{\beta^2 k_y^2\mathcal{H}_{z0}^2}{k^4}\sin^2 k_y y + \mathcal{H}_{z0}^2\cos^2 k_y y\right]dz\,dy \\ &= R_s\left[\frac{\beta^2\mathcal{H}_{z0}^2}{k_y^2}\frac{a}{2} + \frac{\mathcal{H}_{z0}^2 a}{2}\right] \qquad \text{(W)} \end{aligned} \tag{7}$$

The average transmitted power is

$$P_T = \frac{1}{2}\int_{y=0}^{a}\int_{x=0}^{b}R_e[(\mathcal{E}_x)(\mathcal{H}_y^*)]\,dx\,dy \tag{8}$$

$$\begin{aligned} &= \frac{1}{2}\left(\frac{\beta k_y^2\omega\mu\mathcal{H}_{z0}^2}{k_y^4}\right)\int_0^a\int_0^b\sin^2 k_y y\,dx\,dy \\ &= \frac{1}{4}\left(\frac{\beta\omega\mu\mathcal{H}_{z0}^2}{k_y^2}\right)ab \qquad \text{(W)} \end{aligned} \tag{9}$$

Applying (3) to obtain α, we get

$$\begin{aligned} \alpha &= \frac{R_s\mathcal{H}_{z0}^2\left[b + \dfrac{a}{2}\left(1 + \dfrac{\beta^2}{k_y^2}\right)\right]}{2\left[\dfrac{1}{4}\left(\dfrac{\beta\omega\mu\mathcal{H}_{z0}^2}{k_y^2}\right)ab\right]} \\ &= \frac{2R_s}{ab}\left[\frac{k_y^2}{\omega\mu\beta}b + \frac{a}{2}\left(\frac{k_y^2+\beta^2}{k_y^2}\right)\left(\frac{k_y^2}{\omega\mu\beta}\right)\right] \\ &= \frac{R_s\omega^2\mu\epsilon}{b\omega\mu\beta}\left[1 + \frac{2k_y^2 b}{a\omega^2\mu\epsilon}\right] \\ &= \frac{R_s\omega\epsilon}{\beta b}\left[1 + \frac{2bf_c^2}{af^2}\right] \end{aligned} \tag{10}$$

Finally,

$$\alpha = \frac{R_s}{b\sqrt{\dfrac{\mu}{\epsilon}}\sqrt{1 - \dfrac{f_c^2}{f^2}}}\left[1 + \frac{2b}{a}\left(\frac{f_c}{f}\right)^2\right] \qquad (\text{m}^{-1}) \tag{11}$$

where

$$R_s = \frac{1}{\sigma\delta} = \frac{\sqrt{\pi f \mu \sigma}}{\sigma} = \sqrt{\frac{\pi f \mu}{\sigma}} \quad (\Omega)$$

as given in Sec. 11.6. We can further rearrange (11) for more convenience in calculation to

$$\boxed{\alpha = \frac{\sqrt{\pi f \epsilon/\sigma}}{b\sqrt{1-(f_c/f)^2}}\left[1+\frac{2b}{a}\left(\frac{f_c}{f}\right)^2\right] \quad (\text{m}^{-1})} \tag{12}$$

To convert to dB/m, multiply the results from (12) by 8.69, i.e., 20 $\log_{10} e$. Plots of α in dB m^{-1} are given in Fig. 13-6 for an X-band waveguide operating in the TE_{10} mode having a cutoff frequency of 6.55 GHz. The effect of the conductor material (whether copper or aluminum) and the effect of the smaller dimension of the waveguide is clearly demonstrated.

The minimum attenuation for all the plots in Fig. 13-6 occurs at around 14 or 15 GHz, which is above the cutoff frequency for the TE_{20} mode. Thus, one could not take advantage of the minimum attenuation frequency for the TE_{10} mode without risking the propagation of both the TE_{10} and TE_{20} modes and the problems that can develop from double mode operation. This points out a compromise that may have to be made in the choice of waveguide size, minimum attenuation, and cutoff frequencies.

Had it not been for the emergence of low-loss optical fibers for wide-band communications, underground waveguide might have been used for many long-distance communication paths. A very low loss waveguide was designed to propagate the TE_{01}

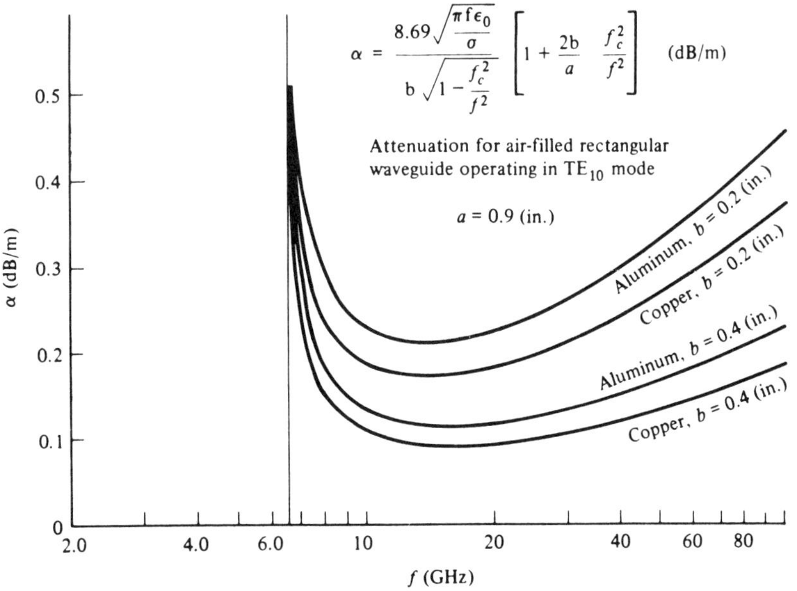

Figure 13-6 Attenuation versus frequency for a rectangular waveguide.

mode in circular waveguide.[1] Attenuation was only about 1 dB km^{-1} in the range 40 to 110 GHz. A remarkable feature of this particular mode is that attenuation decreases with increasing frequency, but it is difficult to control the mode, i.e., to prevent unwanted high loss modes.

Problem 13.5-1 For a 0.9 in. × 0.4 in. copper waveguide, calculate the attenuation in dB per meter at 8.2 GHz and at 12.4 GHz. Check results with Fig. 13-6. [*Note:* Dimensions are in meters and the frequency is in hertz in formula for α.]

13.6 RESONANT CAVITIES

A section of waveguide shorted at each end can support a standing wave pattern just like an ordinary transmission line, provided that the guide is an integral number of $\lambda_g/2$ in length. This is a resonance condition, and the enclosure is referred to as a resonant cavity. A pulse of energy input to the cavity through either a small current loop or a voltage probe will cause the cavity to "ring" at its resonance frequency or frequencies, just as a tuning fork rings after being struck. Resonant cavities are widely used as circuit elements, replacing conventional *LC* circuits at frequencies above 100 MHz or so.

Consider the $\mathscr{E}_x$ component of a TE_{mn} wave propagating in the z direction. With both ends of the guide shorted, $\rho = -1$ at each end, and a standing wave pattern results. Referring to Fig. 13-7,

$$\begin{aligned}\mathscr{E}_x &= \mathscr{E}_x^+ e^{-j\beta z} + \mathscr{E}_x^- e^{j\beta z} \\ &= \mathscr{E}_x^+(e^{-j\beta z} - e^{j\beta z}) \\ &= -2j\mathscr{E}_x^+ \sin \beta z\end{aligned} \tag{1}$$

But

$$\mathscr{E}_x = 0 \quad \text{at } z = 0, c$$

Therefore,

$$\sin \beta z = 0 \quad \text{at } z = 0, c$$

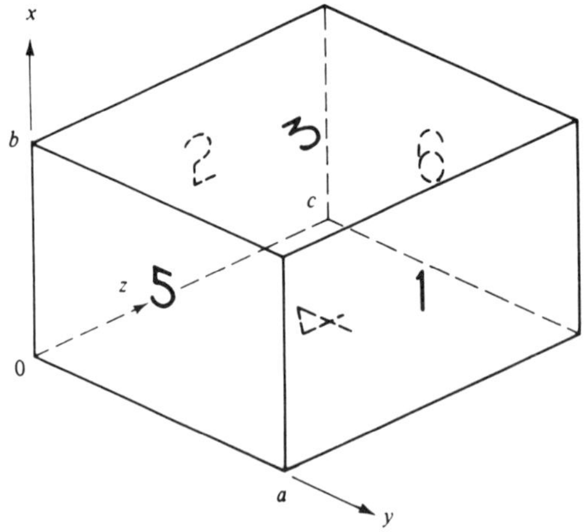

Figure 13-7 Rectangular cavity resonator.

[1] William D. Warters, "Millimeter Waveguide Scores High in Field Test," *Bell Lab. Rec.*, November 1975, pp. 401–408.

or

$$\beta = \frac{p\pi}{c}, \qquad p = 1, 2, 3, \ldots \tag{2}$$

But we have already shown that for a waveguide operating in the TE_{mn} mode,

$$\beta = \sqrt{\omega^2\mu\epsilon - \left(\frac{m^2\pi^2}{a^2} + \frac{n^2\pi^2}{b^2}\right)} \qquad (\mathrm{m}^{-1}) \tag{3}$$

Then, from (2) and (3), we have

$$\omega_r^2\mu\epsilon = \frac{m^2\pi^2}{a^2} + \frac{n^2\pi^2}{b^2} + \frac{p^2\pi^2}{c^2} \tag{4}$$

The frequency for which (4) is true can be called a resonance frequency.

$$f_r = \frac{1}{2\pi\sqrt{\mu\epsilon}}\sqrt{\frac{m^2\pi^2}{a^2} + \frac{n^2\pi^2}{b^2} + \frac{p^2\pi^2}{c^2}}$$

or

$$\boxed{f_{r\mathrm{TE}_{mnp}} = \frac{1}{2\sqrt{\mu\epsilon}}\sqrt{\frac{m^2}{a^2} + \frac{n^2}{b^2} + \frac{p^2}{c^2}} \qquad (\mathrm{Hz})} \tag{5}$$

We refer to the cavity as operating in the TE_{mnp} mode.

The resonant cavity should be viewed as an enclosure with standing waves between all three pairs of walls. Any frequency for which the E and H field structure can satisfy the boundary conditions is a resonance frequency. The assignment of the symbols a, b, c to particular sides of the enclosure is arbitrary.

Example 4

A section of 1.5 in. × 3.0 in. air-filled waveguide 4 in. long is shorted at each end, forming a cavity. Find the three lowest resonance frequencies.

Solution. From (5) we will calculate f_r for the three lowest-order non-trivial modes for TE_{mnp} modes. A trivial mode is where two or more of the subscripts are zero.

$$a = 3.0 \text{ in.} = 0.0762 \text{ m}$$
$$b = 1.5 \text{ in.} = 0.0381 \text{ m}$$
$$c = 4.0 \text{ in.} = 0.1016 \text{ m}$$

Then

$$f_{r\mathrm{TE}101} = 1.5 \times 10^8\sqrt{\frac{1}{(0.0762)^2} + \frac{1}{(0.1016)^2}} = 2.46 \text{ GHz}$$

$$f_{r\mathrm{TE}011} = 1.5 \times 10^8\sqrt{\frac{1}{(0.0381)^2} + \frac{1}{(0.1016)^2}} = 4.205 \text{ GHz}$$

$$f_{r\mathrm{TE}110} = 1.5 \times 10^8\sqrt{\frac{1}{(0.0381)^2} + \frac{1}{(0.0762)^2}} = 4.40 \text{ GHz}$$

The cavity acts like a resonant circuit with a Q that is generally much higher than encountered in ordinary circuits. For an air-filled cavity, the Q is dependent on wall losses, but cavities may also be filled with a dielectric that can add to the total loss and decrease the total Q. Calculation of wall losses is accomplished in the same manner as for waveguides discussed in the previous section.

To calculate the Q of a cavity, both the stored energy and the energy loss per cycle must be determined. Then

$$\boxed{Q = \frac{2\pi(\text{energy stored in cavity})}{\text{energy loss per cycle}}} \tag{6}$$

The total stored energy for the resonant cavity is entirely in the $\overline{\mathscr{E}}$ field when $|\overline{\mathscr{E}}|$ is maximum, thus we can calculate this stored energy by

$$W_{\text{stored}} = \tfrac{1}{2}\epsilon_0 \int_x \int_y \int_z [|\mathscr{E}_x|^2 + |\mathscr{E}_y|^2 + |\mathscr{E}_z|^2]\, dx\, dy\, dz \qquad \text{(J)} \tag{7}$$

and the energy loss per cycle is

$$W_L = T(\tfrac{1}{2})R_s \int_{s(\text{walls})} |\mathscr{H}_{\tan}|^2\, ds \qquad \text{(J)} \tag{8}$$

where T is the period, and where (8) is integrated over all the walls of the cavity as was done for the power loss for the waveguide in the previous section. Consider the TE_{101} mode as an example and use (6), (7), and (8) to derive the formula for the Q of a rectangular waveguide resonant cavity.

First, we must set down the expression for the various fields for the TE_{10} mode modified by the boundary conditions imposed by the shorted ends. For short lengths of guide, we will make the approximations that $\gamma \approx j\beta$. This is to say that the slight loss in the cavity does not significantly affect the distribution of the fields. Then, from (1) and from (13.3-60),

$$\mathscr{E}_x = -2j\mathscr{E}_x^+ \sin\beta z$$
$$= \frac{2\omega\mu \dfrac{\pi}{a}\mathscr{H}_{z0} \sin\dfrac{\pi y}{a} \sin\beta z}{k^2} \tag{9}$$

From (13.3-57) and (10.10-5),

$$\mathscr{H}_z = -j2\mathscr{H}_{z0} \cos\frac{\pi y}{a} \sin\beta z \tag{10}$$

From (13.3-58),

$$\mathscr{H}_y = \frac{2j\beta\dfrac{\pi}{a}\mathscr{H}_{z0}\sin\dfrac{\pi y}{a}\cos\beta z}{k^2} \tag{11}$$

We will now calculate the stored energy.

$$W_{\text{stored}} = \frac{1}{2}\epsilon \int_0^c \int_0^a \int_0^b |\mathscr{E}_x|^2\, dx\, dy\, dz$$
$$= \frac{1}{2}\epsilon \int_0^c \int_0^a \int_0^b \frac{4\omega^2\mu^2\left(\dfrac{\pi}{a}\right)^2 \mathscr{H}_{z0}^2 \sin^2\dfrac{\pi y}{a} \sin^2\beta z\, dx\, dy\, dz}{(\pi/a)^4}$$
$$= \frac{1}{2}\left(\frac{\epsilon\omega^2\mu^2}{\pi^2/a^2}\right)|\mathscr{H}_{z0}|^2 abc$$

$$= \frac{\mu}{2}\left(\frac{\omega^2\mu\epsilon}{\pi^2/a^2}\right)|\mathcal{H}_{z0}|^2 abc$$

$$= \frac{\mu}{2}\left(1 + \frac{a^2}{c^2}\right)|\mathcal{H}_{z0}|^2 abc \qquad \text{(J)} \tag{12}$$

after substituting (4) for $\omega^2\mu\epsilon$ for $m = 1, n = 0, p = 1$. The power loss in the walls is now calculated.

$$P_{1+2} = 2(\tfrac{1}{2}R_s)\int_0^b\int_0^c |\mathcal{H}_z|^2\,dz\,dx$$

$$= R_s\int_0^b\int_0^c 4|\mathcal{H}_{z0}|^2\cos^2\frac{\pi y}{a}\sin^2\beta z\,dz\,dx$$

$$= \frac{4R_s|\mathcal{H}_{z0}|^2}{2}bc = 2R_s|\mathcal{H}_{z0}|^2 bc \qquad \text{(W)} \tag{13}$$

$$P_{3+4} = R_s\int_0^c\int_0^a [|\mathcal{H}_y|^2 + |\mathcal{H}_z|^2]\,dy\,dz$$

$$= R_s\int_0^c\int_0^a\left[\frac{4\left(\frac{\pi^2}{c^2}\right)\frac{\pi^2}{a^2}|\mathcal{H}_{z0}|^2\sin^2\frac{\pi y}{a}\cos^2\frac{\pi z}{c}}{(\pi^2/a^2)^2} + 4|\mathcal{H}_{z0}|^2\cos^2\frac{\pi y}{a}\sin^2\frac{\pi z}{c}\right]dy\,dz$$

$$= 4R_s|\mathcal{H}_{z0}|^2\left[\frac{1}{4}\left(\frac{(\pi/c)^2}{(\pi/a)^2}\right)ac + \frac{1}{4}ac\right] \qquad \text{(W)} \tag{14}$$

$$P_{5+6} = R_s\int_0^b\int_0^a |\mathcal{H}_y|^2\,dy\,dx$$

$$= R_s\int_0^b\int_0^a \frac{4\frac{\pi^2}{c^2}\left(\frac{\pi^2}{a^2}\right)|\mathcal{H}_{z0}|^2\sin^2\frac{\pi y}{a}\cos^2\frac{\pi z}{c}\,dy\,dx}{(\pi^2/a^2)^2}$$

$$= 2R_s\frac{\pi^2}{c^2}\left(\frac{|\mathcal{H}_{z0}|^2}{\pi^2/a^2}\right)ab \qquad \text{(W)} \tag{15}$$

$$W_{L\text{ total}} = TR_s|\mathcal{H}_{z0}|^2\left(2bc + \frac{a^2}{c^2}ac + ac + 2\frac{a^2}{c^2}ab\right) \qquad \text{(J)} \tag{16}$$

and

$$Q_{\text{TE}_{101}} = \frac{2\pi W_{\text{stored}}}{W_{L\text{ total}}}$$

$$= \frac{\pi\mu\left(1 + \frac{a^2}{c^2}\right)\times \text{vol.}}{R_sT\left(2bc + ac + \frac{a^2}{c^2}ac + 2\frac{a^2}{c^2}ab\right)}$$

Now

$$T = \frac{1}{f}, \qquad \pi f\mu\delta = R_s$$

and

$$Q_{\mathrm{TE}_{101}} = \frac{\pi f \mu \left(1 + \dfrac{a^2}{c^2}\right) abc}{\pi f \mu \delta \left(2bc + ac + \dfrac{a^2}{c^2} ac + 2\dfrac{a^2}{c^2} ab\right)}$$

or

$$\boxed{Q_{\mathrm{TE}_{101}} = \frac{\left(1 + \dfrac{a^2}{c^2}\right) \times \text{vol.}}{\delta \left(2bc + ac + \dfrac{a^2}{c^2} ac + 2\dfrac{a^2}{c^2} ab\right)}} \tag{17}$$

If $a = b = c$, then

$$Q_{\mathrm{TE}_{101}} = \frac{2 \times \text{vol.}}{\delta \times \text{surface area}} \tag{18}$$

Problem 13.6-1 A cubical resonant cavity is constructed of copper, $\sigma = 5.8 \times 10^7\ \mathrm{S\,m^{-1}}$. If the edges are 6 cm, find the Q of the air-filled cavity at its resonant frequency in the TE_{101} mode.

Problem 13.6-2 Find the Q at the resonant frequency for an air-filled copper cavity in the TE_{101} mode if the dimensions are 12 cm $\times$ 6 cm $\times$ 12 cm.

13.7 OPTICAL FIBERS (LIGHTGUIDES)

In the study of waveguides, we have studied metal waveguides that may be used up to 100 GHz. Dielectric waveguides are used or proposed for frequencies in the range 100 to 1000 GHz (wavelengths of 3 to 0.3 mm). Dielectric waveguides are not new; the concept dates back to the 1940s. But optical-fiber technology developed independently of dielectric waveguides, even though the permissible modes are the same when guide dimensions are scaled to wavelength.

The fact that light can be guided by a fiber or rod of glass or plastic or even by a stream of liquid has been common knowledge for many decades or even centuries, but optical fibers for communications were first developed in the 1970s, following widespread use of lightguides in flexible probes used for medical applications. Prior to the breakthrough in the 1970s, attenuation to light waves made lightguide fibers impractical for communications over distances on the order of 1 km or greater, but by the end of the decade, attenuation for some types had been reduced to less than 0.5 dB km^{-1}.

The portion of the spectrum being utilized by optical fibers is for wavelengths from about 0.7 to 1.6 μm, or for the red part of the visible spectrum to the infrared. The wavelengths that are popular depend on the availability of LED and laser sources, the availability of sensitive detectors, and the attenuation versus wavelength characteristic of the fiber. Referring to Fig. 13-8, which is an attenuation curve for a hypothetical "typical" fiber, the most used wavelengths were at first the region 0.83 to 0.85 μm in

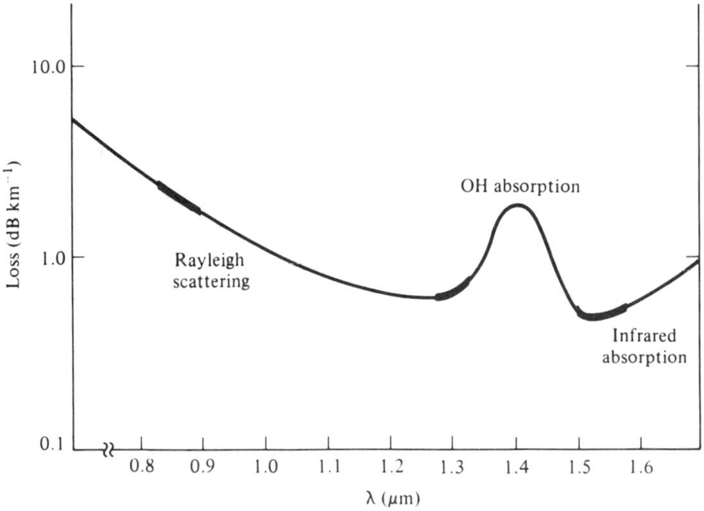

Figure 13-8 Loss versus wavelength for typical optical fiber. Emphasized portions of the attenuation curve represent wavelengths commonly used.

the visible red region. But because of lower attenuation at longer wavelengths, the "windows" at around 1.3 μm and 1.55 μm have become more important.[2,3]

Communications fibers are of three basic designs:

1. Step-index, multimode
2. Graded-index, multimode
3. Step-index, single mode

Referring to Fig. 13-9(a), the step-index construction consists of a glass fiber core on the order of 50 μm in diameter, surrounded by a cladding of lower dielectric constant (lower index of refraction) such that total reflection takes place at the interface between the core and the cladding for all modes accepted. Because the paths are longer for modes having the most reflections, the group velocities are also slower for these modes. The spread in group velocities and the pulse dispersion resulting therefrom places a limit on spacing between pulses for good resolution and limits the maximum data rate.

In the graded-index construction, the index of refraction of the core changes gradually with distance from the center, enabling the paths to bend, as shown in Fig. 13-9(b). Much of the longer path lengths are through lower index of refraction material and hence the increased velocity over this part of the path compensates somewhat for the longer path lengths, and the spread in group velocities between modes is not as great as for the step-index construction.

[2] Stewart E. Miller, "Overview of Telecommunications via Optical Fibers," *Proc. IEEE*, Vol. 68, October 1980, pp. 1173–1174.

[3] Tingye Li, "Structures, Parameters, and Transmission Properties of Optical Fibers," *Proc. IEEE*, Vol. 68, October 1980, pp. 1175–1180.

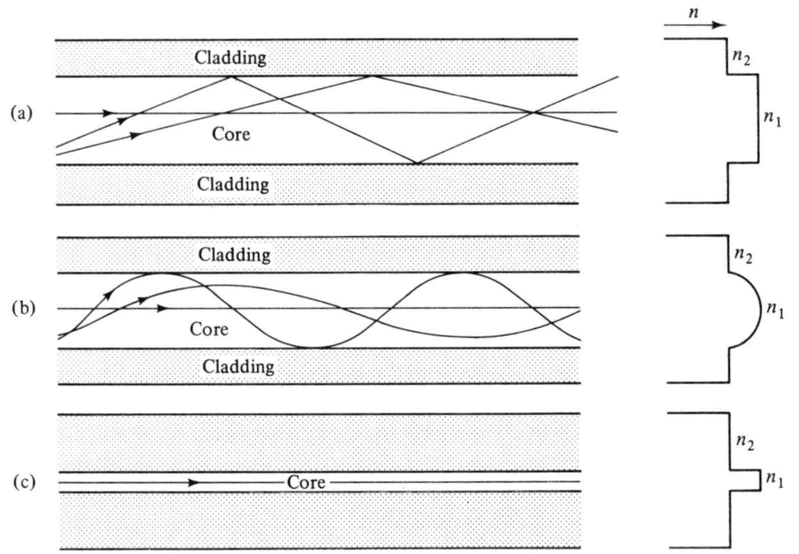

Figure 13-9 (a) Step-index, multimode fiber. (b) Graded-index, multimode fiber. (c) Step-index, single-mode fiber.

For single-mode fibers, Fig. 13-9(c), the core diameter may be as small as 5 μm, and the cladding has only a slightly lower index of refraction than the core. As long as the operating wavelength is longer than the cutoff wavelengths for the higher-order modes, only a single mode, the HE_{11}, will propagate. The advantage of the single-mode fiber is the high data rate made possible by high pulse resolution resulting from a small dispersion with frequency.[4]

Referring to Fig. 13-10 and using (11.10-13), the critical angle for total reflection from the core–cladding interface is

$$\sin \phi_c = \frac{n_2}{n_1} \tag{1}$$

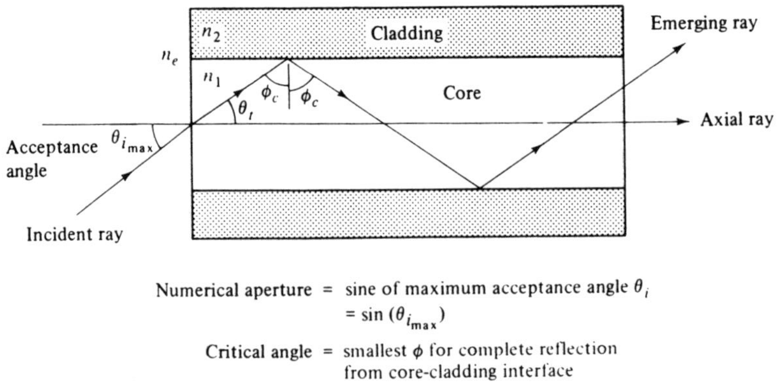

Figure 13-10 Illustration of some basic definitions concerning optical fibers.

[4] Donald B. Keck, "Single Mode Fibers Outperform Multimode Cables," *IEEE Spectrum*, March 1983, pp. 30–37.

where n_1 and n_2 are the indexes of refraction for the core and cladding, respectively. The index of refraction for the medium from which the ray is arriving is n_e. The relationship between the angle of incidence and the angle of transmission into the core (Snell's law) is given by (11.10-14)

$$\frac{\sin \theta_i}{\sin \theta_t} = \sqrt{\frac{\epsilon_1}{\epsilon_e}} = \frac{n_1}{n_e} \tag{2}$$

From the geometry it is clear that $\cos \phi_c = \sin \theta_t$, and substituting (1) into (2) and applying a trigonometry identity yields

$$\begin{aligned} \sin \theta_i &= \frac{n_1}{n_e}\sqrt{1 - \sin^2 \phi_c} = \frac{n_1}{n_e}\sqrt{1 - (n_2/n_1)^2} \\ &= \frac{n_1}{n_e}\sqrt{\frac{(n_1 - n_2)(n_1 + n_2)}{n_1^2}} \approx \frac{n_1}{n_e}\sqrt{\frac{2n_1\Delta}{n_1}} \\ &= \frac{n_1}{n_e}\sqrt{2\Delta} \end{aligned} \tag{3}$$

The angle given by (3) is the maximum incidence angle for which the ray will be accepted, or the *acceptance angle,* and Δ is the approximate fractional difference in index of refraction between the core and the cladding, and is on the order of 0.01 or less. The *numerical aperture* is the sine of the maximum acceptance angle.

The modes for optical fibers are determined in the same manner as for metal waveguides, but because of cylindrical geometry, the radial standing wave patterns result from Bessel functions instead of sinusoids. The reader is referred to another source for details.[5] For single-mode propagation only,

$$V < 2.405 \tag{4}$$

where

$$V = \frac{2\pi a n}{\lambda}\sqrt{2\Delta} \tag{5}$$

where a is the core radius, λ is the wavelength, n is the index of refraction of the core, and Δ is as defined above. Unlike the metal waveguide, the HE_{11} mode in optical fiber does not have a cutoff frequency.

Sources of loss in optical fibers include absorption losses due to impurities in the glass, scattering due to variations in density within the material, Rayleigh scattering, scattering due to minute sharp bends created during the cabling process (called *microbends*), and infrared absorption. The theoretical minimum attenuation, where Rayleigh scattering and infrared absorption are equal, occurs at a wavelength of approximately 1.55 μm. Between 1.3 and 1.55 μm is a peak in the attenuation curve of Fig. 13-8 due to the OH content of the core. Except for the OH peak, the measured attenuation is close to the theoretically expected value. A distance record in the laboratory was established in 1984 by transmitting 2 Gb s^{-1} over a distance of 130 km without using a repeater. The core loss was about 0.19 dB km^{-1} at 1.5 μm.[6]

[5] Michael K. Barnoski, ed., *Fundamentals of Optical Fiber Communications,* Academic Press, Inc., New York, 1981, Ch. 1 by D. B. Keck.

[6] Trudy E. Bell, "Advances in Fiber-Optic Communications," *IEEE Spectrum,* January 1985, pp. 56–57.

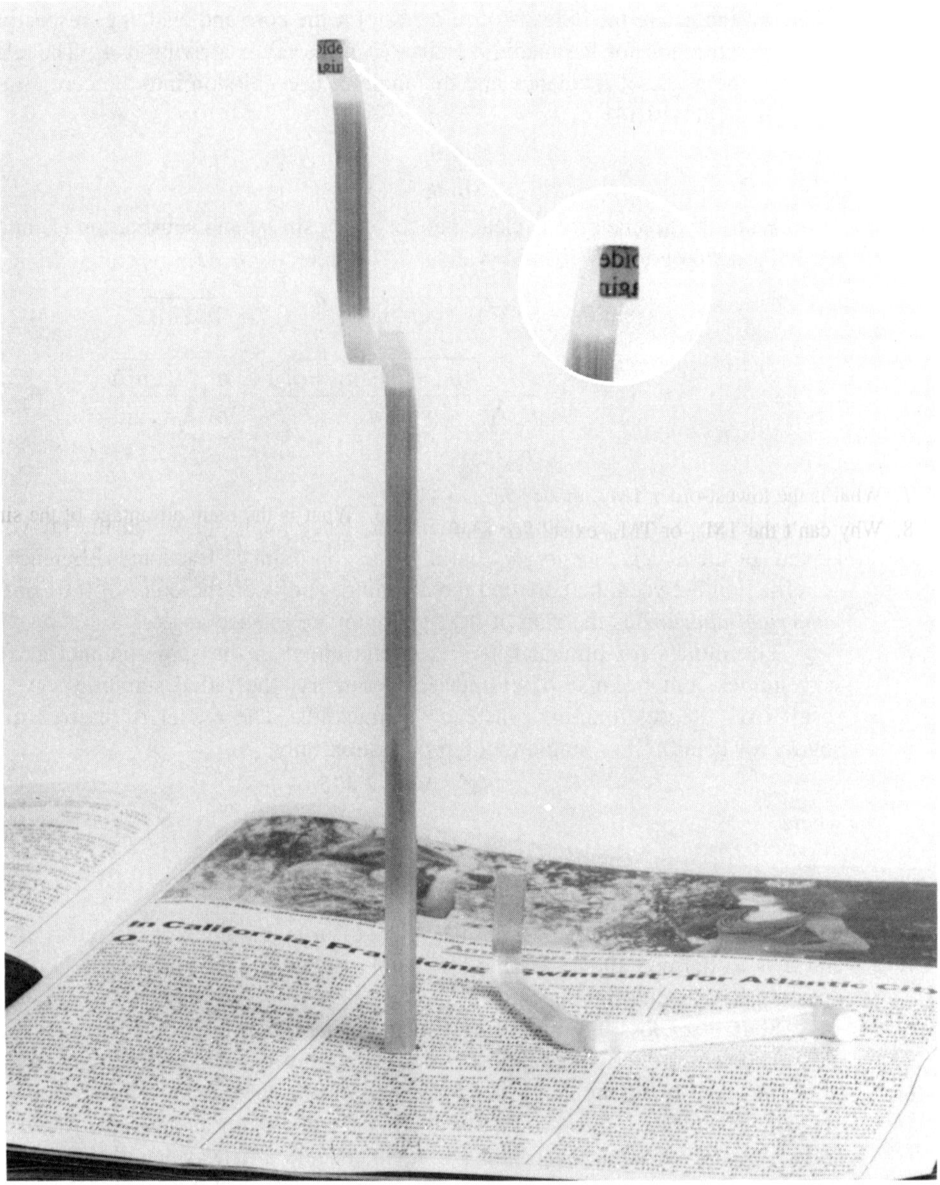

Figure 13-11 Fiber-optic coherent bundle used to transmit an image from one end of the rod to the other. (Courtesy of Schott Fiber Optics.)

Besides being used for communications links, optical fibers are finding wide application in various sensing probes, including acoustic, temperature, electric field, and magnetic field.[7] Optic-fiber bundles or ropes are also being used as image conduits to

[7] T. G. Giallorenzi, J. A. Bucaro, A. Dandridge, G. H. Sigel, Jr., J. H. Cole, S. C. Rashleigh, and R. G. Priest, "Optical Fiber Sensor Technology," *IEEE J. Quantum Electron.*, Vol. QE-18, April 1982, pp. 626–665.

transmit images as indicated in Fig. 13-11. Optical-fiber probes have the obvious advantage that they do not significantly disturb the ambient electric and magnetic fields, and they are not subject to induced emf's from the ambient field.

REVIEW QUESTIONS

1. Is a waveguide a transmission line? *Sec. 13.1*
2. What is meant by a TE mode? *Sec. 13.2*
3. Compare phase velocity, plane wave velocity, and group velocity for a waveguide. *Sec. 13.2*
4. What is meant by cutoff frequency? *Sec. 13.2*
5. What is the lowest-order TE_{mn} mode? *Sec. 13.3*
6. How does β behave near the cutoff frequency? *Sec. 13.3*
7. What is the lowest-order TM_{mn} mode? *Sec. 13.4*
8. Why can't the TM_{01} or TM_{10} exist? *Sec 13.4*
9. What causes attenuation in an air-filled waveguide? How does this attenuation vary with the width of the side walls for a TE_{10} mode? *Sec. 13.5*
10. Can the TE_{100} mode exist in a resonant cavity? *Sec. 13.6*
11. What is the general expression for Q? *Sec. 13.6*
12. What are the three types of optical fibers discussed? *Sec. 13.7*
13. What is the main advantage of the single-mode fiber? *Sec. 13.7*

PROBLEMS

13-1. A planar transmission line is used to "guide" frequencies ranging from dc to 10 GHz. If the spacing between conductors is 2 cm, determine the possible mode or modes of the guided wave over that frequency range, i.e., define the frequency range for each mode.

13-2. For Prob. 13-1, find v_p and v_g for a frequency of: (a) 5 GHz; (b) 9 GHz (all possible modes).

13-3. Find the frequency, relative to the cutoff frequency, for which the waveguide wavelength λ_g is exactly twice the plane wave wavelength λ.

13-4. An air-filled waveguide is carrying energy at a frequency of $2f_c$. For a section of waveguide λ (m) in length, find: (a) the phase shift for a continuous wave (CW); (b) the time delay for short pulses.

13-5. Using material in Sec. 13-3, show that

$$\frac{1}{\lambda^2} = \frac{1}{\lambda_c^2} + \frac{1}{\lambda_g^2}$$

where λ = plane wave wavelength and $k = 2\pi/\lambda_c$.

13-6. Find the wave impedance of a 1.9 cm × 0.95 cm air-filled waveguide operating at 10.6 GHz in (a) the TE_{10} mode; (b) at 18 GHz in the TM_{11} mode.

13-7. (a) On log-log coordinates, plot the phase velocity relative to U as a function of f/f_c from just above the cutoff frequency to $f = 30f_c$. (b) Plot β relative to $\omega\sqrt{\mu\epsilon}$ as a function of f/f_c on the same plot.

13-8. For a 100-ft section of air-filled rectangular waveguide operating in the TE_{10} mode, what is the time delay difference for a 10-GHz pulse and a 7-GHz pulse if the dimensions of the guide are 0.9 in. × 0.4 in.?

13-9. For an air-filled waveguide whose inside dimensions are 1.295 cm. × 0.648 cm, find the cutoff frequencies for the following modes:

$$TE_{10}, TE_{01}, TE_{11}, TE_{20}, TE_{21}, TM_{11}, TM_{21}$$

13-10. From a general expression for β as derived from (13.3-46), show that

$$\frac{d\omega}{d\beta} = \frac{1}{\sqrt{\mu\epsilon}}\sqrt{1 - \left(\frac{f_c}{f}\right)^2} = v_g$$

the group velocity.

13-11. For a 0.622 in. × 0.311 in. aluminum waveguide, calculate the attenuation in decibels per 100 ft for a frequency of 12 GHz for the TE_{10} mode.

13-12. Repeat Prob. 13-11 for a brass waveguide, $\sigma = 2.56 \times 10^7$ S m^{-1}.

13-13. For a 0.510 in. × 0.255 in. brass waveguide, calculate the attenuation in decibels per 100 ft for a frequency of 14 GHz for the TE_{10} mode.

13-14. Repeat Prob. 13.6-2 for a cavity scaled such that the resonance frequency is twice that of Prob. 13.6-2, for: (a) a cavity of aluminum; (b) a cavity of graphite.

13-15. Calculate the acceptance angle for a core having an index of refraction of 1.46, where $\Delta = 0.01$ and the lightguide is in air ($n_e = 1$). Also determine the critical angle ϕ_c referred to in Fig. 13-10.

14

Antennas

14.1 INTRODUCTION

The antenna uses voltage and current from a transmission line or the $\overline{E}$ and $\overline{H}$ fields from a waveguide to "launch" an electromagnetic (EM) wavefront into free space or into the local environment. The antenna acts as a transducer to match the transmission line or waveguide to the medium surrounding the antenna. The launching process is known as *radiation*, and the launching antenna is the transmitting antenna. If a wavefront is intercepted by an antenna, some power is absorbed from the wavefront, and the antenna acts as a receiving antenna.

As we shall see when we study the dipole antenna in some detail, the fields in the immediate vicinity of the antenna include more than just the radiation field, but at a great distance from a transmitting antenna the radiation $\overline{E}$ and $\overline{H}$ fields are the only significant fields. The radiated power spreads as the inverse square of distance from the antenna.

There are many basic types of antenna elements such as the dipole, horn, slot, spiral, long wire, and monopole, and there are also many different types of systems where these elements are arranged in some form of an array, either fixed or electronically controlled. Among the most directive antennas are those used for radio astronomy where 0.1° beamwidth is typical.

The *polarization* of an antenna is the direction of the $\overline{E}$ field for maximum reception by the receiving antenna or the direction of the $\overline{E}$ field transmitted by the antenna. Amplitude-modulated (AM) broadcast band antennas are vertical monopoles or loops and are vertically polarized. Antennas for television and frequency-modulated

(FM) broadcast are horizontally polarized and typically consist of an array of horizontal dipoles. Parabolic dish antennas are used for satellite communication and may use both vertical and horizontal polarization.

14.2 CONCEPT OF GAIN AND BEAMWIDTH

Antenna gain is a measure of the concentration of power density in the radiated wavefront in a given direction. Gain is usually compared to an isotropic antenna, one which radiates equally in all directions and thus has a gain of unity. A rigorous definition of antenna gain is

$$G \triangleq \frac{\dfrac{\text{maximum power radiated}}{\text{unit solid angle}}}{\dfrac{\text{total power delivered to antenna}}{4\pi}} \tag{1}$$

A similar but slightly different definition holds for directivity:

$$G_D \triangleq \frac{\dfrac{\text{maximum power radiated}}{\text{unit solid angle}}}{\dfrac{\text{total power radiated}}{4\pi}} \tag{2}$$

Antenna gain is somewhat less than directivity because of the losses that occur in the antenna.

An ideal antenna would be one for which all the radiated power would be contained within the beam solid angle and for which the losses in the antenna are zero. In this case, the gain and directivity are the same and would be given by

$$G = G_D = \frac{4\pi}{\Omega_B} \tag{3}$$

For practical antennas of moderate to high gain, (3) is a good rule-of-thumb formula, where Ω_B is the beam solid angle (in steradians) measured between 3-dB points.[1] An ideal antenna is represented in Fig. 14-1. All the power is concentrated in a single beam, and the power density is uniform over this beam.

Example 1

An antenna has a gain of 44 dB. Assuming that the main beam of the antenna is circular in cross section, find the beamwidth θ_B.

[1]Actually, (3) tends to give values for gain or directivity that are too high by a factor of 2 or 3-dB when Ω_B is the 3-dB solid angle. In many cases, (3) is more accurate if Ω_B is measured between first nulls on either side of beam center.

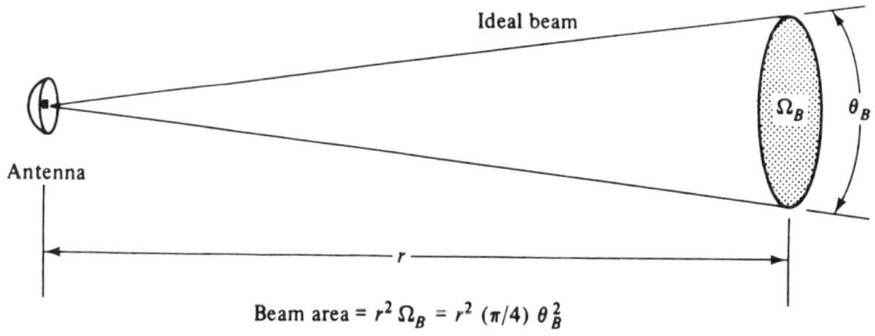

Figure 14-1 Illustration of the concept of a beam solid angle for an ideal antenna.

Solution. The gain is given by

$$G = 10^{44/10} = 2.51 \times 10^4$$

and from (3),

$$\Omega_B = \frac{4\pi}{2.51 \times 10^4} = 5 \times 10^{-4} \text{ steradian (sr)}$$

The beamwidth, assuming a circular cross section, is

$$\theta_B = \sqrt{4/\pi}\,\sqrt{\Omega_B} = 0.025 \text{ rad} = 1.45°$$

that is,

$$\Omega_B = \left(\frac{\pi}{4}\right)\theta_B^2 \tag{4}$$

Just as arc length is measured by the product of the angle subtended by the arc and the radius, the area illuminated by the antenna beam is the product of the beam solid angle and the square of the distance from antenna to area illuminated.

Example 2

A satellite power station in a synchronous orbit is beaming microwave power to earth. If the beamwidth of the transmitting antenna on the satellite is 0.1°, and the distance from the earth's surface is 36,000 km, what is the size of the spot illuminated by the antenna on the earth's surface? Assume a circular spot or beam area.

Solution. The beamwidth in radians is

$$\theta_B = \frac{\pi}{180°} \times 0.1° = 1.745 \times 10^{-3} \text{ rad}$$

and from (4),

$$\Omega_B = \pi/4 \times (1.745 \times 10^{-3})^2$$
$$= 2.39 \times 10^{-6} \text{ sr}$$

Then the area of the spot is $r^2\Omega_B$, or

$$A_{\text{spot}} = 2.39 \times 10^{-6}(36{,}000 \times 10^3)^2 = 3.10 \times 10^9 \text{ m}^2$$

Problem 14.2-1 What is the gain of an antenna, in decibels, that has a beamwidth of (a) 1°; (b) 2°?

Problem 14.2-2 Find the power density, in W m^{-2}, at a distance of 30 km from an antenna that is radiating 5 kW with a directivity of 37 dB.

Problem 14.2-3 What directivity is required for the power density of Prob. 14.2-2 to equal 2.5 mW m^{-2}?

Problem 14.2-4 Convert 10 mW cm^{-2} to W m^{-2}.

14.3 THE ELEMENTAL DIPOLE

A highly directive antenna may have a single radiating element such as a dipole, backed by a parabolic reflecting dish, or it may consist of an array of elements. One of the most commonly used antenna elements is the dipole. It is also the easiest to analyze. In this section we will derive the electric and magnetic fields for a very short dipole. We shall then progress to the more practical half-wave dipole.

Let us consider a very short ($dz \ll \lambda$) length of wire carrying a sinusoidal current, as shown in Fig. 14-2. The approach will be to find the vector magnetic potential (see Sec. 8.9) from the current in the elemental length dz, and from $\overline{A}$ to find $\overline{H}$. Then, using one of Maxwell's curl equations, $\overline{E}$ will be obtained.

The fields at point P are *retarded fields;* i.e., they are due to a source current that existed r/U (s) earlier. Then, using (8.9-16), the retarded vector magnetic potential at point P is

$$\overline{A} = \frac{\mu I_0 \cos \omega\left(t - \dfrac{r}{U}\right) dz}{4\pi r}\,\hat{z}$$

or

$$\boxed{\mathcal{A}_z = \frac{\mu \mathcal{I}_0 e^{-j\beta r}\, dz}{4\pi r} \qquad (\text{Wb m}^{-1})} \tag{1}$$

where $\omega/U = \beta$.

Transforming to spherical coordinates, we find the $\mathcal{A}_r$ and $\mathcal{A}_\theta$ components:

$$\mathcal{A}_r = \mathcal{A}_z \cos\theta = \frac{\mu \mathcal{I}_0 e^{-j\beta r}\, dz \cos\theta}{4\pi r} \tag{2}$$

and

$$\mathcal{A}_\theta = -\mathcal{A}_z \sin\theta = \frac{-\mu \mathcal{I}_0 e^{-j\beta r}\, dz \sin\theta}{4\pi r} \tag{3}$$

$$\mathcal{A}_\phi = 0 \tag{4}$$

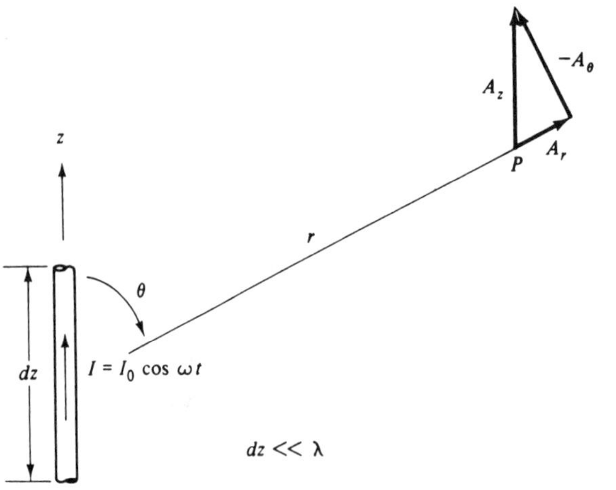

Figure 14-2 Illustration of the elemental or short dipole. The method of connection to the dipole is not shown.

We now find $\overline{\mathscr{H}}$.

$$\overline{\mathscr{H}} = \frac{1}{\mu}\nabla \times \overline{\mathscr{A}} \tag{5}$$

or

$$\mathscr{H}_\phi = \frac{1}{\mu r}\left[\frac{\partial(r\mathscr{A}_\theta)}{\partial r} - \frac{\partial \mathscr{A}_r}{\partial \theta}\right] \tag{6}$$

since

$$\frac{\partial \mathscr{A}_\theta}{\partial \phi} = \frac{\partial \mathscr{A}_r}{\partial \phi} = 0$$

Then

$$\boxed{\begin{aligned}\mathscr{H}_\phi &= \frac{\mathscr{I}_0\,dz}{4\pi r}\left[\frac{\partial}{\partial r}(-e^{-j\beta r}\sin\theta) - \frac{\partial}{\partial\theta}\left(\frac{e^{-j\beta r}}{r}\cos\theta\right)\right] \\ &= \frac{\mathscr{I}_0\,dz\,e^{-j\beta r}\sin\theta}{4\pi}\left(j\frac{\beta}{r} + \frac{1}{r^2}\right) \qquad (\text{A m}^{-1})\end{aligned}} \tag{7}$$

for the total $\overline{\mathscr{H}}$ field from the elemental dipole. To find $\overline{\mathscr{E}}$, we note that

$$\nabla \times \overline{\mathscr{H}} = j\omega\epsilon\overline{\mathscr{E}}$$

or

$$\overline{\mathscr{E}} = \frac{1}{j\omega\epsilon}\nabla \times \overline{\mathscr{H}} \tag{8}$$

and substituting (7) into (8),

$$\overline{\mathscr{E}} = \frac{1}{j\omega\epsilon}\left[\frac{1}{r\sin\theta}\left(\frac{\partial}{\partial\theta}\right)(\mathscr{H}_\phi\sin\theta)\hat{r} - \frac{1}{r}\left(\frac{\partial}{\partial r}\right)(r\mathscr{H}_\phi)\hat{\theta}\right]$$

Then

$$\mathscr{E}_r = \frac{\mathscr{I}_0\,dz\,e^{-j\beta r}}{4\pi j\omega\epsilon}\left(\frac{1}{r\sin\theta}\right)\left(\frac{\partial}{\partial\theta}\right)\left[\sin^2\theta\left(j\frac{\beta}{r} + \frac{1}{r^2}\right)\right]$$

and

$$\boxed{\mathscr{E}_r = \frac{\mathscr{I}_0\,dz\,e^{-j\beta r}\cos\theta}{2\pi\epsilon}\left[\frac{1}{Ur^2} + \frac{1}{j\omega r^3}\right] \qquad (\text{V m}^{-1})} \tag{9}$$

$$\begin{aligned}\mathscr{E}_\theta &= \frac{\mathscr{I}_0\,dz}{4\pi j\omega\epsilon}\left(-\frac{1}{r}\right)\left(\frac{\partial}{\partial r}\right)\left[r\sin\theta e^{-j\beta r}\left(j\frac{\beta}{r} + \frac{1}{r^2}\right)\right] \\ &= \frac{\mathscr{I}_0\,dz\,e^{-j\beta r}\sin\theta}{4\pi j\omega\epsilon}\left[-\frac{\beta^2}{r} + j\frac{\beta}{r^2} + \frac{1}{r^3}\right]\end{aligned}$$

$$\boxed{\mathscr{E}_\theta = \frac{\mathscr{I}_0\,dz\,e^{-j\beta r}\sin\theta}{4\pi}\sqrt{\frac{\mu}{\epsilon}}\left[j\frac{\beta}{r} + \frac{1}{r^2} + \frac{1}{j\beta r^3}\right] \qquad (\text{V m}^{-1})} \tag{10}$$

We note that if we look at only the $1/r$ components of $\mathscr{E}_\theta$ and $\mathscr{H}_\phi$ that

$$\frac{\mathscr{E}_\theta\left(\frac{1}{r}\right)}{\mathscr{H}_\phi\left(\frac{1}{r}\right)} = \sqrt{\frac{\mu}{\epsilon}} \qquad (\Omega) \tag{11}$$

which is the intrinsic impedance of the medium in which the dipole is immersed.

The expressions (7) and (10) contain all the components of the $\overline{\mathscr{E}}$ and the $\overline{\mathscr{H}}$ fields from the elemental dipole, but these are not all radiation fields. By definition of radiated power, the radiated power cannot decrease with distance from the source (we are assuming a lossless medium). Then the radiated power density must vary as $1/r^2$ and

$$\begin{aligned} |\overline{\mathscr{E}}_{\text{rad}} \times \overline{\mathscr{H}}_{\text{rad}}| &= \text{power density radiated} \\ &= \left|\mathscr{E}_\theta\left(\frac{1}{r}\right)\right| \left|\mathscr{H}_\phi\left(\frac{1}{r}\right)\right| \qquad (\text{W m}^{-2}) \end{aligned} \tag{12}$$

Then

$$|\overline{\mathscr{E}}_{\text{rad}}| = \left|\mathscr{E}_\theta\left(\frac{1}{r}\right)\right| = \left|\frac{\mathscr{I}_0\, dz\, e^{-j\beta r}\beta}{4\pi r}\sqrt{\frac{\mu}{\epsilon}}\sin\theta\right| \tag{13}$$

and

$$|\overline{\mathscr{H}}_{\text{rad}}| = \left|\mathscr{H}_\phi\left(\frac{1}{r}\right)\right| = \left|\frac{\mathscr{I}_0\, dz\, e^{-j\beta r}\beta\sin\theta}{4\pi r}\right| \tag{14}$$

and the average power density radiated is

$$\mathscr{P}_{\text{av}} = \frac{1}{2}\left|\mathscr{E}_\theta\left(\frac{1}{r}\right)\right| \left|\mathscr{H}_\phi\left(\frac{1}{r}\right)\right| = \left(\frac{1}{2}\right)\frac{\mathscr{I}_0^2(dz)^2\sin^2\theta\sqrt{\mu/\epsilon}\,\beta^2}{(4\pi)^2r^2} \qquad (\text{W m}^{-2}) \tag{15}$$

and the total average power radiated is, after replacing β by $2\pi/\lambda$,

$$P_{\text{av rad}} = \left(\frac{1}{2}\right)\frac{\mathscr{I}_0^2\, dz^2\sqrt{\mu/\epsilon}}{4r^2\lambda^2}\int_{\phi=0}^{2\pi}\int_{\theta=0}^{\pi}\sin^3\theta\, r^2\, d\theta\, d\phi \tag{16}$$

$$= \frac{\pi\mathscr{I}_0^2(dz)^2\sqrt{\mu/\epsilon}}{3\lambda^2} \qquad (\text{W}) \tag{17}$$

If we define radiation resistance R_{rad} by

$$\boxed{P_{\text{av rad}} = \tfrac{1}{2}\mathscr{I}_0^2 R_{\text{rad}} \qquad (\text{W})} \tag{18}$$

then, from (17),

$$R_{\text{rad}} = \frac{2\mathscr{I}_0^2(dz)^2\sqrt{\frac{\mu}{\epsilon}}\left(\frac{\pi}{3}\right)}{\mathscr{I}_0^2\lambda^2}$$

$$= \frac{2}{3\lambda^2}\pi(dz)^2\sqrt{\frac{\mu}{\epsilon}} \qquad (\Omega) \tag{19}$$

In free space, $\sqrt{\mu/\epsilon} = \sqrt{\mu_0/\epsilon_0} = 120\pi$, and

$$R_{\text{rad}} = \frac{2}{3}(dz)^2\frac{120\pi^2}{\lambda^2} = 80\pi^2\frac{(dz)^2}{\lambda^2} \qquad (\Omega) \tag{20}$$

and

$$\frac{|\mathscr{E}_{\theta\ \mathrm{rad}}|}{|\mathscr{H}_{\phi\ \mathrm{rad}}|} = \sqrt{\frac{\mu_0}{\epsilon_0}} = 120\pi = 377\ \Omega \tag{21}$$

As a receiving antenna, the elemental dipole has an *effective receiving area* A_e, and the power received by the antenna is

$$P_R = P_{\mathrm{rec}} = A_e \mathscr{P}_{\text{incoming wavefront}} \qquad \text{(W)} \tag{22}$$

The voltage induced in the length of wire dz is

$$\mathscr{V} = \mathscr{E}\, dz \qquad \text{(V)} \tag{23}$$

where $\mathscr{E}$ is the electric field strength of the incident wave tangent to the wire.

Referring to Fig. 14-3, for maximum power transfer the antenna must be terminated in an impedance which is the conjugate of the antenna impedance. That is,

$$Z_T = R_{\mathrm{rad}} - jX_A = Z_A^* \tag{24}$$

where X_A is the reactance of the antenna element.

Then the power received by the termination is

$$P_R = \frac{\mathscr{V}_{\mathrm{rms}}^2}{4R_{\mathrm{rad}}} = \frac{\mathscr{E}_{\mathrm{rms}}^2 (dz)^2}{4R_{\mathrm{rad}}}$$

$$= A_e \mathscr{P}_{\mathrm{av}} = A_e \frac{\mathscr{E}_{\mathrm{rms}}^2}{\sqrt{\mu/\epsilon}} \tag{25}$$

Now, since

$$R_{\mathrm{rad}} = \frac{\dfrac{2}{3}\pi (dz)^2 \sqrt{\dfrac{\mu}{\epsilon}}}{\lambda^2}$$

then from (25),

$$A_e = \frac{\sqrt{\dfrac{\mu}{\epsilon}}(dz)^2 \lambda^2}{4\left(\dfrac{2}{3}\pi (dz)^2 \sqrt{\dfrac{\mu}{\epsilon}}\right)} = \frac{3\lambda^2}{8\pi} \qquad (\mathrm{m}^2) \tag{26}$$

for a lossless elemental dipole.[2]

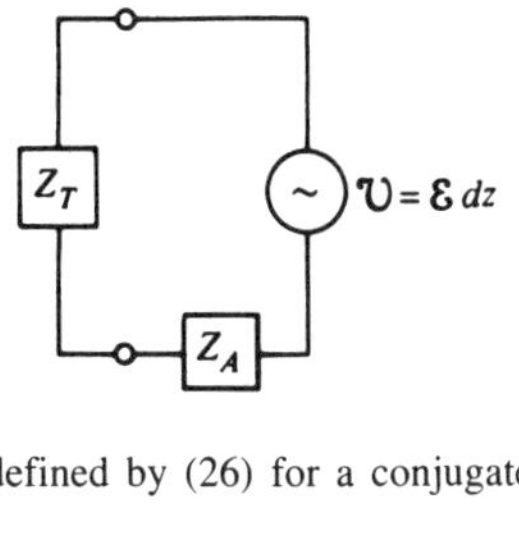

Figure 14-3 Equivalent circuit of the receiving dipole connected to a terminating impedance Z_T.

[2] Note that A_e has been defined by (26) for a conjugate matched load, i.e., A_e is a maximum effective area.

The directivity of the elemental dipole of length dz is computed by substituting (13) and (14) into (14.2-2):

$$G_D = \frac{r^2 1/2(|E_{\theta\,\text{rad}}|\,|\mathcal{H}_{\phi\,\text{rad}}|)_{\max}}{\dfrac{P_{\text{av rad}}}{4\pi}}$$

$$G_D = \frac{\dfrac{r^2}{2}\left(\dfrac{\mathcal{I}_0^2(dz)^2}{(4\pi)^2}\right)\sqrt{\dfrac{\mu}{\epsilon}}\left(\dfrac{2\pi}{\lambda r}\right)^2}{\dfrac{\dfrac{\pi}{3}\mathcal{I}_0^2(dz)^2\sqrt{\dfrac{\mu}{\epsilon}}}{\lambda^2 4\pi}} = \frac{3}{2} \tag{27}$$

Then

$$\frac{G_D}{A_e} = \frac{3/2}{(3/8\pi)\lambda^2} = \frac{4\pi}{\lambda^2}$$

and

$$\boxed{G_D = \frac{4\pi A_e}{\lambda^2}} \tag{28}$$

or

$$\boxed{G = \frac{4\pi A_e}{\lambda^2}} \tag{29}$$

although (28) is strictly true only for a lossless antenna.

Equations (28) and (29) were derived for an elemental dipole, but they happen to be general relations that hold for all antennas. Equation (29) is more generally used, and either the gain or the effective area of an antenna may be defined in terms of the other. A large antenna may have a large effective receiving area because of its size, but the gain of that antenna also depends on the wavlength. A given dish-type antenna may have the same effective receiving area at 5 GHz and at 10 GHz, but the gain at 10 GHz will be four times as great; i.e., to the extent that A_e is the same,

$$\frac{G_{f1}}{G_{f2}} = \frac{\lambda_2^2}{\lambda_1^2} = \frac{f_1^2}{f_2^2} \tag{30}$$

But all of the above, except for (28), (29), and (30), refer to the elemental or short dipole, where the current is assumed constant over the length of the dipole. A more realistic model is a $\lambda/2$ dipole, where the current is assumed to be zero at the ends and maximum at the center or feed point.

Problem 14.3-1 Find the radiation resistance for a 1-m-length elemental dipole radiating at a frequency of 5 MHz.

Problem 14.3-2 Find the radiation resistance of a 0.05 m elemental dipole: (a) at 100 MHz; (b) at 200 MHz.

Problem 14.3-3 A particular 4-ft-diameter dish antenna has a gain of 29 dB at 3 GHz. Find the approximate gain at 12 GHz.

14.4 THE HALF-WAVE DIPOLE

The derivations for the $\overline{E}$ and $\overline{H}$ fields follow the same pattern as for the elemental dipole element. From Fig. 14-4, the z component of the vector magnetic potential is, from Fig. 14-4 and by comparison to (14.3-1),

$$\mathcal{A}_z = \frac{\mu \mathcal{I}_0}{4\pi r}\int_{-\lambda/4}^{+\lambda/4} e^{j\beta z\cos\theta} e^{-j\beta r}\cos\beta z\,dz \tag{1}$$

$$= \frac{\mu \mathcal{I}_0}{4\pi r}\left[e^{j\pi/2\cos\theta}\left(0+\beta\sin\frac{\pi}{2}\right) - e^{-j\pi/2\cos\theta}\left(0-\beta\sin\frac{\pi}{2}\right)\right]\frac{e^{-j\beta r}}{\beta\sin^2\theta}$$

$$= \frac{\mu \mathcal{I}_0 e^{-j\beta r}\cos\left(\frac{\pi}{2}\cos\theta\right)}{2\pi r\beta\sin^2\theta} \tag{2}$$

The spherical components are

$$\mathcal{A}_r = \frac{\mu \mathcal{I}_0 e^{-j\beta r}\cos\left(\frac{\pi}{2}\cos\theta\right)\cos\theta}{2\pi r\beta\sin^2\theta} \tag{3}$$

$$\mathcal{A}_\theta = \frac{-\mu \mathcal{I}_0 e^{-j\beta r}\cos\left(\frac{\pi}{2}\cos\theta\right)}{2\pi r\beta\sin\theta} \tag{4}$$

$$\mathcal{A}_\phi = 0$$

We solve for $\mathcal{H}_\phi$ and $\mathcal{E}_\theta$ in the same manner as we did for the elemental dipole, throwing out all terms that do not vary as $1/r$ and, finally, for the radiation fields,

$$\boxed{\mathcal{H}_{\phi\,\text{rad}} = \frac{j\mathcal{I}_0 e^{-j\beta r}\cos\left(\frac{\pi}{2}\cos\theta\right)}{2\pi r\sin\theta} \qquad (\text{A m}^{-1})} \tag{5}$$

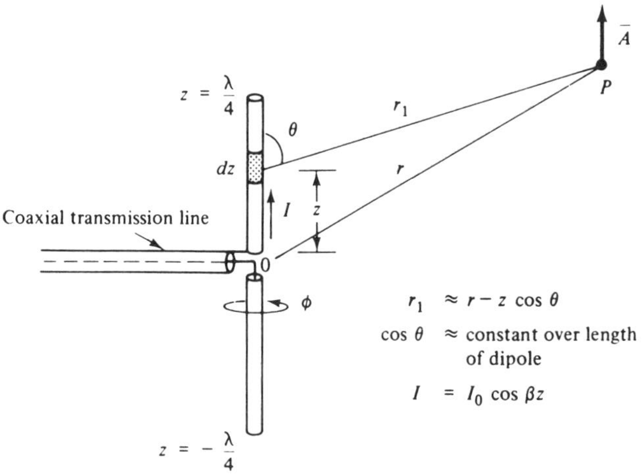

Figure 14-4 Half-wave dipole antenna shown connected to a coaxial transmission line.

and

$$\mathscr{E}_{\theta\,\text{rad}} = \frac{j\beta \mathscr{I}_0 e^{-j\beta r} \cos\left(\frac{\pi}{2}\cos\theta\right)}{2\pi\epsilon\omega r \sin\theta} \qquad (\text{V m}^{-1}) \tag{6}$$

The time average power density radiated is

$$\begin{aligned}\mathscr{P}_{\text{av}}(\theta) &= \frac{\mathscr{I}_0^2 \beta \cos^2\left(\frac{\pi}{2}\cos\theta\right)}{8\pi^2 \epsilon\omega r^2 \sin^2\theta} = \frac{1}{2}|\mathscr{E}_{\theta\,\text{rad}}|\,|\mathscr{H}_{\phi\,\text{rad}}| \\ &= \frac{\mathscr{I}_0^2 \sqrt{\frac{\mu}{\epsilon}} \cos^2\left(\frac{\pi}{2}\cos\theta\right)}{8\pi^2 r^2 \sin^2\theta} \qquad (\text{W m}^{-2})\end{aligned} \tag{7}$$

The average radiated power is

$$\begin{aligned}P_{\text{rad av}} &= \int_0^{2\pi}\int_0^{\pi} \mathscr{P}_{\text{av}}(\theta) r^2 \sin\theta\, d\theta\, d\phi = \frac{2\pi \mathscr{I}_0^2 \sqrt{\frac{\mu}{\epsilon}}}{8\pi^2}\int_0^{\pi} \frac{\cos^2\left(\frac{\pi}{2}\cos\theta\right)}{r^2\sin^2\theta} r^2 \sin\theta\, d\theta \\ &= \frac{\mathscr{I}_0^2\sqrt{\frac{\mu}{\epsilon}}}{4\pi}\int_0^{\pi}\frac{\cos^2\left(\frac{\pi}{2}\cos\theta\right)}{\sin\theta} d\theta \qquad (\text{W})\end{aligned} \tag{8}$$

The integrand in (8) must be solved graphically or numerically to give

$$P_{\text{rad av}} = \frac{1.218\sqrt{\frac{\mu}{\epsilon}}\,\mathscr{I}_0^2}{4\pi} = \frac{1}{2}\mathscr{I}_0^2 R_{\text{rad}} \tag{9}$$

For free space,

$$R_{\text{rad}} = \frac{2(1.218)(120\pi)}{4\pi} = 73\ \Omega \tag{10}$$

Thus, the radiation resistance of a $\lambda/2$ dipole is 73 Ω, but this is not the total impedance of the antenna. There are methods by which the reactive component can be determined, and it would be found to be approximately $j42.5\ \Omega$.[3] The dipole can be made to look purely resistive if it is a little shorter than $\lambda/2$. The 73-Ω radiation resistance of the $\lambda/2$ dipole is the reason for the standard 75-Ω coaxial cable, although it is used for many other applications than the feeding of a $\lambda/2$ dipole.

The directivity of the $\lambda/2$ dipole can be computed, using (7), (9), and (14.2-2) and noting that maximum power radiated per unit solid angle is $r^2\mathscr{P}_{\text{av}}(\theta)$, where $\theta = 90°$.

$$(G_D)_{\lambda/2} = \frac{r^2\left[\left(\frac{1}{2}\right)\frac{\beta\mathscr{I}_0^2}{4\pi^2\epsilon\omega r^2}\right]}{\frac{1.218\sqrt{\frac{\mu_0}{\epsilon_0}}\,\mathscr{I}_0^2}{(4\pi)^2}} = 1.64 \tag{11}$$

[3] R. G. Brown, R. A. Sharpe, and W. L. Hughes, *Lines, Waves, and Antennas*, The Ronald Press Company, New York, 1961, p. 218.

The elemental dipole and the $\lambda/2$ dipole radiation patterns are compared in Fig. 14-5. In either case, the radiation intensity is constant with ϕ and has a maximum at $\theta = 90°$, but the field decreases somewhat more rapidly for angles off the normal for the $\lambda/2$ dipole. The fact that the radiation patterns are similar and that the directivity of the $\lambda/2$ dipole is 1.64 compared to 1.5 for the elemental dipole makes the two antennas appear to be more similar than they really are. For example, using the formula (14.3-20) to compute the radiation resistance of the $\lambda/2$ dipole gives a value of 197 Ω, compared to the actual value of 73 Ω. The formulas for the short elemental dipole apply only when the dipole is much shorter than a half-wavelength.

Example 3

A communication link between two $\lambda/2$ dipole antennas is established with a free space environment. If the transmitter delivers 1 kW of power to the transmitting antenna, how much power will be received by a receiver connected to the receiving dipole 500 km from the transmitter if the frequency is 200 MHz? Assume that the path between dipoles is normal to each dipole; i.e., $\theta = 90°$, and the dipoles are aligned.

Solution

$$G = 1.64 = G_T$$

$$A_e = \frac{G\lambda^2}{4\pi} = \frac{1.64\left(\dfrac{3 \times 10^8}{2 \times 10^8}\right)^2}{4\pi}$$

$$= 0.294 \text{ m}^2 = A_R$$

Then

$$P_R = \frac{P_T G_T A_R}{4\pi r^2} = \text{power received} \qquad \text{(Friis equation)}$$

$$= \frac{(10^3)(1.64)}{4\pi(5 \times 10^5)^2}(0.294)$$

$$= 1.53 \times 10^{-10} \text{ W}$$

Example 4

An incident plane wave at 20 MHz has an rms field strength of 5 mV m^{-1}. How much power would be received by a properly terminated $\lambda/2$ dipole?

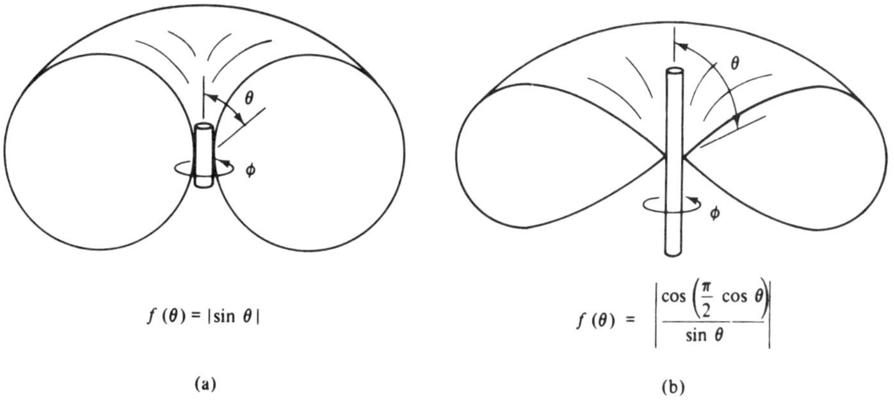

Figure 14-5 (a) Illustration of the variation of the $\overline{E}$ or $\overline{H}$ field radition pattern as a function of θ for the elemental dipole. (b) Same as part (a) for a half-wave dipole.

Solution

$$P_R = A_e \frac{\mathscr{E}_{\text{rms}}^2}{\sqrt{\mu_0/\epsilon_0}} = A_e \frac{\epsilon_{\text{rms}}^2}{120\pi}$$

$$= \frac{\lambda^2 G}{4\pi}\left(\frac{\mathscr{E}_{\text{rms}}^2}{120\pi}\right)$$

$$= \frac{\left(\dfrac{3 \times 10^8}{2 \times 10^7}\right)^2 (1.64)(5 \times 10^{-3})^2}{480\pi^2} = 1.94 \times 10^{-6}\ \text{W}$$

Problem 14.4-1 Repeat Example 3 for a frequency of 100 MHz. Since the gain of a $\lambda/2$ dipole is the same for each problem, why aren't the received signals the same?

Problem 14.4-2 Calculate the received power in Example 3 if the receiving dipole is tilted such that $\theta = 60°$ instead of 90° to the direction of the incident wave.

Problem 14.4-3 Calculate the maximum field strength in volts per meter at a distance of 100 m from a $\lambda/2$ dipole that is radiating 10 W.

14.5 LINEAR ARRAYS

The antenna elements pictured in Fig. 14-6 have radiation patterns that are circular in the plane of the paper. The picture could be the end view of two dipole elements or the top view of two broadcast antenna towers that are d (m) apart. Although each element has a circular pattern in the plane of the paper, together they produce a lobelike radiation pattern, the exact nature of which depends on the spacing between elements in wavelengths.

At a distant point, the path difference from the two elements is

$$\text{path difference} = d \sin \phi \tag{1}$$

and the electrical phase difference ψ is

$$\psi = \frac{2\pi}{\lambda}(d \sin \phi) + \alpha = \beta d \sin \phi + \alpha \tag{2}$$

where α is the phase by which the current to element 2 leads the current to element 1. For the moment, we will consider $\alpha = 0$ (i.e., the elements are fed in phase).

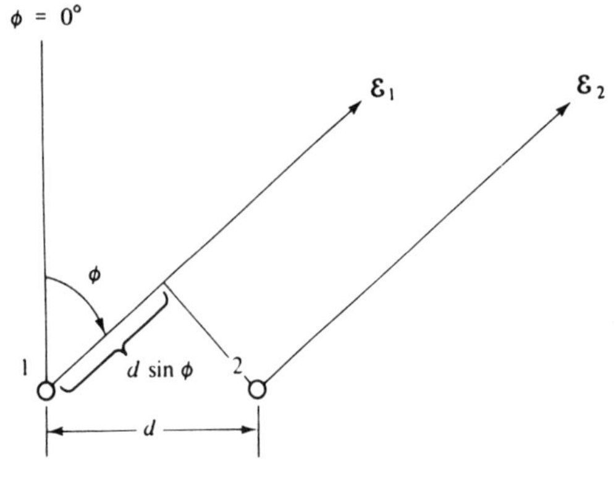

Figure 14-6 Illustration of the path difference from two elements to a distant receiving point, which gives rise to an electrical phase difference.

The electric fields add as shown in Fig. 14-7 for various values of $(2\pi/\lambda)\, d \sin \phi$. The total field strength is

$$|\mathscr{E}_{\text{total}}| = |\mathscr{E}_1|\,|1 + e^{j\psi}| = |\mathscr{E}_1| \left| \frac{\sin \psi}{\sin \dfrac{\psi}{2}} \right| \tag{3}$$

or

$$|\mathscr{E}_{\text{total}}| = 2|\mathscr{E}_1| \left| \cos \frac{\psi}{2} \right| \qquad (\text{V m}^{-1}) \tag{4}$$

where $\mathscr{E}_1$ is the field strength at the distant point due to element 1. Either (3) or (4) apply only in the case of equal amplitude excitation of the elements. Rather than to try to derive formulas that would apply to general cases of different amplitudes and phase, some specific examples will be computed instead.

Example 5

Make a rough polar plot of the field pattern in the plane of symmetry for a pair of dipoles: (a) fed in phase and $\lambda/2$ apart; (b) fed 180° out of phase. Assume $|\mathscr{E}_1| = |\mathscr{E}_2|$.

Solution. (a) In the directions $\phi = 0°$ and $\phi = 180°$, the field is the sum of $|\mathscr{E}_1|$ and $|\mathscr{E}_2|$ and is a maximum since $\mathscr{E}_1$ and $\mathscr{E}_2$ are in phase in those directions. For $\phi = 90°$ and 270°, the path difference is $\lambda/2$ and the two fields cancel, producing a null in those directions. The pattern is sketched in Fig. 14-8(a).

(b) At $\phi = 0°$ and 180°, the fields are 180° out of phase and produce nulls in those directions. At $\phi = 90°$ and 270°, the $\lambda/2$ spacing between elements cancels the 180° phase difference in the elements themselves, and the resultant field is a maximum in those directions. The plot is sketched in Fig. 14-8(b).

Example 6

Find the radiation pattern nulls for two elements separated by 1 wavelength if the current in element 1 leads that of element 2 by 90°. Assume equal amplitude currents.

Solution. The orientation of the elements is as shown in Fig. 14-9. The nulls are located at angles for which $\mathscr{E}_1$ and $\mathscr{E}_2$ are 180° out of phase. That is,

$$\frac{\pi}{2} - (\lambda \sin \phi)\frac{2\pi}{\lambda} = \pm n_{\text{odd}}\pi$$

or

$$\frac{1}{2} - 2 \sin \phi = \pm n_{\text{odd}}$$

or

$$\sin \phi = \frac{\pm n_{\text{odd}} + \frac{1}{2}}{2} = \frac{3}{4}, -\frac{1}{4} \text{ only}$$

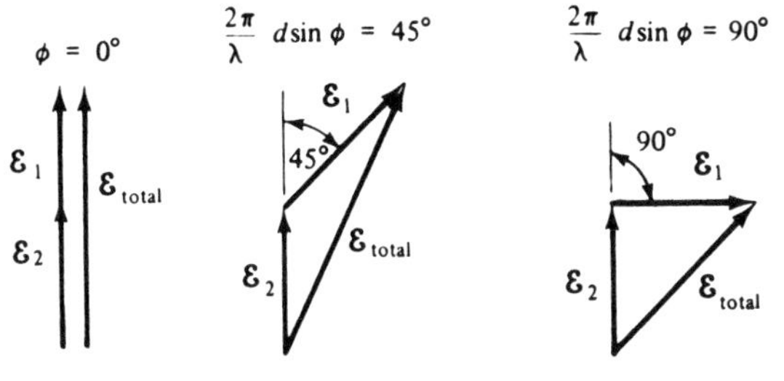

Figure 14-7 Illustration of how the phase difference between two antenna elements affects the resultant field, where $|\mathscr{E}_1| = |\mathscr{E}_2|$.

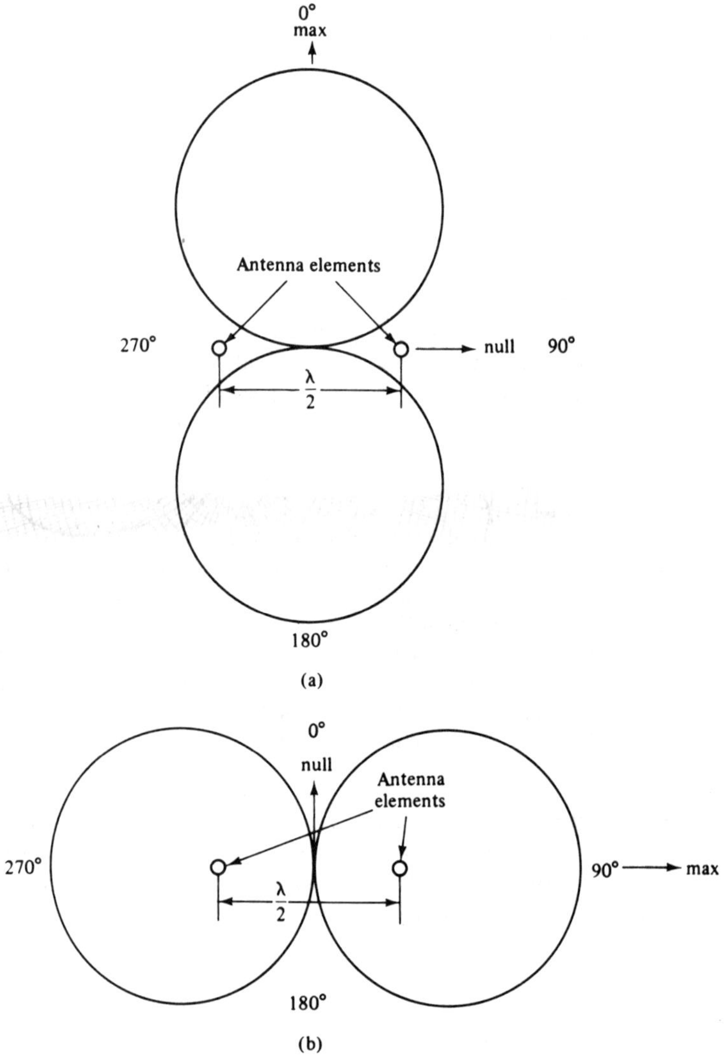

Figure 14-8 Illustration for Example 5.

since

$$|\sin\phi| \leq 1$$

where

$$\sin\phi = \frac{3}{4}, \qquad \phi = 48.59°, 131.41°$$

and for

$$\sin\phi = -\frac{1}{4}, \qquad \phi = -14.48°, 194.48°$$

A complete sketch of the pattern is shown and may be plotted from $f(\phi) = |e^{j\pi/2} + e^{j\psi}|$, where ψ is as given by (2).

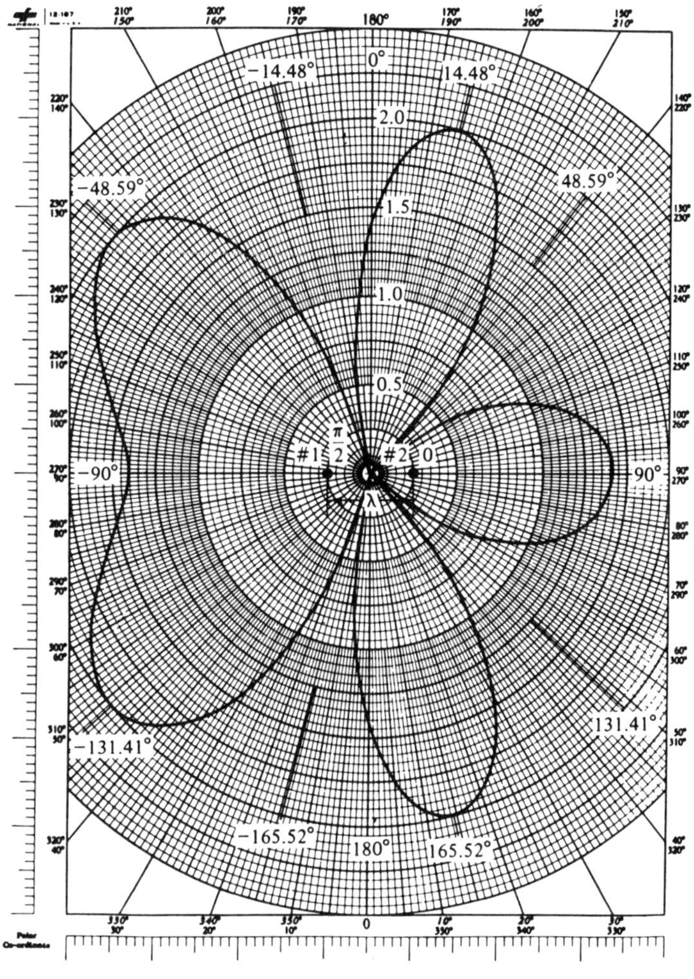

Figure 14-9 Illustration for Example 6.

We shall now look at an N-element linear array. In Fig. 14-10, the N elements are equally spaced and are excited with equal amplitude currents. The total field is

$$|\mathscr{E}_t| = |\mathscr{E}_1|\,|1 + e^{j\psi} + e^{2j\psi} + e^{3j\psi} + \ldots + e^{(N-1)j\psi}| \tag{5}$$

$$|\mathscr{E}_t| = |\mathscr{E}_1| \left| \frac{1 - e^{jN\psi}}{1 - e^{j\psi}} \right|$$

$$= |\mathscr{E}_1| \left| \frac{e^{j(N/2)\psi}}{e^{j\psi/2}} \right| \frac{\sin \dfrac{N\psi}{2}}{\sin \dfrac{\psi}{2}}$$

$$= |\mathscr{E}_1| \frac{\sin \dfrac{N\psi}{2}}{\sin \dfrac{\psi}{2}} \tag{6}$$

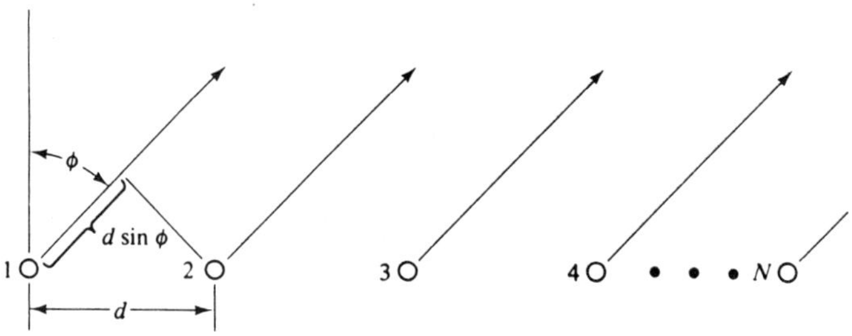

Figure 14-10 An N-element linear array.

The normalized field pattern is

$$F(\psi) = \frac{1}{N}\left|\frac{\sin N\psi/2}{\sin \psi/2}\right| \tag{7}$$

where $\psi = (2\pi/\lambda)d \sin \phi + \alpha$.

To plot (7), it is convenient to first find the nulls of the pattern in order that the remainder of the pattern can be plotted by using a minimum number of points. The nulls (or zeros) of (7) are not just the zeros of the numerator, for both the denominator and numerator have zeros at $\phi = 0$, which is a maximum rather than a null. The zeros of the numerator of (7), for $\alpha = 0$, occur for

$$\frac{N\psi}{2} = 0, \pm\pi, \pm 2\pi, \ldots$$

or

$$\psi = \frac{2}{N}(0, \pm\pi, \pm 2\pi, \ldots)$$

or

$$\sin \phi = \frac{\lambda}{dN}(0, \pm 1, \pm 2, \ldots) \tag{8}$$

The zeros of $\sin \psi/2$ are where

$$\sin \phi = \frac{\lambda}{d}(0, \pm 1, \pm 2, \ldots) \tag{9}$$

Now, let's check $F(\psi)$ at $\phi = 0°$. This gives an indeterminate 0/0 condition, but

$$\lim_{\substack{\psi \to 0 \\ \text{or } \phi \to 0}} \frac{1}{N}\left|\frac{\sin N\psi/2}{\sin \psi/2}\right| = 1 \tag{10}$$

= normalized amplitude at beam center, which is normal to the array of elements

If $\alpha \neq 0$, the maximum of $F(\psi)$ still occurs at $\psi = 0$, but the direction ϕ for maximum $F(\psi)$ is not 0°. Setting

$$\psi = \beta d \sin \phi_0 + \alpha = 0$$

we find that $\alpha = -\beta d \sin \phi_0$, or $\phi_0 = \sin^{-1}(-\alpha/\beta d)$. ϕ_0 is the direction to which the beam is steered by having a successive phase shift α between adjacent elements. In a phased array radar, the beam steering computer computes the necessary α for any given beam direction ϕ_0.

Example 7

Find the beamwidth and location of the first nulls on either side of beam center for a linear array of 80 in-phase elements fed with equal amplitude current and which are $\lambda/2$ apart.

Solution. The beam center is at $\phi = 0$. The first nulls in the pattern on either side of beam center are where the numerator of (10) is zero (except at $\phi = 0$). At the first nulls,

$$\sin\phi = \pm\frac{\lambda}{dN} = \pm\frac{\lambda}{\frac{\lambda}{2}N} = \pm\frac{2}{N} = \pm\frac{2}{80}$$

Then

$$\phi_{\text{first nulls}} = \pm\sin^{-1}\frac{2}{80} = \pm 1.43°$$

To solve for the 3-dB beamwidth, we set $F(\psi) = 0.707$ and solve for ϕ.

$$0.707 = \left(\frac{1}{N}\right)\frac{\sin\dfrac{N\psi_B/2}{2}}{\sin\dfrac{\psi_B/2}{2}} = \left(\frac{1}{N}\right)\frac{\sin N\psi_B/4}{\sin\psi_B/4} \tag{10a}$$

where the main beamwidth extends from $-\phi_B/2$ to $\phi_B/2$ or ψ from $-\psi_B/2$ to $\psi_B/2$, where

$$\psi_B/2 \approx \frac{2\pi}{\lambda}d\sin\phi_B/2$$

Then

$$0.707 = \left(\frac{1}{80}\right)\frac{\sin 20\psi_B}{\sin\psi_B/4}$$

from which

$$\psi_B = 0.0696 = \left(\frac{2\pi}{\lambda}\right)\frac{\lambda}{2}\sin\phi_B = \pi\sin\phi_B$$

Then

$$\phi_B = \sin^{-1}\frac{0.0696}{\pi} = 1.269° = \text{beamwidth}$$

A linear plot of the antenna pattern near beam center is given in Fig. 14-11.

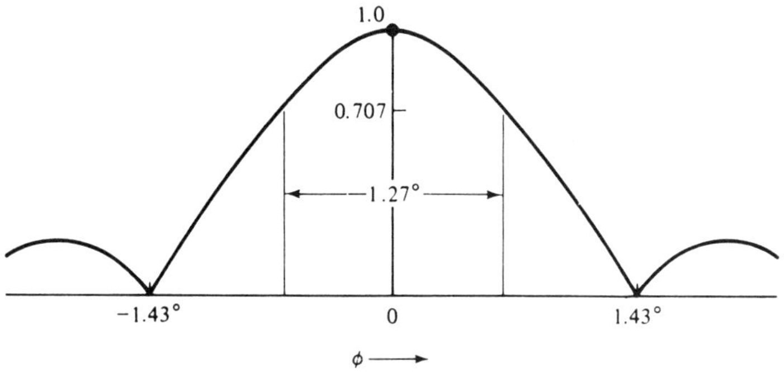

Figure 14-11 Radiation pattern near broadside for the 80-element array of Example 7.

If 80 rows of the 80 elements per row of Example 7 were vertically spaced $\lambda/2$ apart, we would have a vertical planar array of 6400 elements with a beamwidth approximately 1.27° in both the azimuth and elevation directions. In practice, there is some mutual coupling between elements, and because of "tapering" of the antenna, a technique that renders the outer edge elements less effective than the ones in the center and which is done to reduce sidelobe levels, the beamwidth would, in general, be greater than indicated by Example 7. A large phased array radar antenna may have from 5000 to 10,000 antenna elements, each of which has a separate current phase shifter that is electronically controlled for steering the beam center anywhere from the normal to the array to ±60° from normal.

A useful approximate formula for beamwidth, which may be used when the number of elements is very large, is derived by representing the sin $N\psi/4$ and sin $\psi/4$ of (10a) by the first two terms of the sine series each. Then

$$0.707 = \frac{1}{N}\,\frac{\dfrac{N\psi_B}{4} - \dfrac{N^3\psi_B^3}{4^3\cdot 3!} + \cdots}{\dfrac{\psi_B}{4} - \dfrac{\psi_B^3}{4^3\cdot 3!}} = \frac{1 - \dfrac{N^2\psi_B^2}{96}}{1 - \dfrac{\psi_B^2}{96}} \tag{11}$$

Solving for ψ_B^2, we get

$$\psi_B^2 \approx \frac{0.293}{0.0104N^2 - 0.00736} \approx \frac{0.293}{0.0104N^2}$$

for large N.

Then, for large N,

$$\psi_B = \frac{5.307}{N} = \frac{2\pi}{\lambda} d \sin \phi_B \tag{12}$$

Substituting $N = 80$, $d = \lambda/2$ into (12), we have

$$\phi_B = \sin^{-1} \frac{5.307}{80\pi} = 1.21°$$

compared to 1.269° computed from Example 7. A more exact solution to (10a) is

$$\boxed{\psi_B = \frac{5.568}{N}} \tag{13}$$

Problem 14.5-1 Two identical broadcast antenna towers are λ apart and are driven in phase with equal amplitude currents. Find the antenna pattern in the horizontal plane.

Problem 14.5-2 A linear array consists of three isotropic elements that are driven in phase with equal amplitude and are $\lambda/2$ apart. Find the nulls in the antenna pattern in a plane containing the array, and plot the normalized pattern.

Problem 14.5-3 Repeat Prob. 14.5-2 using four elements.

Problem 14.5-4 Find the broadside beamwidth of a linear array of 50 elements driven in phase with equal amplitude currents. Assume $\lambda/2$ spacing.

14.6 APERTURE ANTENNAS

We have seen that a single dipole is not very directive, the $\lambda/2$ dipole having a gain of only 1.64, or 2.15 dB, but from Example 7 a planar array of 80 × 80 elements with $\lambda/2$ spacing could produce a beamwidth of only 1.27° × 1.27°, or a gain of 45 dB, using (14.2-3). A uniformly illuminated aperture antenna is like a limiting case of a planar array of discrete elements with all elements in phase and with equal amplitudes. A single element can be used to illuminate a large reflecting surface that forms the antenna area corresponding to the area of a planar array.

For a linear array of N discrete elements (either a row or column of a planar array), the width w of the array is $(N - 1)d$ but is approximately Nd for large N, and

$$F(\psi) = \frac{1}{N}\left|\frac{\sin\frac{N\psi}{2}}{\sin\frac{\psi}{2}}\right| = \frac{1}{N}\left|\frac{\sin\left(\frac{\pi Nd\sin\phi}{\lambda}\right)}{\sin\left(\frac{\pi d\sin\phi}{\lambda}\right)}\right| \tag{1}$$

and for small ϕ, large N,

$$\lim_{\substack{N\to\infty\\ d\to 0}} F(\psi) = \left|\frac{\sin\frac{N\psi}{2}}{\frac{N\psi}{2}}\right| = \left|\frac{\sin\frac{\pi w\sin\phi}{\lambda}}{\frac{\pi w\sin\phi}{\lambda}}\right| \tag{2}$$

which is like a diffraction pattern from a slot. Equation (2) is the Fourier transform of a rectangular function, normalized to wavelength.

$$f\left(\frac{x}{\lambda}\right) = \begin{cases} 1 & \left(|x| < \frac{w}{2}\right) \\ 0 & \text{otherwise} \end{cases} \tag{3}$$

A rectangular function and its transform are illustrated in Fig. 14-12. The Fourier transform pair are

$$f\left(\frac{x}{\lambda}\right) = \int_{-\infty}^{\infty} F(s)e^{j(2\pi xs/\lambda)}\,ds \tag{4}$$

and

$$F(s) = \int_{-w/2\lambda}^{w/2\lambda} f\left(\frac{x}{\lambda}\right)e^{-j(2\pi xs/\lambda)}\,d\left(\frac{x}{\lambda}\right) \tag{5}$$

Therefore,

$$\lim_{\substack{N\to\infty\\ d\to 0}} F(\psi) = \left|\frac{F(s)}{w/\lambda}\right| = \left|\frac{\sin\pi s'w}{\pi s'w}\right| = |\text{sinc}\ ws'| \tag{6}$$

where $s' = \sin\phi/\lambda$, (not to be confused with area) and $w = Nd$.

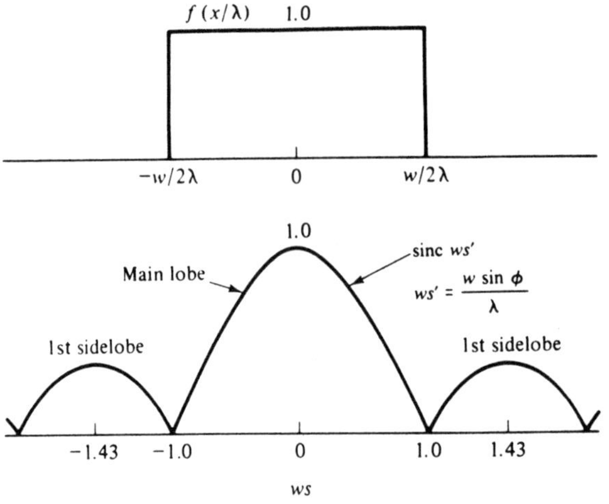

Figure 14-12 Rectangular function and its Fourier transform, corresponding to a uniformly illuminated aperture and its resulting antenna pattern.

From a table of sinc function,[4] we can then determine the beamwidth, location of first nulls, etc. For example, we find that sinc $ws' = 0.707$ at $ws' = \pm 0.443$. Then

$$\frac{(ws')_B}{2} = 0.443 = \frac{w}{\lambda} \sin \frac{\phi_B}{2} \tag{7}$$

Let $N = 80$, $d = \lambda/2$, $w = Nd = 40\lambda$. Then

$$\frac{40\lambda}{\lambda} \sin \frac{\phi_B}{2} = 0.443$$

$$\sin \frac{\phi_B}{2} = 0.0111$$

$$\frac{\phi_B}{2} = 0.6346^\circ$$

$$\phi_B = 1.269^\circ$$

which is the same as that computed from Example 7 for the linear array.

A tapered illumination is illustrated in Fig. 14-13. In general, the illumination at the edges is reduced to near zero. The net effect, and this can be checked using Fourier transforms, is that the amplitudes of the sidelobes are reduced, the beam is slightly broadened, and the gain at the center of the beam is reduced. The main reason for tapering an antenna is to reduce sidelobes. For a uniformly illuminated aperture, the first sidelobes are 0.217 of the main beam center, or they are down 13.3 dB. With suitable tapering, the sidelobe levels can be reduced to −20 dB or less with only 1 or 2 dB loss in antenna gain. For an aperture antenna with typical taper, the effective area A_e is about 60% of the projection of the actual aperture or the *aperture efficiency* is 60%. For a dipole element at the focus of a parabolic dish reflector, as illustrated in Fig. 14-14, the effective area of the overall antenna would be about $0.6(\pi d^2/4)$, where d is the diameter. The dipole reflector part of the feed is to block direct radiation from the radiating dipole and ensure that it is directed to the parabolic reflector instead, from which a plane phase front emerges.

[4] See R. M. Bracewell, *The Fourier Transform and Its Application*, McGraw-Hill Book Company, New York, 1965.

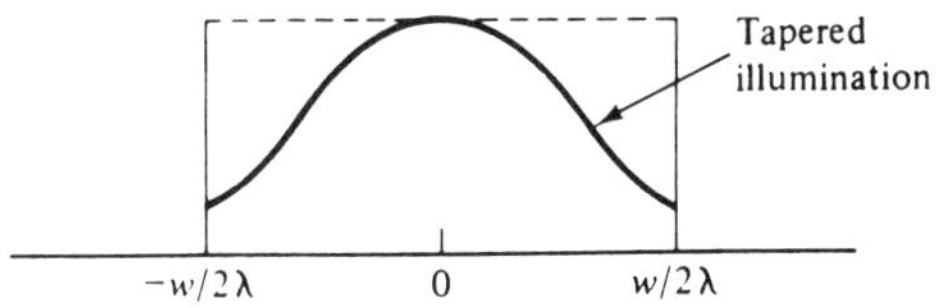

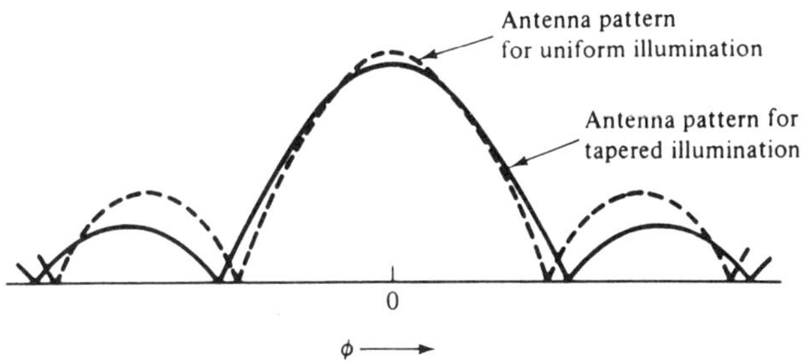

Figure 14-13 Illustration of the effect of tapering an antenna. The dashed lines are for a uniformly illuminated aperture.

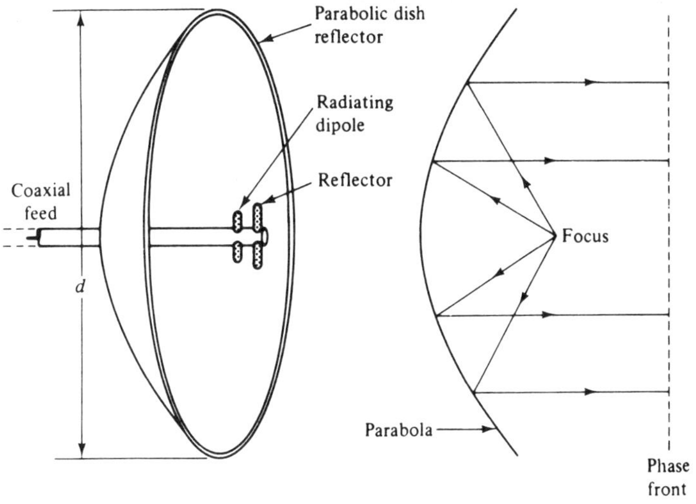

Figure 14-14 Parabolic dish reflector antenna, using a dipole for the feed.

If the illumination were uniform over the total reflecting surface, the beamwidth would be the same as for a planar array with the same area. Actually, the dipole feed has a natural taper, which makes it suited for this application. However, at frequencies above 2 GHz or so, a waveguide horn feed is more frequently used with the parabolic dish reflector, as illustrated in Fig. 14-15. The shape of the horn is designed to illuminate the reflector without too much spillover past the edges, which would create antenna lobes to the rear of the reflector. Antenna gains of 40 to 50 dB or even greater are not uncommon with parabolic dish antennas.

A parabolic dish antenna used to receive television programs via satellite is pictured in Fig. 14-16. Antennas of this type as large as 10 ft in diameter are commonly used.

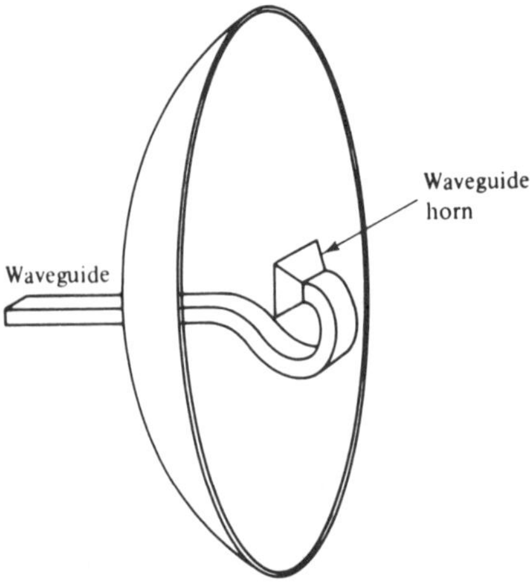

Figure 14-15 Parabolic dish reflector antenna, using a waveguide horn for the feed.

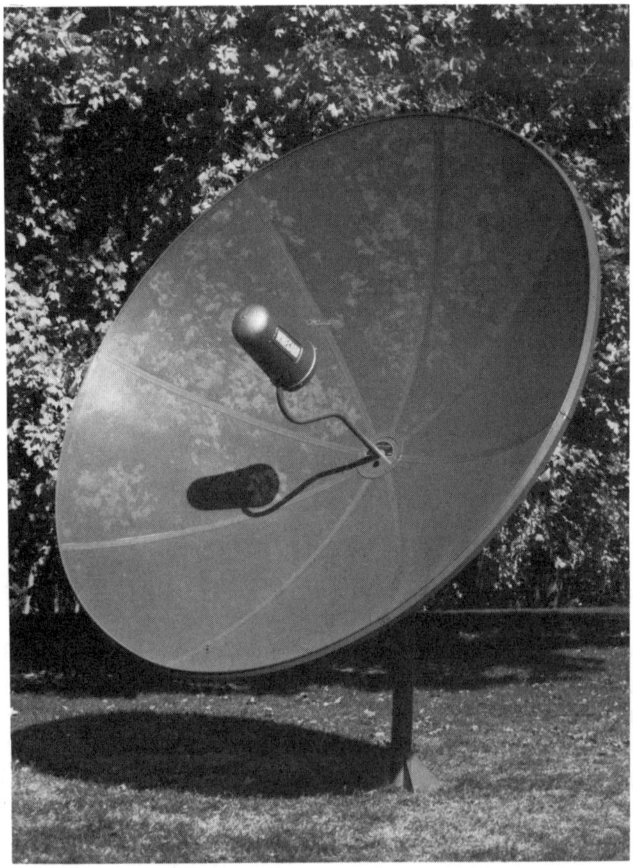

Figure 14-16 Parabolic dish antenna used for reception of television programs via satellite. Both vertical and horizontal feeds are used and frequency conversion is accomplished at the feed.
(Courtesy of Winegard.)

A metal lens antenna, a side view of which is shown in Fig. 14-17, has the feed horn behind the antenna, where it does not create a shadow. The principle of the lens is that the phase velocity in the waveguide is faster than in air. Therefore, the edges of the lens speed up the phase such that the spherical wavefront entering the lens is focused

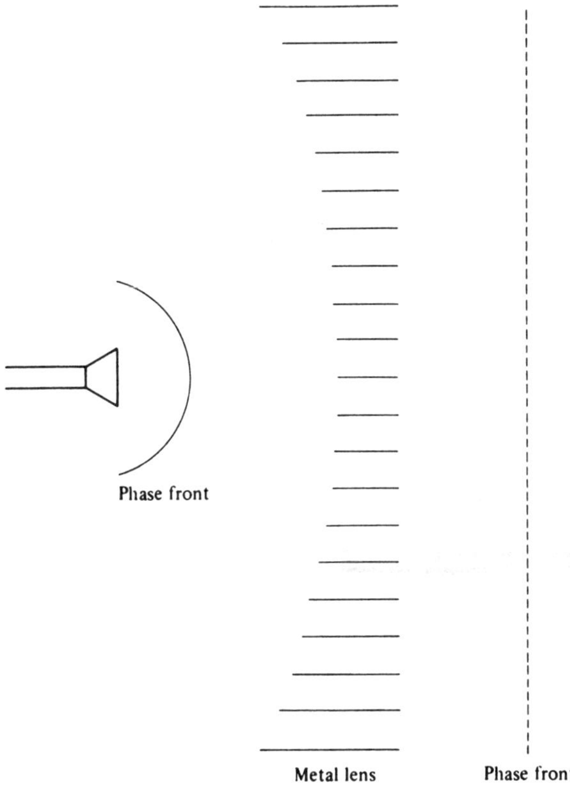

Figure 14-17 Metal waveguide lens antenna.

to a plane wavefront leaving the lens. As a receiving antenna, a plane wavefront arriving at the lens is focused at the feedhorn.

Another type of a spherical lens antenna is the Luneberg lens, depicted in Fig. 14-18. The cross section of the sphere is the aperture area, but the effective area will be dependent on the feedhorn pattern. The sphere is constructed of spherical shells, each of which is a slightly different dielectric constant. A plane wave arriving at the sphere is focused to a point on the opposite side of the sphere, where a horn is located. The advantage of this type of lens is that it permits a large number of feeds to be stacked along the periphery of the lens such that a large angular coverage is obtained with several discrete beams, permitting angular resolution of targets in radar use.

A Luneberg lens receiving antenna, 80 ft in diameter, was part of the acquisition radar for the first antiballistic missile (ABM) system designed in the United States. It was constructed of 40,000 cubes of polystyrene foam, 2 ft on an edge. The blocks of each spherical shell or zone were loaded with a precise weight of aluminum slivers as an artificial dielectric to achieve the desired dielectric constant. Receiving horns were stacked in elevation and the entire antenna rotated for azimuth coverage.

An interesting and ingenious reflecting antenna design is the cassegrain antenna using a twist reflector,[5] the cross section of which is shown in Fig. 14-19. The feedhorn is securely mounted in an opening in the main reflecting dish and is directed toward the

[5] P. W. Hannan, "Microwave Antennas Derived from the Cassegrain Telescope," *IRE Trans. Antennas Propag.*, Vol. AP-9, March 1961, pp. 140–153.

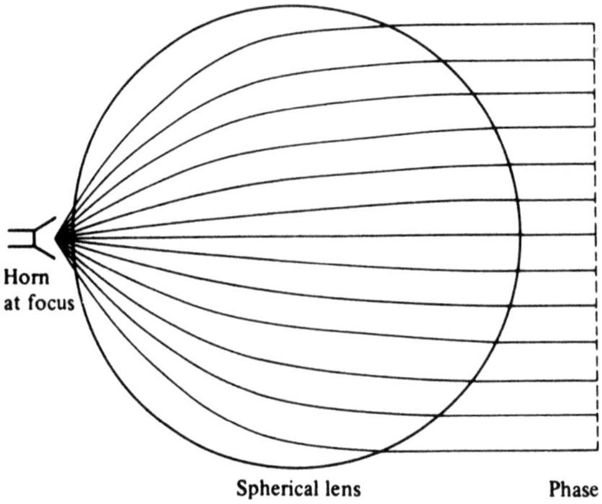

Figure 14-18 Luneberg lens antenna.

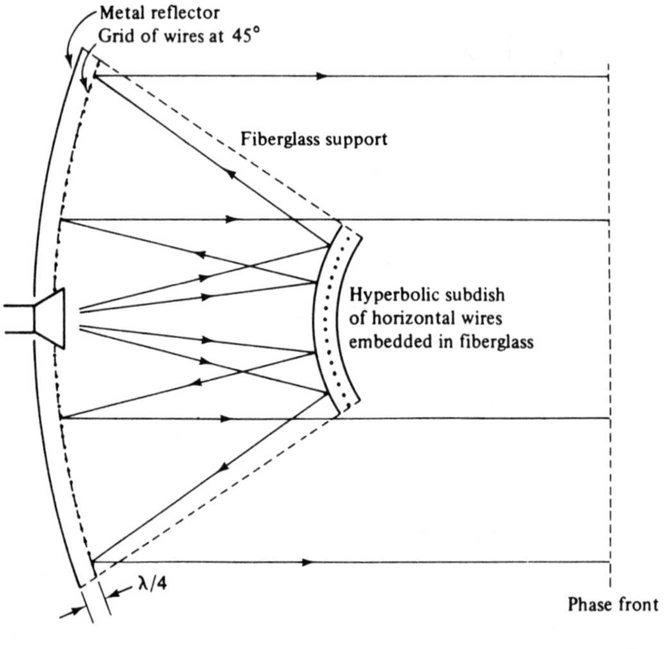

Figure 14-19 Cassegrain antenna, using a twist reflector to rotate the polarization 90°.

subdish reflector. The wave is reflected from the subdish back to the main dish and reflected from the main dish as for the usual parabolic dish antenna.

What is particularly interesting about this antenna is the way the polarization is manipulated. Let's assume horizontal polarization from the feedhorn. Then the subdish, having a hyperbolic shape, has a reflecting surface made up of a grid of horizontal wires. The main dish is covered with a grid of wires at 45° angle and spaced $\lambda/4$ from the metal dish reflecting surface. The polarization is then rotated 90° by this mechanism, as illustrated in Fig. 14-20. A horizontal $\overline{E}$ field E_i is divided into the components

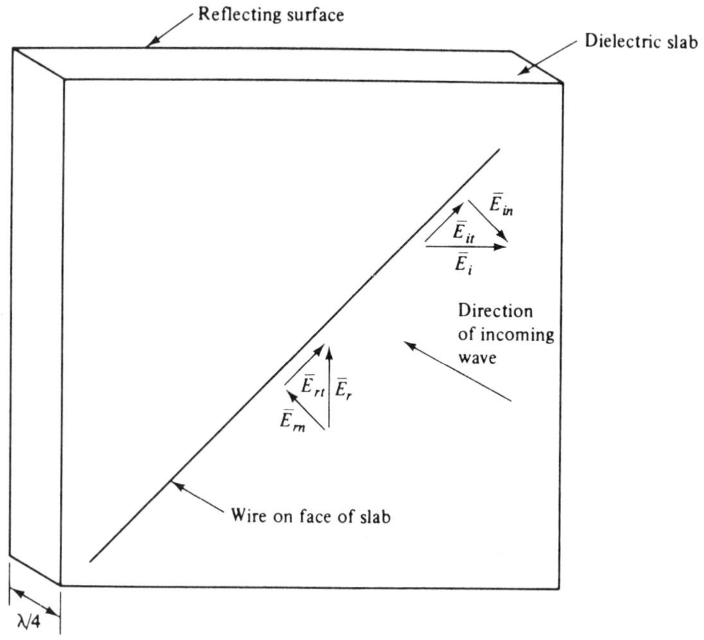

Figure 14-20 Illustration of how the twist reflector works.

E_{it} and E_{in} for the tangential and normal components with respect to the wire. The E_{it} component is reflected from the wire with 180° phase shift, and the normal component passes through the wires to the main dish, where it is also reflected with 180° phase shift. When the two components recombine as they are traveling back toward the subdish, the normal component has traveled altogether $\lambda/2$ farther than the tangential component and is therefore 180° out of phase with respect to the tangential component. Thus, the reflected $\bar{E}$ field E_r is rotated 90° with respect to the incident E_i and is vertically polarized. The horizontal wires of the subdish are transparent to the outgoing vertically polarized wave.

A discussion of the cassegrain antenna would not be complete without mention of the monopulse 4-horn feed used for tracking radars, shown schematically in Fig. 14-21. If this feed were used like that shown in Fig. 14-15, four waveguides would have to be brought out in front of the parabolic reflector and mechanically supported there. This is done with some tracking radars, but a particularly rigid and simple (from the plumbing standpoint) design is where the four-horn cluster is used with the cassegrain twist reflector. The target tracking and missile tracking radars of the ground-to-air Nike Hercules defense system were of this type, and a large number of these antennas were built.

The monopulse feed may be analyzed with the benefit of some knowledge of the hybrid waveguide junction, or "magic tee" as it is often called. Referring to Fig. 14-22, the arms 2 and 3 are considered the input arms. The output of arm 1 is the phasor sum of 2 and 3, and arm 4 is the phasor difference. Then in Fig. 14-21, using four such hybrids, one can obtain from a single pulse of energy (or noise) the elevation and azimuth signals from which dc error voltages can be derived and used to steer the antenna beam

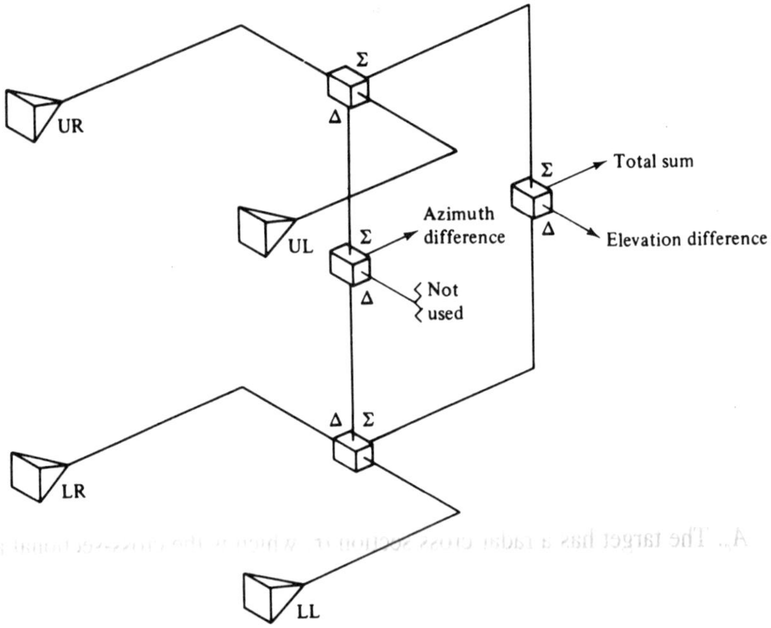

Figure 14-21 Waveguide plumbing for a monopulse tracking-radar antenna feed.

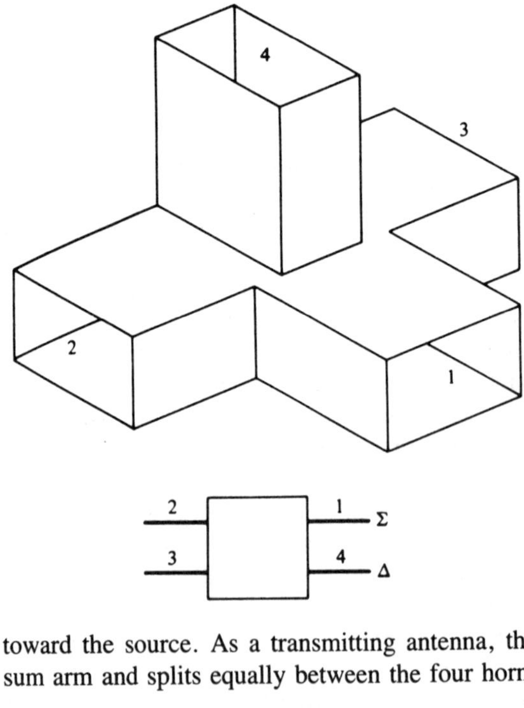

Figure 14-22 Hybrid waveguide junction.

toward the source. As a transmitting antenna, the transmitter is connected to the total sum arm and splits equally between the four horns.

Problem 14.6-1 Find the (a) gain, (b) beam solid angle, and (c) beamwidth for a uniformly illuminated dish antenna whose diameter is 10 ft and which is operating at 12.6 GHz.

Problem 14.6-2 Repeat Prob. 14.6-1 for a tapered illumination such that A_e is only 55% of projected area.

Problem 14.6-3 Assuming an aperture efficiency of 55%, an antenna dish of what diameter is required to produce a 1° beamwidth at 4 GHz?

Problem 14.6-4 What aperture efficiency is required for a 6-ft-diameter dish antenna to have a gain of 37 dB at 4 GHz?

14.7 THE RADAR EQUATION

No application example illustrates antenna concepts such as gain, beamwidth, and effective area as well as the radar equation (or radar range equation), and in view of the ubiquitous nature of radar, it seems appropriate to finish this chapter with a brief look at this application. Figure 14-23 will illustrate.

We shall assume a common antenna for both transmitting and receiving with a transmit-receive (TR) switch isolating the receiver during transmission and disconnecting the transmitter during the receive mode. The antenna has gain G and effective area A_e. The target has a radar cross section σ, which is the cross-sectional area of an equivalent metal sphere. The outgoing transmitted pulse has a power density in its wavefront of

$$\mathcal{P}_{\text{trans}} = \frac{P_T G}{4\pi r^2} \qquad (\text{W m}^{-2}) \tag{1}$$

where r is the distance from the transmitting antenna, and P_T is the transmitter power delivered to the antenna. At the target, at a distance R from the radar, the power density is

$$\mathcal{P}_{\text{target}} = \frac{P_T G}{4\pi R^2} \qquad (\text{W m}^{-2}) \tag{2}$$

The target radar cross section σ is a measure of both receiving area and of reradiating characteristics. For a metal sphere whose diameter is much larger than a wavelength, the radar cross section is $(\pi/4)(\text{dia})^2$. For a flat plate normal to the direction of the radar,

$$\sigma_{\text{flat plate}} = \frac{4\pi A_\sigma^2}{\lambda^2} \qquad (\text{m}^2) \tag{3}$$

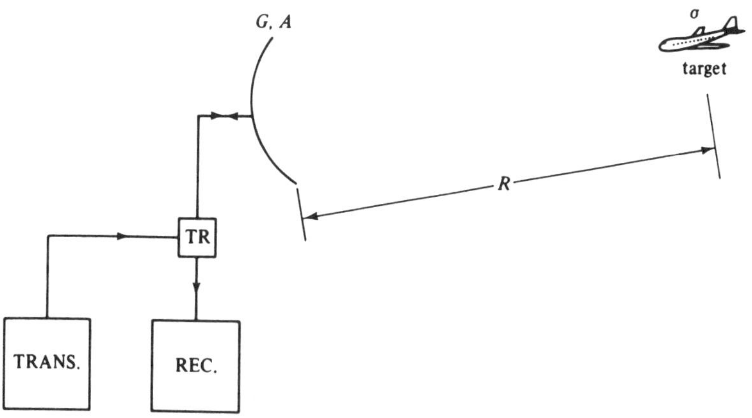

Figure 14-23 Basic radar system.

where A_σ is the area of the plate. The reradiated power density that arrives back at the radar antenna is

$$\mathcal{P}_{\text{rec}} = \frac{P_T G}{4\pi R^2}\left(\frac{\sigma}{4\pi R^2}\right) \qquad (\text{W m}^{-2}) \tag{4}$$

and, finally, the power in watts received by the receiver is

$$P_R = \mathcal{P}_{\text{rec}} A_e = \mathcal{P}_{\text{rec}} \frac{\lambda^2 G}{4\pi} \tag{5}$$

or

$$\boxed{P_R = \frac{P_T G^2 \lambda^2 \sigma}{(4\pi)^3 R^4} \qquad (\text{W})} \tag{6}$$

Using the substitution $G = 4\pi A_e/\lambda^2$ in (6), we have both

$$P_R = \frac{P_T G A_e \sigma}{(4\pi)^2 R^4} \qquad (\text{W}) \tag{7}$$

and

$$P_R = \frac{P_T A_e^2 \sigma}{4\pi \lambda^2 R^4} \qquad (\text{W}) \tag{8}$$

Equations (6), (7), and (8) do not at first glance appear to have the same dependency on wavelength. Perhaps the more accurate picture is gained from (8), which indicates that, for a constant antenna size and target radar cross section, the received signal power would vary directly with the square of frequency. In actual practice, very few if any non-artificial targets have constant cross section with frequency. However, when maximum received power is the main criteria, (8) does suggest large antenna size. In order that the large antenna size not produce too much gain, i.e., too narrow a beamwidth, the wavelength may need to be long. Thus, long-range high-power search radars tend to operate below the GHz region and with antenna diameters on the order of 100 ft. A 500-MHz radar with a 40-dB antenna will receive (ideally) 6 dB more signal than 1-GHz radar with a 40-dB antenna, simply because the 500 MHz antenna is twice the diameter of the 1-GHz antenna and thus has four times the receiving area.

Real targets are usually very complex, being composed of flat surfaces and curved surfaces of various radii such that determination of radar cross section is best done empirically or with scaled models. Radar cross section can also vary many tens of decibels with aspect angle. A small aircraft may have a nose-on radar cross section of 1 m^2 and a broadside cross section of 100 m^2 or more. Although a great effort may be made in some cases to deliberately reduce the radar cross section of an aircraft or other object, the radar cross section cannot be reduced to zero. Birds or even insects have some radar cross section, as do plastic or rubber balloons.

While long-range high-power radars tend toward the long wavelength or low frequencies, compact tracking radars may operate in the 10 to 90 GHz part of the spectrum just to keep the antenna size small. It appears that 100 to 200 GHz will see increased use as well.

The maximum range for a radar for a given target may be computed by rearranging either (6), (7), or (8). If $P_{R\min}$ is defined as the minimum detectable power signal for the receiver, then rearranging (6) gives

$$R_{\max} = \left[\frac{P_T G^2 \lambda^2 \sigma}{(4\pi)^3 P_{R\min}}\right]^{1/4} \quad \text{(m)} \tag{9}$$

Problem 14.7-1 A 9-GHz tracking radar is tracking a 0.02-m^2 target at a range of 100 km. If the gain of the common transmitting and receiving antenna is 45 dB, and the transmitter power is 1 MW, find the received signal P_R in dBm. (A dBm is 1 decibel with respect to 1 milliwatt.)

Problem 14.7-2 For a radar using a common antenna for transmitting and receiving, how much increase in antenna gain is required to double the range capability of the radar?

Problem 14.7-3 What percentage increase in transmitter power is required to double the range capacity of a radar?

Problem 14.7-4 Calculate the maximum range for detecting a $\sigma = 1\ m^2$ target if the radar specifications are:

$$P_T = 1 \text{ MW}$$
$$f = 5 \text{ GHz}$$
$$G = 45 \text{ dB}$$
$$P_{R\min} = -115 \text{ dBm}$$

REVIEW QUESTIONS

1. What is an antenna? *Sec. 14.1*
2. Define antenna gain and beamwidth. *Sec. 14.2*
3. Define solid beam angle. *Sec. 14.2*
4. What is a steradian? *Sec. 14.2*
5. What is a retarded field? *Sec. 14.3*
6. How does radiated field strength vary with distance from the source? *Sec. 14.3*
7. What is the ratio of radiated electric field strength to radiated magnetic field strength? *Sec. 14.3*
8. What is meant by radiation resistance? *Sec. 14.3*
9. What is the effective area of an antenna? *Sec. 14.3*
10. How does the current distribution on a half-wave dipole differ from that for the so-called elemental or *dz* dipole? *Sec. 14.4*
11. What is the term $2\pi/\lambda$? *Secs. 14.3, 14.4*
12. What is a linear array? *Sec. 14.5*
13. How could one electronically "steer" a linear array beam? *Sec. 14.5*
14. Why is an antenna "tapered"? *Sec. 14.6*
15. Why is a parabolic shape used for a dish-type antenna? *Sec. 14.6*
16. How does the received power from a radar target vary with the range to the target? *Sec. 14.7*
17. How does the received power from a flat plate radar target vary with the size of the plate? *Sec. 14.7*

PROBLEMS

14-1. What is the approximate beamwidth of an antenna having a gain of: (a) 30 dB; (b) 45 dB?

14-2. For the elemental dipole, find the distance r for which the radiated $\mathscr{E}_\theta$ field is equal in magnitude to the non-radiated $\mathscr{E}_\theta$ components.

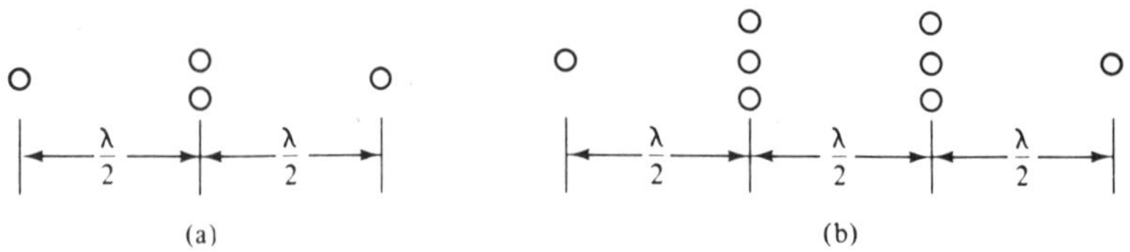

Figure 14-24 Illustration for use with Prob. 14-8.

14-3. Derive the radiation resistance for a $3\lambda/2$ dipole.

14-4. Five dipoles, separated by $\lambda/4$, are fed in phase with equal amplitude currents. In a plane normal to the dipoles, determine the radiation pattern.

14-5. Repeat Prob. 14-4 but with a separation equal to $\lambda/2$ instead of $\lambda/4$.

14-6. Derive (14.5-13). [*Hint:* Assume N large such that $\sin \psi_B/4 \approx \psi_B/4$.]

14-7. Find and sketch the radiation patterns in the plane of the elements for the following pairs of isotropic radiators: (a) separated by $\lambda/4$, fed in phase; (b) separated by $\lambda/4$, fed 45° out of phase; (c) separated by $\lambda/4$, fed 90° out of phase.

14-8. Find and sketch the radiation patterns for the isotropic elements shown in Fig. 14-24. Assume that all elements are fed in phase with equal amplitude currents. [*Hint:* Take a pair at a time and multiply the patterns. These are called *binomial arrays*.]

14-9. An end-fire array is one for which the main beam is parallel to the array of elements. For an array of five elements uniformly spaced, $\lambda/4$ apart, find the successive phase difference between elements to produce an end-fire array, and sketch the pattern for isotropic elements.

14.10. Assuming a typical antenna effective area, calculate the beamwidth in degrees for a 6-ft-diameter dish at: (a) 2.45 GHz; (b) 4 GHz; (c) 12 GHz.

14.11. What is the approximate antenna gain for each part of Prob. 14-10?

14.12. Assuming that due to antenna taper the effective area is 60% actual area, calculate the dish diameter for a radar antenna that has a gain of 45 dB at: (a) 500 MHz; (b) 40 GHz.

14-13. (a) Two isotropic elements are spaced 2λ apart. If they are fed in phase with equal amplitude currents, sketch the field pattern. (b) Now assume these elements are parabolic dishes 2λ in diameter, uniformly illuminated. Sketch the new pattern.

14.14. From a satellite in synchronous orbit, what antenna diameter is required at 2.45 GHz to illuminate a spot 20 miles in diameter on the earth?

14-15. Two identical 4-ft-diameter parabolic dish antennas with typical effective antenna areas are used to transmit and receive at 4 GHz over a 60-mile free space path. What transmitter power is required for a received signal of -90 dBm?

14-16. If the terminals of a receiving antenna are matched to a 50-Ω line, show that the relation between the voltage V_R at the antenna terminals and the plane wave field strength E_R arriving at the antenna is $20 \log_{10} E_R = 20 \log_{10} V_R + 20 \log_{10} (9.73/\lambda\sqrt{G})$, where $20 \log_{10} (9.73/\lambda\sqrt{G})$ is called the *antenna factor,* expressed in dB. V_R is in volts, and E_R is in volts per meter.

14-17. The product $P_T G_T$ in the Friis equation (Example 3) is referred to as the effective isotropic radiated power (EIRP). Show that for a satellite having the same EIRP at 4 GHz as another at 12 GHz, the received power from an earth-mounted receiving antenna of a fixed size would be the same for each frequency. Since the gain of the receiving dish would be nine times as large at 12 GHz, how can this be?

14-18. A radar system has a receiver sensitivity of -115 dBm ($P_{R\min}$ is -115 dBm), and a common transmitting and receiving antenna with a gain of 45 dB. For a 1 m^2 target at 100 miles, what is the transmitter power required for a frequency of 9 GHz?

14-19. Suppose that the radar of Prob. 14-18 has a power P_T of 1 MW. Find the maximum range for a 0.1-m^2 target.

14-20. Compute the phase shift α between successive elements of a phased array radar antenna to steer the beam 20° off array normal if the spacing between elements is 0.5λ. Assume that steering is parallel to a row of elements.

Appendix A

TABLE A-1 ROMAN SYMBOLS AND THEIR UNITS USED IN THE TEXT, LISTED IN ALPHABETICAL ORDER

Symbol	Physical quantity	SI unit	Abbreviation
$\overline{A}$, $\overline{\mathcal{A}}$	Vector magnetic potential	Weber/meter	Wb m^{-1}
A_e	Effective antenna area	Meter^2	m^2
B, $\mathcal{B}$	Magnetic flux density	Tesla	T
C	Capacitance	Farad	F
D, $\mathcal{D}$	Electric flux density	Coulomb/meter^2	C m^{-2}
E, $\mathcal{E}$	Electric field Intensity	Volt/meter	V m^{-1}
f	Frequency	Hertz	Hz
F	Force	Newton	N
G	Conductance	Siemens	S
G	Antenna gain	—	—
H, $\mathcal{H}$	Magnetic field Intensity	Ampere/meter	A m^{-1}
I, $\mathcal{I}$	Current	Ampere	A
J, $\mathcal{J}$	Current density	Ampere/meter^2	A m^{-2}
J_s, $\mathcal{J}_s$	Surface current density	Ampere/meter	A m^{-1}
J_m	Magnetization current density	Ampere/meter^2	A m^{-2}
ℓ	Distance, length	Meter	m
L	Inductance	Henry	H
m	Magnetic dipole moment	Ampere meter^2	A · m^2
m_e	Mass of the electron	Kilogram	kg
M	Mutual inductance	Henry	H
$\overline{M}$	Magnetic polarization vector	Ampere/meter	A m^{-1}
N	Turns	—	—
p	Electric dipole moment	Coulomb-meter	C · m
P	Power	Watt	W
$\overline{P}$	Electric polarization vector	Coulomb/meter^2	C m^{-2}
$\mathcal{P}$	Power density	Watt/meter^2	W m^{-2}
$\mathcal{P}$	Permeance	Weber/ampere	Wb A^{-1}
q, Q	Charge	Coulomb	C
Q	Q of a circuit	—	—
$\mathcal{R}$	Reluctance	Ampere/weber	A Wb^{-1}
R	Resistance	Ohm	Ω
$\overline{R}_p$	Position vector of point P	Meter	m
R_{12}	Distance from point P_1 to point P_2	Meter	m
s	Surface area	Meter^2	m^2
S	Standing wave ratio	—	—
t	Time	Second	s
T	Temperature	Kelvin	K
T	Torque	Newton-meter	N · m
T	Period	Second	s
v	Volume	Meter^3	m^3
v, U	Velocity	Meter/second	m s^{-1}
V, $\mathcal{V}$	Potential difference	Volt	V
W	Work, energy	Joule	J
w_m	Energy density (magnetic)	Joule/meter^3	J m^{-3}
y	Normalized admittance	—	—
Y	Admittance	Siemens	S
z	Normalized impedance	—	—
Z	Impedance	Ohm	Ω

TABLE A-2 GREEK SYMBOLS AND THEIR UNITS USED IN THE TEXT, LISTED IN ALPHABETICAL ORDER

Symbol	Quantity	SI unit	Abbreviation
α	Angle	Radian	rad
α	Attenuation constant	Neper/meter	Np m^{-1}
β	Angle	Radian	rad
β	Phase constant	Radian/meter	rad m^{-1}
γ	Propagation constant	1/meter	m^{-1}
δ	Skin depth	Meter	m
ϵ	Permittivity	Farad/meter	F m^{-1}
η	Intrinsic impedance	Ohm	Ω
θ	Angle	Radian	rad
λ	Wavelength	Meter	m
Λ	Flux linkage	Weber-turn	Wb · N
μ	Permeability	Henry/meter	H m^{-1}
$\mu_{e,h}$	Mobility	Meter^2/volt-second	m^2 V^{-1} s^{-1}
ρ	Reflection coefficient	Dimensionless	—
ρ_ℓ	Line charge density	Coulomb/meter	C m^{-1}
ρ_s	Surface charge density	Coulomb/meter^2	C m^{-2}
ρ_v	Volume charge density	Coulomb/meter^3	C/m^3
σ	Conductivity	Siemens/meter	S m^{-1}
σ	Radar cross section	Meter^2	m^2
τ	Transmission coefficient	Dimensionless	—
τ	Transit time	Second	s
ϕ	Angle	Radian	rad
$\chi_{e,m}$	Susceptibility	—	—
ψ	Electrical phase angle	Radian	rad
Ψ_E	Electric flux	Coulomb	C
Ψ_m	Magnetic flux	Weber	Wb
ω	Radian frequency	Radian/second	rad s^{-1}
Ω_B	Beam solid angle	Steradian	—

TABLE A-3 LIST OF PREFIXES (MULTIPLIERS) USED WITH BASIC SI UNITS, THEIR SYMBOLS, AND MAGNITUDES

Prefix	Symbol	Magnitude	Prefix	Symbol	Magnitude
tera	T	10^{12}	centi	c	10^{-2}
giga	G	10^9	milli	m	10^{-3}
mega	M	10^6	micro	μ	10^{-6}
kilo	k	10^3	nano	n	10^{-9}
hecto	h	10^2	pico	p	10^{-12}
deka	da	10	femto	f	10^{-15}
deci	d	10^{-1}	atto	a	10^{-18}

Appendix B

TABLE B-1 DOT PRODUCTS BETWEEN UNIT VECTORS OF RECTANGULAR, CYLINDRICAL, AND SPHERICAL COORDINATE SYSTEM

	Rectangular			Cylindrical			Spherical		
•	$\hat{x}$	$\hat{y}$	$\hat{z}$	$\hat{r}_c$	$\hat{\phi}$	$\hat{z}$	$\hat{r}_s$	$\hat{\theta}$	$\hat{\phi}$
$\hat{x}$	1	0	0	$\cos\phi$	$-\sin\phi$	0	$\sin\theta\cos\phi$	$\cos\theta\cos\phi$	$-\sin\phi$
$\hat{y}$	0	1	0	$\sin\phi$	$\cos\phi$	0	$\sin\theta\sin\phi$	$\cos\theta\sin\phi$	$\cos\phi$
$\hat{z}$	0	0	1	0	0	1	$\cos\theta$	$-\sin\theta$	0
$\hat{r}_c$	$\cos\phi$	$\sin\phi$	0	1	0	0	$\sin\theta$	$\cos\theta$	0
$\hat{\phi}$	$-\sin\phi$	$\cos\phi$	0	0	1	0	0	0	1
$\hat{z}$	0	0	1	0	0	1	$\cos\theta$	$-\sin\theta$	0
$\hat{r}_s$	$\sin\theta\cos\phi$	$\sin\theta\sin\phi$	$\cos\theta$	$\sin\theta$	0	$\cos\theta$	1	0	0
$\hat{\theta}$	$\cos\theta\cos\phi$	$\cos\theta\sin\phi$	$-\sin\theta$	$\cos\theta$	0	$-\sin\theta$	0	1	0
$\hat{\phi}$	$-\sin\phi$	$\cos\phi$	0	0	1	0	0	0	1

TABLE B-2 RELATIONSHIPS BETWEEN VARIABLES OF RECTANGULAR, CYLINDRICAL, AND SPHERICAL COORDINATE SYSTEMS

	=	Cylindrical	Spherical	Rectangular
Rectangular	x	$r_c \cos \phi$	$r_s \sin \theta \cos \phi$	x
	y	$r_c \sin \phi$	$r_s \sin \theta \sin \phi$	y
	z	z	$r_s \cos \theta$	z
Cylindrical	r_c	r_c	$r_s \sin \theta$	$(x^2 + y^2)^{1/2}$
	ϕ	ϕ	ϕ	$\tan^{-1}\left[\frac{y}{x}\right]$
	z	z	$r_s \cos \theta$	z
Spherical	r_s	$\frac{r_c}{\sin \theta}$	r_s	$(x^2 + y^2 + z^2)^{1/2}$
	θ	$\tan^{-1}\left[\frac{r_c}{z}\right]$	θ	$\tan^{-1}\left[\frac{(x^2 + y^2)^{1/2}}{z}\right]$
	ϕ	ϕ	ϕ	$\tan^{-1}\left[\frac{y}{x}\right]$

TABLE B-3 RELATIONSHIPS BETWEEN SCALAR PROJECTIONS OF VECTORS IN THE RECTANGULAR, CYLINDRICAL, AND SPHERICAL COORDINATE SYSTEMS

=	Cylindrical	Spherical	Rectangular
A_x	$A_{r_c} \cos\phi - A_\phi \sin\phi$	$A_{r_s} \sin\theta \cos\phi + A_\theta \cos\theta \cos\phi - A_\phi \sin\phi$	A_x
A_y	$A_{r_c} \sin\phi + A_\phi \cos\phi$	$A_{r_s} \sin\theta \sin\phi + A_\theta \cos\theta \sin\phi + A_\phi \cos\phi$	A_y
A_z	A_z	$A_{r_s} \cos\theta - A_\theta \sin\theta$	A_z
A_{r_c}	A_{r_c}	$A_{r_s} \sin\theta + A_\theta \cos\theta$	$A_x \cos\phi + A_y \sin\phi$
A_ϕ	A_ϕ	A_ϕ	$-A_x \sin\phi + A_y \cos\phi$
A_z	A_z	$A_{r_s} \cos\theta - A_\theta \sin\theta$	A_z
A_{r_s}	$A_{r_c} \sin\theta + A_z \cos\theta$	A_{r_s}	$A_x \sin\theta \cos\phi + A_y \sin\theta \sin\phi + A_z \cos\theta$
A_θ	$A_{r_c} \cos\theta - A_z \sin\theta$	A_θ	$A_x \cos\theta \cos\phi + A_y \cos\theta \sin\phi - A_z \sin\theta$
A_ϕ	A_ϕ	A_ϕ	$-A_x \sin\phi + A_y \cos\phi$

TABLE B-4 VECTOR IDENTITIES

$\overline{A} \cdot \overline{B} = \overline{B} \cdot \overline{A}$	(1)
$\overline{A} \times \overline{B} = -\overline{B} \times \overline{A}$	(2)
$\overline{A} \cdot (\overline{B} + \overline{C}) = \overline{A} \cdot \overline{B} + \overline{A} \cdot \overline{C}$	(3)
$\overline{A} \times (\overline{B} + \overline{C}) = \overline{A} \times \overline{B} + \overline{A} \times \overline{C}$	(4)
$\overline{A} \cdot (\overline{B} \times \overline{C}) = \overline{B} \cdot (\overline{C} \times \overline{A}) = (\overline{B} \times \overline{C}) \cdot \overline{A} = \overline{C} \cdot (\overline{A} \times \overline{B})$	(5)
$\overline{A} \times (\overline{B} \times \overline{C}) = (\overline{A} \cdot \overline{C})\overline{B} - (\overline{A} \cdot \overline{B})\overline{C}$	(6)
$\nabla(f + g) = \nabla f + \nabla g$	(7)
$\nabla \cdot (\overline{A} + \overline{B}) = \nabla \cdot \overline{A} + \nabla \cdot \overline{B}$	(8)
$\nabla \times (\overline{A} + \overline{B}) = \nabla \times \overline{A} + \nabla \times \overline{B}$	(9)
$\nabla(fg) = f\nabla g + g\nabla f$	(10)
$\nabla \cdot (f\overline{A}) = f(\nabla \cdot \overline{A}) + \overline{A} \cdot \nabla f$	(11)
$\nabla \cdot (\overline{A} \times \overline{B}) = \overline{B} \cdot (\nabla \times \overline{A}) - \overline{A} \cdot (\nabla \times \overline{B})$	(12)
$\nabla \times (f\overline{A}) = \nabla f \times \overline{A} + f\nabla \times \overline{A}$	(13)
$\nabla(\overline{A} \cdot \overline{B}) = (\overline{A} \cdot \nabla)\overline{B} + (\overline{B} \cdot \nabla)\overline{A} + \overline{A} \times (\nabla \times \overline{B}) + \overline{B} \times (\nabla \times \overline{A})$	(14)
$\nabla \times (\overline{A} \times \overline{B}) = \overline{A}(\nabla \cdot \overline{B}) - \overline{B}(\nabla \cdot \overline{A}) + (\overline{B} \cdot \nabla)\overline{A} - (\overline{A} \cdot \nabla)\overline{B}$	(15)
$\nabla \cdot \nabla f = \nabla^2 f$	(16)
$\nabla \times (\nabla f) = 0$	(17)
$\nabla \cdot (\nabla \times \overline{A}) = 0$	(18)
$\nabla \times (\nabla \times \overline{A}) = \nabla(\nabla \cdot \overline{A}) - \nabla^2 \overline{A}$	(19)
$\nabla \times (f \nabla g) = \nabla f \times \nabla g$	(20)
$\oint_s \overline{A} \cdot \overline{ds} = \int_v \nabla \cdot \overline{A}\, dv$ (divergence theorem)	(21)
$\oint_\ell \overline{A} \cdot \overline{d\ell} = \int_s \nabla \times \overline{A} \cdot \overline{ds}$ (Stokes' theorem)	(22)

Suggested References for Further Reading

In addition to those specific references cited in footnotes in the text, the references cited below are general information.

DURNEY, CARL H., and CURTIS C. JOHNSON, *Introduction to Modern Electromagnetics,* McGraw-Hill Book Company, New York, 1969.

HAYT, WILLIAM H., JR., *Engineering Electromagnetics,* 4th ed., McGraw-Hill Book Company, New York, 1981.

HOLT, CHARLES A., *Introduction to Electromagnetic Fields and Waves,* John Wiley & Sons, Inc., New York, 1975.

JOHNK, CARL T. A., *Engineering Electromagnetic Fields and Waves,* John Wiley & Sons, Inc., New York, 1975.

KRAUS, JOHN D., *Electromagnetics,* 3rd ed., McGraw-Hill Book Company, New York, 1984.

LORRAIN, PAUL, and DALE R. CORSON, *Electromagnetic Fields and Waves,* 2nd ed., W. H. Freeman and Company, Publishers, San Francisco, 1970.

PLONUS, MARTIN A., *Applied Electromagnetics,* McGraw-Hill Book Company, New York, 1978.

RAMO, S., J. R. WHINNERY, and T. VAN DUZER, *Fields and Waves in Communication Electronics,* 2nd ed., John Wiley & Sons, Inc., New York, 1984.

RAO, N. NARAYANA, *Elements of Engineering Electromagnetics,* 2nd ed., Prentice-Hall, Inc., Englewood Cliffs, N.J., 1987.

SKILLING, HUGH H., *Fundamentals of Electric Waves,* John Wiley & Sons, Inc., New York, 1948.

SKOLNIK, MERRILL I., *Introduction to Radar Systems,* 2nd ed., McGraw-Hill Book Company, New York, 1980.

SMYTHE, W. R., *Static and Dynamic Electricity,* 3rd ed., McGraw-Hill Book Company, New York, 1968.

WEBER, E., *Electromagnetic Fields,* Vol. 1, John Wiley & Sons, Inc., New York, 1950.

ZAHN, MARKUS, *Electromagnetic Field Theory,* John Wiley & Sons, Inc., New York, 1979.

Answers to Selected End-of-Section Problems

CHAPTER 1

1.4-1 $\overline{R}_p = (+\hat{x}4 - \hat{y}3 + \hat{z}5)$ m, $|\overline{R}_p| = 7.07$ m

1.4-2 $\hat{a}_{\overline{R}_p} = \hat{x}(0.57) - \hat{y}(0.42) + \hat{z}(0.71)$ m

1.4-3 $\hat{a}_{\overline{F}} = -\hat{x}(0.82) + \hat{y}(0.41) + \hat{z}(0.41)$ m

1.6-1 $\overline{F}_1 + \overline{F}_2 = -\hat{x}13 + \hat{y}6 + \hat{z}6$ N

1.7-1 $\overline{R} = \hat{x}t + \hat{y}t^2$ m represents a uniform velocity in the x direction and uniform acceleration in the y direction.

1.7-2 $\overline{P} = -\hat{x} + \hat{y}(3t - 3) - \hat{z}3$

1.7-3 (a) 0, $-\hat{y}(0.05/\sqrt{2})$, $-\hat{y}(0.05)$, $-\hat{y}(0.05/\sqrt{2})$, 0 V m^{-1}; (b) 0, $-\hat{y}(0.05/\sqrt{2})$, $-\hat{y}(0.05)$, 0 V m^{-1}

1.8-1 $\overline{A} \cdot \overline{B} = 0$

1.8-2 $\overline{C} \cdot \overline{A} = 0$, $\overline{C} \cdot \overline{B} = 0$

1.8-4 (a) 21.21; (b) −3.452

1.8-5 First and second determinants differ in sign; first and third are equal.

1.8-6 $\hat{a}_n = \hat{x}(0.408) - \hat{y}(0.816) + \hat{z}(0.408)$

1.8-7 $\overline{T} = \hat{z}20$ N m

1.9-2 (a) $\hat{r}(749.96) + \hat{\phi}(-641.50) + \hat{z}(-27.79)$; (b) 108.32; (c) 76.34°

1.10-2 (a) $\overline{E} = \hat{\theta}(0.370)$ at $(3, \pi/2, 0)$, $\overline{E} = \hat{r}_s(0.741)$ at $(3, 0, 0)$; (b) $\hat{a}_{\overline{E}} = \hat{\theta}$ at $(3, \pi/2, 0)$, $\hat{a}_{\overline{E}} = \hat{r}_s$ at $(3, 0, 0)$

1.11-1 $2\pi \times 10^{-2}$

1.11-2 0

1.12-1 $\overline{E} = (k/r_s^3)(\hat{r}_s\, 2\cos\theta + \hat{\theta}\sin\theta)$ V m^{-1}

1.12-2 $\nabla \cdot \overline{D} = 26$

1.12-3 $\nabla \cdot \overline{E} = 8$

1.13-1 (a) $\overline{A}_{\text{cyl}} = \hat{r}_c r_c + \hat{z}z$; (b) $\overline{A}_{\text{sph}} = \hat{r}_s r_s$

1.13-2 (a) $\overline{A}_{\text{rec}} = \hat{x}\left[\dfrac{x(x^2+y^2)^{3/2} - xy}{x^2+y^2}\right] + \hat{y}\left[\dfrac{y(x^2+y^2)^{3/2} + x^2}{x^2+y^2}\right]$;

(b) $\overline{A}_{\text{sph}} = \hat{r}_s(r_s^2\sin^3\theta) + \hat{\theta}(r_s^2\cos\theta\sin^2\theta) + \hat{\phi}\cos\phi$

CHAPTER 2

2.2-1 (a) $4\pi a k_1$; (b) $2\pi a^2 k_2$; (c) $\dfrac{4\pi a^3 k_3}{3}$; (d) $\pi a^4 k_4$ C

2.2-2 $Q_{\text{en}} = ka^4$ C

2.3-1 (a) $\overline{F}_{Q_t} = \hat{y}(0.1302Q_t)$ N; (b) $\overline{F}_{Q_t} = -\hat{y}(0.047Q_t)$ N

2.3-2 $\overline{F}_{Q_t} = \hat{x}(0.158Q_t)$ N

2.3-3 $\overline{F}_{Q_3} = (\hat{x} + \hat{y})(3.46 \times 10^{-7}$ N$)$

2.3-4 $\overline{F}_{Q_2} = -\hat{x}(1.73 \times 10^{-7}) - \hat{y}(1.73 \times 10^{-7}) - \hat{z}(1.75 \times 10^{-7})$ N

2.4-1 $-\hat{z}(1.27 \times 10^{10})Q$ V m^{-1}

2.4-2 $\hat{z} = -23.74$

2.4-3 $\hat{x}(4.02) - \hat{y}(0.99) + \hat{z}(1.02)$ V m^{-1}

2.5-1 $\dfrac{\hat{z}\rho_\ell a z}{2\epsilon_0(a^2 + z^2)^{3/2}}$

2.5-2 $-\hat{y}11.50 \times 10^3$ V m^{-1}

2.5-3 (a) $\hat{z} = a/\sqrt{2}$; (b) $\overline{E} = \dfrac{\hat{z}\rho_\ell(0.19)}{a\epsilon_0}$ V m^{-1}

2.6-1 $E_z = \rho_s/4\epsilon_0$

2.6-2 $E_z = \rho_s/\epsilon_0$ between sheets; $E = 0$ everywhere else

2.7-1 $\overline{E}_{0.03\text{m}} = \hat{r}_s 4.182 \times 10^{-8}$ V m^{-1}

CHAPTER 3

3.2-1 Fields add between plates and cancel outside plates. Total flux between plates = $\rho_s \times$ area of either plate.

3.2-2 $\rho_s\big|_{r=r_a} = \dfrac{-Q}{2\pi r_a L}$ (C m^{-2})

$\rho_s\big|_{r=r_b} = \dfrac{+Q}{2\pi r_b L}$ (C m^{-2})

3.2-3 $\rho_s\big|_{r=r_a} = \dfrac{+Q}{4\pi r_a^2}$ (C m^{-2})

$\rho_s\big|_{r=r_b} = \dfrac{-Q}{4\pi r_b^2}$ (C m^{-2})

3.3-1 $\Psi_{E\text{ between plates}} = 2 \times 10^{-8}$ C

$D_{\text{between plates}} = \Psi_E/\text{area} = 2 \times 10^{-8}$ C $= |\rho_s|$

3.3-2 $\Psi_{E\text{ total}} = Q = 3.1416 \times 10^{-9}$ C

3.3-3 $\Psi_{E \text{ between spheres}} = 3.69 \times 10^{-10}$ C

3.3-4 $\Psi_{E \text{ outward}} = 0.008\ \mu$C

3.4-1 $\overline{D}_{\text{inside}} = \hat{r}_s r_s \rho_v/3$ (C m^{-2}), $\overline{E}_{\text{inside}} = \hat{r}_s r_s \rho_v/3\epsilon_0$ (V m^{-1}),
$\overline{D}_{\text{outside}} = \hat{r}_s r_0^3 \rho_v/3r_s^2$ (C m^{-2}),
$\overline{E}_{\text{outside}} = \hat{r}_s r_0^3 \rho_v/3\epsilon_0 r_s^2$ (V m^{-1})

3.4-2 $\overline{D}_{\text{inside}} = \hat{r}_c r_c \rho_v/2$ (C m^{-2}), $\overline{E}_{\text{inside}} = \hat{r}_c r_c \rho_v/2\epsilon_0$ (V m^{-1}),
$\overline{D}_{\text{outside}} = \hat{r}_c r_0^2 \rho_v/2r_c$ (C m^{-2}),
$\overline{E}_{\text{outside}} = \hat{r}_c r_0^2 \rho_v/2r_c\epsilon_0$ (V m^{-1})

3.4-3 $\overline{D} = \hat{z}\,\rho_s/2$ (C m^{-2}), $z > 0$; $\overline{E} = \hat{z}\,\rho_s/2\epsilon_0$ (V m^{-1}), $(z > 0)$;
$\overline{D} = -\hat{z}\,\rho_s/2$ (C m^{-2}), $z < 0$; $\overline{E} = -\hat{z}\,\rho_s/2\epsilon_0$ (V m^{-1}), $(z < 0)$

3.5-1 (a) $2Kx$; (b) K; (c) Kz; (d) $4Kr_s$; (e) $-K/r_s^4$; (f) $(2K \sin\theta)/r_s$; (g) $2K\phi/r_s$; (h) $3Kr_c$;
(i) $\dfrac{\cos\phi}{r_s^2 \sin\theta \cos\theta}$ C m^{-3}

3.5-2 (a) 0; (c) 0; (e) ∞; (f) ∞

3.5-3 $\rho_v = (K_1/y) + K_2$ C m^{-3}

3.6-1 (a) $4K\pi b^4$; (b) $4\pi K$; (c) 0; (d) $2\pi KLa^2$

3.6-2 0

CHAPTER 4

4.2-1 (a) -36 pJ; (b) -57 pJ

4.2-2 1.602×10^{-16} J; 1000 electron volts

4.3-1 (a) , (b) 24.75 V

4.3-2 (a) 26.75 V; (b) 24.75 V

4.3-3 (a) 100 V; (b) 0

4.4-1 $a = r_0/2$, $V = \rho_s r_0/\epsilon_0$ V; $a = r_0$, $V = \rho_s r_0/\epsilon_0$; $a = 2r_0$, $V = \rho_s r_0/2\epsilon_0$; $a = 4r_0$, $V = \rho_s r_0/4\epsilon_0$; $a = 8r_0$, $V = \rho_s r_0/8\epsilon$.

4.4-2 $V_z = \dfrac{\rho_\ell r_0}{2\epsilon_0}\left[\dfrac{1}{(r_0^2 + z^2)^{1/2}} - \dfrac{1}{(r_0^2 + 100^2)^{1/2}}\right]$ V

4.5-1 $V = \dfrac{b\rho_\ell}{2\epsilon_0(b^2 + z^2)^{1/2}}$ V; $\overline{E} = \hat{z}\dfrac{b\rho_\ell z}{2\epsilon_0(b^2 + z^2)^{3/2}}$ V m^{-1}

4.5-2 Non-conservative field

4.5-3 (a) $V(r_a) = V_{\text{ref}}$; (b) $r_c = \sqrt{r_b r_a}$; (c) $r_c = \sqrt{r_a^4 r_b}$

4.6-1 -0.2696 J

4.6-2 (a) $W = 2\pi\rho_s^2 r_a^3/\epsilon_0 = Q^2/8\pi\epsilon_0 r_a$ (J); (b) same as part (a); (c) as $r_a \to 0$, $W \to \infty$; (d) for shell $V = \rho_s r_a/\epsilon_0$ when $(0 \le r_s \le r_a)$; also, $V = Q/4\pi\epsilon_0 r_s$ when $(r_s \ge r_a)$, for a point charge $V = Q/4\pi\epsilon_0 r_s$ when $(0 < r_s)$

4.6-3 $\frac{1}{2}\rho_s^2/\epsilon_0$ J m^{-3}. Energy stored is equal to energy required to separate the plates.

CHAPTER 5

5.2-2 $\nabla \cdot \overline{J} = \beta K \sin(\omega t - \beta z)$

5.3-1 $I = 3.64 \times 10^7$ A

5.3-2 (a) $\rho_{v_e} = -1.602 \times 10^7$ C m^{-3}; (b) $U_d = -0.312$ m s^{-1}; (c) $\mu_e = 3.12$ (m^2 V^{-1} s^{-1}); (d) $J = 5 \times 10^6$ A m^{-2}; (e) $\sigma = 5 \times 10^7$ S m^{-1}

5.4-1 $R = 30.88\ \Omega$

5.4-2 $R = \dfrac{4c}{\sigma\pi(r_b^2 - r_a^2)}\ \Omega$

5.4-3 $R = \dfrac{2\ln(r_b/r_a)}{\sigma c\pi}\ \Omega$

5.5-1 (a) $\overline{D}_a = \hat{z}7\epsilon_0$ (C m^{-2}); (b) $\rho_s = 7\epsilon_0$ C m^{-2} for top plate, $-7\epsilon_0$ C m^{-2} for bottom plate

5.5-2 $\rho_s = -3\epsilon_0 E_0 \cos\theta$ C m^{-2}, bottom hemisphere
$= 3\epsilon_0 E_0 \cos\theta$ C m^{-2}, top hemisphere

5.5-3 (a) $\overline{D} = \overline{E} = 0$ inside, $\overline{D} = -\hat{z}10^{-4}/2$ (C m^{-2}) above, $\overline{E} = -\hat{z}10^{-4}/2\epsilon_0$ (V m^{-1}) above, $\overline{D} = +\hat{z}10^{-4}/2$ (C m^{-2}) below, $\overline{E} = +\hat{z}10^{-4}/2\epsilon_0$ (V m^{-1}) below

5.6-1 (a) $\overline{E}$ is directed outward from surface of sphere; (b) $\rho_s = 5\epsilon_0$; (c) $\overline{D} = \hat{r}_s 5\epsilon_0$.

5.6-2 (a) $\sqrt{13}\ \epsilon_0$ C m^{-2}; (b) $(\hat{x}17.7 + \hat{y}26.56) \times 10^{-12}$ C m^{-2}

5.7-1 (a) 0.39 m s^{-1}; (b) 0.19 m s^{-1}; (c) 0.135 m s^{-1}; (d) 0.048 m s^{-1}

5.8-1 (a) $\rho_{sb\,\text{front}} = 1$ C m^{-2}, $\rho_{sb\,\text{back}} = 1$ C m^{-2}, $\rho_{vb} = -1$ C m^{-3}; (b) $\rho_{sb\,\text{front}} = 3$ C m^{-2}, $\rho_{sb\,\text{back}} = -3$ C m^{-2}, $\rho_{vb} = -2x$ C m^{-3}

5.8-2 (a) $Q_{sb} = 8$ C; $Q_{vb} = -8$ C; (b) $Q_{sb} = 0$, $Q_{vb} = 0$

5.8-3 (a) $\overline{E}_i = 0$ in all regions except region 2, $\overline{E}_{i2} = \dfrac{-\hat{r}_s Q}{4\pi\epsilon_0 r_s^2}\left(1 - \dfrac{\epsilon_0}{\epsilon_2}\right)\left(\text{V m}^{-1}\right)$; (b) $\overline{E}_a = \hat{r}_s Q/4\pi\epsilon_0 r_s^2$ (V m^{-1}) in all regions; (c) see Example 11; (d) see Example 11

5.8-4 1/5

5.9-1 (a) $\hat{z}40\epsilon_0$ C m^{-2}; (b) $\hat{z}10$ V m^{-1}; (c) $\hat{z}30\epsilon_0$ C m^{-2}; (d) $\hat{z}30\epsilon_0$; (e) $\hat{z}20\epsilon_0$; (f) 0

5.9-2 (a) $(\hat{x}8\epsilon_0 + \hat{y}12\epsilon_0 - \hat{z}24\epsilon_0)$ C m^{-2}; (b) $(\hat{x}6\epsilon_0 + \hat{y}9\epsilon_0 - \hat{z}18\epsilon_0)$ C m^{-2};
(c) $(\hat{x}8/7 + \hat{y}3 - \hat{z}6)$ V m^{-1}; (d) $(\hat{x}8\epsilon_0 + \hat{y}21\epsilon_0 - \hat{z}42\epsilon_0)$ C m^{-2};
(e) $(\hat{x}48/7\epsilon_0 + \hat{y}18\epsilon_0 - \hat{z}38\epsilon_0)$ C m^{-2}; (f) $0.857\epsilon_0$ C m^{-2}

5.9-3 (a) $\hat{y}20\epsilon_0$ C m^{-2}; (b) $\hat{y}6.67$ V m^{-1}; (c) $\overline{D}_2 = \overline{D}_1$; (d) $\hat{y}15\epsilon_0$ C m^{-2}; (e) $\hat{y}13.33\epsilon_0$ C m^{-2}; (f) $-1.67\epsilon_0$ C m^{-2}

5.10-2 $C = 118$ pF m^{-1}

5.10-3 $C = 0.118$ pF

5.10-4 $C = \epsilon s/\ell$ F

CHAPTER 6

6.2-1 (a) $\rho_v = \nabla \cdot \overline{D} = 0$; (b) $\nabla^2 V = 0$; thus, Laplace's equation is satisfied

6.2-2 (a) $\nabla^2 V = 40 = -\rho_v/\varepsilon$, Laplace's equation is not satisfied; (b) $\nabla^2 V = 0$, except at $r_s = 0$, Laplace's equation is satisfied except at origin; (c) $\nabla^2 V = 2K_2/r_s = -\rho_v/\varepsilon$, Laplace's equation is not satisfied; (d) $\nabla^2 V = 0$, Laplace's equation is satisfied; (e) $\nabla^2 V = K_4/r_c = -\rho_v/\varepsilon$, Laplace's equation is not satisfied.

6.3-1 No, since $V|_{r_c = r_a}$ is not satisfied.

6.3-2 No, solution does not satisfy Laplace's equation.

6.4-1 (a) $V = 10^5 y + 25$ V; (b) $\overline{E} = -\hat{y}10^5$ V m^{-1}; (c) $C = 7.083 \times 10^{-8}$ F m^{-2}

6.4-2 (a) $V = 54.57 \ln r_c + 263.48$ V; (b) $\overline{E} = -\hat{r}_c(54.57/r_c)$ V m^{-1}
(c) $\overline{D} = -r_c(1.933 \times 10^{-9}/r_c)$ C m^{-2}; (d) ρ_s at $r_b = 3.866 \times 10^{-8}$ C m^{-2};
(e) $C = 2.429 \times 10^{-10}$ F m^{-1}

6.4-3 (a) $200\phi/\pi + 100$ V; (b) $-\hat{\phi}200/\pi r_c$ V m^{-1}; (c) $-\hat{\phi}1.127 \times 10^{-9}/r_c$ C m^{-2}; (d) 0.564 nC m^{-2}

6.4-4 (a) $A = V_0/\ln[\tan(\theta_2/2)/\tan(\theta_1/2)]$, $B = [-V_0 \ln \tan(\theta_1/2)]/\ln[\tan(\theta_2/2)/\tan(\theta_1/2)]$; (b) $\overline{E} = \hat{\theta}(-A/r_s \sin\theta)$; (c) $\rho_s|_{\theta_1} = -V_0\epsilon/\{r_s \sin\theta_1 \ln[\tan(\theta_2/2)/\tan(\theta_1/2)]\}$ (C m^{-2})

6.5-1 (a) 25 V; (b) 6.7972 V; (c) 43.1932 V

6.5-2 (a) $V = \sum_{m=1}^{m=\infty} \frac{8V_0(-1)^{(m-1)/2}}{m^2\pi^2 \sinh(m\pi a/b)} \sinh\left(\frac{m\pi x}{b}\right) \sin\left(\frac{m\pi y}{b}\right)$ (V), where m is odd; (b) 32.468 V

6.7-2 -0.428 A

CHAPTER 7

7.2-1 (a) 182 pF m^{-1}; (b) $1.45V_0$ nC m^{-2}; (c) 5.79×10^{-10} C m^{-2}; (d) $1.45V_0$ nC m^{-2}; (e) 5.79×10^{-10} C m^{-2}

7.2-2 (b) 14.84 pF m^{-1}; (c) 1.234 kV m^{-1}; (d) 24.59 nC m^{-2}

7.2-3 (b) $V(a/2, b/2) = 27$ V; (c) region near the gap

7.2-4 (a) 3.25×10^{-7} Ω; (b) $R = 2.87 \times 10^{-7}$ Ω

7.3-1 33.4 V

7.3-2 (a) 42.9 V, 52.7 V, 42.9 V, 18.7 V, 25 V, 18.7 V, 7.1 V, 9.8 V, 7.1 V; (b) 7.1 V, compared to 6.7972 V from Prob. 6.5-1

7.4-1 (a) 51.96 V; (b) $(x, 8.08, 0)$, $r_0 = 6.35$ m

7.4-2 $|\rho_s| = 1.273 \times 10^{-9}\left(\frac{5}{25 + z^2}\right)$ C m^{-2}

7.5-1 8.47×10^8 S

7.5-2 $C = 1500$ pF; $R = 295.13$ Ω

7.6-1 0.133 S m^{-1}

CHAPTER 8

8.3-2 (a) $\hat{\phi}0.0248$ A m^{-1}; (b) $\hat{\phi}0.0124$ A m^{-1}; (c) $\hat{\phi}0.00742$ A m^{-1}

8.3-3 (a) $\overline{H} = \hat{z}0.9I/a$ A m^{-1}; (b) $\overline{H} = [\hat{z}2I(a/2)^2/\pi(z^2 + (a/2)^2(z^2 + 2(a/2)^2)^{1/2}]$ (A m^{-1})

8.3-4 $H = \hat{z}\frac{Ia^2}{2(a^2 + z^2)^{3/2}}$ A m^{-1}

8.3-5 (a) $2499.219I$ A m^{-1}, $2500I$ A m^{-1}, % error = 0.03; (b) 1249.9024 A m^{-1}, $1250I$ A m^{-1}, % error = 0.01

8.3-6 $\overline{H} = \hat{z}\frac{NI}{2\ell}\left[\frac{\ell/2 - b}{[a^2 + (\ell/2 - b)^2]^{1/2}} + \frac{\ell/2 + b}{[a^2 + (\ell/2 + b)^2]^{1/2}}\right]$ (A m^{-1})

8.4-1 Inside, $\overline{H} = \hat{\phi}2r_c^2$ (A m^{-1}); outside, $\overline{H} = \hat{\phi}\frac{2a^3}{r_c}$ (A m^{-1})

8.4-2 $H_{\phi\text{inside}} = 0$; $H_{\phi\text{outside}} = 10^{-7}/2\pi r_c$ A m^{-1}

8.4-3 (a) $\hat{x}4$ μA m^{-1}; (b) 0; (c) 0

8.4-4 (a) $\overline{H} = 0$ $(r_c < a)$, $\overline{H} = \hat{\phi}5/\pi r_c$ $(a < r_c < b)$, $\overline{H} = 0$ $(r_c > b)$

8.4-5 (a) $\overline{H} = \hat{\phi}NI/2\pi r_c$ (A m^{-1}); (b) $\overline{H} = \hat{z}\frac{I}{2a}$; (c) $\overline{H} \approx 0$

8.5-1 (a) 0; (b) 0; (c) $\hat{r}_s K \cot\theta - \hat{\theta}2$; (d) $\hat{z}(10x - 20y)$

8.5-2 (a) $2\pi Ka^2$ A; (b) $2\pi Ka^2$ A; (c) $\pi\, Ka^2$ A

8.5-3 (a) $2\pi K$ A; (b) $2\pi K$ A; (c) πK A

8.5-4 169.6 nA

8.5-5 (a) $\dfrac{wJ_{xo}}{\alpha}(1 - e^{-\alpha d})$ A; (b) $H_y(0) = \dfrac{J_{xo}}{2\alpha}(1 - e^{-\alpha d})$ A m^{-1}, $H_y(d) =$

$\dfrac{-J_{xo}}{2\alpha}(1 - e^{-\alpha d})$ A m^{-1}

8.6-1 (a) $2\pi Ka^2$ A; (b) $2\pi Ka^2$ A; (c) πKa^2 A

8.7-1 $\Psi_m = (\mu_0 I/2\pi)\ln(b/a)$ (Wb)

8.7-2 $\Psi_m = \dfrac{\mu_0 NIs}{\ell}$ (Wb), where $s = h(c - a)$

CHAPTER 9

9.3-2 $\overline{F} = \hat{y}114.28\ \mu$N

9.4-1 $U_z = (-5Qt/m + 10)$ m s^{-1}, $U_x = 3$ m s^{-1}, $U_y = 5$ m s^{-1}

9.4-2 $\overline{U} = \hat{x}\left(3\cos\dfrac{2Qt}{m} + 5\sin\dfrac{2Qt}{m}\right) + \hat{y}\left(5\cos\dfrac{2Qt}{m} - 3\sin\dfrac{2Qt}{m}\right) - \hat{z}\left(\dfrac{5Qt}{m} + 10\right)$

(m s^{-1})

9.5-2 $V_h = -14.86\ \mu$V

9.6-1 $\overline{T} = \hat{z}3.33 \times 10^{-12}$ N m, final torque = 0

9.6-2 (a) $\overline{A} = \dfrac{\hat{\phi}\mu_0 I(d\ell)^2}{4\pi r_s^2}\sin\theta$ (Wb m^{-1}); (b) $\overline{dB} = \mu_0[I(d\ell)^2/4\pi r_s^3](2\hat{r}_s\cos\theta + \hat{\theta}\sin\theta)$ (T)

9.7-1 (a) $\overline{J}_m = \hat{z}3$ (A m^{-2}); (b) $\overline{J}_{sm} = \hat{x}3x$ (A m^{-1}) on top, $\overline{J}_{sm} = -\hat{z}0.3$ (A m^{-1}) on front, $\overline{J}_{sm} = -\hat{x}3x$ (A m^{-1}) on bottom, $\overline{J}_{sm} = 0$ on back; (c) $I_{total} = 0.3$ (A)

9.7-2 (a) $\overline{B} = \hat{\phi}(0.186)$ T; (b) $\overline{M} = \hat{\phi}(1.483 \times 10^5)$ A m^{-1}; (c) $-\hat{z}1.483 \times 10^5$ A m^{-1}; (d) $\overline{J}_m = 0$

9.7-3 If $U_z = 0$, electron will orbit a central point; otherwise, it will spiral about an axis.

9.7-4 $\Delta\omega = q_e B_a/2m_e$ (rad s^{-1})

9.7-5 $\Delta m = -q_e^2 B_a r_c^2/4m_e$ (A m^2)

9.8-1 (a) $B_{gap} = B_{iron} = 1.8$ T; (b) $H_{iron} = 286.48$ A m^{-1}; (c) $H_{gap} = 1.432 \times 10^6$ A m^{-1}

9.8-2 (a) $B_{air} = 3.6 \times 10^{-4}$ T; (b) $H_{air} = 286.48$ A m^{-1}; (c) $H_{iron} = 286.48$ A m^{-1}

9.8-3 (a) $\alpha_1 = 0°$; (b) $\alpha_1 = 0.273°$

9.9-1 (a) $\mathcal{R} = 522{,}800$ (b) $N = 11$ turns

9.9-2 $B_s = 0.81$ T

9.10-1 $L = 3.158 \times 10^{-3}$ H

9.10-2 $L = 0.439$ mH

9.10-3 $L = 6.89 \times 10^{-5}$ H

9.10-4 $L = \mu_0/8$ H m^{-1}

9.10-5 (a) $M_{12} = 0.04$ H; (b) $V_{2rms} = 53.31$ V

9.11-1 $L = (\mu/2\pi)\ln(r_b/r_a)$ H m^{-1}

9.11-2 $F(t) = 1130\cos^2 377t$ N; $F_{av} = 565$ N

CHAPTER 10

10.3-1 $V_1 = 0.8$ V; $V_2 = 0.2$ V

10.3-2 $\Psi_{\text{peak-to-peak}} = 1.061 \times 10^{-3}$ Wb

10.3-3 $s_c = 0.45$ cm^2

10.4-1 Force $= UB^2\ell^2/R$; power $= U^2B^2\ell^2/R$

10.4-2 emf $= (-\ell U_0 B_0 \sin \omega t - \ell y \omega B_0 \cos \omega t)$ (V)

10.4-3 (a) $B = 1.719 \times 10^{-4}$ T; (b) $H = 136.79$ A m^{-1}

10.5-1 $N = 500$ turns

10.5-2 $I_{1_{\text{total}}} = 0.104$ A

10.5-3 18 V; winding would probably burn out quickly!

10.6-1 58.52 A m^{-2} rms at $r_c = 1$ mm; 11.70 A m^{-2} rms at $r_c = 5$ mm

10.6-2 (b) 376.73 Ω

10.7-1 $\frac{1}{2}\sigma E^2$ (W m^{-3})

10.7-2 17.38 W m^{-3}

10.7-3 3.845 W

10.8-1 2.99796×10^8 ms^{-1}

10.9-1 $L = 2.607 \times 10^{-5}$ H

CHAPTER 11

11.3-1 (a) 10 V m^{-1}; (b) 150 MHz; (c) 5.79 m^{-1}; (d) $\hat{z}1.63 \times 10^8$ m s^{-1}; (e) 205 Ω; (f) 0.0488 A m^{-1}; (g) 1.09 m

11.3-2 $E_x = 10$ V m^{-1}, $H_y = -0.0488$ A m^{-1}; both traveling in the negative z direction

11.3-3 2.39×10^8 m s^{-1}, 0.2387 m

11.5-1 10 kHz: $\eta = 0.1258 \angle 45°$ Ω, $\gamma = 0.444 + j0.444$ m^{-1}; 100 kHz: $\eta = 0.3977 \angle 45°$ Ω, $\gamma = 1.404 + j1.404$ m^{-1}; 10 MHz: $\eta = 3.977 \angle 44.745°$ Ω, $\gamma = 13.977 + j14.102$ m^{-1}; 1 GHz: $\eta = 34.37 \angle 24.17°$ Ω, $\gamma = 94.31 + j210.23$ m^{-1}

11.5-2 $\alpha = 0.731$ m^{-1}, $\beta = 292$ m^{-1}, $\lambda_{\text{material}} = 0.0215$ m

11.5-3 11.51 V m^{-1}, $0.0616 \angle -5.2°$ A m^{-1}

11.6-1 At 60 Hz, $\delta = 0.0085$ m; at 6 GHz, $\delta = 8.53 \times 10^{-7}$ m

11.6-2 At 60 Hz, $\delta = 5.227 \times 10^{-4}$ m; at 6 GHz, $\delta = 5.227 \times 10^{-8}$ m

11.6-3 3.03×10^{-11} W m^{-2}

11.7-1 $\eta_1 = 185.77 \angle 30.47°$ (Ω), $\gamma_1 = (21.55 + j36.63)$(m^{-1}), $\eta_2 = 180 \angle 12.11°$ (Ω), $\gamma_2 = (9.21 + j42.89)$(m^{-1}), $\rho = 0.16 \angle -95.71°$, $\tau = 1.0 \angle -9.32°$, $H^+_{y10} = 0.0054E^+_{x10}$, $E^-_{x10} = 0.16E^+_{x10}$, $H^-_{y10} = 0.00086E^+_{x10}$, $E^+_{x20} = 1.00E^+_{x10}$, $H^+_{y20} = 0.00556E^+_{x10}$

11.7-2 $\mathscr{H}^+_{y10}/\mathscr{E}^+_{x10} = 0.0054 \angle -30.47°$, $\mathscr{E}^-_{x10}/\mathscr{E}^+_{x10} = 0.16 \angle -95.71°$, $\mathscr{H}^-_{y10}/\mathscr{E}^+_{x10} = 0.00086 \angle 53.82°$, $\mathscr{E}^+_{x20}/\mathscr{E}^+_{x10} = 1.00 \angle -9.32°$, $\mathscr{H}^+_{y20}/\mathscr{E}^+_{x10} = 0.00556 \angle -21.43°$

11.8-1 $\eta_2 = 243.35$ Ω; $\rho = -0.215$; $S = 1.549$; $\lambda_{\text{air}} = 0.122$ m; $\lambda_{\text{diel}} = 0.079$ m; $\mathscr{P}_{\text{refl}} = 4.6\%$

11.8-3 $\eta_1 = 204$ Ω, $\eta_2 = 377$ Ω, $\rho = 0.298$, $\lambda_2/\lambda_1 = \sqrt{3.4}$, $\mathscr{P}^+_{1_{\text{ave}}} = 24.50$ W m^{-2}, $\mathscr{P}^-_{1_{\text{ave}}} = 2.175$ W m^{-2}, $\mathscr{P}^+_{2_{\text{ave}}} = 22.326$ W m^{-2}

11.8-4 Thickness = 101 nm, $\epsilon'_{r2} = 1.53$

11.8-5 $\epsilon_r = 1.86$; thickness = 6.1 mm

11.8-6 Thickness = 13 mm

11.9-1 At $\lambda/4$ from surface, $E = 200$ V m^{-1}, $H = 0$ A m^{-1}; at $\lambda/2$ from surface, $E = 0$ V m^{-1}, $H = 0.531$ A m^{-1}

11.10-3 $\theta_B = 56.83°$, $\theta_t = 33.17°$

11.10-4 $\theta_c = 40.82°$

11.10-5 $\theta_B = 73.89°$

CHAPTER 12

12.2-2 11.11 Ω

12.4-2 0.344%, 5.36 mV, 8.75 μA

12.4-3 $\rho_L = 0.864 \angle 168.77°$

12.4-4 (a) $516.53 \angle 76.55°$; (b) $725.12 \angle -87.25°$ Ω

12.6-1 0.1263 H

12.8-1 $Z_0 = 41.67$ Ω; $\beta = 10.47$ rad/m

12.8-2 (a) 3.98; (b) 3.37×10^{-8} s

12.8-3 12.57 Ω

12.9-2 (a) $Z_{in} = 30$ Ω, $S = 2.5$; (b) $Z_{in} = 75 \angle -46.40°$ Ω, $S = 2.5$

12.9-3 $S = 1.5$, $Z_{in} = 111.71 \angle -4.17°$ Ω

12.9-4 0.086 λ

12.10-1 $\ell = 1$ m; $Z_0 = 51.23$ Ω

12.10-2 $Z_{in1} = Z_{in2} = 200$ Ω

12.10-3 $Z_{in} = 50 \angle 30°$ Ω

12.11-1 $P_{inc} = 11.25$ W; $P_{refl} = 1.25$ W; $V_{rms} = 25$ V

12.11-2 The 75-Ω line delivers 100 W; the 50-Ω line delivers only 80 W.

12.11-3 $P_L = 3333.33$ W for $S = 2.0$, or 5555.56 W if S were reduced to 1.2.

12.12-1 $S = 2.65$

12.12-2 $Z_{in} = 131.5 - j9$ Ω

12.12-3 $y_L = 0.596 - j0.321$; $S = 1.98$; $y_{in} = 0.55 + j0.25$

12.13-1 (a) $d_1 = 0.292\lambda$, $\ell_1 = 0.114\lambda$; $d_2 = 0.458\lambda$, $\ell_2 = 0.386\lambda$; (b) $S = 3.0$

12.13-2 $\ell_1 = 0.326\lambda$, $\ell_2 = 0.118\lambda$; (b) $S = 3.0$; (c) $S = 2.8$

12.14-1 $f = 1.88$ GHz, $Z_L = 28.5 - j15$ Ω

12.14-2 $f = 1.88$ GHz, $Z_L = 28.5 + j15$ Ω

12.15-1 41.5 miles, $Z_{in} = 2845 + j80.4$ Ω

12.15-2 $S = 1.68$

12.15-3 $Z_{in} = Z_0$

CHAPTER 13

13.2-1 1.5 GHz

13.2-2 $v_g = 2.5 \times 10^{10}$ cm s^{-1}, $v_p = 3.46 \times 10^{10}$ cm s^{-1}

13.3-1 (a) 2.08 GHz $< f <$ 4.17 GHz; (b) 3.15 GHz $< f <$ 6.3 GHz; (c) 7.85 GHz $< f <$ 15.71 GHz

13.3-2 (a) 0.0485 m; (b) 4.37×10^8 m s^{-1}; (c) 130 rad m^{-1}

13.3-3 16.85 GHz

13.4-1 (a) 4.88 GHz; (b) 7.455 GHz; (c) 17.63 GHz

13.4-2

f/f_c	$Z_{TE11}/\sqrt{\mu/\epsilon}$	$Z_{TM11}/\sqrt{\mu/\epsilon}$
1		0
1.1	2.4	0.42
1.2	1.81	0.55
1.4	1.43	0.70
1.6	1.28	0.78
1.8	1.20	0.83
2.0	1.15	0.87
10	1.01	0.99
100	1.00005	0.99995

13.5-1 8.2 GHz, 0.140 dB m^{-1}; 12.4 GHz, 0.097 dB m^{-1}

13.6-1 18,100

13.6-2 19,100

CHAPTER 14

14.2-1 (a) 47.2 dB; (b) 41.2 dB

14.2-2 2.22×10^{-3} W m^{-2}

14.2-3 37.5 dB

14.2-4 100 W m^{-2}

14.3-1 $R_{rad} = 0.219\ \Omega$

14.3-2 (a) 0.219 Ω; (b) 0.877 Ω

14.3-3 41.04 dB

14.4-1 6.13×10^{-10} W; effective area of $\lambda/2$ dipole at 100 MHz is four times the effective area of the $\lambda/2$ dipole at 200 MHz.

14.4-2 1.02×10^{-10} W

14.4-3 0.314 V m^{-1}

14.5-1 Maxima at 0°, 90°, 180°, 270°; nulls at 30°, 150°, 210°, 330°

14.5-2 Main maxima at 0°, 180°; secondary maxima at 90°, 270°; nulls at 41.81°, 138.19°, −41.81°, −138.19°

14.5-3 Main maxima at 0°, 180°; secondary maxima at 47.1°, 132.9°, 227.1°, 312.9°; nulls at 30°, 90°, 150°, 210°, 270°, and 330°

14.5-4 $\phi_B = 2.03°$

14.6-1 (a) $G = 1.62 \times 10^5$; (b) 7.77×10^{-5} sr; (c) 0.57°

14.6-2 (a) $G = 89{,}100$; (b) 1.41×10^{-4} sr; (c) 0.769°

14.6-3 7.38 m

14.6-4 85.3%

14.7-1 $P_R = -99.52$ dBm

14.7-2 To double range, G would have to be increased by a factor of 4.

14.7-3 To double range, P_T would have to be increased by a factor of 16.

14.7-4 $R_{max} = 870$ km or 541 miles

Index

A

D

E

F

G

H

I

J

K

L

Q

R

S

T

U

V

W

X